Guide to Energy Management, Ninth Edition

Guide to Energy Management, Ninth Edition

Editor

Eric Mazzi

Mazzi Consulting, Canada

River Publishers

Routledge
Taylor & Francis Group
NEW YORK AND LONDON

Published 2026 by River Publishers
River Publishers
Broagervej 10, 9260 Gistrup, Denmark
www.riverpublishers.com

Distributed exclusively by Routledge
605 Third Avenue, New York, NY 10017, USA
4 Park Square, Milton Park, Abingdon, Oxon OX14 4RN

Guide to Energy Management, Ninth Edition / by Eric Mazzi.

Routledge is an imprint of the Taylor & Francis Group, an informa business

ISBN 978-87-7022-691-2 (hardback)
ISBN 978-87-4381-379-8 (online)
ISBN 978-87-4381-378-1 (master ebook)
ISBN 978-87-4381-381-1 (ePub)

DOI 10.1201/9788743813781

Contents

Preface

The Guide to Energy Management (GTEM) was written with the intent to raise the bar for how demand side energy and carbon management (DSM) is done in practice. The scope of this book, as described in more detail in Chapter 1, focuses on the demand side ("behind the meter") of energy systems for residential, commercial/institutional, and industrial facilities. While the supply side of energy systems ("front of meter") is also very important, it is generally not in the scope of this book. Similarly, some demand side applications such as transportation (cars, freight, air travel, shipping) are equally important, but not in scope. The GTEM will be most valuable for educators, learners, and practitioners in the U.S. and Canada, and it should also be useful in any higher-income context with modern energy systems. Key themes in the GTEM that make it relatively unique as a DSM resource are as follows:

1. **Climate change is real and the speed and scale of climate action needs to increase.** The current science of climate change is strongly conclusive that achieving net-zero GHGs is a prerequisite for the Earth system to ever realize a stable climate, and that negative emissions will eventually be needed to reverse the cumulative warming that has occurred since pre-industrial times (see Chapter 2). Furthermore, mitigation alone is insufficient because adaptation (see Chapter 8) is now essential (and some argue to consider solar radiation management as well). Today there are many in society who recognize the need to take urgent action to solve climate change, and there are also many who do not. No matter the rationale stated to perpetuate inaction (e.g. those who claim to be dismissive, disengaged, doubtful or cautious – you can search to learn about Yale's "Six Americas" framework), these stated rationales do not change the robust findings of climate science or nullify the damages to human, economic, and ecosystem health being inflicted from a changing climate. The GTEM includes practical guidance for developing and implementing demand-side climate solutions to support an increase in the speed and scale at which solutions are implemented.

2. **All energy systems are built with connected building blocks in order to deliver services.** Developing effective DSM solutions requires an understanding of how human-built energy systems are generally comprised of six connected building blocks (see Chapter 1). The transition to cleaner energy can be envisioned as a need to reconstruct these connected building blocks for demand-side systems (e.g. see Figure 2.9 vs. Figure 2.10). The GTEM is written from the perspective that all solutions should be developed and implemented with careful attention to maintaining, if not improving, delivered energy services (see Chapters 1 and 3).
3. **People are part of energy systems.** Developing effective DSM solutions requires recognition that people are part of the system. This means, for example, that common practice should be to conduct baseline and post-implementation surveys of facility users and O&M staff interviews (see Chapter 3), think carefully about how to best organize an effective energy team (see Chapter 1), and in general to apply proven behavior change methods (see Chapter 7). Most energy and carbon management teams arguably should include individuals with relevant social sciences expertise, to complement the usual engineering, business, economics, finance, and O&M expertise.
4. **DSM will be more effective by embracing and applying clear definitions for key concepts.** In particular, there are four key concepts: Energy management, energy efficiency, energy conservation, and energy savings (see Chapter 1). There is no benefit, and sometimes unintentional harm, by confusing stakeholders by using the terms "efficiency" and "conservation" interchangeably, or by always assuming energy efficiency means energy savings will be realized (yes, it is true that a more energy efficient system may not result in verified energy savings due to rebound effects and eligibility rules for subsidies). Similarly, clear definitions are important in carbon management (see Chapter 2). Obfuscation is unnecessary and can be an enemy of progress.
5. **Integrated solutions are needed.** Successful DSM initiatives are designed to develop integrated solutions which deliver multiple benefits such as energy savings, GHG reduction, climate adaptation, improved energy equity, and/or reduced air pollution exposures while maintaining or improving delivered services (see Figure 3.1). More often than not, solutions which narrowly focus solely on, say, GHG reduction (or only utility bill savings, or only efficiency, or only adaptation) are less

likely to be funded and implemented by comparison to solutions which realistically offer multiple benefits.

The GTEM is written as an educational resource for professional and university programs and for everyday use by practitioners. Each chapter includes at least five practice problems. Solutions to these practice problems are available in a separate GTEM Solutions Manual. Every chapter is written in a manner aiming to be as useful as possible for practitioners who contribute to the assessment, design, implementation, and maintenance of real-world DSM solutions. All chapter authors were challenged with the task to: a) provide the fundamentals of each topic so that someone new to the topic can understand the basics, and b) additionally provide insights useful for DSM veterans. We think that the GTEM delivers on this challenge.

The GTEM is organized by providing key concepts, definitions, and methodologies mostly in the initial chapters. The basics of energy systems and climate change are provided in Chapters 1 and 2, respectively. No matter what one's role is in DSM, you will likely benefit by knowing something about energy and decarbonization studies (Chapter 3), utility rates (Chapter 4), conventional economics (Chapter 5), finance (Chapter 6), human behavior (Chapter 7), resiliency (Chapter 8), commissioning (Chapter 16), and maintenance (Chapter 21). The remaining chapters (9-15 and 17-20) cover essential technologies including: HVAC, heat pumps, building envelope, lighting, electric systems, motors and drives, controls, electric and thermal storage, solar energy, thermal energy networks, and industrial systems. Collectively, these technology-focused chapters cover the fundamentals for technologies and practices needed for most DSM solutions.

There are no significant prerequisites for using the GTEM effectively. However, having practical experience with real-world DSM solutions will help readers gain more from the material. For the technology-focused content that makes use of engineering sciences (this includes electric power, control theory, heat transfer, fluid mechanics, and thermodynamics), it is assumed that readers have completed high school level algebra, chemistry, and physics.

A common question from readers is: what is the most effective way to use this book, to gain the most from it? The GTEM is a **problem-focused resource.** From this perspective, the best starting point is to clearly define what the problem is for which you want to develop a solution. All solutions likely involve technology of some kind, although some solutions may be more people-focused, e.g. to implement a behavior-change initiative such as SEM. Defining the problem means deciding upon the project objectives, and defining the systems and technologies of interest. Section 3.1.2 and Table 3.1 will be helpful. In general, it is recommended that readers find the time to read the first 8 chapters, Chapter 16, and Chapter 21, as this content is likely relevant for most solutions. Then, depending on what systems or technologies are included or excluded from the scope of your project, the technology-focused chapters 9-15 and 17-20 can be chosen. For example, if your project is focused on a specialized industrial process, then Chapter 20 will be most useful. If the focus is a heat pump conversion for a building, then Chapters 9, 10, and 11 are important. If solar PV is planned, then be sure to read Chapter 18, etc.

The GTEM incorporates both English (Inch-Pound) and SI (metric) units throughout. This means that for each chapter where units are used, both systems are included in the chapter content. However, **not every practice problem, example, table or chart will use both sets of units.** The reason for incorporating both systems of units is because practitioners in the U.S. and Canada (as well as other countries like the UK) often encounter both systems of units, and thus practitioners are essentially forced to work with two systems. Richard Feynman, the 20th century Nobel prize winning physicist, complained over 60 years ago of the "idiocy of all the different units which they use for measuring energy[1]" and advocated for using one system of units and one unit for each variable (e.g. energy, mass, momentum, etc.). Unfortunately, this "idiocy" of using multiple units and two systems of units still exists today. It would be preferable if one system of units could be agreed-upon globally, but the reality is that we remain stuck with two systems.

Acknowledgements for contributions to each chapter are provided at the end of each chapter. Additionally, we would like to acknowledge input and guidance provided by Olivier Cappon (Canada) and Ethan Rodgers (U.S.).

We believe it is important to note that the GTEM was written ~100% by humans! Of course, the team of co-authors made use of AI for tasks such as synthesizing information, generating answers to specific questions (which were subsequently verified by the authors), or generating outlines. However, the actual writing of content was not done by AI.

Substantial effort has been made to minimize errors. But expecting zero errors is unrealistic, and so the publisher maintains a website for errata at: https://www.riverpublishers.com/book_details.php?book_id=1532. The

[1]Feynman (1964) "The Character of Physical Law" MIT Press

formula sheet in Appendix A is also freely available https://www.riverpublishers.com/book_details.php?book_id=1532. Any reader who finds significant errors or who wants to provide feedback as to how to help improve the GTEM (we anticipate a 10th edition), please write to the editor/co-author, Eric Mazzi, at mcs@telus.net.

Eric Mazzi, editor/co-author, and the GTEM team of co-authors

List of Figures

List of Tables

List of Abbreviations

AABC	Associated Air Balance Council
AACEI	Association for the Advancement of Cost Engineering – International
AC	Alternating current
ACEEE	American Council for Energy Efficient Economy
ACH	Air changers per hour
ACRS	Accelerated cost recovery system
AEE	Association of Energy Engineers
AHU	Air handling unit
ANSI	The American National Standards Institute
ASHRAE	American Society of Heating, Refrigerating and Air-Conditioning Engineers
ASTM	American Society for Testing and Material
ATB	Annual technology baseline
ATCF	After tax cash flows
BAS	Building automation system
BCA	Building Commissioning Association
BCCB	Building Commissioning Certification Board
BCR	Benefit/cost ratio
BCxA	Building Commissioning Association
BECx	Building envelope commissioning
BECxP	BECx provider
BEV	Battery electric vehicle
BLCC	Building Life Cycle Cost
BOD	Basis of design
BPT	Balance point temperature
BTCF	Before tax cash flows
BTM	Behind-the-Meter
C&I	Commercial and Industrial
CAES	Compressed air energy storage

CBE	Center for the Built Environment
CCF	Certified Commissioning Firms
CCP	Certified Commissioning Professional
CCT	Correlated color temperature
CDD	Cooling degree days
CDR	Carbon dioxide removal
CEM	Certified Energy Manager
CFR	Current facility requirement
CHP	Combined heat and power
COP	Coefficient of performance
CPP	Critical peak pricing
CPUC	California Public Utilities Commission
CRF	Capital recovery factor
CRI	Color rendering index
CSP	Concentrated solar power
CU	Coefficient of utilization
CUSUM	Cumulative sum of differences
Cx	Commissioning
CxA	Commissioning authority
CxP	Commissioning provider
DC	Direct current
DCV	Demand control ventilation
DDC	Direct digital controls
DE	District energy
DER	Distributed energy resource
DHW	Domestic hot water
DLC	Design Lights Consortium
DOAS	Dedicated outdoor air system
DoD	Depth of discharge
DOE	Department of Energy
DR	Demand response
DTT	Double twin-tube
DX	Direction expansion
EBCx	Existing building commissioning
ECI	Energy cost intensity
ECM	Electronically commutated motor
EEM	Energy efficiency measure
EER	Energy efficiency ratio
EISA	Energy Independence and Security Act

EM&V	Evaluation, measurement, and verification
EMIS	Energy management information system
EnMS	Energy management systems
EPDM	Ethylene propylene diene monomer
ER	Ellipsoidal reflector
ERV	Energy recovery ventilator
ESCO	Energy Services Company
EUI	Energy use intensity
EV	Electric vehicle
EVSE	Electric vehicle supply equipment
FDD	Fault detection and diagnostics
FEPA	Federal Energy Policy Act
FIM	Facility improvement measure
GESDB	Global Energy Storage Database
GHG	Greenhouse gas
GWP	Global warming potential
HDPE	High-density polyethylene
HEPA	High-efficiency particulate air
HHV	Higher heating value
HID	High-intensity discharge
HO	High output
HPS	High-pressure sodium
HRV	Heat recovery ventilator
HSPF	Heating seasonal performance factor
HVAC	Heating ventilating and air conditioning
IAQ	Indoor air quality
ICLR	Institute for Catastrophic Loss Reduction
IEA	International Energy Agency
IEC	International Electrotechnical Commission
IEER	Integrated energy efficiency ratio
IEQ	Indoor environmental quality
IES	Illuminating Engineering Society
IIBEC	International Institute of Building Enclosure Consultants
IPLV	Integrated part load value
IPMVP	International Performance Measurement and Verification Protocol
IRL	Incandescent reflector lamp
IRR	Internal rate of return
ISO	International Organization for Standardization

KPI	Key performance indicator
LCC	Lifecycle cost
LDC	Load duration curve
LED	Light-emitting diode
LFP	Lithium iron phosphate
LH	Low-heat
LHV	Lower heating value
LPD	Lighting power density
LPS	Low-pressure sodium
LULUCF	Land use, land use change, and forestry
M&V	Measurement and Verification
MACRS	Modified accelerated cost recovery system
MARR	Minimum attractive rate of return
MBCx	Monitor-based commissioning
MERV	Minimum efficiency reporting value
MLF	Motor load factor
MPP	Maximum power point
MT&R	Monitoring, targeting, and reporting
NCCx	New Construction Building Commissioning
NEBB	National Environmental Balancing Bureau
NEI	Non-energy impact
NEMA	National Electrical Manufacturers Association
NIBS	National Institute of Building Sciences
NLC	Networked lighting control
NOCT	Normal operating cell temperature
NRC	National Research Council of Canada
NREL	National Renewable Energy Laboratory
O&M	Operation and maintenance
OCx	Ongoing commissioning
OLED	Organic light-emitting diode
OPEX	Operational expenditure
OPR	Owner's project requirements
PACE	Property Assessed Clean Energy
PAR	Parabolic aluminized reflector
PCB	Polychlorinated biphenyl
PCIC	Pacific Climate Impacts Consortium
PF	Power factor
PHEV	Plug-in hybrid vehicle
PID	Proportional plus integral plus derivative

PIEVC	Public Infrastructure Engineering Vulnerability Committee
PTAC	Packaged terminal air conditioners
PTHP	Packaged terminal heat pumps
PUC	Public utilities commission
PV	Photovoltaic
RCP	Representative concentration pathway
RCx	Retro-commissioning and re-commissioning
RGB	Red-green-blue
RNG	Renewable natural gas
ROI	Return on investment
RTU	Roof top unit
SBTi	Science Based Target Initiative
SEM	Strategic energy management
SEP	Superior Energy Performance
SHGC	Solar heat gain coefficient
SIR	Savings/investment ratio
SLH	Super-low heat
SMB	Small-to-medium business
SOH	State of health
SPP	Simple payback period
SSL	Solid-state light
TEN	Thermal energy network
TES	Thermal energy storage
THD	Total harmonic distortion
TLED	Tubular LED
TMY	Typical meteorological year
TOU	Time of use
TT	Twin-tube
UBC	University of British Columbia
UHN	University Health Network
UV	Ultraviolet
VAV	Variable air volume
VFD	Variable frequency drive
VHO	Very high output
VLH	Very low heat
VOC	Volatile organic compound
VRF	Variable refrigerant flow
WACC	Weighted average cost of capital
WSHP	Water-source heat pumps

1

Energy Management Basics

Eric Mazzi

Mazzi Consulting, Canada

1.1 A Resource for Practitioners and Learners

The Guide to Energy Management 9th edition, is a resource for energy management practitioners and learners from a wide range of professions and backgrounds. It is written to introduce the basics of demand-side energy and carbon management for technologies and methodologies needed by practicing professionals today. Each chapter is written to help newcomers to energy management understand key concepts, while also providing insights and key solutions so that experienced professionals can be more effective. Probably the most valuable aspect of this book is the breadth of topics covered, which is a necessity for today's professionals to solve real-world energy and carbon management problems.

The scope of this book remains similar to earlier editions, focusing on the demand-side of energy systems, emphasizing end-uses in facilities in the residential, commercial, and industrial sectors. This book also focuses more on retrofits rather than new construction. Regardless, most content applies to both applications. The geographic focus remains on the U.S. and Canada. However, the material in this book is widely applicable in any modern economy across the globe.

One of the realities of energy and carbon management in the U.S. and Canada is that practitioners will encounter both English (imperial) and international system (SI) units. As such, this book is written using both systems of units. Each chapter that includes formulas and calculations provides some examples and practice problems covering both systems of units, although not every calculation uses both systems of units. Appendix A includes common energy management formulas used in this book in both English and SI units, as well as useful unit conversions between the two systems.

For economic analyses and costs, the default in this book is that everything should be assumed to be in U.S. dollar currency, unless stated otherwise.

Last but not least, the Guide to Energy Management focuses on technologies and practices that are proven to be effective and commercially available today. Technologies and practices that are mainly in a research and development phase are generally excluded.

1.2 The Need for Energy Management

Before summarizing why energy management is needed, we should begin by defining energy management. The phrase energy management means different things to different people. In this book, with the chosen focus on the demand-side of energy systems, we define energy management as:

> *The efficient, sustainable, and cost-effective use of energy in end-use systems to deliver the quantity and quality of services that users need and want.*

Energy management is important for many reasons, and there is no single issue driving the need for more effective energy management. We believe it is important for practitioners to be aware of the many reasons why energy management is needed so they can (a) appreciate that effective energy management is important and meaningful work, and (b) be informed advocates of effective policymaking. Six key issues driving the need for better demand-side energy management are summarized below, listed in no particular order of importance.

1. *Energy systems provide valued and essential services for health, safety, economic, and general well-being of people and businesses.* Energy management is not just about cutting consumption and costs. It is about maintaining and improving services while managing consumption and costs. One basic example of an energy service is illumination so that people can read, learn, and earn a living. HVAC systems should provide air that is within a safe and comfortable range of temperature, humidity, and contaminants. Beyond basic health and safety, studies show that worker productivity improves with quality workplace environments such as effective lighting and thermal comfort. In industrial applications, process heating is an example of a service that must support the manufacturing of products that meet quality specifications and support reliable rates of production output. More efficient and effective use of energy can provide multiple benefits to companies including utility bill cost savings,

progress on corporate sustainability goals, improved worker productivity, improved production reliability, and regulatory compliance. One of the earliest pioneers to identify these benefits and the need for energy efficiency was Art Rosenfeld (Rosenfeld, 1999). The importance of energy for businesses is reinforced with this conclusion from a Harvard Business Review article (Winston et al., 2017) "... the majority of firms approach energy as merely a cost to be managed. This is a strategic mistake that overlooks enormous opportunities to reduce risk, improve resilience, and create new value." To quote an assessment for industry in the U.S. (Ungar and Whitlock, 2019), "In addition to cutting emissions and saving money on energy bills, energy efficiency measures can often increase productivity and minimize waste by optimizing processes and equipment, and thus can strengthen competitiveness."

2. *The benefits and costs of energy services in households are distributed unequally and unfairly*. The issue of equity in energy management has rightfully gained traction in recent years. Consider the data in Figure 1.1, which shows disparities in household energy costs for the U.S. The data illustrate that non-whites, low-income, older people, renters, and rural populations generally incur higher burdens of energy costs as a fraction of their income. Energy poverty is often defined as energy cost exceeding 6% of household income. In British Columbia, for example, 272,000 households (15% of total) experience energy poverty, and 17,000 of those households identify as Indigenous (EcoTrust, 2020). It might be tempting to conclude that energy burdens are mainly a result of groups with roughly equal energy use, with some groups having a larger energy burden due to lower income. But research has confirmed that the energy use intensity (consumption per unit floor space) of low-income and non-white households is over 2.5 times greater than high-income and white households (Tong et al., 2021). This means that energy burden is not primarily due to people earning too little money. The energy intensity data indicate that low efficiency homes are probably the biggest factor (e.g., poor insulation, inefficient furnaces, and large air leakages). Effective energy management practices can help alleviate these inequities for vulnerable groups (DOE, 2018).
3. *The production, storage, and distribution of commercial energy results in negative environmental impacts*. Managing the production, transport, and use of energy is critical to the effective management of environmental impacts, the most critical being global climate change. To meet climate change goals and regulations, energy management professionals

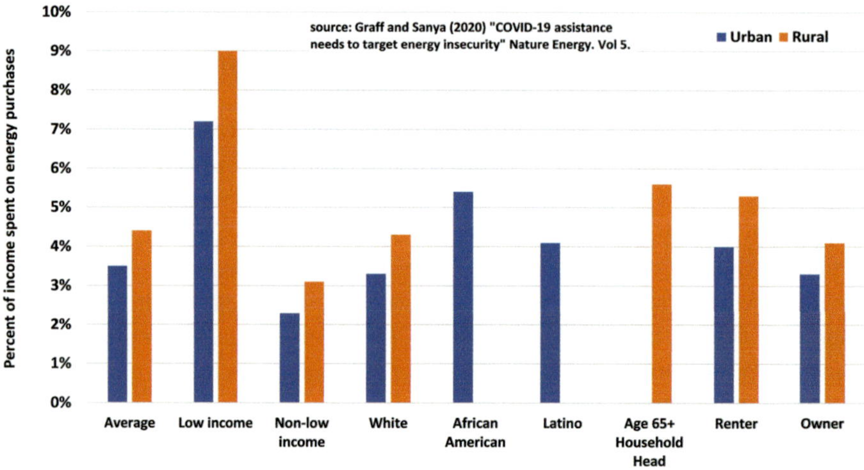

Figure 1.1 Percent of income spent on energy for U.S. households (Graff and Sanya, 2020).

are tasked not only with developing energy-efficient systems but also reducing or eliminating greenhouse gas (GHG) emissions (see Chapter 2). As a direct consequence of climate change, facility energy systems are subjected to extreme events that are increasing in intensity, duration, and frequency. Such events are characterized by more extreme winds, precipitation, and temperatures than have occurred in the past. As a result, facility energy systems need to be designed, operated, and maintained with greater attention to resiliency than ever before (see Chapter 8). Climate change is not the only concern. Fossil-fuelled energy systems contribute to biodiversity losses, ocean acidification, food insecurity, and many other impacts.[1] The combustion of fuels also releases local and regional air pollutants (see Chapter 20), which can be controlled by choices of fuels and end-use technologies.

4. *Americans and Canadians consume, on average, the largest quantities of energy, but we are no better off.* Figures 1.2 and 1.3 provide statistics for energy use per person for a sample of relatively wealthy countries, as well as different metrics of well-being (IEA, 2024; OECD, 2024). One of the key insights from these two charts is that Americans and Canadians are among the highest consumers of energy globally. But are we better off for it? The average employment rates are one indicator that we are not better off. Consider that countries such as the UK, France, and

[1] For a more comprehensive picture of environmental impacts, see the planetary boundaries analysis at www.stockholmresilience.org/research/planetary-boundaries.html.

Japan consume less than half the energy per person, but have roughly equal employment rates. Figure 1.3 shows a similar finding for life satisfaction. One could cite similar statistics for many other measures of welfare such as leisure time, average life expectancy and other health statistics, and access to education – with essentially the same conclusion. These data provide evidence that Americans and Canadians can substantially reduce energy consumption through effective energy management without harm to our economies, health, or general welfare. In fact, many

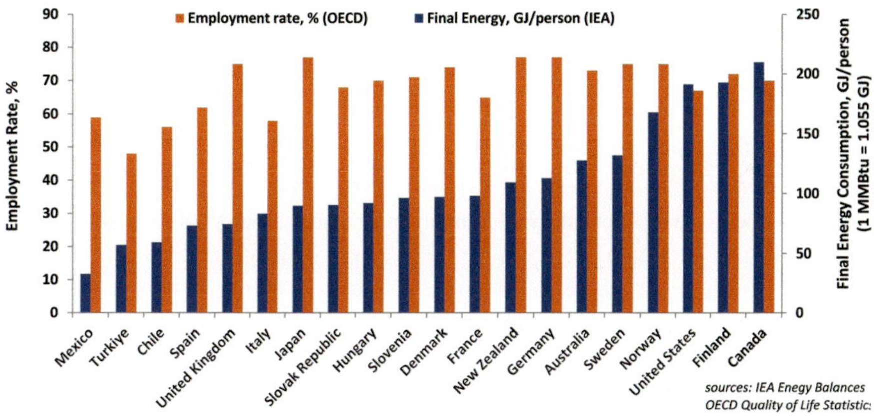

Figure 1.2 Final energy per person and employment rates for a sample of relatively wealthy countries.

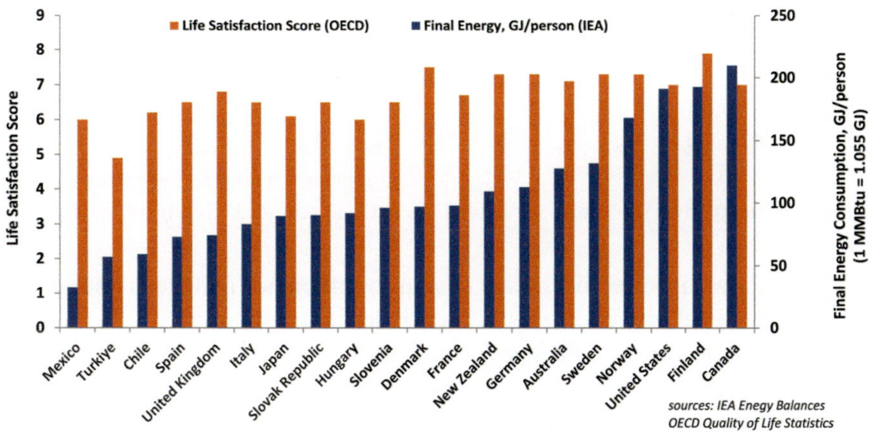

Figure 1.3 Final energy per person and life satisfaction score for a sample of relatively wealthy countries.

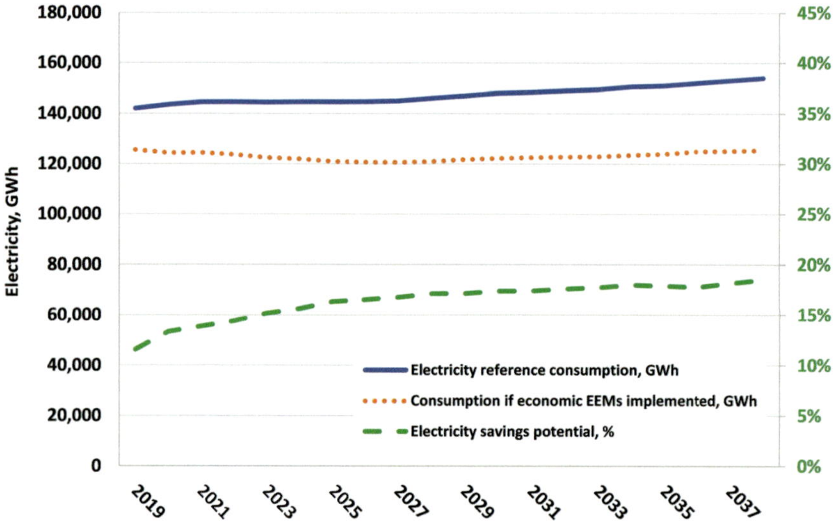

Figure 1.4 Energy savings potential for electricity in Ontario, Canada (Navigant, 2019).

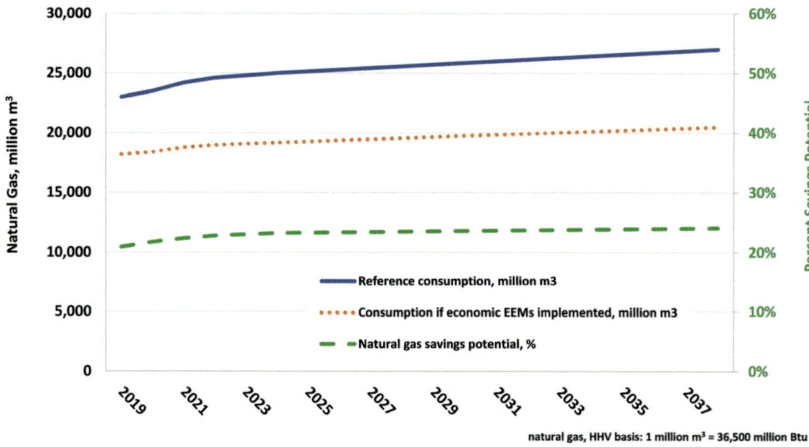

Figure 1.5 Energy savings potential for natural gas in Ontario, Canada (Navigant, 2019).

studies show that effective policy design for efficiency and clean energy can produce net improvements in economics and welfare.[2]

5. *Cost-effective energy efficiency measures (EEMs) often remain unimplemented.* Consider the case of Ontario, Canada as shown for electricity in Figure 1.4, and natural gas in Figure 1.5 (Navigant, 2019).

[2] For example, see www.iea.org/reports/multiple-benefits-of-energy-efficiency.

These are careful estimates of how much energy could be saved annually if all measures were implemented, which are assessed to be both technically viable and cost-effective. These results are from a "conservation potential" study, which is a common analysis performed by energy utilities. The study estimates that over 24% of total natural gas consumption and over 18% of total electricity consumption could be eliminated through cost-effective investments in commercially available technologies and practices, while maintaining economic growth and delivering required services. In the bigger picture, the potential for reducing energy consumption is larger than indicated by these conservation potential studies. Using integrative design (see Chapter 20) by first matching services to needs, then minimizing system losses (e.g., short length and large diameter pipes, thick insulation, advanced windows), followed by efficient if not eliminated mechanical equipment can increase energy savings to 80% or more for many applications (Lovins, 2019; Cullen, 2011). Some of the key reasons why measures remain un-implemented are described in Chapter 6 and Chapter 7.

6. *Timing of electricity consumption in end-use systems needs to be more flexible*. On the supply-side of energy systems, the generation of electricity is increasingly more variable as shares of renewables grow, especially solar and wind. What this means is that there is an increasing need for flexible demand for end-uses of electricity. As an example, Figure 1.6 provides a comparison for a sample summer day versus a winter day for California's grid using 5-minute data (California ISO, 2021). This

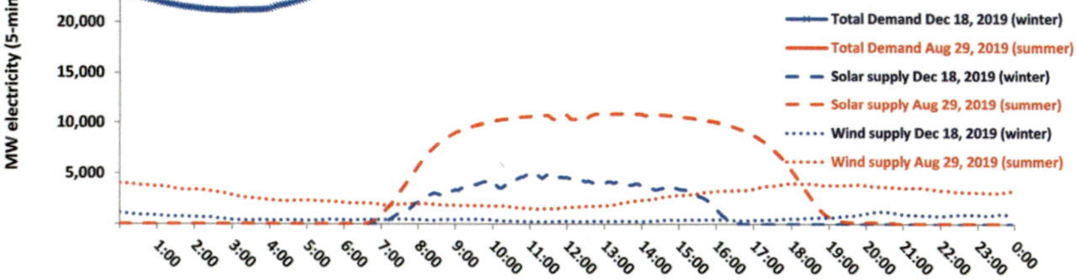

Figure 1.6 Example 24 hours of 5-minute data for electric grid power demand compared to solar and wind supply in California.

24-hour sample dataset illustrates the challenges of matching electricity supply and demand (note that grid demand is the net effect of customer loads and customer-side solar PV generation). Note that solar and wind supplies do not match the timing of demands in either winter or summer. Peak wind and solar supplies do not occur during either the morning or evening peaks in demand. Significant demand persists overnight, when solar is not available (unless solar happens to be coupled with storage) and wind is insufficient. Figure 1.7 provides another timescale, with monthly data on electricity generation for California (EIA, 2021). Note that the variation in supply of renewables is offset by supplies of electricity from natural gas generators. If the goal is to eliminate natural gas (and all fossil-fuelled electricity) from the grid, it would be very challenging for supply to match demand with 100% renewables. These observations do not capture all the key issues, and the situation in California is different from other jurisdictions because every region has unique sources and mix of end-use demands. But the general trend is that the share of variable electricity generation will increase. While part of the solution is on the supply-side with solutions such as increased storage and fossil fuel generators with carbon capture, a key to making carbon-free electricity supplies viable is through flexible demand for end-uses. Effective energy management now requires demand-side solutions that work with the reality of increasingly variable electricity supplies.

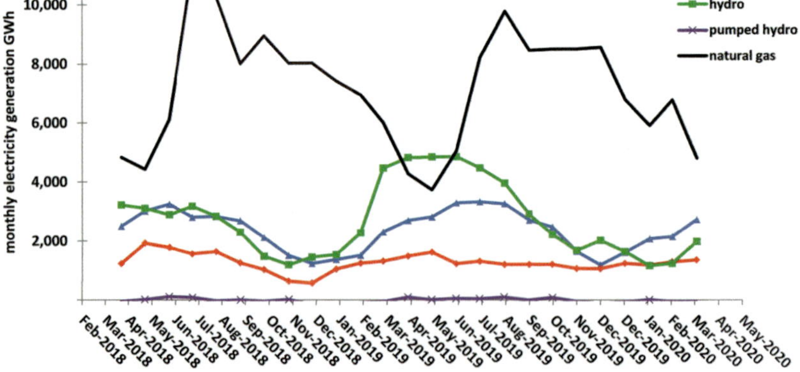

Figure 1.7 Example monthly average data for electricity supply in California.

1.3 Energy System Fundamentals and Key Definitions

In order to be effective, energy management professionals need to clearly understand the basics of energy systems. For example, what is the dividing point between the supply-side and demand-side? How is energy efficiency different from energy conservation? And what is meant by distributed energy resources? Energy management professionals also need to be proficient in energy conversions using both English and SI units. These and other related skills and concepts are addressed here.

1.3.1 Energy forms and units

Physicists tell us that energy is the "ability to do work," which is the ability to exert a force across a distance. Energy comes in different forms such as: electrical energy, chemical energy in combustible fuels, kinetic energy of moving mass such as air or water, electromagnetic radiation such as sunlight, and mechanical energy such as a rotating motor shaft. A critical concept from physics is that there is a natural law called *conservation of energy*. Conservation of energy is based on the physical law that energy is not created or destroyed. Energy may convert from one form to another, such as combustion of natural gas to release heat or the conversion of uranium mass to thermal energy in a nuclear reaction, but the total energy remains the same. For a brief, non-mathematical introduction to the conceptual foundation of energy, readers are referred to chapter 4 of Feynman's Lectures on Physics titled "What is Energy?" (Feynman et al, 1963). For a synthesis of energy use in society, conversions, and history of transitions, readers are referred to Vaclav Smil's writings (Smil, 2016; Smil, 2017).

In the practice of energy management, the law of conservation of energy has very practical implications. Using this law, we can measure energy inputs and outputs for end-use systems in its various forms, and make reliable estimates of energy flows such as purchased energy, energy losses or waste, and the useful energy that produces desired services. In this book, we will make many such practical calculations.

Performing basic calculations is a fundamental skill in energy management. It is not a skill limited to only engineers or physical scientists either. Energy is inherently a numbers game, and all practitioners need to perform basic calculations at various times to be effective. The good news is that a wide range of useful calculations can be made successfully by using introductory algebra and applying high school physics and chemistry. In fact, a key objective of this book is to help practitioners from a wide range of backgrounds to make useful calculations for many common systems (motors, boilers, lighting, thermal storage, heat pumps, etc.).

To begin, it is essential to be able to make unit conversions within and between the English and SI systems. Table 1.1 provides common conversion factors encountered in energy management.

The heating or calorific value of fuels is commonly needed for making everyday calculations in energy management. Table 1.2 provides values for some of the most common fuels, while the appendix includes a more complete table. When dealing with fuel combustion, it is important to distinguish between higher heating values (HHV) and lower heating values (LHV). HHV assumes that water in the exhaust is condensed out as a liquid, while LHV assumes water is exhausted as a vapor (also see Chapter 20). HHV is commonly used to specify the rated efficiency of boilers or HVAC equipment (but the specific basis should always be stated). LHV is commonly used for the rated efficiency of gas turbines and internal combustion engines. It

Table 1.1 Common conversions for English and SI units.

Parameter	Conversion	Common usage or example
Energy	1 kWh = 3,412 Btu	HVAC equipment ratings
Energy	1 GJ =277.78 kWh	Natural gas and electricity common units
Energy	1.055 GJ = 1 MMBtu = 1×10^6 Btu	Energy English-SI conversions
Energy	1 kWh =3.60 MJ	Energy English-SI conversions
Energy	1 Btu = 0.001055 MJ	Energy English-SI conversions
Energy	1 therm = 100,000 Btu	Natural gas in English units
Power	1 hp = 0.7457 kW	Motor power calculations
Power	1 kW = 3,412 Btu/hr	Power English-SI conversions
Power	1 MBH = 1,000 Btu/hr	Boiler ratings in English units
Power	1 ton = 12,000 Btu/hr = 3.517 kW	Thermal power for cooling equipment
Length	1 m = 3.2808 ft	Length English-SI conversions
Area	1 m^2 = 10.76 ft^2	Area English-SI conversions
Volume	1 m^3 = 35.314 ft^3	Volume English-SI conversions
Volume	1 ft^3 = 7.4805 U.S. gal	English unit volume conversions
Volume	1 bbl = 5.615 ft^3	English unit volume conversions
Volume	1 MCF = 1,000 ft^3	Natural gas in English units
Mass	1 lbm = 0.4536 kg	Mass English-SI conversions
Heat transfer	1 W/m^2 °C = 0.1762 Btu/hr-ft^2-°F	Thermal resistance English-SI conversions
Temperature	°C = (°F − 32) /1.8 °F = 1.8 (°C) + 32	Temperature English-SI conversions

Table 1.2 Calorific values of fuels (English and SI units).

Fuel (HHV or LHV)	English units	SI units
Natural gas (HHV)	1,034 Btu/ ft^3 (or use 1,000 Btu for rough approximations)	38.5 MJ/m^3
Natural gas (LHV)	941 Btu/ft^3	35.1 MJ/m^3
Liquified natural gas (HHV)	625,410 Btu/ft^3	23,300 MJ/m^3
Hydrogen (HHV)	60,920 Btu/lbm	141.7 MJ/kg
Crude oil (HHV)	5,722,000 Btu/bbl	45.9 MJ/kg
Gasoline (HHV)	122,700 Btu/gal	34,200 MJ/m^3
#2 fuel oil or diesel (HHV)	138,400 Btu/gal	38,600 MJ/m^3
Propane (HHV)	90,000 Btu/gal	25,100 MJ/m^3
Wet wood, 20% H_2O (HHV)	6,400 Btu/lbm	14.9 MJ/kg
Dry wood, 0% H_2O (HHV)	8,000 Btu/lbm	18.6 MJ/kg
Subbituminous coal (HHV)	10,490 Btu/lbm	24.4 MJ/kg
Landfill gas (HHV)	517 Btu/ft^3	19.3 MJ/m^3
Municipal solid waste (HHV)	5,070 Btu/lbm	MJ/kg

should be noted that calorific values are not exact, because the true value varies with the fuel composition (see Chapter 20). There are always at least minor differences in fuel composition that vary over time and location, but the values in the table are useful for most calculations for the U.S. or Canada. Additional calorific values are provided in the appendix.

1.3.2 Energy system building blocks and connectors

The focus of this book is on the demand-side of energy systems. But what exactly is the demand-side of an energy system, and where is the demarcation between supply and demand? What are the elements of energy systems and how are the elements connected? Before we describe the building blocks of energy systems, it is important to emphasize that practitioners need to understand how energy systems work more than ever. Only a decade or two ago, for example, demand-side professionals needed to know relatively little about supply-side systems because fossil fuel supplies can generally modulate to meet demand. However, those days are disappearing. This is because policy goals globally are targeting massive changes of historical proportion. This is not hyperbole. As introduced above, described further in Chapter 2, and mentioned throughout this book, the overarching goal is to transition energy systems to produce net-zero GHG emissions globally by 2050, if not sooner (and carbon negative soon after). If this goal is to ever be achieved, it means that energy management professionals need to contribute to fundamental reconstruction of energy systems. And to understand how systems need to be reconstructed, one needs to understand how energy systems generally work. The diagram in Figure 1.8 and definitions in Table 1.3 provide a framework to describe all human built (i.e., anthropogenic) energy systems (derived based on Scott, 1994).

Figure 1.8 illustrates how energy systems convert Earth's energy resources to ultimately deliver services. Constructed energy systems begin, at one end, by extracting energy out of Earth's ecosystem. Examples include mining for raw coal, drilling for sour natural gas, harnessing the fluid power of wind, and capturing solar radiation. This energy is transformed into commercial forms, usually described as energy carriers or secondary energy. Modern energy carriers used widely today include electricity, natural gas, pulverized coal, gasoline, and jet fuel. Based on goals to mitigate the effects of climate change, many jurisdictions have adopted policies to reduce, if not eliminate, the use of fossil fuels (see Chapter 2). Low-carbon or zero carbon carriers have gained favor such as electricity from solar or wind resources, electricity from fossil fuels with carbon capture and storage, renewable biofuels (liquids or gases), and hydrogen produced in carbon-neutral processes. Because energy resources are often located far away from end-use systems, energy carriers are delivered to end-use systems by transportation, transmission, and/or distribution systems.

The output side of the diagram in Figure 1.8 illustrates the key purpose of energy systems, which is to provide energy services. Energy services are ultimately delivered by end-use equipment (or service technologies), which convert energy carriers into desired services. *An important concept in energy management is to maintain a focus on reliably delivering the quantity and quality of services that people and businesses need and want, while minimizing energy consumption, cost, and undesirable impacts.* This book

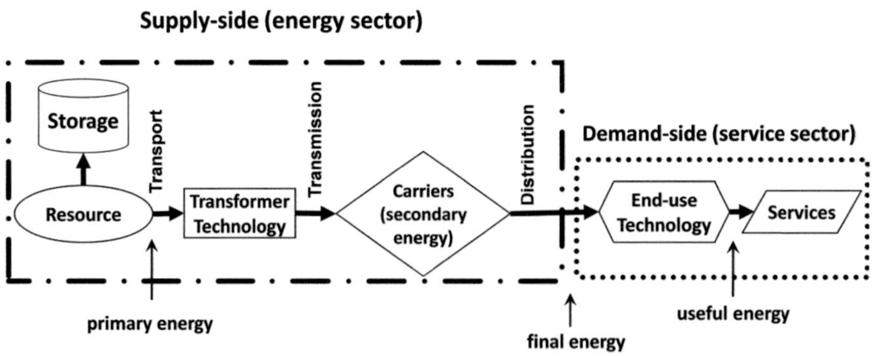

Figure 1.8 Diagram of human built (anthropogenic) energy systems.

Table 1.3 Energy system building blocks and connectors: descriptions and examples.

Element type	Element	Description	Examples
Building block	Resource	Energy forms stored or moving through Earth's ecosystems	Uranium ore, solar radiation, biomass, raw coal, wind, sour natural gas, raw crude oil
	Storage	Energy in a static or stable state that can be held for periods of time, and discharged to help match supply and demand	Hydroelectric reservoir, natural gas in a salt cavern, electric energy in a battery, thermal energy in a hot water tank
	Transformer technology	Converter of energy from one form to another	Nuclear power station, petroleum refinery, solar PV panel, wind turbine, natural gas sweetening plant
	Energy carriers	Energy transformed into commercial forms. Also known as secondary energy	Electricity, wood pellets, hydrogen, sweet natural gas, diesel fuel, steam
	End-use technology	Converts energy carrier into a desired service. Also known as service technology	Gas stove, arc furnace, smart phone, automobile, heat pump, lumber kiln
	Energy services	Desired products and services that require consumption of energy carriers	Cement production, thermal comfort, communication, lumber drying, illumination, entertainment
Connector	Transport systems	Systems to move solid, liquid, or gaseous carriers over short or long distances. Usually ships, trains, or trucks	Super-tanker carrying crude oil, train moving raw coal, truck hauling wood chips
	Transmission	Infrastructure to move energy carriers at high pressure or voltage, typically longer distances	High-voltage conductors, high pressure natural gas pipelines
	Distribution	Infrastructure to transport carriers at low pressure or voltage, typically shorter distances	Low-voltage conductors, thermal energy network hot or chilled water piping, low pressure natural gas pipelines

is written on the premise that keeping this approach in mind is critical to successful energy management.

The framework in Figure 1.8 and Table 1.3 emphasizes three important energy forms along the conversion chain, which are commonly quantified:

1. *Primary energy*[3] is defined as energy forms at or near the point of extraction of resources. Organizations such as the International Energy Agency

[3] This definition of description of primary energy suffices for most situations. For a more detailed definition and analysis methods for primary energy, see chapter 8 of Blok and Nieuwlaar (2021). For primary energy accounting for renewable energies, passive house methods provide another approach. See: https://passipedia.org/basics/energy_and_ecology/primary_energy_renewable_per

and government agencies worldwide commonly produce statistics on primary energy consumption. The precise accounting of primary energy will differ based on the agency, but the general definition is as we present it here. The term "source energy" is sometimes used in energy accounting or benchmarking, which is conceptually the same as primary energy (EPA, 2020; ASHRAE, 2023c; SEP, 2019). In practice, primary energy and source energy are counted differently depending on the source of energy. For example, non-fossil fuel electricity sources such as nuclear, hydroelectricity, or other renewables are counted differently than fossil sources.

2. *Final energy*[4] is defined as energy carriers delivered to end-use systems. In our framework and in general, final energy can be thought of as the metered energy purchased near the point of use such as electricity and natural gas purchased from utility companies, gasoline purchased from a service station, and steam from a district energy system. The term "site energy" is sometimes used in energy accounting or benchmarking, which is essentially equivalent to final energy.
3. *Useful energy* is defined as energy that directly contributes to the services delivered by end-use systems. Examples include thermal energy delivered by HVAC systems, and the embedded energy in manufactured steel materials.

To reinforce the concept of final energy and useful energy, consider the provision of illumination by lighting systems as an example. The approximate breakdown of radiant energy and heat from some common lighting technologies are provided in Table 1.4. For lighting, the total consumption of the luminaire, lamp, and ballast is the final energy (in a facility, one could technically specify the system boundary to also count the energy consumed by transformers, conductors, and lighting controllers as part of the final energy). The visible light delivered is useful energy because it provides illumination. One complication in this example is that, in some circumstances, part or all of the non-visible energy might be considered useful if this energy contributes to space heating as an "interactive effect" (see Chapter 3 for a brief introduction). However, in general, what types of final energy consumed by end-use technologies, which does not directly produce the desired energy

---

[4] This definition of description of final energy suffices for most situations. For facilities that produce, store, and/or sell commercial energy (e.g., electricity or fuels), a more detailed definition and analysis of final energy may be appropriate. See chapter 2 of Blok and Nieuwlaar (2021).

Table 1.4 Illustration of final energy and useful energy for lighting technologies (PNNL, 2007).

	Lighting technology: *final energy* consumption in Watts		
	Incandescent 100 W	**Metal halide 200 W**	**LED 40 W**
Visible light (*useful energy*)	8% (8 W)	27% (54 W)	15%–25% (6–10 W)
Infrared radiation	73% (73 W)	17% (34 W)	0% (0 W)
Ultraviolet radiation	0% (0 W)	19% (38 W)	0% (0 W)
Heat conduction and convection	19% (19 W)	37% (74 W)	75%–85% (30–32 W)
Total	**100% (100 W)**	**100% (200 W)**	**100% (40 W)**

service, is wasted energy. As described in Chapter 12 on lighting, it is critical to understand that the effectiveness of lighting systems (as with all end-use technologies) must consider parameters to characterize both quantity (e.g., lumens) and quality (e.g., color rendering index) of services.

In this section, we have described a framework for understanding energy systems. There is actually one more important element of energy systems that is not explicitly shown in Figure 1.8 or listed in Table 1.3. That element is people! People are part of the energy system in that people make decisions on what services and products are delivered by energy systems, retrofit decisions, decisions on maintenance, operations, etc. The role of people and human behavior in energy systems analysis has been recognized for decades (Schipper *et. al*, 1989). Understanding and applying proven techniques to influence human behavior can result in more effective energy management practices (see Chapter 7).

1.3.3 Key terminology: Defining energy efficiency, conservation, savings, and DERs

All energy management professionals will likely agree that energy efficiency is important. However, in practice, achieving energy efficiency can be challenging – at least in part – because different definitions of efficiency are often assumed and left unclear or undefined. This lack of clarity can create real or perceived barriers to implementing cost-effective measures, lead to measures that fall short of expectations, result in unintended negative consequences, and/or result in avoidable cost overruns. In practice, if one comes across the term "energy efficiency" used by a colleague or in a publication, it is fair to

Table 1.5 Energy efficiency definitions, metrics, and examples.

Energy efficiency category	**Definition**	**Example or metric**
Thermodynamic	first law efficiency (energy or enthalpy) second law efficiency (exergy or entropy)	Input−output efficiency of a steam boiler Isentropic efficiency of a steam turbine
Physical-thermodynamic	"efficiency" = utility/energy "intensity" = energy/utility	Miles per gallon for a car Tons of product per kWh electricity Liters of gasoline per 100 km for a car MMBtu of fuel per number of units produced
Economic-thermodynamic	"efficiency" = energy/economic value "productivity" = economic value/energy national: energy input ($)/economic value	bbl oil equivalent/GDP for a country GDP per mega-ton of oil equivalent for a country Dollars energy expenditure per unit GDP
Economic ($/$) ratios	facility: energy input ($)/economic value	Dollars of energy cost per unit product sales

ask: what definition is being used? Table 1.5 provides a summary of common definitions of energy efficiency (Patterson, 1996).[5] Energy management professionals should expect to encounter and use the different definitions described in the table.

Another practical definition of energy efficiency is provided by ASHRAE in its procedures for energy audits (ASHRAE, 2011):

> "Efficiency refers to providing the same level of service with equipment that uses less energy. Conservation can apply to measures that offer a lower level of service or comfort. However, the terms are often used interchangeably in practice."

Referring to Table 1.5, this definition is a physical-thermodynamic measure of energy efficiency where the "level of service" is a measure of utility. As introduced at the beginning of this chapter (and reinforced below), it is useful and important to measure the level of service in terms of quantity and quality.

ASHRAE's definition of efficiency also introduces a related term, which is "energy conservation." This phrase has a different meaning than the law of conservation of energy from physics as discussed above. As described by ASHRAE, the term conservation is also used to describe measures that

[5] For another perspective on definitions of efficiency, see Lovins (2004).

reduce energy consumption accompanied by a lower level of service (i.e., reduced quantity or lower quality service). While it is true that the terms efficiency and conservation are often used interchangeably, in this book we avoid confusing the concepts. Successful energy management practices avoid compromising the quantity or quality of services delivered. Conservation is a valid strategy in energy management, such as people compromising their optimal comfort by setting their thermostat temperature lower in winter in order to reduce utility costs. However, to maximize effectiveness, energy management professionals should exhaust all energy efficiency measures that meet cost effectiveness criteria before promoting measures that explicitly require sacrifices or hardships by end-users. If reduced consumption results in a compromise in the quantity or quality of service that users need or want, this should be described as energy conservation and not energy efficiency.

Closely related to energy efficiency is the concept of "energy sufficiency," which is defined as follows (www.energysufficiency.org): "When we talk about energy sufficiency we are talking about a situation in which everyone has affordable access to the energy services they need, in which the energy services we want are equitably shared, and in which the environmental limits of the planet are respected." Unlike the definitions and discussion above for energy management and energy efficiency, the concept of energy sufficiency makes a distinction between needs and wants. We do not employ the use of energy sufficiency throughout this book, in particular because it is very challenging for policy makers or energy management professionals to impose distinctions between needs and wants. There are cases where the concept of sufficiency is employed (even if not named), such as the use of inclining block electricity rates where the first-tier quantity of energy is priced cheaper based on a judgment of what is needed, while the higher priced tiers are for quantities estimated to be for end-uses that are arguably wants.

In general, the use of the phrases "energy conservation" or "energy conservation measure" is avoided in this book, except where it is in reference to a measure that results in inferior services being delivered. The phrase energy efficiency is our preferred term, which aligns with ASHRAE's definition of efficiency as stated above. Furthermore, we also use the term *energy efficiency measure* (EEM), which is also used by ASHRAE in its energy audit standard (ASHRAE, 2023b).

To emphasize the distinction between efficiency and conservation and reinforce our definitions above, here is a quote from Amory Lovins, a leading thinker on energy systems (Yale Environment 360, 2008; Lovins, 2004):

> "... don't call it 'conservation.' Lovins, the co-founder and chairman of Rocky Mountain Institute, says that word connotes 'privation, discomfort and curtailment.' By contrast, 'efficiency' means 'doing more and better with less energy and money, but with more brains and technology,' he says."

"Energy savings" is another term that is closely linked to energy efficiency. Energy savings is a concept that is widely reported in energy studies (see Chapter 3), used in the practice of measurement and verification (M&V), and in policy evaluation. In M&V (see Section 1.4.5), energy savings is defined as the difference between pre-installation (baseline) energy and post-installation (reporting period) energy, including energy adjustments to account for differences in conditions between the baseline and reporting period. California's energy policy manual for regulated utilities also uses the term energy savings (CPUC, 2020). For example, California's definition of "gas savings" is:

> "Reduced natural gas usage (or savings) produced by either energy efficiency investments which maintain the same level of end-use service or conservation actions which can reduce energy use by reducing the quantity or quality of the baseline services provided."

Taking into consideration the concepts and industry definitions just described, we provide below a working definition of "energy savings" that is used throughout this book:

> "Energy savings is defined as a reduction in energy consumption for a specified system over a defined period of time due to an intervention, and adjusted for direct rebound effects where applicable. Savings are quantified relative to a baseline system including the baseline technology as well as baseline quantity and quality of energy services. After the intervention, the quantity and quality of services delivered by the new system should equal or exceed the baseline, or otherwise be deemed acceptable (e.g., based on best practices standards, regulations, or stakeholder requirements)."

With this definition, key terms are strategically selected to help practitioners be more effective and avoid common pitfalls that arise when concepts are confused, vaguely described, or key issues are overlooked. We argue that this definition of energy savings is appropriate and useful because:

- The definition is aligned with ASHRAE's definitions for energy assessments, multiple M&V protocols, and leading energy management policies such as in California.
- The term "intervention" is chosen to be inclusive of common equipment retrofits (boilers, lighting, etc.) and behavioral interventions that are described in Chapter 7.
- "Direct rebound effects" (defined in Chapter 7) are included because in some applications, especially industrial systems, rebound effects should not be ignored.
- Explicit mention of a baseline for the quantity and quality of services is included, for reasons discussed above (e.g., ASHRAE and California Public Utilities Commission (CPUC) definitions). Explicit consideration of services in defining energy savings also helps practitioners minimize barriers (discussed further in Chapters 6 and 7).
    - A proposed measure that is estimated to fall short of meeting clearly defined baseline quantity or quality of services is either: (a) not considered to produce valid energy savings, or (b) accepted as a conservation measure with accompanying negative non-energy impacts (NEEP, 2017).
    - A measure resulting in reduced energy consumption while exceeding the baseline service in quantity or quality provides positive non-energy impacts (or non-energy benefits).

Now that we have defined EEMs, energy efficiency, conservation, and energy savings, we will introduce one last commonly used term, which is distributed energy resources (DERs). The term DER is widely used today. But like most terminology, the intended meaning varies across different sources and organizations. In this book, we define demand-side DERs as follows:

> "Demand-side DERs are measures installed behind the meter which reduce final energy consumption, shift the timing of consumption, store energy, switch energy forms, or generate energy – while delivering the quantity and quality of services needed by end-users."

Our definition of DERs includes five types of measures: EEMs, distributed generation technologies, electric and thermal storage systems, fuel switching measures, and demand response technologies. As with most terminology used in energy systems, it is rare to find consensus and consistent definitions for commonly used terminology. Chapter 2 provides further descriptions of distributed generation, storage, fuel switching, and demand response as used in this book.

1.4 Energy Management Approaches and Methods

1.4.1 Energy management systems and programs

Effective energy management for facilities requires pulling together the right people, financial resources, and tools. Generally speaking, all systematic approaches to energy management are similar. Some approaches require formal certification, while some do not. Where formal certification is required, we will refer to these as energy management systems (EnMS). Where formal certification is not necessarily required, we will refer to these as programs.

There are decades of experience with EnMS and programs. Probably the most well-known EnMS is the International Organization for Standardization (ISO) 50001, which is used globally (ISO, 2018). The U.S. DOE also supports a 50001 Ready program that applies the same approach, but does not include formal certification (DOE, 2022). Strategic energy management (SEM) is an energy management program that is widely used in both Canada and the U.S. (Whitlock, 2021; Rogers, 2019). Chapter 7 includes a discussion of SEM.

The basic cycle for effective energy management, which is integral to ISO 50001, is the "plan-do-check-act" approach provided in Figure 1.9.

As the figure illustrates, energy management is a continuous process of planning and organizing, implementing measures, measuring and reporting performance, and making adjustments. From this perspective, an energy manager's job is never done. Facility conditions change over time (e.g., change of occupancy or use, new products produced), new requirements emerge (e.g., more aggressive GHG targets), and/or new solutions reach market maturity (e.g., new technologies).

To create or update an EnMS or program, an important initial task is to define (or redefine) objectives. The actual objectives for an organization can vary widely and should be customized for the organization's unique mandate, circumstances, and culture. A starter list of objectives for an EnMS or program follows:

1. Improve energy efficiency, reduce energy use, and reduce energy costs.
2. Reduce GHG emissions and improve other sustainability metrics.
3. Cultivate good communications to end-users on energy matters.
4. Develop and maintain effective monitoring, reporting, and management strategies for effective energy usage.

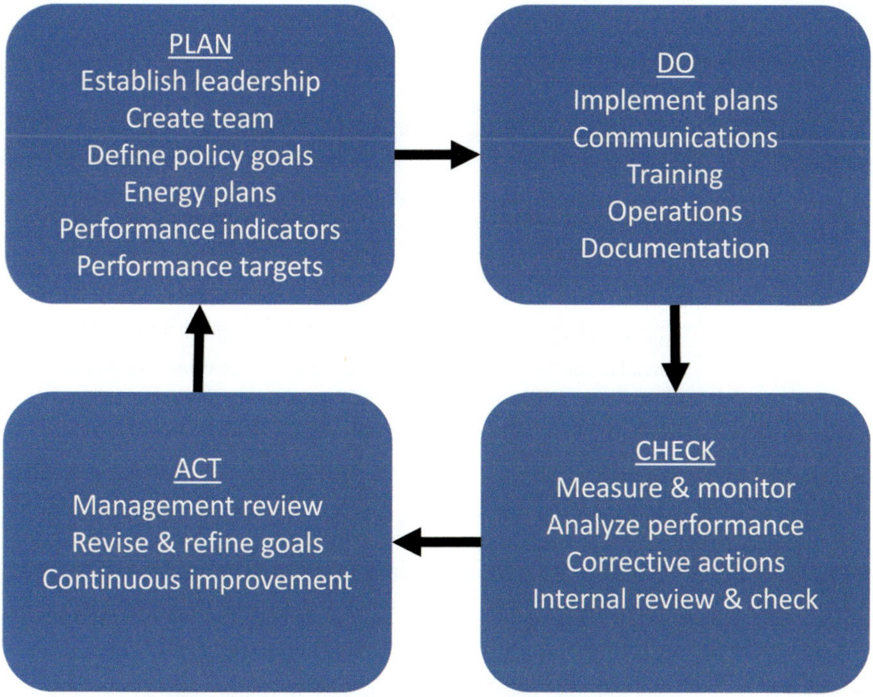

Figure 1.9 Plan-do-check-act energy management cycle.

5. Secure financing for EEMs and other DERs.
6. Cultivate engagement and participation in energy management programs from all users.
7. Improve the reliability and resiliency (see Chapter 8) of end-use systems.

Experience has shown that there are key elements for effective energy management as summarized in Table 1.6.

EnMS and programs have been proven to work for decades. To illustrate the potential, Table 1.7 provides a summary of three case studies.

As illustrated in Table 1.7, the scale of energy and cost savings can vary widely. Some EnMS or programs are designed to be low-cost and focus only on operation and maintenance (O&M) improvements. The Foremost Farms case study in the table is an example of this approach. As well, some behavior change programs (described in Chapter 7) may only focus only on O&M improvements. Other systems or programs are designed to increase the adoption of capital improvements, and therefore count those improvements in reported results. The other two case studies in the table

Table 1.6 Key elements of effective EnMS or program.

Element	Description
Management commitment	An official, documented recognition of the EnMS or program, staff responsibilities, policies, and objectives.
Designated lead person	A designated lead person who is responsible for energy management is critical. This person's job title can vary (e.g., "energy champion" or "energy lead"). Effective energy leaders usually have multiple strengths including technical knowledge, finance/purchasing, and communications. Often this person has limited direct authority over support team members, and achieves results by effectively negotiating a matrix organization.
Support team	Support teams are critical to carry out the many technical and financial functions. Energy management teams are often composed of many individuals who are only partly dedicated to energy management as their job. Contracted or outsourced resources are sometimes utilized as well.
Cost centers	Effective management of energy costs has been found to require dedicated cost centers, and to avoid treating energy purchases as general overhead expenses.
Materials	Material resources are important such as: software tools (e.g., energy management information systems), and instrumentation (e.g., for sub-metering or audits).
Energy assessments	Audits and studies of energy systems (see Chapter 3) identify specific EEMs and other DERs needed to achieve organizational objectives (e.g., energy savings, cost savings, or GHG reduction). These assessments should be updated and refreshed on a periodic basis to stay current with new technologies, current costs, and evolving organizational goals.
Energy performance monitoring	Establish key parameters for energy performance for the facility. The various types of energy efficiency metrics described in Table 1.5 provide an overview of the types of indicators that are commonly used.
Reporting	Systematic reporting of energy monitoring is essential. The mechanism for reporting (e.g., software dashboards, formal annual reports, and email reports) and frequency (e.g., weekly, monthly, or trigger events). Careful consideration of how to visually display actual vs. targeted performance is important.
Training	Selected team members should be afforded training or certification in skills essential to their role. Examples include: energy management, auditing, commissioning, advanced maintenance, M&V, or green finance.

above are examples where capital improvements are explicitly included in the EnMS or program. In practice, all EnMS and programs can contribute to

Table 1.7 EnMS and program case studies.

Organization	EnMS or program	Tools and resources	Results achieved
Foremost Farms (DOE, 2019)	50001 Ready SEM	EMIS visualization for key performance indicators. \$100k investment in sub-meters	3.0% reduction in energy consumption (vs. 2.5% targeted). Reported improvement in product quality and reduced production losses.
Cummins Inc. (DOE, 2015)	Certification to ISO 50001 and Superior Energy Performance 50001 (U.S. DOE certification program)	\$1.41 M total investment in ISO 50001 for nine manufacturing facilities in three countries. Integrated EnMS into existing health, safety, and environment systems.	20% reduction in energy, normalized to work hours, from 2010 to 2015. \$3.54 M annual energy cost savings.
Cree Lighting (DOE, 2020)	50001 Ready SEM	Eleven-member energy management team. Energy management integrated with sustainability management.	From 2018 to 2020, production energy performance improved 18% and paint shop improved 4%.

increased capital investments, but the accounting for energy improvements from capital projects is sometimes counted separately.

1.4.2 Benchmarking

All energy management approaches include performance indicators, such as outlined in Table 1.5. In order to assess sufficient progress using the performance indicators, various benchmarks are used. There are two basic approaches to benchmarking:

1. Compare a facility's energy performance to itself over time. For example, energy use intensity (EUI), defined below, can be compared year-to-year for a school facility over time with the goal of reducing consumption.
2. Compare a facility's performance to similar facilities. For example, a manufacturing facility could be compared to facilities producing the same or similar products using an efficiency metric such as tons of product per kWh of electricity.

Two performance indicators that are commonly used for residential and commercial buildings are introduced here and illustrated with

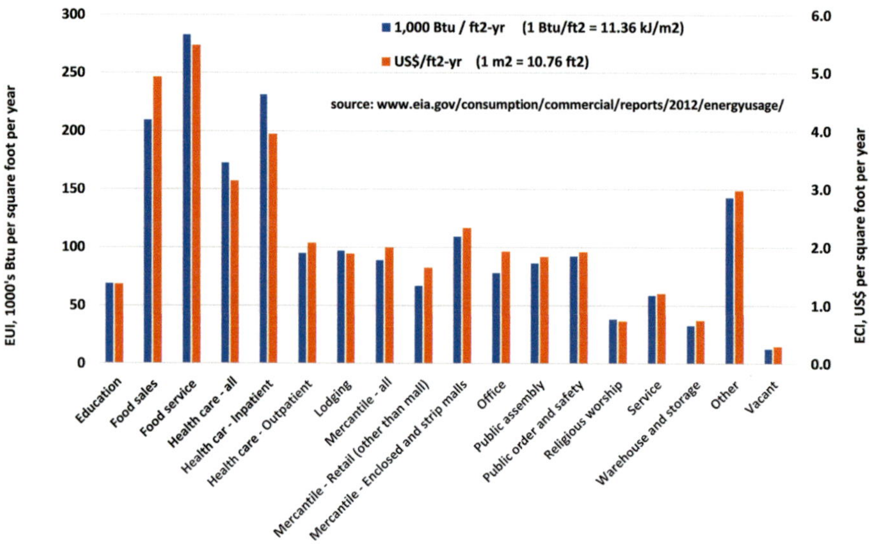

Figure 1.10 Energy use intensity and energy cost intensity for U.S. commercial buildings (2012 Data).

examples: EUI and energy cost intensity (ECI). EUI is defined as annual final energy consumption per unit area of conditioned space. To compute the EUI, all of the energies purchased for a facility are identified, converted to one common unit such as Btu or MJ, summed, and divided by the conditioned floor space. ECI is defined as the total annual energy cost per unit area of conditioned space. Figure 1.10 provides the average data for U.S. commercial buildings, with categories defined by the principal building activity (EIA, 2022).

Example 1.1: Consider an office building with 110,000 ft^2 of floor space, of which 10,000 ft^2 is unconditioned. It uses 1.76 million kWh and 6.5 million cubic feet of natural gas in one year. Estimate the EUI in both English (Btu and ft^2) and SI (MJ and m^2) units. Use Tables 1.1 and 1.2.

Solution: kWh of electricity is converted at 3,412 Btu/kWh and each cubic foot of gas is converted at 1,034 Btu. The total annual energy use is:

$$\begin{aligned}\text{Total energy use} &= (1.76 \times 10^6 \text{kWh})(3{,}412 \text{ Btu/kWh}) \\ &\quad + (6.5 \times 10^6 \text{ ft}^3)(1{,}034 \text{ Btu/ft}^3) \\ &= 6.005 \times 10^9 + 6.721 \times 10^9 \\ &= 12.73 \times 10^9 \text{Btu/yr}\end{aligned}$$

$$\text{In SI units} : (12.73 \times 10^9 \text{ Btu/yr})(0.001055 \text{ MJ/Btu}) \\ = 13,430,000 \text{ MJ/yr}$$

The net, conditioned floor space is 110,000 ft^2 less 10,000 ft^2 unconditioned, which equals 100,000 ft^2. Dividing the total energy use by 100,000 ft^2 gives the EUI:

$$\begin{aligned} \text{EUI} &= (12.73 \times 10^9 \text{Btu/yr})/(100,000 \text{ ft}^2) \\ &= 127,300 \text{ Btu/ft}^2 \text{ yr} \\ \text{In SI units} : \text{EUI} &= (13,430,000 \text{ MJ/yr})/[(100,000 \text{ ft}^2)(\text{m}^2/10.76 \text{ ft}^2)] \\ &= 1,450 \text{ MJ/m}^2 \text{ yr} \end{aligned}$$

Example 1.2: Consider the building in Example 1.1. The annual cost for electricity is \$115,000 and the annual cost for natural gas is \$32,500. Estimate the ECI for this facility.

Solution: The ECI is the total annual energy cost divided by the total area of conditioned floor space.

$$\text{Total energy cost} = \$115,000 + \$32,500 = \$147,500/\text{yr}$$

Dividing this total energy cost by 100,000 ft^2 gives:

$$\begin{aligned} \text{ECI} &= (\$147,500/\text{yr})/(100,000 \text{ ft}^2) \\ &= \$1.48/\text{ft}^2 \text{ yr} \\ \text{In SI units} &: (\$147,500/\text{yr})/[(100,000 \text{ ft}^2)(\text{m}^2/10.76\text{ft}^2)] \\ &= \$15.9/\text{m}^2 \text{ yr} \end{aligned}$$

Based on the average data provided in Figure 1.10, the EUI for the average office building is less than 100,000 Btu/ft^2-yr. Thus, the facility in Example 1.1 consumes greater than the average. There are many underlying reasons (in addition to energy efficiency design features) why this facility could consume more energy such as occupancy, operating hours, climate zone, and magnitude of plug loads. Also from Figure 1.10, the average office building ECI is nearly \$2.0/ft^2-yr (it should be confirmed that facility dollars and reference facility ECI are both based on the same year, as explained in Chapter 5). As such, in Example 1.2, the facility ECI is lower than the average cost for the office category. Other than total energy consumption, the facility ECI can vary based on other factors such as the relative proportion of energy forms (e.g., electricity vs. natural gas), average costs of energy (e.g., regional variations in average prices for electricity), and rate structures (e.g., time of use rates vs. flat rates, see Chapter 4).

Because there are many drivers of energy consumption for buildings and industrial plants, data and tools have been created to perform more sophisticated benchmarking beyond simple ratios such as EUI and ECI. One of the most widely used tools is Energy Star® Portfolio Manager® (used widely in Canada as well as the U.S.), which is freely available from the U.S. Environmental Protection Agency (energystar.gov). These tools utilize energy data and many other variables to produce Energy Star scores ranging from zero (least efficient) to 100% (most efficient). Yet another source of benchmarking for buildings is ASHRAE Standard 100 (Kono, 2021).

For industrial applications, there are two key sources of benchmarking data and tools. DOE's Advanced Manufacturing Office (energy.gov/eere/amo/advanced-manufacturing-office) provides benchmarking support for U.S. industries. Natural Resource Canada's (nrcan.gc.ca) Canadian Industry Program for Energy Conservation includes benchmarking support for industries in Canada.

1.4.3 Monitoring, targeting, and reporting

Energy monitoring, targeting, and reporting (MT&R) is a powerful energy management technique to continuously analyze, control, and report energy savings. The working definitions that commonly apply to MT&R are:

- *Energy monitoring* is the regular collection and analysis of metered energy consumption. Its purpose is to establish a basis of management control, to determine when and why energy consumption is deviating from an established pattern, and to provide a basis for taking corrective action where necessary. Utility meters or system sub-meters can be used, depending on the chosen system boundaries of the MT&R.
- *Targeting* is the establishment of a metric that defines the desired energy performance.
- *Reporting* provides energy management information generated in a form that is meaningful and understandable to decision-makers, enables ongoing control of energy use, helps achieve reduction targets, and provides verification of savings.

MT&R is built around a key statistical technique, described as the cumulative sum of differences (CUSUM). CUSUM is an analysis of the variance between energy consumption predicted by an energy model representing a baseline (i.e., less efficient) system, and the actual measured consumption after efficiency upgrades are implemented. The key steps in an effective MT&R process are:

1. Develop a baseline model using data collected prior to the implementation of EEMs. The baseline model has energy consumption as the explained or dependent variable. One or more explanatory or independent variables are usually identified. Examples of independent variables include production rates (see the case below), degree days (see Chapter 11), or occupancy data.
2. Develop targets for energy reduction.
3. Measure energy consumption and independent variables over time, after EEMs have been implemented (i.e., installed, fully commissioned, and staff trained).
4. For each time interval, calculate the difference between energy predicted by the baseline model, and actual consumption.
5. CUSUM equals the sum of differences from Step 4 for all time intervals after EEMs were implemented.
6. Report consumption and the CUSUM for each time interval.
7. Take corrective action to address variances between the CUSUM and targets.

The achievement of energy cost savings is usually the primary objective of MT&R. As with any energy management practice, there are additional benefits that can be realized if MT&R is effective such as improved occupant or worker productivity, improved product quality, and reduced O&M costs.

1.4.4 Monitoring, targeting, and reporting case study

The CUSUM methodology will be illustrated with a practical case study. Consider a food processing plant that has been upgraded with EEMs. Table 1.8 provides two years of monthly baseline data. Table 1.9 provides one year of monthly data after the EEMs have been installed and commissioned. We will describe the development a baseline model for total energy consumption as a function of production, complete a CUSUM analysis, and discuss options for setting targets and triggering corrective actions.

For this food processing facility and the data provided, the only independent variable available to explain variations in energy (MMBtu) is production (lbm). Monthly time intervals will be used because data are provided as

Table 1.8 MT&R case study – baseline data.

Month	Production lbm	Metered energy MMBtu	Month	Production lbm	Metered energy MMBtu
Jan-2019	39,600	124,891	Jan-2020	40,920	160,823
Feb-2019	21,120	119,458	Feb-2020	43,560	132,753
Mar-2019	13,200	113,940	Mar-2020	50,160	176,224
Apr-2019	44,880	139,089	Apr-2020	10,560	73,803
May-2019	47,520	190,689	May-2020	14,520	95,720
Jun-2019	31,680	115,346	Jun-2020	39,600	117,948
Jul-2019	10,560	94,245	Jul-2020	21,120	109,568
Aug-2019	29,040	109,664	Aug-2020	16,632	80,292
Sep-2019	23,760	119,937	Sep-2020	29,040	125,034
Oct-2019	10,560	79,617	Oct-2020	18,480	84,550
Nov-2019	13,200	99,211	Nov-2020	44,880	157,840
Dec-2019	36,960	119,442	Dec-2020	50,160	134,182

Table 1.9 MT&R case study – post-EEM installation data.

Month	Production lbm	Metered energy MMBtu	Month	Production lbm	Metered energy MMBtu
Feb-2021	21,120	98,524	Aug-2021	10,560	84,905
Mar-2021	15,840	94,630	Sep-2021	29,040	110,510
Apr-2021	13,200	85,649	Oct-2021	23,760	104,051
May-2021	44,880	140,854	Nov-2021	10,560	87,491
Jun-2021	47,520	141,792	Dec-2021	13,200	84,379
Jul-2021	31,680	116,754	Jan-2022	36,960	125,255

monthly totals. A baseline regression model will be developed of this form:

$$\text{monthly MMBtu} = \alpha + \beta(\text{monthly lbm})$$

To develop the baseline model, energy is charted as a function of production as shown in Figure 1.11.

This is a common chart produced with a spreadsheet program, including a line fit plot and regression equation displayed as part of the chart. These types of regression equations are commonly developed using ordinary least squares statistical models, which are described in introductory statistics textbooks (Koop, 2013). For this case study, the regression equation was developed using Microsoft Excel and the REGRESSION function (which is a macro that uses the LINEST function) with summary output shown in Figure 1.12. Numerous statistical software programs are available and will produce the same results as shown below.

For this case study, the regression equation results and parameters used to assess the acceptability or goodness of fit for the regression model follow:

MMBtu = 66,975 + (1.805) (lbm)

p-value for α, the intercept coefficient = 1.8×10^{-8} (accept if $\leq$ 0.1)

p-value for β, the slope coefficient = 1.8×10^{-7} (accept if $\leq$ 0.1)

Cv(RMSE) = Standard error/average metered energy = 13.6% (accept if $\leq$ 20%)

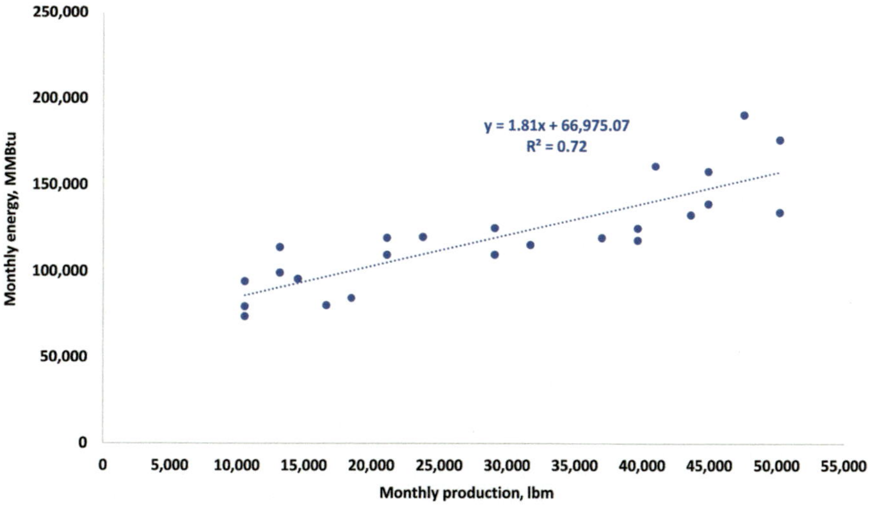

Figure 1.11 MT&R case study – chart of baseline energy vs. production.

Regression Statistics			
Multiple R	0.847203371		
R Square	0.717753552		
Adjusted R Square	0.704924167		
Standard Error	16254.81327	13.6%	Cv(RMSE)
Observations	24		

ANOVA					
	df	*SS*	*MS*	*F*	*Significance F*
Regression	1	14782010773	14782010773	55.94606493	1.76951E-07
Residual	22	5812816994	264218954.3		
Total	23	20594827767			

	Coefficients	*Standard Error*	*t Stat*	*P-value*	*Lower 95%*	*Upper 95%*
Intercept	66975.0655	7798.299511	8.588419233	1.78156E-08	50802.38217	83147.74883
lbm production	1.80538886	0.241371498	7.479710217	1.76951E-07	1.304815011	2.30596271

Figure 1.12 MT&R case study – regression model results.

R^2 = coefficient of determination = 0.72 (accept if $\geq$ 0.5)

Bias error = 0.0000% (accept if $\leq$ 0.005%)

The regression model parameters listed above, as well as the criteria for acceptable goodness of fit, are typical values used for MT&R as well as M&V analyses. The definition of these parameters and criteria for acceptability are not fully described here, and readers are referred to established M&V protocols (ASHRAE, 2023a; IPMVP, 2022) or textbooks (Koop, 2013). In general, the specific parameters chosen to assess goodness of fit and the threshold values for acceptance will vary based on published sources and the judgment of the analyst. For example, the criterion for R^2 varies. A threshold of 0.7 is often used, while other sources cite thresholds that are larger (e.g., 0.9) or smaller (e.g., 0.5).

The results of the CUSUM analysis are provided in Table 1.10 and Figure 1.13. Note that the fourth column of the table is the predicted energy consumption using the regression model. This is equivalent to assuming that the energy consumption would have been the predicted monthly values, if the baseline system had remained in operation. The variance column is the difference between the metered energy of the efficient operation, and the predicted energy if the EEMs had not been implemented. The CUSUM is the cumulative sum of variances, as defined above.

The chart in Figure 1.12 illustrates how CUSUM charts are used to understand and manage energy consumption and costs. Based on the definition of the CUSUM value, energy savings are accumulated over time as the CUSUM

Table 1.10 MT&R case study – CUSUM analysis results.

Month	Production lbm	Metered energy MMBtu	Predicted energy (baseline) MMBtu	Variance MMBtu	CUSUM MMBtu
Feb-2021	21,120	98,524	105,105	−6,581	−6,581
Mar-2021	15,840	94,630	95,572	−942	−7,523
Apr-2021	13,200	85,649	90,806	−5,157	−12,681
May-2021	44,880	140,854	148,001	−7,147	−19,827
Jun-2021	47,520	141,792	152,767	−10,975	−30,803
Jul-2021	31,680	116,754	124,170	−7,416	−38,218
Aug-2021	10,560	84,905	86,040	−1,135	−39,353
Sep-2021	29,040	110,510	119,404	−8,894	−48,247
Oct-2021	23,760	104,051	109,871	−5,820	−54,067
Nov-2021	10,560	87,491	86,040	1,451	−52,616
Dec-2021	13,200	84,379	90,806	−6,427	−59,043
Jan-2022	36,960	125,255	133,702	−8,447	−67,490

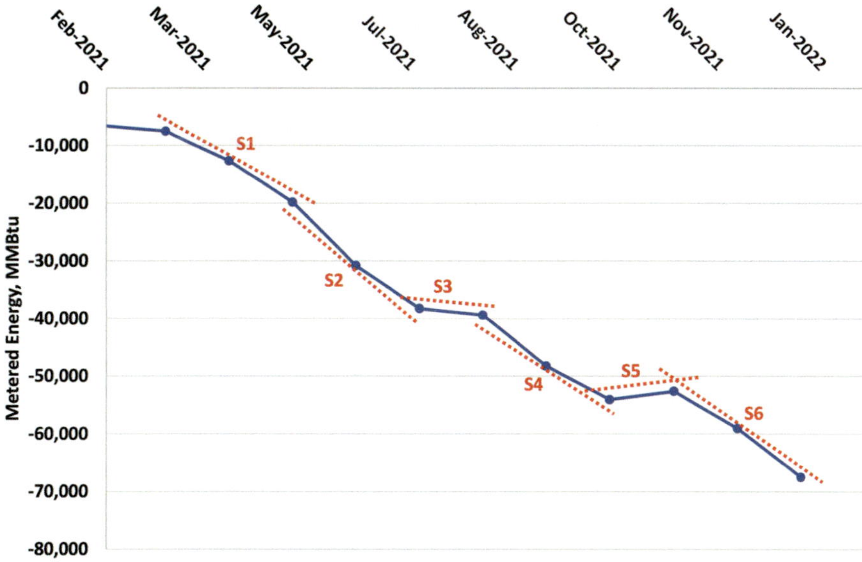

Figure 1.13 MT&R case study – CUSUM results chart.

value is expected to grow increasingly negative for each time interval. A negative sign for CUSUM is the common convention, and can be thought of as a decrease in energy consumption realized due to the EEMs. CUSUM charts are analyzed by observing the slope of the line, illustrated with the dotted red lines in the chart labeled S1–S6. A steep and downward slope is interpreted as increased savings, while a mild slope is interpreted as less energy savings. If the slope turns upward (e.g., S5), this indicates energy consumption has increased relative to what would have occurred with the baseline system.

With an understanding of what CUSUM charts represent (i.e., cumulative energy savings over time), these charts can be effective visual displays to include in reports to relevant stakeholders (e.g., O&M staff, management, facility users). Incorporating CUSUM charts into energy management information systems (EMIS), introduced in, Chapter 15, is one way to effectively use these charts.

CUSUM charts are also useful to establish and track MT&R performance targets. Common strategies for setting performance targets include:

- Target continuous improvement with rules that trigger forensic investigations and corrective actions. One obvious example would be that any flat or upward slope in the CUSUM chart triggers a forensic

investigation. Another rule could be that a decrease in the slope (i.e., less steep) in the CUSUM chart triggers investigation. Yet another example could be to investigate any variance, either positive or negative, which exceeds an absolute value (e.g., +/− 8,000 MMBtu in the case study).
- Because facility energy models typically have a fixed or base load component (i.e., 66,975 MMBtu for case study baseline), and a variable component (i.e., 1.805 MMBtu per lbm of product in the case study baseline), these components can be assessed separately. Additionally, new models of the efficient system could be developed, with targets set to reduce both the base load and variable components of consumption. For example, a new model using the post-EEM data in Table 1.9 produces the following model of consumption:
  MMBtu = 66,731 + (1.589) (lbm).
- New baselines can be established. For example, the first year of post-EEM operation could be modeled as mentioned in the bullet point above. The original baseline could be abandoned, and the first year of post-EEM operation established as the baseline going forward for a new CUSUM model.
- Historical time periods where energy performance was the best or most efficient could be examined in detail to understand the underlying conditions, and then efforts could be made to replicate those conditions. One approach could be to harvest the data by selectively using data from the time intervals with the best performance, create an energy model based on these select data, and then use this model to establish new targets.

1.4.5 Measurement and verification (M&V)

In addition to benchmarking and MT&R, M&V is an established practice to validate and manage energy savings due to investments in energy projects. According to the International Performance and Measurement and Verification Protocol (IPMVP), M&V is defined as (IPMVP, 2022): "… the process of planning, measuring, collecting, and analyzing data to verify and report energy savings resulting from the implementation of energy efficiency measures (EEMs) in one or more facilities."

This section provides a high-level introduction to M&V for EEMs. However, the same M&V methods are also used for other DERs such as electrification loads from fuel switching projects, peak demand reductions due to demand response technologies, or energy production due to solar

PV installations. As this section provides an overview of M&V, readers are referred to the leading protocols for further details (IPMVP, 2022; ASHRAE, 2023a, FEMP, 2015).[6]

The overall goal of any M&V effort is to balance the cost of M&V activity with the targeted level of uncertainty (or accuracy) of results. Common reasons for investing in M&V include:

- To validate energy savings due to EEMs and other DERs.
- To quantify avoided GHG emissions.
- To become eligible for EnMS certification or program funding (e.g., SEM subsidy).
- To improve the performance of energy system upgrades. The process of conducting M&V is proven to contribute to increasing energy and utility bill cost savings, as compared to projects that do not undergo M&V.
- To document credible and accurate utility bill cost savings to validate financial returns for stakeholders (e.g., lenders, facility owners, utilities or agencies providing subsidies, or energy services companies).
- To create data and models that can be utilized for ongoing EnMS or programs to track, maintain, and manage energy cost savings over the long term such as demonstrated above with the MT&R case study.

Figure 1.14 and the basic equation below describe the fundamental M&V concepts and calculations for determining M&V energy savings.

The basic equation for M&V energy savings is:

$$\begin{aligned}\text{Energy savings} = (&\text{Baseline Energy} + / - \text{Energy Adjustments}) \\ &- \text{Reporting Period Energy}\end{aligned}$$

As the figure and basic equation illustrate, energy savings is the difference between adjusted baseline energy and reporting period energy. But what is the difference between baseline energy and adjusted baseline energy? A key issue in M&V is to determine energy consumption for the baseline and EEM systems *for the same operating conditions*, often described as an "apples-to-apples" basis. Conceptually, the goal is to estimate the energy that would have been consumed if the EEM had not been implemented. To assess whether adjustments are needed, an important activity in M&V is to identify all key drivers of energy consumption. If a substantial change for a given variable leads to a substantial change in energy consumption, then this variable is a

---

[6] Additional M&V resources are available at www.bpa.gov/energy-and-services/efficienc y/document-library.

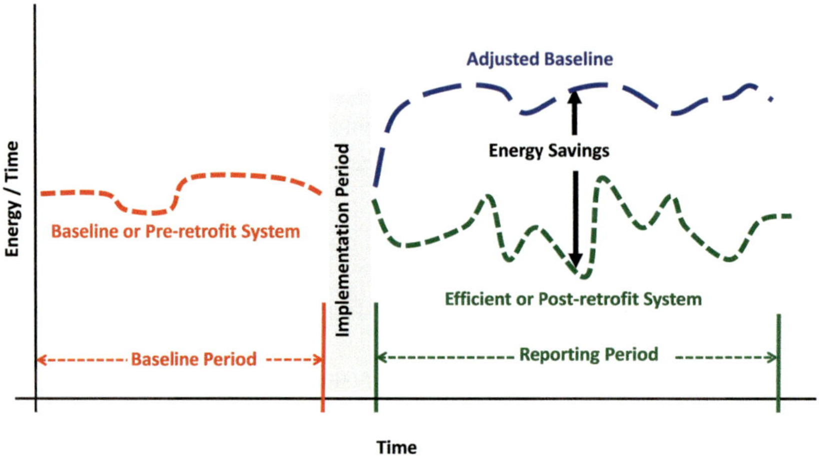

Figure 1.14 Measurement and verification (M&V) of energy savings.

key driver for M&V purposes. How much change is considered substantial is based on the judgment of the responsible M&V professional. Examples of key drivers of energy consumption are occupancy of buildings, outdoor air temperatures or degree days (see Chapter 11), and production rate for an industrial process (e.g., the MT&R case study in Section 1.4.4). If there are no changes in any key drivers of energy, then the adjusted baseline energy would equal the baseline energy, and savings is directly the difference between metered energy for the baseline and reporting periods. However, most real-world projects involve changing conditions such that energy adjustments must be made.

In M&V protocols, adjustments are usually categorized as routine or non-routine. Routine adjustments are for events or variables that are expected to be key drivers, such as production rates, operating hours, or degree days. Non-routine adjustments are for events that are uncommon and not necessarily expected, but significantly affect energy consumption. Examples of non-routine events include changes in floor space, equipment upsets, large changes in ventilation rates (e.g., as occurred during COVID-19), or the unanticipated addition of new equipment.

Note that we have defined M&V energy savings as adjustments to the baseline energy. This is described as the "avoided energy" approach where reporting period conditions are chosen as the reference conditions (in rare cases, the baseline conditions could be the reference, and adjustments made to reporting period energy). Another common alternative to avoided energy is called "normalized energy." With this approach, the reference conditions are neither the baseline nor reporting period, but some other set of conditions such as a standard production rate profile or degree days for a typical meteorological year (ASHRAE, 2021).

In general, there are four options for carrying out M&V as summarized in Table 1.11. This table uses the terminology and concepts as described in the IPMVP (IPMVP, 2022), and is identical or very similar to other M&V protocols (ASHRAE, 2023a; FEMP, 2015; SEP, 2019). In general, the IPMVP is the most widely used protocol, although all the protocols cited provide some unique content. The Superior Energy Performance (SEP) 50001™ Program M&V protocol developed by the U.S. DOE is a robust M&V protocol. A unique aspect of this protocol is that it includes both a top-down regression model methodology that is confirmed by a bottom-up

Table 1.11 Summary of M&V methods based on the IPMVP.

M&V Options	Description	Key issues and applications	Main benefits	Main limitations	Examples
Option A	Retrofit isolation, key parameter measurement	Meter at least one key parameter. More effective for steady state processes.	Validate EEM using metered data to avoid stipulated savings that use no measurements. Low M&V cost.	Greater uncertainty. Not useful for highly transient processes.	Lighting retrofit where either hours of operation or power consumption are metered, but not both.
Option B	Retrofit isolation, all parameter measurement	Identify and meter all key drivers of energy consumption.	Lower uncertainty (greater accuracy). Determine savings for individual EEMs. Works well with highly transient processes.	Higher cost of M&V.	Variable speed drive conversion on fan systems for a commercial building or industrial product line.
Option C	Whole facility	Use whole facility or whole system meters for M&V purposes.	Reduced metering costs if utility meters are used. Option C models often can be used for ongoing CUSUM models.	Uncertainty may be large if savings $<$ 5% of average consumption with hourly or daily data, or if savings $<$ 10% with monthly data. Aggregate savings only, not for individual EEMs.	Residential whole-home retrofit to determine aggregate electricity savings due to multiple EEMs using utility interval data.
Option D	Computer simulation	When computer simulation is essential for M&V.	Validate savings for individual EEMs. Only viable option when baseline energy was not or could not be measured. Models can help estimate routine or non-routine adjustments.	Calibration of simulation models can be challenging. Higher cost of M&V. Requires specialized modeling expertise.	New multi-family residential building where minimum building code is the baseline. Industrial boiler retrofit when baseline fuel consumption was not metered.

check of implemented projects. The SEP 50001^{TM} protocol also includes quantification of both primary (source) energy and site (final) energy.

The top-down energy performance improvement is calculated as the ratio of baseline period to reporting period facility-wide consumption, with adjustments to make the two periods comparable. The bottom-up energy performance improvement is based upon facility-wide energy performance improvement is calculated by aggregating energy savings from individual energy performance improvement actions. The bottom-up analysis serves as a required comparison to the top-down approach.

In summary, Option A and Option B are retrofit isolation methods, Option C uses whole facility or whole system meters, and Option D is when computer simulations are used. The detailed descriptions and tradeoffs in selecting the best M&V option can be found in the M&V protocols, as well as many of the IPMVP application guides (e.g., IPMVP, 2019).

The "V" in M&V is for verification, which is an important task in the M&V process. Verification is not only to confirm that the EEM is installed, but that equipment and systems are fully commissioned and tuned to operate as intended (see Chapter 16). Typically, the M&V reporting period can start after the verification is completed.

To wrap up this introduction to M&V, the important steps and elements of effective M&V efforts are as follows:

1. Establish the overall purpose and specific objectives for conducting M&V, consistent with the contractual needs and (sometimes unwritten) expectations of all stakeholders. Stakeholders should include, but are not necessarily limited to, facility owners and any organization or agency contributing funds or financing.
2. Identify a responsible M&V professional to lead and oversee the M&V project. The person designated as the lead professional could change over the course of a project, but at all times there should be an identified lead M&V professional.
3. Develop an M&V plan following the chosen M&V protocol(s) that is agreeable for all stakeholders. While not strictly essential, the best practice is to complete the M&V plan early in the project prior to committing funds.
4. Delineate measurement boundaries for the baseline and EEM systems.
5. Clearly define the baseline system that includes baseline equipment, baseline time period, baseline operating conditions (e.g., occupancy or production rates), baseline metering and data collection, and data

sources and procedures for establishing adjusted baseline energy consumption.

6. During the implementation period, complete all tasks necessary to operate and maintain the EEM system (e.g., installation, commissioning, and training).
7. Conduct operational verification (the "V" in M&V) prior to initiating the reporting period measurements. It is critical that a member of the M&V team, preferably the assigned M&V lead person, actually visit the site in-person to inspect equipment, observe the operation, and gain insights from facility users and O&M staff. If cost for a site visit is considered prohibitive, a remote visit using live streaming technology may suffice.
8. Complete the M&V analysis and report following the agreed procedures as documented in the M&V plan.
9. Communicate results to stakeholders.

1.5 Summary

In this chapter, we have defined energy management, and clarified the scope of this book as targeting demand-side energy systems for facilities in the U.S. and Canada (with wide application to any high- or middle-income country context). The multitude of issues driving the need for demand-side energy management are summarized, including: the health and economic welfare of facility users and owners, the large magnitude of consumption, inequities in access to energy services and incidence of energy costs, low level of adoption of proven and cost-effective EEMs, environmental protection – especially climate change, and the accelerating need to balance the timing of electricity supplies and end-use demands. To understand how these key issues can be addressed, a framework for understanding all energy systems is introduced, and the demand-side has been demarcated as the systems located on the customer side of (or "behind") the utility meter. Important terminology has been defined including energy efficiency, energy sufficiency, energy conservation, energy savings, and DERs. An overview is provided of the fundamental technical concepts of energy conversions and units. To maximize effectiveness, we present that energy management professionals maintain a focus on reliably delivering the quantity and quality of services that people and businesses need and want, while minimizing energy consumption, cost, and undesirable impacts. To achieve the goals of effective energy management, proven methods are introduced, which describe how to organize, implement, monitor, and validate EEMs and other DERs. These methods include EnMS and programs, MT&R systems, and M&V methods. Understanding the key concepts, definitions, and methodologies introduced in this chapter is a critical prerequisite for all demand-side energy management professionals to be effective.

References

ASHRAE (2011) "Procedures for Commercial Building Energy Audits" (2nd edition).

ASHRAE (2021). ASHRAE Handbook – Fundamentals SI. Chapter 14 "Climatic design information."

ASHRAE (2023a) "Measurement of Energy, Demand, and Water Savings" ASHRAE Guideline 14-2023.

ASHRAE (2023b) "Standard for Commercial Building Energy Audits" ANSI/ASHRAE/ACCA Standard 211-2023.

ASHRAE (2023c) "Standard Method of Evaluating Zero Net Energy and Zero Net Carbon Building Performance" ANSI/ASHRAE Standard 228-2023.

Blok K. and Nieuwlaar E. (2021) "Introduction to Energy Analysis" 3rd edition. Routledge.

California ISO (2021). California Independent System Operator. www.caiso.com/TodaysOutlook.Datafor2019.

CPUC (2020) "Energy efficiency policy manual" (version 6). California Public Utility Commission cpuc.ca.gov

Cullen J., Allwood J., and Borgstein E. (2011) "Reducing Energy Demand: What are the practical limits?" Energy Policy 38 (4), p. 75.

DOE (2018) "Low-Income Household Energy Burden Varies Among States — Efficiency Can Help In All of Them" Office of Energy Efficiency & Renewable Energy. DOE/GO-102018-5122.

DOE (2022). 50001 Ready www.energy.gov/eere/amo/50001-ready-program

DOE (2020). "Energy Department Recognizes Cree Lighting for Achieving 50001 Ready Status" Office of Energy Efficiency & Renewable Energy www.energy.gov/eere/amo/articles/energy-department-recognizes-cree-lighting-achieving-50001-ready-statusPublicationdateDecember21,2020.

DOE (2019) "Foremost Farms—50001 Ready Facility" Better Building. September 25, 2019. https://betterbuildingssolutioncenter.energy.gov/iso-50001/showcase-projects/foremost-farms%E2%80%9450001-ready-facility.

DOE (2015) "Global Energy Management System Implementation: Case Study: Cummins Inc." Better Buildings. https://betterbuildingssolutioncenter.energy.gov/sites/default/files/attachments/Cummins_India_US_UK.pdf

EIA (2021). Electricity data. U.S. Energy Information Administration www.eia.gov/electricity/data/browser.Datafor2018-2020.

EIA (2022) Commercial Buildings Energy Consumption Survey. U.S. Energy Information Administration www.eia.gov/consumption/commercial/reports/2012/energyusage/(note:2012isthemostrecentdataavailableasofMay,2022).

EPA (2020) "Energy Star Portfolio Manager Technical Reference – Source Energy. U.S. Environmental Protection Agency. October, 2020. www.energystar.gov/buildings/tools-and-resources/portfolio-manager-technical-reference-source-energy

FEMP (2015) "M&V Guidelines: Measurement and Verification for Performance-Based Contracts Version 4.0" U.S. DOE Federal Energy Management Program.

IEA (2024). Energy balances. International Energy Agency www.iea.org/data-and-statistics.Datafor2021.

IPMVP (2019) "Uncertainty Assessment for IPMVP" International Performance Measurement and Verification Protocol. EVO 10100 – 1:2019. Efficiency Valuation Organization. www.evo-world.org

IPMVP (2022) "International Performance Measurement and Verification Protocol (IPMVP) Core Concepts" EVO 10000 – 1:2022. Efficiency Valuation Organization. www.evo-world.org

ISO (2018) "ISO 50001:2018. Energy management systems — Requirements with guidance for use" International Organization for Standardization www.iso.org

Kono J. and Kono N. (2021) "Applying ASHRAE Standard 100 to Real-World Building Data" ASHRAE Transactions, V127, Part 1.

Koop G. (2013) "Analysis of Economic Data" 4th edition. Wiley.

Lovins A. (2004) "Energy Efficiency, Taxonomic Overview," Encyclopedia of Energy 2:383-401.

Lovins A. (2018) "How big is the energy efficiency resource?" Env Res Ltr v13.

Navigant (2019) "2019 Integrated Ontario Electricity and Natural Gas Achievable Potential Study" Prepared for Ontario's Independent Electricity System Operator and Ontario Energy Board.

NEEP (2017) "Non-Energy Impacts Approaches and Values: an Examination of the Northeast, Mid-Atlantic, and Beyond" Northeast Energy Efficiency Partnerships. https://neep.org

OECD (2024). OECD Data. Organisation for Economic Co-operation and Development. https://data.oecd.org.Dataforyear2021.

Patterson (1996) "What is energy efficiency? Concepts, indicators, and methodological issues" Energy Policy, 24, no. 5, p377-390.

Rogers E., Whitlock A., and Rohrer K. (2019) "Features and Performance of Energy Management Programs" American Council for an Energy-Efficient Economy, Report IE1901.

Rosenfeld (1999) "The Art of Energy Efficiency: Protecting the Environment with Better Technology" Annual Reviews Energy Environment 24.

Schipper L., Bartlett S., Hawk D., and Vine E. (1989) "Linking life-styles and energy use: a matter of time?" Annual Reviews Energy 14, p.273.

Scott D.S. (1994) *The Energy System.* Int J of Hydrogen Energy Vol. 19, No. 6, pp. 27.

Smil V. (2017) "Energy: A Beginner's Guide" 2nd edition. Oneworld Publications.

Smil V. (2016) "Energy Transitions: Global and National Perspectives" 2nd edition. Praeger

Superior Energy Performance 50001^{TM} (SEP 50001^{TM}). Superior Energy Performance^{TM} Program Measurement & Verification Protocol: 2019. May 6, 2019, Updated October 29, 2019. The Regents of the University of California.

Tong K. *et. al* (2021) "Measuring social equity in urban energy use and interventions using fine-scale data" Proceedings of the National Academy of Sciences of the United States of America, 118, no. 24.

Ungar L. and Whitlock A. (2019) "Energy Efficiency and Corporate Sustainability: Saving Money While Meeting Climate Goals" American Council for an Energy-Efficient Economy, Topic Brief.

Whitlock A., Rightor E., and Hoffmeister A. (2021) "Canadian Strategic Energy Management Market Study" American Council for an Energy-Efficient Economy, Report IE2012.

Winston A., Favalora G., and Healy T. (2017) "Energy Strategy for the C-Suite" Harvard Business Review. January-February.

Yale Environment 360 (November 26, 2008). https://e360.yale.edu/features/amory_lovins_energy_efficiency_is_the_key

Author Biography

Eric Mazzi is a consultant and educator focusing on technology and policy for demand-side energy systems. He specializes in performing independent technical analysis, M&V, policy studies, and educational services. He advocates and contributes to efforts to reduce energy consumption and achieve net-zero GHG emissions, while maintaining or improving energy services and advancing energy equity. His experience includes air quality consulting, utility maintenance, and five years as full-time engineering faculty at the University of British Columbia. He volunteers as a board member for the U.S. Northwest Energy Coalition, and the International Performance M&V Protocol committee. Eric teaches part-time, delivering university and professional certification courses. He is an adjunct professor of energy management at the New York Institute of Technology – Vancouver. He has B.Sc. and M.Sc. degrees in mechanical engineering, and a Ph.D. degree in environmental policy. He is a registered engineer in California and British Columbia.

Acknowledgments

I would like to thank chapter reviewers Ethan Rogers and Jon Feldman for their thoughtful review of drafts of this chapter. I would like to acknowledge the contributions of the authors of the 8th edition of this book, as some parts of this chapter are similar to the 8th edition (in particular, the MT&R case study).

Practice Questions

1. For each metric below, classify it according to the four categories of energy efficiency metrics (see Table 1.5)

    - Thermodynamic
    - Physical-thermodynamic
    - Economic-thermodynamic
    - Economic

    a) Btu purchased per ft^2 of conditioned floor space
    b) GJ consumed per degree day
    c) kWh consumed per unit of production
    d) Btu purchased per sales dollar
    e) Dollars of energy cost per dollar of added product value
    f) MJ purchased per direct labor hour

2. List four (4) examples of end-use (or service) technologies that would commonly be found in a residential property, and describe the final energy consumed, and useful energy delivered by each. One sentence for each example is sufficient.
3. The use of personal space heaters (portable resistance heaters powered by electricity) is often considered an undesirable if not inappropriate end-use technology in offices.
   a) describe a situation or scenario where the use of a personal space heater would be considered to be an *inefficient* use of energy
   b) describe a situation scenario where the use of a personal space heater would be considered to be an *efficient* use of energy
4. As the energy manager for a facility, which tasks can you perform to verify that your energy action plan was maintaining, improving, or otherwise resulting in acceptable levels of thermal comfort?
   a) Measure temperature and relative humidity
   b) Compare to accepted standards for the facility type, such as ASHRAE standards
   c) Survey occupants
   d) Track records of occupant complaints such as hot/cold calls
   e) All of the above
5. A facility has 5,800 m^2 of total space, and 700 m^2 of the total includes unconditioned parking spaces. Calculate the facility energy use intensity if annual consumption is as shown in the table below. Use the lower heating value (LHV) of fuels. Report the results in units of MJ/m^2-yr and also Btu/ft2-yr.

Fuel	**Annual consumption**	**Annual consumption units**	**Annual consumption, MJ**
Natural gas	80,000	m^3	
Dry wood	4,200	kg	
Electricity	335,000	kWh	

6. The baseline regression model for a facility shows that monthly electricity consumption = 3,300 kWh/month + 22.5 (CDD/month), where CDD = cooling degree days. After a cooling system upgrade is completed over the winter, the electricity consumption for the summer months during following year are shown below. Complete the table below to calculate the total electricity savings for the three-month time period.

Month	CDD °C day/month	Actual kWh	Baseline kWh	Savings kWh
June	53	3,900		
July	85	4,200		
August	77	4,150		

7. A single-family home undergoes a retrofit that includes lighting, a new furnace, new windows, more efficient dishwasher and clothes washer, and a new condensing hot water heater. Assuming the goal is to determine the total energy savings (i.e., not critical to determine savings from each individual technology), which International Performance M&V Protocol (IPMVP) option is likely the best fit? State why you chose the specific IPMVP option.
   a) Option A
   b) Option B
   c) Option C
   d) Option D
   e) Insufficient information
8. A university has secured funding to build the "world's most energy-efficient building," which will utilize ground source heat pumps, energy recovery ventilation, low U-values for building envelope, and advanced lighting systems. The university needs to perform M&V of energy savings for the newly constructed building. Which IPMVP option is necessary to utilize for the M&V? (Option A, Option B, Option C, or Option D). Explain why (1–2 sentences will suffice).
9. For the following 12-month CUSUM results, for which month were energy savings *not* achieved?

month	CUSUM, GJ
Jan	-500
Feb	-842
Mar	-1108
Apr	-958
May	-1424
Jun	-1735
Jul	-1810
Aug	-1966
Sep	-2188
Oct	-2654
Nov	-3276
Dec	-3732

a) January
b) April
c) August
d) December
e) Energy savings were achieved for all 12 months

10. The cooling system on an industrial amine process in a natural gas-sweetening plant currently meets cooling requirements during winter, but not summer. The process cools the gas using air-cooled heat exchangers with large, 15-foot diameter axial fans that are connected to the electric motors by belts and pulleys. Average electric power consumption of the fan system is currently 745 kW during the summer. During the summer, the electric motors are loaded to maximum capacity, but the process still runs too hot and experiences equipment failures, which compromises gas quality. The facility energy manager is presented two alternatives by the O&M team:
Alternative #1 (low capital cost): Keep the existing fans (the fan impellers can produce higher flow capacity by increasing speed), but replace existing motors with larger capacity motors to meet the cooling load and operate within acceptable process temperature limits. During summer, the belts and pulleys would be changed to increase fan speed by 30% (in other words, $RPM_{new} = 1.30 \times RPM_{old}$). Total electric power consumption is estimated to increase by $(RPM_{new}/RPM_{old})^{2.4}$ (see Chapter 14 for an explanation).
Alternative #2 (high capital cost): Keep the existing motors, and retrofit with newer, more-efficient fans (i.e., fan housings and impellers). The new fans increase fan efficiency from 23% for existing fans to 42% for the new fans. During the summer, it is estimated that the total power consumption will remain unchanged from the current system. The more efficient fans are estimated to increase air flow enough to operate within acceptable process temperature limits without increasing electric power consumption.

   a) Concisely describe which is the most appropriate baseline and why: the existing system or Alternative #1?
   b) Assume the existing system is the baseline, and Alternative #2 is the EEM. Estimate the average electric power savings (kW) during the summer.
   c) Assume Alternative #1 is the baseline, and Alternative #2 is the EEM. Estimate the average electric power savings (kW) during the summer.

2

Greenhouse Gas Management

Eric Mazzi

Mazzi Consulting, Canada

Almost all energy management initiatives today are directly or indirectly motivated or funded by policies and programs aimed to address climate change. As such, a basic understanding of climate change is critical. The main objective for this chapter is to provide basic methods and strategies to quantify and reduce operational greenhouse gas (GHG) emissions from facilities. Practitioners often need to be able to estimate or measure GHG emissions and develop retrofit measures or new facility designs to reduce GHGs if not completely "decarbonize" the facility. These are the key skills, knowledge, and competencies that demand-side energy management professionals need to be effective.

To begin with, we provide a brief introduction to climate change including some definitions, evidence, and indicators of climate change, major impacts due to climate change, and key facts about human-induced climate change. We then introduce the four general strategies available to solve this massive environmental problem, which are adaptation, solar radiation management, carbon removal, and mitigation. Next, we focus on climate change mitigation for facilities by describing the main GHG pollutants that are emitted from facilities (e.g., from fuel combustion and fugitive fuel and refrigerant emissions) and methodologies to quantify GHG emissions. The remainder of the chapter is focused on the application of distributed energy resources (DERs) to manage and reduce the operational GHG emissions from facilities.

2.1 Overview of Climate Change

2.1.1 What is climate change, and why is it a problem?

Climate is defined as "... the average weather conditions for a particular location and over a long period of time" (WMO, 2022). It is important to understand that climate cannot be understood from the weather occurring today or forecast for next week. Climate is about longer-term averages, such as 30-year averages. Climate is also defined by variations (extremes) in parameters that we commonly use to describe weather such as temperature, humidity, precipitation events, and winds. While climate is proven to change over geologic time scales due to natural events, attribution science for climate change clearly shows that human activity has caused rapid changes in Earth's climate. Provided below is a very brief introduction to climate science. For summaries as well as in-depth understanding of the science of climate change, we refer the readers to these resources:

- The Intergovernmental Panel on Climate Change (IPCC) scientific reports (IPCC, 2021; IPCC, 2022a; IPCC, 2022b)
- World Meteorological Organization's scientific summary (WMO, 2022)
- ASHRAE Handbook's climate change chapter 36 (ASHRAE, 2025).

Fundamentally, climate change is occurring due to modifications to Earth's natural heat balance as illustrated in Figure 2.1. GHGs from human activity and changes in land use by people have altered (and continue to alter) the radiation heat transfer between Earth and outer space (including the sun), which is commonly described as a radiative forcing effect in units of W/m^2.

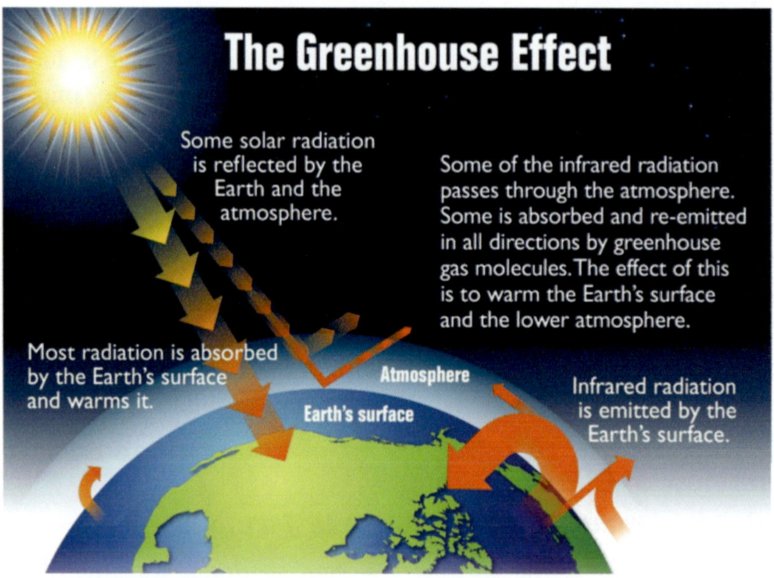

Figure 2.1 Illustration of the greenhouse effect (source: U.S. EPA via www.commons.wikimedia.org).

There are many consequences that result from this altered heat balance, but the main effect is a gradual increase in the mean temperature of the Earth's surface air and sea surfaces.

As described later in Section 2.2 of this chapter, the radiative forcing effects of GHG emissions and land-use changes are quantified in terms of 100-year global warming potential in carbon dioxide equivalents (CO_2e). Global GHG emissions in 2019 were 59 +/− 6.6 $GtCO_2e$, broken down by sectors as shown in Figure 2.2 (IPCC, 2022a).

The evidence of climate change is provided by changes in climate indicators relative to baselines. For example, global mean temperatures are quantified relative to baseline data from the years 1850−1900. GHG levels in the atmosphere are quantified relative to values from the year 1750. Measurements of ocean variables are another critical group of indicators including sea levels, marine heatwaves and cold spells, mean temperatures, and acidification. Yet another group of indicators is described as the cryosphere including sea ice measurements, glaciers, ice sheets in various locations, snow levels, and permafrost. Ongoing measurements for all of these parameters (and

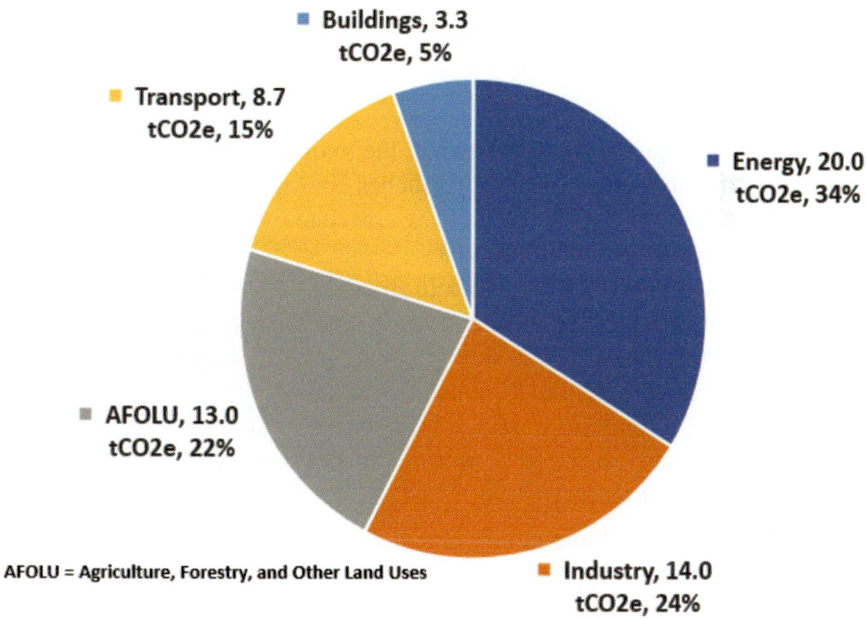

Figure 2.2 Distribution by sector for 2019 global GHG emissions of 59 billion tons of CO_2e.

more) provide a massive body of evidence that climate is rapidly changing and that it is largely attributable to human activity.

The consequences of global climate change are serious and getting worse over time. There are multiple, massive consequences including reduced food security, humanitarian impacts and population displacement, degraded ecosystems (and reduced ecosystem services), and biodiversity loss. It is often not mentioned that global climate change is predicted to result in some populations to be better off (at least for a period time and not necessarily forever) by some measures, for example, realizing increased economic growth or greater crop yields in regions. However, the evidence is clear that the net, aggregate consequences of climate change to all living species are predicted to be negative and catastrophic. To illustrate the scope and severity of climate change impacts, below is a sample list of events that have already occurred as reported for the year 2021 by the World Meteorological Organization (WMO, 2022):

- Heat waves and wildfires: Historic high temperatures recorded in Moscow, British Columbia (with 569 heat-related deaths), Italy, and Finland.
- Cold spells: In 2021, Texas experienced the lowest temperature recorded since 1899 causing nearly 10 million people to live without electric power for over three days; the event is described as the costliest winter storm on record for the U.S.
- Floods: One area of China recorded the most intense rainfall ever recorded in China with 202 mm (8 inches) in one hour and 382 mm (15 inches) over six hours; the event is associated with 380 deaths and US$17.7 billion in economic damages.
- Drought: Lowest precipitation index in Brazil since the 1960s; the driest time period ever on record for the southwest U.S. (by a margin of more than 10%) from January 2020 through August 2021; two-year drought for Madagascar with rainfall 50% below normal.
- Severe storms: Strongest tornado ever on record in the Czech Republic; December tornados in the U.S. were eight times the 20-year average from 1991 to 2010.

A common metric used to quantify global climate change is global mean temperature. As of 2017, the global mean temperature had already risen more than 1.0 °C (1.8 °F) above pre-industrial levels (IPCC, 2018). The IPCC has assessed future scenarios using this same metric. Two common scenarios now analyzed are measures needed to limit global warming to mean temperature increases above pre-industrial levels of: (a) 1.5 °C (2.7 °F) at 50% probability or greater, and (b) 2.0 °C (3.6 °F) at 67% probability or greater. Each scenario is, in turn, associated with a range of indicators and consequences such as the events and impacts described above. For example, IPCC provides a detailed report on the consequences and "reasons for concern" associated with 1.5 °C (IPCC, 2018).

2.1.2 Four basic strategies to solve climate change

There are four basic strategies to solve climate change: mitigation, adaptation, carbon removal, and solar radiation management (IPCC, 2021; IPCC, 2022a; IPCC, 2022b).[1]

IPCC defines *mitigation* of climate change as "A human intervention to reduce the sources or enhance the sinks of greenhouse gases." While IPCC's definition of mitigation includes enhancing sinks, in this book the use of the term mitigation will refer only to reducing the sources of GHGs unless otherwise specified. Enhancing sinks is described below as a separate strategy, common called carbon removal.

IPCC defines *adaptation* for both human and natural systems. Facilities fall under the category of human systems, with this definition provided by IPCC: "In human systems, the process of adjustment to actual or expected climate and its effects, in order to moderate harm or exploit beneficial opportunities." Chapter 8 on resiliency is directly about climate change adaptation for facilities.

Carbon or carbon dioxide removal (CDR), as defined by the IPCC, refers to technologies, practices, and approaches that remove CO_2 from the atmosphere and durably store it (IPCC CDR Factsheet, 2024). For example, the Science Based Target Initiative (SBTi) has a standard for corporations to achieve net-zero GHG emissions, and generally allows for up to 10% of emissions to be offset by carbon removals (SBTi, 2025)[2].

Solar radiation modification (sometimes referred to as solar radiation management) is a strategy to alter the radiation balance of the Earth by injecting particles (e.g., sulfur or other engineered particles) into the upper atmosphere, altering cloud reflectivity, and other similar measures (IPCC, 2022a). This strategy is sometimes referred to as solar geoengineering (Keith,

[1]For an overview of climate change solutions that are less voluminous than the IPCC reports and written for general audiences, see Gates, 2021.

[2]Some literature sources and experts describe CO2 removal from the atmosphere as geoengineering.

2021). The intended outcome of solar radiation management is to induce cooling effects for regions of the Earth. This is an important, complex, and often controversial strategy for solving climate change. While energy management professionals are encouraged to be educated about solar radiation management, in practical terms, this strategy is not within the scope of facility energy management.

For energy management of facilities, reducing GHG sources (i.e., direct and indirect GHG emissions) is the key objective. And the goal or formal policy throughout the U.S. and Canada is to decarbonize 100% of facilities to achieve economy-wide net-zero GHG emissions before 2050 (National Academies of Science, 2021[3]; Canadian Net-Zero Emissions Accountability Act, 2021). For reference, IPCC defines net-zero CO_2 emissions as the "Condition in which anthropogenic carbon dioxide (CO_2) emissions are balanced by anthropogenic CO_2 removals over a specified period." In practice, if the goal of net-zero GHG emissions is to be achieved, then any residual GHG emissions from facility operations will need to be removed using sinks outside the facility boundary, such as CDR methods (Allen, 2022). In general, enhancing sinks and capturing CO_2 from air are challenging due to cost, complexity, ability to verify, and/or ability to ensure permanency of carbon removal (IPCC CDR Factsheet, 2024). For these reasons, achieving zero or very near zero GHG emissions from facilities is critical. Section 2.2 and the remaining sections of this chapter are dedicated to this objective.

2.1.3 GHG trends for the U.S. and Canada

The discussion above on climate science and solutions clearly demonstrates the importance of reducing GHG emissions. Figure 2.3 provides a trend of GHG emissions for the U.S. and Canada. As seen in the data, emissions have declined since the year 2000 in the U.S., but there has been no significant change in Canada (except for a drop in 2020 attributed to COVID-19). The decline in the U.S. has been due to multiple factors, but the largest contributor has been a switch from coal to natural gas for electricity generation. For the year 2019, the U.S. and Canada comprised 12.5% and 1.4% of global emissions, respectively. It may be tempting to consider the U.S. and Canada as having a relatively minor role in solving global climate change. However,

[3] As of the publication date of this book, the U.S. policy has been recinded but the Canadian policy remains in place.

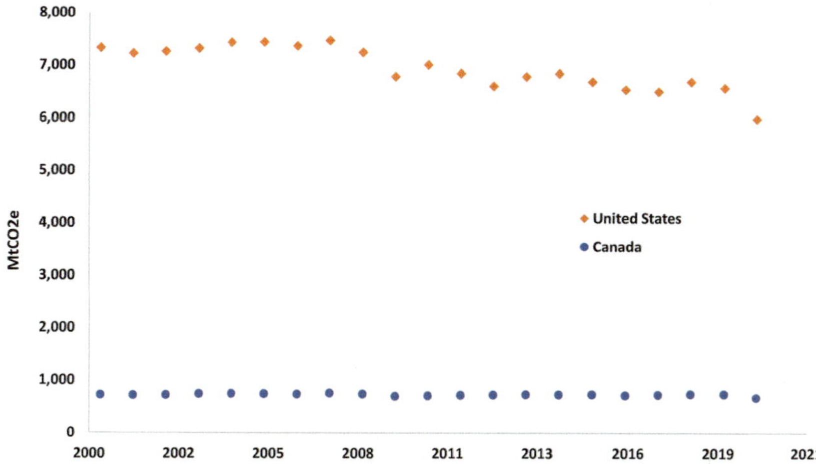

Figure 2.3 Annual trends in GHG emissions for the U.S. and Canada (source UNFCCC, 2022. Excludes land use, land-use change, and forestry).

it is critical to consider that per capita emissions[4] in the U.S. and Canada is about 18 tCO_2e per person in both countries, compared to a global average of less than 7 tCO_2e. Additionally, the U.S. and Canada both contribute a carbon footprint larger than what these statistics indicate through a number of ways including export of hydrocarbon fuels and consumption of goods produced overseas. Another critical fact is that climate change is proportional to cumulative emissions over time, and the U.S. and Canada have contributed about 25% and 2% to current cumulative emissions,[5] respectively. Moreover, the U.S. and Canada exert disproportionate influence on the rest of the world through the development of technologies and policies that can work to either exacerbate or reduce climate change.

A breakdown of GHG emissions by sectors for the U.S. and Canada is provided in Figures 2.4 and 2.5, respectively. In the U.S., electricity, transport, and industry each comprise roughly one-fourth of the total emissions. Buildings and agriculture comprise nearly one-fourth of the remaining total.

[4]Per capita GHG emissions based on 2020 total GHGs excluding land use, land-use change, and forestry (LULUCF) and World Bank population statistics.

[5]https://ourworldindata.org/contributed-most-global-co2
https://www.nature.com/articles/s41597-023-02041-1/figures/1

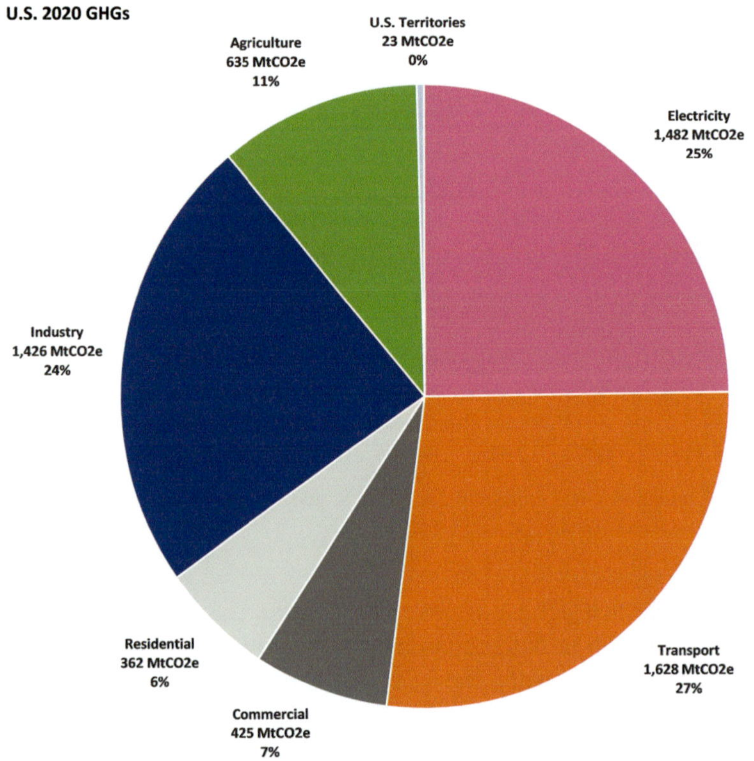

Figure 2.4 2020 GHG emissions for the U.S. (source UNFCCC, 2022. Excludes land use, land-use change, and forestry).

In Canada, the transport and oil and gas sectors each comprise roughly one-fourth of the total. Electricity comprises a relatively small share at 8%, due to the dominance of hydroelectricity and nuclear in the generation mix. Buildings comprise 13% of emissions. Industry (not counting oil and gas industry, reported as a separate sector) comprises 11%, with agriculture, waste, and other sources contributing the remainder.

Because this book focuses on energy management for residential, commercial, and industrial facilities, it is important to emphasize the role of these sectors in GHG management. Buildings and industry account for a larger impact on climate change than indicated in the figures above. Facilities consume by far the largest share of electricity, as well as large quantities of natural gas and other fuels. Additionally, one of the key strategies in climate

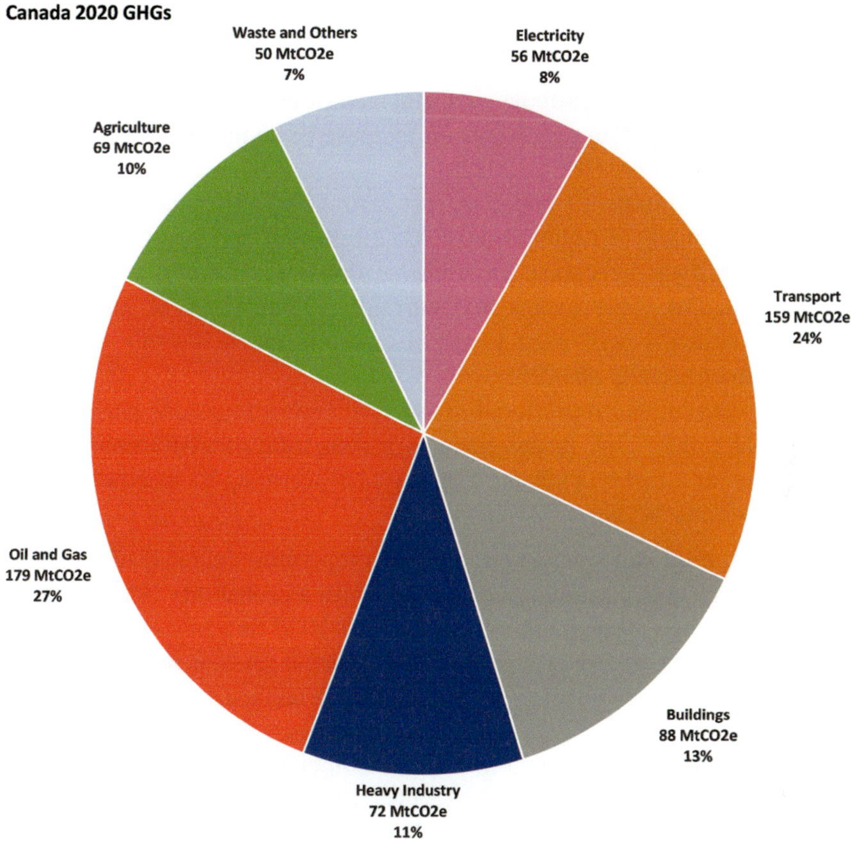

Figure 2.5 2020 GHG emissions for the Canada (source UNFCCC, 2022. Excludes land use, land-use change, and forestry).

policy is to electrify end uses in buildings, industry, and transportation (and a large share of vehicle charging is building electric systems). This means facilities will have an increasing impact on the electricity sector over time. Because of these linkages and policy goals, GHG management for buildings and industrial facilities is critical for solving climate change.

2.1.4 Key facts about climate change science

We conclude this introduction to climate change with some facts based on current science (not necessarily in order of importance):

1. Climate change is the most challenging environmental problem faced by humankind ever, based on the scope, scale, and complexity of the problem.
2. In 2019, global net anthropogenic GHG emissions[6] were estimated at 59 +/– 6.6 $GtCO_2e$. This represents an increase of about 12% since 2010 and 54% since 1990 (IPCC, 2022a).
3. For Earth's climate to stabilize, net-zero GHG emissions must be achieved globally. To quote the IPCC: "Reaching net zero CO_2 emissions globally is necessary for limiting global warming to any level" (IPCC, 2022a). This means that current net emissions of about 59 $GtCO_2e$ annually will need to get to zero to enable Earth's climate system to eventually stabilize.
4. Global warming is proportional to the cumulative emissions since pre-industrial times. This means there is a sum total of GHG emissions over time, or a carbon budget, to limit global warming to specific levels (Matthews and Wynes, 2022):

   - To limit global warming to a mean temperature increase of 1.5 °C (2.7 °F) above pre-industrial levels (with a probability >50%), the remaining carbon budget from the year 2020 onwards is estimated to be 500 $GtCO_2e$ (IPCC 2022a). If current rates of net emissions persist, this means 1.5 °C (2.7 °F) is predicted to occur around 2028.
   - To limit global warming to a mean temperature increase of 2.0 °C (3.6 °F) above pre-industrial levels (with a probability >67%), the remaining carbon budget from the year 2020 onwards is estimated to be 1,150 $GtCO_2e$ (IPCC 2022a). If current rates of net emissions persist, this means 2.0 °C (3.6 °F) is predicted to occur around 2039.

5. While achieving net-zero GHGs globally would stop global mean temperature from increasing, this would not reverse warming, which already occurred, or rapidly stop the inertia of many processes (e.g., sea level rise). For these reasons, adaptation, solar radiation management, and removal of CO_2 from the atmosphere will likely remain as proposed or implemented strategies for generations. However, it would be faulty logic to assume that climate damage is already done and that people should just abandon GHG reductions and rely on adaptation or solar radiation management. It is critical to understand that there is no known

[6]Section 2 of this chapter explains the calculation of CO2e, which is CO2 equivalents.

temperature or emission threshold. Increased global warming and cumulative impacts will continue over time unless net-zero GHGs have been achieved.

2.2 Quantifying GHG Emissions

This section provides an introduction to the methods and data for quantifying GHG emissions for facilities including global warming potential, GHG emission factors, and GHG inventory basics.

2.2.1 Global warming potential of GHG pollutants

Thus far, in this chapter, we have referred to GHGs in units of CO_2e, CO_2, or carbon. In this section, we will define these metrics.

As mentioned in Section 2.1, GHGs refer to gases (or in some cases, aerosols or particles) that alter the radiative forcing and heat balance of the Earth. While there are many GHGs and sources in general, for facility energy management, the most important gases are emitted from fuel combustion or released by fugitive leaks of refrigerants. Because each gas has different radiative forcing effects due to its chemical and physical properties, methods have been developed to quantify global warming effects on a common basis called global warming potential (GWP). Table 2.1 lists the most common GHGs encountered in facility energy management, including common sources and GWP values. The most common sources are fuel combustion, methane leaks, and refrigerant emissions. GWP accounts for the radiative forcing of a gas, and the lifetime of the gas in the atmosphere. Because GWP is a function not only of the gas properties but also of the time horizon of concern, GWP values are commonly provided for different lifetimes. Table 2.1 provides 20-year and 100-year GWP values. In general, 100-year values are the most common metric used in practice. Unless otherwise stated in this book, GWP values are the 100-year values.

To illustrate the importance of GWP, Figure 2.6 (UNFCCC, 2022; IPCC, 2022a) provides the breakdown of GHGs for the U.S. and Canada combined, in units of CO_2e. CO_2 comprises 99.5% of actual emissions vs. 79% when weighted by GWP. Methane comprises 0.47% of actual emissions vs. 7% weighted by GWP. N_2O comprises 0.032% of actual emissions vs. 11% weighted by GWP. Finally, only 0.004% for the other gases accounts for about 3% when weighted by GWP. These facts reinforce the importance of reducing all of the principal GHG pollutants to mitigate climate change.

Table 2.1 Common GHGs, common sources, and global warming potentials (IPCC, 2021; IPCC, 2013).

Chemical symbol or name	Chemical name	Common sources	Atmospheric lifetime	20-year GWP	100-year GWP
CO_2 (R-744)	Carbon dioxide	Fuel combustion	100 years (some fraction > 1,000 years	1	1
CH_4	Methane	Fossil fuel natural gas leaks	12 years	82	29.8
N_2O	Nitrous oxide	Fuel combustion, agriculture	109 years	273	273
SF_6	Sulfur hexafluoride	manufacturing: smelting, high-voltage equipment, photovoltaics	3,200 years	17,500	23,500
R1234ZE ($CF_3CH{=}CHF$)	trans-1,3,3,3-Tetrafluoroprop-1-ene	Heat pumps, refrigeration systems	Days	<1	<1
CHF_3 (HFC-23)	Hydrofluorocarbon	HVAC refrigerant manufacturing	222 years	10,800	12,400
CH_2FCF_3 (HFC-134a)	Hydrofluorocarbon	HVAC refrigerant	14 years	4,144	1,526
CF_4 (PFC-14)	Perfluorocarbons	Manufacturing of aluminum and electronics	50,000 years	5,301	7,380
C_2F_6 (PFC-116)	Perfluorocarbons	Manufacturing of aluminum and electronics	10,000 years	8,210	1,100
NH_3	Ammonia	Food processing and storage, ice-making	2–24 hours	0	0[7]

As noted above, GHG emissions can be reported in terms of actual (i.e., unweighted) emissions, or CO_2e when weighted by GWP factors. Another common metric used is to express GHGs as equivalent mass of carbon emissions. With this metric, actual CO_2 (or sometimes CO_2e) emissions are scaled by the molecular weight of carbon to CO_2, which represents emissions as mass of carbon equivalent. The molecular weights for carbon and CO_2 are 12 and 44, respectively. As an example, 100 kg of CO_2 could be reported as 27.3 kg of carbon (100 kg $\times$ 12/44 $=$ 27.3). Lastly, it should be noted that the use of the term "carbon" is often used generically to mean GHGs, regardless of the units, including the use of that term in this book. However, if calculations or GHG statistics are involved, it is necessary that the specific unit or metric be stated.

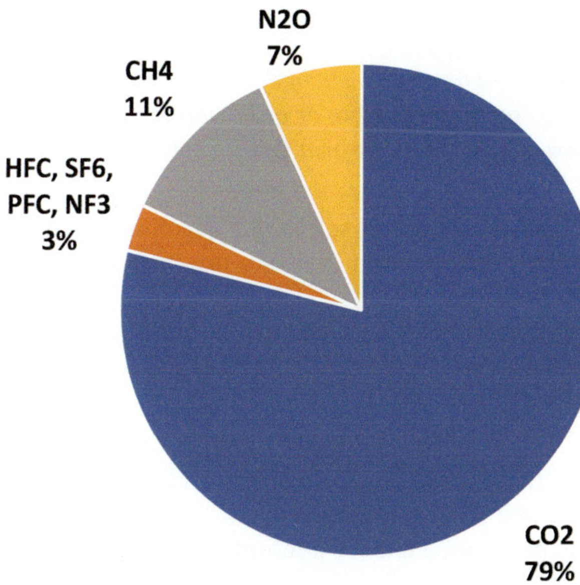

Figure 2.6 Contribution to global warming by pollutant for U.S. and Canada combined based on CO_2 equivalent emissions (CO_2e) using 100-year GWP factors.

2.2.2 Quantifying GHG emissions for facilities

This section provides a concise description of the concepts, methods, and terminology that are essential to quantify GHG emissions as well as to effectively implement and validate GHG reduction projects. The practice of quantifying GHG emissions for facilities is similar to the practice of M&V introduced in Chapter 1. Quantifying GHGs is guided by common protocols, mainly by the International Organization for Standardization (see ISO 14064-1:2018, 0000, ISO 14064-2:2019, 0000, ISO 14064-3:2019). Widely used guidelines, tools, and training are also provided by the Greenhouse Gas Protocol (www.ghgprotocol.org), which is administered by the World Resources Institute and the World Business Council for Sustainable Development.

Quantifying GHGs is conducted by careful selection of system boundaries, and identification of sources, sinks, and reservoirs of GHGs as defined in ISO 14064:

- Sources: *Any physical unit or process that releases a GHG into the atmosphere.*

- Sinks: *Any physical unit or process that removes a GHG from the atmosphere.*
- Reservoirs: *A physical unit or component of the biosphere, geosphere, or hydrosphere with the capability to store or accumulate a GHG removed from the atmosphere by a GHG sink or a GHG captured from a GHG source.*

GHG emissions are then quantified by selecting a quantification methodology, which includes three options: measurement, use of published GHG emission factors, or expert estimates using models.

An important convention in GHG management is to categorize sources based on system boundaries. Table 2.2 provides the three categories for GHG emissions as defined by ISO 14064 and the GHG Protocol.

Two key concepts in quantifying and managing GHGs are:

- *Reversibility* applies to sinks and reservoirs and considers whether stored GHGs could be released (e.g., if a tree plantation is counted as a sink, stored carbon could be released into the atmosphere as CO_2 and CH_4 by wildfire).
- *Leakage* refers to reducing GHGs within the system boundary, which could be offset by release of GHGs outside the system boundary (e.g., use of electricity from solar PV in the afternoon requires the utility to run natural gas peaker plants in idle prior to ramping up to meet late afternoon demand, mitigating some of the carbon reductions from using solar power).

Table 2.2 GHG source categories (www.ghgprotocol.org; IPCC, 2022).

Category	Description	Examples
Scope 1	Direct emissions from within the system boundary	Emissions of CO_2, CH_4, and N_2O from a boiler stack. Fugitive emissions of refrigerant from a heat pump.
Scope 2	Indirect emissions from outside the system boundary associated with purchased electricity, steam, heating, and cooling.	Upstream GHG emissions from the production and transport of electricity or steam, hot water, and chilled water from thermal energy networks.
Scope 3	Other indirect emissions from outside the system boundary	Air travel by company staff. GHG emissions from product manufacturing and transport of materials to a facility through a supply chain. Embodied GHG emissions to construct buildings and factories.

Quantifying GHGs for facilities can involve a substantial amount of time and money, and requires addressing many complexities. However, the process can be summarized concisely as follows:

1. Specify the project boundary.
2. Identify sources, sinks, and reservoirs for both the baseline system and GHG reduction project.
3. Select quantification methodologies (i.e., measurement, emission factors, or expert models).
4. Determine data required.
5. Estimate expected uncertainty (IPCC, 2001).
6. Identify any potential leakage.
7. Identify whether any sinks or reservoirs are reversible.
8. Calculate and report the emissions of GHGs in CO_2e (or some other acceptable metric).

In the management of GHGs for facilities, it is relatively rare to encounter the use of sinks or reservoirs within the facility boundary. Also, leakage is rarely examined at a facility scale. Moreover, Scope 3 emissions are less often included in the accounting GHGs for facilities (although this is changing). In current practice, Scope 1 and Scope 2 sources of GHGs are usually the main focus of GHG management.

2.2.3 Scope 1 GHG emission factors: direct emissions

Emissions of the three principal GHGs from fuel combustion are of primary importance in quantifying and reducing GHG emissions. GHG emissions are a function of the fuel composition (e.g., carbon content, trace elements), combustion equipment design (e.g., gas turbines, steam boilers), and equipment operation (e.g., excess air for combustion, process temperatures). The most common approach is to use published emission factors because direct measurement and modeling are often costly or logistically not feasible. Because emission factors vary by location and application, the specific factors used for a particular project should be clearly specified and agreed upon by project stakeholders. For most purposes, published GHG emission factors, which represent typical fuel composition, common equipment, and average operating conditions, can be used. Table 2.3 provides emission factors for common fuels. Emission factors are provided on both an HHV and LHV basis in international units, and HHV in English units. Additional emission factors for other fuels are provided in the appendix.

Table 2.3 Fuel combustion emission factors for CO_2, CH_4, and N_2O for common fuels (IPCC, 2006; EPA, 2022; WMO, 2022).

Fuel	kg CO_2/ GJ fuel LHV basis	kg CH_4/ GJ fuel LHV basis	kg N_2O/ GJ fuel LHV basis	kg CO_2e/ GJ fuel LHV basis	kg CO_2e/ GJ fuel HHV basis	lbm CO_2e/ therm fuel HHV basis
Natural gas	56.1	0.0010	0.0009	56.4	51.3	11.9
No. 2 oil (diesel)	74.1	0.003	0.0006	74.4	69.5	16.2
RNG	56.1 or 0*	0.0010	0.0009	0.276*	0.248*	0.0578*
Bituminous coal	94.6	0.030	0.0015	104.0	99.8	23.2
Liquified petroleum gas	63.1	0.0050	0.0001	63.3	58.4	13.6
Residential wood (0% H_2O)	82.1 or 0*	0.683	0.0067	22.2*	21.1*	4.90*

* RNG and wood fuel CO_2 emissions are counted as zero on the assumption that carbon in the fuel is sourced from renewable supplies. CO_2e emissions only include CH_4 and N_2O.

Example 2.1: An energy savings project for a boiler in a multifamily building is estimated to reduce natural gas consumption by 100 GJ per month (HHV basis). Estimate the monthly GHG reduction, in units of kg and lbm of CO_2e.

Solution 2.1: To demonstrate the calculation of CO_2e emission factors, the emissions will be calculated using the emission factors for CO_2, CH_4, and N_2O directly. The 100-year GWP values in Table 2.1 and emission factors in Table 2.3 are used. Because emission factors in Table 2.3 are on an LHV basis, first convert the gas consumption to LHV using calorific values in the appendix:
(100 GJ) (49.3 MJ/kg LHV)/(54.2 MJ/kg HHV) = 90.96 GJ on LHV basis.

In international units:
CO_2: (90.96 GJ) (56.1 kg CO_2/GJ) (1 kg CO_2e/ kg CO_2) = 5,102.9 kg CO_2e
CH_4: (90.96 GJ) (0.0010 kg CH_4 /GJ) (29.8 kg CO_2e/ kg CH_4) = 2.7 kg CO_2e
N_2O: (90.96 GJ) (0.0009 kg N_2O/GJ) (273 kg CO_2e/kg N_2O) = 22.3 kg CO_2e
Total GHGs = 5,128 kg CO_2e per month emissions.
Check using HHV CO_2e factor from Table 2.3: (100 GJ/month) (51.3 kg CO_2e/GJ) = 5,130 kg CO_2e.

In English units:
(5,128 kg CO_2e) (lbm/0.4536 kg) = 11,305 lbm CO_2e per month
Check using HHV CO_2e factor from Table 2.3:

Table 2.4 Electricity generation average GHG intensity for U.S. regions for 2020 (EIA, 2022).

North American Electric Reliability Corporation Region (www.nerc.com)	tCO_2e/GWh	$lbmCO_2e/MWh$
Midwest Reliability Organization	449	991
Northeast Power Coordinating Council	229	505
Reliability First Corporation	424	934
SERC Reliability Corporation	395	871
Texas Reliability Entity	407	898
Western Electricity Coordinating Council	326	719
Region Not Reported	643	1,420
US average	**388**	**854**

(100 GJ)(277.78 kWh/GJ)(3,412 Btu/kWh)(therm/100,000 Btu) = 947.8 therms (947.8 therms)(11.9 lbm CO_2e/therm) = 11,300 lbm CO_2e.

2.2.4 Scope 2 GHG emission factors: indirect emissions from purchased energy

Scope 2 emissions account for the indirect GHGs due to purchased energy. Typically, electricity is the main form of energy included in GHG accounting. In conducting GHG analyses, energy management professionals need to use GHG intensity emission factors for purchased electricity. For specific projects, often the stakeholders (e.g., facility owners, funding agency, or electric utility) will specify what values to use. However, there are public sources of these emission factors such as those provided in Table 2.4 for the U.S.[8] and Table 2.5 for Canada. For convenience, average emission factors are provided in both international and English units in both tables.[9]

It is important to emphasize that the GHG intensity values in the tables above represent *average* values for electricity generation for the calendar years as noted. In conducting GHG analyses, these may not be the best values to use. For GHG reduction projects, the *marginal* values would be the most accurate, not the average. Consider a region with a mix of hydroelectricity and natural gas generation. At the margin, it may be the case that natural gas

[8] For GHG intensity factors on a more granular basis, such as by state or generation plant, see www.epa.gov/egrid/data-explorer

[9] GHG intensities based on generation sources published by Canada Energy Regulator for year 2019. GHG intensities in tCO2e/GWh calculated assuming: natural gas = 508, biomass/geothermal = 110, coal & coke = 990, oil = 783. All other generation sources counted as 0 GHGs.

Table 2.5 Electricity generation average GHG intensity for Canadian provinces and territories for 2019 (CER, 2022).

Province or territory	tCO_2e/GWh	$lbmCO_2e/MWh$
Newfoundland and Labrador	29.7	66
Prince Edward Island	4.1	9.1
Nova Scotia	641	1,410
New Brunswick	225	495
Quebec	3.0	6.7
Ontario	41.2	91
Manitoba	0.89	2.0
Alberta	600	1,320
British Columbia	29.7	65
Saskatchewan	614	1,350
Yukon	99.9	220
Northwest Territories	436	961
Nunavut	783	1,730
Canada average	**131**	**289**

generators modulate to meet system peak demand and that electricity savings result in avoided GHGs from natural gas. In this situation, it could mean that avoided GHGs are greater than the average electricity GHG intensity value. On the other hand, consider a region where solar PV is the dominant supply for electricity at specific times of the day or season of the year. Electricity savings during these times may offset solar PV generation and thus represent zero GHG reduction for Scope 2 emissions. In practice, it is challenging to find a reliable (or even any) source of marginal GHG emission intensity values for electricity supplies. However, when accurate marginal values are available, these should be used.

Another consideration is that the GHG intensity of purchased electricity is not solely a function of generation within a region. Imports and exports of electricity from different sources can substantially change either the average or marginal GHG intensity of electricity sold (purchased) within a region.

In conducting decarbonization studies (see Chapter 3), often GHG forecasts are needed for many years going forward for lifecycle analyses. For example, if the expected in-service life for newly electrified industrial equipment is 20 years, then a forecast of electricity GHG intensity would be needed for this time period. The U.S. and Canada have set targets to achieve net-zero GHG for the electricity sector by 2035.[10] As such, using fixed GHG intensity

[10]www.canada.ca/en/environment-climate-change/news/2022/07/government-of-canada-launches-second-round-of-public-engagements-to-establish-the-clean-electricity-regulations.html. Denholm, 2022 for the U.S.

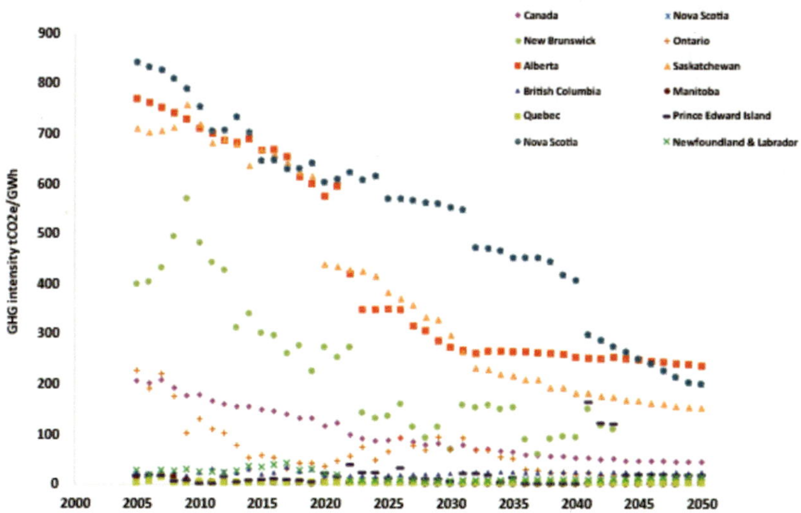

Figure 2.7 Average GHG intensity for electricity for Canada's provinces and territories. Values to 2019 based on data. 2020–2050 values based on Canada's Energy Regulator "Evolving Policies" scenario.

factors based on past or present data will often result in inaccurate GHG lifecycle estimates. Figure 2.7 provides a chart of forecasted GHG intensity for electricity for Canadian provinces and territories to year 2050 based on evolving policies (CER, 2022).[11] This chart illustrates that Canada has much progress to make to achieve the 2035 target. But the chart does provide an example of forecasted GHG intensities that are required to make realistic estimates of lifecycle GHG emissions.

In addition to electricity, Scope 2 GHG emissions apply to other purchased energies including steam, hot water, and chilled water from thermal energy networks (TENs) (Greenhouse Gas Protocol, 2015). The following two examples illustrate the methods to quantify Scope 2 emissions for these sources.

Example 2.2: A facility purchases steam generated and distributed at 80% efficiency (HHV basis) from natural gas. Estimate the Scope 2 GHG emissions per GJ of purchased steam energy.

[11]GHG intensities estimated using forecasted generation sources published by Canada Energy Regulator "Evolving Policies" scenario (https://apps.cer-rec.gc.ca). GHG intensity estimates by source in tCO2e/GWh: natural gas = 508, biomass/geothermal = 110, coal & coke = 990, and oil = 783. All other generation sources are counted as 0 GHGs.

Solution 2.2: One GJ of delivered heating energy requires 1/0.80 = 1.25 GJ of natural gas fuel. Using Table 2.3 emission factor for natural gas on HHV basis: (51.3 kgCO_2e/GJ) 1.25 GJ natural gas/GJ steam = 64.1 kgCO_2e per GJ of purchased steam energy.

Example 2.3: A facility purchases chilled water from a TEN system operating with a system COP of 3.5. Estimate the Scope 2 GHG emissions in lbm CO_2e per MMBtu of purchased cooling energy. Assume the chiller is supplied by electricity with GHG intensity based on the average for the Western Electricity Coordinating Council from Table 2.4.

Solution 2.3: 1 MMBtu of delivered cooling energy requires 1/3.5 = 0.2857 of electricity in units of MMBtu. (0.2857 MMBtu) (1.055 GJ/MMBtu) (0.27778 MWh/GJ) = 0.08373 MWh of electricity. From Table 2.4, the electricity GHG intensity is 719 lbm CO_2e/MWh.

(0.08373 MWh) (719 lbm CO_2e/MWh) = 60.2 lbm CO_2e per MMBtu of chilled water purchased.

Upstream emissions of natural gas are significant, mainly due to methane leaks and emissions from gas processing plant processes. However, indirect emissions from purchased natural gas are *not* counted as Scope 2, and are assigned to Scope 3 for reporting purposes (ICF, 2022).

Nonetheless, for some GHG studies, the upstream emissions from natural gas are quantified, and some utilities are including upstream GHG emissions from natural gas extraction, processing, and transmission in their performance reports. Here are two examples:

- District of Columbia Sustainable Energy Utility reports emission reductions associated with natural gas savings. An emission factor of 11.72 lbmCO_2e/therm (50.4 kgCO_2e/GJ) is used for direct emissions, and 14.25 lbmCO_2e /therm (61.3 kgCO_2e/GJ) including upstream emissions (DCSEU, 2022). This amounts to 2.53 lbmCO_2e/therm (10.9 kgCO_2e/GJ) for indirect emissions.
- FortisBC reports GHG emission reductions associated with natural gas energy savings using a lifecycle (well to burner tip) emission factor of 59.8 kgCO_2e/GJ on an HHV basis (FortisBC, 2022). Using the Scope 1 emission factor of 51.3 kgCO_2e/GJ from Table 2.3, this amounts to 8.5 kgCO_2e/GJ for indirect emissions.

Indirect emissions from RNG have been quantified and are roughly similar to indirect emissions from fossil natural gas, in the range of 8–10 kgCO_2e/GJ (1.86–2.33 lbmCO_2e/therm) (Grubert, 2020).

2.3 Strategies for Reducing GHG for Facilities

In Section 2.1 of this chapter, we provided an overview of climate science, emphasizing the urgency to reduce facility GHG emissions to as close to zero as possible. Section 2.2 provides essential concepts, data, and methods to quantify GHG emissions and emission reductions following common industry practices. With this background information, we now provide a summary of strategies and key concepts to decarbonize demand-side energy systems.

As introduced in Chapter 1, to meet policies and goals aimed to solve climate change, demand-side energy management has evolved to require consideration of multiple distributed energy resources (DERs). This section introduces how each DER (see Chapter 1 for our definition of DERs) can be applied to help achieve decarbonization goals, as well as a discussion of integrating multiple DERs. Because achieving truly zero GHG emissions is not feasible in practice, a brief discussion of counterbalancing residual emissions is provided.

2.3.1 Distributed energy resources to decarbonize end uses

2.3.1.1 Energy efficiency

The need for aggressive GHG reduction is often the first priority for many facility operations, and reducing consumption could be a lesser priority. However, even in cases where decarbonization is the dominant goal, energy efficiency remains as important as ever to achieve net-zero GHGs.

One study for the U.S. illustrates the key role for energy efficiency (Nadel and Ungar, 2019). This study assessed technologies, costs, and policies to achieve aggressive GHG reductions for U.S. facilities and transport systems. The study estimated that by 2050, energy efficient technologies, smart systems, and strategic energy management alone have the potential to achieve reductions of final energy use and GHG emissions by 29% for homes, 41% for commercial buildings, and 15% for industry.[12] These findings illustrate the benefits of combining energy efficient technologies and management practices with fuel switching to achieve large GHG reductions.

As introduced in Chapter 1, electricity is increasingly generated from renewable resources, which enables electrification of end uses as a major decarbonization strategy. Another study examined scenarios to electrify all

[12]Estimated reductions excluded: fuel switching, updated appliance standards, advanced net-zero new building designs, and advanced industrial processes and feedstocks.

building energy use in the U.S. (Buonocore, 2022). The study examined monthly energy data (not just annual averages) and estimated monthly consumption for scenarios to electrify all buildings. The modeled results showed that building electrification would shift utility grids from summer peaking to winter peaking. It also estimated that meeting these electrified building loads with a low-efficiency scenario (compared to a current baseline) would require a 28-fold increase in January wind generation or a 303-fold increase in solar. With a high-efficiency scenario, the January load could be met with a 4.5-fold increase from baseline in wind (i.e., about 84% less increase in capacity by including high efficiency upgrades) or 36-fold increase in solar (i.e., about 88% less increase in capacity by including high efficiency upgrades).

These macro-level studies might seem difficult to relate to specific projects that energy management professionals often work on. A project-specific perspective is provided in Figure 2.8. This figure shows the modeled results of heating load for a prototype home under two scenarios: a standard building envelope and an efficient building envelope (ICF, 2022). The design heating load was modeled as a function of outdoor air temperature. The

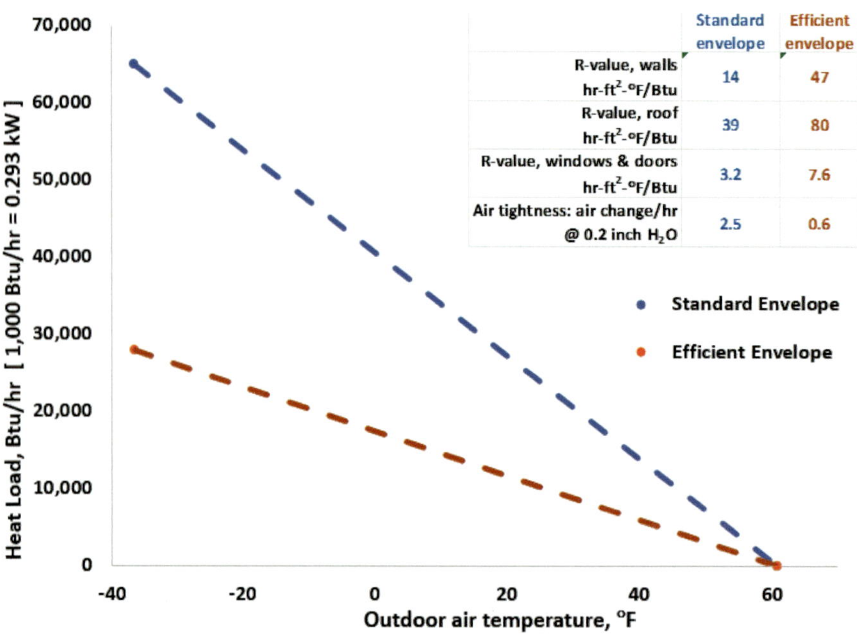

Figure 2.8 Modeled building heat load as function of outdoor air temperature, comparing standard and efficient building envelope designs (IPCC, 2006).

Table 2.6 Hours of life-safety temperatures after a power outage with outdoor temperatures fluctuating around −10 °F (−23 °C)* (RMI, 2020).

House model	**Hours to fall below 40 °F (4.4 °C)**	Walls,R, ft^2-hr-°F/Btu	Attic/roofR, ft^2-hr-°F/Btu	Slab/ foundationR, ft_2-hr-°F/Btu	Windows (15% window to wall ratio) R, ft^2-hr-°F/Btu	Air changes at 50 Pa (0.2-inch H_2O) pressure differential
Typical 1950's home	**8**	None	None	None	0.8	15
Typical 1980's home	**23**	R13 in-frame	R19 blown	8 inch, R5 exterior	1.4	10
Code-compliant 2009 home	**45**	R13 in-frame + R5 continuous	R13 in-frame + R36 blown	4 feet, R10 exterior	2.9	7
Net-zero energy ready home	**61**	R13 in-frame + R10 continuous	R13 in-frame + R36 blown	4 feet, R10 exterior	3.4	3
Passive house	**152**	R13 in-frame + R47 continuous	R35 in-frame + R60 blown	R57, 12 inch foam around 9 inch slab	8.0	0.6

* To convert *R*-values to international units, the conversion factor is: 1 ft^2-hr-°F/Btu = 0.1761 m^2-°C/W.

results show that the heating load could be cut in half across a wide range of temperatures (i.e., range of climate zones) with advanced building envelope upgrades. These design parameters are illustrative only, but do quantify the potential value of investments in upgrading passive systems as described in Chapter 11 (Chapter 20 on industrial also includes some examples of passive system improvements). If the estimates in Figure 2.8 were applied to a heat pump electrification project, the reduction in load would reduce the size and capital cost of the new heat pump by roughly 40%.[13] Additionally, ongoing savings in peak demands and average consumption of electricity would be realized over the life of the project.

The potential benefits of combining energy efficiency upgrades with decarbonization go beyond optimizing electricity grid infrastructure requirements, facility capital costs, and facility operating costs. Consider the results of a modeling study conducted by the Rocky Mountain Institute in Table 2.6 (RMI, 2020). The study modeled a sudden loss of heating systems due to a power outage for several different building envelope designs, while outdoor temperatures fluctuated around −10 °F (−23 °C). The number of hours the indoor temperature stayed above a life-safety level of 40 °F (4.4 °C) was

[13] A 50% decrease in heat pump capacity is assumed to correspond with a 40% decrease in capital cost.

estimated. As the table shows, a home built to typical 1950's standards would dip below 40 °F after about 8 hours, while a passive house design would hold temperatures above 40 °F for 150 hours. It might seem that this hypothetical study may not apply to the real world. However, this study was published about one year before the Texas electric grid blackout occurred in 2021. This outage caused over 10 million customers to survive without power and heating systems for over 72 hours, with temperatures dipping to −25.6 °C (−14 °F), and resulted in 226 deaths (WMO, 2022). The study results in Table 2.6 combined with the actual outage in Texas provide a clear illustration of the resiliency benefits of energy efficient building envelopes. Chapter 8 provides an introduction to resiliency for facilities.

2.3.1.2 Fuel switching

Fuel switching is a strategy to eliminate fossil-fuel consumption by converting end uses to low-/no-carbon energy sources such as renewable electricity, biofuels, or green hydrogen. Common fuel switching strategies employed in decarbonization projects are summarized below.

- *Convert fossil-fueled end uses to electrified equipment with low-/no-carbon electricity, commonly described as electrification.* As noted earlier in this chapter, targets have been set to achieve net-zero GHGs electricity supplies by 2030 for the U.S. and by 2035 for Canada. Converting natural gas-fired space and water heating in buildings, and process heat in industry are common examples of electrification. Chapter 10 provides the basics of electric heat pumps for buildings. For high-temperature or specialized industrial processes, advanced electric heating technologies such as infrared, microwave, and induction heating may be technically viable (Rissman, 2021). In homes, replacing natural gas stoves with induction cooktops is a common substitution.
- *Convert fossil-fueled end uses to low-/no-carbon biofuels such as RNG or biomass.* Building space heating, TENs, combined heat and power (CHP), and industrial processes are examples where low-carbon biofuels may substitute for fossil fuels. One of the benefits of RNG is that it can displace fossil natural gas without the need to replace the end use boilers and heating equipment. Chapter 10 includes a brief introduction to gas-powered heat pumps, which is an application where RNG could be used. Chapter 20 includes a case study for a CHP and district energy TEN system utilizing RNG in the fuel supply mix. Section 2.2 of this chapter

provides typical emission factors for direct and indirect GHG emissions for RNG and some biomass fuels.

- *Convert fossil-fueled end uses to low-/no-carbon hydrogen.* In some applications, such as industrial processes requiring heating temperatures over 400 °C (750 °F), direct electrification of end use equipment may not be technically or economically viable (Ueckerdt, 2021). In these applications, the use of hydrogen may be viable to displace fossil fuels. Two common approaches to decarbonize hydrogen supplies are commonly referred to as "green" and "blue" hydrogen (Rissman, 2021; Uekerdt, 2021). Green hydrogen is produced by electrolysis, splitting hydrogen from H_2O molecules using net-zero GHG electricity. Blue hydrogen is produced from steam reforming of natural gas (mostly CH_4) to split hydrogen and carbon molecules, while employing carbon capture and sequestration to avoid emitting CO_2 to the atmosphere.
- *Replacing gasoline or diesel-fueled vehicles with electric vehicles* is a major fuel switching measure to achieve help decarbonize transport systems. While transport systems are not within the main scope of this book, vehicle charging from building electric systems is rapidly growing. Chapter 13 on electric systems includes a brief introduction to vehicle charging stations.

Referring to the Chapter 1 discussion of services, to increase the chances of a successful fuel switching measure, the quantity and quality of services delivered by the end use system should be maintained at an acceptable (or improved) level. In all of the fuel switching strategies described above, it is important to maintain services at levels acceptable to the end users such as hot water flows and temperatures, cooking performance of stoves, thermal comfort conditions for HVAC, and industrial process heating temperatures.

Future energy prices are a key issue in fuel switching. Table 2.7 provides one forecast of future prices for RNG, green hydrogen, and synthetic natural gas[14] for the state of New York (Nadel, 2022). Relative to expected prices of fossil natural gas, these carbon neutral fuels are projected to range from 6.4 to 16 times more expensive in 2030 and from 6.6 to 8.4 times more expensive by 2050.

[14]Synthetic natural gas is the production of a fuel with roughly the same hydrocarbon composition as natural gas, but produced from carbon captured from air, hydrogen from water, and powered with zero carbon electricity.

Table 2.7 Forecast price of biofuels and fossil natural gas in New York (1 MMBtu = 1.055 GJ).

Fuel		2030	2040	2050
Fossil natural gas	\$US/MMBtu	\$2.5	\$2.5	\$2.5
Biogenic RNG	\$US/MMBtu	\$16.0	\$18.5	\$16.5
	ratio vs. fossil natural gas	6.4	7.4	6.6
Green hydrogen	\$US/MMBtu	\$32	\$21	\$18
	ratio vs. fossil natural gas	13	8.4	7.2
Synthetic natural gas	\$US/MMBtu	\$40	\$26	\$21
	ratio vs. fossil natural gas	16	10	8.4

Similarly, the total cost to produce and transmit electricity to achieve 100% net-zero GHG electricity in the U.S. by 2035 has been modeled (Denholm, 2022). Electricity production cost is projected to increase \$15–\$39 per MWh (35%–91%), compared to a reference scenario not constrained to net-zero GHGs.

Future forecasts of energy prices have often been wrong historically. However, even if these projections are accurate in the rough order of magnitude, the implications are that energy costs for fuel switching may be higher. The simple example below illustrates how facility energy costs can change as a result of fuel switching.

Example 2.4: A commercial building complex is heated with a natural gas heating plant at 90% efficiency (HHV). Consider the year 2035 and compare the energy cost per unit of heat delivered for a fossil natural gas heating plant, an electric heat pump assuming COP = 3.0, and an RNG heating plant. Assume prices of \$2.6/GJ for fossil natural gas, \$16/GJ for RNG (based on Nadel, 2020), and \$61/MWh electricity (based on Denholm, 2022 system cost of \$61 plus 10%). Estimate the energy cost in dollars per unit thermal energy delivered in MWh.

Solution 2.4: This simplified fuel switching example is sometimes referred to as a point-of-use calculation. The three options are calculated as follows:

- Fossil natural gas: (\$2.6/GJ) (GJ/0.27778 MWh) (0.90) = \$8.4 per MWh delivered
- Electric heat pump: (\$67/MWh) (1/3.0) = \$22 per MWh delivered
- RNG: (\$16/GJ) (GJ/0.27778 MWh) (0.90) = \$52 per MWh delivered

In this example, if the electric heat pump option could be combined with a 60% reduction in thermal load through advanced passive system upgrades, then the heat pump energy cost would be about the same as the fossil natural gas as a baseline. This reinforces the importance of combining passive and active energy efficient technologies in decarbonization projects.

2.3.1.3 Distributed generation

In this book, we refer to distributed generation in the more general interpretation as the production of commercial energy at or near end use systems.[15] Driven by policies to decarbonize electricity supplies, the most rapidly growing application of distributed generation for the foreseeable future is expected to be solar PV. Chapter 18 covers solar PV as well as solar thermal systems. Wind power has been dominated by utility scale turbines, which have increased in size and capacity over the past decade (DOE, 2023).[16] While behind the meter wind power has strong economic potential of 919 GW in the U.S. (DOE, 2022), it is currently small capacity and growth rate as compared to solar PV. CHP is another common distributed generation system design, typically defined as the simultaneous production of electricity and thermal energy from one conversion device such as a gas turbine or reciprocating engine. CHP systems are a viable decarbonization technology because CHP typically increases overall system efficiencies, and can be powered by biofuels. TEN systems are yet another common approach to distributed generation. Where TEN thermal production plants are available that utilize low-carbon fuels, this can be an option to help decarbonize facilities. Chapter 19 provides an introduction to TEN systems.

2.3.1.4 Demand response

A common definition of demand response (DR) is "Demand response refers to balancing the demand on power grids by encouraging customers to shift electricity demand to times when electricity is more plentiful or other demand is lower, typically through prices or monetary incentives."[17] Driven by policies to address climate change, two trends are making DR more critical:

---

[15]Many sources define distributed generation as behind-the-meter electricity generation. Some use the term more broadly to include electric or thermal energy (e.g., www.eia.gov/analysis/studies/buildings/dg_storage_chp). Yet other sources define it as near end uses, but not necessarily behind the meter (e.g., www.epa.gov/energy/distributed-generation-electricity-and-its-environmental-impacts).

[16]In the U.S., 56% of front-of-meter, land-based wind turbines were < 2.0 MW in 2011, and 0.2% in 2022. 77% of land-based turbines had rotor diameter <100 m in 2011, and 0% in 2021.

[17]www.iea.org/energy-system/energy-efficiency-and-demand/demand-response. A similar definition is provided by SEE Action, 2020.

increased supply of electricity from variable renewable resources, and electrification. As introduced in Chapter 1, the supply side of electricity systems is transitioning to increase the supplies of renewable sources and decrease generation by fossil fuels. This means that matching electricity supply and demand can no longer be relied upon over the long term by utilizing large underground natural gas storage, large coal piles, or huge oil tanks to power fossil-fueled electric generators.[18] Fuel switching to electrify end uses in facilities is estimated to shift the timing and magnitude of peak electricity demands, as illustrated with the discussion above on building electrification (Buonocore, 2022). In addition to electrifying end uses in buildings, electrification of transportation is yet another factor. For example, one study modeled the requirements to support electrification of all road transport vehicles in British Columbia by 2055 (Keller, 2019). It estimated that the electric grid capacity would need to increase as much as 14 GW, or over 80% of baseline capacity, and in much greater proportion than the increase in annual electricity generation (GWh). DR is a key strategy that can be employed (also storage, efficiency, and distributed generation) to balance electricity supply and demand under conditions of increasing renewable supplies and increased electric peaks resulting from electrified end uses.

Common types of end use equipment involved in DR include heat pumps (see Chapter 10), electric vehicle charging (see Chapter 13), electric water heaters, space cooling systems, household appliances, and data centers. Facility energy managers implement DR motivated by various incentives and penalties (i.e., "carrots and sticks") through the electricity rate design (see Chapter 4), or to take advantage of one-time rebates or subsidies by electric utilities or government agencies. Control systems are critical to the success of DR and can be grouped into behind-the-meter controls used by the facility operator, and more advanced grid-interactive controls. Grid-interactive controls involve the electric utility or a third-party company being provided a degree of control over facility end uses in exchange for compensation provided to the facility owner or customer (Yazdani, 2022).

2.3.1.5 Energy storage

Storage of electric and thermal energy is a key strategy for decarbonization of facilities for the same basic reasons as described for DR. Shifting of

[18]Fossil fuel electricity generation retains a small proportion of generation in some scenarios for some regions, forecasting that carbon capture and storage could become technically and economically viable at large scales.

electric loads can be achieved through control systems connected to the electric equipment using DR technologies, or through storage of electricity in battery systems, or a combination of both DR and storage. Because electric and thermal systems are usually coupled (e.g., electric-driven space heating, process heating, and space cooling), then strategically storing and discharging thermal energy for heating or cooling applications is a viable strategy to accomplish shifts in electricity consumption. In addition to managing the timing of consumption, energy storage can improve facility resiliency by discharging electric or thermal energy in the event of a system outage. Chapter 17 provides an introduction to energy storage.

2.3.2 Integrated demand-side systems for decarbonization

Using the building blocks model of energy systems introduced in Chapter 1, a simplified model of past and present demand-side energy systems is provided in Figure 2.9. These simple end use systems have generally been sufficient because of the ability to rely on centralized energy supplies to deliver the quantity of electricity and fuels when needed. In short, relying on the energy grids has generally been cheap and reliable.

However, as introduced in Chapter 1 and reinforced in this chapter, relying on distant generation resources and energy grids as a cheap and reliable option to power end uses is rapidly changing. Achieving net-zero GHGs will often require deployment of multiple demand-side DERs: energy efficient technologies, fuel switching measures, distributed generation systems, DR technologies, and/or energy storage equipment. While all of these DERs have been used for decades, the speed and scale of deployment needed to achieve net-zero GHG emissions before 2050 is unprecedented. The schematic in

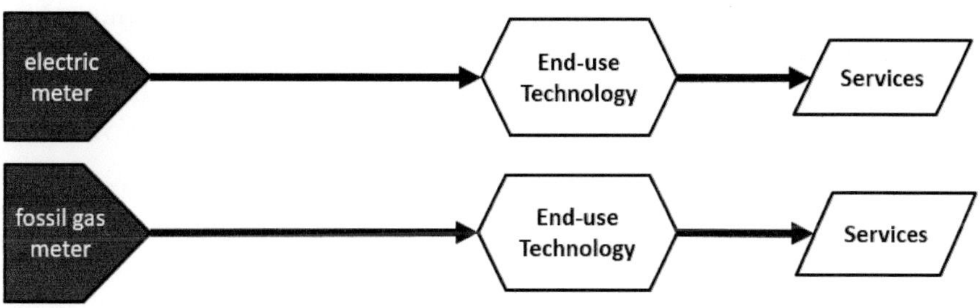

Figure 2.9 Simplified diagram of past and present demand-side energy systems typically fueled by large, remote fossil-fueled energy supplies that modulate to deliver energies when needed to end uses.

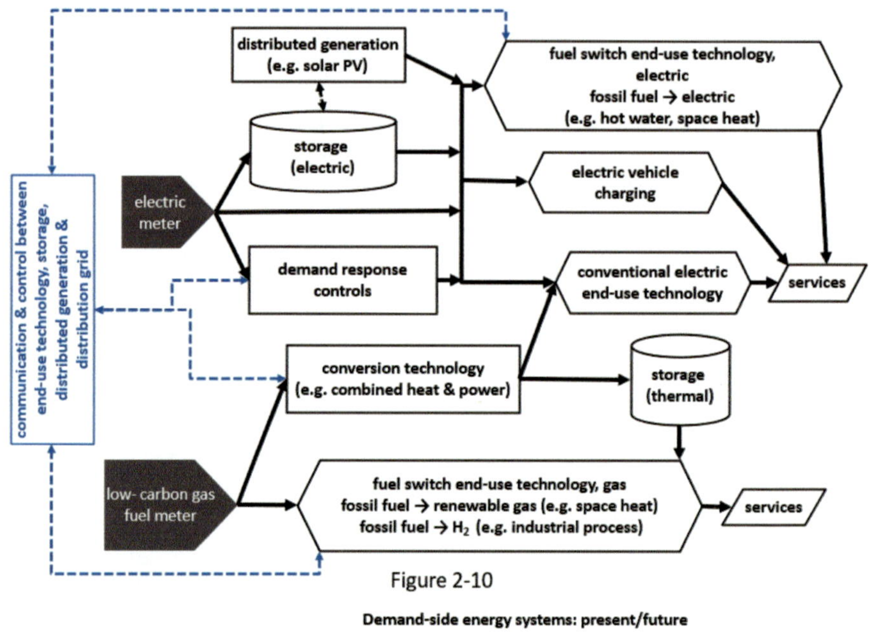

Figure 2.10 Diagram illustrating the complexity of present and future demand-side energy systems required to facilitate renewable energy supplies, achieve net-zero GHGs, and improve resiliency.

Figure 2.10 illustrates how the complexity of behind-the-meter systems is increasing to meet the challenge of decarbonization. Decarbonization is different for new vs. existing facilities. Neither is easy, but retrofits are more costly and difficult than new facility design. "Deep retrofits" is a common phrase used to describe the process for existing facilities. Rocky Mountain Institute provides tools and resources to implement integrative design to adopt aggressive efficiency measures as part of decarbonizing facilities.[19] ASHRAE also has resources for decarbonization of buildings.[20] Of course, not all facilities will need to employ all DERs. But over time, fewer facilities will meet their needs with simple systems shown in Figure 2.9, and a growing number of facilities will resemble the schematic in Figure 2.10.

[19]https://rmi.org/our-work/buildings/deep-retrofit-tools-resources/

[20]https://www.ashrae.org/about/ashrae-decarbonization-related-policies-briefs-and-standards#related

2.3.3 Offsets and carbon removals (CDR) to counterbalance residual emissions

This chapter provides an overview of methods and strategies to decarbonize facilities to help achieve net-zero GHG emissions. However, in practice, it is extremely rare that an organization or a facility can achieve absolute zero GHG emissions. This is because it is nearly impossible to supply a facility all hours of the year with energies that emit absolute zero GHGs. As illustrated in Section 2.2 of this chapter, even regions with the lowest carbon electricity supplies still have non-zero GHG emission intensities. Similarly, biofuels such as RNG or wood fuels contribute GHGs due to CH_4 and N_2O emissions, even if the direct CO_2 emissions are assigned a zero CO_2e emission factor.[21] The goal of real-world decarbonization projects is to reduce GHG emissions attributed to a facility or organization to a practical minimum (or to the limit defined by a chosen standard), usually including Scope 1 and Scope 2 emissions (and sometimes including Scope 3). The remaining GHG emissions are described as residual emissions.

> Two mechanisms to counterbalance or neutralize residual emissions are: 1) offsets and 2) GHG removals (ISO 14068):
>
> 1. Offset: retirement of one carbon credit, where a carbon credit is a tradeable certificate representing one tonne (1,000 kg) of CO2e that is reduced outside the boundary of the organization claiming the credit. The GHG reduction must meet an "additionality" requirement, meaning the reduction is beyond what is required of the provider of the credit.
>
> 2. GHG removal[22]: withdrawal of GHGs from the atmosphere, usually CO2. GHG removal is typically achieved by reforestation, carbon sequestration in soils, enhanced rock weathering, sustainable bioenergy production with CO2 capture and permanent storage, or direct air capture with permanent storage.

---

[21] To determine whether carbon in biofuels is carbon neutral, land-use changes need to be considered. For example, one decarbonization study required that "If biofuels are used, emissions from land use, land-use change,and forestry (LULUCF) must be included in GHG calculations."

[22] Some sources distinguish between GHG removal, which counts removal of any GHG and carbon removal which only counts CO2 removal. In this book, the use of terms "GHG removal" and "carbon removal" should be considered synonymous.

GHG offsets must be verified GHG reduction projects which meet specific criteria, especially the additionality principle incorporated into the definition above. The ISO 14068-1 (2023) standard is commonly used, and it allows for offsetting of residual GHGs. The previously-mentioned SBTi standard (SBTi, 2025) is widely recognized as the leading standard for corporate net zero commitments. SBTi generally requires 90% reduction of GHGs counting Scope 1, 2 and 3, and allows for verified carbon removals (and sometimes offsets) for up to 10% residual GHGs.

2.4 Conclusion and Summary

The science of climate change has shown that global mean temperatures have already risen more than 1.0 °C (1.8 °F) above pre-industrial levels as of 2017, and that warming will continue to increase until net-zero GHGs have been achieved. Global mean temperatures are an indicator of a wide range of impacts and events that result from changes in climate, such as droughts, floods, and extreme precipitation events. The largest contributors to global warming are emissions of CO_2, CH_4 (methane), N_2O (nitrous oxide), and trace gases from refrigerant fugitive leaks. Since these gases have different radiative forcing, a common metric has been developed based on global warming potential (GWP) of gases relative to CO_2. The 100-year GWP is the most common metric, which are CO_2 equivalents (CO_2e) over a 100-year duration. Currently, net GHG emissions are approximately 59 billion metric tons of CO_2e per year. At current global emission rates, a global mean temperature rise of 2.0 °C (3.6 °F) is predicted to occur around 2039. To address climate change, policies and targets have been set globally to achieve net-zero GHGs by 2050 or sooner.

Common GHG accounting principles and methods provide a framework for quantifying and managing emissions. The basic approach to GHG accounting for an organization is to set the facility boundary, and then identify sources, sinks, and reservoirs of GHGs within the boundary. This framework categorizes emissions as Scope 1 (direct emissions), Scope 2 (indirect GHGs from energy consumption), and Scope 3 (everything else not counted in Scope 1 or 2). Currently, most GHG management efforts target Scope 1 and Scope 2, but some also count some or all Scope 3 emissions. For Scope 1, GHG emission factors are commonly used to quantify emissions from fuel combustion. For Scope 2, electricity is usually the largest emission source, which requires quantifying and forecasting the GHG intensity of purchased electricity.

Reducing GHGs to decarbonize existing residential, commercial, and industrial facilities commonly targets operational emissions (i.e., not embodied or Scope 3 GHGs). Reducing or eliminating operational emissions involves the adoption of DERs including energy efficiency, fuel switching, demand response, energy storage, and distributed generation. While energy efficiency alone cannot decarbonize a facility, adoption of aggressive energy efficient technologies and practices are essential to facilitate conversion to low-/no-carbon energies such as renewable electricity, RNG, or green hydrogen. Energy efficiency measures, especially for passive systems such as building envelopes and low-friction industrial piping systems, help reduce the capital costs of decarbonizing facilities, reduce ongoing facility operating costs, reduce costs of upgrading electric grids, and provide resiliency benefits. Due to the variable nature of renewable energies as compared to fossil fuels, energy storage and demand response measures are critical to match electricity supply and demand. Achieving absolute zero facility GHG emissions is challenging. Thus, counterbalancing residual emissions is typically required by investing in either GHG offsets or verified carbon removals.

References

Allen M., Friedlingstein P., Girardin C., Jenkins S., Malhi Y., Mitchell-Larson E., Peters G., and Rajamani L. (2022) "Net Zero: Science, Origins, and Implications" Annual Reviews Environment & Resources 47:849-887.

ASHRAE (2025). ASHRAE Handbook – Fundamentals SI. Chapter 36 "Climate Change."

WMO (2022) "2021 B.C. Best Practices Methodology for Quantifying Greenhouse Gas Emissions" British Columbia Ministry of Environment and Climate Change Strategy.

Bounocore J., Salimifard P., Magavi Z., and Allen J. (2022) "Inefficient Building Electrification Will Require Massive Buildout of Renewable Energy and Seasonal Energy Storage" Scientific Reports 12, 11931.

Canadian Net-Zero Emissions Accountability Act (2021) https://laws-lois.justice.gc.ca/eng/acts/c-19.3/fulltext.html adopted June 29, 2021.

CPUC (2020) "Energy efficiency policy manual" (version 6). California Public Utility Commission cpuc.ca.gov

Denholm P., Brown P., Cole W., Mai T., and Sergi B. (2022) "Examining Supply-Side Options to Achieve 100% Clean Electricity by 2035" National Renewable Energy Laboratory. NREL/TP-6A40-81644.

DOE (2023) "Land-Based Wind Market Report: 2023 Edition" U.S. Department of Energy. National Renewable Energy Laboratory. www.nrel.gov/docs/fy22osti/82519.pdf.

DOE (2022) "Distributed Wind Energy Futures Study" U.S. Department of Energy. Office of Energy Efficiency and Renewable Energy

EIA (2022) "Electricity: emissions by plant and region" U.S. Energy Information Administration www.eia.gov/electricity/data/emissions/

EPA (2022) "AP-42, Fifth Edition Compilation of Air Pollutant Emissions Factors, Volume 1: Stationary Point and Area Sources" U.S. Environmental Protection Agency. www.epa.gov/air-emissions-factors-and-quantification

FortisBC (2022) "FortisBC Energy Inc. Natural Gas Demand-Side Management – 2021 Annual Report."

Gates, B. (2021) "How to Avoid a Climate Disaster: The Solutions We Have and the Breakthroughs We Need" Alfred A. Knopf division of Random House.

GC (2022) "Canada's Greenhouse Gas Offset Credit System" Government of Canada www.canada.ca/en/environment-climate-change/services/climate-change/pricing-pollution-how-it-will-work/output-based-pricing-system/federal-greenhouse-gas-offset-system.html

Greenhouse Gas Protocol (2022) www.ghgprotocol.org. World Resources Institute and World Business Council for Sustainable Development.

Greenhouse Gas Protocol (2015) "GHG Protocol Scope 2 Guidance" https://ghgprotocol.org/scope_2_guidance

Greenhouse Gas Protocol (2013) "Technical Guidance for Calculating Scope 3 Emissions (version 1.0)" https://ghgprotocol.org/scope-3-technical-calculation-guidance

ICF (2022) "Low carbon case studies for residential and commercial space and water heating using air-source or ground-source heat pumps" CEATI Report No. T201700-7073. www.ceati.com.

IPCC (2006) "Guidelines for National Greenhouse Gas Inventories"

IPCC (2013) "Climate Change 2013: The Physical Science Basis. Contribution of Working Group I to the Fifth Assessment Report of the Intergovernmental Panel on Climate Change" [Stocker, T.F., D. Qin, G.-K. Plattner, M. Tignor, S.K. Allen, J. Boschung, A. Nauels, Y. Xia, V. Bex and P.M. Midgley (eds.)]. Cambridge University Press.

IPCC (2018) "Global Warming of 1.5°C. An IPCC Special Report on the impacts of global warming of 1.5°C above pre-industrial levels and related global greenhouse gas emission pathways, in the context of strengthening the global response to the threat of climate change, sustainable development, and efforts to eradicate poverty" [Masson-Delmotte, V., P. Zhai, H.-O. Pörtner, D. Roberts, J. Skea, P.R. Shukla, A. Pirani, W. Moufouma-Okia, C. Péan, R. Pidcock, S. Connors, J.B.R. Matthews, Y. Chen, X. Zhou, M.I. Gomis, E. Lonnoy, T. Maycock, M. Tignor, and T. Waterfield (eds.)]. Cambridge University Press.

IPCC (2021) "Climate Change 2021: The Physical Science Basis. Contribution of Working Group I to the Sixth Assessment Report of the Intergovernmental Panel on Climate Change" [Masson-Delmotte, V., P. Zhai, A. Pirani, S.L. Connors, C. Péan, S. Berger, N. Caud, Y. Chen, L. Goldfarb, M.I. Gomis, M. Huang, K. Leitzell, E. Lonnoy, J.B.R. Matthews, T.K. Maycock, T. Waterfield, O. Yelekçi, R. Yu, and B. Zhou (eds.)]. Cambridge University Press.

IPCC (2022a) "Climate Change 2022: Mitigation of Climate Change. Contribution of Working Group III to the Sixth Assessment Report of the Intergovernmental Panel on Climate Change" [P.R. Shukla, J. Skea, R. Slade, A. Al Khourdajie, R. van Diemen, D. McCollum, M. Pathak, S. Some, P. Vyas, R. Fradera, M. Belkacemi, A. Hasija, G. Lisboa, S. Luz, J. Malley, (eds.)]. Cambridge University Press.

IPCC (2022b) "Climate Change 2022: Impacts, Adaptation, and Vulnerability. Contribution of Working Group II to the Sixth Assessment Report of the Intergovernmental Panel on Climate Change" [H.-O. Pörtner, D.C. Roberts, M. Tignor, E.S. Poloczanska, K. Mintenbeck, A. Alegría, M. Craig, S. Langsdorf, S. Löschke, V. Möller, A. Okem, B. Rama (eds.)]. Cambridge University Press.

IPCC (2001) "Uncertainty - Good Practice Guidance and Uncertainty Management in National Greenhouse Gas Inventories" Chapter 6.

IPCC (2024) Carbon Dioxide Removal Fact Sheet.

ISO 14064-1:2018 Greenhouse gases — Part 1 Specification with guidance at the organization level for quantification and reporting of greenhouse gas emissions and removal. www.iso.org/standard/66453.html

ISO 14064-2:2019 "Greenhouse gases — Part 2: Specification with guidance at the project level for quantification, monitoring and reporting of greenhouse gas emission reductions or removal enhancements" www.iso.org/standard/66454.html

ISO 14064-3:2019 Greenhouse gases — Part 3: Specification with guidance for the verification and validation of greenhouse gas statements www.iso.org/standard/66455.html

ISO 14068-1:2023 "Climate change management – transition to net zero, Part 1: carbon neutrality" www.iso.org/standard/43279.html

Grubert (2020) "At scale, renewable natural gas systems could be climate intensive: the influence of methane feedstock and leakage rates" Env Res Ltr 15, p. 084041.

Keller V., English J., Fernandez J., Wade C., Fowler M., Scholtysik S., Palmer-Wilson K., Donald J., Robertson B., Wild P., Crawford C., and Rowe A. "Electrification of road transportation with utility controlled charging: A case study for British Columbia with a 93% renewable electricity target" Applied Energy 253, 113536.

Langevin J., Satre-Meloy, A., Satchwell A., Hledik R., Olszewski J., Peters K., and Chandra-Putra H. "Demand-side solutions in the US building sector could achieve deep emissions reductions and avoid over $100 billion in power sector costs" (2023) One Earth 6, 1005–1031.

Matthews D.H. and Wynes S. (2022) "Current global efforts are insufficient to limit warming to 1.5°C" Science 376, p.1404.

Nadel S. and Ungar L. (2019) "Halfway There: Energy Efficiency Can Cut Energy Use and Greenhouse Gas Emissions in Half by 2050" American Council for an Energy Efficient Economy Report U1907.

Nadel S. (2022) "What is the Role of Fuels in a Decarbonized Future? What is the Potential Role for Gas Utilities?" Summer Study on Energy Efficiency in Buildings.

National Academies of Sciences, Engineering, and Medicine (2021). Accelerating Decarbonization of the U.S. Energy System. Washington, DC: The National Academies Press.

Rissman et al (2020) "Technologies and policies to decarbonize global industry: Review and assessment of mitigation drivers through 2070" Applied Energy 266, 114848.

Science Based Target Initiative (SBTi) (2025) "SBTi Corporate Net Zero Standard. Version 1.2. March, 2024.

SEE Action (2020). Determining Utility System Value of Demand Flexibility from Grid-Interactive Efficient Buildings. State and Local Energy Efficiency Action Network. Prepared by: Tom Eckman, Lisa Schwartz, and Greg Leventis, Lawrence Berkeley National Laboratory.

Ueckerdt F., Bauer C., Dirnaichner A., Everall J., Sacchi R., and Luderer G. (2021) "Potential and risks of hydrogen-based e-fuels in climate change mitigation" Nature Climate Change https://doi.org/10.1038

UNFCC (2022) https://unfccc.int/ghg-inventories-annex-i-parties/2022.

WMO (2022) "State of the global climate 2021" World Meteorological Organization. https://public.wmo.int/en/our-mandate/climate

Tyndall J. On the absorption and radiation of heat by gases and vapours, and on the physical connection. Philosophical Magazine, 1861; 151:1–36.

Yazdani H., Doostizadeh M., and Farrokh A. (2022) "Unlocking the value of flexibility of behind-the-meter prosumers: An overview of mechanisms to esteemed trends" The Electricity Journal 23, 107126.

Author Biography

Eric Mazzi is a consultant and educator focusing on technology and policy for demand-side energy systems. He specializes in performing independent technical analysis, M&V, policy studies, and educational services. He advocates and contributes to efforts to reduce energy consumption and achieve net-zero GHG emissions, while maintaining or improving energy services and advancing energy equity. His experience includes air quality consulting, utility maintenance, and five years as full-time engineering faculty at the University of British Columbia. He volunteers as a board member for the U.S. Northwest Energy Coalition, and the International Performance M&V Protocol committee. Eric teaches part-time, delivering university and professional certification courses. He is an Adjunct Professor of Energy Management at the New York Institute of Technology – Vancouver. He has B.Sc. and M.Sc. degrees in mechanical engineering, and a Ph.D. degree in environmental policy. He is a registered engineer in California and British Columbia.

Acknowledgments

I would like to thank chapter reviewer Ethan Rogers for his thoughtful review of drafts of this chapter.

Practice Questions:

1. Go to the World Meteorological Organization website and download the latest version of the state of the climate annual report. Summarize in one page total or less using bullet points, short text statements, and/or graphics: (a) the status of key indicators for greenhouse gas levels, global temperatures, ocean, and cryosphere, (b) a sample of extreme weather

and climate events that occurred, and (c) a sample of socio-economic impacts.

2. Search literature for the climate justice ethical issues involved in solar geoengineering, including at least one peer-reviewed source. Summarize briefly, in half a page or less, the major or key climate justice and ethical issues involved if solar geoengineering were to be deployed.
3. Choose one of the following jurisdictions: one Canadian province, one U.S. state, the US as a whole, or Canada as a whole. For your chosen jurisdiction, find the latest government GHG emission inventory and chart the GHG trends for the past 20 years in tons of CO_2e. Comment on (a) how much progress has been made towards net-zero GHG, and (b) starting with the most recent year for which the GHG inventory is available, estimate at what annual percentage rate of reduction (including GHG reduction plus carbon removals) would it take to achieve net-zero GHG by 2050.
4. Consider GHG emissions from the combustion of fossil fuels. For the three principal GHG chemical species (CO_2, CH_4, N_2O), which of these species are the emission rates a strong function of the type of combustion device (e.g., engine vs. boiler vs. combustion turbine), and which of these species are the emission rates mainly dependent on fuel composition?
5. To quantify avoided GHG emissions from electricity savings, the most accurate GHG intensity value to use is:

   (a) average GHG intensity
   (b) minimum GHG intensity
   (c) marginal GHG intensity
   (d) maximum GHG intensity

6. A facility in Canada purchases the following annual energies:
   7,822,000 kWh electricity
   1,805,000 ton-hr chilled water
   24,090 Mlbm steam (1,000s of pounds)
   2,440 GJ natural gas
   GHG intensities are as follows:
   Electricity: 0.030 kgCO_2e/kWh
   Chilled water: 0.023 kgCO_2e/ton-hr
   Steam: 76.6 kgCO_2e/Mlbm
   Estimate the total Scope 2 GHGs in tCO_2e/yr

7. The emission rates (including fuel leaks) for marketable natural gas for a given operation are as shown below. The fuel heating value is 0.03928 GJ/m^3 (HHV basis).
   CO_2: 1,926 gCO_2/m^3
   CH_4: 0.145 g/m^3
   N_2O: 0.033 g/m^3
   Using the GWP values in Table 2.1, estimate the GHG emission factor in $kgCO_2e/GJ$ (HHV basis) using (a) 100-year GWP values, and (b) 20-year GWP values.
8. For diesel fuel consumed in vehicles, Scope 1 emissions can be described as tank to wheels, and indirect (Scope 3) emissions can be described as well to wheels. Given the information listed below, estimate: (a) direct or Scope 1 emission factor in $kgCO_2e/GJ$ and $lbmCO_2e/MMBtu$, (b) indirect or Scope 3 emission factor in $kgCO_2e/GJ$ and $lbmCO_2e/MMBtu$, and (c) total well-to-wheels emission factor also in both SI and English units.
   - Direct emission factor (tank to wheels) is 2,642.88 $kgCO_2e$ per m^3
   - Indirect emission factor (well to tank) is 0.794 $kgCO_2e$ per L
   - Fuel heating value is 0.0426 GJ/L (HHV basis)
9. A natural gas processing plant produces 100 MMSCF/day (360 day/yr) of sales gas. The plant operates off the electric grid and is fueled by hydrofracked (raw) natural gas burned in internal combustion engines to drive compressors and gas turbine generators for ancillary electric loads. The plant consumes 1,168,000 GJ/yr (HHV basis) of raw natural gas to operate. The GHG emission factor for raw natural gas is 55 $kgCO_2e/GJ$. Estimate:
   (a) The annual baseline GHG emissions in tons of CO_2e for the gas processing plant that operates off the electric grid. Include only emissions from the combustion of raw natural gas (i.e., ignore any emissions from maintenance vehicles, methane leaks, or other sources).
   (b) Assume the plant can be converted to 100% electric drives, i.e., replace all internal combustion engines with variable speed-driven electric motors, and replace gas turbine generators with grid electricity. The electricity required to power the facility is 3.50 MWh per MMSCF. Estimate the Scope 1 emissions from the electrified gas plant assuming the grid GHG intensity is 20 $kgCO_2e/MWh$. As

with the baseline GHGs, ignore any emissions from maintenance vehicles, methane leaks, or other sources.

(c) Estimate the annual avoided GHG emissions (tons of CO_2e) by electrifying the gas processing plant.

(d) Plant production of sales gas will ultimately be sold to customers and burned in end use systems (i.e., buildings, manufacturing plants, and perhaps natural gas vehicles). Estimate the annual tons of CO_2e emissions released by the combustion of the 100 MMSCF/day plant production (360 day/yr), assuming the sales gas GHG emission factor is 50 $kgCO_2e/GJ$.

3

Energy and Decarbonization Assessments

Eric Mazzi

Mazzi Consulting, Canada

3.1 Introduction

In Chapters 1 and 2, energy and GHG management systems and practices are introduced. A key task for any energy or carbon management initiative is to conduct facility assessments. In this chapter, we provide an introduction and overview of key issues and best practices for facility assessments. Assessments vary widely in the objectives and scope, and are referred to by different terms such as energy audit, plantwide audit, energy study, energy survey, decarbonization study, net-zero GHG study, net-zero energy study, and more.

In broad terms, the purpose of facility assessments is to analyze measures to optimize energy and peak demands, carbon reduction, or both. Targeted measures could be one or more of the five types we introduced in Chapter 1: energy efficiency, demand response, distributed generation, storage, and fuel switching. In general, assessments can be done for new facilities or retrofit. In this chapter, our emphasis is on retrofits. However, the same issues and practices typically apply to studies of new facility designs. One obvious difference is that new facility designs cannot have site visits and measurements performed, and instead rely on energy modeling and performance data from similar systems and facilities.

Energy and decarbonization assessments often differ as follows:

- Timelines: Energy studies tend to focus on the near term, while decarbonization studies tend to consider the future, long-term vision or plan for the facility.
- Scope: Energy studies tend to identify a narrow scope of upgrades, while decarbonization studies usually are broader in scope. As introduced in Chapter 2 and reinforced later in this chapter and Chapter 8, best

practices for decarbonization studies are to integrate efficiency and adaptation (resiliency) with GHG reduction measures.

When embarking on energy or carbon assessments, it is essential to understand the best practices for identifying measures to improve energy performance and/or reduce GHG emissions. In this introduction, we begin by providing an overview of the optimum process for planning and executing effective energy and decarbonization assessments.

Once objectives are set and assessment projects are launched, the execution of facility assessments is commonly completed in three phases as listed below. After this introduction, there are sections on each of these three phases.

- Phase 1: preparing for facility assessments.
- Phase 2: conducting work on site.
- Phase 3: analysis and reporting.

This chapter concludes with a brief discussion of the next steps after reporting.

3.1.1 Best practice process and sequence for optimum facility assessments

Figure 3.1 provides an overview of the optimum process for performing energy and decarbonization assessments. The correct sequence is important, especially for the first three steps. The first step is to clearly define objectives. This is shown in Box 1 of Figure 3.1, and discussed further in the

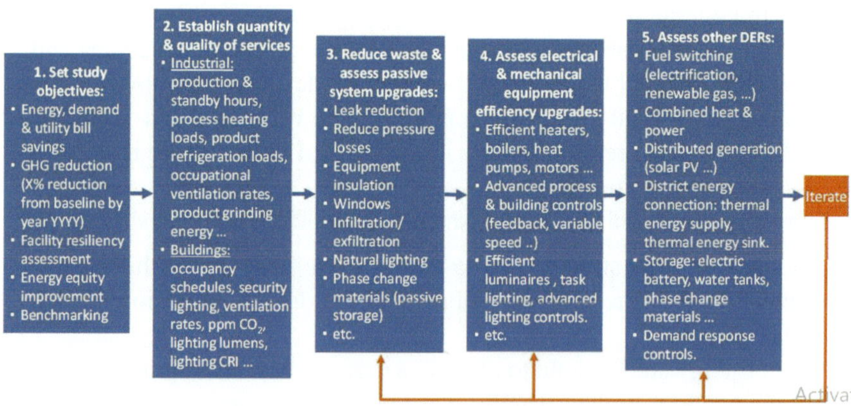

Figure 3.1 Process for effective energy and decarbonization assessments.

next section. As introduced in Chapter 1, it is important to recognize that facility energy systems deliver services that can be described in quantitative and qualitative terms. The best practice is to document services that are expected by facility owners and users. This is shown in Box 2 of the figure. For some facilities, such as homes, it may be relatively straightforward to document heating, cooling, ventilation, lighting, cooking, entertainment, etc. (some homes also function as offices, healthcare setting for people with illnesses, etc.). For industrial processes, the maximum production rate, annual production quantity, and product quality requirements are important.

Examples of services include:

- Ventilation: Quantity is ventilation rate (CFM or L/s). Quality could be measured by maximum indoor CO_2 concentration (see Chapter 9).
- Illumination: Quantity is lumens or footcandles. Quality is measured by parameters such as CRI, lamp color temperature, etc. (see Chapter 12).
- Thermal comfort(ASHRAE Handbook Fundamentals, Chapter 9, 2025): Quantity could be measured in floor space area and/or number of daily occupants. Quality typically targets ranges for dry bulb temperature and relative humidity (see Chapter 9).
- Domestic hot water: Quantity is flow rate (gpm, L/s). Quality could be defined by maximum and minimum temperatures, and response time to achieve desired temperature after opening a valve.
- Manufacturing: Quantity is often the number of production units per day or month. Quality could be parameters such as maximum percent defects, material strength, or other technical performance requirements.

Often the services are not explicitly considered, and it is assumed (without supporting data) that existing systems deliver satisfactory service. However, in practice, this is often not the case. Consider the sample thermal comfort survey response in Figure 3.2. This shows the results from one question on heating from a 10-question online survey conducted for occupants of a 40,000 ft^2 commercial building located in a cold climate (over 6,000 °C-days/year HDD). This survey was conducted as part of an ASHRAE Level 2 energy audit. The responses show that only about half of respondents considered the heating satisfactory. Moreover, an open-ended question was included in the 10-question survey, which included the following comments:

> "Air conditioning in the summer time is too cold! Must be a way of finding the right balance so that everyone is happy."

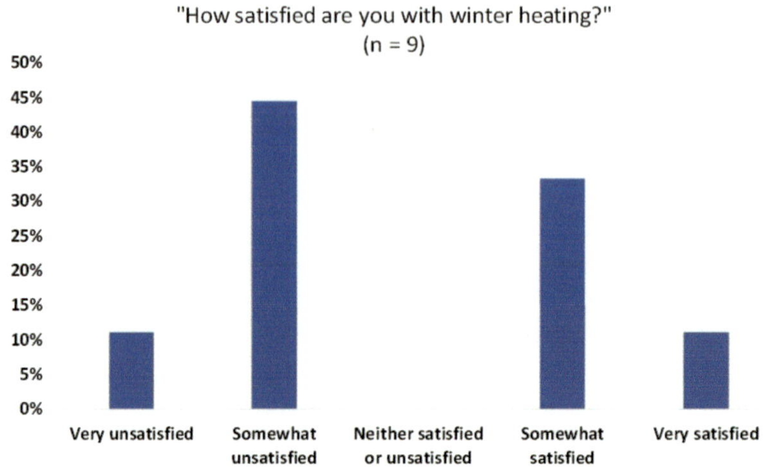

Figure 3.2 Sample response from a thermal comfort survey for a 40,000 ft^2 commercial building.

> "The basement office I work in is cold all the time and it's summer. I can't imagine how it's going to feel come winter."
>
> "Temperature is hard to regulate for the seasons. For example too cold in the winter and too hot in the summer."
>
> "Better cooling needed in server room."

While it might seem limiting to obtain only nine responses, it is important to note that nine responses are better than zero! The field work for this audit was intentionally conducted off-hours during the summer, in order to minimize disruption to occupants. If the survey had not been conducted, then the assessment team would likely have assumed that the thermal comfort was satisfactory, in the absence of any contrary evidence. Feedback such as provided by this example provides an opportunity to improve both thermal comfort and HVAC system efficiency (e.g., note the survey comments about overcooling in summer). Moreover, having documented evidence of occupant dissatisfaction provides added motivation for recommended upgrades to actually be funded and implemented. The point of this example is to encourage energy assessment teams to invest some time in confirming services required and the level of satisfaction with the services. No matter the outcome of this effort, it is worthwhile. In cases where users are highly satisfied, then the assessment team can focus its efforts on ensuring existing service levels are maintained or improved. Further reinforcement of the value of conducting surveys and interviews of facility users is provided in Chapter 7. Appendix H provides a sample survey questionnaire for a building. Additional examples of facility surveys (e.g., thermal comfort) can be found from many sources, such as ASHRAE, LEED, and others.

Next, shown in Box 3 of Figure 3.1, is to assess practical opportunities to reduce waste and upgrade passive systems. It is important that this step is completed prior to examining mechanical and electrical equipment upgrades. Skipping this step or making only superficial efforts to examine waste reduction and passive system improvements increases the risks of overspending capital dollars on mechanical and electrical equipment. Examples of waste reduction include leak management for compressed air systems, eliminating unnecessary recirculation for industrial pumping systems, reducing the number of leaking steam traps, and reducing the number or sharp angles or bends in duct and piping retrofits. Several chapters in this book provide approaches to reducing waste, especially Chapters 9 and 20–21.

After objectives are clearly defined, service requirements documented, and waste and passive systems have been assessed, then this is the appropriate time to examine capital-intensive upgrades to mechanical and electrical systems. This is shown in Box 4 of Figure 3.1. This is the step in the energy assessment where analysis of efficiency measures is completed such as described in Chapters 9 and 10 (HVAC and heat pumps), Chapters 12–15 (lighting, motors, controls, and industrial measures (Chapter 20).

For most facility assessments today, the objective is seldom limited to energy efficiency measures and utility cost savings, and decarbonization is a major objective. This leads to the task identified in Box 5 of Figure 3.1, which is to identify and analyze other DERs. This potentially includes storage (see Chapter 17), distributed generation (see Chapters 18 and 20), fuel switching (see Chapters 10, 19, and 20), and demand response (e.g., through controls, see Chapter 15, using storage, see Chapter 17).

Finally, it is important to optimize proposed measures (often bundling measures together) by iterating as shown in Figure 3.1. This could involve, as examples, examining tradeoffs to investing more in advanced building envelope upgrades to reduce the size of a heat pump, added thermal storage to reduce the rated capacity of cooling systems, optimizing industrial piping (i.e., shorter, larger diameter pipes), and more. Typically, once the objectives and service requirements are set (Boxes 1 and 2 in Figure 3.1), these do not change. But in rare cases, objectives and service requirements could change as well. The iteration exercise can also target improvements in non-energy impacts such as increased reliability for manufactured products, reduced maintenance costs (see Chapter 21), and improved resiliency (see Chapter 8).

3.1.2 Setting objectives

Now that the best practices for the overall process of conducting energy and decarbonization assessments have been described, let us discuss in more detail the important first step, which is to set objectives. Table 3.1 provides a list of common objectives with illustrative examples. Facility assessments could target one or multiple objectives as listed in the table.

Table 3.1 Common objectives for facility assessments.

Objectives	Illustrative example
Reduce facility utility bill costs	• Identify all measures with a positive NPV using a discount rate of 6% • Reduce total utility costs by 3.0% each year
Improve energy efficiency of a specific system or the whole facility	• Reduce lighting system consumption by 18% from a defined baseline • Deep energy retrofit with minimum 40% reduced consumption from a defined baseline for all purchased energy • Achieve net-zero energy by a target year applying a specific standard (e.g., https://sciencebasedtargets.org)
Reduce peak electricity demand and/or shift timing of electric demand	• Reduce monthly peak kVA by a specific percentage or target value • Minimize energy consumption during the peak time period under a time-of-use rate
GHG reduction	• 50% GHG reduction from 2007 baseline by 2030 • Net-zero GHG by 2045, including Scope 1 and 2 emissions
Improve facility resilience (climate adaptation)	• Design for maintenance of life-safety indoor temperatures under specific extreme cold or hot climate events (e.g., 100-year heat wave) with loss of electricity for up to 72 hours. Add cooling to existing facilities where active cooling did not previously exist • Upgrade systems to maintain indoor air quality levels to specified levels (e.g., less than 10 $\mu g/m^3$ 1-hour average $PM_{2.5}$) during wildfire events when outdoor particulate levels are high (e.g., up to 200 $\mu g/m^3$ 1-hour average $PM_{2.5}$)
Benchmarking	• Reduce weather-normalized EUI by 3% each year • Attain the top 15 percentile status for similar facilities applying an energy consumption, energy cost, or GHG emission metric
Energy poverty reduction	• Reduce annual utility bill costs to 6% or less of household income • Improve indoor air quality and thermal comfort • Engage landlords or property managers in energy assessments and help identify funding sources to improve building deficiencies
Facility rehabilitation and life extension	• assess equipment refurbishment vs. replacement • condition assessment and remaining life estimate • lifecycle economic analysis

It is important to think through the objectives of a proposed facility assessment, because the objectives can have major implications for the scope and cost of the assessment. Key issues to consider while defining objectives are listed below. Completing all the relevant tasks in the list below will help ensure that the scope and resources for the assessment are realistic so that the defined objectives can be achieved with the available funding for the assessment.

- ☐ Clearly define systems included in (or excluded from) the study scope, and system boundaries.
- ☐ Define baseline year, energy consumption, and GHG emission rates. As described in Chapter 2, it is especially important to consider forecasted GHG intensity of electricity supplies.
- ☐ Establish clear targets for GHG, energy, and/or peak demand reductions and targeted timelines for achieving targets. Define which GHG emissions are included: Scope 1, 2, and 3 (see Chapter 2).
- ☐ Document financial criteria to be applied such as minimum rate of return, discount rate, lifecycle cost per ton of carbon abated, carbon fees (e.g., as adopted by some municipalities such as Vancouver, BC and New York City) and maximum payback period (see Chapter 5).
- ☐ Document utility rate structure and costs (current and forecast), and marginal (not just average) rates to be applied to reductions in energy and demand (see Chapter 4).
- ☐ Confirm available financial resources for assessments: utility subsidies, grants, Energy Services Company (ESCO) funding, or owner operating fund.
- ☐ Confirm potential financial resources for capital upgrades: loans, utility, or government subsidies (rebates), ESCO funding, or owner capital (see Chapter 6).
- ☐ Document applicable requirements for the assessments. Examples include ASHRAE Standard 211 (2023) for energy audits (specifying Level 1, 2, or 3), an energy utility's industrial assessment guideline, or a municipality's net-zero GHG study guideline.

3.1.3 Key references and resources for completing energy assessments

Before we discuss details to prepare for assessments, complete on-site work, and perform analysis and reporting, it should be emphasized that this chapter will only highlight key tasks and practices. For energy and decarbonization assessments of buildings, below is a list of some key resources that may provide more detailed guidance:

- ASHRAE (2023) "Standard 211-2023 Standard for Commercial Building Energy Audits" provides a widely used standard for defining Level 1, 2, and 3 energy audit scopes. The appendix to the standard includes a link to spreadsheet templates such as equipment inventory, condition rating, energy use breakdown, and space function.
- ISO (2025) Standard 50002 "Energy audits – Requirements with guidance for use" is also a widely used standard for energy audits.
- ASHRAE (2011) "Procedures for Commercial Building Energy Audits" second edition[1] provides a comprehensive guideline for energy audits for buildings. The book includes online resources including audit forms, checklists, formulas, and other spreadsheet templates (https://xp20.ashrae.org/PCBEA/PCBEA_Supplemental_Files.html).
- Doty, S. (2021) "Commercial Energy Auditing Reference Handbook" (4th edition) has over 900 pages of detailed guidance for conducting energy audits for buildings.
- ASHRAE has a website with Decarbonization Related Policies, Briefs, and Standards for buildings (www.ashrae.org/about/ashrae-decarbonization-related-policies-briefs-and-standards#related).
- U.S. DOE Better Buildings Program (https://betterbuildingssolutioncenter.energy.gov/carbon-hub/erp).

For energy and decarbonization assessments of industrial plants, below is a list of some key resources:

- The Canadian Industry Partnership for Energy Conservation provides many resources for industrial energy and carbon reduction, including a 280-page "Energy Savings Toolbox – An Energy Audit Manual and Tool."[2]

[1]The forward to ASHRAE 211-2018 states that a new edition for "Procedures for Commercial Building Energy Audits" had already begun in 2018. As of the publication date of this book, the 2nd edition is still the latest version.

[2]The link for CIPEC is https://www.nrcan.gc.ca/energy-efficiency/energy-efficiency-for-industry/canadian-industry-program-energy-conservation-cipec/20341 and the guidebook at www.nrcan.gc.ca/sites/nrcan/files/oee/pdf/publications/infosource/pub/cipec/energyauditmanualandtool.pdf

- The U.S. DOE Industrial Assessment Center includes many resources for industrial energy and decarbonization such as case studies, a guidebook on implementing measures, and a link to software tool at www.energy.gov/eere/amo/software-tools.
- Bonneville Power Authority (BPA, 2017) "Industrial Audit Guidebook: a Guidebook for Performing Walk-through Energy Audits of Industrial Facilities" is a useful resource for industrial walk-through audits.
- The "Industrial Energy Systems Handbook (1st edition)" (Williams, 2023) provides guidance on industrial energy audits in Chapter 6.

3.2 Phase 1: Preparing for Facility Assessments

Effective preparation activities will help get the most benefit from an energy or GHG assessment at the minimum cost. Listed below are key tasks to perform as preparation for the on-site work. It should be noted that some of the items listed below could be shifted to be done during the on-site time or even afterwards. However, the most effective assessments would complete the items listed below *before* arriving on-site. Having a basic understanding of the facility design and equipment, operation, benchmarking energy performance, and targeted measures will help make the best use of the on-site time. Effective preparation reduces the chances that something critical will be missed during the on-site time.

☐ Obtain facility design information and activity data (e.g., occupancy and operating schedules for a commercial building; production shifts and production data for a manufacturing plant). Obtain data in the shortest time interval available such as hourly production for a manufacturing plant or daily occupancy for a building. Obtain copies of facility layout drawings and equipment lists as available (e.g., motor lists, natural gas equipment, cooling units). In some cases, remote access to building automation systems can be provided.

☐ In some cases, facility condition assessments (e.g., per ASTM E2018-15) have been done in the past, or are coordinated to be conducted with energy audits or GHG studies. Condition assessments provide valuable information such as: equipment specifications and rated performance, comprehensive and coded condition assessments, current economic value, remaining life and targeted year for replacement (capital upgrade plans), and replacement value. Recent and accurate condition assessment reports, including capital plans, provide valuable information to confirm the facility baseline scenario, and to develop the most economical upgrade plans by timing upgrades (e.g., fuel switching) to coincide with equipment end-of-life timelines.

- ☐ Collect historical energy purchases, ideally for three years, but typically a minimum of 12 months is required. Obtain data in the shortest time interval available (e.g., hourly electricity data for a building, 15-minute electricity data for an industrial pump system, and/or daily data for natural gas for a building).
- ☐ Establish a measurement plan and secure audit instrumentation. This could be as simple as a list of spot readings to record with portable instruments, or a detailed plan to conduct continuous measurements for a number of days or weeks. For example, ASHRAE Level 3 audits require either energy modeling or two weeks of measurements (ASHRAE, 2023). Audits of compressed air systems typically involve installing flow, temperature, and pressure instruments to collect hourly (or shorter interval) data for one week.
- ☐ Perform utility data analysis and create charts including: Time series, energy vs. OAT, and benchmarking. Figure 3.3 provides an example of an energy vs. OAT chart, which was used to determine a balance point temperature (BPT) of 70 °F. The figure includes a regression model based on HDD using 70 °F as the reference temperature. Chapter 1 includes a discussion of benchmarking.
- ☐ Develop a preliminary list of measures for energy and/or carbon reduction. Good practice is to provide the list to facility management and O&M staff for comment. This helps confirm the measures are consistent with the owner's preferences and future plans.
- ☐ Obtain and review previously completed energy and GHG studies, as well as energy modeling. Reports from previous studies and modeling projects can provide a good understanding of the facility design and energy performance (the more recent the better).
- ☐ Obtain and review previously completed commissioning reports (see Chapter 16) and M&V reports, if available.
- ☐ Confirm site safety protocols and personal protective equipment required. Some facilities, especially industrial plants, may require that online safety modules be completed with proof of completion provided before entering the gate of the facility.

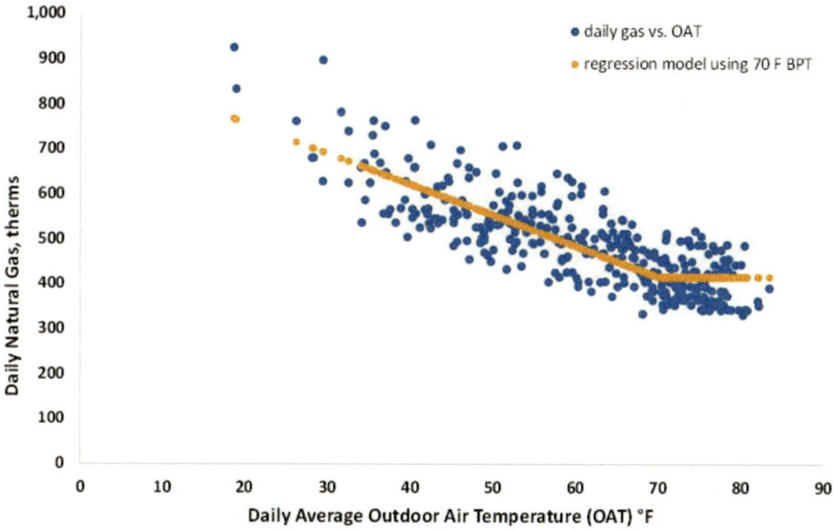

Figure 3.3 Sample chart of daily natural gas consumption vs. OAT.

3.3 Phase 2: Conducting Work On-Site

The on-site work should begin with a kickoff meeting involving, at a minimum, the lead person for the assessment field team and the lead contact at the facility. The facility contact is often an O&M manager or energy management staff member who functions as the owner's representative. The initial meeting is done to confirm the facility assessment objectives, schedule of on-site activities, safety protocols, and other important details. An initial walk-through tour of the facility is typically performed after the site kickoff meeting. This initial tour would be to confirm the major systems to be inspected during the on-site work and introductions to key contacts (e.g., lead staff for specific areas or processes in a facility). On-site work can vary from a few hours to a few weeks, depending on the scope of the assessment. Below is a high-level checklist of key tasks to complete as part of the on-site work.

- ☐ Conduct a site safety briefing to review safety hazards and required procedures.
- ☐ Gather data on all equipment considered significant including nameplate data, fuel sources, and equipment counts.
- ☐ Take spot reading measurements using portable instruments.
- ☐ Set up temporary instrumentation according to the measurement plan.

- ☐ Obtain facility operating data from whatever systems are available such as building automation system screenshots and/or data downloads, industrial data historians, and/or hardcopy logs and reports.
- ☐ With prior permission, take pictures and video clips of key operating events and equipment nameplate data.
- ☐ Perform thermal imaging, if this is part of the assessment scope.
- ☐ Interview key staff such as operations, maintenance, and (with prior permission and considering potential disruption) occupants and facility users.
- ☐ Check out with the designated facility contact before departing the site.

3.3.1 Common facility assessment tools and instruments

The site work phase of energy and carbon assessments always includes using various tools and instruments to collect data. Following is a list of commonly needed tools for energy audits:

- Tape measures and digital distance meters: A 25-foot tape measure is useful to check relatively short dimensions of facility design dimensions of interest such as doors, windows, and the distances between pieces of equipment. Digital distance meters are useful for taking measurements of long dimensions (often inaccessible without scaffolding) such as floor to ceiling heights, lengths of ducting or piping, and building outside dimensions.
- Lightmeter: A portable lightmeter that reads in footcandles or lumens allows direct analysis of lighting systems and comparison with recommended light levels (see Chapter 12). Measuring the illumination allows the auditor to recommend increases or decreases in lighting levels to match user needs.
- Thermometers: Several thermometers may be needed to measure temperatures in offices and other worker areas, and to measure the temperature of process equipment. Knowing process temperatures allows the auditor to determine process equipment efficiencies, and also to identify waste heat sources for potential heat recovery retrofit. Electronic thermometers with interchangeable probes are available to measure temperatures for different systems. Some common types include an immersion probe, a surface temperature probe, and a radiation-shielded probe for measuring true air temperature. Infrared thermometers and thermographic equipment are also available. For example, an infrared

thermometer can measure temperatures of steam lines that are inaccessible without a ladder.

- Relative humidity: Portable electronic humidity meters (often including dry bulb temperature sensors) are useful to measure relative humidity in rooms. Spot checks can be made to compare local measurements to HVAC setpoints and to confirm whether targeted thermal comfort conditions are being met.
- Wattmeter/power factor meter: A portable hand-held wattmeter and power factor meter is used for determining the power consumption and power factor of individual motors and other inductive devices, and the load factors of motors. This meter typically has a clamp-on feature that allows an easy and safe connection to the current-carrying conductor, and has probes for voltage connections. As previously noted, electrical measurements should only be conducted by qualified personnel, typically based on the system voltage.
- Combustion analyzer: Combustion analyzers are portable devices used to measure the combustion efficiency of furnaces, boilers, or other fuel burning equipment. Digital combustion analysis equipment performs the measurements and reads out combustion efficiency (also see Chapter 20).
- Ultrasonic leak detector: Compressed air leak detectors are electronic ultrasonic receivers that are tuned precisely to the frequency of the hissing sound of an air leak. Multi-purpose leak detectors are available, which can also be used to test steam traps. The sound pattern will allow a determination of whether the steam trap is functioning properly and whether it is stuck open (i.e., leaking live steam) or stuck shut (i.e., not removing condensate and non-condensable gases).
- Airflow measurement devices: Airflow measurement devices can be used to identify problems with HVAC air duct flows as well as combustion air or exhaust flows. Typical airflow measuring devices include a velometer, pitot tube, an anemometer, or an airflow hood.
- Blower door attachment: Building or structure tightness can be measured with a blower door attachment. This device is frequently used in buildings to determine the air leakage rate or the number of air changes per hour in the facility at a standard pressure differential.
- Indoor air quality (IAQ): Cost-effective IAQ instrumentation is now widely available. Pre-retrofit (baseline) IAQ monitoring can identify air quality issues that may be addressed in the facility upgrade measures. At a minimum, pre-retrofit IAQ data provides evidence of the baseline

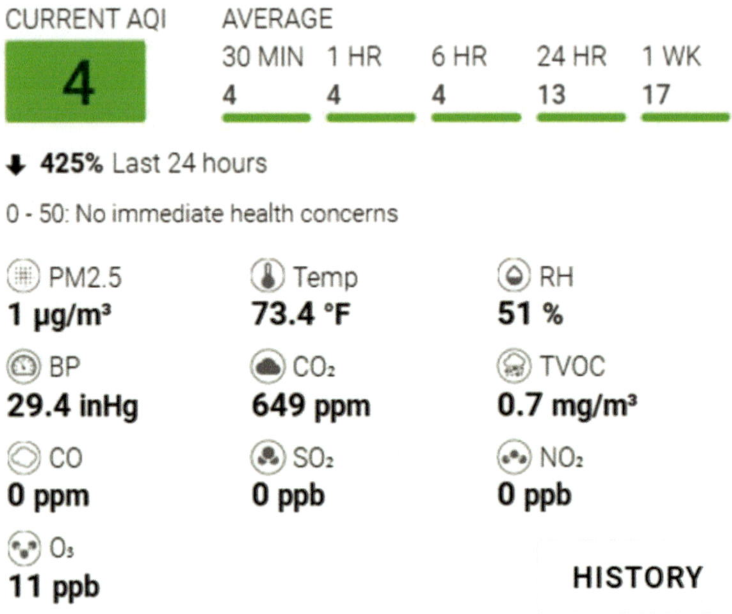

Figure 3.4 Indoor air quality dashboard.

air quality that should be maintained or improved. Figure 3.4 provides a sample of an online IAQ dashboard using an instrument purchased for US$1,100 in 2022.

- Smoke generator: A smoke generator can be used in residences, offices, and other buildings to find air infiltration and leakage around doors, windows, ducts, and other structural features. Care must be taken to ensure that personnel exposure limits for the chemical smoke are maintained and/or protective equipment is used such as breathing apparatus.
- On/off loggers and timers: In many cases, the operating time of equipment such as motors and luminaires is highly uncertain. Inaccurate estimates of operating hours for key equipment can lead to large over- or under-estimates of baseline energy consumption. Field confirmation of operating hours alone can greatly improve the accuracy of energy savings estimates. Low-cost loggers are commercially available, such as devices that sense indoor light level changes, motor on/off status, and room occupancy. Figure 3.5 provides an example of an on/off logger in use for a residential kitchen exhaust fan (part of a heat recovery

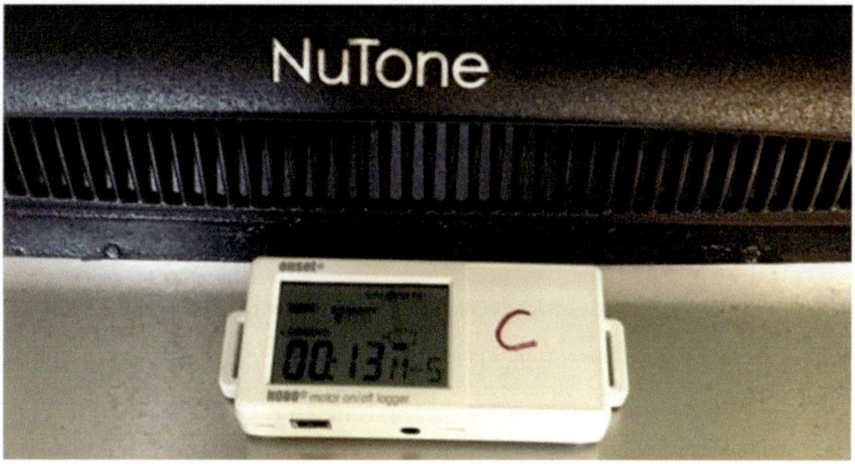

Figure 3.5 Motor on/off logger being used to record operating time for a residential kitchen exhaust fan.

ventilation and indoor air quality study). Even if logging is done for relatively short time periods (e.g., hours, days, or a few weeks) or on a sample of equipment (e.g., sample of luminaires), measured data will always provide superior accuracy compared to estimates, assumptions, or verbal reports by users regarding operating time.

3.3.2 Safety considerations

Safety is a critical part of any facility assessment. The field team should have essential knowledge and training regarding safety equipment and procedures. Table 3.2 provides a basic checklist of safety requirements to consider during the planning phase of facility assessments.

Before starting the facility tour, the field team should be thoroughly briefed on any specialized safety equipment and procedures for the facility, such as issuance of safety permits. Workers should be cautious when examining any operating piece of equipment, especially those with open drive shafts, belts, gears, or any form of rotating machinery. Field staff should coordinate and communicate with the equipment operator or supervisor before inspecting equipment and systems. If necessary, the inspection may need to be scheduled when the machine or device is idle or shutdown in order to conduct the inspection safely.

Table 3.2 Safety requirements to consider for on-site facility assessments.

Risk type	Precaution
Personal protective equipment to limit exposure to hazards	• High visibility vest • Steel toe shoes or boots • Safety glasses • Hearing protection plugs or earmuffs to reduce sound to 30 decibels • Long sleeve, fire resistant clothing • Work gloves
Electrical	• Work only on circuits for which you have the training and qualifications. Arrange for qualified electricians or certified technicians as needed • Never work on live circuits
Respiratory	• Test for safe oxygen levels (20.9% by volume) where risk of oxygen deficiency exists • Test for contaminants depending on the risk (e.g., CO, H_2S) When necessary, wear a full-face respirator mask with adequate particle and/or carbon filter
Confined space	• Use two-person confined space procedures (lookout person, communication devices, etc.) for spaces assessed as confined space hazards based on a safety risk assessment
Heat stress	• Assess for heat stress risks (typically if work environment is 100 °F or 38 °C or above) and adopt appropriate procedures (e.g., planned breaks, extra water, portable air cooling)
Fall protection	• Use fall protection systems when working at heights above safety threshold as defined by the local safety authority (e.g., 6 feet or 2 meters)
Working alone	• Avoid working alone in facilities, or arrange for procedures as needed with the facility safety staff (e.g., communication and check in protocol)

3.3.3 System inspections and data collection for common systems

The audit team should tour and inspect all systems in the facility, which are included in the assessment scope. Field staff should examine the operational patterns and equipment usage, collect detailed data on the facility structure, as well as on all energy-using equipment. As each of these systems is examined, the following questions could be asked:

1. What functions and services does this system deliver (consider quantity and quality)?
2. What is the energy consumption of this system, as well as peak demand magnitude and timing?
3. What are the indications that this system is working properly?
4. If this system is not working, how can it be restored to good working condition?
5. How can the energy cost of this system be reduced, without compromising performance?
6. What are the maintenance requirements for this equipment?

7. What options are there to upgrade or replace this system to reduce or eliminate GHG emissions?
8. Who has direct responsibility for maintaining and improving the operation and energy efficiency of this system?

With the above questions in mind, below is a list of nine common systems to be inspected for most facility assessments. The resources described in Section 3.1.3 include links to many template forms and spreadsheets that could be used to inspect the systems listed below.

1. Building envelope: The building envelope includes all building components that are directly exposed to the outside environment. The components of the building envelope to be inspected include: outside doors, windows, walls, the roof, and, in some cases, the floor. The inspection should record information on the insulation levels in the various parts of the facility using thermal imaging, the condition of the roof and walls, the location and size of any leaks or holes, and the location and size of any door or windows that open from conditioned to unconditioned space.
2. Space and process heating boilers and steam distribution system: Steam or hot water boilers typically consume the majority of natural gas (or other fuels) in a facility. Inspections typically include the major components of the steam or hot water generator (burners, fans, fuel supply, etc.), distribution piping insulation, steam traps, insulation on valves and expansion joints, and end-use heat exchangers.
3. HVAC system: All HVAC equipment should be inventoried. Generally, it should not be assumed that installed equipment matches record drawings. Images of nameplates should be collected to confirm equipment data. The equipment should be inspected to determine the condition of the heat exchanger coils, the air filters, and the insulation on the refrigerant lines. Air velocity measurements may also be made and recorded to assess operating efficiencies or to discover conditioned air leaks.
4. Electrical supply system: This system consists of transformers, wiring, switches, and fuses – all the components needed to enable electricity to move from the utility-owned wires at the facility boundary to its point of use within the facility. Nameplate data of key equipment should be recorded. Thermal imaging can be used to detect hot spots on electrical connections, which could indicate opportunities to improve safety, reliability, and reduce energy losses.

5. Lighting system, including all lights, windows, and adjacent surfaces: The functions of this system are to provide sufficient illumination for work tasks, to assist in building and area security, and for general facility aesthetics. Making a detailed inventory of all lighting systems is important, if lighting is part of the scope. Data should be recorded on numbers of each type of light fixture and lamp, the wattages of the lamps, and the hours of operation of each group of lights.
6. Domestic hot water system: The hot water system distributes hot water for washing, for use in industrial cleaning, and for use in kitchens. Its main components are hot water heaters (natural gas or electric), storage tanks, piping, and faucets or nozzles. All water heaters should be examined, and data recorded on their type, size, age, model number, electrical characteristics, and fuel use. What the hot water is used for, how much is used, and the timing of hot water demand should all be noted. The temperature of the hot water at the point of use can be measured and recorded.
7. Compressed air distribution system: Air compressors and the air distribution system provide motive power for tools and some machinery, and often provide air to operate the heating, ventilating, and air conditioning system. Inspections should look for signs of leaks, opportunities to switch from compressed air (see Chapter 21), and document system pressure.
8. Motors: Common practice is to inventory all electric motors over a chosen threshold (e.g., over 2 hp). Typical data to record includes motor size, use, age, model number, estimated hours of operation, and ideally the power factor at multiple load points. Measurement of voltages, currents, power factors, and shaft speed (if accessible and safe) may be appropriate for larger motors.
9. Manufacturing systems (also see Chapter 20): Industrial systems can usually be divided into two categories: (a) specialized manufacturing equipment, and (b) common ancillary systems. Industrial facilities will have highly specialized process and production operations and machines, depending on the products that are manufactured. Examples include electrolytic processes, chemical reactors, high-temperature furnaces and kilns, pulp refiners, product drying equipment, and many others. Common ancillary systems include boilers and steam distribution, compressed air, water supply, wastewater treatment, motors and controls, industrial lighting, and refrigeration. Data on types, sizes,

capacities, fuel use, electrical characteristics, age, and operating hours should be recorded for all of this equipment.

3.3.4 Virtual energy audits

During the COVID pandemic, virtual energy auditing techniques were developed and pilot tested (e.g., see https://neep.org/remotely-virtual-energy-audit-solution). Due to public health restrictions on travel and the ability to visit sites in person, conventional energy auditing practices were not possible. Virtual audits can be similar to conventional ASHRAE audits in many ways, with the key difference being there is no in-person site visit. Instead of the in-person site visit, the auditor may engage with the facility users or O&M staff to conduct a live, remote site visit using specialized software and smart phones (e.g., see https://smartenergy.illinois.edu/virtual-energy-assessments). In some cases, there is not even a virtual site visit, and energy analytics is performed using utility data and non-intrusive load monitoring techniques using smart meter data (Schirmer, 2023). In other cases, data for remote audits can include public data sources such as municipal records or public internet mapping tools. In theory, virtual audits could produce the same study outcomes as described for conventional assessments as described in this chapter. In practice, there can be significant shortcomings due to the lack of first-hand examination of the facility such as: missing measurements of available space for new equipment, no condition assessment of existing equipment, absence of submetering data, ability to analyze only electricity data (not other fuels), and inability to engage directly with facility users. Nonetheless, virtual auditing techniques can be appropriate where funding is limited for conducting assessments or where privacy concerns prohibit visiting a facility.

3.4 Phase 3: Identifying Measures, Analyses, and Reporting

3.4.1 Identifying energy and decarbonization measures

The best practice sequence for identifying and analyzing measures is described in Figure 3.1. As described in Section 3.2, the best practice also includes identifying potential measures during the preparation phase. One reason to identify measures proactively is that the range of measures to be considered will have a significant impact on the scope and cost of the assessment. Also, if the assessment team waits until after leaving the site to propose specific measures, then the opportunity to collect critical data needed for specific measures may have been missed. Yet another reason is that facility owners or O&M staff may have preferences for what measures they will or will not accept. For example, some organizations may prefer 100% electrification as a decarbonization strategy, while others may prefer diversified fuel types. Throughout the assessment process, including on-site and during the analysis phases, new or modified measures should be examined. As a general rule, the greatest effort should be devoted to analyzing measures that show the greatest energy savings or GHG reduction, and the least effort to those with the smallest improvement potential.

In terms of generating ideas for specific measures to meet assessment objectives, Chapter 7 on human behavior provides additional guidance on working with people and organizations. For specific technologies, Chapters 8–21 should be reviewed. Additionally, the resources listed in Section 3.1.3 provide volumes of information on technologies, lists of potential measures, and case studies.

3.4.2 Accuracy requirements for energy savings, GHG reduction, and cost estimates

The accuracy requirements for estimates of energy savings, GHG reduction, and implementation costs are important factors that affect the scope and cost of facility assessments. However, the accuracy requirements are typically not specified in standards and guidelines. For example, ASHRAE 211 does not state specific accuracy requirements for energy savings estimates or implementation costs. While it is clear that Level 3 audits are the most accurate and Level 1 the least, the specific targets are not stated. Accuracy requirements are often stated by funding organizations (e.g., utility or government loan programs), or otherwise agreed upon between the facility assessment team (often a consultant) and the facility owner.

Typical accuracy requirements are stated as a simple plus or minus percentage, such as energy savings accuracy of +/−10% and cost estimate accuracy of +/−15%. However, best practices use asymmetrical accuracy requirements. For example, the Certified Energy Manager^{TM} (CEM) training materials provide accuracy targets as summarized in Table 3.2 (CEM, 2022). Note that energy savings accuracies are all negative values, and implementation cost accuracies are positive. This is because financial risks increase when energy savings are less than estimates and when actual costs

Table 3.3 Example of accuracy requirements for energy audits.

	ASHRAE Level 1	ASHRAE Level 2	ASHRAE Level 3
Energy savings	−30% to −50%	−15% to −25%	−5% to −10%
Implementation costs	+30% to +50%	+15% to +25%	+5% to +10%

exceed estimates. Similarly, in Chapter 5, Table 5-1 summarizes standard cost estimate accuracy requirements that are asymmetrical (AACEI, 2020).

In general, the accuracy requirements for energy, GHGs, and costs should be agreed upon early in the planning of an assessment because it will affect the scope and cost of the assessment. For example, highly accurate energy savings estimates may require substantially more field measurements. Highly accurate implementation cost estimates may require that basic concept designs be worked out involving multiple engineering disciplines, and solicitation of budget quotes from vendors.

3.4.3 Accounting for interactive effects

In general there are two definitions of interactive effects used in energy management:

1. In the practice of M&V, the IPMVP (2022) defines interactive effects as interactions between systems:

   > "Any energy impacts created by an Energy Efficiency Measure that occur outside of the Measurement Boundary of the EEM and are not captured in the energy measurements."

   A common example is lighting EEM interactions with heating or cooling systems.
2. ASRHAE 211-2023 defines interactive effects as interactions between multiple measures:

   > "the change in resultant energy savings estimates or actual energy savings due to analyzing or implementing multiple EEMs that impact one another."
   >
   > A common example of multiple measure interactions is to analyze a lighting control EEM and a lighting fixture EEM for a given lighting system, where the net energy savings would be less than the sum of each EEM independently.

In energy assessments, the best practice is to include interactive effects if they are significant in terms of affecting the energy and/or utility bill savings.

What is "significant" is in the judgment of the energy analyst or auditor or as prescribed by a funding stakeholder.

3.4.4 Report outline and content

The next step in the facility assessment process is to prepare a report that details the results of the energy and/or GHG analyses. The length and detail of the report will vary depending on the objectives, applicable standards (e.g., ASHRAE 211, Level 3 vs. Level 1), and facility owner requirements. Additionally, if financial support is provided by an energy utility (e.g., custom incentive) or government agency, these organizations may have specific analysis and reporting requirements (e.g., specific software models to be used). If expectations about the report organization or content are not fully clear, then a good practice is to provide a table of contents to relevant parties in advance of completing the full report. A sample outline for an ASHRAE Level 2 energy assessment report is provided in Figure 3.6. Examples and best practices for various types of facility assessments can also be found in the resources listed in Section 3.1.3.

Following is a list of best practices for writing effective facility assessment reports. These practices are supported, in part, by research for residential energy audits (Sussman et al., 2019). But the findings should be generally applicable to any energy or decarbonization assessment.

1. Summarize all key findings in an executive summary, aiming for five pages or less.
2. Compare current or baseline performance to the expected performance (energy, GHG emissions, and financial) if recommended measures were implemented.
3. If non-energy impacts are projected to be realized, then clearly state these expected benefits with supporting data and evidence.
4. Make an effort to empathize and understand the concerns and needs of facility owners and users, and then prioritize recommendations showing how these concerns and needs could be met.
5. Minimize technical jargon, and write narrative summaries using simple and clear language appropriate for the targeted audience (e.g., eighth grade reading level is recommended for home audits).
6. Use clear visuals (images, diagrams, and charts) and uncluttered tables while aiming that all features of visuals accurately convey data and meaning without unintentionally distorting reality (Tufte, 2006).

Executive Summary
1. Introduction
    1.1 Background and Audit Personnel
    1.2 Deliverables
    1.3 On Site Audit and Data Collection Activities
    1.4 Audit Methods, Metrics, Data, Assumptions, and Accuracy
2. Facility Description
    2.1 Facility Description and Use
    2.2 Mechanical and Electrical Equipment
    2.3 Thermal Imaging
    2.4 Occupant Survey
3. Utility Data, GHGs, and Energy Benchmarking
    3.1 Energy Consumption and Degree Days
    3.2 Energy Use and Breakdown
    3.3 Energy Costs
    3.4 GHG Emissions
    3.5 Energy Consumption and Cost Benchmarking
4. Energy Efficiency Measures
    4.1 Electricity EEMs
    4.1.1 Lighting Fixture Retrofit
    4.1.2 Lighting Controls Retrofit
    4.1.3 HVAC Optimization
    4.1.4 Variable Frequency Drive Conversion
    4.1.5 Strategic Energy Management
    4.1.6 Recommended Bundle of EEMs
    4.2 Natural Gas EEMs
    4.2.1 HVAC Optimization
    4.2.2 Install Condensing Hydronic Boilers
    4.2.3 Install Condensing MAU Units
    4.2.4 Hot Water Blankets
    4.2.5 Strategic Energy Management
    4.2.6 Recommended Bundle of EEMs
    4.3 Measurement and Verification
5. Conclusions and Recommendations
6. References

Appendix 1: Building user survey comments
Appendix 2: Electric and gas utility Site ID's, meter numbers, and data
Appendix 3: Detailed electrical and mechanical analyses
Appendix 4: Detailed lighting analysis
Appendix 5: Option C M&V Plan

Figure 3.6 Sample table of contents for an ASHRAE Level 2 energy audit report.

7. Review findings with facility stakeholders who have decision-making authority to implement recommendations. Ideally, conduct the review in person and at the facility.

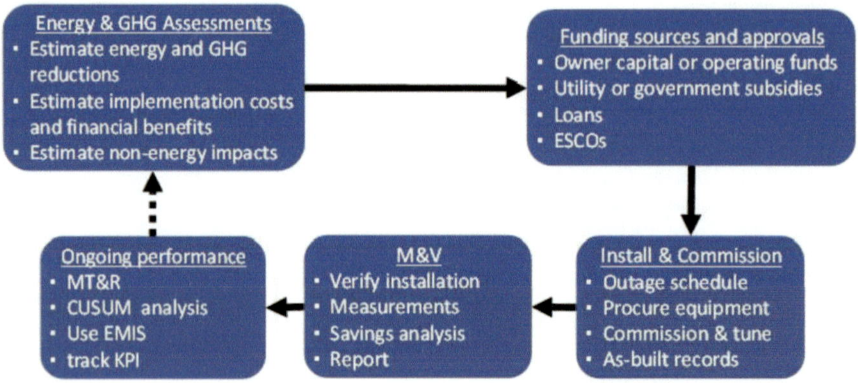

Figure 3.7 Flowchart illustrating the cycle for implementing facility assessment recommendations.

3.5 Implementing Measures and Monitoring Performance

After reports have been completed and the most attractive measures are targeted for implementation, the next steps are illustrated in Figure 3.7. Ideally, these steps as illustrated in the figure are carried out through an ongoing energy management program, as outlined in Chapter 1. The first step is to identify and secure funding sources for targeted measures. These sources could be owner funds, various subsidies such as utility rebates, loans, or energy services companies (ESCOs). With funding secured, plans for procuring equipment and installing measures are carried out. Commissioning and tuning are a critical part of the implementation (see Chapter 16). If the measures are behavior based, then apply the best practices from Chapter 7. As described in Chapter 1, M&V is often carried out on large projects, or smaller projects on a sampled basis. Because the performance of measures cannot be assumed to persist indefinitely, ongoing performance monitoring using various tools such as CUSUM are good practices. Lastly, it is important to note that ongoing monitoring over the long term will eventually reveal the need to update a facility assessment to identify and implement new measures. This is shown with the dashed line in the figure.

3.6 Summary

In this chapter, we have introduced the best practices for carrying out energy and decarbonization assessments. Assessments vary widely in the objectives and scope, and are referred to by different terms such as energy audit, plantwide audit, energy study, energy survey, decarbonization study, net-zero GHG study, net-zero energy study, and more. In practice, many facility assessments are done with limited effort to define objectives and service requirements, and often jump directly to analyzing capital-intensive upgrades for mechanical and electrical equipment. A key concept is that assessments should begin by carefully defining objectives, setting service requirements for energy systems, and analyzing the end uses first. The sequence is important. Analyzing options to eliminate waste (e.g., avoidable leakage and pressure losses in air and water systems) and upgrade passive systems (e.g., building envelope, industrial pump system leak reduction) should be completed before investigating capital-intensive upgrades.

Best practices are discussed for the three common phases of facility assessments. The first phase is preparation. The preparation phase is often the "make or break" opportunity for a highly successful assessment project. During the preparation, professionals are encouraged to: invest the time to define objectives, consider safety issues, obtain as much data and design information on the facility systems as possible, conduct preliminary utility bill analyses and benchmarking, develop measurement plans, and develop an initial list of targeted measures to achieve the energy or decarbonization objectives. For real-world projects, the ideal set of preparation tasks is not always feasible. However, the more of these tasks that can be done before the site work, the more likely the on-site work will collect needed information in a cost-effective manner. The second phase is to complete on-site work, which is discussed with a heavy emphasis on safety and careful consideration of facility users and O&M staff needs. Site work could range from a few hours to a few weeks, depending on the scope of the assessment and the measurement needs. Best practices for analysis and reporting are introduced. In particular, one aspect is to be clear about accuracy targets for estimates of energy savings, GHG reductions, and implementation costs. Common practice is to express these accuracies as a simple plus or minus percentages, but the best practice is often to use asymmetrical targets. Best practices for reporting are introduced. In particular, the most effective reports provide concise and clear summaries, emphasize both financial and non-energy benefits for facility users and owners, make effective use of visual information, and present the results in person to decision-makers. Finally, a short discussion is provided describing effective steps to implement recommended measures, including M&V and long-term performance monitoring.

References

ASHRAE (2011) "Procedures for Commercial Building Energy Audits" second edition.

ASHRAE (2023) "Standard 211-2023 Standard for Commercial Building Energy Audits".

ASHRAE (2023) Guideline 14-2023: Measurement of Energy, Demand, and Water Savings.

ASHRAE Handbook Fundamentals SI (2025) – "Thermal Comfort" Chapter 9

BPA (2017) Bonneville Power Authority "Industrial Audit Guidebook: a Guidebook for Performing Walk-through Energy Audits of Industrial Facilities" www.bpa.gov/-/media/Aep/energy-efficiency/industrial/Industrial-files/audit-guide.pdf (downloaded December 24, 2022).

CEM (2022) Certified Energy Manager^{TM} international training materials. Association of Energy Engineers.

Doty, S. (2021) "Commercial Energy Auditing Reference Handbook" 4th edition. . River Publishers.

ISO (2025) Standard 50002 "Energy audits — Requirements with guidance for use." International Standards Organization.

Williams, A. (2025) "Energy Management Handboo" River Publishers

Schirmer P. and Mporas I. (2023) "Non-Intrusive Load Monitoring: A Review" IEEE Transactions on Smart Grid, Vol. 14, No. 1.

Sussman R., Chikumbo M. and Miller N. (2019) "After the audit: improving residential energy efficiency assessment reports" American Council for an Energy Efficiency Economy. Report B1901.

Tufte E. (2006) "Beautiful Evidence" Graphics Press.

Williams, A. (2023) "Industrial Energy Systems Handbook" (1st edition). River Publishers

Author Biography

Eric Mazzi is a consultant and educator focusing on technology and policy for demand-side energy systems. He specializes in performing independent technical analysis, M&V, policy studies, and educational services. He advocates and contributes to efforts to reduce energy consumption and achieve net-zero GHG emissions, while maintaining or improving energy services and advancing energy equity. His experience includes air quality consulting, utility maintenance, and five years as full-time engineering faculty at the

University of British Columbia. He volunteers as a board member for the U.S. Northwest Energy Coalition, and the International Performance M&V Protocol Committee. Eric teaches part-time, delivering university and professional certification courses. He is an Adjunct Professor of Energy Management at the New York Institute of Technology – Vancouver. He has B.Sc. and M.Sc. degrees in mechanical engineering, and a Ph.D. degree in environmental policy. He is a registered engineer in California and British Columbia.

Acknowledgments

I would like to thank Dawn-Marie Barreira for her thorough review of this chapter, and helpful recommendations for improvements in content.

Practice Questions:

1. You are tasked to complete a site visit to a natural gas processing plant, located in a remote, rural location. The purpose is to conduct a plantwide audit for energy savings, and a screening analysis for decarbonization measures to reduce GHGs by 25%. Describe what tasks and activities you would complete for the preparation phase of the energy and decarbonization assessment (i.e., before visiting the site).
2. A 10-storey multi-family residential building was built in 1972, is owned by one company, and contains 110 rental units. It is located in ASHRAE climate zone 6, with 3,000 heating degree days (°C units). It is a cast-in-place concrete structure with single pane windows and glass balcony doors in all suites. All suites have electric stoves with hot water baseboard heating. All facility heating and domestic hot water is supplied by one conventional hydronic boiler rated at 2,000 MBH. Ventilation is provided by two rooftop makeup air units, equipped with heating coils, which deliver outdoor air to corridors at each floor. Lighting is mostly fluorescent tube in common spaces, with a mix of lamps in suites ranging from incandescent to modern LEDs. Each suite has its own electric meter, and the common facilities have one electric meter and one gas meter. Suite ventilation is achieved by corridor air entering through the door undercuts. Common spaces in the building include a gym, pool, and laundry. An ASHRAE Level 3 energy audit is to be performed with the stated objective of reducing energy consumption by 30%. Develop the basic elements of a measurement and modeling plan for the study.

3. Describe the key differences between an energy assessment versus a decarbonization assessment, in terms of the general scope and time horizons for the studies.
4. The administrative building for an Indigenous community has four storeys and about 40,000 square feet, and is located in a remote northern location in Canada. The facility is undergoing a decarbonization assessment with plans to implement the upgrades in the next two years. You are tasked to develop a plan for an indoor environmental quality assessment for the facility to be completed in parallel with the decarbonization study and construction project. It is anticipated that the key upgrades will be to replace natural gas space and hot water heating with electric heat pump and electric boiler systems, new triple pane windows, air leakage sealing, LED lighting, and new electrically heated ventilation air systems. Describe the key elements of your indoor environmental plan. Consider what data you would collect, what methods and instrumentation you might use, and the timing of your activities.
5. An ASHRAE Level 3 energy audit is to be performed for a university academic building that contains classrooms and laboratories. Describe what accuracy requirements you would recommend for (a) the energy savings estimates, and (b) the cost estimation estimates. State the rationale for your recommendations.
6. A low-income, three-storey social housing multi-family building (24 units) is located in the southeast U.S. The building was built 50 years ago with wood frame construction, natural gas stoves, common space laundry facility, individual suite electric metering, central natural gas heating (hot water radiators) and domestic hot water, mostly compact fluorescent lighting, and rooftop ventilation units for the corridors. About one-fourth of the occupants have window-mounted air conditioners. There is government funding available to conduct a complete energy and decarbonization assessment for the building, with the objective of achieving net-zero GHG counting Scope 1 and Scope 2 emissions, and improving resiliency. Outline the key tasks, methods, and data collection that you would plan on completing for the preparation (before going to the site) and the on-site phases of your energy and decarbonization assessment. State what unique issues might exist for this facility, as compared to a newer multi-family housing unit located in a high income neighborhood.

4

Energy Billing Rates

Eric Mazzi

Mazzi Consulting, Canada

4.1 Introduction

Energy billing rates are critical in energy management. The billing rates are what translate the consumption of purchased energies into costs, so that economic analyses can be performed and energy budgets can be managed. Utility costs traditionally have been considered overhead costs for organizations, and seen only by accounting staff. Although energy management practices have advanced over decades, this is still often the case. The best practice for financial and energy management is to actively track and manage energy purchases. This is increasingly important as facilities undergo decarbonization and energy costs rise.

The purpose of this chapter is to introduce the basic concepts and energy rate structures for electricity, natural gas, and Thermal Energy Networks (TEN)[1]. An overview of energy suppliers is provided, followed by a description of the common components of rate structures. It is important to emphasize that some components of utility bills are fixed and some are variable. As well, some components are based on total consumption or energy (e.g., kWh of electricity) and some components based on peak demands (e.g., peak 30-minute kVA of electricity) or timing of consumption such as time of use (TOU) rates. Common rate structures and examples are presented for electricity, natural gas, and TEN. Example rates presented in this chapter are not necessarily current in the dollar values stated, but are based on prevailing rates for the years 2021–2023.

The importance of energy rates can be understood by considering the five types of distributed energy resources (DERs) introduced in Chapters 1 and 2.

[1]See Chapter 19 introduction for an explanation for the use of the term Thermal Energy Network vs. District Energy

- For energy efficiency projects, it is critical to understand the concept of marginal cost, which is the incremental cost or savings by adding or avoiding one unit of energy to the energy bill. If energy efficiency investments are made for electric-driven or natural gas-fired equipment, the fixed components of these utility costs will not change. Some or all of the variable components will change, which is captured in marginal costs.
- Demand response usually involves electricity, although some natural gas rates include demand charges. With demand response, it is important to understand how shifting the timing and magnitude of consumption translates into changes in utility costs. How the utility costs change from shifting demand will depend on how demand charges are built into the rate structure, and whether the rate includes TOU or other time-variant pricing.
- Energy storage – depending on whether it is thermal, electric, or both – can potentially effect changes in both the energy and demand components of utility bills. As with demand response, assessments of energy storage systems on utility bills need to consider the quantity and timing of stored energy, which purchased energies are affected by storage (e.g., electricity, gas, and/or TEN supply), and how these shifts translate to changes in utility costs.
- Distributed generation such as solar PV and combined heat and power (CHP) can significantly transform energy costs, for better or worse. The electricity generated in a facility that is not exported will reduce purchased electricity at the marginal rate, similar to electric efficiency improvements. Generated electricity can also shift peak demands and timing of power drawn from the grid, and needs to be analyzed similar to demand response. If the electric tariff allows for exporting of power into the grid, then the value of exported power needs to be analyzed using the specific terms of the net metering (or net billing) rate structure. Adding CHP systems to facilities always involves an increase in the purchase of a fuel to supply the CHP system (e.g., natural gas, renewable natural gas, or biomass), but will also offset purchased electricity and energies used to supply thermal energy to a facility. For example, the recovered heat from a CHP unit could offset electric heating, natural gas boilers, or purchased hot water from a TEN utility.
- Fuel switching measures, such as electrifying end uses or switching from fossil natural gas to low carbon fuels, can also fundamentally change utility costs in a similar fashion to distributed generation. Fuel switching

will significantly increase the quantity and timing of consumption for one energy (e.g., electricity for electrification), and significantly decrease if not eliminate another energy (e.g., natural gas).

In all cases, it is important for energy management professionals to understand the rate structure and specific pricing terms. This is necessary to analyze rate structures and costs, and estimate the potential economic benefits or impacts of implementing DERs.

An overview of energy suppliers is provided. Then, common rate structures, cost elements, and key concepts needed to assess the impact of DERs on utility costs are presented in the remainder of this chapter.

4.2 Energy Suppliers

Before introducing common utility billing rate structures and concepts, it is helpful to review some basic types of energy utility business models (Chaudhari, 2021; Price, 2021; Lazar, 2020).

Utility companies are typically owned by investors (i.e., profitmaking companies) or are publicly owned. For example, many provincial government-owned electric utilities exist in Canada, some federally owned utilities exist in the U.S., and there are many municipality-owned utilities. In the U.S. and Canada, most electric, natural gas, and TEN utilities operate with oversight by a public utilities commission (PUC), which mandates that utilities work within strict regulations. In a regulated market, energy utilities have exclusive rights to supply and distribute energy within a specific geographic service territory. In a deregulated market, supply and distribution are separated. The common situation in deregulated markets is that transmission and distribution are regulated monopolies, but customers can purchase their supplies (i.e., electricity or natural gas) from competing independent firms. These independent supply firms are usually described as marketers (e.g., "power marketers" or "gas marketers"). One of the practical issues for facilities operating in deregulated markets is that there may be more than one utility bill for a facility, such as one bill from independent suppliers, and another from the regulated utility operating the transmission and distribution systems.

Another common business model for energy utilities is a cooperative or "coop," which operates under the oversight of a local board, rather than a PUC. Coops are typically owned by the customers and members, are non-profit organizations, and often operate in rural areas. In the U.S., coop service territories cover 56% of the land area (www.electric.coop/electric-cooperative-fact-sheet). Some natural gas and TEN utilities also operate with the coop model. Typically, coops have more flexibility to set rates and change rate structures as compared to utilities operating under PUCs.

4.3 Common Elements of Energy Rates

In general, there are many components of energy bills that could potentially apply. Following are descriptions of the most common elements of energy bills.

- *Energy or consumption charges* are for the total quantity of metered energy consumed for the billing period. This covers electricity (kWh), natural gas (GJ, Dekatherms, MCF), hot water (GJ, MMBtu), chilled water (ton-hours), or other energies (e.g., biomass, steam, RNG). Typically, but not always, the consumption charge is the largest cost for a utility bill.
- *Demand charges* are for the peak demand of electricity (kW or kVA) and natural gas (e.g., GJ or therms), averaged over a specified period of time. For electricity, 15-minute or 30-minute averages are common. Where natural gas peak demand is part of the rate structure, daily peaks are more common. Demand ratchets are sometimes used for commercial or industrial rates. Ratchet charges are typically based on the maximum demand over a 12-month period. For a given billing period, a common ratchet demand charge would be based on the larger of the current period demand, or a percentage of the highest demand over the previous 11 months.
- *Power factor* charges apply to electricity rates based on a minimum threshold for the power factor at the customer meter, such as 0.90 (see Chapter 13 for an explanation of power factor).
- *Transmission and distribution* charges are for the movement or delivery of energy from the production site to the customer's facility. These charges are typically priced per unit of energy consumption (e.g., per kWh for electricity, or per GJ of natural gas).
- *Storage* costs are sometimes used for natural gas based on the portion of natural gas that was kept in storage. Sometimes storage costs are included with transmission and distribution as part of "delivery" charges. Electricity storage on the utility side of the meter (called "front of

meter") is becoming more common, and these storage costs are typically incorporated into transmission and distribution charges.

- *Fuel adjustment* charges are sometimes used by electric utilities with thermal power generation as well as TEN utilities. This charge is to adjust the price of electricity or thermal energy based on changes in the price of fuels for energy production (e.g., natural gas market rate), which are higher or lower than the fuel cost used to determine the base rate for energy.
- *Customer fees* cover utility costs such as administration, metering, and billing.
- *Taxes* such as state or provincial sales tax almost always apply, although some customers have exemptions. Taxes are multipliers of the total, pre-tax utility bill (sometimes excluding specific costs). Carbon taxes sometimes apply based on the GHG emissions intensity of the energy.
- *Riders* are additional charges to cover regulatory requirements such as environmental charges, energy efficiency programs, or government clean energy research funding.
- *Late payment fees* can be significant and are avoidable. The best practice is to pay utility bills on time. Some utilities offer discounts for early or prompt payment of bills.

4.4 Electricity Rates

4.4.1 Common electricity rate structures

Common types of electric rates structures are summarized in Table 4.1 (Price, 2021; ASHRAE, 2019). These rates are applied to different customer classes. In general, utilities offer many types of customer classes. Some of the most common include:

- Residential, sometimes with two or more options to choose from such as TOU, block rates, or net metering.
- Commercial or general service, often segmented into small, medium, and large (e.g., based on voltage, peak demand, or whether transformation is provided by the utility or customer).
- Industrial, often segmented based on connection voltage or peak demand.
- Other specialized classes, such as lighting.

Table 4.1 Common rate structures for electricity.

Rate structure	Description	Comments
Flat rates	Constant unit price of energy at any time and any quantity.	The simplest rate design.
Block rates	Energy pricing is in tiers that could be inclining (price increases with quantity) or declining.	Declining block pricing does not incentivize energy efficiency. Inclining block typically incentivizes energy efficiency. Variable block options exist.
TOU	Defined peak periods (on-peak) with higher unit pricing, and one or more off-peak periods with lower unit pricing.	Typically peak demand charges would not be combined with TOU pricing.
Critical peak charges	Critical peak pricing: utility increases price during designated peak events. Critical peak rebate: utility rebates the customer for reducing consumption during designated peak events compared to their historical baseline.	Critical peak charges or rebates could be combined with other rate designs. Requires smart meters.
Real-time pricing	Electricity prices vary based on market prices, and customers receive communications with forecast prices (e.g., day ahead).	Requires smart meters. Pricing information must be clear and readily available to customers.
Curtailment and interruptible	Customer agreement allows utility to curtail part or all of their load, with specified time period of advance notice (e.g., 30 minutes to hours).	Customers receive compensation in exchange for flexibility to curtail load.
Lifeline rate	Lower rate based on income eligibility.	Designed to help low-income customers afford basic energy services. One option is discounted price for a baseline quantity of energy each (or % discount on whole bill) for billing period.
Variable peak pricing	TOU rate structure where one of the defined time periods is priced using some form of real-time pricing (e.g., TOU rate where the price of the afternoon time period changes daily).	Requires smart meters. Pricing information must be clear and readily available to customers.

Table 4.1 *Continued.*

Rate structure	Description	Comments
Net metering[1]	Customers accrue kWh credits. No direct purchase of energy by the utility.	For customers with behind-the-meter electricity generation. Requires smart meters.
Net billing	Customers are compensated for electricity supplied to the grid	For customers with behind-the-meter electricity generation. Requires smart meters. Compensation for power supplied to the grid can be more, less, or equal to the price paid for purchased power.
Buy all, sell all	Generated electricity is metered separately from the facility consumption, and all generated energy is purchased by the utility at a contract rate.	For customers with behind-the-meter electricity generation. Compensation for power supplied to the grid can be more, less, or equal to the price paid for purchased power.
Seasonal rates	Rate structure where energy prices vary by season, such as winter, summer, and/or shoulder seasons.	Can be combined with TOU, block, or real-time pricing. Net billing rates can also vary seasonally.

[1]A summary of net metering policies for the U.S. can be found at: www.dsireusa.org/resources/detailed-summary-maps/net-metering-policies-2/. Also see Zinaman, 2017.

4.4.2 Residential electricity rate examples

This section provides examples of electric utility rates to illustrate the range of rate structures and options. Sales taxes generally apply to all utility bills (except in cases where customers have exemptions) but are not shown in these examples. It should be noted that the rate examples provided here were representative of rate offerings and prices in the years 2021–2023. Utility rate offerings and price terms can change as frequently as every year.

An example of a residential rate for a multi-family customer is provided in Table 4.2, based on Commonwealth Edison rates for basic electric service.[2] Commonwealth Edison is a large, regulated utility based in Illinois. Notice that these residential rates do not include any demand charges, which is often the case for residential customers.

[2]www.comed.com/MyAccount/MyBillUsage/pages/currentratestariffs.aspx.L Prices for each component are illustrative and not necessarily current rates.

Table 4.2 Commonwealth Edison basic electric service for multi-family residential without electric space heat.

Supply charges	**Price**	**Comment**
Electricity supply charge	$0.09738/kWh	Summer rate shown (June to September). Non-summer months have lower rate.
Transmission services charge	$0.01360/kWh	
Purchased electricity adjustment	$1.37	Adjustment to cover utility supply cost variances.
Deliver charges		
Customer charge	$8.06/month	This charge is slightly higher for multi-family with electric space heating.
Standard metering service charge	$3.10/month	
Distribution facilities charge	$0.03246/kWh	This rate is lower for multi-family with electric space heating.
Illinois electricity distribution tax charge	$0.00132/kWh	
Taxes, riders, and fees		
Environmental cost recover	$0.00053/kWh	
Renewable portfolio standard	$0.00502/kWh	
Zero emission standard	$0.00195/kWh	
Energy efficiency programs	$0.00233/kWh	
Energy transition assistance	$0.00082/kWh	
Franchise cost	Percentage times pre-tax cost	Minor charge to cover franchise agreement between utility and municipality.
Municipal tax	Variable charge multiplier of the kWhs delivered	Applicable municipal taxes.

Mid-Carolina Electric Cooperative is an electric coop located in the southeastern U.S. with a residential rate structure as shown in Table 4.3.[3] A relatively unique feature of this rate structure is that there is a demand charge, based on the peak utility hours by season.

Mid-Carolina also offers a residential net metering rate[4] (e.g., for customers with solar PV). The rate is the same as Table 4.3 for the account charge

[3] www.mcecoop.com/member-benefits/rate-structure

[4] https://www.mcecoop.com/media/yb1djw5b/rate-rider-net-metering-rider-jan-2021.pdf

Table 4.3 Mid-Carolina Electric Cooperative residential rate.

Fixed charges	**Price**	**Comment**
Account charge	\$0.95/day	Covers administration, metering, and other utility costs.
Energy charges		
Flat rate	\$0.0515/kWh	Plus a monthly adjustment factor to cover variable utility costs. Historically has ranged from \$0.00 to \$0.0047 per kWh.
Demand charges*		
Single highest hour during the on-peak time period: Winter on-peak season is November 1–March 31, 6:00–9:00 am Summer on-peak season is April 1–October 31, 4:00–7:00 pm	\$12.00/kW	Plus a monthly adjustment factor Historically has ranged from \$0.00 to \$0.062 per kW.
Riders		
Subdivision lighting	\$3.24/month	Not applicable to all customers
Franchise fee, city right of way	\$4.17/month	Not applicable to all customers

* Customers on net metering do not pay the per kW charge, and replace it with:
(Net Peak Period kWh)(\$0.3440/kWh)

and energy charge. Each kWh of electricity exported to the grid is credited to the customer's account (i.e., valued at the purchase price). Energy charges bottom out at zero each billing period, and any additional credits are carried forward to future billing periods (i.e., the utility does not pay out dollars if energy supplied to the grid exceeds energy drawn from the grid). With net metering, demand charges are not per kW as shown in Table 4.3. Instead, customer demand charges are based on the net energy consumed during the on-peak time periods for winter and summer with the following formula:

$$\text{Peak Charge} = (\text{Net Peak Period kWh})(\$0.3440/\text{kWh})$$

Fort Collins, Colorado presents an example of a municipal utility rate. The utility offers time-of-day pricing[5] to residential customers. The rate structure is provided in Table 4.4, which is a hybrid of TOU, inclining block rate, and seasonal rates. For customers with solar PV (or other self-generation), the utility offers net billing credits as shown in Table 4.5 (the negative sign indicates a credit to the customer). Note that the net billing credits are similar

[5]https://www.fcgov.com/utilities/residential/rates/electric/#cb-33260-4651-0

Table 4.4 Fort Collins' time-of-day pricing for customers with non-electric space heating.

Fixed charges	**Price**	**Comment**
Customer monthly charge	$10.65/month	Covers administration, metering, and other utility costs.
Energy charges		
Summer: May–September On-peak Monday–Friday 2:00–7:00pm	$0.2715/kWh	
Summer: May–September Off-peak	$0.0754/kWh	All other hours, weekends, and major holidays.
Non-summer: October–April On-peak Monday–Friday 5:00–9:00pm	$0.2503/kWh	
Non-summer: October–April Off-peak	$0.0754/kWh	All other hours, weekends, and major holidays.
Tier charge		
Additional charge for each kWh used above 700 kWh in a month	$0.0265/kWh	Tier charge represents an inclining block.

Table 4.5 Fort Collins' net billing solar PV credits for customers with non-electric space heating.

Energy charges		
Summer: May–September On-peak Monday–Friday 2:00–7:00pm	– $0.2510/kWh	
Summer: May–September Off-peak	– $0.0636/kWh	All other hours, weekends, and major holidays.
Non-summer: October–April On-peak Monday–Friday 5:00–9:00pm	– $0.2179/kWh	
Non-summer: October–April Off-peak	– $0.0636/kWh	All other hours, weekends, and major holidays.

in value, but slightly lower than the rates that customers pay for power drawn from the grid.

4.4.3 Commercial and industrial electricity rate examples

Commercial and industrial class customers are non-residential customers such as retail businesses, healthcare facilities, educational facilities, industrial

Table 4.6 Manitoba Hydro's general service medium rate (Canadian currency).

Fixed charges	Price	Comment
Basic monthly charge	CAD$33.69/month	Covers administration, metering, and other utility costs.
Energy charges		
First 11,000 kWh	CAD$0.09485/kWh	This is a declining block rate.
Next 8,500 kWh	CAD$0.07277/kWh	
Balance of kWh	CAD$0.04492/kWh	
Demand charges		
First 50 kVA measured demand	No charge	
Balance of measured demand	CAD$11.52/kVA	Note that kVA captures power factor
Billed demand	Greatest of: (1) measured kVA, (2) 25% of contract kVA, or (3) 25% of highest measured kVA in the previous 12 months.	

plants, and many others. Typically, these customer classes are segmented with many optional rate schedules, depending on voltage and peak power demand. Common voltage categories are secondary voltage (typically 600 V or less, but sometimes up to several kV), primary voltage (above secondary voltage, and typically below 69 kV), or transmission (often 69 kV or higher).[6]

Manitoba Hydro is a large, crown corporation (i.e., government owned) located in the prairies of Canada. It is an integrated utility providing electricity and natural gas to the province. Table 4.6 shows an example of a commercial rate with the utility's general service medium rate class.[7] This rate illustrates some common rate design features. One observation is that the energy charge is a declining block rate, which does not incentivize energy efficiency. Note that demand charges are based on kVA, not kW, which means that the demand rate captures power factor effects (see Chapter 13). Also note that the billed demand is not necessarily the measured demand, but the greatest of three different kVA values. The third option is the billed demand, which is 25% of the highest measured kVA for the previous 12 months, and is an example of a demand ratchet clause.

[6] These voltage ranges are not standardized, and can vary widely from one utility to another. Some utilities include a sub-transmission voltage category, which is between primary and transmission.

[7] www.hydro.mb.ca/accounts_and_services/rates/commercial_rates/#e-mgs.\Customers \with\utility-owned\transformer\and\billing\demand\exceeding\200\kVA.

Table 4.7 Sacramento Municipal Utility District commercial and industrial time-of-day rate (secondary voltage < 12 kV, 300–499 kW).

Fixed charges		**Price**	**Comment**
Non-summer (October–May)	System Infrastructure Charge	\$428.35/month per meter	
Summer (June–September)	System Infrastructure Charge	\$428.35/month per meter	Charge is the same Summer vs. non-summer
Energy charges			
Non-summer (October–May)	Peak: weekdays 4:00–9:00pm, excluding holidays	\$0.1236/kWh	
	Off-peak saver: every day 9:00 am–4:00 pm, including holidays	\$0.1000/kWh	
	Off-peak: all other hours, including holidays	\$0.0990/kWh	
Summer (June–September)	Peak: weekdays 4:00–9:00 pm, excluding holidays	\$0.2195/kWh	Note much higher peak energy price in summer
	Off-peak: all other hours, including holidays	\$0.1333/kWh	
Demand charges			
Non-summer (October–May)	Site Infrastructure Charge	\$4.597/kW	12 month maximum 15-minute peak kW
Summer (June–September)	Site Infrastructure Charge	\$4.597/kW	12 month maximum 15-minute peak kW
	Summer peak charge	\$9.877/kW	Monthly maximum 15-minute kW

Another example of a commercial/industrial rate is provided in Table 4-7, based on Sacramento Municipal Utility District.[8] This rate structure combines seasonal (summer vs. non-summer) and TOU pricing. It also splits peak charges into two components, one based on annual peak kW plus a summer peak based on monthly peak kW.

An example of a large, industrial customer rate schedule is illustrated by Hydro Quebec's Rate L.[9] The rate requires customers to have a minimum billing demand of 5,000 kW. The basic rate is simple: CAD\$13.779 per kW

[8] www.smud.org/en/Rate-Information/Business-rates

[9] www.hydroquebec.com/data/documents-donnees/pdf/electricity-rates.pdf?v=20230401\

for monthly demand, and a flat rate of CAD$0.03503 per kWh for energy. Billing demand is the larger of (a) contract demand or (b) maximum power demand during a given month. Maximum power demand is determined based on 15-minute metered data for real power (kW) and apparent power (kVA), and is the larger of the highest metered kW value or 95% of the highest metered kVA value. The full rate schedule includes a number of variables, such as an optimization charge if the maximum power demand exceeds 110% of the contract demand.

4.4.4 Examples of electric billing rates for special applications

We will conclude this section on electric utility rates with a few examples of special billing rate applications.

Power factor was defined in Chapter 13 and is an important parameter for electric utilities. As such, power factor clauses are often built into billing rates. One example of this is DTE Energy, based in Detroit. The utility has a rate class D11 for primary supply customers, which includes a Power Factor Clause.[10] The terms of this rate specify that the facility power factor must be at least 0.70 (or 70%). If it falls below 0.70, then the customer is required to install corrective equipment (e.g., capacitors, synchronous motors). The rate includes substantial cost penalties for power factor less than 0.85 according to the following schedule:

- 0.850 and higher, no penalty.
- 0.800–0.849, penalty of 1% times monthly billing amount for energy (kWh).
- 0.750–0.799, penalty of 2% times monthly billing amount for energy (kWh).
- 0.700–0.749, penalty of 3% times monthly billing amount for energy (kWh).

The previous section highlighted examples of net metering, which most often is used for behind-the-meter solar PV installations (see Chapter 18). Some utilities have special rates to purchase power from combined heat and power (see Chapter 20) or other small power generation systems, which meet legal "qualified facility" criteria. One example is Northern Indiana Public Service Company, which has a specific rate rider for CHP as listed

[10] www.dteenergy.com/us/en/business/service-request/pricing/rate-options.html

Table 4.8 Northern Indiana Public Service Company Rate Rider 778 for qualified CHP.

Summer (May–September)	On-peak: Monday–Saturday, 8:00 am–11:00 pm	$0.03709/kWh
	Off-peak: Monday–Saturday, 11:00 pm–8:00 am All day Sunday Designated holidays, 24 hours	$0.02460/kWh
Winter (October–April)	On-peak: Monday–Saturday, 8:00 am–11:00 pm	$0.03692/kWh
	Off-peak: Monday–Friday, 11:00 pm–8:00 am All day Saturday and Sunday Designated holidays, 24 hours	$0.02942/kWh

in Table 4.8.[11] This example illustrates that utilities value customer supply to the grid differently based on day of the week, time of day, and season.

BC Hydro is a crown corporation (i.e., government owned) supplying electricity for most of the province. As noted in Chapters 1 and 2, fuel switching is a common strategy employed to achieve climate policy goals. British Columbia's electricity supply has a relatively low GHG intensity (see Chapter 2, Table 2-5) and the utility is mandated to incentivize its customers to switch fossil-fueled end uses to grid electricity. BC Hydro offers its industrial transmission customers (69 kV connection), based on meeting fuel switching eligibility criteria, the option of fuel switching Rate 1895.[12] This rate was linked to the provinces climate mitigation goals to increase the uptake of low-carbon grid electricity. Rate 1895 offers substantial discounts to the standard transmission rate for seven years as follows:

- 20% discount the first five years
- 13% discount in year six
- 7% discount in year seven

Rural communities that are not connected to the main transmission and distribution systems represent another special application. Typically, these communities are supplied with diesel-electric generation facilities, which

[11] www.nipsco.com/docs/librariesprovider11/rates-and-tariffs/electric-rates/current-rates/table-of-contents/rider-778.pdf?sfvrsn=5

[12] https://app.bchydro.com/accounts-billing/rates-energy-use/electricity-rates/electrification-rates.html

is expensive. One example is SaskPower's residential Rate E04 for customers supplied with diesel electricity generation.[13] The rate is an inclining block structure as follows: CAD$29.99 basic monthly charge, CAD$0.14895 for the first 650 kWh, CAD$0.58148 for energy over 650 kWh, and CAD$0.010125/kWh federal carbon charge.

Yet another special electric rate tariff, which is rapidly becoming more common, is for electric vehicles (EVs) (Tiedemann and Sulyma, 2023). For example, Excel Energy in Minnesota has multiple electric vehicle charging rate options for residential customers.[14] One option is for customers to rent a Level 2 charger, pay a monthly fee of $16.48 (includes rental cost), and pay for electricity applying a TOU and seasonal rate schedule. The charger controls are set so that vehicles automatically charge from midnight to 6:00 am every day at a rate or $0.028 per kWh, while customers can still charge at other times at higher rates.

4.5 Natural Gas Rates

Natural gas rates are generally less complex than electrical rates. Rate structures typically include the three common elements: fixed charges to cover costs such as metering and billing, delivery charges for transport and storage, and energy charges. Some rate schedules for larger consumers include demand charges.

Enbridge Gas is based in Ontario, Canada and is a regulated utility. Their residential natural gas rates[15] are provided as typical examples. Table 4.9 shows the rate schedule for customers purchasing supply from Enbridge (Rate 1), and Table 4.10 for customers purchasing gas supply from a marketer (Rate M1). These are declining block rates. Note that Canada has a carbon tax system that applies either a provincial or federal carbon tax, based on the Scope 1 emissions for the natural gas (see Chapter 2).

Piedmont Natural Gas is a regulated natural gas utility operating in the southeast U.S., and owned by Duke Energy. An example of a natural gas rate is provided in Table 4.11[16] for the utility's firm Large General Service Rate 203 in South Carolina. This service is for customers with daily consumption of at least 50 dekatherms per day (50 MMBtu/day or 53 GJ/day). Note that

[13] www.saskpower.com/Accounts/Power-Rates/Power-Supply-Rates

[14] https://ev.xcelenergy.com/ev-charging-programs

[15] www.enbridgegas.com/residential/my-account/rates

[16] www.piedmontng.com/our-company/about-piedmont/our-rates

Table 4.9 Example residential natural gas Rate 1 based on Enbridge Gas (supply included).

Fixed charges		
CAD$22.88/month		
Delivery and energy* charges		
Delivery: first 30 m^3	CAD$0.120499/$m^3$	
Delivery: next 55 m^3	CAD$0.113754/$m^3$	
Delivery: next 85 m^3	CAD$0.108472/$m^3$	
Delivery: over 170 m^3	CAD$0.104535/$m^3$	
Facility carbon charge	CAD$0.0079/$m^3$	Carbon charge for Enbridge facilities
Transportation to Enbridge	CAD$0.042937/$m^3$	
Federal carbon charge	CAD$0.123900/$m^3$	Canadian federal carbon tax
Gas supply charge	CAD$0.132318/$m^3$	Cost to Enbridge with 0% markup
Cost adjustment	CAD$0.013259/$m^3$	Monthly variations in Enbridge's supply, transport, and delivery costs

* 1 m^3 = 0.03514 MCF = 0.0385 GJ (HHV basis) = 0.0368 MMBtu = 0.368 therms

Table 4.10 Example residential natural gas rate based on Enbridge Gas (supply from a marketer).

Fixed charges		
CAD$23.98/month		
Delivery and energy* charges		
Delivery: first 100 m^3	CAD$0.073881/$m^3$	
Delivery: next 150 m^3	CAD$0.070757/$m^3$	
Delivery: over 250 m^3	CAD$0.062692/$m^3$	
Facility carbon charge	CAD$0.0162/$m^3$	Carbon charge for Enbridge facilities
Federal carbon charge	CAD$0.123900/$m^3$	Canadian federal carbon tax
Gas supply charge	Not applicable	Customer supply is from marketer
Cost adjustment	CAD$0.013259/$m^3$	Monthly variations in supply, transport, and delivery

* 1 m^3 = 0.03514 MCF = 0.0385 GJ (HHV basis) = 0.0368 MMBtu = 0.368 therms

there is a demand charge for customers under this rate. The energy charges are a declining block rate.

Piedmont Natural Gas offers a companion Rate 204 (also 50 dekatherms per day), which is for interruptible service. With the interruptible option, the customer's supply can be curtailed with notice from the utility. The tariff requires that customers on interruptible rates be equipped with an alternative

Table 4.11 Piedmont Natural Gas Large General Service Rate 203 for firm gas.

Fixed charges		
$250.00/month		
Demand charges		
$1.2500/therm	Demand is based on the highest daily therm measured from the previous winter period from November 1 to March 31.	
Energy charges:	November to March per them	April to October per therm
First 15,000 therms	$0.45502	$0.38400
Next 15,000 therms	$0.39901	$0.34650
Next 75,000 therms	$0.36150	$0.32649
Next 165,000 therms	$0.31449	$0.29699
Next 330,000 therms	$0.26148	$0.26255
Over 600,000 therms	$0.24898	$0.24898

* 1 therm = 100,000 Btu = 0.1055 GJ

fuel supply that is kept in useable condition. Diesel fuel tanks to supply dual-fuel boilers are a common example of an alternative fuel supply. There is no specific curtailment period stated in the tariff, but natural gas curtailments typically last from hours to a few days.

Large industrial customer rate schedules for natural gas are typically similar in structure to the residential and commercial examples above. Large natural gas users are much more likely to purchase gas supply from marketers, and so the rate with the local gas utility would include fixed charges and delivery charges per unit energy transported. Gas marketers are paid separately by the customer.

Renewable natural gas (RNG) was introduced in Chapter 2. Many utilities offer RNG to their customers to help meet carbon reduction goals. When customers purchase RNG, their gas bill is typically the same as their usual natural gas bill, except the natural gas supply cost is replaced with the RNG rate. Customers pay a premium for RNG. Prices vary by utility, based on the utility supply chain for RNG production. In Canada, the price for RNG is roughly CAD$20 per GJ (roughly 3–4 times the market price for natural gas in 2023).[17] In the U.S., RNG prices are in a similar range of $20 per MMBtu. The production cost of RNG ranges from $7 to $44 per MMBtu, depending on feedstocks and other variables (ICF, 2020). Future prices are forecast to

[17] www.cga.ca/energy-magazine-post/canadas-growing-renewable-natural-gas-supply/

Table 4.12 Saint Paul, Minnesota district cooling (chilled water) rate schedule.

Demand charges	
$30.85 per ton monthly	Customer demand is determined each year, and is the same charge for every month. It is based on the customer's highest 1-hour demand (tons*) from June to September of the previous two 12-month periods.
Energy charges:	
Monthly ton-hours	$0.097/ton-hour Multiply by fuel adjustment charge Fuel adjustment charge was zero for 2022
Other charges	
City franchise fee	3.51965%

* 1 ton of delivered thermal power = 12,000 Btu/hr = 3.517 kW. 1 ton-hr = 12,000 Btu.

range from $14 to $18 per MMBtu by 2030, but forecasts are highly uncertain (Nadel, 2022).

4.6 Thermal Energy Network Rates

The use of TEN systems is growing (see Chapter 19). For example, in the U.S., there are 86,000 commercial buildings connected to district heating, and 55,000 connected to district cooling (EIA, 2023a). In this section are presented examples of TEN rate schedules for common cooling and heating TEN systems.

An example of a district cooling rate is provided in Table 4.12, based on District Energy St. Paul's schedule.[18] The operation is a non-profit utility developed as a public−private partnership with the city. Chilled water is distributed via subsurface piping and metered at the customer facilities.

The hot water district energy utility in Jamestown, New York provides an example of district heating providing hot water rate schedule[19] as shown in Table 4.13.

The city of Vancouver, British Columbia owns and operates a neighborhood energy utility. The utility utilizes heat pumps to extract heat from sewage, supplemented with natural gas boilers, to generate hot water that is distributed through subsurface piping to multi-family residential and commercial buildings. The utility's hot water district heating rate schedule is

[18]www.districtenergy.com/wp-content/uploads/2020/09/Cooling-Brochure-2021.pdf. Updated to 2022.

[19]ww.jamestownbpu.com/DocumentCenter/View/120/Schedule-of-District-Heat-Rates-PDF

Table 4.13 Jamestown, New York district heating (hot water) rate schedule.

Basic service and meter charge	
Basic charge	$5.95 per month
Btu meter	Range of $13.05 for 1 inch Btu meter, to $43.28 monthly for 6 inch Btu meter
Energy charges:	
Class 2 – Small	$7.11 per MMBtu, plus $3.00 per MMBtu* fuel adjustment charge[20] for all consumption.
Class 3 – Large	$5.63 per MMBtu, plus $3.00 per MMBtu* fuel adjustment charge for all consumption, or minimum of 300 MMBtu/month, whichever is greater.
Class 6 – Special	$3.90 per MMBtu, plus $3.00 per MMBtu* fuel adjustment charge for all consumption, minimum of 20,000 MMBtu per calendar year.

* 1 MMBtu = 1.055 GJ

Table 4.14 Vancouver, British Columbia district heating (hot water) rate schedule.

Capacity charge	
Class 1 (residential within development plan area)	CAD$0.613 per m^2 net floor area (100% of building for residential) per month
Class 2 (residential outside development plan area) and Class 3 (non-residential)	CAD$9.225 per kW peak demand monthly
Energy charges:	
Class 1 (residential within development plan area)	CAD$57.446 per MWh*
Class 2 (residential outside development plan area) and Class 3 (non-residential)	CAD$57.446 per MWh*

* 1 MWh = 3.60 GJ = 3.412 MMBtu.

provided in Table 4.14. [21] Note that the Class 1 rate has a demand charge based on net floor space, which means this is a fixed charge to the customer.

Consolidated Edison (ConEd) is a large, regulated utility serving New York City with electricity, natural gas, and steam. Steam is distributed to customers at average pressures of 125 psig or higher. Table 4.15 provides a summary of ConEd's Classification 1 rate schedule for selling steam. Note the pattern of increasing unit cost up to the first 50 Mlb, then decreasing unit cost, and then increasing again as consumption increases. This is an example of a variable block rate. The rate schedule includes a fuel adjustment, which is determined annually. The rate is set using a base cost of fuel, and then the

[21] https://council.vancouver.ca/20211201/documents/1e.pdf

Table 4.15 Con Edison New York district heating (steam) general service (Classification 1).

Customer charge	
Basic charge	\$1,259.52 per month
Energy charges:	
First 20 Mlb*	\$15.903 per Mlb
Next 30 Mlb	\$50.129 per Mlb
Next 950 Mlb	\$39.126 per Mlb
Over 1,000 Mlb	\$37.652 per Mlb

* 1 Mlb = 1,000 lbm. Assume 140 psia steam supplied and heat is utilized to discharge condensate at 200 °F. Heat used = 1,193.4 – 168.1 = 1,025.3 Btu/lbm. Each Mlb of steam = 1.025 MMBtu = 1.082 GJ.

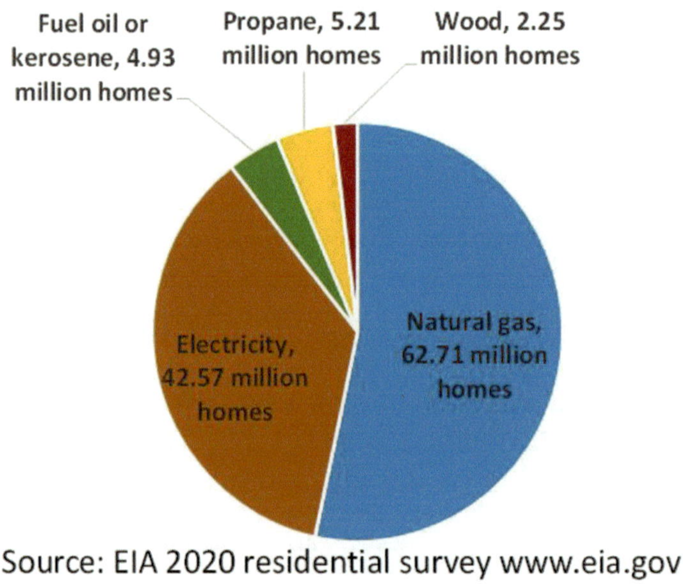

Figure 4.1 Residential heating energy by source for U.S. homes in 2020.

fuel adjustment is assessed based on actual fuel costs to determine the fuel adjustment in dollars per Mlb of steam.

4.7 Other Fuels

Energy other than electricity, natural gas, and TEN are used in certain circumstances. For example, in the U.S., there are over 12 million homes heated with energy sources other than electricity and natural gas, as shown in Figure 4.1 (EIA, 2023b). In Canada, 10% of residential energy use (not counting transportation) is due to the combined total of heating oil, coal, propane, and wood (NRCan, 2023). Similarly, commercial buildings use other fuels in many cases. In the U.S., industrial[22] facilities use large quantities of petroleum, as well as biofuels and coal. Fuels such as propane, kerosene, and fuel oil are typically purchased in batches. These energy purchases are often made on irregular time intervals, and the energy bills can be more difficult to compile such as for completing energy audits. Wood fuel for residential use may be purchased, or in some cases harvested locally.

4.8 Summary

This chapter has provided an overview of energy rates, focusing on electricity, natural gas, and TEN. Most often, energy suppliers are regulated energy utilities, but many are also coop or municipal utilities. In completing energy or decarbonization studies, the largest value of avoided energy or GHG emissions is the utility bill cost savings. As such, understanding utility rate structures is critical for effective energy and carbon management. The common elements of energy bills include fixed components to cover metering and other costs, variable components for energy and demand charges, and rate riders for a variety of other purposes. Essentially, all facilities use electricity, and electric rates are the most complex and diverse. Common electric rate designs include fixed rates, time-variant rates such as TOU and seasonal rates (Badtke-Berkow), and tiered rates (e.g., inclining block). For customers with self-generation that can export to the grid, such as solar PV or CHP, net metering or net billing rates are common. Natural gas rates are similar to electricity, but generally less complex and involve fixed components, energy charges (which can be time-variant and/or blocks), as well as various riders. This chapter includes examples of TEN rates for purchases of steam, hot water, or chilled water. Finally, there are other energies such as fuel oils, wood, coal, other biofuels, and propane, which are used in facilities. For example, there are millions of households in the U.S. and Canada heated by fuels other than electricity or natural gas.

---

[22] In EIA datasets, "industrial" data includes manufacturing, mining, construction, and agriculture.

References

"ASHRAE Handbook – HVAC Applications" Chapter 57 "Electrical Considerations."

Badtke-Berkow M., Centore M., Mohlin K., and Spiller B. (2015) "A Primer on Time-Variant Electricity Pricing" Environmental Defense Fund.

Chaudhari S., Sundaramoorthy S., Price C., Botts A., and Wenning T. (2021) "Understanding Your Utility Bills: Natural Gas" Oak Ridge National Laboratory

U.S. Energy Information Administration (EIA) (2023a) "2018 Commercial Building Energy Consumption Survey Data" www.eia.gov/consumption/commercial/data/2018/#b19-b22

U.S. Energy Information Administration (EIA) (2023b) "2020 Residential Energy Consumption Survey Data" www.eia.gov/consumption/residential/data/2020/#sh

U.S. Energy Information Administration (EIA) (2022) "Monthly Energy Review, Table 2.4, April 2022" www.eia.gov/energyexplained/use-of-energy/industry.php

ICF (2020) "Study on the Use of Biofuels (Renewable Natural Gas) in the Greater Washington, D.C. Metropolitan Area" report for Washington Gas Light Company.

Lazar J., Chernick P., and Marcus W. (2020) "Electric Cost Allocation for a New Era: A Manual" Regulatory Assistance Project.

Nadel S. (2022) "What is the Role of Fuels in a Decarbonized Future? What is the Potential Role for Gas Utilities?" ACEEE Summer Study on Energy Efficiency in Buildings.

Natural Resources Canada (2023) "National Energy Use Database - Residential Secondary Energy Use (Final Demand) by Energy Source and End Use" https://oee.nrcan.gc.ca

Price C., Botts A., Sundaramoorthy S., Chaudhari S., and Wenning T. (2021) "Understanding Your Utility Bills: Electricity" Oak Ridge National Laboratory

Tiedemann K. and Sulyma I. (2023) "Literature Review of Evaluations of EV Charging Station Rates & EV Technology Incentive Programs" CEATI Report T211700-7074 13.

Zinaman O., Aznar A., Linvill C., Darghouth N. Dubbeling T., and Bianco E. (2017) "Grid-Connected Distributed Generation: Compensation Mechanism Basics" NREL.

Author Biography

Eric Mazzi is a consultant and educator focusing on technology and policy for demand-side energy systems. He specializes in performing independent technical analysis, M&V, policy studies, and educational services. He advocates and contributes to efforts to reduce energy consumption and achieve net-zero GHG emissions, while maintaining or improving energy services and advancing energy equity. His experience includes air quality consulting, utility maintenance, and five years as full-time engineering faculty at the University of British Columbia. He volunteers as a board member for the U.S. Northwest Energy Coalition, and the International Performance M&V Protocol committee. Eric teaches part-time, delivering university and professional certification courses. He is an Adjunct Professor of Energy Management at the New York Institute of Technology – Vancouver. He has B.Sc. and M.Sc. degrees in mechanical engineering, and a Ph.D. degree in environmental policy. He is a registered engineer in California and British Columbia.

Acknowledgments

This chapter benefited from a peer review completed by Ken Tiedemann.

Practice Problems:

1. A company is under Hydro Quebec's Rate L (see Section 4.4.3) with a contract demand of 5,500 kW. For a given month, the largest metered (15-minute) demand was 6,000 kW and the power factor was constant at 0.92 for the whole month. What is the billed demand for this month?
2. ConEd's steam rate schedule is provided in Table 4.15. Note that steam is sold on a mass basis (per pound), and that customers have control over how much thermal energy is extracted from each pound of steam. Assume steam is supplied at 140 psia. Calculate the amount of heat recovered per pound of steam if (a) customers utilize only the latent heat of supplied steam; (b) customers utilize all the heat to discharge condensate at 150 °F. Hint: you may need to review steam fundamentals in Chapter 20.
3. The terms for an electricity rate are defined below:
   Energy consumption charge:
   $0.08/kWh first 50,000 kWh monthly
   $0.12/kWh over 50,000 kWh monthly

$100/month customer charge
$25/kW-month peak demand charged only from 2:00 pm to 6:00 pm
The customer's total consumption ranges from 60,000 to 80,000 kWh each month during a typical year. You find an idling motor that consumes 10 kW, which can be turned off from 12 noon to 6:00 pm daily, for 365 days per year. Calculate the annual cost savings of turning off this motor.

4. A facility's billing rate for electricity has a 65% ratchet clause. The utility ratchet clause requires that the billed demand for any month is 65% of the highest on-peak season maximum demand of the previous 12 months or the actual demand, whichever is greater.
The rate schedule demand charges are $13.27 per kW for the peak season months of June to September, and $4.82 per kW for all other months.
The facility's actual, metered demand is shown in the table below. The metered demand is in units of kW.

Month	2021 Demand, kW	2022 Demand, kW
January	505	495
February	520	550
March	605	580
April	625	600
May	590	610
June	890	900
July	1,150	1,100
August	965	1,000
September	905	900
October	615	600
November	515	500
December	510	515

Calculate: (a) 2022 demand charges if there was no ratchet clause, (b) 2022 demand charges with the ratchet clause, and (c) the increase in charges due to the ratchet clause.

5. A university campus has a combined heat and power plant that generates 2,000 kW of electricity from an internal combustion engine. The electric utility purchases the total output of the generator (less ancillary loads) at a contract rate that is different from the rate schedule

that applies to university purchased power. What rate design is this, and why?

6. Consider the following three electric rate designs and what equity issues might exist, in consideration of low-income residential customers and renters: inclining block rates, time of use rates, and net metering rates. Hint: consider customers who are low-income or tenants.

5

Economic Analysis and Lifecycle Costing

Eric Mazzi

Mazzi Consulting, Canada

5.1 Introduction

Once an energy efficiency measure (EEM) or other distributed energy resource (DER) project has been identified, the energy manager must determine the cost-effectiveness of the project in order to recommend it to decisionmakers (e.g. owner, manager) for implementation. If a group of measures has been identified, then they should be ranked on some economic basis, with the most cost-effective ones to be implemented first. There are many measures of cost-effectiveness, and organizations typically use their own unique procedures and criteria to make investment decisions. The basic elements of cost-effectiveness analysis are discussed in this chapter and some of the common techniques or measures of cost-effectiveness are presented. The emphasis in this chapter is on techniques that use the time value of money to calculate the lifecycle cost (LCC) of projects. LCC is the total cost of purchasing and operating a piece of energy consuming equipment or process over its entire lifespan.

The chapter begins by discussing the various types of costs that a project can incur. Sections 5.2–5.4 present fundamental economic analysis methods. Sections 5.5–5.7 develop and discuss basic methods to perform discounted cash flow analysis to account for the time value of money. Section 5.8 presents decision-making based on the time value of money. Sections 5.9 and 5.10 present fundamental LCC concepts and demonstrate the ease with which they can be applied in evaluating capital investments for energy-related projects. Incorporating taxes and inflation into an economic analysis is discussed in Sections 5.11 and 5.12, and LCC software is introduced in Section 5.13. Chapter 6 provides concepts and methods financing projects.

The material presented in this chapter includes only the basics of economics and LCC. This basic material as well as more advanced economics and costing material are covered in Newnan (2020) and Farr and Faber (2019), or any current engineering economics textbook.

5.2 Costs

Organizations incur various types of costs. These are generally classified into two broad categories: expenses and capital investments. Expenses are the routine, ongoing costs that are necessary for conducting business or operations. Capital investments have four important characteristics. First, they are relatively large, ranging from thousands to millions of dollars, depending on the size of the organization. Second, the benefits of a capital investment are returned over the lifetime of the investment, which is typically several years. Third, capital investments are relatively irreversible. After the initial investment has been made, altering the project significantly, or terminating the project, has substantial (usually negative) implications. Fourth, capital investments can have significant tax implications, depending on the choice of financing methods.

5.2.1 Categories of costs for energy project investments

The costs incurred for a project investment can almost always be placed in one of the following categories: acquisition, utilization, and disposal. Acquisition costs are the initial, or first, costs that are necessary to prepare the investment for service. These costs include the purchase price, installation costs, training costs, engineering work, and permits.

Methods for estimating acquisition costs range from using past experience to obtaining more precise estimates from vendors, catalogs, and databases. RS Means (www.rsmeans.com) is one commonly used source of data for estimating costs. In terms of cost estimation methodologies, the Association for the Advancement of Cost Engineering – International (AACEI) provides many publications on producing reliable cost estimates (www.aacei.org). The desired accuracy of the estimate depends on many factors and is ultimately determined by the stakeholders with a financial interest in a project (e.g., owners, utilities, or agencies providing subsidies, energy services companies, and lenders). AACEI's guideline for process industries provides a general guideline that has been found to be useful for determining the required accuracy of cost estimates for energy management projects, as summarized in Table 5.1 AACEI, (2020). Examples of applying AACEI's cost classes to energy projects would be to require plant-wide energy audits to meet Class 5 accuracy, and projects receiving utility incentive funds to meet Class 3 accuracy.

Table 5.1 Cost classification and accuracy guidelines for process industries AACEI, (2020).

Estimate class	Level of project definition	Typical purpose	Estimation methodology	Expected range of accuracy
Class 5	0%–2%	Concept screening (e.g., Level 1 energy audit)	Capacity factored, parametric model	Low: −20% to −50% High: +30% to +100%
Class 4	1%–15%	Study or feasibility (e.g., Level 2 energy audit)	Equipment factored or parametric models (may include budget quotes)	Low: −15% to −30% High: +20% to +50%
Class 3	10%–40%	Budget authorization or control (e.g., Level 3 energy audit)	Semi-detailed unit costs with assembly level line items (may include budget quotes)	Low: −10% to −20% High: +10% to +30%
Class 2	30%–75%	Control or bid/tender	Detailed unit cost with forced detailed take-off	Low: −5% to −15% High: +5% to +20%
Class 1	65%–100%	Check estimate or bid/tender	Detailed unit cost with detailed take-off	Low: −3% to −10% High: +3% to +15%

Utilization costs are those required on a routine basis for operating and maintaining the investment. These include purchased energy, maintenance, and operating labor. Utilization costs can be direct or indirect. Direct costs include labor and materials for routine repair and maintenance. Indirect costs are the costs not directly attributable to the project but necessary for conducting business. Energy costs are often classified as indirect costs, but good energy management practice is to classify energy costs as direct costs (see Chapter 1). Understanding energy costs and estimating reductions in energy costs requires a clear and complete understanding of the rate structures, as introduced in Chapter 4. Indirect costs are also often referred to as "overhead." Examples of indirect costs include salaries for staff personnel, janitorial costs, and facility leasing costs. Estimates for utilization costs can be obtained from databases or professional experience.

Disposal costs are those incurred (or recovered) when the project has reached the end of its useful life. These are costs required to retire or remove the asset. These costs are often referred to as the salvage value if the project has a positive worth at the end of its lifetime. The actual salvage value may depend on many factors, including how well the asset was maintained and the market for used equipment. Estimates for disposal costs/benefits are often obtained from experienced judgments and current values of such used equipment.

5.2.2 Cash flow diagrams and tables

It is often helpful to present costs visually. This can be accomplished with a tool called a cash flow diagram. Alternatively, a cash flow table could be used. A cash flow diagram is a pictorial display of the costs and revenues associated with a project. An interest rate or discount rate can also be shown. A common convention for cash flow diagrams is that costs are represented by arrows pointing down, while revenues are represented by arrows pointing up. The time periods for the costs dictate the horizontal scale for the diagram. Although some costs occur at different points in time, most of the time, an end-of-year approach is sufficient. This chapter will use end-of-year cash flows for all analyses.

Figure 5.1 provides an example of a cash flow diagram to visualize the costs and benefits of purchasing a heat pump. The costs and benefits associated with the heat pump are:

- The heat pump costs $10,000 initially
- The heat pump saves $2,500 per year in energy costs for 20 years
- The maintenance costs are $500 per year for 20 years

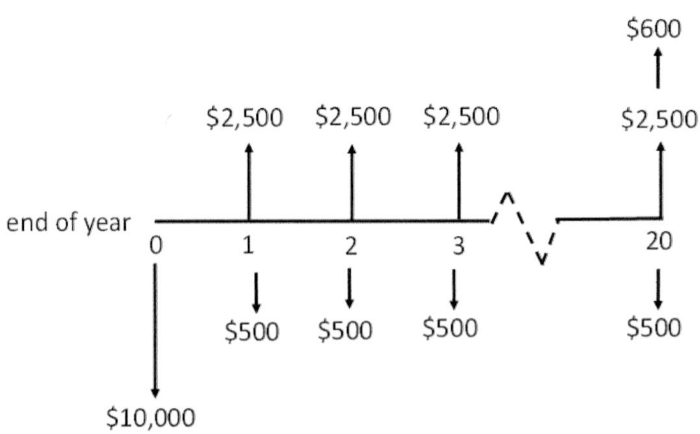

Figure 5.1 Cash flow diagram

Table 5.2 Cash flow in tabular form.

End of year	Cash flow
0	–\$10,000
1–19	\$2,500–\$500 = \$2,000
20	\$2,500–\$500 + \$600 = \$2,600

- The estimated salvage value is \$600 at the end of 20 years

Costs can also be listed in a tabular format as illustrated in Table 5.2.

5.3 Simple Payback Period Cost Analysis

One of the most commonly used cost analysis methodologies is the simple payback period (SPP) analysis (sometimes described as "payback period" analysis). The SPP determines the number of years required to recover an initial investment through project returns. SPP does not account for the time value of money. The formula is:

$$\text{SPP} = (\text{Initial cost})/(\text{Annual savings}) \tag{5.1}$$

Example 5.1:

The heat pump discussed above has an initial cost of \$10,000, an energy savings of \$2,500 per year (ignoring the salvage value), and a maintenance cost of \$500 per year. Thus, the net annual savings is \$2,000. Therefore, its SPP would be (\$10,000)/(\$2,000/yr) = 5 years.

The advantage of the SPP is its simplicity, and it is easily understood by workers and management. It does provide a rough measure of the worth of a project. The primary disadvantages are: (1) the methodology does not consider the time value of money; and (2) the methodology does not consider any of the costs or benefits of the investment after the payback period. No specific lifetime estimate of the project is required, but it is assumed that the lifetime is longer than the SPP. These limitations mean the SPP tends to favor shorter-lived projects. This is a bias that is often economically unjustified. However, using the SPP in conjunction with one of the discounted cash flow methodologies discussed later in this chapter can provide a better understanding of an investment's worth.

5.4 Economic Analysis using the Time Value of Money: Discounted Cash Flow Analysis

5.4.1 Introduction

Most people understand that money changes value over time. If a person is offered \$1,000 today, or \$1,000 in one year, almost everyone would choose the \$1,000 today. The decision-maker clearly places higher value on money today. This decision occurs for a variety of reasons, including how much the person needs the money, but a primary factor is the recognition that \$1,000 today is worth more than \$1,000 will be worth in one year.

The change in worth is due to two primary factors: interest (opportunity cost) and inflation. Interest is the return earned on money when someone else uses it. If \$1,000 were deposited in a bank or invested in a financial instrument (stock, mutual fund, bond, etc.), then in one year, an amount greater than \$1,000 would be available for withdrawal. Inflation is a decrease in the purchasing power of money. In other words, \$1,000 will buy more at a store today than it will in one year. Inflation is discussed in greater detail in Section 5.12 and will not be incorporated into our economic analyses until that time. The effects of both interest and inflation are important to consider in a full economic analysis of energy projects.

Energy management expenditures are typically justified in terms of avoided energy costs (which could include environmental costs, such as a carbon tax). Expenses typically come at the beginning of a project, while savings/benefits occur later. The sums of money for costs and revenues that need to be compared are paid or received at different points in time. Because money changes value over time, these sums of money, or cash flows, *should not be directly added together unless they occur at the same point in time*. In order to correctly add these cash flows at different times, we need to reduce them to a common basis through the use of the interest rate, also called the discount rate. This method of treatment is called discounted cash flow analysis, and first requires us to understand the mathematics of interest and discounting.

5.4.2 The mathematics of interest and discounting

Two factors affect the calculation of interest: the amount and the timing of the cash flows. The basic formula for calculating interest is:

$$F_n = P + I_n, \tag{5.2}$$

where:

F_n = a future cash flow of money at the end of the nth year
P = present cash flow of money
I_n = interest accumulated over n years
n = number of years between P and F

Interest is stated as a percentage rate that is to be paid for the use of the money for a time period. Although interest rates can have many different compounding periods (e.g., days or months), this chapter will be limited to yearly interest, and end of year cash flows, for simplicity. The term "discount rate" is used to reflect the fact that future sums of money must be discounted by the interest rate to find their initial values or present values. The discount rate is specific to an organization or company. It must be supplied by the organization's accounting department or management and is usually known for evaluating investments in new business opportunities or new facilities. This interest rate is often referred to as the minimum attractive rate of return (MARR). In general, private companies (especially fast-growing companies) will use a higher discount rate, and government agencies will use a lower discount rate. For example, an information technology company might use 35% as the MARR, while a municipality might use 4%. However, there are always exceptions, and the best practice is to confirm the applicable discount rate for the specific organization involved.

5.4.3 Simple interest

There are two primary types of interest: simple and compound. Simple interest is earned (charged) only on the original principal amount, and paid once, at the end of the time period. The formula is:

$$I = (P)(n)(i),$$

where:

I = the interest accumulated over n years
P = the original principal amount
n = the number of interest periods (often measured in years)
i = the interest rate per period

The general formula for the total amount owed or due at the end of a loan (investment) period of n years when using simple interest is:

$$\begin{aligned} F_n &= P + I \\ &= P + (P)(n)(i) \\ &= P[1 + (n)(i)]. \end{aligned} \tag{5.3}$$

Example 5.2:

The ABC Corporation wants to borrow $10,000 for five years at a simple interest rate of 18%/year. How much interest would be owed on the loan?

$$\begin{aligned} I &= (P)(n)(i) \\ &= (\$10,000)(5 years)(0.18/year) \\ &= \$9,000. \end{aligned}$$

The total amount owed at the end of the five years would be the principal, $10,000 plus the interest, $9,000, for a total of $19,000.

5.4.4 Compound interest

Interest that is earned (charged) on the accumulated interest as well as the original principal amount is said to be compounded. In other words, the interest that accumulates at the end of the first interest period is added to the original principal amount to form a new principal amount due at the end of the next period. The power (or penalty) of compound interest is illustrated in Table 5.3.

The effect of compound interest is substantial. When using simple interest, the amount owed at the end of the loan period for the principal amount of $10,000 was $19,000. With compound interest, the amount owed at the end of the same loan period is $22,877. Clearly, a borrower would prefer simple interest, but a lender would prefer compound interest. Since compound interest is used far more commonly in practice than simple interest, the remainder of this chapter uses compound interest.

Formulas for calculating compound interest have been derived for several patterns of cash flows. These formulas are discussed later in this chapter.

Table 5.3 Compound interest.

Year	*A* Amount owed at the beginning of the year	$B = i \times A$ Interest owed at the end of the year	$C = A + B$ Total amount owed at the end of the year
1	P $10,000	$P \times i$ $10,000×0.18 = $1,800	$P + P \times i$ $11,800
2	$P + P \times i$ $11,800	$(P + P \times i) \times i$ $11,800 × 0.18 = $2,124	$13,924
3	$13,924	$13,924×0.18 = $2,506	$16,430
4	$16,430	$16,430×0.18 = $2,957	$19,387
5	$19,387	$19,387×0.18 = $3,490	$22,877

5.5 Discounted Cash Flows: Basics and Single Sum Analyses

Once we understand the effects of interest or discounting from Section 5.4, we can proceed with our goal of finding a method that correctly allows the addition of two cash flows occurring at different points in time. The general approach used is to reduce cash flows occurring at different times to a common basis through the use of the interest rate or discount rate. This method of treatment is called discounted cash flow analysis, and it is a fundamental approach that is necessary to correctly account for costs and savings that occur at different times.

This section presents the formulas and compound interest factors used to convert a sum of money from its value in one time period to its corresponding value in another time period. Unless stated otherwise in this chapter, interest is assumed to be compounded annually, and cash flows are assumed to occur at the end of the year. The following notation will be used:

$i =$ Annual interest rate (or discount rate; also known as the MARR)

$n =$ The number of annual interest periods (in our case, the number of years)

$P =$ Present value (or present worth)

$A =$ Single payment in a series of n equal annual payments

$F =$ Future value (or future worth)

The following important points apply in the use of discounted cash flow analysis:

- The end of one year is the beginning of the next.
- P is at the beginning of a year at a time regarded as being the present.
- F is at the end of the nth year from a time regarded as being the present.
- A (an annual payment) occurs at the end of each year of the period under consideration.
- When P and A are involved, the first A of the series occurs one year after P.
- When F and A are involved, the last A of the series occurs simultaneously with F.

5.5.1 Single sum, future worth

The first task is to determine how to convert a single sum of money from a present amount to a future amount. This analysis is illustrated by asking the question “If I borrow \$5,000 at 10% interest for five years, how much will

I owe at the end of the five years?" In this example, the present amount, *P*, and the interest rate are known. The unknown is the future amount, *F*. The formula for finding *F* is:

$$F = P(1 + i)^n \tag{5.4}$$

The term $(1 + i)^n$ is one of the six compound interest factors that are commonly used and tabulated. It is often described as the single sum, future worth factor. The term is also known as the factor (*F*/*P*,*i*,*n*), which is read: "find *F*, given *P*, at *i*% for *n* periods" (years in our case). It is also sometimes referred to as the "*F* given *P*" factor. Both formulas and the compound interest factors obtained from tables will be presented in this chapter, although the tables will be used for most computations. All compound interest factors are in the same format: (*X*/*Y*,*i*,*n*), where:

X = unknown quantity – read as "Find *X*"
Y = known quantity – read as "Given *Y*"
i = known interest rate – read as "At *i*"
n = known number of time periods (usually years) – read as "For *n* periods of time"

Figure 5.2 provides a summary of the formulas, factors, and sample cash flows for various combinations of finding *P*, *F*, and *A*.

Single Payment 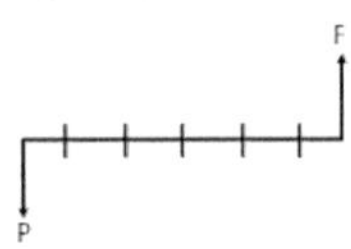	To Find *F* Given *P*	(*F/P,i,n*)	$F = P(1+i)^n$
	To Find *P* Given *F*	(*P/F, i,n*)	$P = F(1+i)^{-n}$
Uniform Series 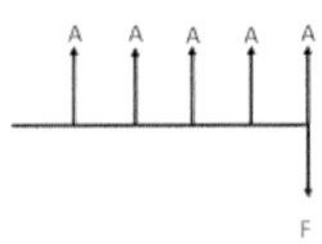	To Find *F* Given *A*	(*F/A, i, n*)	$F = A\left[\frac{(1+i)^n - 1}{i}\right]$
	To Find *A* Given *F*	(*A/F,i,n*)	$A = F\left[\frac{i}{(1+i)^n - 1}\right]$
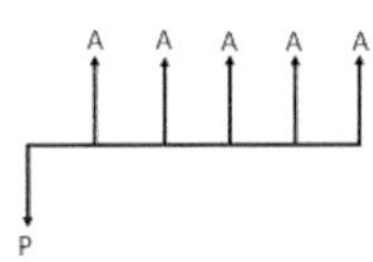	To Find *A* Given *P*	(*A/P,i,n*)	$A = P\left[\frac{i(1+i)^n}{(1+i)^n - 1}\right]$
	To Find *P* Given *A*	(*P/A,i,n*)	$P = A\left[\frac{(1+i)^n - 1}{i(1+i)^n}\right]$

Figure 5.2 Formulas and factors for P, F, and A.

Although we are only discussing yearly interest and yearly time periods in this chapter, it is important to note that other interest periods can occur, such as quarterly. If this is the case, the interest period and the time period must match.

Since the compound interest factors are used extensively in discounted cash flow analysis, numerical tables have been developed for various combinations of i and n. See Table 5.4 for an illustration of the table for $i = 10\%$,

Table 5.4 Interest (discount) rate factors for i = 10%.

n	To Find F Given P $(F\|P,i\%,n)$	To Find P Given F $(P\|F,i\%,n)$	To Find F Given A $(F\|A,i\%,n)$	To Find A Given F $(A\|F,i\%,n)$	To Find P Given A $(P\|A,i\%,n)$	To Find A Given P $(A\|P,i\%,n)$
1	1.1000	0.9091	1.0000	1.0000	0.9091	1.1000
2	1.2100	0.8264	2.1000	0.4762	1.7355	0.5762
3	1.3310	0.7513	3.3100	0.3021	2.4869	0.4021
4	1.4641	0.6830	4.6410	0.2155	3.1699	0.3155
5	1.6105	0.6209	6.1051	0.1638	3.7908	0.2638
6	1.7716	0.5645	7.7156	0.1296	4.3553	0.2296
7	1.9487	0.5132	9.4872	0.1054	4.8684	0.2054
8	2.1436	0.4665	11.4359	0.0874	5.3349	0.1874
9	2.3579	0.4241	13.5795	0.0736	5.7590	0.1736
10	2.5937	0.3855	15.9374	0.0627	6.1446	0.1627
11	2.8531	0.3505	18.5312	0.0540	5.4951	0.1540
12	3.1384	0.3186	21.3843	0.0468	6.8137	0.1468
13	3.4523	0.2897	24.5227	0.0408	7.1034	0.1408
14	3.7975	0.2633	27.9750	0.0357	7.3667	0.1357
15	4.1772	0.2394	31.7725	0.0315	7.6061	0.1315
16	4.5950	0.2176	35.9497	0.0278	7.8237	0.1278
17	5.0545	0.1978	40.5447	0.0247	8.0216	0.1247
18	5.5599	0.1799	45.5992	0.0219	8.2014	0.1219
19	6.1159	0.1635	51.1591	0.0195	8.3649	0.1195
20	6.7275	0.1486	57.2750	0.0175	8.5136	0.1175
21	7.4002	0.1351	64.0025	0.0156	8.6487	0.1156
22	8.1403	0.1228	71.4027	0.0140	8.7715	0.1140
23	8.9543	0.1117	79.5430	0.0126	8.8832	0.1126
24	9.8497	0.1015	88.4973	0.0113	8.9847	0.1113
25	10.8347	0.0923	98.3471	0.0102	9.0770	0.1102
26	11.9182	0.0839	109.1818	9.159E-03	9.1609	0.1092
27	13.1100	0.0763	121.0999	8.258E-03	9.2372	0.1083
28	14.4210	0.0693	134.2099	7.451E-03	9.3066	0.1075
29	15.8831	0.0630	148.6309	6.728E-03	9.3696	0.1067
30	17.4494	0.0573	164.4940	6.079E-03	9.4269	0.1061

with n ranging from 1 to 30 years. This table will be used in the examples in this section. The appendix contains tables for various interest rates ranging from 1% to 100%. In practice, spreadsheets or calculator functions are commonly used.

Example 5.3:

If \$5,000 was deposited in an account that paid 10% interest annually, how much would be in the account at the end of five years?

For this problem,
$P = \$5{,}000$
$i = 10\%$
$n = 5$ years
$F = ?$
Using eqn (5.4),
$F = 5{,}000(1 + 0.10)^5 = \$8{,}053$
or using the F/P factor and Table 5.4, with $i = 10\%$ and $n = 5$,
$F = 5{,}000(F/P, 10\%,5) = 5{,}000(1.6105) = \$8{,}053.$

5.5.2 Single sum, present worth

The single sum, present worth formula, or single payment, present worth formula, is:

$$P = F(1 + i)^{-n}. \tag{5.5}$$

The term $(1 + i)^{-n}$ is called the present worth factor, $(P/F,i,n)$, which is read: find P, given F, at i% for n periods (years).

Example 5.4:

An energy manager expects a boiler to last seven years, and he/she expects it will cost about \$150,000 to replace at that time. How much money should the company deposit today in an account paying 10% per year in order to have \$150,000 available in seven years?

For this problem:
$F = \$150{,}000$
$i = 10\%$
$n = 7$ years

$P = ?$
Using eqn (5.5),
$P = \$150{,}000(1.1)^{-7} = \$76{,}974.$
Using the present worth factor (P/F,10%,7) and Figure 5.2,
$P = \$150{,}000(P/F,10\%,7) = \$150{,}000(0.5132) = \$76{,}980.$
The results are different due to number rounding, which is not significant.

5.6 Discounted Cash Flows: Uniform Series

The next type of cash flow pattern to understand is a uniform series of costs or savings. A uniform series is a cash flow pattern with cash flows that are of the same magnitude, occurring at the end of several consecutive periods. We will be using years for our periods. Examples of uniform series of costs are loan payments (e.g., for a house or car), or any other type of regular payment. Most of our energy management projects produce a uniform series of savings in future energy costs.

Four types of "conversions" will be addressed:

1. Given a present amount, find the equivalent uniform series. This is known as find A, given P, and is denoted by the factor (A/P, i, n).
2. Given a future amount, find the equivalent uniform series. This is known as find A, given F, and is denoted by the factor (A/F, i, n).
3. Given a uniform series, find the equivalent present worth. This is known as find P, given A, and is denoted by the factor (P/A, i, n).
4. Given a uniform series, find the equivalent future worth. This is known as find F, given A, and is denoted by the factor (F/A, i, n).

Formulas for all series conversions are given in Figure 5.2. Uniform series factors at 10% interest are included in Table 5.4. Additional compound interest factors will be used in the examples below, since the series formulas are more complex than earlier formulas. Examples of each type of series conversion follow.

Example 5.5:

An energy manager has \$5,000 available today to purchase a high efficiency air conditioner with a life of six years. He/she would like to know what energy cost savings would be needed each year to justify this project if the company MARR is 10%.

$P = \$5{,}000$
$i = 10\%$

$n = 6$
$A = ?$
This problem is the first type of conversion: find A, given P. Therefore:
$A = \$5{,}000(A/P, 10\%, 6)$
$= \$5{,}000(0.2296)$
$= \$1{,}148.$

Thus, if the air conditioner produces annual energy cost savings of \$1,148 or greater, the company will earn its MARR at the least. If the savings is greater than \$1,148, then the actual rate of return will be greater than 10%.

Example 5.6:

A heat pump is expected to produce energy cost savings of \$1,500 per year over a lifetime of 20 years. What is the equivalent present sum or present worth for this series of cash flows, if the company MARR is 10%?
$A = \$1{,}500$
$i = 10\%$
$n = 20$
$P = ?$
This is the find P, given A problem.
$P = \$1{,}500(P/A, 10\%, 20) = \$1{,}500(8.5136) = \$12{,}770.$

Thus, \$12,770 is the worth in today's dollars of the series of annual savings of \$1,500 over 20 years.

Example 5.7:

A company needs to begin saving money for the new boiler in Example 5.4. The company will make a deposit each year for seven years to a savings account paying 10% annually. How large should the annual deposits be if they want to have \$150,000 in the bank in seven years?
$F = \$150{,}000$
$i = 10\%$
$n = 7$
$A = ?$
This problem is characterized as a find A, given F problem.
$A = \$150{,}000(A/F, 10\%, 7)$
$= \$150{,}000(0.1054)$
$= \$15{,}810.$

Example 5.8:

A high efficiency lighting project for a company is saving $10,000 per year in energy costs. If that $10,000 a year is deposited into an energy management savings account paying 10%, how much money will be available in five years to use to invest in another EEM?

$A = \$10{,}000$

$i = 10\%$

$n = 5$

$F = ?$

This is the find F, given A problem.

$F = \$10{,}000\ (F/A,\ 10\%,\ 5)$

$= \$10{,}000\ (6.1051)$

$= \$61{,}051.$

By placing the $10,000 a year energy cost savings into an account that earns 10% interest, the company will have $61,051 to spend on a new project in five years.

5.7 Cost Analysis Methodology using Discounted Cash Flows

It is important to have a methodology to follow when performing a discounted cash flow analysis. This chapter recommends the following methodology for performing economic analyses using discounted cash flows:

1. Define the alternatives. State the problem and list all feasible solutions or alternatives that have been selected for economic analysis.

2. Estimate the relevant costs. Each alternative from Step 1 is defined in terms of its cash flows. Vital information includes the amount, timing, and direction (benefit or cost) of each cash flow.

3. Analyze the alternatives. Identify the most cost-effective alternative by analyzing each alternative using the discounted cash flow methodologies described later in this chapter.

4. Perform sensitivity analyses. Since the analysis above is generally based on estimated costs, these costs can be varied, depending on the uncertainty of the estimates, to see if the uncertainties have a pronounced effect.

The three parameters needed for the above methodology are: (1) an estimate of the cash flows, (2) an estimate of the interest rate or discount rate, and (3) an estimate of the life of the project.

5.7.1 Estimation of cash flows

Sources of data for estimates of the cash flows of acquisition costs, utilization costs, and disposal costs were discussed in Section 5.2.1. Purchase cost estimates are generally obtained from vendor quotes, cost databases, or based on past experience (e.g., bid or actual costs from similar projects completed in the past). Operating and maintenance cost (and savings) estimates can also be based on past project experience, from analyses performed by the energy manager (or maintenance managers), or possibly obtained from vendors. Salvage value estimates are often based on experience, or communication with vendors.

5.7.2 Interest rate

The interest rate (discount rate or MARR) is a company-specific or organization-specific value. It must be supplied by the company or agency, accounting department, and is usually the value the company uses for evaluating investments. The magnitude of the discount rate depends on the source of capital that will be used to finance the project. If money must be borrowed (debt capital), the discount rate is likely to be higher than if company funds (equity capital) are to be used.

5.7.3 Lifetime

Estimates for the lifetime of equipment are often difficult to obtain, but can usually be found by contacting vendors, industry-specific organizations, or using estimates based on experience. For example, lifetimes for energy-using equipment for buildings might be found in the ASHRAE (http://weblegacy.ashrae.org/publicdatabase/).

The expected lifetime of equipment (sometimes called "technical lifetime") may not always be the best or appropriate determinant of a project lifetime used in an economic analysis. In economic analysis, a project is sometimes referred to as the "study period." In demand-side energy management, project lifetime might be established based on "effective useful life," "persistence," or other parameters Gold and Nowak, (2019).

5.7.4 Equivalence of cash flows

Once the concept of converting money from one time to another is understood, then a method is needed to compare sums of money that are paid or received at different points in time. To accomplish this, cash flows are "moved around" in time using a discount rate, or MARR, and then compared. If two cash flows have the same present value, future value, or annual worth, they are said to be equivalent. In economic analysis, "equivalence" means "the state of being equal in value." Normally, the concept of equivalence is applied to two or more cash flows. The choice of time period is arbitrary. In other words, if two sums of money with the same MARR are found to be equivalent at one point in time, they will be equivalent at any other point in time.

Example 5.9:

Is the sum \$1,000 today equivalent to \$1,331 three years from today at 10% interest?

Solution:

Compare the values at $t = 0$ (using present value or present worth):

$$\begin{aligned} PV(1) &= \$1,000 \\ PV(2) &= \$1,331(P/F, 10\%, 3) = \$1,331(0.7513) = \$1,000. \end{aligned}$$

Therefore, they are equivalent at 10%.

Making decisions between cash flow profiles for different projects where the cash flows occur at different times requires using the equivalence of cash flows to make comparisons.

5.8 Cost-Effectiveness Measures using Discounted Cash Flows

The goal of this section is to present methods that help energy managers, or other decision-makers, determine whether a project is economically feasible or cost-effective. Multiple methodologies exist for deciding whether a project is worth pursuing. Section 5.3 discussed the simple payback period. Because this method does not consider the time value of money, SPP should be used in combination with a method that explicitly considers the time value of money.

We cover five economic decision-making methods that consider the time value of money in this section: present worth, future worth, annual worth, benefit/cost ratio or savings to investment ratio, and internal rate of return. We present a single example throughout this section so that the reader can see the consistency of decisions that are made using these methodologies.

Example 5.10:

A single zone heating unit is being used in a small office building. A variable air volume system retrofit can be purchased and installed for a cost of \$100,000. The retrofit system is estimated to save 450,000 kWh per year for its economic life of 10 years. The company uses an MARR of 10%. If the company pays \$0.06 per kWh for electricity, and the system will have a salvage value of \$500 at the end of its life, should the new system be purchased?

The cash flow table for this example is shown below, in tabular form.

End of year	Cash flow
0	−\$100,000
1	450,000 kWh (\$0.06/kWh) = \$27,000
2	\$27,000
3	\$27,000
4	\$27,000
5	\$27,000
6	\$27,000
7	\$27,000
8	\$27,000
9	\$27,000
10	\$27,000 + 500 = \$27,500

5.8.1 Present worth, future worth, and annual worth

A present worth comparison converts all the cash flows to a present worth value; a future worth comparison converts all the cash flows to a future value at a common future time (usually at the end of the study period, in this case End of Year 10); and an annual worth analysis converts all the cash flows to a uniform annual series over the study period (in this case, 10 years).

There are three different cash flow patterns in Example 5.10: the initial cost, the annual savings, and the salvage value. Each of these is converted using the compound interest factors discussed earlier. Costs carry a negative sign, while benefits carry a positive sign.

The decision rules for present worth, future worth, and annual worth are the same: a positive value indicates the project is economically feasible and cost-effective for a given MARR. It is important to note that the three measures are economically equivalent. This means that if the annual worth value were converted to either future or present worth, the same numerical value would be obtained. The reader is encouraged to verify this statement using the factors. The ultimate result is that only one of these three measures needs to be calculated. However, different organizations and different people have their own preferences for which of these metrics to use. Since all three are in fairly common use, each of the three methods is discussed below.

The first method finds the present worth of the project using the *P/A* factor to find the equivalent cash flow of a series of $27,000 annual savings for 10 years with a 10% MARR, and the *P/F* factor to discount the $500 salvage value in year 10 to the present – the start of year 1. The project cost of $100,000 is a present value since it must be paid at the start of the project.

$$\begin{aligned} \text{PW} &= -\$100,000 + \$27,000(P/A, 10\%, 10) + \$500(P/F, 10\%, 10) \\ &= -\$100,000 + \$27,000(6.1446) + \$500(0.3855) \\ &= \$66,097. \end{aligned}$$

Since the present value of the benefits of the project exceeds the present value of the cost by $66,097, this is a cost-effective project.

The second method finds the future worth by calculating the equivalent cash flows at the end of year 10. The *F/P* factor is used to move the $100,000 cost to a value at the end of year 10, and the *F/A* factor is used to find the equivalent future value of the series of $27,000 annual savings over the project's 10-year life. The $500 salvage value is already based at the end of year 10, so it can be added in directly.

$$\begin{aligned} \text{FW} &= -\$100,000(F/P, 10\%, 10) + \$27,000(F/A, 10\%, 10) + \$500 \\ &= -\$100,000(2.594) + \$27,000(15.937) + \$500 \\ &= \$171,399. \end{aligned}$$

The positive future worth of the project also confirms that it is a cost-effective measure.

The third method reduces the initial cost and the salvage value to equivalent uniform annual amounts over the life of the project. The *A/P* factor is used for the cost and the *A/F* factor is used for the salvage value.

$$\begin{aligned} \text{AW} &= -\$100,000(A/P, 10\%, 10) + \$27,000 + \$500(A/F, 10\%, 10) \\ &= -\$100,000(0.1627) + \$27,000 + \$500(0.0627) \\ &= \$10,751. \end{aligned}$$

Even after discounting the savings at a 10% MARR, there is still a net savings of $10,751 per year from implementing this new system.

Since the present, future, and annual worth values are all positive, this indicates the project should be pursued, from a strictly economic standpoint.The resulting value of a present, future, or annual worth analysis also reveals information on the IRR as follows:

- a negative value indicates the IRR is less than the MARR
- a value of zero indicates the IRR = MARR
- a positive value indicates the IRR is greater than the MARR

As will be discussed in Chapters 6 and 7, there may be other reasons why a project with a positive present, future, or annual worth would not be implemented (i.e. barriers).

As will be discussed in Chapters 6 and 7, there may be other reasons why this project would not be implemented.

5.8.2 Benefit/cost ratio, or savings/investment ratio

A benefit/cost ratio (BCR), also known as a savings/investment ratio (SIR), calculates the present worth of all benefits, then calculates the present worth of all costs, and takes the ratio of the two sums. The BCR or SIR is another alternative economic decision-making criterion. It is favored by some organizations including parts of the U.S. Government and Military, especially the Army Corps of Engineers. Energy utilities and government agencies also use BCR analysis for funding decisions.

A careful definition of "benefits" and "costs" is important, as well as determining which parties are the recipient of benefits and which incur the costs (funding sponsors could also be the parties reaping the benefits, or not). In general, benefit−cost analysis can be much more complex and readers are referred to Newnan (2020) and NESP (2020) for further study of this topic. In this chapter, we present a simplified approach to benefit−cost analysis. Benefits are defined to mean all the advantages, less any disadvantages, to the project users. Costs are defined to mean all costs, less any savings, that will be incurred by the funder(s). With these definitions, a salvage value would reduce the costs for the funders(s), rather than increase the benefits for the user.

$$\text{BCR} = \text{Net benefits}/\text{Net costs}. \tag{5.6}$$

A benefit/cost ratio greater than 1 is necessary for the project to be cost-effective. For our example, the only cost is the initial cost, which already occurs at the start of the project, so it is already a present worth. The salvage value of $500 would serve to reduce the costs. One benefit occurs, the annual savings. Note that the ratio is explicitly defined as benefits divided by costs, so the sign on both values in the benefit/cost ratio will be positive.

The calculations required for this example are:

$$\begin{aligned}\text{PW of Annual savings} &= \$27,000(P/A, 10\%, 10)\\ &= \$27,000(6.1446) = \$165,904.\\ \text{PW of salvage value} &= \$500(P/F, 10\%, 10) = \$500(0.3856) = \$193\end{aligned}$$

$$\begin{aligned}\text{PW of total costs} &= \$100,000 - \$193 = \$99,807\\ \text{BCR} &= \$165,904/\$99,807 = 1.66.\end{aligned}$$

Since the benefit/cost ratio is greater than 1, the project is economically attractive. Some organizations set a low value of the discount rate or MARR, and then require a minimum BCR or SIR of 1.25−1.5 to identify a project as cost-effective.

5.8.3 Internal rate of return

A final method in this section is to base a project decision on the internal rate of return or IRR of the project. IRR is defined to be that value of the interest rate or discount rate that makes the present worth of the costs of a project equal to the present worth of the benefits of the project. If the computed IRR is equal to or greater than the MARR for an organization, the project is cost-effective. Many private organizations prefer the IRR method since it produces a rate of return to compare to their MARR that they have already established.

Continuing with the estimates in Example 5.10, the equation to solve for IRR in this case is:

$$\$100,000 = \$27,000(P/A, i, 10) + \$500(P/F, i, 10).$$

The value of i that makes this equation balanced is called the IRR, and can be found using a financial calculator, a computer program, or by trial and error with our standard tables. Since the present worth of the salvage value is going to be quite small, we can start by solving:

$$\begin{aligned}\$100,000 &= \$27,000(P/A, i, n)\\ (P/A, i, 10) &= 100,000/(27,000) = 3.7037.\end{aligned}$$

The next step is to look through the tables of *P/A* factors with $n = 10$ to find the value of the *P/A* factor closest to 3.7037. For $i = 25\%$, the *P/A* factor is 3.5705; for $i = 20\%$, the *P/A* factor is 4.1925. Thus, the value of *i*, or IRR, is approximately 24%. This is quite close to the correct answer.

The actual solution for IRR is found using an economic analysis computer program and is 23.8%. Since the company has an MARR of 10%, the 23.8% rate of return indicates this project is an excellent investment.

5.9 Lifecycle Costing

The LCC for a project or a piece of equipment is its total cost of purchase and operation over its entire service life. This total cost includes the costs of acquisition, operation including energy costs, maintenance and disposal. Most of these costs occur at some future time beyond the purchase date and must be analyzed using the time value of money. Thus, the discounted cash flow analysis presented in Sections 5.4–5.8 form a fundamental part of the LCC analysis.

Many businesses and organizations still use economic analysis methods that do not include all costs, and do not use the time value of money. The SPP method is commonly used by businesses and other organizations, but it is not a lifecycle cost analysis method. Some organizations still make purchase decisions based on lowest initial costs and do not consider the operating and maintenance costs at all. Costs that occur after the project or equipment are purchased and installed are ignored in these cases. This omission can create a very inaccurate view of the economic viability of a project in many cases. Use of LCC can lead to more rational purchase decisions and can lead businesses to higher profits.

Fortunately, many government agencies and private companies alike recognize the value of LCC when making investment decisions for EEMs or other DER projects. Thus, it is important for all energy management professionals to understand lifecycle costing.

5.10 LCC Decision-Making among Multiple Alternatives

Most lifecycle analyses involve choosing between more than one alternative. Therefore, a simple methodology for making valid economic comparisons between alternatives is necessary. This section uses the present worth method for this comparison. Several other methodologies exist, including future worth, annual worth, internal rate of return, and benefit/cost. Future worth and annual worth are economically equivalent to present worth and can be interchanged if desired. Internal rate of return and benefit/cost methods require an "incremental" approach to make valid comparisons between projects. See Farr and Faber (2019) and Newnan (2020) for further explanation. In addition, multiple solutions for the value of IRR can occur depending on the form of the cash flow. Since present worth is considered the standard against which other methods are judged, present worth is recommended.

In the U.S., the Federal Energy Management Program prescribes the use of a present value-based measure of worth in performing LCC analyses. All cash flows must be brought to a single, present-day baseline. In addition, the maximum study period for federal energy efficiency projects is 25 years from the date of occupancy. Therefore, the study period should be the lesser of 25 years, the life of the building, or the life of the system. In any case, the study period must be the same for all alternatives under consideration.

The LCC method using present worth, when applied properly, allows the analyst to compare projects with different cash flows occurring at different times. The present worth of the total costs of each alternative is calculated, and the alternative with the lowest LCC is selected. If multiple projects can be selected, then the projects can be ranked by LCC and the project with the next lowest LCC can also be selected.

The present worth analysis in lifecycle costing has two important requirements. First, when comparing multiple alternatives, different pieces of equipment may have different lifetimes. In this case, a common "study period" must be chosen so that all alternatives are considered over the same timeline. There are four choices for this common timeline. They are: shortest life, longest life, least common multiple of the two lives, and an arbitrary choice.

If shortest life is used, a salvage value must be estimated for the longer-lived alternatives. If longest life is used, shorter lived alternatives are assumed to be repeatable. If the least common multiple life is used, projects with lives of three years and five years would be compared over a 15-year study period. Both projects would have to be repeatable. An arbitrary choice results in possibly having to assume both repeatability, and a salvage value. The critical point is that all projects must be compared over the same time horizon.

The second important requirement of the analysis is that the interest rate used (MARR) must be the same for all alternatives.

End of Year	Alternative 1	Alternative 2
0	\$30,000	\$25,000
1	\$1,000 + \$6,000 = \$7,000	\$500 + \$10,000 = \$10,500
2	\$7,000	\$10,500
3	\$7,000	\$10,500
4	\$7,000	\$10,500
5	\$7,000	\$10,500
6	\$7,000	\$10,500
7	\$7,000	\$10,500
8	\$7,000	\$10,500
9	\$7,000	\$10,500
10	\$7,000	\$10,500

Note: In an LCC analysis, all cash flows are costs, and the signs are ignored.

Example 5.11:

An energy efficient air compressor is proposed by a vendor. The compressor will cost \$30,000 installed and will require \$1,000 worth of maintenance each year for its life of 10 years. Energy costs will be \$6,000 per year. A standard air compressor will cost \$25,000 and will require \$500 worth of maintenance each year. Its energy costs will be \$10,000 per year. If your company uses an MARR of 10%, would you invest in the energy efficient air compressor?

Alternative 1: Energy efficient air compressor
Alternative 2: Standard air compressor
Cash flows:

Analysis:

$$\begin{aligned}
\text{LCC(Alternative1)} &= \$30,000 + \$7,000(P/A, 10, 10) \\
&= \$30,000 + \$7,000(6.1446) \\
&= \$73,012. \\
\text{LCC(Alternative2)} &= \$25,000 + \$10,500(P/A, 10, 10) \\
&= \$25,000 + \$10,500(6.1446) \\
&= \$89,518.
\end{aligned}$$

The decision rule for LCC analysis is to choose the alternative with the lowest LCC. Since Alternative 1 has the lowest LCC, it should be chosen.

5.11 Taxes and Depreciation

Depreciation and taxes can have a significant effect on the lifecycle analysis of energy projects and should be considered for all large-scale projects for organizations that pay taxes. Depreciation is not a "cash flow," but it is considered a business expense by the government, and, therefore, lowers taxable income. The effect is that taxes are reduced, and taxes are a cash flow.

Tax laws and regulations are complex and intricate. A detailed treatment of tax considerations as they apply to lifecycle analysis is beyond the scope of this chapter, and generally requires the assistance of a tax professional. A high-level summary of important considerations follows. This discussion uses the context of the U.S. considering federal corporate income taxes to illustrate the taxation concepts. It should be noted that the principles discussed can apply to tax considerations in other countries as well.

Depreciation is a recognition that assets decrease in value over time. Depreciation deductions are an attempt to account for this reduction in value. These deductions are called depreciation allowances. To be depreciable, an asset must meet three primary conditions: (1) it must be held by the business for the purpose of producing income, (2) it must wear out, or be consumed in the course of its use, and (3) it must have a life longer than one year.

Table 5.5 MACRS property classes.

Property class	Example assets
3-year property	Special material handling devices and special tools for manufacturing
5-year property	Automobiles; light and heavy trucks; computers, copiers
7-year property	Office furniture, fixtures, property that does not fall in other classes
10-year property	Petroleum refining assets, assets used in manufacturing of certain food products
15-year property	Telephone distribution equipment; municipal water and sewage treatment plants
20-year property	Municipal sewers, utility transmission lines and poles
27.5-year residential rental property	Apartment buildings and rental houses
31.5-year non-residential real property	Business buildings

Table 5.6 MACRS percentages for three-, five-, and seven-year property.

Year	Three-year	Five-year	Seven-year
1	33.33%	20%	14.29%
2	44.45%	32%	24.49%
3	14.81%	19.2%	17.49%
4	7.41%	11.52%	12.49%
5		11.52%	8.93%
6		5.76%	8.92%
7			8.93%
8			4.46%

Multiple methods for calculating depreciation have been used in accounting. These include straight line, sum of the year digits, declining balance, and the accelerated cost recovery system (ACRS). The depreciation method used depends on when an asset was placed in service. Assets placed in service before 1981 use straight line, sum of the year's digits, or declining balance. Assets placed in service between 1981 and 1986 use the accelerated cost recovery system (ACRS). Assets placed in service after 1986 use the modified accelerated cost recovery system (MACRS). This chapter will discuss the MACRS depreciation methodology.

MACRS depreciation allowances are a function of: (1) the asset's "property class" (a life defined by the government), and (2) the asset's "cost basis" (purchase price plus any costs required to place the asset into service). Note that the salvage value is not used when calculating depreciation using MACRS. Also, the property class "life" may have little relationship to the actual anticipated life of the asset.

Eight property classes are defined for assets that are depreciable under MACRS. Table 5.5 shows the eight classes and gives some examples of properties that fall into each class. Professional tax guidance is recommended to determine the MACRS property class for a specific asset.

Tables for calculating the MACRS allowances are published by the U.S. government. These percentages are based on declining balance, switching to straight line, for most property classes. The real property classes are depreciated using straight line. The depreciation percentages for three-year, five-year, and seven-year property are shown in Table 5.6.

Example 5.12:

Calculate the depreciation allowance in year 4 for an asset that is a five-year MACRS property and has a cost basis of $10,000.

$$\begin{aligned} \text{Depreciation} &= \text{CostBasis} \times \text{MACRS\%} \\ &= \$10,000 \times 0.1152 = \$1,152. \end{aligned}$$

U.S. federal taxes are determined based on a tax rate multiplied by a taxable income. The tax rate depends on the corporation's income range and has varied from 15% to 39% of taxable income (current rates may differ; this range is presented to illustrate calculations that take into account taxes). Taxable income is calculated by subtracting allowable deductions from gross income. Allowable deductions include salaries and wages, materials, interest payments, and depreciation, as well as other "costs of doing business."

Tax-paying companies may want to use after tax cash flows (ATCF) when comparing alternatives. The calculation of taxes owed, and the ATCF, requires knowledge of:

- Before tax cash flows (BTCF), the net project cash flows before tax considerations.
- Total loan payments incurred by the project, including a breakdown of principal and interest components of the payments.
- Total bond payments incurred by the project, including a breakdown of the redemption and interest components of the payments.
- Depreciation deductions attributable to the project.

Once the above information has been determined, the ATCF is calculated in the following way:

Taxable income = BTCF – Loan interest – Bond interest – Depreciation

Income Taxes = (Taxable income) (Tax Rate)

ATCF = BTCF – Total loan payments – Total bond payments – Income taxes

At the tax-paying company's choice, all economic evaluation procedures (e.g. present worth, SIR, IRR) may then be applied to the ATCF.

5.12 Inflation

One of the fundamental principles supporting LCC analysis is the recognition that money changes value over time. There are two primary reasons for this. First is the "opportunity cost," or the interest loss that is incurred when dollars are not invested. The second reason is inflation. Inflation is the term for the loss in purchasing power of a dollar over time and should be accounted for in any lifecycle analysis.

Several additional terms must be defined before the term inflation can be explained clearly, including:

Constant dollars – constant dollars reflect the purchasing power (not face value) of the cash flow. These dollars are generally stated in relation to purchasing power at some base year, for example, 2020 dollars.

Current dollars – these are the out-of-pocket dollars that will actually change hands at any point in time. In other words, if a person visits a store in 10 years, current dollars will be the amount of cash paid at that time. However, the value of the items in the cart if purchased today (e.g., 2026 dollars) would be the constant dollar amount.

Inflation rate (f) – Rate typically published by the government agencies (e.g., based on the consumer price index in Canada or U.S.).

Real interest rate (j) – The amount of real growth in the earning power of money. This is also known as the inflation-free interest rate.

Market interest rate (i) – Opportunity to earn as reflected by the rates of interest available in finance and business. This rate contains both the inflation effect and the real earning power effect. This is the interest rate quoted by banks and other financial institutions.

The mathematical relationship between market interest rate, the inflation rate, and the real interest rate is:

$$i = f + j + (f)(j). \tag{5.7}$$

The key to proper analysis under inflation is to match the type of cash flow with the proper interest rate. If cash flows are estimated in current dollars, there are two effects contained in the cash flows: inflation, and real earning power. Therefore, the market interest rate, which includes both effects, is used to discount the cash flows. If cash flows are estimated in constant dollars, they do not contain inflation effects, and the real (inflation-free) interest rate should be used.

Type of dollars	**Interest rate to use**
Current (actual)	i = combined (market) interest rate
Constant worth	j = inflation-free interest rate

A current cash flow can be converted to a constant worth cash flow by removing the inflation effect. This is the same principle as discounting any cash flow and finding the present worth. The equation to be used is:

$$\text{Constant \$} = (\text{Current \$})/(1+f)^n, \tag{5.8}$$

where f is the inflation rate and n is the number of years between the "base year" (usually the present), and the time the money is to be spent. In the same way, constant dollars can be converted to current dollars by multiplying the constant dollars by the inflation effect.

$$\text{Current \$} = (\text{Constant \$})(1 + f)^n. \tag{5.9}$$

It should be noted that most energy commodities, such as electricity, inflate at a different rate than the overall inflation rate quoted by the government. This rate is called the escalation rate, k. In this case, equations (5.7) to (5.9) can still be used, with k substituted for the inflation rate, f. Energy escalation rates may be quoted by region, end-use sector, and fuel type. Also, agencies or utilities that contribute funding to EEM or other DER projects may specify the escalation rate to assume in project economic analyses.

Example 5.13:

A company has energy costs of \$25,000 a year for the next three years. The cost of energy is subject to a 5.8% escalation rate. The company's real discount rate is 4%. Find the present worth of the energy costs using (a) constant dollars and (b) current (actual) dollars.

(a) Constant dollar analysis: The constant dollar cash flow is \$25,000/yr for three years. Using the real interest rate of 4% to discount the constant dollar cash flow, we obtain:

$$\begin{aligned} \text{PW (constant \$)} &= \$25,000(P/F, 4\%, 1) + \$25,000(P/F, 4\%, 2) \\ &\quad + \$25,000(P/F, 4\%, 3) \\ &= \$25,000(1.04)^{-1} + \$25,00(1.04)^{-2} \\ &\quad + \$25,000(1.04)^{-3} \\ &= \$24,038 + \$23,114 + \$22,225 = \textit{\$69,377}. \end{aligned}$$

(b) Current dollar analysis: Two quantities must be calculated before the current (actual) dollar analysis can be performed: the current (actual) dollar cash flow, and the combined interest rate, which will be used to discount the current (actual) dollar cash flow.

Current (actual) dollars = Constant dollars (1 + escalation rate)^n

$$\begin{aligned} \text{Current (actual)\$(yr1)} &= \$25,000(1.058)^1 = \$26,450 \\ \text{Current (actual)\$(yr2)} &= \$25,000(1.058)^2 = \$27,984 \\ \text{Current (actual)\$(yr3)} &= \$25,000(1.058)^3 = \$29,607. \end{aligned}$$

Combined interest rate = 0.04 + 0.058 + (0.04)(0.058) = 0.10032 or 10.032%

$$\begin{aligned}\text{PW (current or actual\$)} &= \$26,450(P/F, 10.032\%, 1)\\ &\quad+\$27,984(P/F, 10.032\%, 2) + \$29,607(P/F, 10.032\%, 3)\\ &= \$26,450(1.10032)^{-1} + \$27,984(1.10032)^{-2} + \$29,607(1.10032)^{-3}\\ &= \$24,039 + \$23,114 + \$22,225 = \mathit{\$69,378}.\end{aligned}$$

Note that the present worth obtained using either method is the same (within rounding errors). This will always be true if the method is properly applied. Thus, it does not matter whether a constant dollar or current (actual) dollar analysis is used in this analysis.

5.13 Lifecycle Costing Software

The advent of the computer age has significantly helped the practitioner to perform lifecycle cost analysis. Spreadsheets have built-in functions that calculate economic values automatically. Users should read the definition of any built-in function carefully in order to make certain it is performing the desired calculation.

A common lifecycle costing software program is "Building Life Cycle Cost" (BLCC), which is distributed through the U.S. National Institute of Standards and Technology (nist.gov). This package was written for governmental agencies that must perform lifecycle cost analyses that conform to government standards.

BLCC was designed to provide an economic analysis of capital investments that reduce future costs. It focuses on energy efficiency in buildings. Potential users include federal, state, and local governments, as well as private sector users.

BLCC is menu-driven. Data requirements fall under three categories: general assumptions, capital investment data, and operation-related data as summarized in Table 5.7.

BLCC provides two report types: one for individual project alternatives, and another for comparative analysis. The individual project report provides an input data listing, an LCC analysis, both detailed and summary, and a yearly cash flow analysis. The comparative analysis provides a listing of all LCCs for all alternatives, with the lowest LCC flagged, as well as comparative economic measures.

The software provides an efficient and effective method to perform the last step in the analysis methodology, a sensitivity analysis. Sensitivity analysis refers to determining how "sensitive" a solution is to changes in the inputs and assumptions. Since many of the cash flows and interest rates are based on estimates, these estimates can be varied easily to see if minor changes affect the final results.

5.14 Summary

This chapter has discussed the fundamental methods needed to perform basic economic evaluations and LCC assessments for individuals and organizations to make decisions to invest in EEMs or other DER projects. Although the focus was on energy-related projects, the concepts can be applied to any investment. The methodology provides a sound basis to analyze projects and make "rational" decisions based on cost-effectiveness, such as SPP, BCR, or LCC metrics. These methods are essential for all energy management professionals to develop proficiency in economic analysis. However, we will learn in Chapters 6 and 7 that basic economics and rational cost considerations are necessary, but not sufficient, to explain many real-world decision-making outcomes. The next two chapters present important concepts and tools to complement the methods presented in this chapter.

References

AACEI Recommended Practice No. 18R-97 (2020) "Cost Estimate Classification System as Applied in Engineering, Procurement, and Construction for the Process Industries" Association for the Advancement of Cost Engineering, International aacei.org.

Farr, J. and Faber, I. (2019) "Engineering Economics of Life Cycle Cost Analysis". Boca Raton: CRC Press, DOI:10.1201/9780429466304.

Table 5.7 Data requirements for Building Life Cycle Cost software.

General assumptions	Capital investment data	Operation-related data
Analysis type	Initial cost	Annual recurring operating and maintenance costs
Base data	Capital replacements	Non-annual recurring operation and maintenance costs
Service data	Residual values	Energy consumption and cost data
Study period	Escalation rates	Discount rate

Gold R. and Nowak S. (2019) "Energy Efficiency Over Time: Measuring and Valuing Lifetime Energy Savings in Policy and Planning" American Council for an Energy Efficiency Economy Report U1902 www.aceee.org.

NESP (2020) "National Standard Practice Manual For Benefit-Cost Analysis of Distributed Energy Resources" National Energy Screening Project (NESP) www.nationalenergyscreeningproject.org/national-standard-practice-manual/.

Newnan (2020) "Engineering Economic Analysis" (14th edition) Oxford University Press.

Acknowledgments

This chapter was derived from the 8th edition version written by Camille DeYong, with only minor changes made by Eric Mazzi. One significant change was the addition of AACEI cost accuracy guidelines. Another significant change was that the sections on energy financing options were removed, because this 9th edition now has a new material on financing in Chapter 6. Minor text edits and changes to examples were made throughout the chapter. New, updated references are provided.

Practice Problems

1. A fan is driven by a 400 hp (2,300 V, three-phase) electric motor and operates with a power factor of 0.87. Average motor amperage drawn is 50 amps. The fan operates 6,500 hours each year. Electricity costs \$0.085 per kWh. The fan can be upgraded at a capital cost of \$28,000 to save 20% of baseline energy consumed. Hint: From Chapter 13, Electric power $= (3)^{1/2} \times$ Voltage $\times$ Current $\times$ Power Factor

   a. Calculate the annual energy and energy cost savings.
   b. Estimate the simple payback period (years) based on energy savings.
   c. Estimate the net present value of the project over a 10-year lifetime, using an 8% interest rate.

2. A facility uses a 10 hp motor 16 hours per day, 5 days per week, and 50 weeks per year. This motor is 90% efficient, and it is near the end of its useful life. The company is considering buying a new ultra-premium efficiency motor (97% efficient) to replace the old one instead of buying

a premium efficiency motor (94% efficient). Assume the motor operates at 100% of rated load. The ultra-premium efficiency motor costs $1,500 more than the premium efficiency model, and should have a 15-year life. The company pays $15.00/kW per month and $0.15/kWh. The company has set a discount rate of 12% for their use in comparing projects. Determine the SPP, IRR, and BCR for this project.

3. The annual savings in electricity for an EEM is 400,000 kWh/yr valued at $0.11 per kWh. The capital cost of the EEM is $375,000. How much of a utility incentive is required to achieve a 1.5-year simple payback period?
4. A company must repair their main air conditioning system, and they are considering two alternatives.

   - Purchase a new refrigerant compressor for $20,000 that will have a future salvage value of $2,000 at the end of its 15-year life. The new compressor will save the company $6,500 per year in electricity costs.
   - Purchase a high efficiency heat pump for $28,000 that will have a future salvage value of $3,000 at the end of their 15-year useful life. The heat pumps will save $8,500 per year in electricity costs.

   The company's discount rate is 18%. Compare the two projects on a lifecycle cost basis, and advise which project would be recommended.
5. A six-year project has the following cash flows:

   - Purchase and installation cost $100,000
   - Maintenance per year $10,000
   - Energy saving per year $45,000
   - Salvage value $20,000

   Assume that the minimum attractive rate of return is 12%/year. The corporate tax rate is 34%. Assume the equipment has a five-year life for tax purposes (MACRS percentages for "five-year" property). Calculate the net present value for this project.
6. An energy efficiency upgrade will cost $110,000 to install and will save an estimated $31,000 per year in energy costs over a 15-year life. The project has $2,100 per year in additional maintenance costs. The company's tax rate is 26%. Calculate the annual after tax cash flow for the first 10 years if it depreciated on a straight-line basis.

6

Financing Energy Management Opportunities

Thor Jensen

Halitra, Canada

6.1 Introduction

Energy managers routinely plan capital budgets, procure building services, and apply for grants or incentives. Increasingly, the pressures to manage energy costs and reduce GHG emissions lead energy managers to seek external financing. External financing can supplement internal budgets to support investments in the energy transition. A growing number of financing sources target measures to reduce energy or GHGs, sometimes with government support. This chapter provides a resource for energy managers looking for appropriate financing sources.

This chapter is organized as follows. Section 6.2 introduces finance and begins with describing common sources of finance in Section 6.2.1. Section 6.2.2 notes that organizations with low-cost capital still require high rate-of-return thresholds for projects. The organizations weighted average cost of capital, risks, and opportunity cost of capital explain why organizations seek high returns. Debt financing instruments for commercial and institutional energy efficiency and related projects are then reviewed.

Section 6.3 lists common reasons why organizations reject energy efficiency measures (EEMs) and other distributed energy resources[1] (DERs) that offer compelling returns. In some cases, financing can determine whether management proceeds or passes on an investment. Examples will show when financing can improve project returns. Readers will be able to identify

[1]As introduced in Chapter 1, DERs include: energy efficiency measures (EEMs), energy storage (e.g., batteries), distributed generation (e.g., solar panels), fuel switching, and demand response.

appropriate funding sources and discuss their limitations. Summary tables in Section 6.4 list common barriers by asset class and financing intervention. The chapter concludes with practice questions and additional resources.

The economics in Chapter 5 is a pre-requisite reading. This chapter assumes an understanding of the time value of money, discounting, minimum acceptable rates of return, net present value, and internal rate of return calculations.

6.2 Introduction to Finance

The three learning objectives for Section 6.2 are to:

- Describe financing sources and instruments
- Describe the advantages and disadvantages of green financing sources

6.2.1 Financing sources

Like any discipline, finance comes with its own jargon. Developing a financial vocabulary will aid in understanding and evaluating financing options. This chapter begins by differentiating common financing types, instruments, providers, and enhancements.

To fund their investments, organizations may draw on their cash savings or they may seek external financing. The two primary types of financing are equity and debt capital. Debt financing must be repaid, and it includes all types of loans and bonds. In contrast, equity investors have no repayment obligation but receive an ownership position in the organization. Hybrid financing types convert from debt to equity under certain conditions. Table 6.1 summarizes common financing types, instruments, and enhancements catering to DERs.

The three main financing types include subcategories of financial instruments.[2] In short, financial instruments are contracts with monetary rights.[3] Investors buy, or banks lend, debt instruments for rights to an interest-bearing asset. Common equity investors exchange cash for the right to a share of ownership in the company, for example. Preferred shares might include special rights for voting or dividend payments.

[2]In Table 6.1, the financial instruments split into public and private markets. Public financial instruments trade on public capital markets, such as the New York Stock Exchange or NASDAQ. Loans originate in private markets and trade with less frequency.

[3]Financing instruments also include derivatives, such as options and futures. These advanced concepts are not covered in this chapter.

Table 6.1 Summary of common financing types, instruments, sources, and enhancements.

Financing type	Financing instrument		Financing providers/sources	Financing enhancement
	Public market	Private market		
Debt	Bonds, **Green Bonds,** Corporate Bonds, Project Bonds.	**Senior Secured Loan, Property Tax Secured Loan, Utility Bill Secured and Unsecured Loans, Unsecured Loan,** Subordinated Loan.	Revolving Funds, Energy Savings Companies, Endowments, Specialized Lenders, Utilities, Commercial Banks, Retail Lenders, Green Banks, Governments, Public/Private Debt Investors.	**Energy Savings Guarantee, Financial Savings Guarantee, Recoverable Expense, Transferable Loan, Energy Services (off-balance sheet).** Interest Rate Buydowns Credit Enhancements Risk Mitigants.
Hybrid	Convertible bonds.	Mezzanine financing.	Hybrid debt/equity funds.	
Equity	Common shares, preferred equity.	Unlisted shares, private equity.	Real estate investment trusts, limited partnerships infrastructure funds, special purpose vehicles, public/private equity investors.	

Note: Bolded text refers to financial instruments and enhancements discussed in this chapter. Source: OECD , 2015).

This chapter focuses on debt financing instruments. Debt instruments, such as loans, are often used to finance facility improvements. There is a tremendous variety of loan types. For simplicity, here loans are distinguished by security and seniority:

- **Secured vs. unsecured loans:** Secured loans give the lender a right to claim some or all of the borrower's assets as collateral. If the borrower fails to pay, then the lender can seize or sell the assets as collateral. Unsecured loans have no claim to assets or collateral. They are lent based on the credit worthiness of the project or borrower.
- **Senior vs. junior or subordinated loans:** Senior lenders receive the right to first payment in exchange for capital. Payment priority becomes valuable when an organization is in distress. An organization would pay senior lenders before junior and subordinate lenders.

Financing providers provide the capital for the financial instruments. A financing provider can supply capital through one or several financial instruments. A revolving fund could offer both secured and unsecured loans, while an energy savings company may focus on providing on one compelling loan targeting school districts, for example.

Financing enhancements include special features to make DERs more attractive. They are closely linked to the financing provider and instrument. These enhancements may overcome common barriers preventing the implementation of measures. For example, energy savings guarantees and loan transferability address different, but common, problems. The second half of this chapter treats these financing enhancements in detail.

6.2.2 Sources of finance dedicated to DERs

When the cost of internal funds is high, it may be worthwhile to pursue financing. External financing sources may be less costly and offer enhancements favorable to DERs. This section will describe six debt financing instruments, their providers, and common enhancements, including:

- Green bonds
- Revolving funds
- Utility bill financing
- Property tax financing
- Energy performance contracts
- Energy services

6.2.3 Green bonds

Bonds are public debt financing instruments. They provide investors a series of interest (coupon) payments followed by loan repayment (principal repayment often due at maturity). Investors purchase bonds for portfolio income and diversification, trading them on public markets.

Green bonds are simply bonds with proceeds earmarked for green projects. Governments and firms both issue green bonds. Government agencies use bonds to fund many public works and corporations to finance their growth and operations. Increasingly, banks and real estate investment trusts also issue green bonds. Bonds can allow for borrowing large amounts at a low cost over a long-time horizon. Bond issuances tend to be large but can fund many smaller projects if grouped together.

There are significant time and contracting costs associated with creating a green bond (OECD , 2017). As a result, most green bonds will be issued in amounts in the tens or even hundreds of millions of dollars. For governments, bond issuances may involve legislative or voter approval. For firms, a bond issuance typically requires board and management approval.

6.2.4 Revolving funds

Revolving funds provide dedicated sources of capital for DER projects. Revolving loans supply a consistent, recurring source of funding exclusively for DER projects (McCabe , 2019). This helps projects overcome the opportunity cost problem that arises when competing against other projects for funding.

Revolving funds can be both internal and external funds. Internal funds see the parent organization borrow for its separate divisions and departments. External revolving funds allow for multiple, independent organizations to access the revolving fund. An external revolving fund may allow the private sector to borrow at low government rates.

In some cases, a green bond can capitalize a revolving fund. In the case of an external fund, the financing instrument is usually a secured or unsecured loan. Energy savings help offset the loan cost, and once loans are repaid, the funds can be lent to new projects.

The challenge with revolving funds is the slow rate of loan repayment. It may take some time to recover the capital lent to long-term projects. They also require time and resources to set up and manage.

6.2.5 Utility bill financing

Utility bill financing is common among energy utilities throughout the world. Utilities invest in energy efficiency and renewables to adhere to regulatory mandates. Utilities may prefer to invest in energy efficiency than to increase power supply. Utility bill financing allows for loans to be repaid as part of the customer's utility bill. There is significant variation in financing programs, but most fall into one of two categories:

- On-bill financing: The utility uses its own capital to invest in building improvements. It then collects the loan as part of its monthly utility bill.
- On-bill repayment: The utility partners with a lender to supply capital. The utility collects the loan payments on the bill and then transfers the funds to the lender.

Secured and unsecured loans are common to both program types. The key difference is whether the utility or a third-party provides loan capital. Governments may also offer capital and incentives to encourage utility participation.

Businesses rarely default on their utility payments, which provide excellent security for lenders (Bianco et al. 2020). The advantage to borrowers is the ability to borrow capital with generous terms. If energy savings exceed loan repayments, then the project can provide immediate cost savings. In some cases, the loan is passed through to tenants or transferred to future owners. This helps overcome split incentive and holding period concerns, discussed in the next section. However, the passing on loans to tenants or future buyers is a negotiation.

Utility bill financing is common, but the eligible measures will vary over time, region, and policy mandate. Energy managers should check with their local utilities to understand current offerings. Offerings may have a narrower focus that aligns with the utility's interests.

6.2.6 Property tax financing

Property tax financing allows for local governments to fund building improvements like other public works. Local governments pass enabling legislation so that buildings can repay the cost of building improvements on their property taxes. Water conservation measures and safety improvements, such as seismic upgrades, may be eligible. Also known as property assessed clean energy or a local improvement charge, this financing is available in half of the United States and a growing number of Canadian provinces.

Property tax financing is a type of secured loan. If the building defaults on its taxes, the local government would place a lien against the building. Property taxes provide excellent security, so lenders extend generous terms to borrowers. Fixed rates of 20 years or longer are possible. Property tax financing may even have payment priority above the senior lender. Thus, borrowers can access capital at attractive interest rates over a long term.

Having the loan repaid through property taxes offers other advantages. If the tenant pays their share of property taxes, then building owners might pass through the loan to tenants. If the energy cost savings exceed loan payments, the tenant will benefit. It is also possible to transfer the loan upon sale. Like utility bill financing, this will be a negotiation between the tenants and future buyers. Some tenants may prohibit paying for capital improvements in their lease.

Property tax financing requires government buy in. Governments may have to seed loan capital, and it may take years to establish a local program. The effort to receive property tax financing can be considerable (Leventis et al. , 2018). Property tax financing requires mortgage lender consent, adding months to loan approvals. Property tax financing projects loans tend be large as a result.

6.2.7 Energy performance contracts

Energy performance contracts pay for capital improvements supported by energy cost savings. There is a significant variety in contract types, but most involve a level of guaranteed energy cost savings (U.S. DOE, 2022). Organizations can finance the upgrades and then pay a fixed price for a guarantee, like insurance. Or organizations may hire an energy savings company to design, install, and finance the work. In the latter arrangement, the company earns their profit and recovers their costs from the energy cost savings. After the contract term, the borrower owns the equipment outright.

Governments often enter energy performance contracts to renew their buildings. Governments can have budgetary constraints leading to excessive deferred maintenance. If energy savings exceed costs, then equipment replacement may be budget neutral. Energy savings companies can also provide significant technical and project management expertise.

Energy saving guarantees are not free. Besides the cost of improvements, borrowers should expect to pay for the guarantee and the company's profit. Nor does an energy savings guarantee equate to financial savings. Most guarantees do not include changes in energy prices or building occupancy, either of which would have financial implications. Due to their size and complexity, most energy performance contracts target large project sizes.

6.2.8 Energy services contracts

Energy services contracts are based on the delivery of a level of comfort, utility cost savings, or energy performance. The service company has wide discretion in how to achieve those goals. The contract may specify the delivery of therms (or GJ) of hot water or lumens of light, for example, without specifying the type of equipment or form of energy. In regions with time-of-use pricing, for example, energy services can increase the value of distributed energy resources.

In all cases, the emphasis is on providing an ongoing service rather than financing equipment. A growing number of companies offer energy services for buildings. Some common types of energy services are:

- **Energy service agreements:** Like an energy performance contract, the service provider uses energy cost savings to recover their costs. Unlike energy performance contracts, a separate entity owns and finances the equipment. The building owner then subscribes to the energy service.
- **Metered energy service agreements:** Like an energy service agreement, a service company owns the equipment. As part of metered energy

service agreements, the service company pays the building owners' utility bill. If the service provider pays the utility bills, it allows for a level of guaranteed energy and financial savings.

- **Lighting as a service:** An independent company provides lighting services. They have wide discretion in providing lighting technologies, using controls, daylighting, and scheduling. The service provider may also guarantee a level of energy savings.
- **Hot water as a service:** A company contracts with the building owner to deliver therms (or GJs) of hot water at an agreed-upon pricing schedule. The company has discretion in terms of investment in capital equipment and operations.

Organizations facing borrowing restrictions may also prefer services over equipment ownership. The government may face legislation prohibiting borrowing, for example (U.S. DOE, 2022). In these cases, contracting for energy services avoids adding a liability to the balance sheet. The cost of equipment and services reduce the operating budget instead.

An energy service may be more expensive than owning the equipment. Owning equipment comes with interest and depreciation tax shields and omits the service provider profits. In some cases, services may reduce operating income (profitability) relative to ownership. The value to borrowers depends on the quality of services rather than its ability to avoid borrowing constraints.

6.3 The Role of Financing in Implementing DERs

In order to receive funding, DER projects must often provide a high return on investment. It is common for minimum acceptable rates of return of 20% or higher to be applied, or 3–4 years' simple paybacks. In some cases, projects that offer high returns on investment are still rejected. This section will review why investment hurdle rates are high, some common barriers to investment, and discuss when financing can help overcome those barriers. Questions addressed in this section include:

- What is the organization's weighted average cost of capital?
- Why are some projects with high returns still rejected?
- When can financing overcome common barriers to investment?

6.3.1 The weighted average cost of capital

For most organizations, financing will include a combination of equity and debt. Equity investors seek high rates of return to compensate for placing their

Table 6.2 Weighted average cost of capital of an organization.

Capital stack	**Payment priority**	**(A) Amount ($M)**	**(B) Cost before tax**	**(C) Cost after tax**[1]	**(D) WACC**[2] **($M)**
Equity	4	40%	15%	15%	
Preferred shares	3	10%	12%	12%	
Subordinated debt	2	10%	7%	5.25%	
Senior debt	1	40%	5%	3.75%	
Total		**100%**			**(D) 9.23%**

1) After Tax = Cost of Borrowing (1-Tax Rate). Applies only to debt.
2) Weighted Average Cost of Capital (D) = Σ $(A) \times (C)/\Sigma$ (A)

capital at risk.[4] Debt financing is less costly but is subject to repayment. The sources of funds, and their combined cost, partially explain the high return requirements.

The total corporate cost of capital for an organization is its weighted average cost of capital (WACC). To calculate the WACC, multiply the share of each funding source by its cost. Table 6.2 shows the WACC calculation for an organization with four sources of capital and a corporate tax rate of 25%. In this example, the organization with a 15% cost of equity has a WACC of 9.23% due to the low cost of debt financing.

There is also a tax advantage to borrowing, further lowering its cost. Borrowing lowers taxable income by deducting interest expenses paid. To calculate the cost of borrowing, multiply the interest rate by (1 – Corporate Tax Rate). The cost to an organization borrowing funds at a 5% interest rate drops to 3.75% with a 25% corporate tax rate, for example.

From an investment perspective, the WACC is the organization's required rate of return or minimum acceptable rate of return. In the long run, an organization must target projects providing returns greater than its WACC. Otherwise, organizations will fail to create value for equity investors and may struggle to meet its loan repayment obligations.

In practice, an organization will target rates of return greater than its WACC. Organizations target higher rates of return because of risk and opportunity costs. Projects involve risks so that target returns must make allowances

[4]Calculating the cost of equity, or in some cases the equity market premium, is a firm specific exercise (Saalmuller, L. , 2022). Here we assume management provides the cost of equity.

for losses. Competition with other projects for funding further raises return thresholds. Thus, the WACC can be considered the floor when allocating capital. The need to meet all funding requirements and provide value to equity investors raises the rate even higher.

6.3.2 Barriers to DER investment

Management may still reject DER projects with positive net present values and high internal rates of return. Often, the reason can be traced to common barriers interfering with the project's financial return. This section will describe five common barriers to DER investment and propose when they can be resolved by financing instruments. The five common barriers addressed here are:[5]

- Performance risk
- Project scale
- Split incentives
- Borrowing limitations
- Holding period concern

The barriers will be discussed using a simple example of a rooftop unit (RTU) replacement, with a cost of $120,000. The examples will include necessary calculations. However, motivated readers can calculate the loan and interest payments using the methods outlined in Chapter 5 on economics, or they may choose to use computer spreadsheets for calculations (such as Microsoft Excel PMT, IPMT, and PPMT functions).

6.3.2.1 Performance risk

DER projects are subject to performance risks. Performance risk increases with the project's technical complexity, size, and the organization's budgetary constraints. With greater performance risk comes wider forecasts in project costs and benefits. The greater the uncertainty in forecasted energy savings, the riskier the project becomes. As a result, management may lower the project's expected income to reflect its risk.

One method of resolving the performance risk is to pay for a performance guarantee from an energy service company. The energy guarantee typically includes a level of energy savings or equipment performance. Energy service

[5]See Chapter 7 to learn more about barriers to investing in projects that appear to be economically attractive.

companies often bundle one or more phases of project design, build, finance, and operation. Consider the following example.

> *The energy manager finds considerable variance in energy savings when evaluating the DER project. Depending on the quality of roof repair and installation, energy savings may be $270,000 or $220,000 annually. At a 4% cost of borrowing, this is an 18% or 10% IRR over ten years, respectively. Management is unwilling to proceed with a project that might return only a 10% IRR. The energy manager can also partner with an engineering firm to guarantee the energy savings. The engineers would model the risk and determine an insurable level of savings. The cost of the guarantee increases total project costs lowering the project return to 15% IRR. The 15% IRR is acceptable for management, helping to advance the project today.*

Like insurance, performance guarantees are not free. The energy service company will receive compensation for accepting performance risks. Management will weigh the benefit of performance certainty against the cost of insurance. Energy managers must also carefully negotiate risk-sharing in the performance guarantee. Energy savings guarantees tend not to include risks outside the guarantor's control. If energy prices rise, for example, it is possible to still pay higher operating costs.

6.3.2.2 Scale and transaction costs

A DER may offer significant returns on investment, but if the dollar amount is too small, it may not attract attention from management. Further, the "hidden" costs of organizing, planning, and implementing projects can be high (or at least perceived to be high). These transaction costs can make external financing untenable for small projects.

Project size can make some sources of external financing untenable. For example, performance contracts can take half a year or longer to negotiate. For this reason, energy performance contracts often target larger government contracts. Other financing types targeting DERs at the scale of one building or project can entail significant effort as well. Commercial property tax loans have average project sizes over $1M. Property tax financing can entail significant effort to attain, such as receiving permission from the senior lenders to add this type of financing to the capital stack.

Financing solutions can be designed to lower the cost facing potential borrowers. Some revolving loan facilities offer smaller loan sizes, providing a recurring source of capital that is accessible for smaller DER projects. Some utility financing programs have developed rapid approval for simple DER measures financed by unsecured loans. Utility programs often post their financing terms, their eligibility criteria, and the average time to approve a loan to facilitate borrowers. Consider the following example of how project scale winnows down the list of suitable financing option.

> *When considering external financing for a RTU upgrade, the energy manager notes that the total project cost is $120,000. Given its relatively small budget, the energy manager decides this project is unsuitable for a performance contract. Based on the available offerings, the energy manager considers two other sources of financing, property tax financing or an unsecured loan product offered by a revolving fund. The project is still on the small side for property tax financing. The energy manager could defer the project to the next year and bundle the project with a lighting retrofit. The lighting retrofit would double the project size, making it worthwhile to apply for property-tax financing. Alternatively, the energy manager could apply for an unsecured loan. At $120,000, this is in the typical range for a revolving fund, making it suitable if the energy manager decides to install the RTU this year.*

The scale barrier presents a persistent challenge for smaller, independently owned buildings. Not only do small buildings have a lower ceiling on energy savings, if they are independently owned then staffing is also a concern. With fewer staff, the cost of searching and applying for financing can be disproportionately high in small buildings.

6.3.2.3 Split incentives

Split incentive barriers emerge when lease or rental contracts separate energy savings from the capital investment. In a triple net lease, for example, the tenant pays a charge for rent, property taxes, and the tenant pays utility bills directly to the utility company. In this lease structure, the tenant directly benefits from greater investments in DER. If the building owner pays for the DER, the financial impact will be negative from the building owner's perspective.

If utility bills and property taxes are not directly paid by tenants, one remedy is to recover the cost of DER investments from the tenant. Landlords can pass through higher maintenance charges, utility bills, or property taxes.

This is one benefit of using property tax and utility bill financing mechanisms. If energy cost savings exceed the increase in property taxes or utility charges, then it will have a positive impact on the tenant's cash flow. Consider the following example:

The same building owner is considering whether to invest in an RTU upgrade. In this case, the building is a retail center with multiple tenants all served by triple net lease. The manager considers three scenarios: A) an investment with no tenant recoveries; B) an investment with tenant recoveries over five years; and C) an investment with tenant recoveries over ten years. It is possible to borrow at the same 4% interest rate over five- or ten-year periods. These scenarios are quantified in Table 6.3.

As shown in scenario A, the year 1 impact on the landlord's cash-flow is negative. The tenant reaps the full benefits from the project. In scenario B, the landlord is cash flow neutral, but the short-term financing horizon has made the tenant worse off. In scenario C, the longer-term financing arrangement lowers the annual charge. As a result, the tenant is still cash flow positive, and the landlord is cash flow neutral. Finding similar win–win opportunities can help increase the likelihood of DER implementation.

Even when DERs offer benefits to landlords and tenants, it may not be possible to directly recover costs. All leases are a negotiation, and tenants may object to an increase in costs. Some leases may outright prohibit the pass through capital costs. In these instances, the landlord may prefer to recover DER costs through higher rents. Efficient buildings are more competitive, allowing landlords to indirectly recover DER costs through higher rents.

Table 6.3 Year 1 impact on cash flows.

Tenants	**A. None**	**B. 5 years**	**C. 10 years**
Energy savings	$27,000	$27,000	$ 27,000
Loan payment	0	−$27,717	−$15,541
Net impact	$27,000	(−$717)	$11,459

Building owner	**A. None**	**B. 5 years**	**C. 10 years**
Energy savings	0	0	0
Loan payment	−$27,717		
Net impact	($27,717)	0	0

6.3.2.4 Balance sheet constraints

Restrictions on borrowing are common among private investors and governments alike. Among commercial properties, senior lenders may require their approval on any additional borrowing. Avoiding excessive leverage helps protects the senior lender's interests. Governments may also restrict borrowing of their various agencies and departments through legislation. For example, balanced budget legislation will restrict government borrowing outside the annual budget.

Borrowing restrictions are sometimes referred to as balance sheet constraints. This is because borrowing restrictions prohibit the increase in liabilities on the balance sheet. Capital equipment is a long-term asset on the balance sheet of an organization. If purchased using debt, then a corresponding loan would appear under long-term liabilities.

If an organization faces restrictions on borrowing, then external financing is moot. However, contracting for energy services may still be possible. Energy services would see the whole cost of the DER project deducted as an operating cost. If there is no associated long-term liability, the service may be considered "off balance sheet." As a rule, value of services provided should outweigh its accounting treatment.

Energy services may offer preferable accounting treatment, but they could cost more than traditional financing. Equipment ownership comes with the benefit of interest and depreciation tax shields for the borrower. Further, energy services will deduct their entire cost from the operating budget, including an additional service charge.

Energy managers should always compare the cost of energy service contracts to traditional financing. If energy services cost more than the traditional finance cost, this cost should be weighed against its value. In some cases, energy services may provide superior energy management services, which justifies the greater cost.

6.3.2.5 Holding period concern

Many DER investments can have long investment time horizons. Deep renovations might have a payback period greater than 20 years and building owners may seek financing spread over this duration. The advantage of a longer loan term is smaller principal repayments. Smaller principal payments increase total interest costs but may allow for energy cost savings to exceed borrowing costs creating a positive impact on cashflow.

If a property is likely to be sold in the near term, then ownership may be hesitant to take on long-term project financing. This dilemma is known as the hold barrier, referring to the short holding period return of some commercial properties.

Some financing mechanisms allow the loan to be transferred to the future owner. Loan transferability is associated with property tax financing and some utility bill financing programs. In both cases, loan terms can stretch to 20 years or longer with the intention of the future owner assuming the loan payment schedule.

Whether loan transferability overcomes the hold barrier is subject to debate. Loan transfer is possible, but it may negatively affect the asset sales price. Normally, buildings are sold free and clear of any previous financial obligations. If financial obligations do carry forward, as higher property taxes or utility fixed charges, then they will be factored into the sales price. The future loan obligations will reduce the income available to investors, and the buyer will have to approve the loan transfer with their own senior lenders.

6.4 Summary

To summarize financing, Table 6.4 shows how the six barriers discussed in this chapter apply by asset class as well as suitable financing products. As shown in Table 6.4, it is possible for multiple barriers to afflict the same property type. Table 6.5 shows that financing products can provide multiple benefits but that no financing product offers a resolution to all barriers. Consider utility bill financing as an example. Utility bill financing commonly offers long-term loans and low interest rates; it is sometimes recoverable from tenants to overcome the split incentive barrier and depending on the program could be transferred to the new owner to overcome the holding period barrier. Utility bill financing is sometimes paired with sophisticated technical expertise to overcome operational challenges and can overcome scale problems through replication over many projects. However, utility bill financing is not off-balance sheet (appearing as an asset and a liability on the balance sheet).

Financing products are continually evolving (and so is their accounting treatment). Therefore, energy managers should always consult with their finance departments about new products.

Table 6.4 Barriers to investment by property type.

	Budgetary priorities	*Split incentives*	*Holding period concern*	*Borrowing restrictions*	*Operational challenges*	*Scale and trans-action cost*
Office and *retail*[1]	●	●	○	○	○	○
Multifamily	●	●	○	○	○	○
Industrial	●	-	-	○	●	○
Hospitality	●	-	○	○	●	○
Public sector[1]	●	-	-	○	○	○
Notes	1) Municipal, university, schools, and healthcare buildings.					
Legend	● Often ○ Sometimes – Uncommon					

Table 6.5 Possible financial interventions to overcome investment barriers.

	Budgetary priorities	*Split incentives*	*Holding period concern*	*Borrowing restrictions*	*Energy saving uncertainty*	*Scale and Transaction Cost*
Possible resolution	Long-term, low-interest rates	Recoverable expenses	Loan transferability	Energy services	Guarantee/expertise	Project Replicability
Revolving funds (internal)	•	-	-	-	○	○
Green loans from banks	•	-	-	-	○	○
Utility bill finance[1]	•	○	○	○	○	○
Property tax finance	•	○	○	○	○	○
Energy performance Contract	○	-	-	○	•	○
ESA/MESA[2]	○	○	-	•	•	○
Notes	On-bill financing and on-bill repayment. 2) Energy service agreements, Managed energy service agreements.					
Legend	•Often ○ Sometimes – Uncommon					

References

Bianco, V., Marchitto, A., Jankovic, I., (2020), *Overview of On-Bill Building Energy Renovation* Schemes, RenoBill. Accessed January 20, 2022 at www.renonbill.eu.

Leventis, Greg, Schwartz, Lisa, Kramer, Chris, and Deason, Jeff. Lessons in Commercial PACE Leadership: The Path from Legislation to Launch. United States: N. p., 2018. Web. doi:10.2172/1425438.

McCabe, M.J. (2019). Driving Investment in High-Performance Commercial Buildings. In: Walker, T., Krosinsky, C., Hasan, L.N., Kibsey, S.D. (eds) Sustainable Real Estate. Palgrave Studies in Sustainable Business In Association with Future Earth. Palgrave Macmillan, Cham. https://doi.org/10.1007/978-3-319-94565-1_11

OECD (2015), Mapping Channels to Institutional Investment in Sustainable Energy, Green Finance and Investment, OECD Publishing, Paris, https://doi.org/10.1787/9789264224582-en.

OECD (2017), *Mobilising Bond Marke*ts for a Low-Carbon Transition, Green Finance and Investment, OECD Publishing, Paris, https://doi.org/10.1787/9789264272323-en.

Saalmuller, L., (2022). *Cost of Capital: What It Is and How to Calculate It,* Harvard Business School Online. Accessed June 1, 2022 at https://online.hbs.edu/blog/post/cost-of-capital.

U.S. Department of Energy, (2020). What is an Energy Savings Performance Contract, Better Buildings Financing Navigator, Accessed January 15, 2022 at https://betterbuildingssolutioncenter.energy.gov

Bibliography and Resources

Online resources for learning more about financing options are widely available. The resources below were consulted in writing this chapter and can provide detailed treatment of the financing options discussed in this chapter.

- Office of Energy Efficiency & Renewable Energy, U.S. Department of Energy, Accessed Online January 23, 2022: https://www.energy.gov/eere/slsc/establish-financing
- Better Buildings Financing Navigator, U.S. Department of Energy, Accessed Online January 23, 2022: https://betterbuildingssolutioncenter.energy.gov/financing-navigator

- National Association of State Energy Officials, NASEO. Accessed Online January 23, 2022: https://www.naseo.org/issues/energy-financing/resources
- Centre on Green Finance and Investment, Organization for Economic Co-operation and Development. Accessed Online January 23, 2022: https://www.oecd.org/cgfi

Author Biography

Thor Jensen (Ph.D., CFA) works at the intersection of buildings, climate policy, and investment. His doctoral research examined the role of incentives and financing on the adoption of low-carbon technologies in buildings. Thor is a chartered financial analyst, having worked in private equity, pursuing direct investments in real estate and clean technology ventures. He currently advises investors, governments, and companies on low-carbon investment strategies. An entrepreneur, he founded his own property-tech company in 2022 with the vision of decarbonizing commercial buildings.

Acknowledgments

I would like to thank Bill Demopoulos and Adam Rosenfeld for their generous comments on earlier drafts, and to Eric Mazzi for his lively engagement and patience.

Practice Questions

1. Marcus Young, an energy manager, learns that his employer, Superstore International, recently issued a $400M long-term corporate bond yielding 3%. Marcus' supervisor states that it will make it much easier to approve DERs, which offer returns at twice as high as the bond yield. Describe why Marcus' supervisor may be incorrect with respect to his assumption on the availability of debt capital for DERs.

   Solution: Marcus' supervisor is comparing the cost of debt to the project returns. The relevant hurdle rates would be the weighted average cost of capital for the organization, which may be significantly higher than 3%.
2. Some of Superstore International's stores are in regions that offer Property Assessed Clean Energy (PACE) financing. Marcus determines that

the financing is suitable for several projects at company-owned stores. Marcus' supervisor agrees but wonders what would happen should some buildings be sold as part of a regional consolidation play. Marcus responds that should management decide to sell the building, then the loan would simply be transferred to the future buyer without consequence to Superstore International. Is Marcus correct with respect to loan transferability?

Solution: While it is possible to transfer the loan, it will certainly factor into the sales price during negotiations with the future buyer. Marcus should highlight this potential event to management.

3. Electricity prices have been volatile and rising; therefore, Marcus is debating adopting a state-of-the art rooftop unit package for air conditioning. Marcus is concerned about the price premium, which is only justified if the rooftop units achieve the forecast electricity savings. A local engineering firm offers an energy savings guarantee based on previous energy use at the site. Marcus thinks the energy saving guarantee is reasonably priced but wonders if there are other risks he should consider. State two additional factors Marcus should consider prior to proceeding.

   Solution: (a) Energy savings guarantees may or may not include changes in electricity price. The units could perform as expected, but if prices spike then the store may still see their operating expenses increase. (b) Energy savings guarantees based on historical performance may not carry into the future. Changes in building use or occupancy may invalidate baseline assumptions.

4. Marcus is frustrated after project fail to secure internal financing for a DER, which offered a 10% internal rate of return. Marcus is curious whether the project was rejected for failing to exceed the firms' weighted average cost of capital. After reaching out to a friend in the corporate finance department, Marcus learns the firm is financed with 60% debt at an average borrowing of 4%, and 40% equity at an equity cost of 20%. The corporate tax rate is 30%. Was the project rejected for failing to exceed the firm's weighted average cost of capital?

   Solution: No, the project's internal rate of return exceeded the firm's weighted average cost of capital is 9.68%. Using the formula in Table 6.2: $9.68\% = (0.6) \times (0.04)(1 - 0.3) + (0.4)(0.2)$. Marcus might want to investigate other reasons for the project's rejection, including competition with projects offering even greater returns or technical risks, for example. The weighted average cost of capital is only the floor for project returns.

5. Marcus has been tasked by reviewing a recently acquired portfolio of multi-residential portfolio of buildings. The buildings were constructed during the 1970s and refurbished during the early 2000s. Marcus notes that the buildings are more energy intense than average, and after reviewing a number of energy audits is able to find numerous energy savings opportunities. Marcus' manager reviews the recommendations and rejects most of the capital-intensive opportunities on the grounds that "energy savings go to the tenants so won't impact the companies' bottom line and besides borrowing is off the table." What are the barriers Marcus' managers has referred to?

   Solution: Marcus' manager has referred indicated barriers of split incentives and borrowing restrictions. Split incentives are common in multi-residential buildings where tenants pay their own utility bills. Borrowing restrictions occur when properties have borrowed the maximum their loan covenants permit.

7

Human Behavior and Facility Energy Management

Kady Cowan

Sustainability Studio, Canada

7.1 Introduction: Importance of People and Behavior in Energy Management

Why focus on behavior to save energy? Because people are not going away. Many energy professionals have stories about capital-intensive high-tech energy management projects that do not perform as expected over the long term. They often seek expert guidance to uncover why. Sometimes it is due to the "human side," such as know-how, culture, permission, values, trust, and relationships. Often, these human aspects have not been sufficiently understood or integrated into the project from the outset. Energy professionals in charge of these projects and programs usually do not have training or experience in the high-potential, low-cost realm of human-focused energy management. This chapter will turn a spotlight on human behavior, and how behavior change interventions and programs can help tap into new energy management opportunities that are attractive to facility and business leaders in any sector.

Increasingly, published findings from residential, commercial, institutional, and industrial sectors show that the energy savings from well-designed, implemented, and monitored human-focused energy management programs can save between 3% and 25% (Smith, 2020). Dr. Reuven Sussman, Director of the Behavior, Health & Human Dimensions Program at the American Council for Energy Efficient Economy (ACEEE), has commented in personal communication: "Ignore the people at your peril." Although human-focused energy management can be complementary to traditional energy management practices, we find that the energy-saving potential of addressing the interactions of people, technology, and organizational processes is often discounted or ignored. Here, we would like to demonstrate how even a little attention on the human aspect will help amplify achievable energy savings, avoid energy saving decay over time, and reveal new energy-saving opportunities for the future.

7.1.1 Typical energy management patterns

It is commonly accepted that economic benefits are a primary reason why energy management is important. The assumption is that if cost-effective measures can be identified to save energy and reduce costs in facilities, then decision-makers will act to implement these measures and achieve the cost savings. But how often is this true? Consider these hypothetical but very common examples.

7.1.2 Common energy waste scenarios

1. You are a consultant to an industrial operation. You have identified a pump retrofit project that offers a rate of return of 15% (larger than the company's profit threshold), but your client fails to implement your recommendations.
2. A school facility received a heavy envelope upgrade including wall and roof insulation and new high-performance windows. However, facility energy consumption does not decrease.
3. As the operations manager of a hospital facility, you have determined that the hot water thermostat for washing laundry could be reduced by four degrees, saving thousands of dollars yearly while meeting all hygiene requirements. Yet the cleaning staff decline to act on your recommendation.
4. A major restaurant chain does not train employees to save energy, so they continue to waste energy by keeping freezer doors propped open, putting hot food in the refrigerators, and relying on an outdated HVAC temperature control system.
5. A government subsidy program offers free energy audits and cash rebates to commercial businesses for high-efficiency water heaters. After 6 months of intensive advertising and outreach, hardly any businesses have signed up.
6. A small factory owner had a controls upgrade implemented on a heating ventilating and air conditioning (HVAC) system and expects to see reduced utility bills that do not occur.

Is there something wrong with the people or technologies in these examples? No, there is nothing wrong with either. The decision-makers and employees are exhibiting normal human behavior, which is not always economically rational, at times works against our own best interest, and has difficultly valuing the future. However, most energy-saving interventions are based on the dominant neoclassical economic theory of *Homo Economicus* (perfectly rational economic humans), which incompletely accounts for natural human behavior and often results in measures with attractive financial benefits to "sit on the shelf." The name for this phenomenon in energy management is the "energy efficiency gap." This gap is defined as the difference between energy savings that are estimated to be cost-effective and those that are actually implemented.

Figure 7.1 presents a theoretical illustration of the energy efficiency gap for a facility. Over time, a facility delivers various energy services, such as manufacturing processes, thermal comfort, entertainment, and illumination,

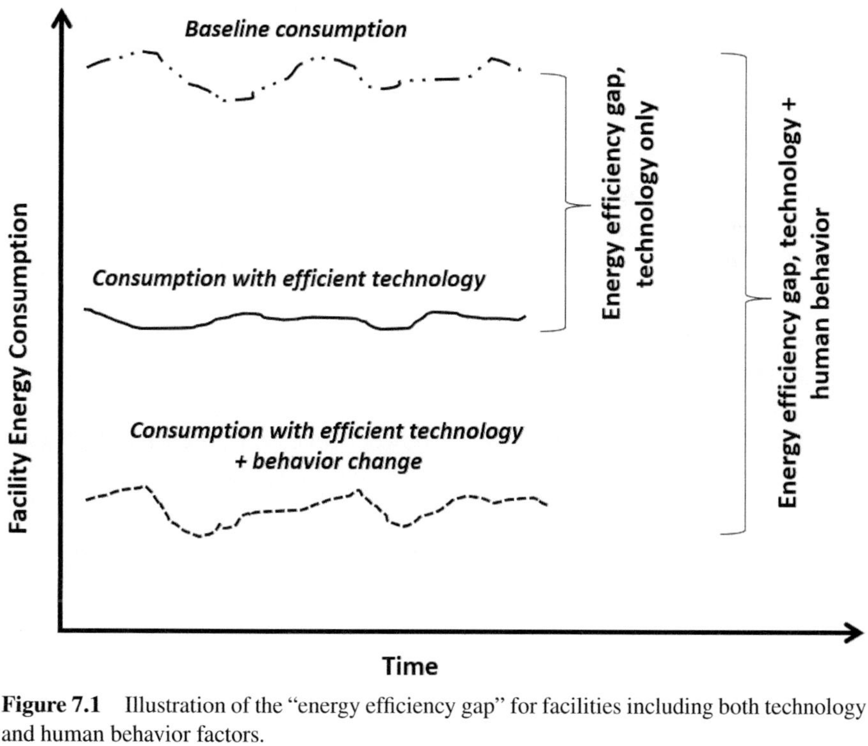

Figure 7.1 Illustration of the "energy efficiency gap" for facilities including both technology and human behavior factors.

and consumes a certain quantity of energy. The efficiency gap traditionally describes energy savings that could be achieved, while delivering the same services, with cost-effective and efficient technology alone. Increasingly, experts recognize that consumption could be reduced even further with a cost-effective combination of technology and behavior change. In other words, what is meant by the "energy efficiency gap" is expanding in the field of energy management to include not only technology utilization but also the influence of human behavior.

It is important to understand that human behavior can work in two directions when it comes to energy consumption. Later in this chapter, we introduce various behavior change actions, and programs that have proven to help reduce energy consumption beyond what is achievable with technology alone. However, some human behaviors erode energy savings, with the result being that the adoption of efficient technologies does not lower energy consumption as much as predicted or, in extreme cases, might even increase consumption. Examples of such behaviors include improper technology commissioning, poor maintenance practices, and rebound effects;[1] the increased use of services once they are made more energy efficient. *Technologies alone do not guarantee savings. The final objective is not implemented technology, it is realized energy savings.*

In order to understand why the energy efficiency gap exists, and what can be done to narrow it, conventional economics and engineering science do not offer enough help. Instead, they can contribute to the problem when they misunderstand, or have an unrealistic understanding of human behavior. Rational economics and engineering treat energy efficiency as a complicated issue that has a solution. They often use the *deficit model*, working on the premise that something is missing, and that if you provide more information or more financial incentive, you will get behavior change. Instead, energy management is a complex issue, which does not have one best solution, may not be resolved with computational power alone, and often requires a combination of interventions based on contextual understanding and often multi-stakeholder collaboration.

The key is recognizing that a facility energy system includes not only the physical and technological systems and organizational processes but

---

[1] "Rebound Effect" – is an increase in energy efficiency that leads to eroded energy savings through a behavioral response, for example buying a fuel-efficient car and driving more miles each year.

also the people. People function in complex social systems, in addition to their interaction with the built and mechanical infrastructure. These people include the individuals and groups who live and/or work in the facility, or participate in the design, construction, operation, maintenance, and other activities occurring within the facility. The facility energy system includes all the physical–mechanical–human elements where each part affects the others because they are interdependent. Furthermore, the actions of the people can have a very discernible impact on facility energy performance. The people involved with the life of a facility make energy management decisions that may appear irrational according to conventional economics and engineering ways of thinking. However, the people are the reason facilities were created in the first place; they are part of the overall system. Their understanding and preferences about facility operations need to inform energy management solutions for long-term results.

7.1.3 Social science methods

Not all information can be reliably and effectively put into numbers. Social sciences such as psychology, sociology, and anthropology use qualitative data to understand *why* something happens, not just what happens. In energy management, qualitative data helps us to understand the circumstances, identify leverage points for change, describe how well a behavior change program is working for specific audiences, and/or how the program might be improved over time. Four common qualitative methods are used: exploratory surveys, interviews and storytelling, focus groups, and observation (ethnography). Qualitative methods can use more open-ended questions to gather facts and perspectives that may otherwise not be discovered; for example, if a behavior program has a positive or negative effect on other factors beyond energy consumption, such as work productivity or sense of teamwork. Using qualitative methods can help reveal these perspectives. In addition, critical information such as barriers to energy-saving behavioral actions can be uncovered with qualitative data-gathering. Shifting focus to include people, their motivations, and the quality of their interactions and relationships with energy-efficient technology, processes, and other people deepens the understanding of energy management opportunities. The remainder of this chapter will provide an overview of the core concepts, real-world examples, and practical techniques for integrating human-focused energy management into practice.

7.2 Defining Behavior, Behavior Change, Actions, and Programs

Energy management professionals will benefit from becoming familiar with general concepts of behavior and behavior change, including why a human-focused approach matters, how it works in practice, and what tools and resources help to make a difference. Exploring the connection between behavior and facility energy management can improve energy savings, as well as minimize building complications.

7.2.1 What is behavior and behavior change?

The broadest definition of behavior is the response of an individual, group, or species to its environment. With this definition, it becomes clearer how all-encompassing behavior can be and, thus, how behavior can be an important clue about the environment and circumstances. In personal communication with Mithra Moezzi, Ph.D.,[2] she says: "Circumstances convey a system-oriented view of things and acknowledge that people do the things the way they do because of the situation they are in." This is especially true in a workplace setting. *As energy professionals, it is helpful to know our task is not to change the people, but to change the environment or wider circumstances, so that the desired behavior may follow*. Behavior change in this context refers to changes in human actions, which were directly or indirectly influenced by a variety of programs and interventions. The term for this work is *energy behavior*; practices (e.g., habits, protocols, skills) and structures (e.g., policy, training, equipment) that affect the way individuals and groups interact with energy systems and services.[3] Incorporating behavior and behavior change into all energy management opens up opportunities to learn about your circumstances and discover ways to change them; so additional energy savings are realized.

7.2.2 Energy ecosystem: physical–mechanical–human interdependence

Another helpful term for understanding energy behavior in the wider context is *energy ecosystem*. Like ecosystems in nature, the whole is greater

---

[2] Mithra Moezzi, Ph.D., Anthropological Folkloristics, California Energy Commission, Energy Research Development.

[3] Definition from Rotmann & Mourik (2013, p.107). https://userstcp.org/wp-content/\uploads/2019/12/26.Task24_Phase1_eceee_paper_2013_1-183-13_Rotmann.pdf

than the sum of its parts. Energy ecosystems account for quantity and *quality* of interactions, relationships, and coordination between the interdependent physical–mechanical–human elements. A mindset that appreciates the whole helps us understand, and then value, the diversity and responsiveness required to produce a healthy and productive energy ecosystem. Recognizing the roles, skills, and experiences of human actors in the facility energy ecosystem expands our view of the system overall, and includes anyone who is impacted, involved, or who cares about facility performance. Human actors may include facility owners and administrators, boards of directors, senior management, engineering designers, facility and equipment operators, facility occupants, technical service providers, equipment vendors, community representatives, as well as the utility companies that power the facility. Knowing who is affecting energy outcomes at all levels allows us to coordinate their efforts to achieve results (Rotmann, 2016).

7.2.3 Energy behavior actions versus programs

There is an important distinction between actions and the programs that lead to these actions. Programs are the various techniques and organizational rules that are intended to change the environment and influence people to increase the occurrence of desired, energy-saving actions. They may include but are by no means limited to: a policy review process, training and education, facility audits, access to incentives, social norms, and feedback. Energy behavior action is the act of carrying out the tasks that save energy and can occur at any stage of energy management. For example:

- *Purchasing*, ongoing and repetitive spending such as OPEX (operational expenditure) focused on low-cost technologies such as light bulbs and faucets. When purchasing protocols specify energy-efficient light bulbs or low flow faucets that is an energy behavior.
- *Investment*, which may be one-off or time-bound larger scale spending requiring CAPEX (capital expenditure). For example, buying a variable speed gas compressor versus single speed, and choosing condensing water heaters versus conventional appliances and retrofit projects.
- *Installation and commissioning*, by verifying set-points and ensuring components and sub-components are sized accurately.
- *Ongoing operations*, by utilizing enhanced control schemes such as temperature setback and matching equipment schedules to run time needs or turning off after hours.
- *Maintenance, cleaning, and repair*, by changing HVAC filters, cleaning coils, tuning up heating and cooling equipment, and tagging steam leaks.

- *End use by activity substitution*, such as wearing a sweater versus turning up the thermostat; arranging furniture strategically; sweeping sawdust instead of blowing it away using compressed air.

The reason the distinction between actions and behavior change programs is important is that the desired actions do not occur as often as we would like without some program or intervention designed to influence people's behavior. The idea that sharing information about energy-efficient actions is enough to get people to actually carry them out is generally false. You may have experience putting up a sign or sticker to "turn off" some equipment, with limited results. Access to information such as the sign or sticker alone is not enough to result in desired behaviors consistently. Without the intention to turn off, the reminder (sign or sticker) is practically invisible to people. This is not because people are foolish or like to waste energy, but because up to 95% of our energy use is habitual or routine, and habits are the hardest to break.[4] If you approach energy management with the expectation that access to information alone will increase desired behaviors, you can expect to be perpetually frustrated. People do not always make choices in their best interest. Next, let us identify some typical behavioral programs and actions that are proven to save energy; later in this chapter, we will introduce some techniques that help influence people to carry out the targeted energy-saving actions.

7.2.4 Role of SEM – strategic energy management programs

One approach that values the role of humans in the energy ecosystem is strategic energy management (SEM). SEM is based on the ISO 50001 Energy Management Systems standard and requires companies and organizations to concentrate on systematically reducing energy waste by focusing on the details of their energy ecosystem. Most SEM techniques focus on building energy teams, continuous improvement, training and engagement, and operations and maintenance. SEM programs can be very useful to move from entirely technology-focused energy management to more holistic energy management, producing up to 23% in electricity and natural gas savings in commercial and industrial buildings (Stern, 1992).

One of the longest running and most successful SEM programs with a specific focus on behavior change is in British Columbia, Canada. It

[4] Outlined in DEFRA (2011). Habits, Routines and Sustainable Lifestyles. Summary report. 71 pp.

is a partnership between BC Hydro, Fortis BC, and Prism Engineering. Since 2007, these utilities have been running workplace energy-efficiency campaigns emphasizing the cohort model, providing workshops, ongoing coaching, incentive funding, step-by-step instructions, and deployable communication tools for participants. Program participants have demonstrated that successful behavior change campaigns can achieve between 2% and 10% energy savings depending on the end-use, building and sector type (Smith, 2020). Despite the known opportunities of behavior change, in general, energy behavior remains a small component of traditional SEM programs. For the curious energy professional, supplemental techniques, like those described throughout this chapter, are necessary to achieve the full potential of human-focused energy management programs.

7.2.5 Behavioral actions proven to save energy in every sector

Listing all the potential behavioral actions proven to save energy in residential, commercial, institutional, and industrial facilities is impossible. Not only because there are too many but because they keep improving and evolving. Provided here are a sample from one published field experiment, one scenario analysis, and two case studies.

Residential: Results of a widely cited "behavioral wedges" study published in the U.S. Proceedings of National Academy of Sciences studied residential weatherization (weather stripping), equipment upgrades (e.g., water heaters), maintenance (e.g., HVAC filter replacement), adjustments (e.g., selecting hot/cold water temperature), and daily behaviors (e.g., use of a clothesline) and estimated that these "low or zero cost" actions could save energy in U.S. households that would reduce greenhouse gas emissions by 123 million metric tons, which is greater than emissions of the entire country of France (Dietz, 2009). Although energy behaviors may seem like small stuff, when they are coordinated across large groups and systems, the impacts can often surpass our own expectations.

Commercial: A 2015 modeling study titled *Behavior-Based Energy Saving Opportunities in Commercial Buildings: Estimates for Four U.S. Cities* (Ehrhardt-Martinez, 2015) found that commercial buildings such as office, education, food service, and retail facilities could achieve at least 7% energy savings by applying various combinations of 91 specific actions in nine categories: space heating, space cooling, ventilation, water heating, lighting, cooking, refrigeration, office equipment, and computers.

University Health Network in Toronto

Going back to the early aught's (2000s), the Energy and Environment team with the leadership of Ed Rubinstein made it a priority to integrate physical, mechanical, and human elements into all sustainability policy and programs. An early win for UHN was installing LED lamps across all campuses before LEDs were the industry standard. The success of this human-focused initiative is described in the 8th edition of this book and was based on tactics of listening to electricians, sharing dollar saving data with finance, and cultivating trust and then investment in new technology with small tests and trials in high-profile spaces like the office of the CEO. Building on that success, UHN is demonstrating energy behavior leadership again by scaling up large applications of district energy.

By helping all actors in the energy ecosystem make sense of energy behavior challenges, UHN is getting faster board-level approval for larger and larger energy projects, improving contract negotiations with energy vendors, and is proving that permanent staff focused on energy is critical. The Energy and Environment Team support facilities staff, who are often overwhelmed and tend to operate in more reactive modes (Cowan, 2017). Dedicated energy staff have the capacity to take the complementary longer, strategic view. This combination of skills and methods is the basis of UHN's overall energy behavior strategy. In general, UHN expands their perspective and sources of input to identify the broadest array of challenges and problems. Then they narrow their focus to prioritize issues and develop specific solutions that meet their needs. Once they are happy with the results, they return to an expanded view to see if the solution can be applied more broadly where there are similar circumstances across the organization. *This approach is discussed in more detail in Section 7.4.1, Behavior Program Design Phases: Discover, Define, Develop, Deliver, Evolve.*

An example of how this approach works at UHN is from former energy manager Chad Berndt. His focus on a wide range of facilities showed that one building was not performing as expected. After narrowing his focus and investigating the issues and energy data in detail, he determined one particular chiller was the problem. Returning again to the expanded view with the team he brought together, including building operators, finance, and energy system vendors, many potential solutions were considered. This view revealed that a district energy system based on deep lake cooling[1] would improve the building energy performance as well as meet financial, climate resilience, and reliability criteria most effectively. Once the process was seen to work well at one site, district cooling was applied at two additional hospitals with more applications in the works.

UHN's appreciation for complexity, using simple methods and focusing on the people (decision-makers, building operators, and occupants), has had an incredible impact on both energy and behavior. The first deep lake cooling project will save $22 million over 20 years, reduce water consumption by 65 million liters, save 7 million kWh, and reduce greenhouse gas emissions by 63% and simple payback of 4.2 years. Not only that, the system provides 30% more capacity for new equipment and renovations (extremely frequent in healthcare) and better control of cooling in the hospital. What is particularly exceptional about this story is that the people involved were listened to, their concerns were addressed, and, as a result, district energy, which had been turned down by the board several times in the past, is now being considered at every UHN campus. This example is illustrative of how mechanical solutions rely on human behavior along the way to meet their full potential.

Institutional: One of the first organizations in North America to publicly embrace human behavior in facility energy management is University Health Network (UHN) in Toronto, Canada.

Industrial: Staff at SierraPine Composite Solutions, a manufacturer of fiberboard and particleboard in Oregon, did not think they would find any operational or maintenance energy savings in their facility. Once they started working with the Energy Trust of Oregon, who helped them create an energy team, provided targeted energy training, and completed facility walkthroughs and audits, they found that improving two 6,000-horsepower refiners, the plates at the heart of the company's production, they were able to cut annual energy costs by about $400,000 with no impact on product quality. By focusing on the people, they found additional energy savings by shutting down underutilized equipment and adjusting equipment schedules to minimize idle time. Overall SierraPine saved $588,700 or 12% of their annual energy costs (Northwest SEM Collaborative, 2021), well beyond any of the staff's expectations.

What is common between these stories is that by tapping into the existing human potential on site, energy projects are more successful and lead to additional energy-saving opportunities.

7.2.6 Specific targets: audience and behavior

One of the key attributes of behavior change is that the audience and associated behavioral actions that you target must be specific and clearly defined. Seldom, if ever, can we expect 100% uptake of energy-saving actions by all energy users, but the more specific you can get, the greater the likelihood that uptake will be sufficient. One useful way to categorize behavioral actions is to group them into one-time actions that are performed more rarely and more consciously (e.g., buying a boiler), and habitual actions that are performed more frequently and often less consciously (Mourik et al., 2015) (e.g., turning off equipment). Additionally, Dr. Sea Rotmann,[5] one of the leading international researchers on energy behavior, has found that energy behavior actions typically fall into the following categories:

- Financial decisions (e.g., investment vs. purchasing)
- Technological changes (e.g., hardware vs. software)
- Curtailment behaviors (e.g., habitual vs. one-off energy efficiency efforts)

[5] More work by Dr. Sea Rotmann is on her website. https://sustainableenergyadvice.org

- Equipment-related (e.g., maintenance/cleaning vs. repair)
- Policy interventions (e.g., rules/processes vs. regulation/legislation)
- Communication (e.g., training/education vs. mass marketing or top-down information)

One reason to clearly define and categorize energy behavior actions is that it helps to determine who your target audience should be. Often different actors from different parts of the organization are responsible for different actions. This is true in households, commercial buildings, as well as industrial facilities, where different individuals or unique staff roles have responsibility for purchasing, installation, operation, and maintenance of equipment. Using the energy "behavioral socio-ecosystem" or energy ecosystem for short accounts for the interaction between all people, the infrastructure, and technology in the environment (or circumstances). With this perspective, all actors and departments should be accounted for in energy management decisions because they all affect final energy performance. But for energy behavior to be successful, aligning target behaviors with the target audience is the first step (Rotmann et al., 2020).

7.2.7 Specific targets: tactics and interventions

A fully developed energy behavior program is one where the context and circumstances for change are fully understood. This understanding helps to reveal points in your energy ecosystem where a targeted tactic called an intervention can be designed and implemented to influence the practice (e.g., habits, protocols, skills) and structures (e.g., policy, training, equipment) that will lead to energy-saving actions and additional co-benefits (also called non-energy impacts, NEI) such as: improved comfort, longer equipment life, or better air quality. There is an almost endless list of possibilities and past experiences that help guide the choice about intervention design. In this section, we will describe the most common human behavior and facility management interventions. A short synopsis of *Key Behavior Tactics for the Energy Sector* (Illume, 2015) provides additional intervention options for consideration.

Local energy champions use social influence to affect change. Individual champions influence others' behavior by what they say and do. Energy champions set an example of social norms, which are opportunities for others to compare themselves within their own community to learn what is possible, normal, and approved of (Sussman et al., 2016). Energy champions often, but not always, have a personal interest in energy management. Energy champions are usually well-liked or respected within their department or community, and are outgoing or interested and willing to help others. Finally, energy champions help others make sense of energy management by listening, sharing, educating, and connecting. This works because local champions provide information as a familiar and trusted person, in a style that is relevant to the audience. Energy champions impart an implicit social pressure because others tend to care about the energy champion as an individual and are interested in preserving their social connection (e.g., friendship). Thus, individuals who do not necessarily care a lot about energy savings may decide to take energy-saving actions more willingly because it is important to someone who is important to them, the energy champion. The ability to identify internal champions is a sure sign of success as local energy champions are especially important because they act as a route for rolling out subsequent behavioral interventions. Internal champions can lead energy management for their department, help to identify new target energy behaviors, and keep energy management as a priority for all actors across the organization.

Energy feedback interventions are effective when recipients understand the information being presented to them. Energy feedback in real time through (sub)metering, building automation systems, smart technology, web portals, or social media can encourage behavior change if the recipient of the information already knows what action to take as a result of the feedback. More utilities now provide feedback in real time with access to smart meter data, various tariffs, and rate structures like time of use, or demand response and other flexible demand incentives (see Chapter 4). Additionally, companies have access to powerful software tools and models to support energy management programs. Delayed feedback such as monthly utility bills or annual performance evaluation is still useful to recipients because they encourage reflection and learning that links previous actions to energy outcomes. For example, a building operator who adjusts set-points in the summer may be able to learn how much energy that action saved through utility bill feedback. However, delayed feedback often relies on someone remembering what happened sometime in the past and therefore is not always as beneficial as real-time feedback. Feedback also improves energy literacy skills such as interpreting energy data. Energy feedback is most effective when it is personalized, provided by a trusted source, and is provided as close to the action as possible (e.g., real-time).

> Key Behavior Tactics for Energy Management
>
> *Commitments* influence behavior by asking participants to pledge (preferably publicly) either in a written or verbal format to adopt an energy-saving or carbon reduction action. This can include a wide variety of actions such as: signing up to formal programs such as ISO 50001, turning down the thermostats over night, or meeting with a manager to talk about energy savings at work.
>
> *Appeals* influence behavior by providing access to repetitive, consistent messages to participants over time, which serve as a reminder when the participant is ready to take action.
>
> *In-person interaction,* as opposed to digital communication, influences behavior by sending a program representative, e.g., an energy advisor or auditor, to speak with a participant face-to-face, to provide reminders, feedback, appeals, or education.
>
> *Rewards or gifts* such as incentive dollars or subsidies influence behavior by providing certain extrinsic benefits (coming from outside oneself) in exchange for completion of desired actions. Extrinsic benefits are useful to get people's attention at first, and run the risk that once the reward or gift is gone, so is the behavior. In general, it is better to strive for intrinsic motivation (coming from within oneself), such as a sense of accomplishment, for behavior change to persist.
>
> *Social norms* influence behavior by comparing participants to other members of their community. Usually focused on energy savings or attitudes about energy savings, individuals can self-determine if they are "keeping up with the pack." Home energy reports delivered on monthly bills are one of the most popular examples of social norms. Other examples include competitions and benchmarking.
>
> *Celebration and recognition* are simple and cost-effective means to build and keep the momentum for programs going. Providing meaningful recognition to the team, audience and decision-makers can go a long way to preserve trust, build relationships, and cultivate engagement. Celebration and recognition need not be fancy; being celebrated and recognized by those you respect in a way that is personally meaningful is often enough.

7.3 Energy Behavior Successes

Where there is proven energy behavior change, there is often an energy manager or champion at work. Behind all three of the following success stories was a talented and ambitious energy manager who would not take "no" for an answer.

7.3.1 Unifying fragmented building automation systems, City of Markham

The City of Markham in Ontario, Canada, with the guidance of their award-winning energy manager Amanda Martin and BAS Coordinator Aaron Cheung,[6] led a multi-year project to unify and improve the City's fragmented collection of building automation systems (BASs). Like many large organizations, the City of Markham had BASs from many different vendors, and multiple vintages, effectively resulting in dozens of different types of systems.

Their approach started with the simplification of operations by creating an updated BAS design specification document that mandated a reduction in the diversity of BASs. The first major application of this new standard was the retrofit of a community center. As a result, this City is saving 260,000 kWh/year electricity and 90,000 m^3/year natural gas and they are just getting started. The equipment replacement process started with the following steps:

- An energy audit and subsequently incorporating audit improvement findings into the BAS design.
- Using the new standardized BAS design specification.
- Unifying on-site BAS into one centralized, web-based system.
- Re-evaluating equipment to prevent like-for-like replacement, which encouraged higher-efficiency, and right-sized equipment replacements.
- Applying for utility incentives to offset incremental capital cost.
- Finally, completing post-project monitoring and verification.

At first, people were not that receptive to installing energy-efficient equipment. Amanda used a multi-prong approach to engage them and get them on side. She invested the time to visit other facilities with upgraded equipment, and she talked with contractors and staff to help them feel comfortable with the proposed changes. In her role as energy manager, she acted as an internal consultant where she was able to identify the external expertise required, apply for incentives, and help the facilities staff get better products. This effort enabled facilities staff to focus on linking BAS operations with energy expertise for the long term. Because of her consistent approach and success, the City of Markham now has facility staff regularly asking for energy efficiency in all electrical and mechanical replacement projects.

---

[6] For more information watch City of Markham 2018 Energy Manager of the Year video, www.youtube.com/watch?v=uKTdSS-PU1o

7.3.2 User-centered design, Price Fitness Center at the Presidio of Monterey

The gym at the Price Fitness Center has ample windows for natural light and yet the lights were kept on all day. The facility's energy manager knew there had to be a better way. If the lights could be turned off, they would eliminate a 5-kW load and save $2,500/year.[7] The problem was, it was impossible for gym users to find the light switches because they were located 40 feet away in a staff break room.

To facilitate energy savings, the switches were moved closer to the gym entrance, and clearly labeled so that users would know which lights corresponded to which courts. Staff were required to keep lights off as often as possible and prompts (stickers) were placed on the switches as reminders to "conserve energy." The results of this simple but effective project confirmed that relocating the switches reduced energy use by 24%. There was an additional 6% energy saving after the reminders were added to the switches. Now, gym users always have ample light to shoot hoops without any energy waste.[8]

7.3.3 Make energy management visible to all levels of the organization, New Afton

Energy efficiency projects often result in more than just energy savings. At New Gold New Afton mine in Kamloops, British Columbia, their previous energy specialist, Andrew Cooper, led the mine through ISO 50001, Energy Management Systems certification. He began the process from the ground up, working with operators to implement a number of energy projects to demonstrate the effectiveness of energy management in cost savings, the potential to increase production, and other NEIs. This approach was beneficial in making the case to senior leadership about the value of pursuing ISO certification.

Achieving ISO 50001 certification from a trusted third party has helped demonstrate that the mine is serious about energy management. The focus on strategic energy management (SEM) has helped to realize significant cost savings as well as production improvements. As a result of the energy performance improvement initiatives, New Afton is realizing cumulative energy savings in excess of 10% of its annual energy consumption. Energy efficiency

[7] Printed with permission from U.S. Army Corps of Engineers, Construction Engineering Research Lab, Department of Defence. Website not publicly available.

[8] Source: www.milsuite.mil/university/energy_resilience_culture

projects like a floatation blower improvement also increased productivity because it improved airflow on the production circuit. They achieved better safety, because operators do not need to manually adjust valves anymore and reduced environmental health impacts, such as equipment noise reduction. In addition to the energy efficiency and production benefits, the mine is also paying lower energy bills, which are typically a mining company's second biggest expense (after labor).[9]

7.4 Energy Management Traps: Cognitive Biases and Psychological Barriers

This chapter about the role of human behavior in energy management acknowledges that people act the way they do because of the situation they are in, the experiences they have had, their beliefs and values, as well as subconscious habits. To better understand the circumstances of people, it is important to be familiar with some of the psychological theories that explain participation in, or aversion to, energy management actions. We acknowledge that this topic may be especially difficult for energy managers who often do not have training, skills, or practical experience in applied psychology. The intention of this section is to introduce some of the most common behavioral science concepts to ease energy managers into familiarizing themselves with the language, and how the most common cognitive biases and psychological barriers play out in energy management practice.

7.4.1 Cognitive biases, making sense of the world around us

Cognitive biases are the fundamental thought patterns that help us understand and judge the world around us. Research has shown that cognitive biases play a key role in people misunderstanding how much energy they use, and poorly estimating how much can be saved through common actions (Attari et al., 2010, Dietz, 2010). Understanding common cognitive biases provides insight for practitioners to be more effective in applying behavior change to improve energy management. It is helpful to understand your own cognitive biases when designing and implementing behavior change programs, as well as biases of the target audience. Qualitative social science approaches from behavioral economics, psychology, and sociology help practitioners notice

---

[9] Source: www.prismengineering.com/other_resources/case-study-for-new-gold-new-afton-mine-iso-50001\

Table 7.1 Cognitive biases – how they are relevant to energy management.

Cognitive bias description	Relevance to energy management	Example
Status quo bias: Preference is given to environments and situations that stay the way they already are.	Reluctance to adopt new technologies that can save energy and money. Established behaviors do not change without a significant incentive.	Procurement officer insists on double pane windows when higher-performance windows are available.
Framing: People make different decisions, depending on how information is presented to them.	The same facts presented in different ways can impact the adoption of energy-saving behaviors.	Sharing statistics about how many people *are* turning off lights at work is more encouraging than how many *are not* turning off lights.
Loss aversion: People are affected more by a loss than a gain. The fear of losing money is greater than the joy of gaining money.	Presenting energy-efficient opportunities as protecting what you have already got may increase adoption.	Presenting a heat pump purchase as preventing a $500 annual loss induces more buyers than presenting it as a chance to profit $500 per year.

patterns in individuals and groups that reveal non-obvious cognitive biases that may get in the way of effective or persistent behavior change.

At any time, there can be dozens of cognitive biases[10] at work in the background shaping our preferences, aversions, and perceived risks. In Table 7.1, we define just a few of the most common cognitive biases involved in energy management decisions.

7.4.2 Psychological barriers to adoption of energy-efficient actions and programs

The reader may begin to notice that there is overlap between cognitive biases and psychological barriers. Some psychological barriers are a direct result of cognitive biases, such as the status quo bias (preference to preserve what is currently in place), and framing (how information is presented). In other cases, psychological barriers that prevent people from taking energy-efficient actions may arise from other circumstances such as: perceived risks, ideologies, mistrust, limited thinking about the problem, lack of energy literacy (Sussman et al., 2016), and learned helplessness.

[10] For more examples please review the cognitive bias codex https://commons.wikimedia.org/wiki/File:Cognitive_bias_codex_en.svg\

For many energy management professionals, familiarity with these concepts helps with understanding why cost-effective and energy-efficient interventions often remain "on the shelf." On first glance, some barriers, such as access to capital, may appear to be structural. In fact, this is precisely the opportunity for an energy manager to make progress. If a decision-maker is not acting on an energy efficiency investment that appears to be cost effective, is it because the person truly has no access to the capital? Or are there psychological barriers at work?

A comprehensive review of all potential barriers and detailed descriptions are beyond the scope of this chapter. The reader can refer to articles by lead authors Sorrell and Thollander in the bibliography for further details. In Table 7.2, we define just a few of the most common psychological barriers involved in energy management decisions.

Table 7.2 Common psychological barriers to the adoption of energy-saving technologies and behavioral actions.

Psychological barrier description	Relevance to energy management	Example
Learned helplessness: People expect that nothing they do will change the circumstance because every effort they made in the past was stalled or prevented.	Everyone is aware of major energy loss from equipment but seems like nothing ever gets done about it.	Maintenance teams stop reporting steam leaks because whenever they have in the past, nothing was ever done.
Diversity aversion (heterogeneity): Preference for keeping solutions the same across different circumstances.	Interventions that work well in one facility with one specific audience do not necessarily scale up, as expected, in other similar facilities.	Due to various factors such as different activities or preferences of building occupants, interventions may need to be customized or tailored to get the same results across various facilities.
[3pt] **Hyperbolic discounting:** Smaller, immediate rewards are selected over larger, delayed ones.	The reward of energy savings in the future is perceived as less valuable than the instant gratification of the lowest first cost.	Purchasing a less expensive and less energy-efficient boiler today is more appealing to decision-makers because it is cognitively more difficult to value energy and dollar savings that will occur in the future.

7.4.3 Examples of common real-world biases and barriers

All actors involved in energy management will benefit from being aware of common biases and barriers. In many cases, there are market barriers such as split incentives or hidden costs that can compound or strengthen the impact of psychological barriers. Being familiar with psychological and non-psychological barriers so that they can be reduced or eliminated assists to increase realized energy savings. All of the major biases and barriers of demand-side energy management policies and practices directly or indirectly involve human behavior. Let us return to those common, hypothetical examples in the introduction of this chapter and list some potential biases and barriers that might be at work.

1. You are a consultant to an industrial operation. You have identified a pump retrofit project that offers a rate of return of 15% (larger than the company's profit threshold), but your client fails to implement your recommendations.
   - **Risk aversion**: There might be a maintenance manager who is concerned that the new pumps may not perform well during system start-up. The risk may be real or perceived. In any event, this barrier should be addressed – this is why qualitative methods like focus groups or interviews or site visits with operators or users are very useful to find out "why" something is not happening.
   - **Hidden cost**: The buyer may have past experience with a pump vendor where the installation costs were double what was proposed because of delays or added training requirements.
2. A school facility received a heavy envelope upgrade including wall and roof insulation, and new high-performance windows. However, facility energy consumption did not decrease as expected.
   - **Rebound**: Unbeknownst to the energy manager, the building operator increased the space heating temperature settings after the retrofit in order to reduce the number of complaints about cold rooms in winter.
   - **Split incentives**: The teachers in the school (who do not pay the utility bills) perceived the ventilation to be inadequate after the upgrade, and began leaving windows open much more often.
3. As the operations manager of a hospital facility, you have determined that the hot water thermostat for washing laundry could be reduced by four degrees, saving thousands of dollars yearly while meeting all hygiene requirements. Yet the cleaning staff declined to act on your recommendation.

- **Trust and credibility**: The cleaning staff hold chronic distrust of management due to a variety of issues unrelated to energy, and therefore do not believe it would save as much money as stated.
- **Inertia/competing priorities**: Operator staff have been happy with the performance of the laundry service, and they have other priorities.

4. A major restaurant chain does not train employees to save energy, so they continue to waste energy by keeping freezer doors propped open, putting hot food in the refrigerators, and relying on an outdated HVAC temperature control system.
   - **Form of information:** People generally prefer information that is simple, clear, specific, and personal.
   - **Status quo bias**: Established behaviors do not change without a significant incentive.
5. A government subsidy program offers free energy audits and cash rebates to commercial businesses for high-efficiency water heaters. After six months of intensive advertising and outreach, hardly any businesses have signed up.
   - **Framing**: The marketing was presented in a complicated fashion that did not make the business proposition and call to action clear.
   - **Principal-agent relationship**: Facility owners have authority to purchase water heaters, but they delegate all facility decisions to the property management firms that do not control capital purchases, and they pass on utility costs averaged into the leases.
6. A small factory owner had a controls upgrade implemented on a HVAC system and expects to see reduced utility bills that do not occur.
   - **Diversity (heterogeneity)**: The controls upgrade was standard for an institutional building, and was not customized for this factory where the office space cooling loads include substantial waste heat from large furnaces within the factory.

7.5 Introduction to Behavior Change Program Design: Seeing Patterns in Energy Management Behavior

Implementing programs that acknowledge and account for human behavior is a relatively new dimension of energy management. Most of the theory and a majority of the programs developed to date have been aimed at the residential sector and the individual energy behaviors residents engage in at home. In fact, commercial, institutional, and industrial sectors (and especially small-to-medium businesses, SMBs) have been underserved, under-researched, and considered hard-to-reach with regard to energy behavior applications (Rotmann et al., 2020). Accounting for patterns in energy management behavior, if done effectively, can produce greater energy savings that persist for the long term compared to when human behavior is ignored. This section will provide energy management professionals with some of the basic steps required to observe energy management patterns and roll out energy behavior programs and interventions in the workplace. The guidance in this section is complementary to the overall energy management program design presented in Chapter 1.

Four principles of behavior change help to frame the program design process[11] (Karlin et al., 2021). Being comfortable with them will make program development and implementation smoother, and increase the likelihood of success.[12]

1. Let context be your guide. There is no one best way or one-size-fits-all approach for behavior change. Solutions emerge from your specific organizational context, which requires being open to new perspectives, listening, noticing, and learning. Taking the time to observe and understand the circumstances, relationships, drivers, and barriers in your energy ecosystem results in appropriate design and successful programs.
2. Focus on designing a continuous improvement process and not for one particular outcome. This will set the direction you want, but not limit how you get there. Designing for continuous improvement incorporates a feedback cycle and willingness to change.
3. Make participants feel valued by designing solutions collaboratively. This creates trust and results in desired behaviors that persist for the long term.
4. Change takes time, so please be patient. Take the time to observe the system and get to know the people in order to better execute principle #1.

---

[11] https://seechangeinstitute.com/wp-content/uploads/2021/08/Building-Blocks-of-Behavior-Change.pdf

[12] The first known training course for energy behavior and facility energy management is described in the paper by Rotmann and Karlin (2020) in the bibliography.

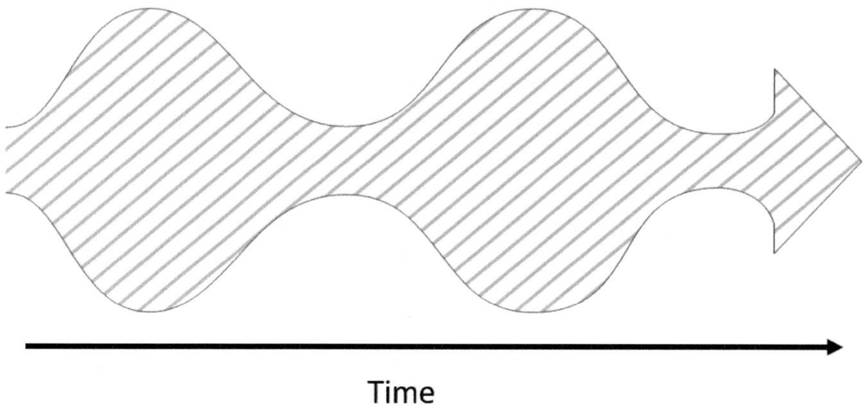

Figure 7.2 Wave technique: Expanding and narrowing focus to clarify challenge and solution.

7.5.1 Behavior program design phases: discover, define, develop, deliver, and evolve[13]

Energy management is a complex process that emerges as a result of human interaction with physical and mechanical energy systems and organizational structures. Breaking behavioral program into phases helps to reveal what may be clear within these complex systems. The approach presented here based on British Design Council's discover, define, develop, and deliver model provides an overview of the steps involved in behavioral program definition and execution and describes success beyond energy savings alone. In addition to the four phases in the model, we have added the evolve phase. Each has unique characteristics, but all comply with the simple wave technique pictured in Figure 7.2, which helps to visualize the process over time as one of emergence. Expanding possibilities and options by casting a wide net, followed by a period of convergence, narrowing possibilities, refining options, and focusing on a specific target like first described in the UHN example in Section 7.1.5. This process is repeated as many times as necessary to gain the insight and clarity needed to define the right problem and execute the most appropriate and durable solutions.

- **Discover:** Scan to understand the needs and motivations of people in your energy ecosystem. In this phase, you will focus on the unique needs

[13]Based on the double diamond design framework from the design council, www.designcouncil.org.uk

and circumstances to ensure you are working on the right thing. Do not assume you know what is possible until you start including others in your facility. Do the research, including quantitative and qualitative aspects. Use observation, surveys, interviews, audits, utility data, and sub-meters. Decide on a shared goal and or objectives.

- **Define:** Focus by narrowing down what you have learned into themes and patterns. Identify key individuals who want to work with you. Decide together what must happen for the program to be successful. Select your target audience, and determine how they can be included. Clarify your target behavior(s) by focusing on high impact actions or easy wins with as few barriers as possible. Identify your target tactics based on which interventions are likely to work best. Determine your target metrics, such that they are credible and meaningful to all participants as well as the overall organization.
- **Develop:** Gather baseline measurements including metrics beyond energy such as staff engagement, reduced occupant complaints, reduced emergency service calls, reduced maintenance costs, more investment in energy projects, and number of senior leaders asking about energy. Start to experiment with prototypes (impermanent and inexpensive tests), where you can quickly refine your ideas into tangible interventions based on feedback. In this phase, you are striving to "fail fast," and to land on potential solutions or a proof of concept that you can expand on in the next phase.
- **Deliver:** Action starts with implementation. Monitor what is working and what needs to be changed by providing feedback and asking for feedback from all actors. At this point, new target behaviors that are adopted and persist are a clue that you have changed your energy ecosystem in the direction you were intending. Continue to provide support for the early adopters and make time to talk about the program as often as possible. Monitor co-benefits such as reputation of the program, clear communication channels between staff, routine procedures being in place to retain savings, and extended equipment lifetime. Keep an eye out for potential costs or unexpected barriers that may arise, e.g., more onus and pressure on the energy champions, rebound, greenwash, more red tape, or lack of persistence.
- **Evolve:** In order to build on your success, it is critical to look back without judgment to separate the signals from the noise by identifying who is enthusiastic about the program, and if you have achieved systemic or long-term change. Gather quantitative and qualitative measurements

at regular intervals to help confirm if your program is moving in the right direction and providing the insight you need for improvement. This is a chance to check assumptions and expectations again, make changes, set new targets, recruit new actors, and share results with new audiences.

7.5.2 Skills and experience that support large-scale change

Techniques described by Westervelt in 2014 will help program managers focus on the interaction and interdependence of the physical–mechanical–human elements of facility energy management. In brief, they include forming a team, discussing objectives, barriers, drivers, and priorities, building relationships, facilitating communications, and assisting in the establishment of new behaviors.

- **Establish an energy management team:** Multiple stakeholders with unique mindsets and skill sets affect facility operations and pull in competing ways of knowing. To coordinate and harmonize the many disjointed objectives of the people involved in the life of the facility, we recommend forming a broad-spectrum team (Cowan, 2019) as a means of corralling and focusing efforts, and filling in gaps of understanding among stakeholders. By establishing shared objectives, actors agree on the operational success that they are willing to support.
- **Discuss objectives, barriers, drivers, and priorities:** Within the team, discussing the objectives, barriers, drivers, and priorities will uncover why people do what they do and what tools, skills, and resources each distinct actor can bring to the table, and who can shed the necessary light on workable solutions. Some prime barriers to action are heavy workloads, lack of site-specific operator training, discomfort with hierarchies, responding to user complaints and comfort issues, and cash flow limitations. A thoughtful review of the full context of the facility energy system, including the perspectives, needs, perceptions, and desires of the target audience, allows us to meet people where they are. This approach appropriately positions us to focus them where they want to be, in a way they are comfortable with and accept.
- **Develop relationships:** Developing trusting relationships is essential to obtain a full view of the drivers and constraints impacting people at work. Trusting relationships help when we are pulling together

multi-stakeholder groups and need to influence, persuade, or convince people to implement recommendations. People will not share their real concerns and the real obstacles to energy performance until they trust that you will do them no harm and that you have something of value to offer. The more vulnerable or fearful they are (e.g., of losing their jobs), the more distrustful and risk averse they will be. People will not act on recommendations unless they believe that you are suggesting prudent changes, with accurate calculations, and have their best interests in mind. Useful methods in relationship building include: listening before reflexively trying fix problems, assuming a constructive viewpoint, reading the non-verbal communication, framing interactions positively, and being respectful, non-threatening, and helpful.

- **Facilitate communication:** Facilitating communication means offering support so that people can discuss what is happening and making it simple to connect with each other so that during that discussion what is heard is what was actually meant. Facilitated communication entails bringing people together, setting up easy online tools or virtual connections, and clarifying shared language. Clarifying language involves verifying understandings of words and using words carefully, both by distilling words to make them more accessible to busy participants and by providing language so that they can describe the energy management efforts to others. Storytelling and the use of narratives is a particularly useful, overarching "language" that can help bridge gaps and translate between actors from very diverse and sometimes conflicting positions (Moezzi et al., 2017; Rotmann, 2017).
- **Support new behaviors:** Persistent energy savings do not happen by accident. Persistence in savings requires understanding and participation from many diverse *people*, such as operations staff and C-Suite decision-makers. Other supports include: coaching, accurate, and accessible *data* collection and reporting tools, standards, protocols, and *structures* both material, such as incentives, and cultural, such as capital approval timelines, to ensure performance gains do not evaporate.

Working on the human side of energy management is challenging, but also highly rewarding. The challenges present another layer of the detective-like work of systems optimization. Addressing the human behavior component of energy systems is a fundamental piece of overall energy sustainability.

7.5.3 Start somewhere

Complementary to these recommendations are several theories of change that are well documented in the academic and popular literature to enable energy behavior change at the individual, social, and organizational level. Further information on the tools introduced here can be found in the bibliography for this chapter.

Community-based social marketing is one of the most common starting points and is grounded in the discipline of environmental psychology. The author of this approach, Doug McKenzie-Mohr, proposes eight tools to save energy: commitment, prompts, norms, social diffusion, goods and services, communication, financial incentives, and convenience.

More recently, the "User-Centred Energy Systems" Technology Collaboration Programme[14] by the International Energy Agency (IEA) developed the demand-side management Task 24 Behavior Changer Framework (Rotmann, 2016). This international research collaboration was developed to tackle real-world energy management challenges. Applying this framework is useful to hone in very clearly on the exact behavior that needs to be changed and who exactly the target audience is. The behavior changer framework goes through the top interventions, relationships, mandates and tools, and identifies their individual and collective potential. In a new research collaboration focusing on hard-to-reach energy users in residential and commercial sectors, this framework has been expanded on and incorporated into a "Building Blocks of Behavior Change" process (Karlin et al., 2021), which combines design thinking with robust social science methodologies (Rotmann, 2021). Additionally, practical options for learning more about human behavior and energy management are available in the published literature listed in the bibliography. Seek out similar sources to learn about the latest research and real-life findings. Invest some of your time and available training funds into workshops and courses. There are likely accessible options near you in the form of post-secondary courses or professional workshops. Also try to attend conferences or webinars (such as those run by the ACEEE or Users Academy[15]) to listen to case studies and meet like-minded colleagues.[16] Last, but not least, experiment with some of the various concepts and tools, like surveying and prototyping, in your own work.

[14] See https://userstcp.org

[15] https://userstcp.org/academy

[16] One conference that is particularly noteworthy is the Behavior, Energy, and Climate Change Conference held annually in the U.S. (www.beccconference.org).

7.6 Real-World Examples of Energy Behavior Change

Examples of large-scale real-world human-focused energy management, beyond the residential sector, are still limited in the academic and popular literature. At the residential level, rigorous and large-scale studies like randomized controlled trials[17] are often used to confirm which actions lead to energy savings and how much energy is saved. On the workplace side, randomized control trials are often not feasible because the number of similar facilities that can be grouped and compared is far lower than large numbers of households that are more homogeneous. In addition, for commercial operations, there is a vast diversity of building types, business uses (in many different sectors), specialized energy equipment and infrastructure, and a wide variety of roles and decision-makers within organizations, which drastically increases the complexity of any comparative investigation. The success of any energy management intervention relies on the ability to quantify and attribute savings to specific interventions. For quantifying savings, the International Performance Measurement and Verification (M&V) Protocol (IPMVP) is the industry standard (see Chapter 1). M&V is typically used to determine savings attributable to time-bound technical or mechanical interventions.

There are three key challenges with M&V for behavior change interventions:

1. Energy savings from behavioral interventions often occur gradually and with irregular patterns over weeks to months. Changes in energy use are typically not specific events like common equipment retrofits.
2. There are often investments in other projects or programs that overlap behavioral interventions. This confounds the ability to attribute savings to a behavior change program. Because behavior often has a complementary or contribution effect to energy savings (or is indeed the underexplored barrier to success), the effort to assess attribution of savings can be greater than the effort to implement the energy-saving interventions.
3. Behavior changes can affect multiple systems, and it is often not feasible *a priori* to know which systems might be optimized by staff (or when). For example, in commercial buildings, behavior change interventions

[17] Randomized control trials are accomplished by randomly allocating subjects to two or more groups, treating them differently, and then comparing them with respect to a measured response.

for O&M (operations and maintenance) staff can effect changes in space heating, cooling, plug loads, lighting, or hot water systems.

Because of the above characteristics of behavior change programs, IPMVP Option C (whole facility performance) is usually the best M&V approach. Option C was utilized for a hospital case study described below.

7.6.1 Energy behavior program at a large American hospital system[18]

In this section, we describe an energy behavior program that was the first known attempt in the commercial/institutional sector to spread human-focused energy management across one of the largest not-for-profit healthcare systems in the U.S. The hospital launched their energy management program in 2012 at 12 of the system's 40 hospitals where retro-commissioning was the leading technical intervention. In 2015, a human-focused behavioral approach was integrated to complement the traditional energy management practices already in place. The program creator (and author of this chapter) Kady Cowan knew that there were limited resources to help guide the effort of validating and verifying results from behavior-based programs in workplace settings. To help ensure that the program would be effective and verifiable, Kady convened 13 energy experts from around the world to act as advisors and external evaluators for this initiative. This is the story of how we came to learn that human-focused energy management can have big returns in energy savings and organizational learning.

The award-winning *Energy Behavior Program* was conceived to amplify energy savings that were previously achieved, to avoid energy-saving decay over time, and to tap into new energy-saving potential quickly. After forming an *Energy Behavior Program* design team with representatives from the hospital energy ecosystem, discussing objectives and drivers with external and internal experts, building relationships up, down and across the organization, facilitating communication with our target audience – facility operators, mechanics and building engineers – we were ready to support new behaviors with six interventions at six pilot facilities. After considering several high impact behaviors that we could target, we decided to initially focus on reducing overrides in the BAS.

[18] The name and location of this program have been anonymized at the request of the hospital.

- In the *Discover Phase*, we collected energy hero and horror stories (Rotmann, 2017), completed interviews and focus groups, and spent significant time meeting people and listening to their concerns and ideas. These activities were the foundation of a comprehensive qualitative energy culture baseline report, which revealed the circumstances, structures, and practices in place that were common among staff and between sites.
- In the *Define Phase*, we started to describe patterns of behavior that helped us to organize our attention, narrow our focus on what really matters, and to clarify how we might intervene in the system.
- In the *Develop Phase*, we mapped information and data flows so that we could visualize how the actors and structures in the energy ecosystem were connected, or not. Finally, we collected and tracked utility bill information, energy use intensity, overrides and manual BAS adjustments, maintenance requests, and comfort complaint calls. With this foundational work complete, we started to deliver the program.
- In the *Deliver Phase*, we focused on six interventions to help shape the circumstances so that energy-saving behavior would follow. The first six interventions were:

    1. Conduct in-person energy literacy and basic energy management training (2 hours a week for 5 weeks) for building operators, maintenance, and engineering teams.
    2. Make site-specific energy data visible to the facilities workforce.
    3. Select and support a site-based energy champion.
    4. Develop a standard process flow for hot/cold call response.
    5. Document a protocol for adjustments in the BAS.
    6. Promote dialogue between occupants and facilities staff with regard to energy savings.

As a result of our multi-year, zero capital-cost effort, we verified energy savings in five out of six pilot facilities. Using a conventional IPMVP Option C M&V approach, baseline models were developed for electricity and natural gas as a function of outdoor air temperature. The realized energy savings are provided in Table 7.3 for the initial 12 months following implementation of the first interventions.

7.6.2 Stories, circumstances, practice, patterns

To start, we collected stories from all actors in the *Energy Behavior Program* ecosystem. These insights made the organizational energy culture visible by defining system boundaries, and gave us time to check assumptions. Seeking

Table 7.3 Energy savings for Energy Connect pilot facilities in 2018.[19]

Pilot site	Electricity savings	Gas savings
Rural Hospital	3.1%	6.7%
Acute Hospital	−2.0%	−11.4%
Medical office 1	8.6%	N/A (100% electric)
Medical office 2	1.5%	6.6%
Emergency care facility	8.7%	24.3%
Data and call center	12.2%	N/A (100% electric)

to understand the circumstances at the universal, organizational, social, and individual levels revealed a clearer picture of the practices and patterns that hold energy-wasting behavior in place. From this vantage point, we could determine the best points in the system to intervene to create durable and appropriate behavior changes that supported energy-saving objectives. Practices from the stories are summarized in Table 7.4, followed by a sample of the themes and patterns that emerged from the qualitative research phase.

The clearest themes to emerge from the research are outlined in Table 7.5. Each had multiple direct and indirect impacts on the circumstances that can invite or repel behavior change. Included here are a few of the impacts we observed from qualitative research. Human behavior and facility energy management is no more difficult than traditional energy management, but for many energy managers it is new, and thus perceived as intimidating or risky. None of us are perfect the first time we try something new. But adopting behavior change is feasible with encouragement from peers and bosses, expanded timelines, connection with like-minded others, and access to learning the low-cost and durable energy-saving opportunities of working with people. Energy behavior should be complementary to traditional energy management and contributes in obvious (e.g., improving energy-saving know-how) and non-obvious ways (e.g., pride in accomplishments) to energy savings and organizational improvement. Many of the non-obvious impacts go undetected because qualitative impacts often show up at unpredictable times and are rarely measured as part of energy efficiency programs. The business case for doing this work is extremely strong, especially if you are prepared to look for non-obvious results at various points throughout the life of your project or program. The success of the *Energy Behavior Program* came because we made time for those with high impact energy decisions at

[19]Published with permission from the program evaluator. Except for the Acute Hospital, all M&V savings met the recommended minimum uncertainty requirements per ASHRAE M&V Guideline 14-2023. The Acute Hospital site had major air handling unit upgrades occurring coincident with the behavioral program, which explained the findings of negative energy savings.

Table 7.4 Practices and patterns in place at the beginning of the energy behavior program.

Circumstance level	Practices and patterns
Universal	• Limited or inappropriate thinking about large/complex problems. • Buildings morph over time and operations become difficult. • Willingness to fix is more acceptable than paying higher first costs. • Culture and policy structures that support the status quo need to be overcome.
Organizational	• Energy not previously considered a priority. • The skills to run complex buildings are not valued highly enough. • Employees are not free agents in the organizational hierarchy. • Building operators and facilities staff often describe their days as "Juggling hand grenades and putting out fires."
Social	• Lack of social cohesion and hierarchy makes change hard to spread. • Unclear and inconsistent instructions about energy and cost savings. • Contradictions between what is said and what is done lead to an attitude that nothing will ever change.
Individual	• People fear getting in trouble for doing things differently. • There is no time for staff to think about or plan for the future. • People worry that energy action will take too much time. • Habits, motivations, values, skills, and psychological barriers not fully understood.

their fingertips, building operators, and maintenance staff. What the *Energy Behavior Program* took advantage of that other previous efforts have failed to fully appreciate is that the day-to-day decisions of building and facility teams have a large impact on energy consumption. It is not just executive decision-makers who can move the dial on energy efficiency; it is the staff who are delivering energy services and controlling processes directly. Getting to know these actors is time well spent, especially as we come to expect as an industry that our patterns of energy management produce fewer reactive or "band aid" outcomes and instead more anticipatory or repeatable outcomes.

After the successful pilot year of the *Energy Behavior Program* at six facilities, the program was adopted at all 40 hospitals in the health system and dozens of healthcare facilities. One of the building engineers who became a site-based energy champion was able to accomplish such good work in his buildings that the corporate head of office (from out of state) came to meet him to learn more about how he was making change happen. What they saw were energy savings increasing year over year because of the

Table 7.5 Themes, impacts, and remedies from energy behavior program.

Theme	Direct impact	Indirect impact	Remedy
Permission to save energy at work.	People do not turn equipment off because they are not sure if they should.	Very few people know they are allowed to save energy at work.	Make energy part of everyone's job, talk about energy often, support staff to act.
Short-termism dominates planning horizons.	Facilities and maintenance time is generally spent reacting to issues and complaints.	Everything is framed in the short term and underlying longer term issues are overlooked.	Management delegates time, resources, and praise for longer term energy efficiency projects.
Connectivity between the human/mechanical and human/human energy systems.	Each individual sees themselves as caring about energy issues at work more than others.	Individuals do not perceive support from peers and bosses to save energy.	Create opportunities to visibly demonstrate caring about energy efficiency among peers.
Value of learning and skill development.	Staff not hired for energy expertise.	Time to notice and learn from wins or mistakes not included.	Support champion, energy leaders, training, and network building.

interest and expertise of the facilities team; maintenance and engineering staff participating in energy-saving projects during the building or renovation design phase, not just once buildings were open; more time and resources being dedicated to energy efficiency; and, critically, during recruitment, by giving preference to applicants with energy management experience.

The building engineer in this example was just as busy as all the others. But he made an effort to identify and build relationships with enthusiastic collaborators from across the organization to support his own facility team's energy projects; he was clear about his data needs and what learning was required, and he built on past successes. As a result, he helped expand the understanding of the energy ecosystem, the physical, mechanical, and human system interaction, and tapped into the massive energy-saving capacity that was available.

7.6.3 Army energy culture, U.S. Army Corps of Engineers

A group of researchers with the U.S. Army Corps of Engineers are investigating ways to add social science insights to energy management efforts to increase energy resilience through the efficiency of Department of Defense facilities. Although only a limited number of studies have been published that specifically focus on energy behavior related to military facilities, the team found numerous attempts from multiple levels of the organization to improve facility performance by including the human aspect of energy use (Westervelt, 2020).

At the highest levels, military leaders recognize the need to establish a pervasive culture of efficiency and official guidance calls for coupling behavioral and operational change to technology implementation efforts for optimal results. Goals for energy reductions and requirements for ongoing energy auditing and retro-commissioning provide the performance objectives and methods for identifying efficiency opportunities. Furthermore, authority for hiring an energy manager at installations and requirements for training federal facility operators establish roles and rules that increase efficiency.

Compulsory installation of energy meters in new construction was established to enable monitoring. Staff were trained to set up web-based energy data to inform decision-making for achieving energy reduction targets, and allow production of mock billing to provide building users who do not receive utility bills with the necessary feedback to understand their energy use profile. Authority to retain some of the monetary savings of efficiency projects for use at the installations is providing appropriate incentive for change. At the energy manager and facility operator level, many creative individuals have devised methods to further energy efficiency within limited budgets.

A program to train building operators in retro-commissioning techniques is developing needed skills in equipment testing and analysis to identify root causes of inefficiency and resolve building operational faults. In instances where automated controls for facility equipment (such as lighting and HVAC) are cost prohibitive, relocating hard-to-find light switches and thermostats to accessible locations with signage of operational instructions and expectations (to turn off when not in use) has deputized building occupants to support efficiency at minimal cost. Most locations have some level of energy awareness outreach and award programs.

Although these efforts to increase energy efficiency are laudable, there is still opportunity for refinement, more widespread utilization, and integration across multiple stakeholders. To further move the Department of Defense toward efficiency, the U.S. Army Corps of Engineers combined best practices from energy management, change management, and complementary military processes into a step-by-step *human-centered efficiency process*, which calls out considerations and insights collected from academia, industry, and many levels of the federal workforce to holistically coordinate the use of technology by people for optimal efficiency. It pulls in social science methods for influencing people to act in desirable ways at an individual level, as well as procedures for embedding business processes at an organizational level for sustained change.

The *human-centered efficiency process* calls for continual improvement with adaptive problem-solving and shared ownership of the outcomes. It provides methods for determining the current status of the energy culture through interviews, focus groups, and surveys. It gives a sample description of an aspirational future state and many ideas for bridging the gaps in efficiency by supporting people. Data gathered thus far shows opportunities to increase understanding of institutional processes, to invest in facility repair and expanded staffing, and to further explore local conditions and develop solutions with local personnel that harmonize the actions of different groups. An online repository for knowledge sharing has been launched to support federal employees in their human-centered efficiency efforts.

7.7 Performance of Energy Behavior Programs

Evaluation, measurement, and verification[20] (EM&V) are general techniques that often apply to large-scale programs generally at a state, provincial, or national level. EM&V techniques include both quantitative and qualitative elements and are also useful at the facility level or systems within a facility scale. In this section, we focus on the facility scale, or smaller. There are a wide variety of evaluation tools and evaluation types. We have chosen to focus on practical tools and methods that we believe are most often accessible to facility and energy managers, given their practical constraints of time, money, and expertise. There are a wide variety of research methods and research designs from the social and physical sciences that require longer time frames, bigger budgets, and access to greater levels of expertise than is needed for everyday facility energy management.

7.7.1 Quantifying energy savings: A critical outcome of behavior programs

Quantifying energy savings due to behavior-based interventions at the facility scale (or smaller) can be approached similarly to technology-based interventions. Quantitative tools such as M&V and ongoing monitoring, targeting,

[20] In general, approximately equivalent terminology found in other literature is "monitoring and evaluation."

and reporting (MT&R) introduced in Chapter 1 can be utilized. With any quantitative effort, it is critical to be clear what intervention or action is being evaluated, to plan ahead for collection of baseline data, and consider the system boundary for the interventions. In practice, quantifying energy savings can be complicated due to several factors such as multiple interventions occurring in series or in parallel, different system boundaries, and non-routine changes in facility use or activity.

Consider this hypothetical, yet common, situation where a facility has implemented some energy-efficient technology, and then sometime later implements a behavior change program. In this example, energy consumption creeps up over time, which could occur due to a variety of reasons such as equipment degrading, wasteful rebound effects, changes in use or service levels or operating hours, staff turnover, or normal variation in operating conditions, such as climate or seasonal changes. In this example, it could be that the behavior change program was applicable to multiple (or even all) systems in the facility, while the efficient technology was for one component only. Thus, each intervention would have different start times, different persistence time periods, different physical system boundaries, and potentially different organizational boundaries as well. If other factors change such as activity (for example, occupancy in a building, or product quality in a plant), operating hours, or environmental conditions (like weather patterns), then quantifying savings will require adequate data on these variables so that the appropriate baseline adjustments can be performed, and net energy-saving attributions can be estimated. In this example, there is one technology intervention and one behavior change intervention, but in real-life situations there can be multiple interventions of both types, occurring simultaneously. Anticipating these potential complexities and planning M&E during the early stage of behavior program development is important. In particular, it is important to document what "proof" is most vital to participants, decision-makers, and any funding actor, and to develop an M&E plan designed to provide this proof. Thoughtfully planned and implemented M&E helps energy managers adequately prepare and execute rigorous, quantitative validation of savings that is feasible within budget and time constraints.

7.7.2 Establishing attribution and behavior program impact

In energy behavior programs, it is almost always important to collect data on variables beyond kWh. Gathering evidence about your program through process, outcome, or narrative evaluation helps determine the extent to which savings can be attributed to the intervention. Two particular types of variables are performance and impact measures. Performance measures are focused on program activity, such as the number of people receiving a survey request, the number of pamphlets distributed, or the number of participants in a focus group. These variables are useful in order to understand the reach and cost of the behavior program. Impact measures are variables that more directly quantify the actual participation of actors in the program. This could include quantifiable measures such as the number of survey responses completed, the number of people requesting further information, or the number of product samples taken.

Quantifying these variables may be necessary to establish attribution, that is, to provide evidence that the behavior change program is responsible for the actions that saved energy. For example, if there is evidence in a school facility that lights are turned out more frequently at night, or if purchasing staff in an industrial plant begin ordering motors that exceed premium efficiency standards, how can you be sure these actions are due to your behavior change program? Performance and impact measures of the behavior change program (often based on qualitative data) help provide evidence that attribute actions to the energy behavior program. From this vantage point, the exercise of quantifying energy savings can be carried out with data to support that savings are actually attributable to the program.

7.7.3 Ensuring and qualifying non-energy impacts

A good behavior change program will be carefully designed to maintain or improve energy services, and to avoid (unintentionally) imposing negative impacts on services. For example, any behavioral actions targeted for HVAC systems should be specifically designed to ensure thermal comfort is maintained or improved by building users (confirmed with baseline data collection), while tracking metrics to avoid degrading service levels. For industrial facilities, a behavior program design should pay careful attention to worker safety, reliability of production, and product quality. Approached in this way, energy behavior programs will be designed to deliver NEI that can sometimes be even better indicators of success than energy cost savings. Tracking of key metrics on the quality and quantity of services can confirm that you now have more of what you want from your energy systems, and provide a sure sign of success. As the expectations in energy management change, we will come to expect as an industry that our patterns of energy management produce fewer perverse outcomes such as: operation disruption, reactivity, short sightedness, blame, isolation, status quo bias, "band aid" solutions, and spending to fix. Careful planning and execution of a behavior change program can instead lead to more positive outcomes including: smooth and reliable operations, preparedness for unexpected events, process improvement, access to learning, encouragement from peers and managers, acting in uncertainty, spending to optimize lifecycle costs, and a preventative mindset. Your program will have unique opportunities to realize NEIs, and these should be deliberately identified and measured as part of the M&E. In closing this discussion, some less common NEIs are listed here as a guide.

- Build a sense of care about energy management.
- Stabilize working relationships and improve team communication.
- Fund and implement more large and complex energy efficiency projects.
- Increase rate of investment in energy-saving programs.
- People have the right skills to participate in energy management.
- Energy-saving goals are related to core business.
- Acceptance of "new" ways of doing things.
- Increase focus on energy-saving practices and behaviors.
- Maintain current programs and savings with little upkeep.
- Write policies and procedures that enable energy savings.
- Energy-saving behaviors and practices do not revert back over time.
- Application of energy management is consistent across the organization.
- Energy-saving successes and energy champions are celebrated.

7.8 Summary and First Steps

What has been described in this chapter is usually referred to as "soft skills." But this work is neither soft nor easy; it is essential, entails persistent effort and hard work, and requires specialized knowledge and skills in applied social sciences. In this chapter, we have delved deeper into these essential skills to describe why human behavior is relevant in everyday practice. We have defined *energy behavior*, and *energy ecosystems*, identified target audiences, behaviors, tactics, and metrics, and presented examples of how behavioral actions have been proven to save energy. If you are one of the many humans involved in energy management, it is always important to keep in mind the difference between actions that save energy, and the interventions and programs designed to make these actions occur more frequently. If behavioral actions proven to save energy were readily implemented by simply sharing information, then this chapter would be unnecessary. We believe that practitioners can be significantly more effective if they are supplied with the additional knowledge, tools, and techniques of behavioral sciences to complement engineering and conventional economic approaches.

We have introduced the knowledge and tools of human behavior in two sections. In the section on biases and barriers, we have provided the reader an overview of common cognitive biases and psychological barriers that are important in order to understand many of the non-obvious barriers to realizing energy savings. In the section on program design, we have introduced a variety of programs and techniques that are likely to help increase the occurrence of energy-saving behavioral actions. You will no doubt notice that there are areas of overlap between biases, barriers, and programs. Examples include the extensive team-forming aspects of programs to help overcome credibility and trust barriers, and offset status-quo biases.

It is important to emphasize that this chapter provides only an introduction to the knowledge, tools, and techniques of behavior change in facility energy management. There is much more to learn and apply based on what is already well-supported by research and practical experience, and new knowledge and tools are constantly emerging. We suggest that maintaining a curious mind and a humble attitude will help. By now, you know enough to start somewhere, even if it is imperfect at first. We encourage you to integrate human behavior into all of your facility energy management strategy and tactics, and hope that you find it as rewarding as we do.

References

ASHRAE (2023) Measurement of Energy, Demand, and Water Savings. ASHRAE Guideline 14-2023.

Attari, S.Z., Dekay, M.L., Davidson, C. I., Bruine, W., & Bruin, D. (2010). Public perceptions of energy consumption and savings. Proceedings of the National Academy of Sciences of the United States of America, 107(37), 16054–16059.

Backlund, S., Thollander, P., Palm, J., & Ottosson, M. (2012). Extending the energy efficiency gap. Energy Policy, 51, 392–396. doi:10.1016/j.enpol.2012.08.042

Cowan, K. (2019) Energy Efficiency Landmark Case Study: Energy Connect. In Tools of Change. https://www.toolsofchange.com/en/case-studies/detail/718/

Cowan K. Sussman, R., Rotmann, S., Cox, R. (2017) Designing a Behaviour Change Programme for Hospital Facilities Staff. International Energy

Agency. Energy Technology Initiative on Demand-Side Management Technologies and Programmes https://userstcp.org/wp-content/uploads/2019/12/30.Task24_Phase2-Subtask-11_CHS-case-study_FONTS.pdf

Cowan, K. (2016). Designing Energy Behaviour – Leading Change. In Summer Study on Energy Efficiency in Buildings. Pacific Grove, USA. American Council for an Energy-Efficient Economy.

Darnton, A. (2008). GSR Behavior Change Knowledge Review. Practical Guide: An overview of behavior change models and their uses. Government Social Research Unit.

Dietz, T. (2010). Narrowing the US energy efficiency gap. Proceedings of the National Academy of Sciences of the United States of America, 107(37), 16007–16008.

Dietz, T., Gardner, G., Gilligan, J., Stern, P., & Vandenbergh, M. (2009). Household actions can provide a behavioral wedge to rapidly reduce U.S. carbon emissions. Proceedings of the National Academy of Sciences of the United States of America, 106(97), 18452–18456.

Ehrhardt-Martinez, K. (2015). Behavior-based Energy Saving Opportunities in Commercial Buildings: Estimates for Four U.S. Cities. In 2016 ACEEE Summer Study on Energy Efficiency in Buildings.

Gillingham, K. (2013). The Rebound Effect is Overplayed. Nature, 493, 475.

Jaffe, A.B., & Stavins, R.N. (1994). The energy-efficiency gap What does it mean? Energy Policy, 22(10), 804–810. doi:10.1016/0301-4215(94)90138-4

Illume Advising, Indicia Consulting & E. Vine (2015). Energy Efficiency Behavioral Programs: Literature Review, Benchmarking Analysis, and Evaluation Guidelines.

IPMVP (2022) "International Performance Measurement and Verification Protocol (IPMVP?) Core Concepts" EVO 10000 – 1:2022. Efficiency Valuation Organization. www.evo-world.org

Karlin, B., Forster, H., Chapman, D., Sheats, J., and Rotmann, S. (2021). The Building Blocks of Behavior Change: A Scientific Approach to Optimizing Impact. See Change Institute.

Klotz, L. (2010). Cognitive biases in energy decisions during the planning, design, and construction of commercial buildings in the United States: an analytical framework and research needs. Energy Efficiency, 4(2), 271–284. doi:10.1007/s12053-010-9089-z

Lutzenhiser, L. (1993). Social and behavioral aspects of energy use. Annual Reviews Energy and Environment, 18, 247–289.

Maxwell, D., Owen, P., McAndrew, L., Muehmel, K., Neubauer, A., (2011). "Addressing the Rebound Effect" a report for the European Commission DG Environment.

Mazzi, E. (2019). "Perspective on M&V behavioral change programs in commercial and industrial facilities." *M&V Focus Issue #4.* Efficiency Valuation Organization (EVO).

McKenzie-Mohr, D. (2011). Fostering Sustainable Behavior. New Society Publishers.

McKenzie-Mohr, D., Lee, N., Shultz, P.W., & Kotler, P. (2012). Social Marketing to Protect the Environment: What Works. SAGE Publications.

Moezzi, M., Janda, K., Rotmann, S. (2017). Using stories, narratives, and storytelling in energy and climate change research. Energy Research & Social Science. Volume 31, Pages 1-10.

Mourik, R.M., van Summeren, L.F.M., Breukers, S., Rotmann, S. (2015). Did you behave as we designed you to? IEA DSM Task 24, Energy Technology Network.

Northwest Strategic Energy Management Collaborative (2021). "Case Studies." *SEMHUB* https://semhub.com/case-studies

Rotmann, S. (2016). How to Create a 'Magic Carpet' for Behaviour Change. BEHAVE 2016, 4th European Conference on Behaviour and Energy Efficiency. https://userstcp.org/wp-content/uploads/2019/12/36.Task24_Phase2_Rotmann-BEHAVE-2016.pdf

Rotmann, S. (2017). "Once upon a time..." Eliciting energy and behaviour change stories using a fairy tale story spine. Energy Research & Social Science. Volume 31, Pages 303-310.

Rotmann, S., Mundaca, L., Castaño-Rosa, R., O'Sullivan, K., Ambrose, A., Marchand, R., Chester, M., Karlin, B., K. Ashby, Butler, D., and J. Chambers (2020). Hard-to-Reach Energy Users: A Literature Review. Prepared for User-Centred Energy Systems TCP - HTR Task. Published by SEA - Sustainable Energy Advice Ltd: Wellington. 252pp. ISBN: 978-0-473-64983-8

Rotmann, S. and Karlin, B. (2020). Training commercial energy users in behavior change: A case study. ACEEE Summer Study on Energy Efficiency in Buildings: Monterey. https://doi.org/10.47568/3CP104

Schiller, S.R., Skumatz, L., Carvallo, J.P., Burdette, J., Klucher, M. (2016). Evaluating and Quantifying the Non-Energy Impacts of Energy Efficiency. Berkeley Lab. Electricity Markets and Policy.

Smith, S., Seo, P. (2020). A Brief History of Behaviour Change Demand Side Management Programs - Supporting Staff Engagement in the Commercial Sector. ACEEE Summer Study on Energy Efficiency in Buildings.

Sorrell, S., Schleich, J., Scott, S., O'Malley, E., Fergal, T., Boede, U., Ostertag, K., Radgen, P. (2000). Reducing Barriers to Energy Efficiency in Public and Private Organisations. University of Sussex: SPRU Environment & Energy.

Stern, P. (1992). What psychology knows about energy conservation. American Psychologist, 47(10), 1224–1232. doi:10.1037/0003-066X.47.10.1224

Sussman, R., Gifford, R., Abrahamse, W. (2016). Social Mobilization: How to Encourage Action on Climate Change. Pacific Institute for Climate Solutions.

Sussman, R., Rotmann, S., Cowan, K., Cox, R. (2017). Designing A Behaviour Change Program for Hospital Facilities Staff. International Energy Agency: Energy Technology Initiative on Demand Side Management Technologies and Programs.

Sweeney, D. (2009). Show me the change: a review of evaluation methods for residential sustainability behavior change projects.

Thollander, P., Palm, J., & Rohdin, P. (2010). Categorizing Barriers to Energy Efficiency: An Interdisciplinary Perspective. In Energy Efficiency (pp. 49–62).

Westervelt, E. (2014). The Human Side of RCx. Strategic Planning for Energy and the Environment, 34(3).

Westervelt, E., Loechl, P., Clark, S., DuPont, C. (2020). Enhancing Army Energy Culture with Behavioral Approaches. U.S. Army Corps of Engineers, Engineer Research and Development Center.

Williamson, C. (2015). Estimating Impacts from Behavioral Programs Using Experimental and Quasi-Experimental Design. Applied Energy Group. Load Analysis Membership Services Report #LAMS-0011.

Author Biography

Kady Cowan is Founder of the Sustainability Studio. She invites organizations and institutions to collaborate in the Studio where they have time to learn and practice combining sustainability, environmental protection, energy and climate safety, and a wholistic ecosystem perspective into both strategic and everyday activities at work. Kady facilitates the adoption of positive environmental behaviors while integrating ecological decision-making in complex work settings. Kady's focus on learning, simple prototyping, appropriate scaling, and evaluation of dynamic low-carbon projects and programs leads to the protection of environmental and human health, climate resilience, financial savings, and conservation of natural and energy resources. Kady has been proud to serve as a climate, environment, and sustainability leader for local, national, and international organizations including the Canadian Coalition for Green Health Care, Doctors Without Borders (MSF), Independent Electricity System Operator (IESO) in Ontario, Atrium Health in North Carolina, and University Health Network (UHN) in Toronto.

Acknowledgments

It is a delight to update this chapter, which first appeared in the 2016 edition of Guide to Energy Management. It is only great because of the contributions and editorial support from Dr. Eric Mazzi, Ms. Eileen Westervelt, Dr. Sea Rotmann, and Dr. Andrew Knox. I want to especially acknowledge all of the energy managers who are achieving the best energy savings possible in their facility, industry, and sector by including human behavior in their projects. I am particularly grateful to all the energy managers who shared their stories with me and are featured in this chapter.

Practice Questions

1. What is the energy efficiency gap?
   A) Delivery of energy services to some parts of the facility.
   B) The difference between energy savings that are estimated to be cost-effective, and those that are actually implemented.
   C) The adoption of efficient technologies does not lower energy consumption as much as predicted or, in extreme cases, might even increase consumption.
   D) Unrealized reductions in energy consumption, which could be achieved with a cost-effective combination of technology and human behavior change.
   E) Both B and D
2. Interventions such as energy feedback are most likely to result in behavior change when accompanied by:

A) A system of penalties to reinforce implementation of behavior change.
B) Knowledge of the environmental and financial benefits of the behavior change.
C) Reminders at the physical point of use/decision.
D) A trusted source provides a personalized message as close to the action as possible (e.g., real-time).
E) Tracking systems that show people how frequently they did and did not implement energy-saving actions.

3. When well-designed human-focused energy management programs are implemented and monitored, they can typically save:
   A) 3%–25% of a facility or organization's energy consumption.
   B) 10%–20% of a facility or organization's energy consumption.
   C) 2%–3% of a facility or organization's energy consumption
   D) 3%–10% of a facility or organization's energy consumption
4. What are the four principles of behavior change that inform program design?
   A) Placement of a visible suggestion box, hyperbolic discounting, building trust and credibility, and framing the issue positively.
   B) Be patient, make participants feel valued, focus on continuous improvement, and let context be your guide.
   C) Friendship, respect, encouragement, and tolerance.
   D) Supporting technically minded energy champions, interpreting energy data, celebration and recognition, and evaluation.
5. What is the term for the energy management work that leads to changes in human actions based on direct or indirect programs and interactions and provide one example of how it might be applied?
6. Describe two tactics (specific actions) that University Health Network used to integrate physical, mechanical, and human elements into all energy and environment policy and programs.
7. Why are communication and relationship building important skills for realizing maximum energy savings at a facility? Please give three reasons.

8

Resilience

Remi Charron[1] and Stefan Storey[2]

[1]Innovation Investment Advisor, National Research Council of Canada, Canada
[2]Sensible Building Science, Canada

8.1 Introduction

Chapter 2 provides an introduction to the challenges we face with climate change. Many of this book's chapters cover strategies to mitigate or reduce the amount of climate change that will occur. However, regardless of how much mitigation efforts take place around the world, the climate system has already changed considerably, and changes will continue for the coming decades based on the amount of greenhouse gases that have been emitted to date. The impacts of this changing climate are leading to bigger and more frequent natural disasters. This has already led to an increase in human health impacts (e.g., premature deaths from extreme heat) as well as a significant growth in the damage and destruction of buildings and other infrastructure. This chapter will discuss how facilities can be made more resilient in an effort to protect people and buildings from our changing climate. The specific learning objectives for this chapter are as follows:

- Understanding the imperative of resiliency planning
- Differentiating between climate and weather
- Impacts of climate-change-driven hazard events
- Contextualizing vulnerability
- Gain familiarity with the meaning and application of a "vulnerability gap"
- Identifying potential design strategies that can help mitigate risk in a building

One of the best financial measures that we have utilized to protect assets from damage has been insurance. Over the years, the insurance industry has

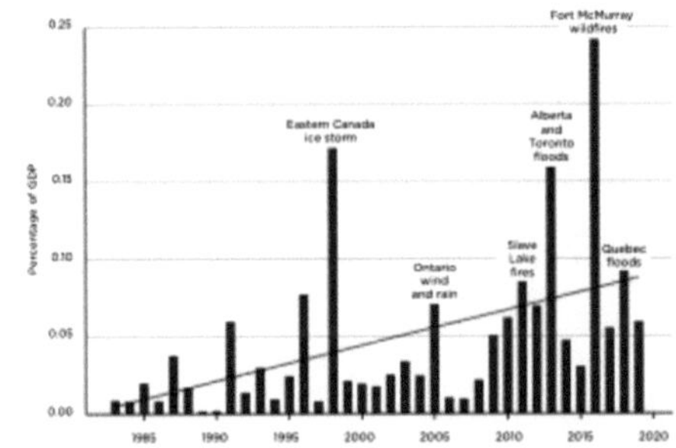

Climate Mitigation

Mitigation is defined by the IPCC as the human intervention to remove or abate GHG sources or to sequester or store GHG out of the atmosphere for a predetermined period of time. Most of the Guide to Energy Management is focused on reducing emissions; in contrast, this chapter is focused on adaptation and resilience with particular attention to urban systems. The future success of global mitigation efforts is hard to predict. Adaptation and resiliency are therefore a response to the inevitable changes that are in the pipeline, motivating engineers, architects, and urban planners to consider how to reduce risk to both human health and infrastructure damage.

Figure 8.1 Catastrophic insured losses in Canada (1983–2019) as a share of GDP. Source: Council of Canadian Academies, 2022. Building a Resilient Canada, Ottawa (ON): The Expert Panel on Disaster Resilience in a Changing Climate, Council of Canadian Academies.

been faced with rising losses from climate-related events. Figure 8.1 shows catastrophic losses increasing over time and includes a spectrum of climate-enhanced extreme weather events driving costs upwards. This has led to rising insurance premiums, as well as the industry refusing certain coverage in areas more likely to be hit with disaster (e.g., forest fires and flooding).

Given the rising financial burden placed on the insurance industry, it may not be enough to rely on them to protect assets. There is a need to make the buildings themselves more resilient to prevent losses in the first place. The good news is that there is a net benefit of investing in disaster mitigation. The Institute for Catastrophic Loss Reduction (ICLR) conducted a review of cost–benefit ratios of resiliency measures being applied to core public infrastructure guidelines and codes (ICLR, 2020). The ICLR reported high return benefit-to-cost ratios at 6:1 (Wildland Urban Interface Guideline & Standard), 11:1 (National Guidelines for the Flood Resilience of Buildings), and 9:1 (Canadian Highway Bridge Design Code).

Not all stakeholders may agree on the cause of climate change. When working with stakeholders that do not adhere to the science of climate change, it is useful to leverage the statistics of climate-related damages that are not contestable. Humanity is building more infrastructure in harm's way and economic damages arising from events is measurably increasing. Resiliency planning must proceed, independently of personal "belief" in climate change.

8.1.1 Key challenges to infrastructure

Assessing the risk of climate change should be treated formally and under a regime that addresses uncertainty. Risk reduction is a complex field of study and is based on probabilistic dynamics between systems and variables. For this chapter, risk is considered from an applied point of view, and using the basic principles as defined by ISO 31000. An advanced study of risk mathematics is beyond the scope of this guide.

Opportunities for cost-effective risk reduction can be divided between existing and new infrastructure. Strategies for the remediation of old infrastructure require an entirely different approach compared to new development. Urban infrastructure is long-lived and climate impacts beyond 2050 will be felt by most infrastructures built today. Considering an annual replacement rate of under 5% means that a significant proportion of existing buildings will experience events that are far outside the original designs.

Most existing infrastructure was built with building codes and planning bylaws implemented before the impacts of climate change were known.

1. Previous regulations were often prescriptive and implemented without adequate enforcement. This has led to crises such as sick building syndrome starting in the late 1980s, through "leaky condo" errors later in the 1990s.
2. Building codes have been developed using the assumption that historical data represents a baseline for future performance. Most energy codes are based on the assumption that the past informs both the present and the future. Climate change means this assumption is now obsolete.
3. Most building regulations were designed to protect people rather than assets. For example, fire and seismic codes are specified so that buildings are sufficiently robust to ensure occupants can safely vacate within timescales of minutes or hours. These work well for short-duration impacts, but the threat of climate events are broader and threaten human health and infrastructure on large regional scales over much longer timeframes. For example, rising sea level can impact an entire littoral or flood plain zone, with a duration of decades.

In terms of planning for the future, the concept of a "historical" climate is no longer sufficient, which means that codes and regulations are often ineffective at risk reduction. The intent of this chapter is to provide information and best practices to reduce risk relative to dynamic climate change, to evaluate infrastructure exposure, to identify vulnerabilities, and to design remedial measures that improve system resilience. The guidelines provided in this chapter are therefore delivered from a practitioner's perspective and introduce key aspects and themes for professionals intending to support the design and implementation of resiliency practices.

8.1.2 Definitions: comparing resilience and adaptation

Practitioners tend to use the terms adaptability and resilience interchangeably. There are minor conceptual differences in terms of framing, timescale, and scope (Nelson and Adger, 2007; Leggett, 2021):

- Adaptation often refers to an ability to respond dynamically to current climate change events. Adaptation measures usually involve making changes to existing systems that are inadequately responding to current extremes.
- Resiliency is oriented to future climate change where system robustness, reliability, and rapid recovery from disaster are key outcomes.

The two approaches are synergistic in that adaptation and resilience serve to reduce the risk of harm, but their deployment may be semantically different. For example, reducing risk for existing buildings located in a flood plain could deploy adaptability measures such as introducing vegetated swales to attenuate runoff, whereas a resiliency measure may include physical changes such as flood defense berms or even land-use changes to anticipate long-term hazards. For the purposes of this chapter, which is focused on applied solutions and risk reduction, the two terms will be used interchangeably. The key aspect of success is focused on the importance of context for the design and deployment of resiliency strategies. Just as in the field of public policy, context is profoundly important for matching solutions to real-world limitations and capabilities.

8.1.3 Defining vulnerability: exposure, sensitivity, and adaptive capacity

The main factors that determine how vulnerable an asset is include exposure, sensitivity to the potential impacts, and the adaptive capacity to absorb or accommodate potential harm. Any of these three drivers may be considered for resiliency planning, as shown in Figure 8.2. Even when exposure and sensitivity cannot be fully managed, "vulnerability" can be reduced by improving adaptive capacity (reactive and proactive), hence reducing risk. Bear in mind that risk can only be managed and never fully eliminated.

While future climate may cause an increased exposure to a particular type of climate event, this may not be of concern if infrastructure has low sensitivity to a specific stressor. For example, an existing building may be located in a floodplain, but is found to be constructed on an elevated platform, which may have reduced sensitivity to internal flooding. Though the building

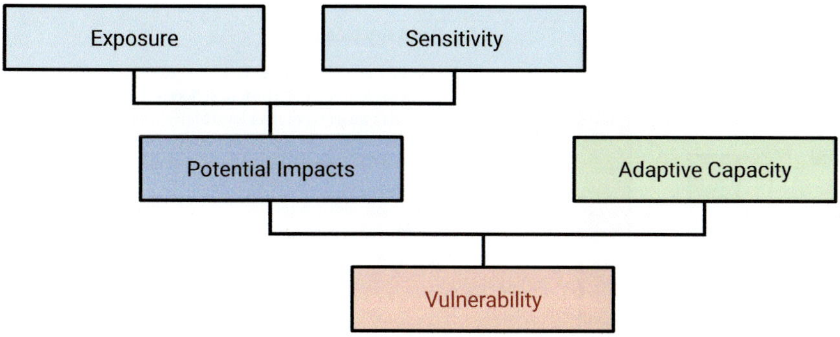

Figure 8.2 Framework for vulnerability assessment (IPCC 2007).

may not be sensitive, occupants within the building can still be impacted if the building is no longer reachable by road or is cut off from utilities.

Evaluating exposure and sensitivity can be informed by insight from historical data (temperature, floods). However, a full evaluation requires an assessment of future frequency, scale, severity, and impacts of climate events driving increasing exposure of assets to degradation. New infrastructure design can reduce both exposure and sensitivity considering 10-, 20-, and 50-year (or longer) recurrence intervals for climate events. Existing infrastructure has fewer paths for exposure reduction. Nonetheless, sensitivity reduction can be gained from retrofits and periodic improvements in asset robustness on a component-by-component basis.

The final step is an assessment of adaptive capacity. Capacity is often defined at an organization level as the ability to respond to both long-term and short-term stress events using physical metrics such as existing system ability to accommodate events. For example, the existing capability of HVAC equipment to provide cooling services for a 100-year heat event. However, capacity can also be defined on human metrics that might include human health, emergency response, access to amenities, and other determinants of human health and safety.

Another relevant concept related to a vulnerability assessment is that of the "vulnerability gap." Most infrastructure starts life with an inherent capacity to resist harmful impacts or "loading." However, as systems age, or become subject to enhanced impact vectors, their capacity may be insufficient to prevent system failures. The vulnerability "gap" may become apparent when forecast loads exceed existing capacity, as seen in Figure 8.3. A system

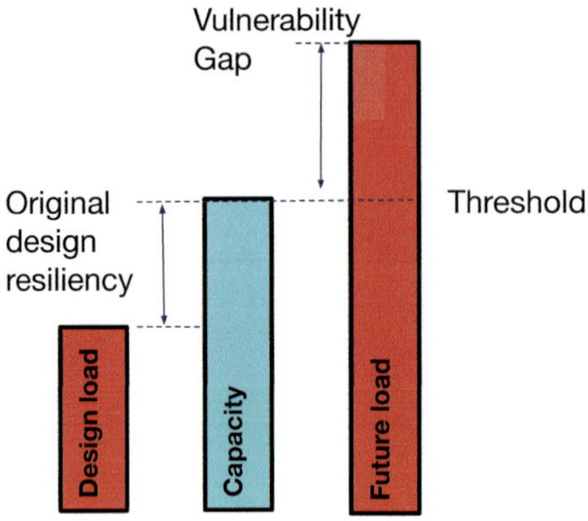

Resiliency Example

Consider an existing river levee designed to contain storm surges for typical historical floods. Typically, a rainfall event causes a river level to rise without harm until it reaches the top of the levee. Over-spillage results in fast erosion of the levee wall and collapses suddenly. The point of spillage marks the capacity and its exceedance results in catastrophic flooding across a flood plain and potentially irreversible damage to building infrastructure.

Figure 8.3 Vulnerability gap as measured relative to design and future loading. Source: Adapted from Engineers Canada PIEVC Protocol.

is resilient if capacity is greater than the forecast load. A system is vulnerable if capacity is less than the forecast load; hence, there is potential of failure. The point at which loading exceeds capacity is called a "threshold." For simple systems, there may be a single threshold, and for most real-world complicated systems, there may be multiple thresholds.

Design and retrofit for resilient systems depends on assessing loading (or exposure as defined earlier), and capacity (analogous to sensitivity). Understanding the level of exposure and sensitivity reduces vulnerability, and thereby manages the system-level risk.

8.2 Risk Assessment

Retrofitting a building for the explicit purpose of making it resilient can be an expensive proposition. As with energy efficiency, the best time to implement resilience measures in existing buildings is in coordination with other building upgrades. For example, adding exterior shading during a comprehensive cladding and window renewal project can reduce overall project costs as cranes and/or scaffolding would already be in place, and would make detailing and sealing of any penetrations much easier to accommodate (Nelson and Adger, 2007). It would also make the overall look of the project more cohesive. Failing to consider and implement energy efficiency and resiliency upgrades when the building is being renewed is a lost opportunity. If climate change or its associated regulations make an upgrade required at a future date, it would undoubtedly be more expensive than if done in conjunction with other work. Conducting a formal risk assessment on a building can be a great way to develop a plan for current and future building renewal retrofit projects.

Risk can be considered as the level of uncertainty against objectives. If the objective is to have resilient infrastructure, risk must be considered as a product of the likelihood of a harmful event and the consequence to an asset. These two variables, likelihood and consequence, are independent of each other. Each variable must be evaluated separately.

ISO 31000 provides a framework on how risk assessments should be conducted. The ISO framework has been used as the backbone of many specialized risk assessment systems such as The Public Infrastructure Engineering Vulnerability Committee (PIEVC) assessment method (https://pievc.ca/protocol).

In terms of risk assessment, shown in the core of the ISO 31000 framework, the three key steps are risk identification, risk analysis, and risk evaluation.

Risk identification:

a. Infrastructure response: Each building component and climate parameter interaction are considered by engineers and expert stakeholders.
b. Preliminary review: A screening analysis eliminates components that are not of interest or obviously low risk.
   Risk analysis:
c. Likelihood scoring: Climate-driven events are scored on a sliding scale (for example, 0–5) for each climate parameter.
d. Consequence scoring: Potential of damage and thresholds are modeled by a technical evaluation. Asset managers and stakeholders then discuss the impacts of the interactions between the climate parameter and infrastructure and to create consequence score.
e. Consequences and likelihood scores are evaluated either individually or collectively in a matrix to evaluate total risk scores. Risk is computed for each combination of likelihood and consequence, as defined by the equation: $Risk = Likelihood \times Consequence$. The results of each combination are used to populate a matrix, as shown in the example contained in Figure 8.4.

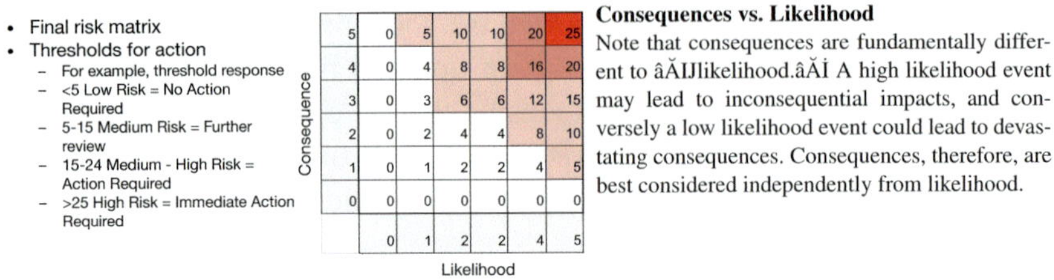

Figure 8.4 Vulnerability gap as measured relative to design and future loading. Source: Adapted from Engineers Canada PIEVC Protocol.

Risk evaluation:

Risk evaluation should be guided by what type of assessment outcomes are required. Outcomes can include many different infrastructure metrics such as damage functions, insurability, durability, harm to human health, and wellbeing. Alternately, outcomes may be guided by institutional key performance indicators (KPIs). Depending on desired outcomes, a risk evaluation reveals three types of risks:

- Acceptable: Risk is deemed to be sufficiently low such that climate adaptation efforts are not justified.
- Tolerable: Risk reduction is required. Adaptive risk strategies are required for reducing residual risk to a level acceptable by asset managers.
- Intolerable: Feasible or affordable adaptation strategies are impossible or unavailable. Intolerable risks require specialized attention outside of the bounds of this chapter.

Bear in mind that outcomes that involve human safety will typically have low levels of acceptable residual risk.

A full discussion of tolerability and risk perception are outside the scope of this chapter. However, it must be noted that in complex projects, these factors become influential. If decision-makers are indifferent, there will be no action taken, regardless of the outcome of a risk assessment. For further guidance, review ISO 31000.

For decision-makers, the following categories describe actions:

- Protect: Large-scale reactive strategy to shield infrastructure via physical means (e.g., fire perimeter, flood wall, shading structures)
- Accommodate: Adaptive strategies at a smaller scale (e.g., permeable surfaces, ground-source heat pump to shed seasonal thermal energy fluxes)

- Top-down strategic action to relocate or abandon assets in-place (e.g., relocating communities away from a flood zone)
- Avoid: Prevent any infrastructure development in at-risk areas. Can be embedded in land-use plans, legal permitting process (e.g., building prohibition in 200-year flood plains).

A combination of the above responses is possible. Selecting the most appropriate risk treatment option(s) involves balancing the potential benefits and achievement of objectives against costs, effort, or disadvantages of implementation.

Risk treatment options are not necessarily mutually exclusive or appropriate in all circumstances. Options for treating risk may involve one or more of the following:

- Avoiding the risk by deciding not to start or continue with the activity that gives rise to the risk;
- Accepting an increased risk in order to pursue an opportunity;
- Removing the risk source;
- Changing the likelihood;
- Changing the consequences;
- Sharing the risk (e.g., through contracts, buying insurance);
- Retaining the risk by informed decision.

Evaluators of a risk assessment should communicate that risks can never be fully eliminated. At best risk can be effectively managed. Ultimately, they must determine which risks are already acceptable or at least tolerable with reduction strategies. The question of "who decides" is important to any situation, and an inclusive process will help ensure buy-in, which is critical to advancing to implementation.

Ideally, a robust risk assessment is key to choosing the right resiliency approach. But analysts should consider that even a high-quality risk assessment will fail when risk perceptions are out of alignment with actual risk.

8.3 Risk Mitigation

This section discusses specific risks that can impact buildings along with concrete actions to mitigate these risks. Risks will vary by location and can be assessed using the risk assessment framework presented in the previous section. In the LEED v4 rating system, one point is provided for assessing the different risks and up to two points for implementing solutions to reduce the building's exposure to the most likely risks. The potential hazards that are identified in LEED's assessment include:

- Flooding
- Sea-level rise and storm surge
- Wildfire
- Hail
- Hurricanes and high winds
- Tornados
- Earthquakes
- Tsunami
- Drought
- Landslides and unstable soils
- Extreme heat
- Winter storms

There are additional risks that should be considered, such as, but not limited to:

- Poor outdoor air quality caused by wildfire smoke
- Global pandemics requiring new ventilation and filtration strategies
- Loss of permafrost in Northern regions

Almost all of the potential risks that buildings are exposed to are increasing as a result of climate change. As discussed earlier in the chapter, in a risk assessment, one should not simply rely on historical data. Instead, future projections should be used as much as possible.

In addition to the risks posed to a building from distinct events, climate change is also introducing a longer-term chronic risk associated with the changing climate. Building designs are often geared toward the local climate. If over years, the climate shifts to being hotter, dryer, or wetter, exposed to more frequent higher winds, etc., a building that was not designed for these conditions might face premature failures.

Another risk that should be assessed is power outages. Power outages can last from minutes to days, and can drag into weeks and even months after larger events. Often power outages are the result of some of the other risks already discussed. Texas had several days of power outages when Winter Storm Uri struck in February 2021. Its power system was not resilient as its gas, electricity, and water systems are inter-linked; so failures in one of them can lead to failures in others (Leggett, 2021). It could be that the building is resilient to the elements, but if the local power grid is not, it will still have an impact on the building. This is assessed separately to other risks in LEED v4, where an additional two points are available for passive survivability and backup power during disruptions.

8.3.1 Concurrent events and cascading impacts

Concurrent events can happen that will impact how resilient a building is. For example, wildfire smoke typically occurs in the summer, and can be coincident with heat waves. In a building that relies on the use of windows and natural ventilation for its primary source of cooling, the poor outdoor air quality can become an issue. Cascading impacts are when one event or failure leads to another impact or failure. A power outage is often a cascading impact of a natural disaster or extreme weather event. A building dealing with the concurrent wildfire smoke and heat wave with mechanical cooling and ventilation may be meeting the needs of the occupants, but the forest fires and/or heat wave can lead to power outages as a cascading impact. In California, people have had to deal with forest fires, extreme heat, smoke, and rolling power outages happening concurrently across the state on a regular basis. In Haiti, two days after being struck by a magnitude 7.2 earthquake on August 14, 2021, the area was hit by Tropical Depression Grace, causing extensive flooding.

The remaining parts of the chapter will talk about solutions, but will be limited to items related to HVAC, although this can include passive measures to reduce loads.

8.3.2 Resilience to extreme heat

Depending on the region, climates are becoming more extreme and with longer periods of excessive heat. This means that buildings that previously might not have required cooling to keep occupants comfortable may be overheating. In this case, the forecast load on an HVAC system will be larger than the existing loading capability, creating a vulnerability gap. Buildings with cooling systems might find that the system no longer provides adequate cooling during heat waves, or simply that it is getting unaffordable to maintain comfortable conditions. With the extreme heat conditions that are becoming more common, we need to consider both how to keep occupants comfortable, and how to maintain safe temperatures suitable for occupant health. We need to not only consider recent heat waves but also plan for even hotter ones that are likely coming as a result of climate change. Although a mechanical system can help achieve thermal comfort, we have seen many instances where heat waves are too much for our strained electrical grids, and power outages occur. This can be considered a threshold of system capability. What passive measures can be added that will help keep the building within safe operating limits in the event that a power outage occurs at the same time as a heat wave? These are all points that will be discussed in this section.

8.3.3 Thermal comfort

A number of different thermal comfort indices have been developed to help assess whether occupants would be comfortable in a building. Given that thermal comfort can be subjective, many of these indices try to find conditions that would be comfortable for a maximum number of occupants. These indices typically take into account personal factors that can impact an individual's comfort (e.g., age, gender, clothing, and activity levels) as well as various environmental factors (air temperature, relative humidity, air speed, operative temperature, and mean radiant temperature). There are some indices that only apply to un-airconditioned or naturally ventilated buildings. These often apply adaptive thermal comfort models, which assume occupants are more tolerant of hotter temperatures when it is hot outside; thus, temperature limits vary with outdoor temperatures.

It is outside the scope of this chapter to provide detailed information on thermal comfort indices. The reader is encouraged to review the various indices and to find the one most applicable to their building and region. These include:

- ASHRAE Standard 55[1]Thermal Environmental Conditions for Human Occupancy
- CIBSE TM59[2] Design Methodology for the Assessment of Overheating Risk in Homes (2017)
- TM52[3] The Limits of Thermal Comfort: Avoiding Overheating in European Buildings
- ISO 7730:2005[4] Ergonomics of the Thermal Environment

---

[1]www.ashrae.org/technical-resources/bookstore/standard-55-thermal-environmental-conditions-for-human-occupancy

[2]www.cibse.org/knowledge-research/knowledge-portal/technical-memorandum-59-design-methodology-for-the-assessment-of-overheating-risk-in-homes

[3]www.cibse.org/knowledge-research/knowledge-portal/tm52-the-limits-of-thermal-comfort-avoiding-overheating-in-european-buildings

[4]www.iso.org/obp/ui#iso:std:iso:7730:ed-3:v1:en

- BS EN 16798-1:2019[5] Energy Performance of Buildings. Ventilation for buildings indoor environmental input parameters for design and assessment of energy performance of buildings addressing indoor air quality, thermal environment, lighting, and acoustics.

The Center for the Built Environment (CBE) has developed a free Thermal Comfort Tool to assist users in verifying compliance with ASHRAE Standard 55 and EN 16798 (Tartarini, 2020). WELL certification from the International WELL Building Institute^{TM} includes a Thermal Comfort category that has a requirement that the majority of building users find the thermal environment acceptable, with additional points for elements such as zoning, individualized control, humidity control, etc.

8.3.4 Health and safety

Not only do we need to consider traditional "thermal comfort" metrics, we also need to start considering how hot is too hot in terms of health and increased risk of mortality. Being too hot can cause a number of heat-related illnesses, such as heat rashes, cramps, and heat stroke. Extreme heat can also worsen pre-existing conditions, such as cardiovascular and respiratory diseases, which lead to stroke, and increase susceptibility to infectious diseases (Busby et al., 2021). All of these conditions lead to people dying because of heat waves.

Figure 8.5 shows the results of a study that estimated the current and projected heat-wave-related deaths for Canada. In the 49 years from 1971 to 2020, an estimated 14,700 deaths (or 300/year) were associated with heat waves. Depending on population growth, global climate mitigation achievements, and the level of adaptation measures that are adopted, this rate is projected to increase by roughly 45%–455%. This stresses the importance of both mitigation and adaptation efforts. Depending on the representative concentration pathway (RCP) scenario (Busby et al., 2021), we see that adaptation measures can reduce annual mortality by roughly 100%–450%.

ISO 7933:2023 provides guidance on maximum allowable exposure times of occupants in hot conditions that would result in no expected physical harm to individuals in good health (ISO, 2023). The Thermal Comfort Tool

[5] www.en-standard.eu/bs-en-16798-1-2019-tc-tracked-changes-energy-performance-of-buildings-ventilation-for-buildings-indoor-environmental-input-parameters-for-design-and-assessment-of-energy-performance-of-buildings-addressing-indoor-air-quality-thermal-environment-lighting-a

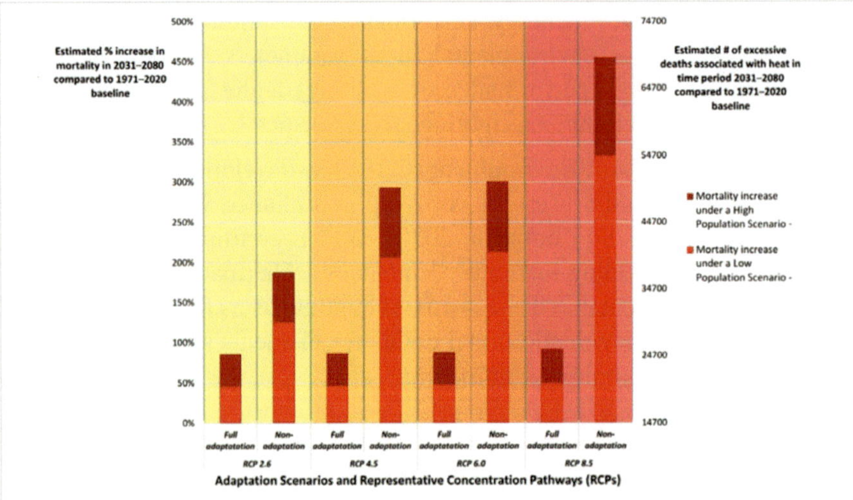

SOURCE: Guo et al., 2018

Figure 8.5 Estimation of the current and projected heat-wave-related deaths for Canada, in the 49 years from 1971 to 2020 (Guo et al., 2018).

developed by CBE provides this maximum allowable exposure time for a variety of different input parameters:

- Air temperature
- Mean radiant temperature
- Air speed
- Relative humidity
- Metabolic rate
- Clothing level

In terms of lowering exposure to heat events, short duration heat waves can be ameliorated by considering temporary changes to these parameters. However, safeguarding comfort for long duration events may require retreating to a dedicated cooled area of a building.

8.3.5 Modeling with future weather files

Modeling with future climate files enables an evaluation of the "consequence" of climate-driven extremes. When it comes to reducing and evaluating the risk of overheating, an important element is to consider how the building will maintain health and comfort as the climate warms. A number of organizations have created future weather files that can be used in energy modeling software. For example:

- *Climatedata.ca*[6] includes future weather files for various locations in Canada developed by the Pacific Climate Impacts Consortium (PCIC) and the National Research Council of Canada (NRC). These files are being referred to in some building codes to verify that minimal comfort standards will continue to be met into the future.
- *CCWeatherGen*[7] Climate Change Weather File Generator is a software tool that was developed to enable users to generate climate change weather files for the UK, ready for use in building performance simulation programs.
- *Meteonorm Software*[8] allows users to generate climate data in various formats. It also includes three IPCC scenarios with projections to the year 2100.

When modeling a system, the behavior of the building under historical or existing climate conditions can be considered a base-case performance. This represents a snapshot of the existing capacity of the system to provide cooling services. The future climate files are testing for projected loading and allow a modeler to identify thresholds when an HVAC system will fail to protect occupants. Verify with your local building authorities if future weather data files are available for your climate. There are a number of factors that can impact future weather data files, including:

Climate change model and emission scenarios: Climate models are developed assuming different projected GHG levels. The results of the global models can be fed into other climate models to develop more granularity to get more regional or localized projections. The global climate models, GHG emission scenarios, and regional analysis methods are being systematically updated as the science evolves. Therefore, depending on the global model used, the emission scenarios, or the regional analysis performed, the resulting projected climate can vary.

Generating future weather estimates: Once the level of warming has been assessed with the climate change modeling, these estimates are then used to create the hourly weather files used in energy modeling software. There are several approaches developed to construct weather files, with a morphing technique being the most well-known (Jiang et al., 2019). Morphing uses

[6] https://climatedata.ca/climate-data-in-action-future-weather-files

[7] https://energy.soton.ac.uk/ccweathergen

[8] https://meteonorm.com/en

existing weather files based on historical data and makes adjustments using climate modeling results. Based on the climate data and methodology used to generate the files, the resulting weather estimates can vary.

Timeframe: Future weather files can be generated for different future time periods. These can be generated for the current period, 2020s, as well as into the future 2050s, 2080s, and 2100.

A guideline document developed by BC Housing suggests using the worst-case emissions scenario (RCP 8.5) and the 2050 timeframe to assess the performance of the building with future climate (BC Housing, 2020). Given that the climate science and the tools are still evolving, simply doing an analysis with future climate scenarios and considering design strategies that will allow the building to maintain comfort as the climate warms is a good starting point.

8.3.6 Mechanical cooling

The consequences of any given climate event are important to consider for risk management. In the case of mechanical cooling, consequences will be directly felt by occupants but also exerted on building systems. A study commissioned by the University of British Columbia examined the impacts of climate change on the thermal comfort of multifamily residential buildings in and around Vancouver (RDH, 2020). The objective was to identify cost-effective design measures that would maintain thermal comfort under future climate conditions for new and existing low- and high-rise residential buildings. The study found that even in the moderate climate of Vancouver, some type of mechanical cooling will be required to maintain thermal comfort as the climate warms.

The need for a mechanical cooling system is amplified when trying to maintain thermal comfort during a wildfire smoke event, which prompts occupants to keep windows closed. The region got a glimpse into the future during a particularly bad heat wave in June 2021, where hundreds died with the abnormally hot temperatures. The study found that for some buildings that kept cooling loads down, thermal comfort could be maintained to a reasonable level with partial cooling added to the ventilation system. The ventilation system would need to include a high efficiency heat recovery ventilator (HRV – with boost and bypass modes), plus a cooling coil downstream of the HRV. For centralized systems, distribution ducts could be oversized during design to allow additional capacity for cooling in the future. The study found that sizing to 150%–160% of capacity enabled boost airflow and additional cooled air to be circulated when needed.

For most climates, and for buildings with large window areas, full mechanical cooling will be needed to maintain thermal comfort. Passive design measures such as adding exterior shading and using low solar heat gain windows will help to lower cooling loads and energy consumption. As discussed in Chapters 2 and 10, building electrification, building envelope improvements, and the use of heat pumps for space heating is a key measure in climate change mitigation. The benefit is that with most heat pump installations, the same system can also be used for mechanical cooling. The use of heat pumps is thus effective at both mitigation and adaptation. Refer to Chapters 9 (HVAC), 10 (Heat Pumps), and Chapter 11 (Building Envelope) for more details on heat pumps and other HVAC systems.

One aspect that should be considered when designing a mechanical cooling system is the future climate. How will the system be able to accommodate the growing cooling demand? Do you install an oversized variable capacity system now that will be able to meet future demands, or do you design a system that can easily be upgraded with additional cooling capacity in the future? How much investment in the building envelope should be made to reduce cooling loads and provide increased hours of life safety in the event of a power outage (e.g. see Table 2.6), possibly cascaded with a heat wave? There may be multiple strategies that could work. What is required is to iterate with the design until an optimized solution is found that achieves all project objectives at the lowest life cycle cost (see Figure 3.1).

8.3.7 Fans to mitigate overheating

Although air-conditioning is common and one of the main options in addressing extreme heat, it is not affordable in many countries with hot climates, or to existing building owners that currently do not have a cooling system, and it is not the most sustainable option (Lehrer, 2021). The proliferation of air conditioning adds strain on electrical grids, creating high peak loads during heat waves that can lead to power outages, which can further exacerbate heat-related health impacts.

The research team at Berkley's Centre for the Building Environment (CBE) has been studying the benefits of fans. They have found fans to be beneficial even in buildings with AC systems. In one of their field studies, they installed fans in houses in California where summer temperatures regularly exceed 100 °F (38 °C). During their study, the AC failed and repairs took months, yet surveys found occupants were satisfied with their comfort despite indoor temperature rising to the mid-80s °F (~29 °C). The cooling effect of the ceiling fans kept the spaces usable.

In an effort to encourage the effective use of fans, the CBE has developed a Fan Tool to help evaluate the cooling effects of different fans in a building (Center for Built Environment, 2019). This is in addition to the CBE Comfort Tool (RDH, 2020) that gives users an indication of when fans are beneficial at eliminating or reducing the level of heat strain at different indoor conditions.

8.3.8 Passive cooling

Although mechanical cooling systems and fans will likely play an important role in keeping buildings comfortable and habitable into the future, the implementation of passive cooling measures is critical. As discussed, it is not uncommon for power outages to occur concurrently with heat waves. If this happens, passive cooling measures can help keep indoor temperatures within safe operating limits (Sun et al., 2021). Even when the power is on, passive cooling helps reduce the size of mechanical cooling systems, and reduce overall building energy consumption.

In existing buildings, the key element is to reduce the amount of solar gains coming into the building. Reducing solar gain can be achieved by either replacing windows with some with lower solar heat gain coefficient (SHGC) or by introducing shading. Exterior shading is much more effective than internal shades. They can reduce the risk of overheating by more than 90% compared to a reduction of only 11%–30% for internal reflective blinds (Laouadi et al., 2021). Other strategies can include introducing more operable windows, and mitigating the heat island effect by using reflective/white surfaces on the facade and roof, and planting trees and other vegetation, on and around the building.

There have been many studies and guides developed to improve the resilience of buildings. Given that warming temperatures are central to climate change impacts, many of the guides have a focus on limiting overheating. To get more specific design guidance, readers are encouraged to examine the resources listed in the bibliography at the end of this chapter.

8.3.9 Resilience to smoke events

Climate change has been leading to more intense forest fires, with this trend expected to continue into the future. With large forest fires burning, large regions can be blanketed with thick smoke. People are generally told to limit the time outside to reduce their exposure to this smoke. However, without proper ventilation and filtration systems, indoor air can also be unhealthy. Measurements taken indoors in commercial buildings during the Australian bushfires found that recommended critical thresholds were often exceeded, with peaks up to 12 times higher than recommended for PM10 and 24 times higher for PM2.5 (Brambilla et al., 2021).

Buildings that rely primarily on operable windows for cooling and fresh air can be particularly impacted. Closing windows can reduce particulate matter but can lead to overheating and other indoor air quality issues such as elevated CO_2 concentrations. Having an airtight building envelope is useful in reducing the amount of particulate matter that infiltrates into the building. The amount of infiltration can be lower when a mechanical ventilation system is operating and keeping a building pressurized. Monitoring in Australia found that commercial buildings that reduced or stopped ventilation overnight to conserve energy experienced spikes in indoor particulate matter from increased infiltration.

In buildings with ventilation systems, filtration of air with high efficiency filters is required to remove smoke particulates. To effectively remove smoke, a MERV 13 filter or better is required (ASHRAE, 2021). Using higher efficiency filters can lead to reduced airflow or higher fan energy consumption. To limit the cost and impacts associated with using these filters, they can be reserved to be deployed only during wildfire smoke events where conventional filters are used on a regular basis. Ventilation with filtration requires occupants to keep windows closed. This could prove problematic in buildings that do not have mechanical cooling systems. Bear in mind that not all HVAC systems are capable of generating the high-pressure gradients required to drive air across a MERV 13 filter-bank. Even if a high-pressure gradient is maintained, there may be other consequences such as pressure damage to older ducting systems.

There are a number of buildings that do not have the proper ventilation and filtration system to provide clean air throughout the building during wildfire smoke events. Alternatively, a building can have a good ventilation and filtration system, but a leaky building envelope could still lead to higher indoor particulate concentrations during a wildfire smoke event. If high particulate concentrations are found indoors during wildfire smoke events, portable air cleaners rated with a high clean air delivery rate for smoke could help improve air quality. Portable HEPA filters have been shown to reduce indoor PM2.5 concentrations by 60% on average (Harriman and Brennan, 2019). In areas prone to wildfires, one strategy could be to acquire one or more portable air cleaners and deploy them during wildfire smoke events.

In summary, passive solutions and mechanical solutions provide a pathway to lower risk to occupants. Depending on a building design, age, and condition, a combination of mechanical and passive strategies may be deployed, along with backup portable HEPA solutions.

8.3.10 Resilience to flooding

The first step in evaluating the flooding risk is to get the latest flood projection levels for your area that take into consideration our changing climate. This chapter focuses on mechanical equipment and will not cover the building structural and durability considerations of flooding. Following Hurricane Sandy in the US, many buildings did not incur substantial damage to their structure, but did experience damage to the building's systems such as furnaces, boilers, water heaters, and electrical panels (City of New York, 2014). The inundation of critical systems including mechanical, electrical, plumbing, elevators, emergency power systems, fuel tanks, sump pumps, communication, and alarm systems can cause a building to be unusable for weeks or even months.

The Intact Centre on Climate Adaptation in Canada has developed guidance for the retrofit of commercial buildings to make them more resilient to flooding (Moudrak and Feltmate, 2019). The following summary, shown in Table 8.1, provides a list of their recommendations in terms of plans and procedures, equipment and supplies, and major retrofits. As with other plans, one of the best things that can be done is to elevate critical systems above expected flood levels. This should also include locating penetrations through the building envelope for utilities above flood levels (EGBC, 2020). The Intact Centre guidance document provides more detailed information, including more guidance on elevators.

8.3.11 Resilience to power outages

Larger buildings generally are required to supply emergency power for a short duration ($\sim$2 hours) to meet minimum life safety requirements related to occupant evacuation and firefighter access. However, in residential buildings, it would be preferable if occupants did not need to evacuate for a power outage. In this case, providing some backup power to meet non-life safety requirements could be important in maintaining occupant well-being (e.g., water supply, heating, elevators), such that they can remain in their building safely and with a degree of comfort. The City of Toronto has developed a voluntary Minimum Backup Power Guidelines for MURBS that would ensure occupants a minimum level of comfort for at least 72 hours, without

Table 8.1 List of recommendations and procedures for resilient retrofits (Moudrak and Feltmate, 2019).

Plans and procedures	Equipment and supplies	Major retrofits
Emergency operations center: Designated space is available for operations staff to use during an emergency.	**Backup power generation:** Onsite backup power is available to operate emergency systems, including sump-pumps for 24–72 hours.	**Elevate and flood proofing equipment:** Critical electrical, HVAC, and communications equipment as well as server rooms are located above expected flood levels. If not possible, equipment is flood proofed.
Emergency supply contracts: Contracts are in place to have generator fuel supply replenished during emergencies.	**Emergency lighting:** Battery operated emergency lighting in mechanical and electrical rooms as well as to assist building egress.	**Isolating electrical circuits:** All floors situated below flood levels (e.g.. parkades) have their own isolated electrical circuits.
	Elevator water sensors: Sensors are in place to prevent elevator from going down to flooded levels.	**Electrical panel upgrade:** Electrical panels are equipped for remote access such that breakers can be shut down from elsewhere.
	Backwater valves: Backwater vales are installed to prevent backflow from storm and sanitary sewer pipes.	

requiring refueling generators (City of Toronto, 2016). The guideline includes a number of recommendations:

- At least one elevator, in addition to the firefighter's designated elevator, should be backed up for resident use during a power outage.
- Sump pumps in basements should be backed up to ensure that any water entering below grade, such as during a rainstorm, can be pumped out during a power outage.
- Backing up booster pumps is necessary to provide adequate water supply to units, especially units located above the fourth floor. Water for drinking, washing, and flushing toilets is essential during extended power outages.
- Ensuring hot water circulation is an important aspect of asset management as it would help preserve the integrity of pipes during power

outages in the winter. Additionally, it would ensure greater occupant comfort during sustained power outages by making showering and dishwashing easier.

- Units in MURBs with radiant heating, which tend to be older apartment buildings, can be provided with space heating during a 72 hour power outage at a low incremental cost because it only requires providing backup power to the boiler and pumps. However, units with heating systems like fan coils or heat pumps – more typical of newer condominiums – are unlikely to have space heating during a 72 hour power outage because each individual heating system would require backup power. The additional electrical distribution costs to each unit tend to make this approach cost-prohibitive.
- Common refuge areas are shared spaces where residents can gather to stay warm, charge cell phones, and perhaps prepare a meal, for example. This is a key resilience solution in situations where it is cost-prohibitive to provide heating (and cooling) to individual units.
- When considering adding a new generator or replacing an existing one in a building without separated loads, coordinating with other electrical work can help reduce electrical distribution costs, for both life safety and non-life-safety loads.
- If at-grade spaces are anticipated to host important community services, consider designing the backup power system to also provide backup power to this part of the building.
- Use natural gas generators, especially in tall buildings.
- Design load management systems with load selection capability.

Historically, generators were the sole option to provide backup power. However, as the cost of solar energy and backup battery systems continues to decrease, they can potentially serve as the backup system, or as a hybrid system to reduce the required size and use of a traditional generator. A 2018 NREL study found that depending on utility rate structures, as solar and battery costs decline, that solar with battery storage became economical in 10 of the 17 markets they studied (McLaren et al., 2018). In order to boost the economics of using battery storage, they should be used for more than just one function (e.g., demand-charge reduction, time-of-use optimization) (Fitzgerald et al, 2015). If batteries are the sole source of backup power, it would severely limit their use for other purposes in order to always have a minimum amount of power available in the case of a power outage. If used for other functions, it can be paired with a local generator to provide backup power at multiple scales, ranging from second-to-second power quality maintenance for industrial operations to daily backup for residential customers. As with other new technologies, there are several regulatory barriers that may restrict the use or increase the cost of deploying batteries for multiple uses within a building.

8.4 Summary

Throughout this chapter, a framework for risk evaluation and assessment is provided, along with examples for risk reduction strategies for building systems. For new buildings, there are many opportunities to design for climate change, attainable at small increases in cost. Deep retrofits also provide opportunities, although such retrofits can be more technically challenging and costly. In either case, there is a burden of responsibility to future-proof our built environment and replacing like-for-like is no longer an option. In terms of risk management, the challenge is even greater because we cannot change the likelihood of a climate event; it is predetermined by our global RCP pathway. This means that for the risk equation, where risk is a product of likelihood and consequence, we can only provide measures to reduce the harm of consequences. This is a key aspect of climate change and ISO 31000. For conventional risk management, both likelihood and consequence can be reduced and managed. For climate change, we can only control the extent of consequences. With "likelihood" being entirely dependent on our global success or failure to mitigate GHG emissions at source, managing consequences may well be our last resort.

Given that managing "consequences" of climate change is the only way for building owners and designers to effectively manage the increasing level of risk, the importance of building resilience will continue to grow. There are many emerging pathways that resilience is being adopted into buildings, including by voluntary approaches such as the RELi rating system (https://gbci.org/reli), or through regulations by increasing requirements in building codes and standards. The extensive bibliography presented at the end of the chapter includes just a fraction of the resources on resilience.

References

ASHRAE (2021) "Filtration/Disinfection" Available: www.ashrae.org/technical-resources/filtration-disinfection.

BC Housing (2020) "Builder Insight 19 - Modelling the Future Climate in Passively Cooled Buildings," BC Housing, Burnaby.

Brambilla A., Gocer O., Candido C., Gocer K., and Sangiorgio M. (2021) "Can commercial buildings cope with Australian bushfires? An IAQ analysis," Building & Cities, vol. 2, no. 1, pp. 593-598.

Busby J.W., Baker K., Bazilian M., Gilbert A., Grubert E., Rai V., Rhodes J., Shidore S., Smith C. and Webber M. (2021) "Cascading risks: Understanding the 2021 winter blackout in Texas," Energy Research & Social Science, vol. 77.

Center for Built Environment (2019) "CBE Fan Tool." Available: https://centerforthebuiltenvironment.github.io/fan-tool/.

City of New York (2014) "Coastal Climate Resiliency - Retrofitting Buildings for Flood Risk," Department of City Planning.

City of Toronto (2016) "Minimum Backup Power Guidelines for MURBs - Voluntary Performance Standards for Existing and New Buildings."

EGBC (2020) "Practice Advisory - Electrical Engineering Considerations in Flood-resilient Design of Buildings - Version 1.0," Engineers & Geoscientists British Columbia.

Fitzgerald, G., J. Mandel, J. Morris and H. Touati, "The Economics of Battery Energy Storage: How multi-use, customer-sited batteries deliver the most services and value to customers and the grid," Rocky Mountain Institute, 2015.

Guo Y. et al. (2018) "Quantifying excess deaths related to heatwaves under climate change scenarios: A multicountry time series modelling study," PLoS Medicine, vol. 15, no. 7, 2018.

Harriman L. and Brennan T. (2019) "New Guidance for Residential Air Cleaners" ASHRAE Journal, September.

ICLR (2020) "Estimating the Benefits of Climate Resilient Buildings and Core Public Infrastructure," Institute of Catastrophic Loss Reduction.

International Organization for Standardization (2023). "ISO 7933:2023 Ergonomics of the thermal environment — Analytical determination and interpretation of heat stress using calculation of the predicted heat strain."

Jiang, X. Liu, E. Czarnecki, C. Zhang, 2019. Hourly weather data projection due to climate change for impact assessment on building and infrastructure. Sustainable Cities and Society 50

Laouadi A., Bartko M., Gaur A., and Lacasse M. (2021) "Climate resilience buildings: guideline for management of overheating risk in residential buildings," National Research Council of Canada. Construction, Ottawa.

Leggett J. (2021), "Climate Change: Defining Adaptation and Resilience, with Implications for Policy," Congressional Research Service, Washington DC.

Lehrer D. (2021) "Alternatives to Air Conditioning: Fans as a Path to Resilience," Available: https://cbe.berkeley.edu/centerline/alternatives-to-ac-fans-for-resilience.

Liu, E., Czamecki C., and Zhang C. (2019) "Hourly weather data projection due to climate change for impact assessment on building and infrastructure," Sustainable Cities and Society, vol. 50.

Moudrak N. and Feltmate B. (2019) "Ahead of the Storm - Developing Flod-Resilience Guidance for Canada's Commercial Real Estate," Intact Centre on Climate Adaptation, University of Waterloo.

McLaren, J., K. Anderson, N. Laws, P. Gagnon, N. DiOrio and X. Li, "Indentifying Critical Factors in the Cost-effectiveness of Solar and Battery Storage in Commercial Buildings," NREL - National Renewable Energy Laboratory, 2018.

Nelson D. and Adger W.K. (2007), "Adaptation to Environmental Change: Contributions of a Resilience Framework," Annual Review of Environment and Resources, vol. 32, no. 1, pp. 395-419.

RDH Building Science (2020) "Designing Climate Resilient Multifamily Buildings," University of British Columbia, Vancouver.

Sun K., Zhang W., Zeng Z., Levinson R., Wei M. and Hong T., (2021) "Passive cooling designs to improve heat resilience of homes in underserved and vulnerable communities." Energy and Buildings, vol. 252.

Tartarini F., Schiavon S., Cheung T., and Hoyt T. (2020) "CBE Thermal Comfort Tool," Center for the Built Environment. Available: https://comfort.cbe.berkeley.edu/.

Bibliography

- American Institute of Architects, Resilience Design Toolkit, 2023.
- American Institute of Architects, Resilient Project Process Guide, 2022.
- ASHRAE British Columbia Chapter, Building Sustainability & Resilience Guide, 2022.
- ASHRAE Guideline 29-2019 – Guideline for the Risk Management of Public Health and Safety in Buildings
- Attia, S., et. Al., Resilient cooling of buildings to protect against heat waves and power outages: Key concepts and definition, *Energy and Buildings*, Volume 239, 2021
- BC Housing, BC Energy Step Code Design Guide Supplement S3 on Overheating and Air Quality, 2019.

- Climate Data Online, National Centers for Environmental Information, www.ncdc.noaa.gov/cdo-web/
- Climate Resilience Guidelines for BC Health Facility Planning & Design October 2020. Original data source: HealthADAPT, 2020
- Climdex indices, www.climdex.org
- Coastal Climate Resiliency – Resilient Retail, Department of City Planning City of New York, New York, 2016.
- Collection of PIEVC Assessments, PIEVC Program
- CSA S478:19 Durability in buildings, 2019
- CSA PLUS 4011.1:19 Technical Guide: Design and construction considerations for foundations in permafrost regions
- CSA S504:19 Fire resilient planning for northern communities
- CSA Z800 Guideline on basement flood protection and risk reduction
- CSA/W204: Flood Resilient Design of New Residential Communities
- European Commission, EU-level technical guidance on adapting buildings to climate change – Best Practice Guide, 2023.
- European Commission, EU-level technical guidance on adapting buildings to climate change, 2023.
- FEMA, Building Codes Save: A Nationwide Study Losses Avoided as a Result of Adopting Hazard-Resistant Building Codes, 2020
- Health Canada (2020) "Reducing Urban Heat Islands to Protect Health in Canada - An Introduction for Public Health Professionals," Government of Canada, Ottawa.
- INVESTING IN CANADA'S FUTURE: The Cost of Climate Adaptation at the Local Level. *The Federation of Canadian Municipalities (FCM) and Insurance Bureau of Canada (IBC),* 2020, http://assets.ibc.ca/Documents/Disaster/The-Cost-of-Climate-Adaptation-Report-EN.pdf
- International Code Council (ICC), Global Building Resilience Guidelines, 2022.
- International Code Council (ICC), Resilience Toolkit.
- IPCC Climate Change 2014: Impacts, Adaptation, and Vulnerability. Part A Chapter 6 and Chapter 19
- ISO 14090:2019 Adaptation to climate change — Principles, requirements and guidelines (preview here)
- ISO 14091:2021 Adaptation to climate change — Guidelines on vulnerability, impacts and risk assessment (preview available here)

- Kesik, T., O'Brien, L., Peters, T.Enhancing the Liveability and Resilience of Multi-Unit Residential Buildings (MURBs) Version 2.0, 2019
- Kesik, T., O'Brien L., Thermal Resilience Design Guide, University of Toronto, 2019
- LEED V4, Assessment and Planning for Resilience_and Design for Enhanced Resilience, US Green Building Council
- Lepage, R., Climate Change Resilience for Buildings, RDH Building Science and BC Housing, 2021.
- NOAA, Implementing the Steps to Resilience A PRACTITIONER'S GUIDE, 2022
- NYC Mayor's Office of Climate and Environmental Justice, Climate Resiliency Design Guidelines – Version 4.1, May 2022.
- Persily, A., Emmerich, S. Technical Note 1882 Indoor Environmental Issues in Disaster Resilience, NIST. 2015
- NIST, Community Resilience Planning Guide for Buildings and Infrastructure Systems, 2015
- RELi^{TM} 2.0 Rating Guidelines for Resilient Design and Construction, USGBC, 2021
- Standards in Action: Building a Climate-Resilient Future, Standards Council of Canada, 2021
- Sandink, D., Robinson, B., Dale, N., Okrutny, P., Practical guidance for private-side drainage systems to reduce basement flood risk: addressing critical information gaps. National Research Council of Canada, 2021.
- Urban Land Institute, Resilient Retrofits – Climate Upgrades for Existing Buildings, 2022.
- World Bank, BUILDING REGULATION FOR RESILIENCE, Managing Risks for Safer Cities, 2015.
- World Green Building Council, Climate Change Resilience in the Built Environment – Principles for adapting to a changing climate, 2022.
- Wright, K., Whitehouse, K., Curti, J., Voluntary Resilience Standards: An Assessment of the Emerging Market for Resilience in the Built Environment (Kresge, and Barr Foundations), 2017
- Zhang, C., et al., Resilient cooling strategies- A critical review and qualitative assessment. *Energy and Buildings* **251,** 2021.

Author Biography

Stefan Storey is an educator and innovator with a focus on energy and resource management. He teaches courses in energy policy, adaptation, entrepreneurship, and related subjects at New York Institute of Technology and the University of British Columbia. Stefan also leads Sensible Building Science, a data science technology company dedicated to optimizing building controls for efficiency and occupant comfort. His areas of expertise include data science, environmental assessment, automated systems, energy modeling, risk analysis, financial planning, and design for resiliency. He is a dedicated champion for his students and an advocate for universal and accessible education. He believes open-access knowledge is essential for empowering learners everywhere.

Remi Charron is a professional engineer with a Ph.D. degree in net-zero energy buildings. He has spent the last 12 years as an independent consultant for the provincial (British Columbia) and Canadian governments on projects related to making housing more energy-efficient and resilient. From 2016 to 2024, he was Associate Dean and Professor for a Master's of Energy Management at the New York Institute of Technology's Vancouver campus. In 2023, he published *Plan for a Turbulent Future*, a guidebook helping readers build personal resilience to climate change impacts. He recently started a role with the Canadian government, helping support innovation in the construction industry. He lives in Pemberton, British Columbia, where he and his wife run a small family farm.

Practice Questions

1. Which of the following can be considered climate events rather than climate processes?
   - A. Monsoon floods
   - B. Hurricanes and typhoons
   - C. Desertification
   - D. Water scarcity
   - E. Food insecurity
   - F. Salinization of agricultural land
2. In terms of Risk = Consequence × Likelihood, which of the following statements are correct.

A. Very low likelihood and high consequences are never a significant risk
B. Very high likelihood and low consequences may be a very low and manageable risk
C. Very high likelihood and very high consequences can be a low risk
D. Medium likelihood and medium risk may need further review

3. Risk assessment typically involves multiple experts and decision-makers around the table. Which of the following is not usually invited?

   A. Facility owners and managers
   B. Debt manager from a banking authority
   C. Health, safety, and emergency responders
   D. Community representatives
   E. Climate expert

4. Which of these statements is not true?

   A. Small changes in climate event magnitude can result in significant increases in damage or other impacts
   B. There is only one threshold for each infrastructure asset
   C. A system is vulnerable when loads exceed capacity
   D. An asset is resilient if capacity exceeds loads, but may become vulnerable if expected loads will be higher in the future

5. In terms of the difference between weather and climate, which of the following statements are applicable to weather?

   A. It rained 20 mm in the morning
   B. It has not rained this month yet
   C. Usually rainfall amounts in March will exceed 80 mm
   D. This April is predicted to be warmer than normal
   E. It has never exceeded 38 °C in the month of April

6. You are a resiliency planner working for a small-town municipality on coastal British Columbia. You are tasked with creating a resiliency plan for a community building for a time horizon out to 2050. Describe the steps you would take to build your plan. Consider:

   A. How would you assess risk in terms of consequence and likelihood?
   B. Which stakeholders would you include in your resiliency plan?
   C. What considerations would be important to them?
   D. How would you reduce risk?

7. Using the CBE Comfort Tool in the Fans & Heat tab (Tartarini, 2020), what air speed would be required to keep you from having any heat strain while reading this book with typical summer indoor clothing when there is an operative temperature of 44 °C and 25% relative humidity?

9

Heading, Ventilation, and Air Conditioning

Mike Reimer

Firstlight Energy Solutions, Canada

9.1 Introduction

Heating, ventilation, and air conditioning (HVAC) systems are pivotal in maintaining indoor environmental quality, encompassing temperature control, humidity regulation, and air purity. Considering that people spend up to 90% of their time indoors,[1] well-designed and maintained HVAC systems facilitate comfortable and healthy indoor environments for occupants. For industrial applications, HVAC systems maintain specific indoor climate conditions for processes and products. By controlling the indoor climate, HVAC systems help to optimize human health, productivity, and well-being, highlighting their indispensable role in modern building design and operation.

However, HVAC systems are associated with significant energy consumption and greenhouse gas (GHG) emissions. According to the International Energy Agency (IEA), space and water heating account for nearly 45% of total energy use in buildings worldwide.[2] This substantial energy demand has a considerable environmental impact, contributing to approximately 80% of direct GHG emissions from buildings.[3] As the effects of climate change intensify, the importance of HVAC systems for maintaining occupant safety and well-being grows. Yet, the impact on energy use and climate change cannot be overlooked. This juxtaposition highlights the urgent need to implement more energy-efficient and sustainable HVAC solutions for both new and

[1] The Inside Story: A Guide to Indoor Air Quality | US EPA

[2] World Energy Outlook 2023 – Analysis - IEA

[3] World Energy Outlook 2023 – Analysis - IEA

existing buildings to mitigate climate impacts and ensure occupant health and well-being.

This chapter aims to equip energy management professionals with a foundational knowledge of HVAC systems and their role in maintaining energy-efficient and healthy indoor environments. This chapter focuses on the following key areas:

1. **The purpose of HVAC systems.** Providing a clear understanding of the purpose of HVAC systems and emphasizing their role in maintaining indoor air quality, thermal comfort, and overall well-being for people.
2. **Energy principles.** Outlining fundamental principles that govern the energy processes in HVAC systems, and helping energy managers grasp the core concepts needed for effective system analysis and optimization.
3. **Heating and cooling loads.** Providing an overview of the factors driving building heating and cooling loads, directly impacting HVAC system sizing and overall performance. This section will highlight key load contributors and their influence on HVAC systems.
4. **Ventilation systems.** Reviewing common components of ventilation systems and discussing strategies for optimizing energy efficiency without compromising indoor air quality.
5. **Heating and cooling systems.** Providing an overview of different heating and cooling systems, including central and decentralized configurations. This section also covers distribution and terminal systems, including strategies to improve energy performance.

9.2 Indoor Environmental Quality

Indoor environmental quality (IEQ) is a crucial aspect of building design and operation that directly impacts occupants' well-being, productivity, and health. A growing body of research underscores the clear association between high-quality indoor environments and improved productivity, reduced absenteeism and presenteeism, enhanced academic learning and test scores, and an overall sense of well-being. This relationship highlights the importance of good IEQ for individual health and the broader societal and economic benefits it brings. As indoor spaces are where people spend approximately 90% of their time, the significance of IEQ cannot be overstated.

Along with building envelope performance, healthy indoor materials, and lighting systems, HVAC systems play a vital role in providing a comfortable and healthy indoor environment, addressing several key pillars of IEQ:

- *Air quality* is a primary concern in IEQ, with indoor air often containing a mix of pollutants such as chemicals, biological agents, and particulate matter that can negatively impact health and comfort. Effective ventilation and air filtration are essential for maintaining low levels of indoor pollutants. These systems ensure an adequate supply of outdoor air, diluting and removing contaminants from indoor spaces to safeguard occupant health.
- *Thermal comfort* is another critical aspect of IEQ, directly influenced by the ability of an HVAC system to control air temperature, humidity, and circulation within a space. Factors such as radiant temperature and airspeed also significantly affect how individuals perceive their thermal environment. Proper design and operation of HVAC systems are vital in achieving a balance to meet the diverse comfort preferences of building occupants.
- *Lighting and daylighting* contribute significantly to visual comfort and the regulation of circadian rhythms, affecting both physical and psychological well-being. While not directly managed by HVAC systems, integrating building systems to support access to natural light while controlling heat gain and loss is essential to holistic IEQ strategies. Appropriate artificial lighting solutions that avoid glare and discomfort are also vital in maintaining optimal conditions throughout indoor spaces. Refer to Chapter 12 for more information on lighting systems.
- *Acoustics and vibration* is the aspect of IEQ that addresses the impact of sound and vibration on an occupant's ability to concentrate, communicate, and feel comfortable within a space. When poorly designed or operated, HVAC systems can contribute to noise levels through mechanical sounds or airflow noise, detracting from acoustic quality. Therefore, careful design to minimize noise and vibration transmission and selecting quiet equipment are crucial in preserving an indoor environment conducive to productivity and well-being.

9.2.1 Thermal comfort

ASHRAE Standard 55 defines thermal comfort as the condition of mind that expresses satisfaction with the thermal environment and defines an acceptable thermal environment as one which more than 80% of occupants find acceptable. Thermal comfort is complex, and environmental and personal factors influence subjective perception, as illustrated in Figure 9.1.

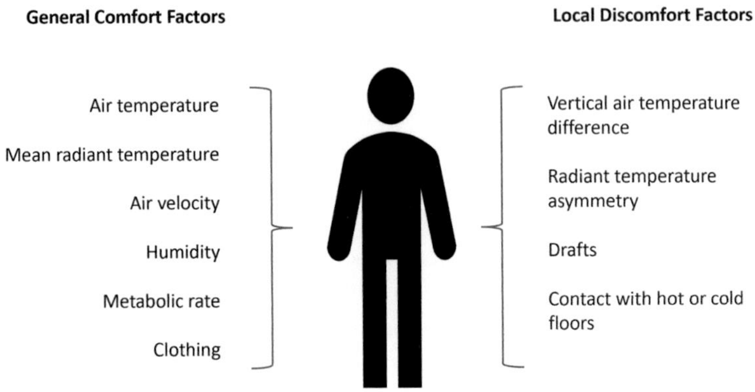

Figure 9.1 Factors affecting thermal comfort.

9.2.1.1 General comfort factors

General comfort factors encompass the overall conditions affecting the thermal sensation of an individual in a given space. These include both environmental and personal factors.

Environmental factors contributing to general thermal comfort include air temperature, radiant temperature, air speed, and humidity. Air temperature is the average air temperature of the occupied space. In contrast, radiant temperature is the average temperature of all surfaces in the occupant's line of sight, affecting heat loss or gain through radiation. Air speed influences the body's convective heat loss or gain, and humidity affects the body's ability to lose heat through sweat and evaporation.

Personal factors also play a significant role in thermal comfort, including a person's metabolic rate and clothing. The metabolic rate reflects the body's internal heat production, varying by activity level, body mass, age, and gender. At the same time, clothing insulation represents the resistance to heat flow provided by the occupant's attire.

9.2.1.2 Localized discomfort factors

Localized thermal discomfort factors may cause discomfort in specific areas of the body despite the overall thermal conditions being within an acceptable range. Localized discomfort conditions will overshadow general comfort conditions, often the source of thermal comfort complaints within a space. These include vertical air temperatures between the feed and head, radiant temperature asymmetry caused by hot or cold surfaces such as windows, localized cooling of the body caused by drafts, and extremes in floor temperature causing discomfort to the feet.

9.2.1.3 Assessing thermal comfort

ASHRAE Standard 55 and ISO 7730 are widely accepted thermal comfort standards used to assess thermal comfort within the built environment. EN 16798 is also often used with the European Union to evaluate thermal comfort. The Center for the Built Environment (CBE) Thermal Comfort Tool is a free online tool for assessing thermal comfort compliance with all three standards: CBE Thermal Comfort Tool for ASHRAE-55 (https://comfort.cbe.berkeley.edu).

9.2.1.4 Additional references

- ANSI/ASHRAE Standard 55 (2023). *Thermal Environmental Conditions for Human Occupancy*. Atlanta, GA: American Society of Heating, Refrigerating and Air-Conditioning Engineers.
- Standard 55 User's Manual (2016). Atlanta, GA: American Society of Heating, Refrigerating and Air-Conditioning Engineers.
- ISO 7730 (2025). Ergonomics of the thermal environment. Analytical determination and interpretation of thermal comfort using calculation of predicted mean vote (PMV) and predicted percentage of dissatisfied (PPD) indices and local thermal comfort criteria. Geneva, Switzerland: International Organization for Standardization (ISO).
- CBE Thermal Comfort Tool: CBE Thermal Comfort Tool for ASHRAE-55 (berkeley.edu). Centre for the Built Environment, Berkeley University of California.

9.2.2 Indoor air quality

Indoor air quality (IAQ) significantly impacts human health, comfort, and productivity. The effects of poor IAQ include short-term irritations, such as eye and throat discomfort, and long-term health issues, such as respiratory diseases and heart problems. Immediate effects, including symptoms like colds or other viral diseases, are often treatable. Conversely, long-term exposure to polluted indoor air can lead to severe health conditions, including chronic respiratory disease, heart disease, and even lung cancer.[4] Understanding these impacts is crucial for mitigating risks associated with indoor pollution. Table 20.3 in Chapter 20 provides a concise summary of common air pollutants, impacts, and control methods.

Improving IAQ involves a combination of strategies that focus on removing or reducing pollutants directly at their source, diluting contaminants

[4] www.epa.gov/indoor-air-quality-iaq

through ventilation, and removing particles and gases from the air using filtration. Below is a summary of the common methods to enhance IAQ:

1. **Source control.** This is considered the most effective way to improve indoor air quality. By eliminating individual sources of pollution or reducing emissions, the overall air quality can be significantly enhanced without additional measures.
2. **Ventilation.** Increasing the amount of outdoor air entering the indoor space helps to dilute or displace the concentrations of indoor air pollutants. Additional ventilation can be supplied through natural methods such as opening windows and doors or mechanically through HVAC systems. The key is bringing in fresh air and expelling polluted indoor air.
3. **Filtration.** Air filtration systems and air cleaners are used to remove particles and, in some cases, gaseous pollutants from indoor air. High-efficiency particulate air (HEPA) filters, for instance, are very effective at capturing particulate matter such as dust, pollen, and mold spores. Some systems also include activated carbon or other adsorbents that help reduce gaseous contaminants. How air is distributed into a space is important because air patterns impact the effectiveness of delivering air into the occupied zone.
4. **Air cleaning.** In addition to traditional filtration, various air cleaning technologies are available that target specific pollutants. These include UV germicidal irradiation, which can deactivate airborne pathogens, and photocatalytic oxidation, which can break down volatile organic compounds (VOCs) into less harmful substances.

Both ventilation and air filtration impact the energy efficiency of HVAC systems. Ventilation air requires heating, cooling, and dehumidification throughout the year to ensure the air being introduced into the space is suitable to maintain thermal comfort. Filtration, however, impacts the fan energy required to move ventilation air throughout the building. Section 9.6 of this chapter discusses strategies to improve ventilation system energy efficiency.

9.2.2.1 Ventilation standards

ASHRAE Standard 62.1 "Ventilation for Acceptable Indoor Quality" defines the minimum outdoor air ventilation rates and local exhaust rates to provide acceptable indoor air quality to minimize adverse effects on health. The standard provides three procedures for determining minimum ventilation rates. These include: (1) The Ventilation Rate Procedure, which is prescriptive and most commonly used; (2) The Indoor Air Quality Procedure, which is performance-based and requires an analysis of contaminant sources and their impact on IAQ; and (3) the Natural Ventilation Procedure, which includes prescriptive or engineered system design procedures for naturally ventilated buildings.

Discussion of these procedures is beyond the scope of this chapter. Refer to the additional resources for more information. Table 9.1, adapted from

Table 9.1 Check table for minimum ventilation volumes (adapted from ASHRAE 62.1-2019 Table L-1).

Occupancy Category	Combined Outdoor Air Rate L/s-m^2	Combined Outdoor Air Rate cfm/ft^2	Occupancy Category	Combined Outdoor Air Rate L/s-m^2	Combined Outdoor Air Rate cfm/ft^2
Educational Facilities			**Office Buildings**		
Art classroom	2.53	0.51	Breakrooms	2.47	0.49
Classrooms (ages 5 through 8)	2.47	0.49	Main entry lobbies	0.73	0.15
Classrooms (ages 9 plus)	3.13	0.63	Office space	0.57	0.11
Computer lab	2.47	0.49	Reception areas	1.40	0.28
Lecture classroom	3.69	0.73	**Public Assembly Spaces**		
Lecture hall (fixed seats)	8.00	1.58	Auditorium seating area	5.40	1.08
Libraries	1.13	0.23	Courtrooms	2.73	0.55
Multiuse assembly	5.47	1.08	Legislative chambers	2.07	0.41
Food and Beverage Service			Libraries	1.13	0.23
Cafeteria/fast-food dining	6.27	1.24	Lobbies	5.40	1.08
Restaurant dining rooms	4.75	0.94	Museums/galleries	2.43	0.48
General			Places of religious worship	4.40	0.88
Break rooms	1.23	0.25	**Retail**		
Coffee stations	1.07	0.21	Sales	1.56	0.31
Conference/meeting	2.07	0.41	Mall common areas	2.43	0.48
Corridors	0.40	0.08	Supermarkets	0.81	0.16
Hotels, Motels, Resorts, Dormitories			**Sports and Entertainment**		
Barracks sleeping areas	1.07	0.21	Casinos	7.28	1.44
Bedroom/living room	0.73	0.15	Dance floors	13.73	2.75
Lobbies/prefunction	1.92	0.38	Game arcades	2.21	0.44
Multipurpose assembly	4.40	0.88	Gym, sports arena (play area)	2.13	0.43
Miscellaneous Spaces			Health club/aerobics room	5.73	1.15
Banks	1.16	0.23	Health club/weight rooms	1.73	0.35
Computer (not printing)	0.53	0.11	Spectator areas	8.00	1.58
General manufacturing (excludes heavy industrial and processess using chemicals)	1.67	0.33	Stages, studios	5.07	1.01
Shipping/receiving	0.93	0.19	Swimming (pool and deck)	3.20	0.64
Sorting, packing, light assembly	1.15	0.23	**Transient Residential**		
Warehouses	0.47	0.09	Dwelling unit	0.50	0.10

source: Adapted from ASHRAE 62.1-2019, Appendix L, Table L-1

Note: Not intended to be used for design

ASHRAE 62.1, provides a list of typical minimum outdoor ventilation rates by occupancy category and can be used as a quick reference for back-of-the-envelope calculations when assessing ventilation impacts for various types of facilities.

Additional codes and standards typically govern industrial and healthcare applications. For example, ASHRAE Standard 170 governs ventilation requirements for healthcare facilities in the U.S., and a similar standard, CSA Z371.2, defines ventilation requirements for healthcare facilities in Canada. The applicable code or standard must be referenced when assessing energy efficiency improvements for ventilation systems.

9.2.2.2 Additional resources

- ANSI/ASHRAE Standard 62.1 (2019). *Ventilation for Acceptable Indoor Air Quality.* Atlanta, GA: American Society of Heating, Refrigerating, and Air-Conditioning Engineers.
- ANSI/ASHRAE Standard 170 (2017). *Ventilation of Health Care Facilities.* Atlanta, GA: American Society of Heating, Refrigerating, and Air-Conditioning Engineers.
- Standard 62.1 User's Manual. Atlanta, GA: American Society of Heating, Refrigerating, and Air-Conditioning Engineers.
- *Indoor Air Quality Guide: Best Practices for Design, Construction, and Commissioning* (2009). Atlanta, GA: American Society of Heating, Refrigerating, and Air-Conditioning Engineers.
- CAN/CSA Z317.2 (2019). Special requirements for heating, ventilation, and air-conditioning (HVAC) systems in health care facilities. Mississauga, Ontario: Canadian Standards Association (CSA Group).
- Industrial Ventilation: A Manual or Recommended Practice for Design, 31st Edition; American Conference of Governmental Industrial Hygienists (ACGIH).
- Indoor Air Quality (IAQ). https://www.epa.gov/indoor-air-quality-iaq United States Environmental Protection Agency (EPA).

9.3 Basic First Principles Related to HVAC Systems

A basic understanding of first principles related to energy systems is essential for analyzing the energy efficiency of HVAC systems. Key concepts include thermal energy efficiency, which assesses how effectively HVAC systems convert energy into useful heating or cooling, and the concept of energy quality, which refers to the suitability of energy for specific applications.

Additional critical principles include sensible and latent heat processes, psychrometric relationships for moist air, and the basic methods of heat transfer – conduction, convection, and radiation – all of which play a vital role in HVAC system analysis and design.

9.3.1 Energy/thermal efficiency

Thermal systems are commonly characterized by their thermal efficiency, often referred to as 1st law efficiency. The first law of thermodynamics (the conservation of energy) states that energy cannot be created or destroyed within a closed system but must be conserved. This concept is illustrated in Figure 9.2. For systems with no internal energy storage, a common assumption for HVAC energy conversion systems such as heat pumps, chillers, and boilers, the first law reduces to [Energy Input] = [Energy Output]. In its simplest form, thermal efficiency measures the amount of useful energy output from the systems divided by the energy input.

For combustion-based systems, the thermal efficiency is always less than 100% due to heat loss in flue gases, radiation and conversion losses, and water blow-down required for water treatment purposes (see Figure 9.3). Thermal efficiency is one efficiency term used to characterize combustion-based heating systems, and it is important to avoid confusing it with the other terms to ensure heating system options are compared on a level-playing field. The three common efficiencies (including thermal efficiency) are detailed in the sidebar below.

Refer to Chapter 20 for more information on combustion-based heating systems.

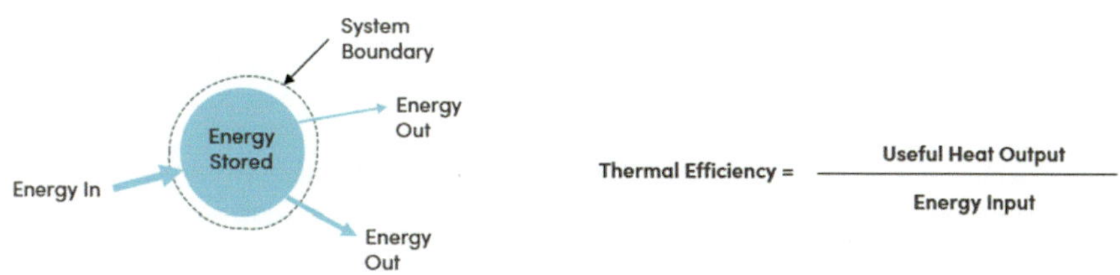

[Energy Input] – [Energy Output] = [Increase in Energy Stored]

For no energy storage:

[Energy Input] = [Energy Output]

Figure 9.2 Conservation of energy and thermal efficiency.

Electric resistance heaters, including baseboard heaters, domestic water heaters, and heating boilers, are characterized by similar efficiencies. Because

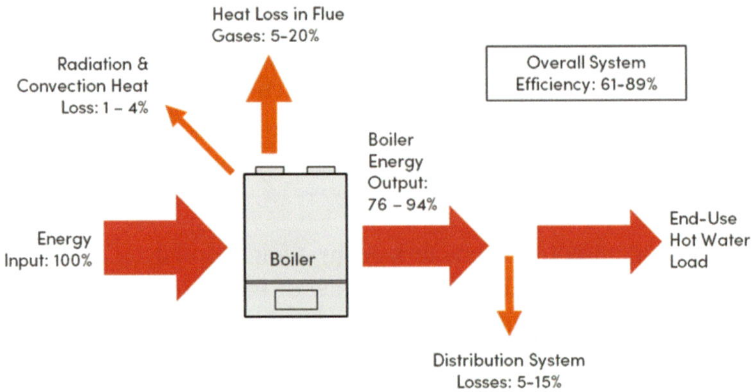

Figure 9.3 Energy balance and thermal efficiency of combustion-based heating systems.

there is no combustion, virtually all the heat input for an electric heater is converted to useful heat. The thermal (overall) efficiency of electric domestic water heaters and boilers is generally 92%–96%,[5] with losses attributed to jacket losses or heat loss from blow-down.

SIDEBAR: COMBUSTION-BASED EFFICIENCIES

Combustion efficiency. A measure of the combustion efficiency only and does not account for thermal losses except losses through the flue gas. Combustion efficiency will always be higher than thermal (overall) efficiency and seasonal efficiency.

Thermal (overall) efficiency. A measure of the boiler efficiency based on the useful heating delivered to the heat transfer media (air or water). Thermal losses include flue gas heat, radiant and convective jacket losses to the surrounding room, water blow-down required for chemical treatment, and energy losses from equipment cycling on and off.

Seasonal efficiency. A measure of the operating efficiency achievable over a heating season at various loads. Accounts for efficiency gains or losses due to part-load operation, including heat losses when the equipment is cycled on and off. The control method (on/off vs. modulating operation) can significantly impact seasonal efficiency.

[5]ASHRAE. 2020. ASHRAE Handbook – HVAC Systems and Equipment, Chapter 32 Boilers.

Air conditioners, water chillers, and heat pumps operate fundamentally differently from combustion-based heating or electric-resistance heating systems, using input energy to move thermal energy rather than solely converting the input energy into heat. The performance of these systems is typically quantified by the coefficient of performance (COP), which is the beneficial heat absorbed (in cooling mode) or heat rejected (in heating mode) divided by the energy input to drive the refrigeration cycle (refer to Figure 9.4). Because the COP is influenced by useful energy output, perspective is essential when defining a COP. For cooling applications, the COP is defined by the amount of heat the unit removed from a space or process, indicated as the "Cooling Effect" in Figure 9.4. For heating applications, the heat output from the process is the "Heating Effect." Finally, for processes that can use both sides of the cycle simultaneously, the COP is defined by the total cooling effect plus the heating effect produced by the refrigeration cycle. Refer to Chapter 10 (Heat Pumps) for more information.

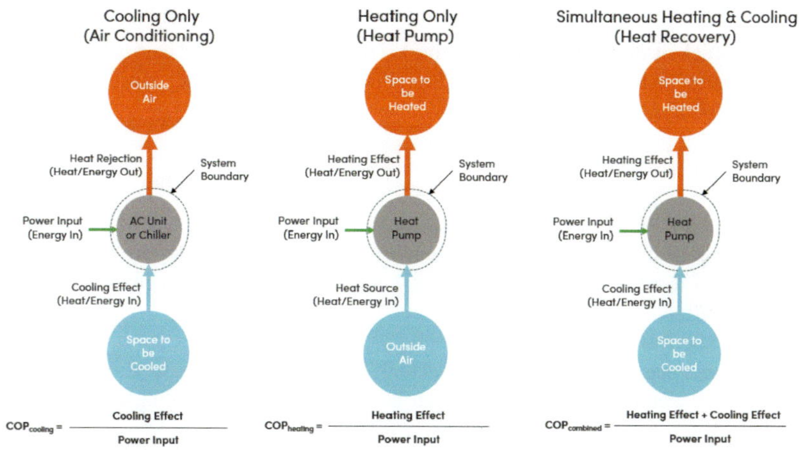

Figure 9.4 Air conditioning and heat pump efficiency.

SIDEBAR: EFFICIENCY TERMS FOR AC, CHILLERS, AND HEAT PUMPS

Numerous terms are used to characterize air conditioner, chiller, and heat pump efficiencies. It is necessary to compare like efficiency terms when assessing equipment. For this discussion, it is important to remember a few useful unit conversions:

1 ton (refrigeration) = 12,000 Btu/hr = 3.517 kW
1 kW = 3,412 Btu/hr

Energy efficiency metrics are used to define the efficiency of an air conditioning unit, chiller, or heat pump at a specific operating condition.

EER (energy efficiency ratio). The ratio of the net cooling capacity (Btu/hr) to the total rate of electric input (Watts) under designated operating conditions. It is commonly used for unitary air conditioners and heat pumps (in cooling mode).

$$\text{EER (Btu/Wh)} = \frac{\text{Cooling Output (Btu/hr)}}{\text{Total Energy Usage (W)}}.$$

kW/ton. The ratio of energy input (kW) to the rate of cooling delivered (tons). It is commonly used for air-cooled and water-cooled chillers.

$$\text{kW/ton} = \frac{\text{Power Input (kW)}}{\text{Cooling Capacity (tons)}}.$$

COP$_C$ (Coefficient of performance, cooling). The ratio of the cooling rate delivered to the energy input rate in consistent units (Btu/hr per Btu/hr or W per W). It is commonly used for heat pumps and chillers.

$$\text{COP}_\text{C} = \frac{\text{Cooling Output (W)}}{\text{Electrical Input (W)}} = \frac{\text{Cooling Output (Btu/hr)}}{\text{Electrical Input (Btu/hr)}}.$$

COP$_H$ (Coefficient of performance, heating). The ratio of the rate of heat delivered to the rate of energy input in consistent units (Btu/hr per Btu/hr or W per W). It is commonly used for heat pumps.

$$\text{COP}_\text{H} = \frac{\text{Heating Output (W)}}{\text{Electrical Input (W)}} = \frac{\text{Heating Output (Btu/hr)}}{\text{Electrical Input (Btu/hr)}}.$$

A *figure of merit* characterizes equipment performance over a range of operating conditions and is intended to reflect actual usage patterns and seasonal variations. They cannot be directly compared with energy efficiency metrics when comparing the efficiency of different equipment options.

HSPF (heating seasonal performance factor) is the ratio of heating output (in Btu) to the total electric energy input (in Watt-hours) over a typical heating season. It is primarily used for small unitary heat pumps.

IEER (integrated energy efficiency ratio). A weighted average of the part-load EER efficiency based on weighted operation at various load capacities. It is primarily used for commercial unitary air conditioners.

IPLV (Integrated part load value). A weighted average efficiency for a chiller operating at various capacities, reflecting typical annual usage. It is primarily used as a performance metric for commercial air-cooled and water-cooled chillers.

SCOP (Seasonable coefficient of performance). The ratio of heating or cooling output to energy input over an entire cooling or heating season.

SEER (Seasonal energy efficiency ratio). The ratio of cooling output (in Btu) to the total electric energy input (in Watt-hours) over a typical cooling season. It is primarily used for small unitary air conditioners.

Example 9-1

A large room in a commercial office building has a heat production rate of 600,000 Btu per hour from lights, equipment, people, and heat flow from the outside. How many tons of air conditioning is required to remove this heat?

Solution

Using the conversion factor between Btu/hr to tons:
X tons = (600,000 Btu/hr)/(12,000 Btu/hr/ton) = 50 tons

Example 9-2

At test conditions, a roof-top air conditioner provides 240,000 Btu/hr of cooling with an electric input of 20,000 W. What is the EER of the unit?

Solution

Using the definition of EER:
EER = (240,000 Btu/hr)/(20,000W) = (12 Btu/hr)/W or 12 Btu/Wh

Example 9-3

An air conditioner removes 240,000 Btu/hr of heat and has an input power of 23 kW. What is the COP of the unit?

Solution

First, find the Btu/hr by converting the 23 kW to Btu/hr. Since 1 kWh = 3,412 Btu, we can divide each side by hours, h, and get 1 kW = 3,412 Btu/hr. So, 23 kW in Btu/hr = (23 kW) $\times$ (3412 Btu/hr) = 78,476 Btu/hr.

Now, we can solve for the COP:
COP = (240,000 Btu/hr)/(78,476 Btu/hr) = 3.06.
Remember, COP has no units.

Example 9-4

A 20-ton air conditioner has a full load power draw of 25 kW. What is the kW/ton rating?

Solution

Just solve for the given kW divided by the given tons.

kW/ton = 25 kW/20 tons = 1.25 kW/ton.

Example 9-5

What is the COP of an AC system with an EER of 12 Btu/Wh?

Solution

COP = (12 Btu/Wh)/(3.412 Btu/Wh) = 3.517 (no units).

Example 9-6

What is the EER of an AC system with a COP 5?

Solution

EER = 5 $\times$ 3.412 Btu/Wh = 17.1 Btu/Wh.

Example 9-7

What is the kW/ton for an AC system with an EER of 12?

Solution

kW/ton = input/output = (1 Wh/12 Btu) $\times$ (12,000 Btu/ton hr) $\times$ (kW/1000 W) = 1 kW/tonkW/ton = 12/12 = 1 kW/ton.

Example 9-8

What is the kW/ton for an AC system with a COP of 3.517?

Solution

kW/ton = input/output = (1/3.517) x (12,000 Btu/ton hr) x (kWh/3,412 Btu) = 1.0 kW/tonkW/ton = 3.517/3.517 = 1 kW/ton.

9.3.2 Energy quality (exergy or availability)

Exergy is a measure of the quality or usefulness of energy[6] in terms of its potential to perform work. Unlike energy, which is conserved, exergy is consumed due to inefficiencies and irreversibilities in real-life systems. Any time entropy is produced in a process, exergy is destroyed. Not all energy sources are equal in terms of their usefulness for a given process. We inherently understand this concept in many areas of everyday life but often fail to apply this concept within building systems and when deciding on suitable forms of energy supply, particularly for heating systems.

A simple example of this concept is the process of heating water with electricity. While the total energy remains the same – energy is conserved – the "quality" of the energy is diminished once it is converted into heat.

[6] Exergy is a term more commonly used in the EU. In North America, the term Availability is often used.

Initially, electricity has high quality (i.e. exergy or availability) because it is useful for a large number of applications such as running motors, powering electronics, lighting, etc. However, after the energy is converted, the resulting hot water can only serve a limited range of purposes, such as washing or space heating. Reversing the process would require an additional input of energy, demonstrating that what is ultimately lost is the availability of the original energy form.

For example, we understand these concepts through the heating of water using electricity. We can heat up water using an electric element, but we cannot reverse this process. In this process, the energy may not be lost, but "something" has been lost since we cannot obtain the original condition without an input of energy. What is lost is the quality of the energy. With electricity, the energy was available for a broad range of applications. Once this energy is converted to hot water, the energy has limited availability for further use, as the hot water has limited applications.

While a detailed discussion on exergy is beyond the scope of this chapter, the ability to quantify energy quality is a valuable tool in making resource-responsible decisions for HVAC systems.

The quality of a form of energy can be expressed using the exergy factor (often called the quality factor). The exergy factor is the ratio of exergy per unit of energy:

$$\text{Exergy Factor} = \frac{\text{Exergy}}{\text{Energy}}.$$

For thermal energy, the exergy factor depends on the temperature at which heat transfer occurs (T) compared to a reference temperature state (T○), which is often taken at ambient temperature:

$$\text{Exergy Factor} = \left(1 - \frac{T_o}{T}\right).$$

Understanding the quality factor of various energy sources and demands allows us to make informed decisions on matching the most appropriate energy source for a given energy demand. By matching low-exergy factor sources to low-exergy factor demands, we can ensure high-exergy energy sources are reserved for high-exergy demands, minimizing exergy destruction. Figure 9.5 illustrates this concept.

Table 9.2 Exergy factors of various forms of energy (adapted from[7]).

Energy form	Exergy factor (exergy/energy)
Mechanical energy	1
Potential energy	1
Electrical energy	1
Nuclear energy	0.95
Solar radiation (sunlight)	0.93
Chemical exergy: Oil, petroleum products	1.04
Chemical exergy: natural gas	0.94
Chemical exergy: coal	1.03
Chemical exergy: wood	1.05
Heat at 1,600 °C	0.85
Heat at 1,000 °C	0.78
Heat at 200 °C	0.41
Heat at 100 °C	0.25
Heat at 80 °C (common for hot-water heating)	0.21
Heat at 60 °C (common for domestic hot water)	0.17
Heat at 20 °C (common room temperature)	0.05
Note: For heat, T_0 = 5 °C (278 K)	

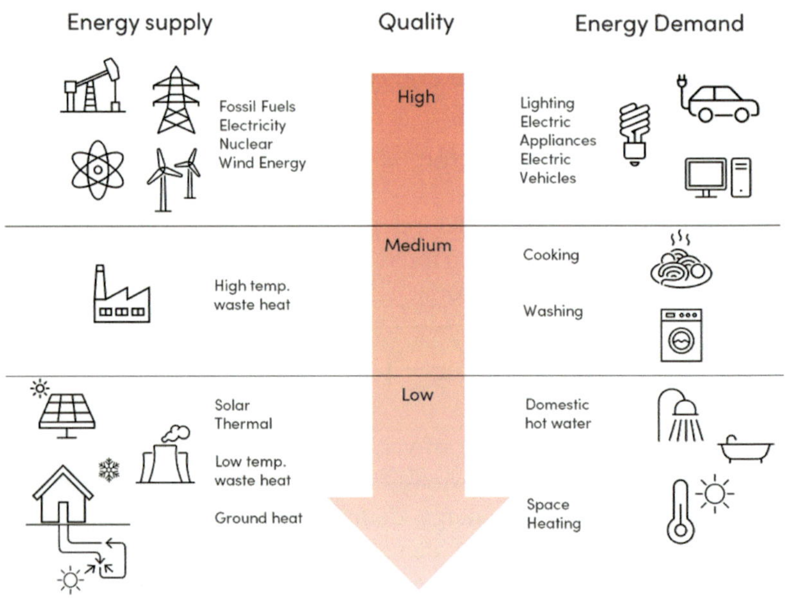

Figure 9.5 Matching demand side and supply side exergy factors.

[7] Exergy Guidebook for Building Professionals. Sabine Jansen. Delft University of Technology, October 2014, Delft.

The exergy factor required to satisfy building heating loads is very low. Nevertheless, high-quality energy sources, such as fossil fuel and electric resistance heating, are commonly used to satisfy these demands, leading to misuse of energy sources that are better reserved for high-exergy loads.

The term exergy efficiency can also be used to compare the suitability of various energy sources. Exergy efficiency can be calculated by multiplying energy efficiency by the ratio of the exergy factor of the demand and the source.

$$\Psi\ (\text{exergy efficiency}) = \text{ energy efficiency } \times \frac{\text{exergy factor}_{\text{demand}}}{\text{exergy factor}_{\text{source}}}.$$

Example 9-9

Calculate the exergy efficiency of a natural-gas condensing boiler with an efficiency of 90%, producing water at 60 °C.

Solution

Using the exergy factors for thermal energy at 60 °C and for natural gas gives:

$$(\text{exergy efficiency}) = 0.90 \times \frac{0.17}{0.94} = 0.163 \text{ or } 16.3\%.$$

Example 9-10

Calculate the exergy efficiency of an electric heat pump with a COP of 2.5, producing water at 60 °C.

Solution

$$\Psi\ (\text{exergy efficiency}) = 2.5 \times \frac{0.17}{1} = 0.425 \text{ or } 42.5\%.$$

9.3.3 Sensible and latent heat

Sensible heat refers to the heat added or removed from a substance resulting from a temperature change with no associated phase or state change. The amount of sensible heat gained or lost by a substance can be quantified as expressed by

$$\dot{Q}_s = \rho \times c_p \times \dot{V} \times \Delta T$$

where:

$\dot{Q}_s$ = sensible heat transfer rate

ρ = density of media

c_p = specific heat of media

$\dot{V}$ = volume flow rate of media
ΔT = temperature differential of media

For air, sensible heat refers to changing the air's temperature without altering the amount of water vapor present. For example, when air is heated by passing over a heating coil, only its temperature is affected, and this process is purely a transfer of sensible heat. For air at standard conditions with a density of 1.2 kg/m^3 (0.075 lb/ft^3) and a specific heat of 1.006 KJ/kg-°C (0.24 Btu/lb-°F), the sensible heat entering or leaving an airstream is expressed by

SI unit	English units
$\dot{Q}_s = 1.2 \times \dot{V} \times \Delta T$	$\dot{Q}_s = 1.08 \times \dot{V} \times \Delta T$
where: $\dot{Q}_s$ = sensible heat transfer rate, W $\dot{V}$ = volume flow rate of air, L/s ΔT = temperature differential of media, °C	where: $\dot{Q}_s$ = sensible heat transfer rate, Btu/hr $\dot{V}$ = volume flow rate of air, CFM ΔT = temperature differential of media, °F

For liquid water, sensible heat is associated with changing the temperature of water, but always keeping it in the liquid state. Thus, if water is heated to increase or cooled to decrease its temperature but remains in the liquid state, this is sensible heat transfer. For water at standard conditions with a density of 1,000 kg/m^3 (62.4 lbm/ft^3) and specific heat of 4.184 kJ/kg-°C (1 Btu/lbm-°F), the heat transferred to or from water is expressed by

SI unit	English units
$\dot{Q}_s = 4.2 \times \dot{V} \times \Delta T$	$\dot{Q}_s = 500 \times \dot{V} \times \Delta T$
where: $\dot{Q}_s$ = heat transfer rate, W $\dot{V}$ = volume flow rate of fluid, L/s ΔT= temperature differential of media, °C	where: $\dot{Q}_s$ = heat transfer rate, Btu/hr $\dot{V}$ = volume flow rate of fluid, GPM ΔT = temperature differential of media, °F

Latent heat, on the other hand, is the heat absorbed or released by a substance during a change in its physical state (e.g. from liquid to vapor or vice versa, freezing or thawing of water) that occurs without a temperature change.

Latent heat is involved when air is humidified or dehumidified; for example, water vapor is added or removed. Water vapor can be added to the air by spraying steam or atomized water droplets into an airstream, and water vapor can be removed from the air by cooling the air below its dewpoint temperature. The quantity of latent heat added or removed from air is expressed by the following equation:

$$\dot{Q}_l = \rho \times \dot{V} \times \Delta W$$

where:

$\dot{Q}_l =$ latent heat transfer rate

$\rho =$ density of air

$\dot{\mathcal{V}} =$ volume flow rate of media

$\Delta W =$ humidity ratio differential between leaving and entering conditions

For air at standard conditions with a density of 1.2 kg/m^3 (0.075 lbm/ft^3) and a latent heat of vaporization of 2,450 kJ/kg (1,060 Btu/lbm), the latent heat entering or leaving an airstream is expressed by

SI unit	English units
$\dot{Q}_l = 2960 \times \dot{\mathcal{V}} \times \Delta W$ where: $\dot{Q}_l =$ latent heat transfer rate, W $\dot{\mathcal{V}} =$ volume flow rate of air, L/s $\Delta W =$ humidity ratio differential, kg_w/kg_a	$\dot{Q}_l = 4780 \times \dot{\mathcal{V}} \times \Delta W$ where: $\dot{Q}_l =$ latent heat transfer rate, Btu/hr $\dot{\mathcal{V}} =$ volume flow rate of air, CFM $\Delta W =$ humidity ratio differential, lbm_w/lbm_a

For water and refrigerants, latent heat is involved when a state change is involved. Latent heat is added to the fluid when it changes state from a liquid to a vapor (evaporates or boils), and heat is removed when it changes state from a vapor to a liquid (condenses).

9.3.4 Psychrometrics

Psychrometrics is the study of air and its properties, specifically temperature, humidity, and energy content. It helps us understand how moisture interacts with air and its temperature. Engineers use psychrometrics to design efficient and comfortable indoor environments. This understanding allows them to quantify energy exchange in various HVAC processes, including heating, cooling, humidification, and dehumidification.

The core concepts of psychrometrics are best described and visualized using a psychrometric chart (see Figure 9.6). Engineers use this chart (or software) to visualize and quantify the properties of air in HVAC processes quickly. Two psychrometric properties of air must be known to quantify the other characteristics. The psychrometric properties of air are outlined below.

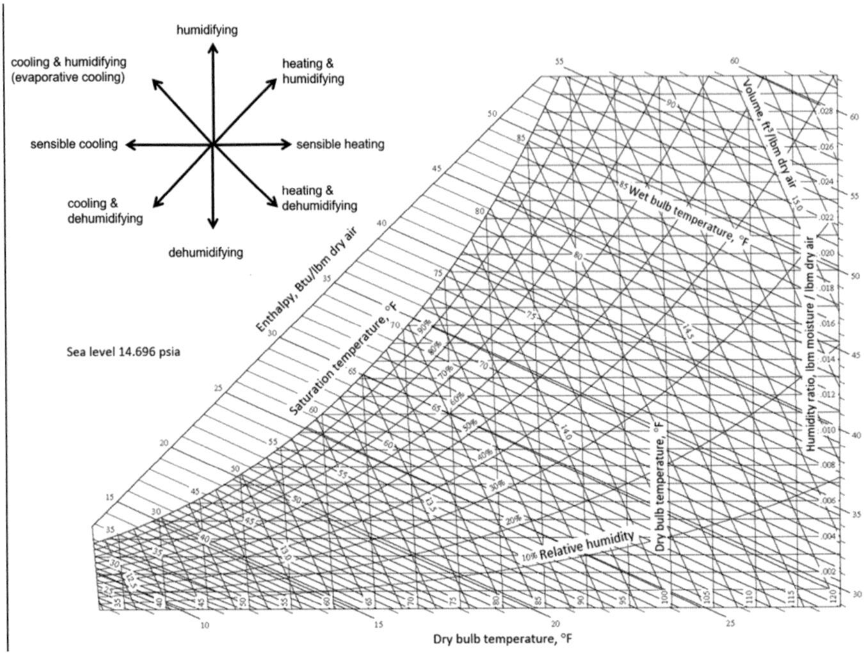

Figure 9.6 Psychrometric chart.

9.3.4.1 Dry-bulb temperature (air temperature)

Dry-bulb temperature is the temperature of air measured with a regular thermometer; this is commonly referred to as air temperature. The dry-bulb temperature of air is not affected by moisture or humidity. The dry-bulb temperature is represented on the horizontal axis of the psychrometric chart.

9.3.4.2 Absolute humidity (humidity ratio and dew point)

Absolute humidity measures the amount of water vapor in a given air volume. It is typically measured as a mass ratio of water to air (i.e. kg of water per kg of air). It represents the total moisture content independent of air temperature and is one way to express humidity.

Absolute humidity is represented on the psychrometric chart's vertical axis, with the air's water content increasing from the bottom to the top. The absolute humidity is typically indicated using one of two air properties:

1. **Humidity ratio.** Also known as specific humidity, the humidity ratio represents the ratio of the mass of water vapor present in a unit mass of dry air. It is typically measured in kilograms of water vapor per kilogram

of air (kgw/kga), grains of moisture per pound of air (gr/lbm), or pounds of water vapor per pound of air (lbmw/lbma). The humidity ratio is a fundamental parameter in psychrometrics, along with the dry-bulb temperature.

2. **Dew point.** Dew point is the temperature at which air becomes saturated with moisture, leading to condensation. It is the temperature to which air must be cooled to reach saturation. As the air cools below the dew point temperature, it cannot hold all its moisture, forming dew, fog, or frost.

9.3.4.3 Relative humidity and saturation point

Relative humidity is the ratio of moisture in the air to the maximum amount of moisture the air can hold at a specific dry-bulb temperature, expressed as a percentage. It is relative because the amount of moisture air can hold depends on the dry-bulb temperature.

The saturation point of air is the condition where air holds the maximum amount of water vapor possible at a given temperature and pressure without condensation. The relative humidity is 100% because the air is fully saturated with moisture. At this point, the dry-bulb temperature is equal to the dew point.

The curved line on the left side of the psychrometric chart is called the saturation curve. This curve connects the horizontal and vertical axes, representing the conditions where the air is fully saturated with moisture. The point where the saturation curve intercepts both axes signifies a relative humidity of 100%, indicating complete air saturation with water vapor. The curve's shape is due to warmer air being capable of holding more moisture before reaching saturation compared with colder air.

The subsequent curved lines below the saturation curve represent different relative humidity levels. These lines indicate various moisture levels in the air corresponding to specific temperatures. Each curved line represents a different relative humidity, indicating how much moisture the air holds compared to its maximum capacity at various temperatures.

9.3.4.4 Enthalpy

Enthalpy, in the context of psychrometrics, refers to the total thermal energy content of the air, combining both its sensible heat (related to its temperature) and latent heat (associated with its moisture content). It is a measure that accounts for the thermal energy required to change the air's temperature and the energy associated with the phase change of moisture (evaporation or condensation) in the air.

On a psychrometric chart, enthalpy is represented by the diagonal lines that slope downwards from left to right. These lines are called constant enthalpy lines. They help visualize the total energy content of air at different conditions, combining both sensible and latent heat components.

These lines allow engineers and HVAC professionals to analyze and assess the air's total energy content as it changes in temperature and moisture content. The change in enthalpy determines the energy requirements to heat, cool, dehumidify, or humidify the air, directly impacting the energy requirements for HVAC systems.

9.3.4.5 Psychrometric processes

Psychrometrics can characterize several common HVAC processes, including heating, cooling, humidification, and dehumidification. This is useful for sizing HVAC equipment to maintain indoor environmental quality and to quantify the energy requirements for such processes. Figure 9.6 provides a visual summary of the air conditioning processes overlaid on a psychrometric chart.

Example 9-11: Sensible Heating

8,000 CFM of supply air at 60 °F and 50% relative humidity passes over a heating coil and leaves the coil at 90 °F. Given standard air conditions, calculate the heat transfer rate of the system. Plot the process on a psychrometric chart. What is the relative humidity of the leaving air?

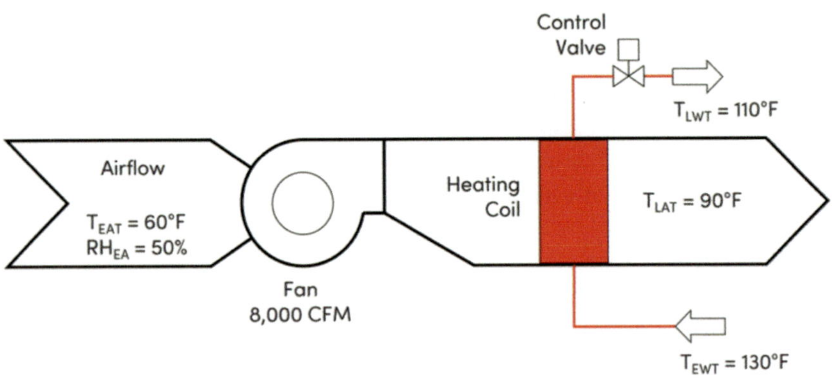

Solution

Standard air conditions apply; hence, the heat transfer rate from the coil to the air can be calculated as follows:

$$\dot{Q}_s = 1.08 \times 8{,}000 \text{ CFM } \times (90\,°\text{F} - 60\,°\text{F}) = 259{,}200 \text{ Btu/hr.}$$

This process can be plotted on the psychrometric chart as illustrated below:

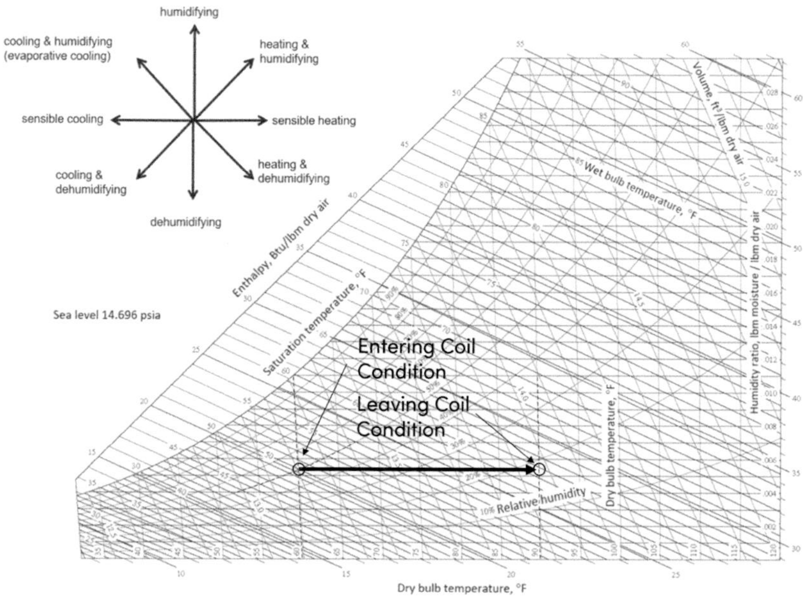

The leaving air's relative humidity can be estimated at 19% by plotting the process on a psychrometric chart.

Example 9-12: Heating Coil Flow Rate

The heating coil in the previous example is supplied with 130 °F heating water, and the heating water temperature leaves the coil at 110 °F. Determine the water flow rate required assuming standard water conditions.

Solution

The heat transfer rate added to the air equals the heat transfer rate lost from the water. As such, the flow rate can be calculated as follows:

$$\dot{\mathcal{V}} = \frac{\dot{Q}_s}{500 \times (\Delta T)}$$

$$\dot{\mathcal{V}} = \frac{(259,200 \text{ Btu/hr})}{500 \times (130\ ^\circ\text{F} - 110\ ^\circ\text{F})} = 25.9 \text{ GPM}.$$

Example 9-13: Cooling and Dehumidification

8,000 CFM of supply air at 90 °F dry-bulb temperature and 70 °F wet-bulb temperature passes over a cooling coil and leaves the coil at 55 °F at saturated conditions (i.e. a relative humidity of 100%). Given standard air conditions, calculate the sensible, latent, and total heat transfer rate of the system. Plot the process on a psychrometric chart. What is the total cooling load, in tons?

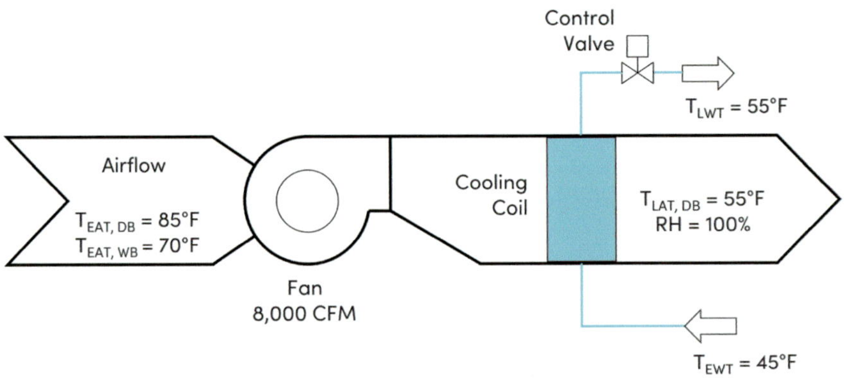

Solution

Standard air conditions apply; hence, the sensible heat transfer rate from the air to the coil (i.e. sensible cooling of the air) is calculated as follows:

$$\dot{Q}_s = 1.08 \times 8,000 \text{ CFM} \times (85\ ^\circ\text{F} - 55\ ^\circ\text{F}) = 259,200 \text{ Btu/hr}.$$

To calculate the latent heat transfer rate, the humidity ratio of the entering and leaving air conditions must be determined using the psychrometric chart. Using the intersection of 85 °F dry-bulb temperature and 70 °F wet-bulb temperature gives an entering air humidity ratio of 0.0123 lbw/lba. Using the intersection of 55 °F at the saturation curve gives a leaving air humidity ratio of 0.092 lbw/lba. Hence, the latent heat transfer rate from the air to the coil is calculated as follows:

$$\dot{Q}_l = 4,780 \times 8,000 \text{ CFM} \times \left(0.0123\frac{\text{lbm}_\text{w}}{\text{lbm}_\text{a}} - 0.092\ \frac{\text{lbm}_\text{w}}{\text{lbm}_\text{a}}\right) = 117,644 \text{ Btu/hr}.$$

The total heat transfer rate (i.e. cooling load on the air) is the sum of the sensible and latent air transfer rates:

$$\dot{Q}_T = 259,200 \text{ Btu/hr } + 177,644 \text{ Btu/hr} = 376,844 \text{ Btu/hr}.$$

This estimate can be cross-checked using enthalpy values from the psychrometric chart: $\dot{Q}_T = \dot{m}\,\Delta \text{h} = 0.0752\text{lbm/ft}^3 \times 8{,}000 \text{ ft}^3\text{/min} \times 60 \text{ min/hr}$ $\times$ (34 Btu/lbm - 23 Btu/lbm) = 397,056 Btu/hr which is approximately the same result as above.

Converting this to tons of cooling gives

$$\dot{Q}_T = \frac{376,844 \text{ Btu/hr}}{12,000 \ \frac{\text{Btu/hr}}{\text{Tons}}} = 31.4 \text{ tons}.$$

This process can be plotted on the psychrometric chart as illustrated below:

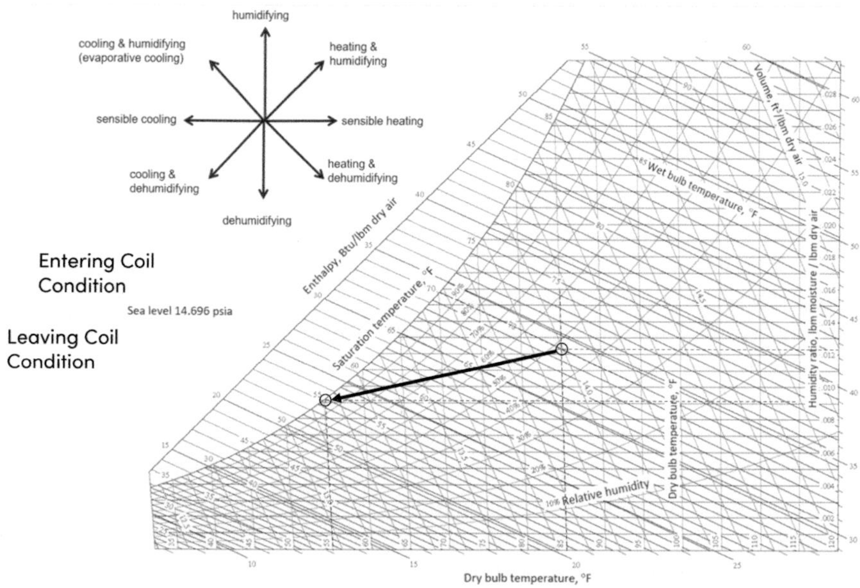

9.3.5 Heat transfer

Heat is transferred through three separate modes:

- **Conduction.** In a solid material, heat flows from a hot region to a cooler region when there is a temperature difference.

- **Convection.** In liquids or gases, heat flows from hot to cold regions due to a change in density.
- **Radiation.** Heat transfers directly between two separate bodies due to wave motion and temperature difference between two bodies.

Refer to Chapter 11 (Building Envelope) and the formula sheet in the appendix for a brief overview of heat transfer.

9.3.6 Additional resources

- ASHRAE Fundamentals 2025, Chapter1 – Psychrometrics.
- ASHRAE Fundamentals 2025, Chapter 4 – Heat Transfer.
- Exergy Guidebook for Building Professionals: How to use exergy to analyze and develop energy systems for the built environment. Sabine Jensen, Delft University of Technology, October 2014, Delft.

9.4 Heating, Cooling, and Ventilation Loads

The design of an HVAC system starts with estimating the heating and cooling loads of the building the system serves. Peak (i.e. design) heating and cooling loads determine the size of heating and cooling systems, ventilation systems, piping and ductwork distribution, and every other component of the HVAC system. Annual heating and cooling loads are used to model the expected performance of HVAC systems to characterize building energy consumption and make informed choices regarding equipment configurations to maximize efficiency during part load conditions.

The magnitude of loads can significantly affect both the first cost and ongoing operating costs of the building. Excessive heat gain and losses through the building envelope also negatively impact occupant comfort, particularly in extreme climates. Reducing loads should be the first priority for new construction and energy retrofits before implementing EEMs or replacing mechanical equipment (refer to Chapter 3, Figure 3.1). Understanding the components that make up heating and cooling loads is important to prioritize the highest-impact interventions to lower heating and cooling loads.

Heating and cooling loads are composed of many influencing factors, including external factors such as solar radiation; heat gains and losses through the building envelope, including walls, windows, floors, and roof; infiltration of outside air through cracks in the building's construction; ventilation air required to maintain indoor air quality, and internal heat gains from occupants, lighting, and equipment. Figure 9.8 illustrates the external and internal factors influencing heating and cooling loads. Each factor is briefly touched upon in this section.

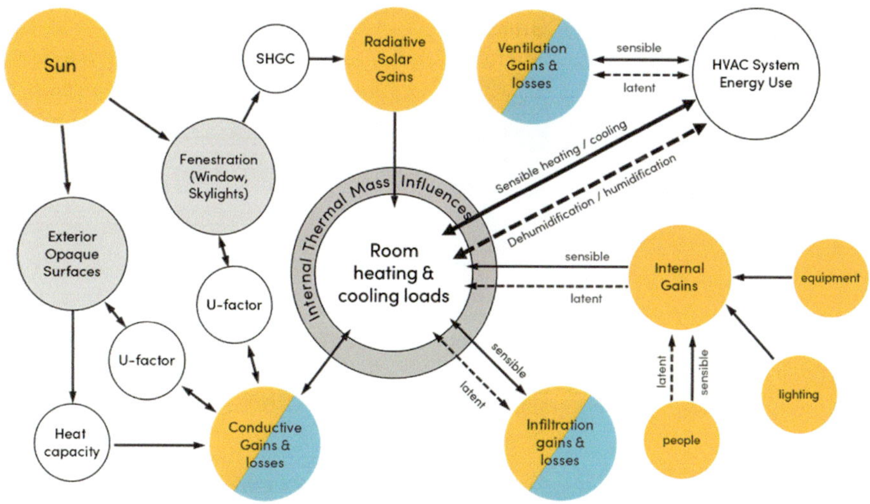

Figure 9.8 Factors Impacting heating and cooling loads (adapted from Advanced Design Guide for Small to Medium Office Buildings Achieving 50% Energy Savings Toward a Net Zero Energy Building, ASHRAE, 2011).

9.4.1 Climate influences gains and losses

The surrounding climate and building envelope heavily influence buildings' heat gains and losses. Components such as walls, windows, doors, roofs, and floors allow heat to flow into or out of the building due to temperature differences between indoor and outdoor air. Solar energy also plays a large role in heat gains, either transmitting directly into the space through windows or skylights or by being absorbed by exterior surfaces, which transfer heat into the building as they warm up.

Heat is transferred inside the building primarily through radiation and convection. Convection directly impacts the air temperature, leading to instantaneous heating or cooling loads. At the same time, radiant heat is absorbed by surfaces and later released, causing a delay between the time of heat gain and load on the HVAC system.

9.4.1.1 Radiative solar gains

Solar energy plays a significant role in both peak cooling loads and the annual heating and cooling requirements of a building. The overall impact of solar gain depends on various factors, including the building's orientation, aspect ratio, location, and the materials used in its construction. During colder months, solar gains can be beneficial by providing passive heat, reducing the

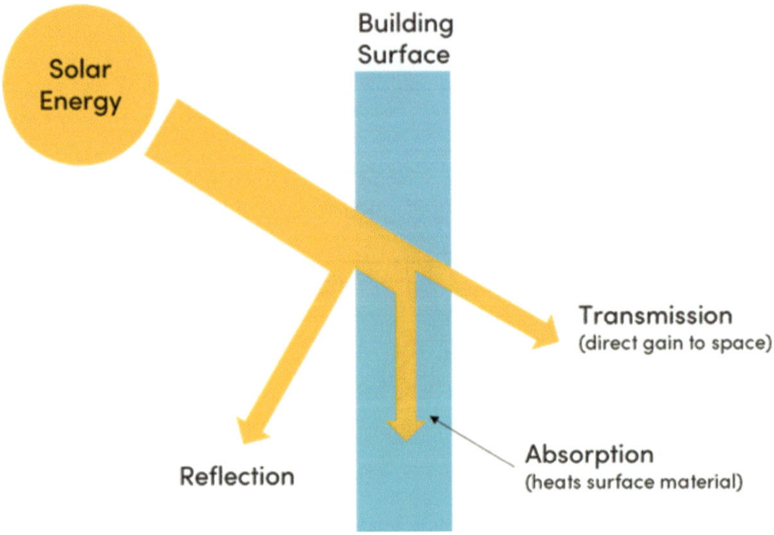

Figure 9.9 What happens when solar energy hits a building surface.

need for mechanical heat. However, during warmer months, excessive solar gains can lead to overheating, increasing the demand for mechanical cooling. Direct solar radiation can also affect occupant comfort, making it challenging to maintain acceptable indoor conditions.

Effective management of solar gains requires balancing the benefit of passive heating in winter and natural lighting with the need to prevent overheating. Several strategies can be employed to optimize solar gains, such as appropriately sizing and orienting windows, using reflective or insulating materials for opaque surfaces, and incorporating shading devices like overhangs, louvers, or vegetation. Figure 9.9 illustrates the effect of solar radiation on a surface, showing how part of the energy is reflected, part is absorbed, and – if the surface is transparent – part is transmitted directly into the building.

The surface properties of the materials making up the building envelope play a crucial role in determining the amount of radiation that is reflected, absorbed, or transmitted. Surfaces should have high reflectivity, low absorption, and low transmittance to minimize unwanted heat gains.

Solar Gain Through Transparent Building Elements

Fenestration refers to the light-transmitting components of a building envelope, such as windows, skylights, and glazed doors. These elements are often the most significant contributors to external heat gains and losses, as they

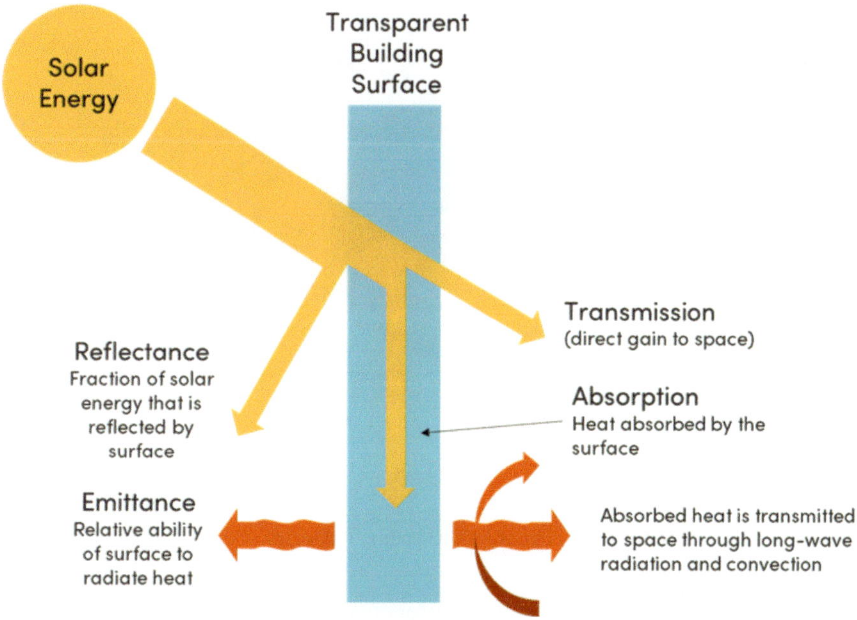

Figure 9.10 Solar heat gain through a transparent surface.

typically occupy a substantial portion of the building's surface area and have poorer thermal properties than opaque building materials. However, well-designed fenestration is essential for providing natural daylighting and views, contributing to indoor environmental quality.

Figure 9.10 illustrates the processes of solar heat gain through transparent materials. The primary mechanism of solar heat gain through fenestration is transmission, where radiant energy passes through the glass and into the interior space. This energy is absorbed by surfaces such as floors, walls, and furniture, raising their temperatures and subsequently increasing indoor air temperatures. Glass absorbs solar energy, which heats the glass and releases heat into and out of the building through radiation and convection.

The solar radiation entering a building through a window is measured by the solar heat gain coefficient (SHGC), or the g-value in Europe. Both metrics quantify the fraction of solar radiation transmitted through a window assembly, with a lower SHGC (or g-value) indicating less solar heat gain. An SHGC of 1 means all solar radiation is transmitted, while a value of 0 means no solar energy enters.

Shading devices, such as blinds, curtains, overhangs, or louvers, can help control solar heat gains, particularly in the summer. External shading is more

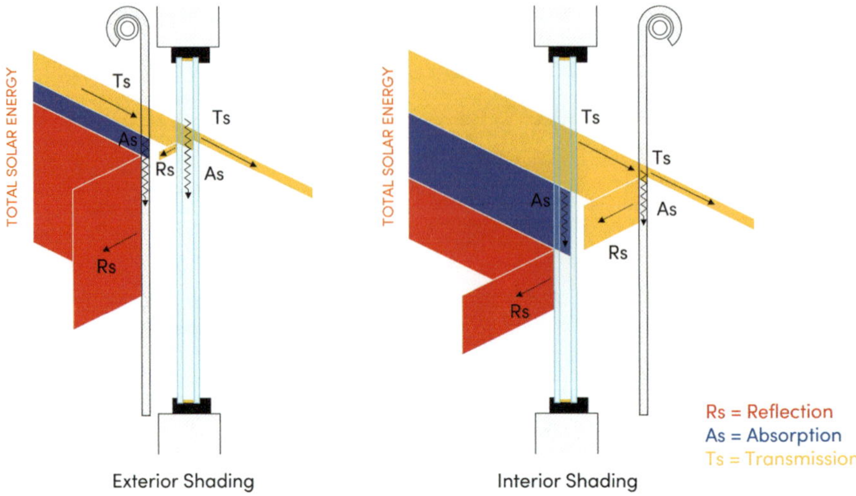

Figure 9.11 Fenestration shading strategies.

effective at reducing solar heat gain as it prevents both the transmittance and absorption of solar energy by the glass (see Figure 9.11).

Solar Gain on Opaque Surfaces

Solar radiation also impacts opaque building surfaces, such as walls and roofs (see Figure 9.12). When solar energy strikes these surfaces, it is absorbed, raising the temperature of the materials. This heat is then conducted through the surface and transferred to indoor spaces through radiation and convection. The material and color of these surfaces influence the extent of heat absorption. Dark-colored, low-reflectivity materials absorb more solar radiation, leading to more significant heat gains, while light-colored or reflective surfaces reduce heat absorption and lower heat gains.

The thermal mass of the materials is another critical factor. High-mass materials like concrete absorb and store heat during the day, releasing it gradually over time, which helps stabilize indoor temperatures by slowing the heat transfer rate. In contrast, low-mass materials, such as steel-framed walls, release absorbed heat more quickly, causing more significant fluctuations in indoor temperatures. The role of thermal mass is explored further in Section 9.4.3.

Indirect Solar Gains and Reflections

Indirect solar gain can also increase cooling loads as reflections from surrounding surfaces, such as nearby buildings, roads, or water, can intensify the

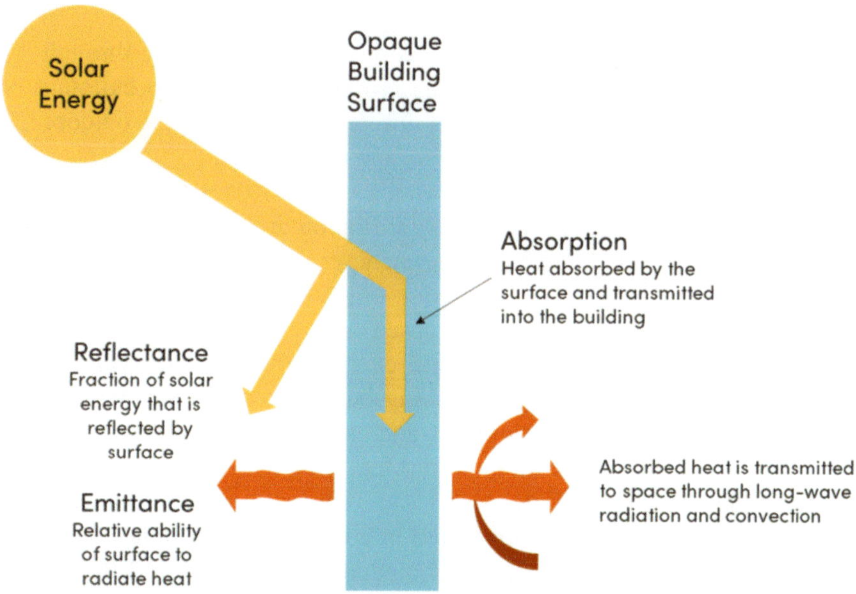

Figure 9.12 Solar heat gain on opaque surface.

solar radiation a building receives. This effect is more pronounced in urban areas, where reflective materials like glass facades and concrete amplify solar heat gains, contributing to the urban heat island effect.

Additionally, solar heating of the surrounding environment can raise outdoor air temperatures, increasing the temperature difference between indoor and outdoor spaces. This results in greater heat transfer into the building through conduction, further increasing cooling demands.

9.4.1.2 Conductive gains and losses

Heat gain from radiative solar energy absorbed by building materials and the temperature difference between the outdoor and inside air result in conduction through the walls, windows, and other exterior building surfaces. The amount of heat transfer through a material depends on its thermal conductance (or U-value). Adding insulation to walls and roofs can significantly reduce unwanted heat gain or losses through opaque building elements. For buildings with high window-to-wall ratios, improving the thermal performance of window elements by reducing the U-factor results in the most significant improvements to overall building envelope performance. Thermal conductance, insulation, and heat transfer calculations through opaque assemblies are discussed in detail in Chapter 11.

9.4.1.3 Infiltration gains and losses

Infiltration is the uncontrolled flow of outdoor air into a building through cracks, gaps, and openings in the building envelope and through building entrances. This flow is driven by pressure differentials between the outdoor and indoor environments caused by the wind and stack effect.

In large buildings, the stack effect – where warm air rises and creates a pressure difference – can induce significant infiltration on lower floors. Maintaining a slight positive building pressure whenever possible is important to minimize infiltration.

It is challenging to predict infiltration based on first principles accurately. Infiltration rates for load calculations and energy modeling are typically chosen based on empirical data. For existing buildings, in situ air leakage tests can be used to quantify air tightness and estimate infiltration (e.g. see ASTM E3158-18 Standard Test Method for Measuring the Air Leakage Rate of a Large or Multizone Building).

Infiltration rates are typically measured in air changers per hour (ACH) or as air volume per unit area of wall area (e.g., CFM/ft^2 or L/s-m^2). Figure 9.13 summarizes typically expected air leakage rates, including performance requirements of various building standards. For consistency, all rates are based on a pressure differential of 75 Pa (0.3 inches w.g.). The ASHRAE Advanced Energy Design Guides recommends a target leakage rate of less than 0.25 CFM/ft^2 at 75 Pa in climate zones 0–6 and no higher than 0.15 CFM/ft^2 at 75 Pa in climate zones 7 and 8.[8]

The volume flow rate ($\dot{V}$) of air infiltration can be calculated from the ACH if the volume of the space (V) is known using the following equations:

SI units	English units
$\dot{V} = \text{ACH} \times \frac{V}{3,600}$	$\dot{V} = \text{ACH} \times \frac{V}{60}$
where: $\dot{V} =$ volume flow rate of air, L/s V = space volume, m^3	where: $\dot{V} =$ volume flow rate of air, CFM V = space volume, ft^3

With the infiltration airflow determined, psychrometric equations can be used to estimate the sensible and latent energy transfer associated with infiltration.

[8] ASHRAE, Achieving Zero Energy: Advanced Energy Design Guide for Small to Medium Office Buildings, 2019.

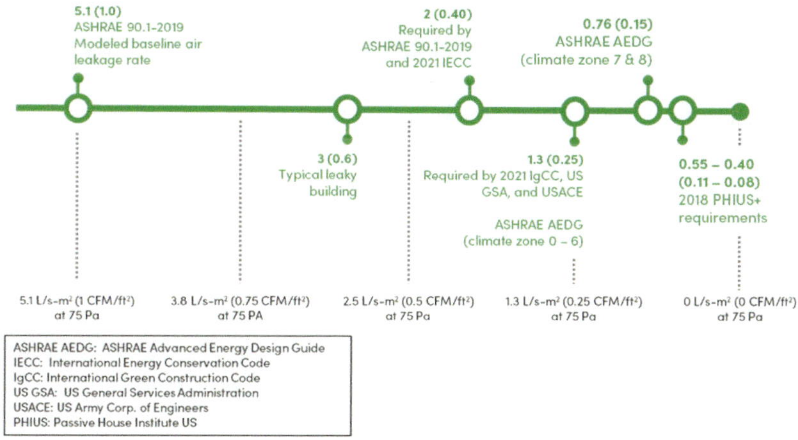

Figure 9.13 Representative infiltration rates by different codes and standards (adapted from Measured Air Leakage Requirements in Context, Daniel Overbey).

Improving air tightness in existing buildings can be challenging without substantial investment in upgrading exterior envelope components. For new construction and envelope retrofits, infiltration can be minimized by providing continuous air barriers as part of the building envelope. Additionally, the following strategies should be considered to reduce infiltration:

- **Seal gaps and cracks.** Inspect the building envelope for openings in walls, roofs, foundations, windows, and doors. Use caulking and weather-stripping to seal these gaps, preventing air leakage.
- **Enhance window and door seals.** Ensure all windows and doors are correctly installed and sealed. Inspect gaps when windows and doors are shut to ensure proper sealing.
- **Use vestibules and revolving doors:** Vestibules and revolving doors at main entrances create buffer zones, reducing the direct exchange of indoor and outdoor air when doors are opened.
- **Implement air curtains.** Air curtains minimize infiltration when doors are open by creating a high-velocity air barrier across the opening.
- **Maintain proper building pressurization.** A slight positive pressure inside the building prevents air from infiltrating through openings.
- **Seal ductwork.** Ensure all ductwork is properly sealed, especially if ducts pass through unconditioned spaces. Air leakage from ductwork can cause pressure imbalances that increase infiltration.

- **Commissioning and testing.** Building air leakage tests, such as blower door tests, help quantify the building infiltration rates and identify weak points for infiltration. The results of the tests allow a targeted approach to reduce infiltration most effectively.

9.4.1.4 Ventilation loads

Ventilation loads are associated with the intentional and controlled introduction of outdoor air into a building through the HVAC system. Unlike infiltration, which involves uncontrolled air leakage, ventilation is a controlled process designed to maintain acceptable indoor air quality by supplying conditioned and filtered outside air.

Ventilation loads consist of both sensible (temperature-related) and latent (moisture-related) heat gains or losses. Loads are calculated using the same sensible and latent heat equations used for infiltration loads, considering the differences in temperature and humidity between the indoor and outdoor air. Ventilation rates are determined in accordance with relevant ventilation codes and standards (see Section 9.2.2).

Effective strategies for reducing loads associated with ventilation include demand-controlled ventilation to minimize overventilation during partial occupancy and air-to-air heat recovery, which reclaims energy from exhaust air to precondition incoming outdoor air. Further details on demand control ventilation and energy recovery are discussed in Section 9.6.

9.4.2 Internal gains

Internal heat gains are sources of heat generated within a building that contribute to cooling loads on the HVAC system. These gains typically fall into the following categories: occupants, interior lighting, equipment, and processes. While interior lighting generates sensible heat gains, occupants and some equipment generate both sensible and latent heat gains.

9.4.2.1 Equipment

Equipment within a building contributes significantly to the internal heat gains, affecting both the sensible (and sometimes latent) loads on an HVAC system. These gains generally arise due to inefficiencies in converting electrical power into useful work, with the excess energy released as heat.

Equipment gains can originate from various sources, including:

- **Information technology (IT) equipment.** Servers, computers, and network devices produce sensible heat due to their continuous operation. The rate of heat gain is proportional to the rate of data processing.
- **Office equipment and appliances.** Printers, copiers, and other machinery contribute sensible heat gains in office environments.
- **Kitchen appliances.** Cooking equipment emits sensible heat and often moisture into the air, adding a latent heat component that increases indoor humidity.
- **Electric motors.** Electric motors are widely used in processes and HVAC systems and generate heat based on their efficiency and operational load.
- **Industrial and process equipment.** Machinery used in manufacturing or laboratory settings can produce substantial heat gains and make up a high proportion of cooling loads in these settings.
- **Specialized equipment or processes.** Certain facilities may have unique internal gains (or losses), such as latent heat gain due to evaporation from pool surfaces in aquatic centers or the cooling effect from ice surfaces in skating rinks.

Understanding the magnitude and frequency of equipment heat loads, particularly in industrial settings, can be helpful for identifying opportunities for heat recovery.

Accurate estimation of equipment heat gains involves considering both the operational characteristics of the equipment and the usage patterns within the space. For example, office equipment nameplate ratings often indicate the maximum power usage, not the typical operational consumption. Actual heat gains are usually in the range of 25%–30% of the nameplate value. Empirical data for heat gain from common types of equipment is provided in Chapter 18 of the 2021 ASHRAE Fundamentals handbook.

The diversity of equipment use also dramatically affects overall heat gains. The diversity factor is the ratio of the peak combined electrical load to the sum of the individual equipment loads within a space. It accounts for the likelihood that not all equipment operates at peak capacity simultaneously.

Diversity factors can vary widely, and the occupancy type for a given space must be considered. For example, an office building for accountants and lawyers will have less diversity than an office for engineers or sales staff, where staff often travel for work.

Estimating Heat Gain from Electric Motors

Electric motors generate heat due to inefficiencies in converting electrical energy into mechanical work. The instantaneous heat gain from an electrical motor can be calculated using the equations below. Refer to Chapter 14 for more information on motors and drives. The equations below include several factors that impact the heat gain:

- *Motor efficiency* (η_m) represents how effectively the motor converts electrical input energy into mechanical energy. Motor efficiency varies with motor size and design. Motors less than 3 hp in size typically have an efficiency between 65% and 80%, whereas larger motors typically have efficiencies between 85% and 96%.
- *Motor-use factor* (F_{UM}) accounts for intermittent motor operation. A motor running continuously has a factor of 1.0, while one that operates intermittently will have a lower factor.
- *Motor-load factor (MLF)* is the ratio of the actual load on the motor to its rated capacity.

The time-average heat gain from an electric motor located within a conditioned space can be calculated using the following equation:

SI units	English units
$Q_{\mathrm{em}} = \frac{1{,}000 \times P \times F_{\mathrm{UM}} \times \mathrm{MLF}}{\eta_m}$ where: Q_{em} = Heat gain from the motor (Watts) P = Motor output power rating (kW) η_m = Motor efficiency (decimal fraction < 1.0) F_{UM} = Motor-use factor (decimal fraction ≤ 1.0) F_{LM} = Motor-load Factor (decimal faction ≤ 1.0) 1,000 = Conversion factor (Watts per kilowatt)	$Q_{\mathrm{em}} = \frac{2{,}545 \times P \times F_{\mathrm{UM}} \times \mathrm{MLF}}{\eta_m}$ where: Q_{em} = Heat gain from the motor (Btu/hr) P = Motor output power rating (hp) η_m = Motor efficiency (decimal fraction $<$ 1.0) F_{UM} = Motor-use factor (decimal fraction ≤ 1.0) MLF = Motor-load factor (decimal faction ≤ 1.0) 2,545 = Conversion factor (Btu/hr per hp)

Only the mechanical work (which becomes heat) within the space is considered for motors located outside of the space:

SI units	English units
$Q_{\mathrm{em}} = 1{,}000 \times P \times F_{\mathrm{UM}} \times \mathrm{MLF}$	$Q_{\mathrm{em}} = 2{,}545 \times P \times F_{\mathrm{UM}} \times \mathrm{MLF}$

For conditions where the motor is inside the space, but the drive equipment is outside the space, only the motor's efficiency contributes to the heat gain within the space.

SI units	English units
$Q_{\text{em}} = 1{,}000 \times P \times \left(\frac{1.0-m}{m}\right) \times F_{\text{UM}} \times$ MLF	$Q_{\text{em}} = 2{,}545 \times P \times \left(\frac{1.0-m}{m}\right) \times F_{\text{UM}} \times$ MLF

9.4.2.2 Lighting

All input energy from lighting is eventually converted to heat within the building. The amount of heat generated depends on the required illumination levels and the type of light fixture. Table 9.3 summarizes common lighting types, including efficiency and typical lifespan. The luminous efficacy measures how effectively a light source converts input energy into visible light; the higher the value, the more efficient the light source. Refer to Chapter 12 for more information on lighting systems.

Table 9.3 Efficiency and lifespan comparison of common types of lighting.

Type	Luminous efficacy (Lumen/Watt)	Relative energy savings	Typical lifespan
Incandescent bulb	10–17	60 W (–)	Up to 2,000 hours
Halogen bulb	15–30	42 W (30%)	Up to 2,500 hours
Fluorescent	50–100	15 W (75%)	5,000–20,000 hours
LED	80–150	12 W (80%)	25,000–50,000 hours

Energy codes typically mandate a maximum lighting power density (LPD) for lighting systems based on the type of space or building they serve. The LPD defines the maximum allowable input power for lighting per unit floor area, expressed in Watts/ft^2 or Watts/m^2. Table 9.4 summarizes the maximum LPD by the building type specified in past versions of ASHRAE 90.1. Over the last few decades, advances in lighting technology have resulted in significant reductions in energy use and corresponding heat gain associated with lighting, which are reflected in the drastic cutbacks in allowable LPD, as illustrated in the table.

The information in Table 9.4 can help set reasonable targets for lighting retrofits and can also be used for preliminary energy savings estimates.

Table 9.4 ASHRAE 90.1 maximum LPD by building type based on vintage.

Building area type	ASHRAE 90.1-1999	ASHRAE 90.1-2010	ASHRAE 90.1-2019
Automotive facility	1.5	0.82	0.75
Convention center	1.4	1.08	0.64
Courthouse	1.4	1.05	0.79
Dining: bar lounge/leisure	1.5	0.99	0.80
Dining: cafeteria/fast food	1.8	0.90	0.76
Dining: family	1.9	0.89	0.71
Dormitory	1.5	0.61	0.53
Exercise center	1.4	0.88	0.72
Fire station	1.3	0.71	0.56
Gymnasium	1.7	1.00	0.76
Healthcare clinic	1.6	0.87	0.81
Hospital	1.6	1.21	0.96
Hotel	1.7	1.00	0.56
Library	1.5	1.18	0.83
Manufacturing facility	2.2	1.11	0.82
Motel	2.0	0.88	0.56
Motion picture theatre	1.6	0.83	0.44
Multifamily	1.0	0.60	0.45
Museum	1.6	1.06	0.55
Office	1.3	0.90	0.64
Parking garage	0.3	0.25	0.18
Penitentiary	1.2	0.97	0.69
Performing arts theatre	1.5	1.39	0.84
Police station	1.3	0.96	0.66
Post office	1.6	0.87	0.65
Religious building	2.2	1.05	0.67
Retail	1.9	1.40	0.84
School/university	1.5	0.99	0.72
Sports arena	1.5	0.78	0.76
Town hall	1.4	0.92	0.69
Transportation	1.2	0.77	0.50
Warehouse	1.2	0.66	0.45
Workshop	1.7	1.20	0.91

Sources:

(1) **ASHRAE/IESNA Standard 90.1-1999 (I-P), Table 9.3.1.1.**

(2) **ANSI/ASHRAE/IES Standard 90.1-2010 (I-P), Table 9.5.1.**

(3) **ANSI/ASHRAE/IES Standard 90.1-2019 (I-P), Table 9.5.1.**

To calculate the heat added to the space from lighting, remember that each kWh of electricity equals 3,412 Btu. Therefore, the heat produced each hour is found by simple unit conversion from kWh to Btu. Note that the input power must include power consumption for the complete light fixture, including any ballasts connected to the lights. Refer to Chapter 12 for more information on lighting systems.

Example 9-14: Heat Gain from Lighting

For air conditioned facilities, a lighting retrofit will provide energy savings through reduced lighting energy and additional savings based on the hours the air conditioning is needed. This same retrofit will result in a heating energy increase based on the number of hours that heating is required.

A 10,000-ft^2 office has been retrofitted with new lighting, which reduces the lighting power density from 1.3 W/ft^2 to 0.7 W/ft^2. A packaged rooftop unit serves the office with a cooling EER of 10 and a heating efficiency of 80%. Estimate:

1. The cooling load reduction resulted from the lighting retrofit in tons.
2. The annual reduction in lighting energy use, assuming the lighting operates 3,500 hours per year.
3. The annual reduction in air conditioner energy use, assuming the equivalent full load hours of the air conditioning system are 1,200 hours.

Solution

Step 1: Determine the Energy Savings of the Lighting Retrofit

Pre-Retrofit Lighting Power = 10,000 ft^2 $\times$ 1.3 W/ft^2 = 13,000 Watts
Post-Retrofit Lighting Power = 10,000 ft^2 $\times$ 0.7 W/ft^2 = 7,000 Watts
Lighting power savings = 13,000 W–7,000 W = 6,000 W (or 6 kW)

Step 2: Determine the Peak Cooling Load Reduction

Cooling load reduction equals the reduction in lighting power converted to tons:

6,000 W $\times$ 3.412 Btu/hr per W $\times$ 1 ton per 12,000 Btu/hr = 1.71 tons (20,472 Btu/hr)

Step 3: Determine the Annual Lighting Energy Savings

Annual lighting energy savings = 6 kW $\times$ 3,500 hours per year = 21,000 kWh per year

Step 4: Determine the Annual Reduction in Energy Associated with Air Conditioning

Annual air conditioning energy savings = (20,472 Btu/hr $\times$ 1,200 h)/(10 Btu/Wh $\times$ 1000 Wh/kWh) = 2,457 kWh per year.

9.4.2.3 Occupants

Occupants contribute both sensible and latent heat gain, which depends mainly on their metabolic activity. Age, sex, muscle mass, and physical activity influence an individual's metabolic rate. Table 9.5 and Table 9.6 provide typical heat gain factors adjusted for an average percentage of men, women, and children representative of the activity listed.

Sensible heat gain consists of convective and radiant components. The space's thermal mass mainly influences the conversion rate of the radiant portion to a cooling load. In contrast, the convective sensible gain and latent gain are typically considered instantaneous loads.

Table 9.5 Rates of occupant gain based on activity level – English units (adapted from ASHRAE Handbook – Fundamentals (2021), Chapter 18, Table 1).

Activity	Total heat (Btu/hr)		Sensible heat Btu/hr	Latent heat Btu/hr	% Sensible heat that is radiant	
	Adult male	**Normalized, M/F²**			**Low Velocity**	**High Velocity**
Seated, at rest	390	350	245	105	60	27
Seated, very light work	450	400	245	155		
Moderately active office work	475	450	250	200		
Standing, light work; walking	550	450	250	200	58	38
Light bench work	800	750	275	475		
Walking 3 mph; light machine work	1,000	1,000	375	625		
Heavy work	1,500	1,450	580	870	54	19
Athletics	2,000	1,800	710	1,090		

Notes:
1. Values are based on 75 °F dry-bulb room temperature.
2. Normalized values are based on a representative percentage of adult male, adult female and children for the activity listed and assumes that an adult female heat gain is 85% of an adult male, and heat gain from a child is 75% of an adult male.
3. Values are approximate based from ASHRAE 2021 Fundamentals, Chapter 9, Table 6.

Table 9.6 Rates of occupant gain based on activity level – SI units (adapted from ASHRAE Handbook – Fundamental (2021), Chapter 18, Table 1).

Activity	Total heat (Btu/hr)		Sensible heat Watts	Latent heat Watts	% Sensible heat that is radiant	
	Adult male	Normalized, M/F[2]			Low Velocity	High Velocity
Seated, at rest	114	103	72	31	60	27
Seated, very light work	132	117	72	45		
Moderately active office work	139	132	73	59		
Standing, light work; walking	161	132	73	59	58	38
Light bench work	235	220	81	139		
Walking 3 mph; light machine work	293	293	110	183		
Heavy work	440	425	170	255	54	19
Athletics	587	528	208	320		

Notes:
1. Values are based on 23.9 °C dry-bulb room temperature.
2. Normalized values are based on a representative percentage of adult male, adult female, and children for the activity listed and assumes that an adult female's heat gain is 85% of an adult male, and heat gain from a child is 75% of an adult male.
3. Values are approximate based from ASHRAE 2021 Fundamentals, Chapter 9, Table 6.

Example 9-15: Heat Gain from Occupants

An open office space consists of 30 occupants at design conditions. Estimate the sensible and latent heat gain from the occupants, assuming a normal distribution of male and female occupants. Assume the degree of activity is "Moderately active office work.

Solution

Step 1: Determine Heat Gain Per Person

For the degree of activity, use "Moderately active office work." Sensible and latent heat gains for the adjusted M/F group are:

Sensible heat per person = 250 Btu/hr per person

Latent heat per person = 200 Btu/hr per person

Step 2: Calculate Sensible Heat Gain

Total Sensible Heat Gain = 30 people × 250 Btu/hr per person = 7,500 Btu/hr

Step 3: Calculation of Latent Heat Gain
Total Latent Heat Gain = 30 people × 200 Btu/hr per person = 6,000 Btu/hr

9.4.3 Internal thermal mass influences

Heating and cooling loads are the energy needed to heat or cool a space to maintain a comfortable indoor temperature. Heat gains or losses are immediate changes that directly affect these loads. However, the building's thermal mass – the ability of materials such as walls, floors, and furniture to absorb, store, and release heat – plays a significant role in moderating these changes.

This buffering effect means that actual heating and cooling loads fluctuate more gradually and are generally lower than the sum of the immediate heat gains and losses. Figure 9.14 demonstrates how this dampening effect stabilizes indoor temperature. The key takeaway is that increasing thermal mass in a space can help moderate indoor temperature, particularly by reducing peak cooling loads.

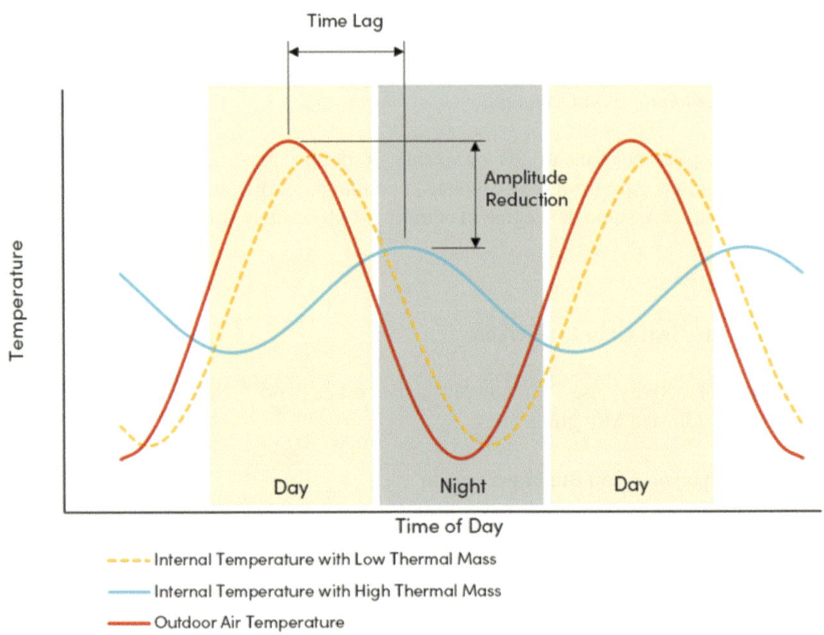

Figure 9.14 Thermal mass influence of heating and cooling loads (source: FirstLight Energy Solutions).

9.5 Load Analysis and Equipment Right-Sizing

Load analysis is a cornerstone of HVAC system design and optimization. A comprehensive understanding of a building's heating and cooling demands can be understood using load profiles and duration curves. This insight allows for the right sizing of equipment, enhancing efficiency, reducing costs, and improving occupant comfort. Embracing data-driven approaches, particularly in retrofit projects, further refine this process, ensuring that HVAC systems can efficiently meet the actual needs of the building without unnecessary costs associated with oversizing.

Benefits of a load analysis include:

- **Informed decision-making.** Enables data-driven choices in equipment selection and system design.
- **Enhanced system performance.** Leads to HVAC systems that are better matched to building demands, improving overall performance.
- **Energy savings.** Identifies opportunities for load reduction and efficiency improvements.
- **Peak load management**. Helps implement strategies to reduce peak demand charges and alleviate strain on the electrical grid.
- **Thermal energy storage opportunities.** Highlights the potential for TES systems to shift loads and improve system flexibility.

9.5.1 Design heating and cooling loads

Detailed heating and cooling load calculations are inherently complex due to the multitude of factors influencing them, including:

- *Variable heat gains and losses* that stem from multiple sources discussed in Section 9.4. These sources vary based on environmental factors, building schedules, and occupancy patterns.
- *Solar radiation exposure* that varies with the time of day, season, and building orientation, impacting loads throughout the day.
- *Thermal mass effects* that characterize the ability of building materials to absorb and store heat, affecting the timing and magnitude of heating and cooling loads. Thermal mass effects interact with sensible forms of heat gain and loss, impacting the loads on HAC systems. Failure to account for thermal mass effects can lead to unnecessary oversizing of HVAC equipment.

Given this complexity, commercial software and spreadsheets are commonly employed to perform detailed load calculations. Standard approaches include the radiant time series and the heat balance method. The heat balance method is often utilized in commercial software due to its comprehensive handling of heat transfer processes.

9.5.2 Data-driven approach in retrofit projects

For retrofit projects, leveraging actual building performance data and historical utility and O&M records can enhance the accuracy of load analysis. This data-driven approach offers several benefits:

1. Minimize guesswork by reducing reliance on assumptions inherent in modeling approaches.
2. Identify effective load reduction strategies by helping to identify problem areas of performance within a building and pinpointing where energy efficiency improvements will have the most impact.
3. Reduce oversizing risks by using historical data to determine actual peak loads.
4. Maximize return on investment by helping to identify which capital expenditures yield the most significant energy savings or GHG emission reduction potential.

Common data sources include: utility data, building automation system (BAS) trend logs, energy sub-metering, maintenance trouble calls, facility user surveys, energy audit reports, and O&M staff interviews. Utility data provides historical energy consumption, peak demand, and energy costs for a building, which are helpful for quantifying energy use and GHG emission profiles. BAS and sub-metering information offer granular data on operating temperatures, pressures, flow rates, and equipment runtimes. Sub-metering is particularly useful for monitoring specific subsystems like lighting, plug loads, or individual HVAC systems Facility user surveys and O&M staff interviews provide valuable insights as to whether HVAC services are over- or under-performing relative to expectations. It is critical to conduct surveys and interviews both before (baseline) and post-retrofit to ascertain whether services are worsened, maintained, or improved as a result of a retrofit.

9.5.3 Annual load profiles

Annual load profiles (e.g. Figure 9.15) represent the heating and cooling demand over a year, providing insights into seasonal fluctuations and how loads change with outdoor temperature and occupancy patterns. It is particularly useful for identifying periods of simultaneous heating and cooling demands that present opportunities for heat recovery. Heat recovery reduces heating loads by reclaiming heat from cooling loads – which would

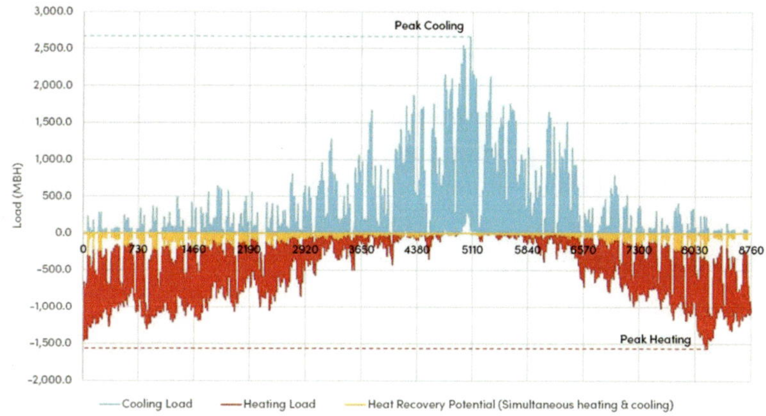

Figure 9.15 Example annual load profile for heating and cooling demands.

otherwise be rejected to the atmosphere – and utilizing it for space heating, absorption cooling, and domestic hot-water needs.

Annual load profiles also help evaluate the potential for storing excess thermal energy to reduce peak heating or cooling demands. Combining thermal energy storage with heat recovery can be a powerful strategy to lower

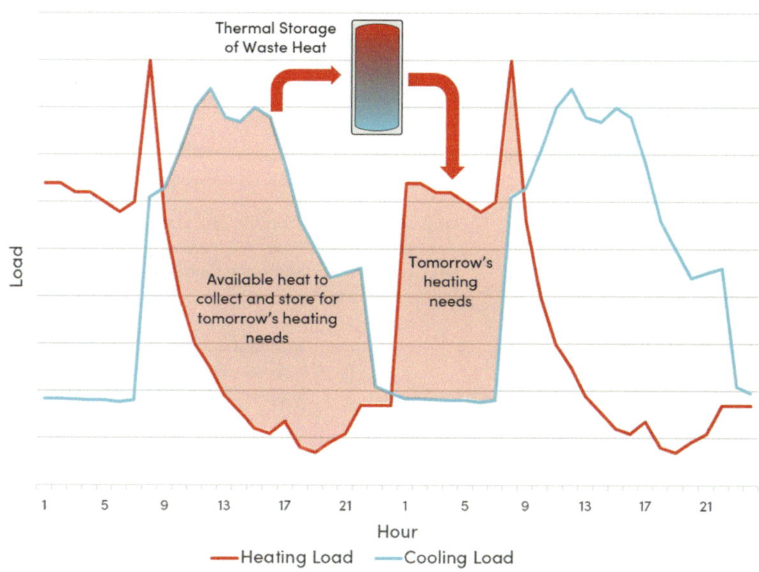

Figure 9.16 Example design day load profile illustrating thermal storage potential.

heating system carbon emissions and peak loads. Figure 9.16 provides an example snapshot of two days within the annual load profile of Figure 9.15, illustrating the potential benefits of thermal energy storage coupled with heat recovery to save excess heat from afternoon cooling loads to meet the next day's morning warm-up heating needs. Refer to Chapter 17 for more information on thermal energy storage and Chapter 20 for heat recovery.

9.5.4 Load duration curves

A load duration curve (LDC) is a graphical representation that ranks system demand from highest to lowest over a period of time (typically a year), illustrating how frequently each level of demand occurs. Figure 9.17 presents an example of an LDC for heating and cooling systems. LDCs are useful for analyzing peak versus part-load conditions and understanding how the system operates at various levels. This is valuable for optimizing equipment selection by informing decisions on equipment efficiency at part-load conditions and understanding equipment turn-down requirements.

LDCs are also useful when investigating hybrid system configurations that combine differential fuel sources for improved efficiency. For buildings that operate primarily at part-load, a high-efficiency heating source sized at less than 50% of the peak load may be capable of supplying 70–90% of the

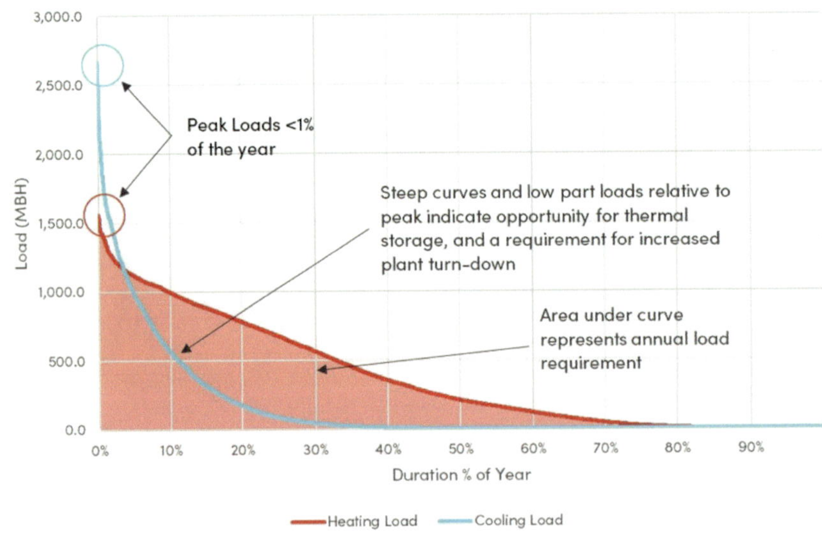

Figure 9.17 Example load duration curve for heating and cooling demand.

annual heating energy. An existing or less costly peaking system can then be relied upon to meet peak heating requirements with a relatively minor impact on annual heating energy consumption.

Key insights from load duration curves:

- **Peak loads occur infrequently.** Both peak heating and cooling loads may occur for a small number of hours annually.
- **Systems operate mostly at part-load.** Most HVAC systems often run at less than a certain percentage (e.g. 50%) of peak load for most of the year.
- **Area under the curve.** It represents the total annual load requirement, visually quantifying energy need. This allows for a quick assessment of the energy savings impact of high-efficiency systems sized for less than peak load.
- **Distinguishing base load vs. variable load.** Understanding the proportion of constant (base) load versus variable load aids in prioritizing load reduction strategies.

9.5.5 Equipment right sizing

Properly sizing HVAC equipment is important for achieving optimal energy efficiency, comfort, and cost-effectiveness in building systems. Right-sizing ensures that equipment capacity matches the actual heating and cooling demands of a building, as determined through detailed analysis. Oversized equipment operates inefficiently, resulting in increased energy consumption, higher operating costs, and unnecessary capital expenditure. Conversely, undersized equipment may fail to maintain desired indoor conditions, compromising occupant comfort.

The tools discussed above enable informed decisions about equipment selection and sizing. These tools help address the following key equipment sizing considerations:

- **Part-load efficiency.** Equipment should operate efficiently at the load levels most commonly experienced. High part-load efficiencies ensure energy savings since buildings often run under partial loads.
- **Modulating capabilities.** Systems with good turn-down ratios adjust output to match varying demands. Modulating equipment, like variable-speed compressors and fans, provides precise control without frequent cycling.
- **Hybrid systems.** Combining different equipment types, such as a primary heat pump with a secondary boiler, optimizes efficiency across

load ranges. High-efficiency equipment can be sized to meet the base loads, while supplemental equipment handles short-duration peaks.

Right-sizing avoids the inefficiencies of oversized equipment operating at low loads. Oversized systems tend to short-cycle, wasting energy and increasing wear, which leads to higher maintenance costs and reduced lifespan. Appropriately sized equipment runs longer at optimal efficiency, enhancing reliability and reducing energy consumption.

Additionally, right-sizing lowers capital costs since smaller equipment is less expensive to purchase and install. Operating costs decrease due to reduced energy usage and less frequent maintenance. Investing in equipment that precisely meets building needs yields a better return on investment and lowers total ownership costs.

9.6 Ventilation Systems

Ventilation systems are essential components of HVAC systems, primarily responsible for ensuring that indoor spaces receive adequate fresh air to maintain healthy air quality. Proper ventilation removes contaminants, controls humidity, and provides a comfortable environment for occupants by diluting and displacing indoor pollutants.

Ventilation can be classified into two main categories:

- **Natural ventilation.** This approach relies on passive means such as windows, vents, and other controlled openings to facilitate air exchange between indoor and outdoor environments. Natural ventilation utilizes natural forces like wind pressure and the stack effect to move air through the building without mechanical assistance. While it can be energy-efficient and cost-effective, natural ventilation offers limited control over airflow rates and may not be sufficient in tightly sealed or densely occupied buildings.
- **Mechanical ventilation.** Mechanical ventilation uses fans and ductwork to control air movement within a building. Air handling units (AHUs) are commonly employed in these systems to filter, condition, and distribute air. Mechanical ventilation provides precise control over airflow, temperature, humidity, and filtration, ensuring consistent indoor air quality regardless of external conditions.

Air can serve a dual purpose by controlling space temperatures and providing ventilation. Systems that deliver both conditioned air (for heating or cooling) and fresh outdoor air through a single distribution network are commonly referred to as *all-air systems*. These systems integrate ventilation and thermal conditioning, supplying air that meets both the ventilation requirements and the thermal comfort needs of occupants.

Alternatively, a *dedicated outdoor air system (DOAS)* is specifically designed to handle only the outdoor air ventilation requirements of a building. DOAS operates independently from the primary heating and cooling systems, directly supplying fresh, conditioned outdoor air to occupied spaces or other air distribution systems. The thermal conditioning of spaces is managed by separate systems such as fan coil units, radiant panels, or variable refrigerant flow (VRF) systems.

Table 9.7 provides a comparison highlighting the key features of each system.

Table 9.7 Comparison of all-air vs. DOAS.

Feature	All-air system	DOAS
Ventilation and conditioning	Combines ventilation and thermal conditioning in one integrated system.	Separates ventilation from heating and cooling, where the DOAS provides ventilation and thermal conditioning is provided by another system.
Efficiency	May be less efficient due to potential over-ventilation or simultaneous heating and cooling, which can lead to energy waste.	Offers improved efficiency by optimizing ventilation rates and often incorporates energy recovery technologies to precondition incoming air.
System complexity	Involves a single system, which can simplify installation but requires larger ductwork to handle combined airflow. Control systems can be more complex due to the integration of multiple functions.	Utilizes two separate systems, increasing mechanical complexity but allowing for smaller ductwork dedicated to ventilation. Controls can be simpler because each system manages a specific function.
Humidity control	Managing both temperature and humidity simultaneously can be challenging, potentially leading to less effective dehumidification.	Provides better humidity control by conditioning ventilation air separately, allowing for precise dehumidification or humidification of ventilation air.
Ductwork	Requires larger ducts to accommodate the total airflow needed for ventilation and conditioning, which can impact building design and space utilization.	Uses smaller ducts for ventilation air, with separate ductwork or piping for thermal conditioning systems, offering more flexibility in design and space utilization.

9.6.1 Air handling unit components and energy-saving strategies

Air handling units (AHUs) are integral to all-air and dedicated outdoor air systems (DOASs), serving as the heart of mechanical ventilation by conditioning and circulating air throughout a building. This section will examine the typical components of air handling units – including fans, filters, heating and cooling coils, and strategies associated with each. Figure 9.18 illustrates an example air handling unit labeled for clarity, providing a visual guide of each component.

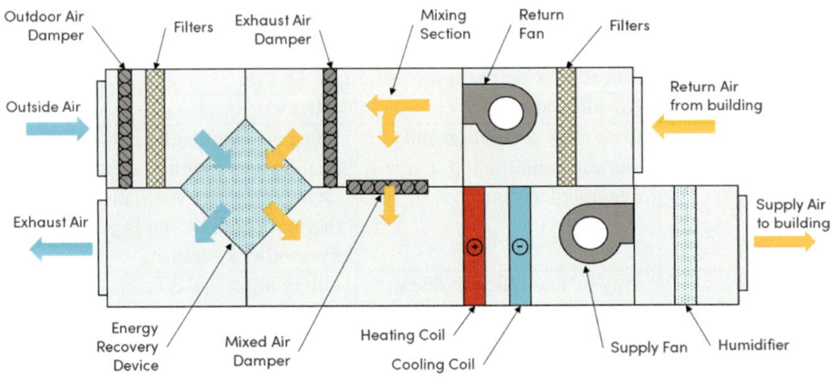

Figure 9.18 Air handling unit common components.

9.6.1.1 Unit casing

The casing is the external housing of the air handling unit that encloses all the components. It protects the internal components from environmental factors, provides insulation to reduce thermal losses, and directs air through components.

Efficiency Strategies:

- **Test and seal the air handling unit to minimize air leakage.** Ensure that the AHU casing is adequately sealed to prevent unwanted air leakage, which can significantly reduce system efficiency. Air leakage should not exceed 1% of the design airflow when tested at 1.5 times the total static pressure rating of the AHU.
- **High-performance insulation to reduce heat loss.** Apply high-quality insulation to the AHU casing and internal components to minimize thermal losses and prevent condensation. Use insulation materials with

a minimum thermal resistance (R-value) appropriate for the application – typically R-8 or higher for AHU casings. Insulation should be non-combustible, moisture-resistant, and meet or exceed ASHRAE 62.1 requirements for resistance to mold growth and erosion.

- **Consider larger casing sizes to reduce air velocity.** Standard AHUs are often selected with air velocities around 2.5 m/s (500 fpm). Reducing air velocity by increasing the cross-sectional area of the unit lowers the pressure drop across internal components, improving energy efficiency. Consider designing AHUs with reduced air velocities of 1.75–2 m/s (350–400 fpm) for new construction or retrofit projects.
- **Ensure proper trapping of condensate and casing drains.** Ensure drains from AHUs have adequate depth traps or mechanical air traps to prevent air leakage through the drain lines.

9.6.1.2 Fans (supply, return, and relief)

Fans are responsible for the movement and circulation of air through an HVAC system, ensuring that conditioned air reaches the occupied spaces, returning air to the AHU for processing, and relieving air from the building to maintain building pressure relationships. Fans generate the necessary pressure to overcome the resistance of components within the AHU, ductwork, and other devices included in the air distribution system.

Supply fans deliver conditioned air from the AHU to the occupied spaces within a building. This air has typically been filtered, heated or cooled, and possibly humidified to meet indoor comfort requirements.

Return fans draw air from occupied space back to the AHU for reconditioning and recirculation. They are common in air handlers in large buildings with extensive ductwork to overcome the resistance in return air ducts, ensuring air from the buildings gets back to the AHU.

Relief or exhaust fans discharge indoor air to the outside, necessary when outdoor air is introduced into the system for ventilation or economizer cooling. They help to maintain appropriate building pressure levels by ensuring the amount of air leaving the building balances the amount of air entering it.

Efficiency Strategies:

Fan energy can account for 30%–40% of building HVAC system energy. Refer to Section 7.2 for additional strategies to reduce fan energy consumption.

- Use *premium efficiency motors* to reduce electrical energy use associated with fan systems. Use inverter-duty motors for variable speed applications to ensure compatibility with variable frequency drives.
- Use *direct-drive fans* instead of belt-drive fans whenever possible. Direct-drive configurations connect the motor directly to the fan, eliminating energy losses associated with belts and sheaves. This also reduces maintenance requirements by removing the need for belt replacements and adjustments.
- Install *variable speed fan motors* by implementing variable frequency drives (VFDs) or electronically commutated motors (ECMs) to modulate fan speed based on real-time demand (also see Chapter 14). Variable speed fans are typically controlled to maintain a static pressure setpoint within the distribution system.
- Implement *static pressure resets* in the building control system by resetting the duct static pressure setpoint based on zone demands. Demand-based static pressure resets can reduce fan energy consumption by up to 50% in variable air volume (VAV) systems (also see Chapter 15).

9.6.1.3 Mixed air section and dampers

The mixing section combines outdoor air and return air from the building before undergoing filtration, heating, cooling, and redistribution back into the building. For all-air systems, the mixing section plays a vital role in maintaining indoor air quality by controlling the proportion of outdoor air supplied. It is generally composed of the following components:

- Outside air dampers regulate the amount of outdoor air entering the system from the outside.
- Return air dampers control the flow of air returning from the occupied spaces to be recirculated.
- Exhaust or relief air dampers allow excess return air to exit the buildings, maintaining proper pressure balancing.
- Mixing plenums are the chambers where the outdoor air and return air are blended before being further conditioned by heating and cooling coils.

For DOAS systems, the mixing section is typically not included. It is replaced by separate outdoor air intake and exhaust air discharge sections, each with a motorized damper that closes when the AHU is shut down.

Efficiency Strategies:

- Utilize *low-leakage dampers* designed with tight seals and precision construction to minimize air leakage when closed.
- Regularly *inspect damper seating and seals* to ensure dampers fully close and that seals are intact and functioning correctly (also see Chapter 21).
- Utilize *air-side economizers* to reduce mechanical cooling energy by leveraging cool outdoor air to meet building cooling loads during appropriate conditions.
- Implement *demand control ventilation (DCV)* to adjust ventilation rates based on real-time occupancy levels detected through CO_2 sensors or occupancy sensors to reduce heating and cooling energy required to condition ventilation air.
- Optimize *mixing damper sequencing* by implementing control strategies that coordinate the operation of outdoor, return, and exhaust dampers to minimize pressure drop through the mixing section. Refer to ASHRAE Guideline 36 for details.
- *Avoid stratification* in the mixing plenum, which can reduce coil efficiency and lead to uneven heating or cooling of air. For cold climates, baffles or air blenders may be required to ensure adequate mixing of airstreams.

9.6.1.4 Air filtration and cleaners

Particulate air filters and air cleaners are essential components in HVAC systems, removing particulate matter from the air to protect equipment and maintain indoor air quality by eliminating dust, allergens, and pollutants. The effectiveness of a filter is characterized by its MERV rating (minimum efficiency reporting value) or ISO 16890 rating.

Air cleaning technologies extend beyond mechanical filtration. Electronic air cleaners use electrically charged materials to capture particles, while UV-C light systems inactivate airborne pathogens such as bacteria and viruses by disrupting their DNA. Activated carbon filters are also commonly used to remove gases and odors through adsorption.

Efficiency Measures:

- **Replace filters more frequently.** Filters should be replaced well before they reach the maximum pressure drop recommended by the manufacturer. Regular replacement ensures optimal airflow and maintains energy efficiency.

- **Monitor pressure drop across filters.** Implementing pressure drop monitoring across filter banks through the control system allows for automated filter change alerts based on actual filter conditions, minimizing the risk of filters not being changed at appropriate intervals.
- **Eliminate pre-filters.** Pre-filters often add unnecessary pressure drop and maintenance costs without significantly extending the life of primary filters due to their low dust capture ratings. Some manufacturers offer extended surface area filters designed for long life without the need for pre-filters.
- **Consider air cleaners as alternatives to particulate filters.** Electronic air cleaners can be effective alternatives to traditional particulate filters. Electronic air cleaners often provide similar MERV ratings with reduced air pressure drops.
- **Install UV-C systems within AHUs.** UV-C systems can enhance air filtration by improving indoor air quality and maintaining the cleanliness of air coils. They prevent biofilm buildup, reduce pressure drop across coils, and improve heat transfer efficiency.
- **Use extended surface area filters.** Extended surface filters have lower pressure drops than conventional panel filters, improving airflow and reducing energy consumption.

9.6.1.5 Energy recovery

Energy recovery devices transfer otherwise wasted heat or cooling energy from the exhaust air. Generally, this energy is transferred into the incoming outdoor air to reduce the energy required for conditioning. However, this energy source can also be reclaimed through heat pump systems and used for different systems throughout the building (e.g. domestic hot-water heating). Energy recovery is discussed in more detail in Section 9.6.2.

9.6.1.6 Heating and cooling coils

Heating and cooling coils are the primary interface between air handling systems and the heating and cooling plants. They regulate indoor air temperature and humidity by conditioning the supply air delivered to the building.

Cooling coils contain chilled water or refrigerant that absorbs heat from the incoming air. Heat is transferred into the chilled fluid as warm air passes over the coil, lowering the air temperature. This process also causes moisture in the air to condense on the coil surface, dehumidifying the air and helping control indoor humidity levels.

Heating coils utilize hot water, steam, or electric resistance elements to raise the air temperature. As air passes over the heated element, it absorbs heat, increasing its temperature. Heating can also be achieved directly through fossil fuels by using a furnace. The hot combustion gases from the burner pass through a metal chamber, which is a heat exchanger. Air is heated as it passes over the heat exchanger.

Efficiency Strategies:

- *Select coils for reduced face velocity* to reduce air pressure drop. Selecting coils for a lower face velocity of around 2 m/s (400 fpm) instead of the typical 2.5 (500 fpm) reduces air pressure drop across the coil, which lowers fan energy. It is also one method to increase coil surface area, increasing heating and cooling plant performance.
- *Increase coil surface area* by selecting coils with larger surface areas, such as deeper coils or those with more fins per inch. Increased surface area improves the heat exchange between the air and the coil fluid, improving the efficiency of associated heating and cooling systems.
- Using a *switchover coil for heating and cooling* reduces pressure drop by utilizing a single coil to provide heating and cooling functions. This approach is attractive in retrofit applications to reduce heating water temperatures by repurposing the existing chilled-water cooling coil as a switchover coil. This helps reduce heating water temperatures because the chilled-water cooling coil typically has a lot larger surface area due to increase coil rows.
- Incorporate a *coil bypass damper* to reduce air pressure drop by bypassing air around the coil when not required.
- Implement a *demand-based supply air temperature setpoint reset* in the building control system to adjust the supply air temperature setpoint based on real-time zone heating and cooling demands. In cooling mode, the supply air temperature can be reset upward, maximizing the potential for free cooling through airside economizers. In heating mode, this prevents unnecessary overheating of supply air, reducing energy use by the heating system.
- Add a *heat transfer additive* to the hot-water and chilled-water systems to reduce the surface tension of the fluid, allowing it to make better contact with the coil tubes. This increased contact improves the heat transfer efficiency, reducing the energy consumption of the associated heating and cooling plants. Empirical studies indicate up to 15% energy savings, depending on the specific system and operating conditions.

- Ensure *regular coil cleaning and maintenance* to prevent dust, dirt, and biofilm buildup on coil surfaces, which impede heat transfer and increase air resistance. Installing UV-C systems within the coil sections can aid in maintaining cleaning coil surfaces and lower the frequency of manual coil cleanings.

9.6.1.7 Humidification

Humidifiers are integrated into air handling units (AHUs) to add moisture to the supply air, maintaining appropriate humidity levels within a building. They are ubiquitous in healthcare facilities and buildings in very cold climates, where maintaining indoor humidity is crucial for comfort and health.

Humidifiers typically operate based on one of the following principles:

- **Steam humidification (isothermal).** These humidifiers generate steam using boiler-generated steam or heating water with electrodes or resistive elements. The steam is introduced directly into the airstream, increasing humidity without significantly altering air temperature.
- **Atomization (adiabatic humidification).** This method involves atomizing water into fine droplets or mist. The tiny water particles evaporate into the airstream without additional heat, utilizing the air's thermal energy and slightly cooling the air.
- **Evaporative humidification (adiabatic).** Evaporative humidifiers pass air over a water-saturated medium. As the air moves across the wet surface, water evaporates into the airstream, increasing humidity and producing a cooling effect due to heat absorption during evaporation.

Efficiency Strategies:

- **Demand-based humidity setpoint reset.** Adjust the supply air humidity setpoints in the building control system based on the specific requirements of different zones within the building. By resetting humidity levels according to actual demand, the system avoids over-humidifying spaces, which reduces unnecessary energy consumption associated with excess moisture production.
- **Implement adiabatic humidification methods.** Adiabatic humidification techniques, such as atomization or evaporative systems, can lead to significant energy savings – often over 90% compared to isothermal steam humidification. Since adiabatic methods do not require heating water to generate steam, they consume considerably less power. However, it is essential to ensure proper water treatment when using

adiabatic systems to prevent issues like mineral buildup or microbial growth.

9.6.2 Air-to-air energy recovery

Energy recovery is essential to modern HVAC systems to reduce the energy required to condition outdoor air. Air-to-air energy recovery systems can reduce heating and cooling loads associated with conditioning outdoor air by capturing energy from one airstream and transferring it to another.

Air-to-air energy recovery provides a mechanism for recovering sensible (temperature) and latent (moisture) heat between two airstreams. Heat is recovered from warmer exhaust air and transferred to the incoming outdoor air when heating is required. In cooling, heat and moisture are extracted from the warmer and more humid outdoor air intake and rejected into the exhaust stream, effectively pre-cooling and dehumidifying the incoming air before further conditioning.

Energy recovery devices can be categorized based on their function:

- **Sensible heat recovery devices.** These transfer only sensible heat and are commonly referred to as heat recovery ventilators (HRVs).
- **Total energy recovery devices.** These transfer heat and moisture are commonly referred to as energy recovery ventilators (ERVs), enthalpy recovery devices, or total energy recovery devices.

9.6.2.1 Estimating energy recovery savings

The effectiveness of an air-to-air energy recovery device is a measure of its ability to transfer heat (and moisture) from one airstream to another. The overall effectiveness can be expressed as

$$\text{Effectiness, } \varepsilon = \frac{\text{Actual Energy Recovered}}{\text{Maximum Energy Available for Recovery}}.$$

The effectiveness of an energy recovery device can be broken down into sensible and latent components:

- Sensible effectiveness refers to the ability of the device to transfer sensible heat, measured as a percentage of the temperature difference between incoming outdoor air and exhaust air.
- Latent effectiveness refers to the ability of the device to transfer moisture, measured as a percentage of the difference between the humidity ratio of the incoming outdoor air and exhaust air.

Energy savings are calculated based on the reduced heating or cooling load achieved through energy recovery. For example, suppose an energy recovery system with 70% sensible and 50% latent effectiveness is installed. In that case, 70% of the temperature differential and 50% of the moisture content between the exhaust and supply air is transferred.

9.6.2.2 Frost control and frosting

When outdoor air temperatures fall below freezing, heat transfer from the warm exhaust air stream to the incoming outdoor air stream can cool the exhaust air below its dew point. This causes water vapour to condense on the heat transfer surface. If the surface temperature is also below freezing, the condensate freezes, forming frost. As frost accumulates, it restricts airflow and degrades heat transfer effectiveness, necessitating a defrost or frost-control strategy to maintain system performance.

The outdoor air temperature at which frost begins to form, known as the frost threshold, depends on the exhaust air temperature and humidity, the type of recovery device (sensible-only versus total energy recovery), and the heat recovery effect.

Various frost prevention and control strategies are used to mitigate this issue; however, most reduce overall energy recovery performance. These strategies include:

Frost prevention. Frost prevention strategies maintain continuous ventilation and prevent frost buildup on the energy recovery device by limiting heat recovery effectiveness to keep the exhaust air temperature above the frost threshold.

- **Outdoor air preheating.** Preheat frost control is a preventative strategy to prevent frost formation while maintaining continuous ventilation. With this strategy, the outdoor air is preheated sufficiently to prevent the exhaust air temperature from dropping below the frost threshold.
- **Face and bypass dampers.** Face and bypass dampers bypass a portion of outdoor air around the energy recovery device to reduce the amount of energy recovered and keep the exhaust air temperature above the frost threshold. Continuous ventilation is maintained with a downstream heating coil providing supplemental heat to maintain the supply air temperature.
- **Energy recovery device modulation.** For heat and enthalpy wheels, the wheel's rotation speed is lowered to reduce the heat recovery effectiveness of the device, preventing the exhaust air from dropping

below the frost threshold. Run-around loops can also utilize this strategy by reducing the circulating fluid flow rate.

Defrost. Defrost strategies are intermittently used after frost accumulation on the energy recovery device. Defrost strategies are common for small stand-alone HRVs and ERVs in residential and light commercial applications.

- **Recirculation defrost.** This method shuts down the exhaust fan, closes the outdoor air damper, and uses the supply fan to recirculate return air through the energy recovery device to melt frost accumulation. This process disrupts ventilation but prevents the building from being driven into negative pressure.
- **Exhaust-only defrost.** Exhaust-only defrost stops introducing ventilation air into the building and runs the exhaust fan to warm up the heat recovery device to melt frost accumulation. This process disrupts ventilation and drives the building into negative pressure.

9.6.2.3 Types of passive energy recovery

Several air-to-air energy recovery systems are available, each with unique characteristics, benefits, and applications.

Fixed plate (sensible only) and membrane plate (sensible and latent) heat exchangers	**Description:**
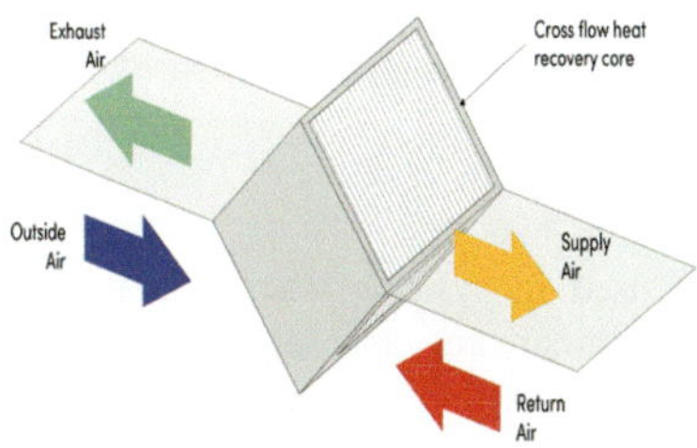	A fixed plate or membrane plate heat exchanger consists of thin plates or membranes separating the exhaust and supply airstreams. Heat is transferred through the plates without mixing the two airstreams. Membrane plate heat exchangers also facilitate moisture transfer for latent heat recovery. **Effectiveness:** Sensible: 50%–75% Latent: 25%–60% (membrane plate) **Applications:** Common for smaller heat and energy recovery systems such as residential and commercial applications. **Benefits:** Low maintenance, no moving parts.

<table>
<tr>
<td>

Heat (sensible only) and enthalpy (sensible and latent) wheels

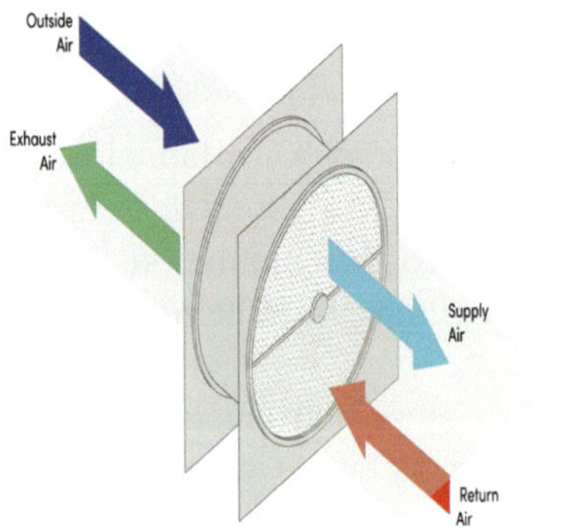

</td>
<td>

Description:
Heat and enthalpy wheels are rotating devices with a matrix of materials that absorb heat and moisture from the exhaust stream and transfer it to the incoming outdoor air as the wheel spins.

Effectiveness:
Sensible: 65%–80%
Latent: 50%–80% (enthalpy wheel)

Applications:
Common in large buildings with high ventilation rates, such as hospitals and large commercial buildings.

Benefits:
High efficiency for sensible and latent (enthalpy wheels) heat recovery, and compact footprint for larger airflow capacities.

</td>
</tr>
<tr>
<td>

Run-around loop (sensible only)

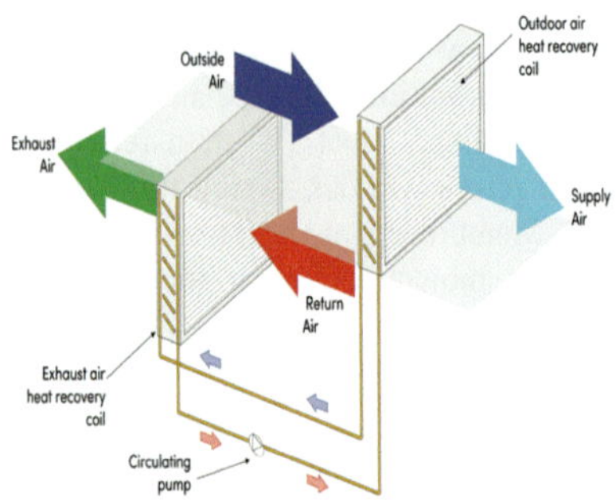

</td>
<td>

A run-around loop uses two air coils connected by a closed loop of fluid (typically a water/glycol mixture) to facilitate sensible-only heat transfer between the two airstreams. A pump is used to circulate the water between the two coils.

Effectiveness: Sensible: 45%–65%
Latent: 0%

Applications:
Typically used in systems where exhaust and supply airstreams are separated and in applications with cross-contamination concerns like laboratories or hospitals.

Benefits:
Flexibility in design, no cross-contamination risk, suitable for retrofits in buildings where supply and exhaust duct systems are separated.

</td>
</tr>
</table>

Heat pipe (sensible only) 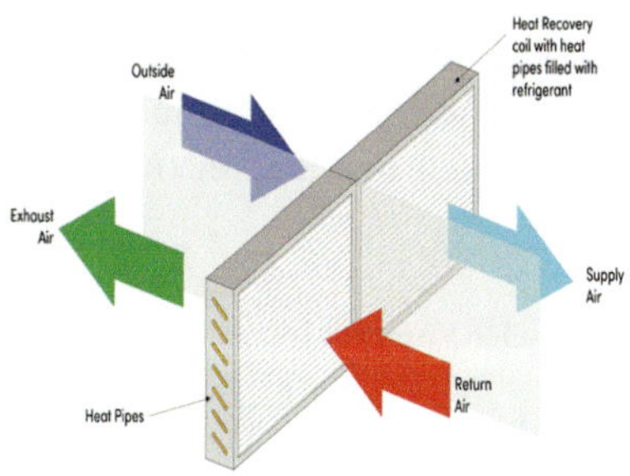 	**Description:** A heat pipe consists of several sealed tubes filled with a refrigerant as the working fluid. Heat is absorbed from the warm airstream, evaporating the refrigerant, which travels to the cold end of the pipe and releases its heat into the other cooler airstream as it condenses. **Effectiveness:** Sensible: 40%–60% Latent: 0% **Applications:** Useful for applications where cross-contamination between airstreams is a concern. **Benefits:** Low maintenance requirements and reliable.
Dual core heat exchanger (sensible and latent) 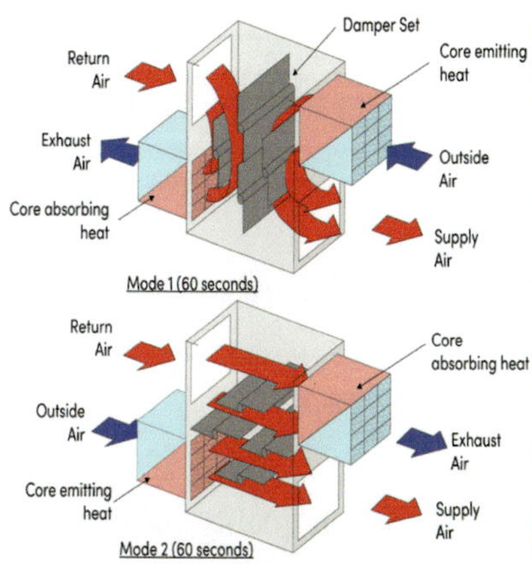 	**Description:** A dual-core energy recovery system uses two separate heat exchangers (cores) and dampers to transfer heat between the exhaust and supply air. The dampers control which airstream is directed over a core, with the airstreams being switched every 60–90 seconds. Exhaust air heats one of the cores while supply air depletes the other core of its heat. After a short period, the flow is reversed, and the depleted core is charged with exhaust air, while the freshly charged core is depleted with supply air. **Effectiveness:** Sensible: 85%–90% Latent: 50%–70% **Applications:** Useful in cold climates where frost formation is a significant concern. **Benefits:** It allows for continuous energy recovery and prevents frosting without needing bypass systems or defrost cycles. It does not suffer from a defrost penalty associated with other forms of energy recovery.

HRVs can substantially improve IAQ as illustrated in Table 9.8. These are the results for the retrofit of a counterflow HRV installed in a detached home equipped with a natural gas heating furnace, hot water tank, and cooking stove. The HRV recovered more than 80% of the sensible heat from the exhausted stale air. The concentration of five pollutants decreased due to the HRV. Ozone did not decrease, but it did remain much lower than outdoor (background) levels.

Table 9.8 Measured indoor air quality improvements due to HRV retrofit for a single family home

description	**start**	**end**	CO_2 ppm	NO_2 ppb	PM2.5 $\mu g/m^3$	VOC mg/m^3	co ppm	O_3 ppb
outdoor (background air quality)	2023-May-11	2023-Jul-13	428	14.72	2.30	1.27	0.26	9.16
indoor baseline (no HRV)	2023-Nov-06	2024-Jan-31	963	22.06	6.53	0.74	1.34	1.55
indoor HRV in operation	2024-Feb-17	2024-Sep-18	545	18.25	1.79	0.51	0.12	3.82
outdoor (background air quality)	2024-Sep-18	2024-Oct-19	433	18.15	2.21	1.38	0.01	10.94
difference: HRV vs. baseline			-418	-3.82	-4.75	-0.23	-1.22	2.27
difference: HRV vs. outdoor			115	1.81	-0.47	-0.82	-0.01	-6.23

- See Table 20.3 for a summary description of air pollutants.
- ppm = parts per million; ppb = parts per billion; $\mu g = 10^{-6}$grams; mg $= 10^{-3}$ grams
- Measurements made using an AirAssure model 8144-6.
- Indoor O_3 levels likely increased baseline vs. with HRV due to injection of outdoor air, but overall indoor levels remained well below outdoor concentrations.

9.6.3 Demand control ventilation (DCV)

Demand controlled ventilation (DCV) optimizes indoor air quality and energy efficiency by dynamically adjusting ventilation rates based on real-time occupancy and pollutant levels. Unlike fixed ventilation, DCV systems modulate airflow (often based on space CO_2 concentrations) to deliver ventilation when and where it is needed.

Key components of DCV systems include:

1. **Sensors.** Sensors are relied upon to monitor indoor air quality conditions. Carbon dioxide (CO_2) and, increasingly, particulate matter sensors continuously monitor indoor air quality. Occupancy sensors are also commonly integrated to detect space occupancy.

2. **Controller.** A building automation system or standalone controller processes sensor data, determining ventilation rates based on predefined thresholds. For instance, elevated CO_2 levels beyond a specific concentration will trigger increased airflow.
3. **Modulating equipment.** Variable-speed fans and dampers, adjusted by the controller, regulate outdoor air intake to meet specific ventilation requirements.

There are several methods to implement DCV, each tailored to different building types and occupancy patterns:

1. *CO_2-based DCV* is the most common type of DCV, where CO_2 sensors monitor carbon dioxide levels in the air. As occupancy increases, CO_2 levels rise, triggering increased ventilation to dilute indoor pollutants. Conversely, when occupancy decreases, ventilation rates are reduced to save energy.
2. *Occupancy schedule-based DCV* involves adjusting the ventilation rate based on the building or space's anticipated occupancy schedule. This approach can be used, with caution, for spaces with well-defined and predictable occupancy patterns, such as cafeterias, lecture halls, and theatres.
3. *Occupancy sensor-based DCV* relies on occupancy sensors to detect the presence of people in a space and determine when ventilation is required. When occupancy is detected, ventilation is provided at design capacity; when spaces are vacant, ventilation is reduced to background levels or wholly shut off where possible.
4. *Occupant counter-based DCV systems* rely on sensors to estimate space occupancy, such as infrared beams, camera systems, thermal imaging, or Wi-Fi tracking, to count occupants in real-time and adjust ventilation to match the calculated ventilation requirement. These systems are complex to implement and have not been widely adopted due to the complexity of technology. Venues such as arenas or theatres that involve ticket sales or turnstiles are examples of where counter-based DCV could be utilized.

9.7 Heating and Air Conditioning Systems

Heating and cooling systems can be broken into three fundamental components: the energy source (heating or cooling source), the energy distribution system, and the space or process interface (terminal system). Each element plays a vital role in meeting the thermal comfort needs of a building or process, and each significantly impacts the overall efficiency of an HVAC system.

The *energy source* is the system element responsible for generating the heating or cooling necessary to meet the building loads. Energy sources can vary widely, including boilers, furnaces, heat pump condensers, heat recovery chillers, and solar panels for heating; chillers, cooling towers, direction expansion (DX) units, and heat pump evaporators for cooling. The choice of energy source is often dictated by climate, energy availability, cost, and environmental considerations.

The *energy distribution system* transports the thermal energy between the energy source and the occupied spaces or processes within the building. Air, water, and refrigerants are common energy transport media with unique advantages and limitations. The distribution system includes all necessary infrastructure for energy transport, including ductwork, piping, fans, and pumps.

The final component, the *space or load interface*, represents the demand for heating or cooling within a space or process. This demand originates from various sources such as occupants, equipment, lighting, and the building envelope. The form of the interface is dependent on the transport media. For air, the interface is a direct connection with room air through diffusers or grilles. In contrast, a fluid-to-air heat exchanger facilitates thermal energy transfer when using water or refrigerant.

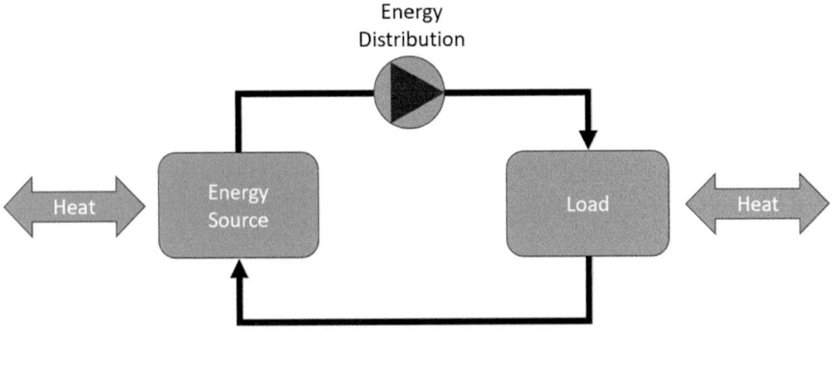

Figure 9.19 Thermal energy system fundamental components.

9.7.1 Heating and cooling sources

Energy sources generate heating and cooling that is necessary to meet building loads and maintain indoor comfort. The selection of fuel sources – fossil fuels, electricity, or renewable energy – plays a significant role in determining the operational costs for a building and carbon emissions profile.

In many buildings, particularly larger facilities, there is often a need to provide both heating and cooling simultaneously for different zones, presenting an opportunity for energy efficiency through heat recovery, where waste heat from cooling can be repurposed for heating demands elsewhere in the building. Recovering and reusing thermal energy from one process to support another enhances overall system efficiency and is a cornerstone of modern HVAC system operation.

Depending on the building type and HVAC system configuration, heating and cooling sources can be centralized or decentralized. Large buildings typically use centralized plants that generate heating and cooling in a single location, distributing it throughout the building. In contrast, smaller buildings often employ decentralized systems like packaged rooftop units, where individual units are distributed throughout the building. Each decentralized unit contains all the necessary components for heating and cooling.

Subsequent sections explore the distinction between central and decentralized equipment, followed by strategies to improve energy efficiency across these systems.

9.7.1.1 Central systems

Boilers  (Credit: First Light Energy Solutions)	Boilers generate hot water or steam for heating purposes. Combustion (natural gas, oil, or biomass) or electric resistance (electricity) heat water in a closed system. Condensing boilers are more efficient than conventional (non-condensing) boilers by condensing flue gases but require a low return water temperature to facilitate condensing. Refer to Chapter 20 for more information on boilers.

Chillers  (Credit: First Light Energy Solutions)	Chillers remove heat from water using a refrigeration cycle, producing chilled water for air conditioning. Air-cooled chillers reject heat to the ambient air using fans, whereas water-cooled chillers reject heat to a cooling tower for more efficient heat rejection.
Heat recovery chillers and water-to-water heat Pumps  (Credit: First Light Energy Solutions)	Heat recovery chillers are specialized chillers that simultaneously produce chilled water for cooling and hot water for heating. They use heat extracted during the cooling process for heating instead of rejecting it to the ambient air. Water-to-water heat pumps operate similarly but have a reversing valve that facilitates reversing the direction of heat flow. Refer to Chapter 10 for more information on heat pumps.
Air-to-water heat pumps  (Credit: First Light Energy Solutions)	Air-to-water heat pumps transfer heat from the ambient air to a water loop to provide heating or cooling using a refrigeration cycle. In heating mode, they extract heat from the outdoor air and transfer it into the building's hot-water loop. The process is reversed in cooling, and heat from the chilled-water production is rejected to the ambient air. Refer to Chapter 10 for more information on air-to-water heat pumps.

Cooling towers  (Credit: First Light Energy Solutions)	Cooling towers reject waste heat from water-cooled chillers or other processes to the atmosphere. Water is cooled through evaporative cooling, where a portion of the water evaporates, lowering the temperature of the remaining water.
Thermal energy storage (TES)  (Credit: Trane)	Thermal energy storage (TES) systems store thermal energy (typically chilled water, ice, or hot water) during off-peak periods for use during peak demand periods to provide heating or cooling. TES systems can reduce peak energy loads by shifting energy consumption to times when electricity is cheaper. TES is advantageous when combined with heat recovery systems as waste heat can be stored for later use, enabling asynchronous energy recovery. Chapter 17 explores thermal energy storage systems in more detail.

9.7.1.2 Decentralized (local) systems

Packaged rooftop units (RTU)  (Credit: First Light Energy Solutions)	Packaged rooftop units (RTUs) combine all components of an HVAC system (compressor, condenser, evaporator, fan, filters, and heating section) into a single unit that is typically installed on the roof of a building. A/C rooftop units have a refrigeration system that provides cooling, with optional gas or electric heating. Heat pump RTUs can provide heating and cooling by reversing the refrigeration cycle. They are typically equipped with supplemental heating in colder climates. Heat pumps are further discussed in Chapter 10.
Split systems  (Credit: First Light Energy Solutions)	Split systems have two main components: an outdoor unit (containing the compressor and condenser) and an indoor unit (containing the evaporator and fan). Refrigerant lines connect the two units, and the system provides either cooling or heating and cooling (if equipped with a heat pump).
Packaged terminal air conditioners (PTAC) and heat pumps (PTHP)  (Credit: First Light Energy Solutions)	PTAC and PTHP units are self-contained heating and cooling systems installed through a wall or window. They operate independently, providing heating and cooling for individual rooms or small spaces.

Terminal water-source heat pumps (WSHP) 	Water-source heat pumps transfer heat between a water loop and the indoor space, providing either heating or cooling. In heating mode, the heat pump extracts heat from the water loop and transfers it to the indoor space, while in cooling mode, it rejects heat to the water loop. The temperature of the water loop is typically maintained at ambient temperatures between 10°C (50°F) up to 30°C (86°F). Excess heat is rejected from this loop often using a cooling tower or fluid cooler. Heat can be added through any combination of heating source from boilers to air-to-water heat pumps. The use of a geo-exchange ground-loop is often used with this type of system. WSHPs are further discussed in Chapter 10.
Variable refrigerant flow (VRF) systems  (Credit: First Light Energy Solutions)	VRF systems use refrigerant as the cooling and heating medium, circulating between an outdoor unit and multiple indoor units. The system can modulate refrigerant flow to each indoor unit, allowing for simultaneous heating and cooling in different zones. VRF systems are further discussed in Chapter 10.
Furnaces 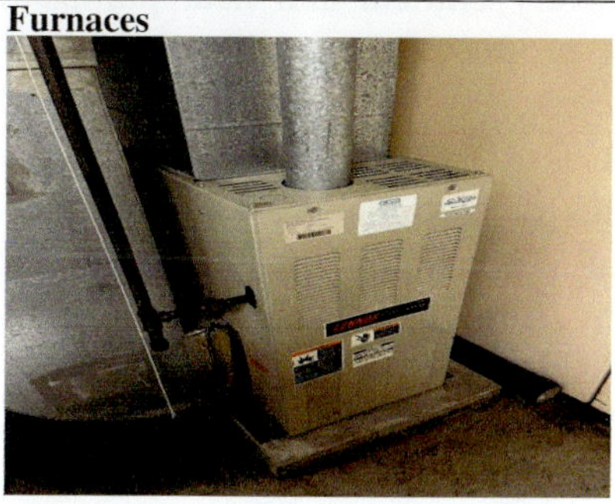 (Credit: First Light Energy Solutions)	Furnaces provide space heating by delivering warm air through ductwork. Heat is typically generated through combustion (natural gas, propane, oil) or electric resistance heating.

9.7.1.3 Energy efficiency strategies

Applicable energy efficiency strategies depend on the type and application of heating and cooling technology. Below is a summary of some common methods that should be considered for heating and cooling sources.

1. **Implement heat pump technology:**
   - Heat pumps offer higher operating efficiencies than traditional heating equipment and eliminate on-site combustion. Heat pumps offer the ability to provide heating and cooling within one piece of equipment. Heat pumps and applications are discussed in further detail in Chapter 10.
   - For regions with low-GHG electrical grids, heat pumps are the technology of choice for reducing energy use and carbon emissions.
2. **Simplify system configurations:**
   - Overly complex systems are more challenging to operate and maintain and are more prone to underperforming in practice due to these complexities. Implement systems as simply as possible wherever possible (without compromising performance).
3. **Select high-efficiency equipment:**
   - Choose equipment that exceeds the minimum efficiency requirements listed in applicable energy codes (e.g. ASHRAE 90.1).
   - Choose modulating compressors, fans, and pumps: modulating equipment allows for precise control based on demand and improves efficiency during part-load conditions.
4. **Energy efficiency strategies for hot-water and chilled-water systems:**
   - **Utilize heat recovery.** Heat recovery chillers and heat pumps capture waste heat from cooling processes to provide heating, improving overall system efficiency.
   - **Leverage thermal energy storage.** Use thermal energy storage for hydronic-based systems to shift energy consumption from peak to off-peak periods, reducing demand. Thermal energy storage also improves heat recovery potential by storing heat for later use.
   - **Design systems around variable flow.** Variable flow hydronic systems adjust water flow based on demand, improving energy efficiency by reducing pump power requirements.

- **Maximize turn-down capability with modular equipment.** Use a larger number of smaller boilers or chillers to achieve better turn-down ratios, improving efficiency at low loads and offering redundancy without overcapacity.

5. **Energy efficiency strategies for boilers: (see Chapter 20)**
   - **Flue gas condensation.** Use condensing technology to capture waste heat from flue gases for combustion-based boilers to improve overall system efficiency.
   - **Combustion air and O_2 control.** Implement oxygen trim control using real time monitoring or excess air to improve combustion efficiency.

6. **Energy efficiency strategies for chillers and heat pumps:**
   - **Lower compressor lift.** Implement demand-based temperature reset strategies to increase chilled-water temperature setpoints and lower heating water temperature setpoints, reducing compressor lift to increase equipment efficiency. "Lift" refers to the temperature difference the compressor must overcome between the heat source and heat sink. Higher lifts require more energy to move the same amount of heat.
   - **Lower condenser water temperature.** In cooling applications, reduce the condenser water temperature to decrease chiller lift. Refer to cooling tower strategies.
   - **Variable speed compressors.** Install VFDs on compressors to improve capacity control and part-load efficiency.
   - **Refrigerant charge.** Ensure proper refrigerant charge to maintain system performance and efficiency.
   - **Water treatment.** Maintain an aggressive water treatment program to prevent scaling and fouling, which degrade heat transfer efficiency.

7. **Energy efficiency strategies for cooling towers:**
   - **Lower approach temperatures.** Select cooling towers with a lower approach temperature between the leaving water temperature and ambient wet-bulb temperature to improve overall efficiency by reducing condenser water temperatures.
   - **Condenser water temperature reset.** Implement this control strategy to lower condenser water temperature during mild outdoor conditions, improving chiller efficiency.

- **Variable speed fans.** Install VFDs on cooling tower fans to match fan speed with demand, reducing power use.
- **Select induced draft, axial fan cooling towers.** This type of cooling tower configuration is preferred over forced draft and centrifugal towers to reduce fan energy consumption.
- **Water-side economizer.** For systems without air-side economizer operation, consider implementing a water-side economizer to provide "free cooling" through the chilled-water system during cooler outdoor air conditions.

9.7.2 Energy distribution systems

The purpose of an HVAC energy distribution system is to transport thermal energy between heating and cooling sources and loads throughout the building. Unlike small packaged systems, common in residential and smaller commercial applications, most HVAC systems rely on an energy distribution network. These systems transport heat to maintain occupant comfort.

Energy distribution systems can use different media – air, water, or refrigerant – to carry thermal energy. Each medium has distinct characteristics that affect system design and performance.

9.7.2.1 Energy distribution media

Energy is typically distributed using three primary media (or energy carriers – see Chapter 1): air, water, or refrigerant. Each medium has specific properties that determine its appropriate application in a building's HVAC system (see Table 9.9 and Figure 9.20).

Air

Air is the most used medium in HVAC distribution systems, particularly in North America, because it can serve multiple purposes simultaneously, including heating, cooling, and ventilation. However, due to its low specific

Table 9.9 Comparison of distribution media.

Medium	Energy density	Space efficiency	Environmental impact	Distribution efficiency	Heat loss
Air	Low	Poor	Low (but energy-intensive)	Low	High
Water	High	Good	Low	Good	Low
Refrigerant	Very high	Excellent	High (GWP potential)	Excellent	Low

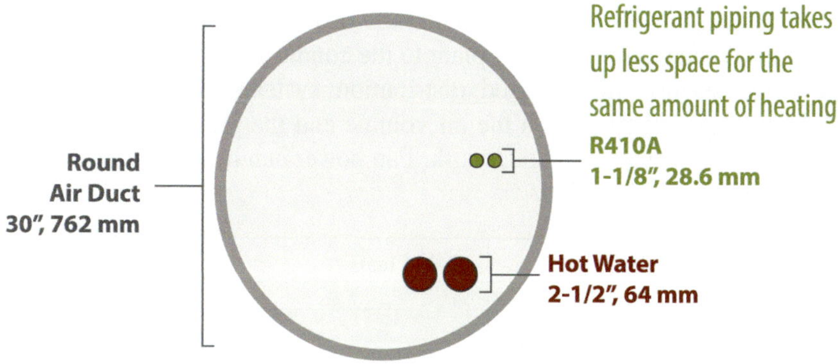

Figure 9.20 Comparison of air, water, and refrigerant distribution sizes required to meet the same load (source: NREL, Decarbonizing Building Thermal Systems: A How-to-Guide for Heat Pump Systems and Beyond).

heat capacity, air is the least efficient medium for transporting thermal energy, resulting in large ductwork sizes and higher distribution energy use than water or refrigerants.

Water

Water has a high volumetric heat capacity (Btu/lbm ft^3 or or kJ/kg m^3), approximately 3,500 times higher than air, allowing it to transport significant amounts of thermal energy in a small cross-sectional area, making water-based distribution systems, such as chilled-water or hot-water systems, more space- and energy-efficient. Moving water requires less energy than moving air, improving energy efficiency in many applications, particularly for larger buildings. Water is the media of choice for extensive systems, including Thermal Energy Network energy systems (see Chapter 19).

Refrigerants

Refrigerants are used in vapor-compression systems to transport energy through phase change, which makes them highly efficient at moving large quantities of thermal energy relative to their mass flow rate. These systems are commonly used in direct expansion (DX) systems, where refrigerant is distributed directly between the plant and zones. However, refrigerant leaks pose significant environmental risks and sometimes risks to human health, and detecting small leaks can be difficult.

9.7.2.2 Moving thermal energy: fans, pumps, and compressors

Transporting thermal energy within buildings involves moving the chosen heat transfer media from the HVAC plant to the conditioned spaces.

Fans are essential in air-based distribution systems. The fan power required to move air depends on the air volume and the pressure needed to overcome friction losses in the ductwork. Fan power can be estimated using the following equation:

SI units	English units
$P = \frac{\dot{v} \times \Delta TP}{\eta_p \times \eta_m \times \eta_t \times \eta_f}$	$P = \frac{\dot{v} \times \Delta TP}{6{,}343 \times \eta_p \times \eta_m \times \eta_t \times \eta_f}$
where: P = Motor output power rating (kW) $\dot{v}$ = volume flow rate of air, L/s ΔTP = Total fan pressure rise, Pascal η_p = Power supply (VFD) efficiency (decimal fraction < 1.0) η_m = Motor efficiency (decimal fraction < 1.0) η_t = Mechanical transmission (belts, pulley, bearings) efficiency (decimal fraction < 1.0) η_f = Fan aerodynamic efficiency (decimal fraction < 1.0)	where: P = Motor output power rating (hp) $\dot{v}$ = volume flow rate of air, ft^3/min ΔTP = Total fan pressure rise, inches of water η_p = Power supply (VFD) efficiency (decimal fraction < 1.0) η_m = Motor efficiency (decimal fraction < 1.0) η_t = Mechanical transmission (belts, pulley, bearings) efficiency (decimal fraction < 1.0) η_f = Fan aerodynamic efficiency (decimal fraction < 1.0) 6,343 = Conversion factor

Power supply efficiency is applicable for fans equipped with a variable speed drive. The controller efficiency is the ratio of the drive's electrical power output to its electrical input. Losses typically result from electronic switching, heat, and auxiliary circuits (e.g. cooling fans). Typical power supply efficiencies range from 90% to 98%.

Motor efficiency is the ratio of the mechanical power delivered at the motor shaft to the electrical power input to the motor. Typical efficiencies vary from 85% to 96%, with larger motors generally performing at higher efficiencies.

Transmission efficiency accounts for losses in belts, pulleys, and fan bearings. Depending on the drive type and belt quality, friction, bending resistance, and slippage can lead to losses of 2%–5%. In-field efficiency can deteriorate by up to an additional 5% due to installation quality. Overall, belt-drive configurations commonly achieve 90%–96% efficiency. Direct-drive fans, which have no belts or pulleys, are typically considered close to 100% efficiency.

Fan efficiency is the ratio of air power output (useful airflow and pressure delivered) to the mechanical power at the fan shaft. Fan efficiency varies widely based on fan type, design, and operating conditions, usually ranging from 40% to 85%. Lower-capacity fans typically have lower efficiencies between 50% and 60%.

Pumps are used in hydronic systems to move heated or chilled water between the plant and zone loads. Pump power can be estimated using the following equation:

SI units	English units
$P = \frac{\text{sg} \times \dot{v} \times H_P}{1{,}000 \times \eta_p \times \eta_m \times \eta_t \times \eta_f}$	$P = \frac{\text{sg} \times \dot{v} \times H_P}{3{,}956 \times \eta_p \times \eta_m \times \eta_t \times \eta_f}$
where: P = Pump power, kW sg = Specific gravity of fluid (1 for water) $\dot{v}$ = Volume flow rate of fluid, L/s H_p = Fluid head added by pump, m η_p = Power supply (VFD) efficiency (decimal fraction < 1.0) η_m = Motor efficiency (decimal fraction < 1.0) η_t = Mechanical transmission (belts, pulley, bearings) efficiency (decimal fraction < 1.0) η_f = Pump hydraulic efficiency (decimal fraction < 1.0) 1,000 = Conversion factor	where: P = Pump power, hp sg = Specific gravity of fluid (1 for water) $\dot{v}$ = Volume flow rate of fluid, U.S. gal/min H_p = Fluid head added by pump, ft η_p = Power supply (VFD) efficiency (decimal fraction < 1.0) η_m = Motor efficiency (decimal fraction < 1.0) η_t = Mechanical transmission (belts, pulley, bearings) efficiency (decimal fraction < 1.0) η_f = Pump hydraulic efficiency (decimal fraction < 1.0) 3,956 = Conversion factor

Pump hydraulic efficiency is the ratio of air power output (useful fluid flow and pressure delivered) to the mechanical power at the pump shaft. Pump efficiency varies widely based on pump type, design, and operating conditions, usually ranging from 40% to 85%. Lower-capacity fans typically have lower efficiencies between 40% and 60%.

Chapter 20 includes additional content on fan and pump efficiency.

9.7.2.3 Energy use in distribution systems

The energy required to distribute thermal energy accounts for a significant portion of total HVAC energy consumption, particularly in large commercial buildings. According to data from the U.S. Department of Energy (DOE), HVAC distribution systems – including fans and pumps – consume approximately 30%–40% of total HVAC energy. Energy is expended to power the equipment, move the heat transfer medium, and overcome air leakage and thermal losses.

Fan energy alone can account for nearly 20%–30% of total HVAC energy, whereas pump energy is a lower proportion, typically between 10% and 15%. Figure 9.21 provides a breakdown of commercial buildings' typical electrical energy consumption distribution systems.

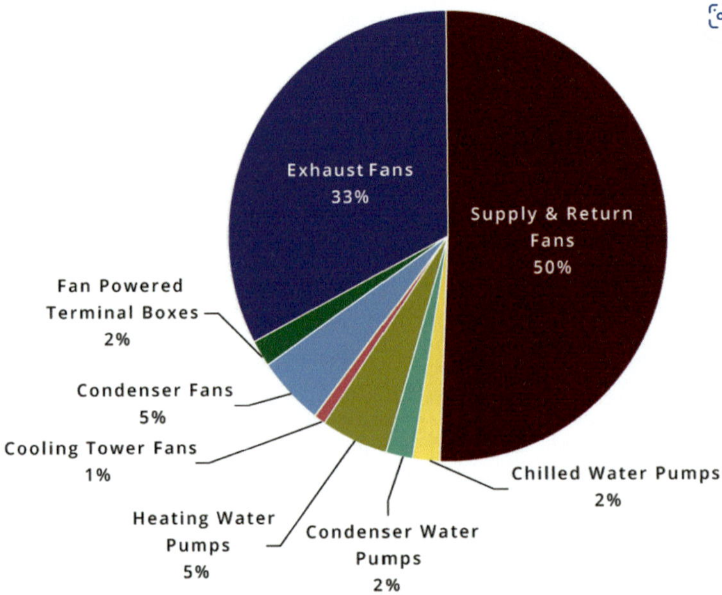

Figure 9.21 Breakdown of typical distribution energy (data source: Energy Consumption Characteristics of Commercial Building HVAC Systems, U.S. Department of Energy).

9.7.2.4 Energy efficiency strategies

As summarized below, several strategies can be employed to reduce energy consumption associated with distribution systems.

- **Design for distribution efficiency.** Establishing a target distribution efficiency is useful when designing distribution systems. It is also a helpful metric when monitoring and troubleshooting existing systems to provide a benchmark for energy improvements. For example, ASHRAE

90.1 includes fan system power limits and efficiency benchmarks that can be used for targets to meet or improve.

- **Minimize friction losses.** The HVAC designer greatly influences the friction losses within a system. Friction losses should be minimized wherever possible and include the following strategies:
    - Reduce distribution system lengths whenever possible.
    - Size ductwork and piping for low-pressure drops.
    - Optimize ductwork and piping layouts to reduce changes of direction and specify the use of low-pressure drop transitions and fittings.
- **Vary flow rates based on requirements.** Design new systems for variable flow and convert existing systems to variable flow wherever possible. Variable flow systems reduce fan and pump energy by adjusting flow rates to meet demand. In addition to saving fan and pump energy consumption, variable flow air systems reduce heating and cooling energy by reducing reheat energy, and variable flow hydronic systems improve the system temperature differential, increasing heating plant efficiency.
- **Implement demand-based pressure reset.** Variable speed fans and pumps are typically controlled to maintain a pressure setpoint within the distribution system. Resetting the pressure setpoint based on zone equipment feedback in the building control system can help further reduce energy consumption associated with fans and pumps.
- **Select high-efficiency fans and pumps.** The type of fans and pumps selected contributes to the mechanical efficiency of the system. Consider multiple types of fans and pumps for applications, analyzing the efficiency of the selected equipment.
- **Insulate distribution systems.** All distribution systems should be insulated to minimize parasitic thermal energy losses. Energy codes typically specify the thermal insulation performance requirements for duct and piping systems.
- **Reduce duct leakage.** Air leakage can result in significant energy losses, as conditioned air is lost to unconditioned spaces and requires fans to move more air to ensure zones receive sufficient conditioned air to maintain comfort conditions. Duct systems should be sealed to SMACNA (www.smacna.org) Class A standards, and systems with static pressures above 750 Pa (3 in. w.g.) should be leak tested.

9.7.3 Terminal systems

The interface with the occupied space is where thermal energy and moisture are either added to or removed from the space to maintain occupant comfort.

For air distribution and decentralized systems, energy exchange typically occurs by directly mixing conditioned supply air within the space, typically through air diffusers or grilles.

Energy is exchanged via heat exchangers within terminal units for water and refrigerant distribution systems. The heat exchangers are typically water- or refrigerant-to-air coils but can also take the form of panels or piping embedded within the building structure.

Terminal units significantly impact overall HVAC system efficiency as their sizing determines the heating and cooling temperatures required from central plant equipment. For example, sizing terminal unit heat exchangers for high surface areas can enable lower supply temperatures for heating or higher temperatures for cooling, improving the central plant's efficiency. For water-based systems, selecting heat exchangers for higher temperature differentials can reduce the flow rate required to deliver the same amount of heat, reducing distribution energy use.

Terminal systems are where occupants interface with the HVAC system. As occupants adjust temperature setpoints or airflow controls, their behavior can significantly influence energy use. Strategies that engage and inform occupants of how their behavior impacts building energy use are essential to reduce consumption patterns (see Chapter 7).

For central systems, terminal systems are typically composed of terminal units, where energy (hot water, chilled water, or conditioned air) is supplied from the central plant to various spaces in a building. For decentralized systems, the terminal system is often integrated directly into the unit or is connected to the space through small air distribution systems with diffusers or grilles.

9.7.3.1 Terminal units

Terminal units are small components typically associated with central systems that contain a heating coil, cooling coil, modulating damper, or a combination of these to control the temperature of a single zone within the building. This section provides a brief description of common terminal units.

CAV reheat (constant air volume reheat)  (Credit: First Light Energy Solutions)	A CAV reheat system supplies a constant volume of air to each space. The air handling unit sets the supply air temperature low enough to meet the highest cooling demand among all zones. To prevent overcooling in other zones served by the same air handler, each zone has a reheat coil in its terminal unit to raise the supply air temperature as needed. Because this approach involves both cooling and subsequent reheating, it is very energy intensive. Consequently, most energy codes prohibit this type of terminal unit.
VAV reheat (variable air volume reheat)  (Credit: Price Industries)	In a VAV reheat system, a VAV box damper varies the supply air volume based on the space's cooling demand. The air is reheated at the terminal unit when additional heating is needed. This system is more efficient than CAV reheat by reducing airflow when heat is required to reduce the amount of reheat energy required.
Fan-powered VAV (FPVAV)  (Credit: Price Industries)	This system combines a VAV box with a fan at the terminal unit. The fan can draw air from the ceiling plenum (return air) and mix it with supply air, reducing the need for reheating by using warm return air. FPVAV units are commonly used for perimeter zones because they can make use of recirculated return air to provide heating, which can be particularly useful to provide setback heating without the need to run the main air handling system.

Dual duct VAV 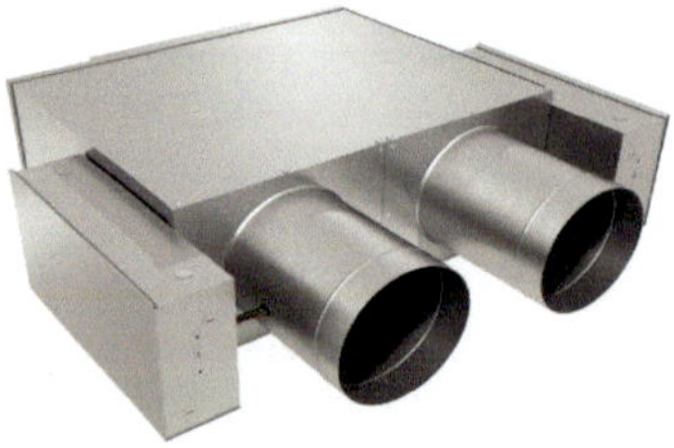 (Credit: Price Industries)	A dual-duct VAV system supplies two airstreams, one hot and one cold, mixed at the terminal unit to achieve the desired temperature. The air volume is variable to match the load, providing good control and flexibility but with high energy consumption due to simultaneous heating and cooling periods when airstreams are mixed.
Fan coil units  (Credit: Price Industries)	Fan coil units use a fan to blow recirculated air over a heating or cooling coil to maintain the space temperature setpoint. Fan coil units provide a high level of individual zone control without reheating previously cooled air.
Radiant panels  (Credit: Price Industries)	Radiant panels are typically installed in ceilings or walls and provide heating or cooling through radiant heat transfer. Hot or cold water flows through pipes embedded in the panels. They are energy efficient and offer quiet operation, but their capacity is limited to the available surface area of the wall or ceiling. Cooling applications require careful control of space humidity to prevent condensation.

Radiant slabs  (Credit: First light Energy Solutions)	Similar to radiant panels, radiant slabs consist of piping embedded in floors or ceilings, which circulate heated or chilled water. Because the piping is embedded within the building, slab systems leverage the thermal mass of the building to reduce HVAC loads. However, they are slow to respond to temperature changes and are best suited in applications with highly efficient building envelopes to reduce transient loads.
Chilled beams  (Credit: Price Industries)	Chilled beams use convection to cool or heat a room. Passive chilled beams use natural convection as warm air rises and passes over chilled-water coils. Active chilled beams use ventilation air to induce room air through the beam. These energy- and space-efficient systems require specialized humidity control to avoid condensation.
Induction units  (Credit: Price Industries)	Induction units use high-velocity air supplied from a central air handling unit to induce room air through a coil for heating or cooling. They offer individual zone control but can be noisy due to the high air velocities.

Radiators and convectors 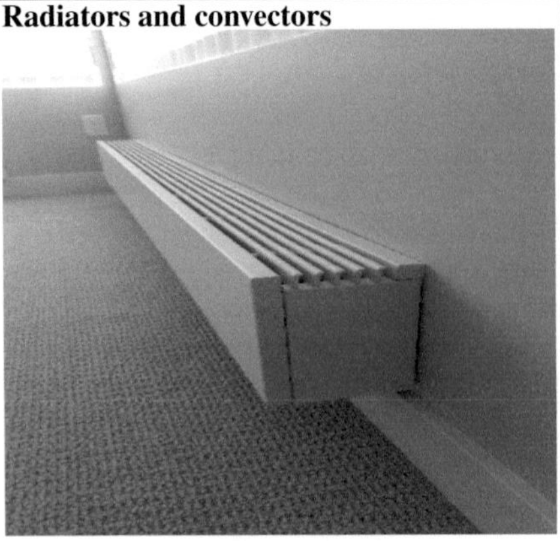 (Credit: Jaga Canada)	Radiators and convectors provide heat by circulating hot water or steam through metal surfaces that transfer heat to the room. Radiators primarily transfer heat through radiation, while convectors use fins to enhance natural convection.

9.7.3.2 Energy efficiency strategies

Below are key energy efficiency strategies to consider for terminal systems and zone controls:

- **Prevent simultaneous heating and cooling within a zone.** Simultaneous heating within a thermal zone is one of the most wasteful operating conditions, often occurring when control systems are improperly coordinated. Ensure that terminal equipment within a thermal zone works in coordination to prevent energy waste.
- **Avoid constant volume reheat.** CAV reheat systems are inefficient and should be avoided in new construction. Consider retrofitting with VAV systems to reduce airflow and reheat energy requirements when encountered in existing buildings.
- **Minimize reheat whenever possible.** Reheating cooled air for temperature control is energy intensive. Select system types that do not require reheating for zone temperature control to minimize reheat. For VAV systems, implement advanced control sequences such as dual-maximum control logic specified in ASHRAE Guideline 36 to minimize reheat.
- **Increase heating and cooling coil sizes.** Large coil surfaces improve heat exchange efficiency, allowing for lower temperature differences between the supply medium (water or refrigerant) and the space. Coil surface areas can be increased by providing more coil rows, increasing the cross-sectional area, or a combination of both. These large surface areas enable lower heating water temperatures and higher chilled-water temperatures to deliver the same amount of energy, which increases the efficiency of upstream equipment, such as boilers, chillers, and heat

pumps. Care must be taken to ensure existing fan systems can handle increases pressure drops of adding additional rows to coils.

- **Select hydronic coils for large temperature differentials.** Large coils typically facilitate higher temperature differentials between the supply and return water temperatures of heating and cooling coils. Higher temperature differentials reduce the water flow rate required to deliver the same amount of energy, reducing distribution energy requirements and smaller pumps.
- **Consider switchover coils for heating and cooling.** A single coil in a switchover arrangement reduces air pressure drop. It also ensures the maximum heat transfer surface area is available for heating and cooling. Consider this approach in retrofit applications where cooling coils (which typically have a greater surface area than heating coils) within terminal units can be repurposed as switchover coils to reduce the required heating water temperatures in heating mode.
- **Increase radiant panel surface area.** Similar to coils, increasing the surface area of radiant systems allows lower heating water and higher chilled-water temperatures to deliver the same energy output.
- **Utilize low-temperature rated convectors.** Low-temperature convectors are specifically manufactured to operate at lower heating water temperatures, increasing heating plant efficiency. Some options include small low-voltage fans to enhance convection across the convector.
- **Utilize ECM fans.** Utilizing direct-drive electronically commutated motors (ECMs) for terminal units such as fan coils and fan-powered VAV boxes. These motors provide inherent speed control and operate at higher efficiencies than similar-sized induction motors (also see Chapter 14).
- **Integrate terminal unit controls with building automation systems (BAS).** Integrating terminal unit controls, including space temperature sensors, with a central BAS allows for real-time monitoring and fine-tuning of HVAC performance. This information enables demand-based reset strategies to reset pressure setpoints and supply water temperatures based on actual zone demand to maximize energy savings. This information is also helpful in identifying problem zones or terminal units that may negatively impact the larger system (also see Chapter 15).
- **Use advanced zoning control strategies.** Advanced zone control strategies, such as those included in ASHRAE Guideline 36, utilize control logic that has been field tested and shown to reduce energy consumption while providing high thermal comfort.

- **Implement dual setpoint temperature sensors.** Dual setpoint control utilizes separate setpoints for heating and cooling. The HVAC system will not heat or cool the space if the room temperature is within the two set points.
- **Implement night setback for zones.** Night setback strategies involve reducing heating and raising cooling setpoints during unoccupied hours, preventing unnecessary energy consumption when spaces do not require full conditioning.
- **Implement zone-based optimal start routines.** Zone-based optimal start routines predict when the system should begin heating or cooling to achieve the desired setpoint when occupants arrive. By anticipating thermal loads, the system can operate more efficiently, reducing unnecessary operation during unoccupied periods.
- **Implement occupied standby controls.** Occupied standby controls relax temperature setpoints within a zone using occupancy sensors to detect when the space is vacant. This approach reduces unnecessary conditioning without compromising occupant comfort.

9.8 Controls

The primary function of HVAC control systems is to regulate indoor environmental conditions while ensuring energy-efficient operation. Building control systems are a critical component of HVAC systems, monitoring and adjusting equipment such as boilers, chillers, air handling units, control dampers, control valves, and terminal devices in response to varying conditions to maintain comfortable indoor environments.

Effective control systems adjust equipment operation based on real-time occupancy, changing outdoor conditions, and building loads to deliver energy while minimizing consumption. Control systems also monitor equipment performance and alert operators to maintenance needs or equipment failures.

Modern HVAC systems typically employ building automation systems (BAS) to integrate the monitoring and control of HVAC, lighting, security, and other building systems into a centralized platform. BAS use direct digital controls (DDC) to automate and control HVAC system operations, often with remote access capabilities and sophisticated algorithms for energy management.

Further information on control systems can be found in Chapter 15.

9.9 Summary

This chapter provided an overview of HVAC systems, focusing on their role in maintaining healthy indoor environments while optimizing energy performance. The chapter began by outlining the impact HVAC systems have in maintaining acceptable indoor environmental quality, which includes both thermal comfort and indoor air quality.

Fundamental principles were introduced to aid in understanding the energy processes and characterization of efficiency as applicable to HVAC systems. Of particular relevance to HVAC was the introduction of Psychrometrics. Building on these fundamentals, an overview of heating and cooling loads was provided, explaining factors that influence system sizing and energy use and highlighting the impact that duration and variations of these loads has on system right-sizing and overall system efficiency.

Ventilation systems were explored in detail, including descriptions of common components and strategies to improve energy efficiency. Finally heating and cooling systems, including both centralized and decentralized systems, were covered. Strategies for improving the energy performance of thermal systems were outlined, focusing on efficient equipment selection, system design considerations, and control methods to reduce energy consumption.

References

ASHRAE. 2025. ASHRAE handbook – Fundamentals, Chapter 1 Psychrometrics.

ASHRAE. 2025. ASHRAE Handbook – Fundamentals, Chapter 2 Thermodynamic and Refrigeration Cycles.

ASHRAE. 2025. ASHRAE Handbook – Fundamentals, Chapter 4 Heat Transfer.

ASHRAE. 2025. ASHRAE Handbook – Fundamentals, Chapter 10 Indoor Environmental Health.

ASHRAE. 2025. ASHRAE Handbook – Fundamentals, Chapter 18 Nonresidential Cooling and Heating Load Calculations.

ASHRAE. 2024. ASHRAE Handbook – HVAC Systems and Equipment, Chapter 26 Air-to-Air Energy Recovery.

ASHRAE. 2024. ASHRAE Handbook – HVAC Systems and Equipment, Chapter 32 Boilers.

ANSI/ASHRAE Standard 55 (2023). Thermal Environmental Conditions for Human Occupancy. Atlanta, GA: American Society of Heating, Refrigerating and Air-Conditioning Engineers.

ASHRAE Standard 55 User's Manual (2016). Atlanta, GA: American Society of Heating, Refrigerating and Air-Conditioning Engineers.

ANSI/ASHRAE Standard 62.1 (2019). Ventilation for Acceptable Indoor Air Quality. Atlanta, GA: American Society of Heating, Refrigerating, and Air-Conditioning Engineers.

Advanced Design Guide for Small to Medium Office Buildings Achieving 50% Energy Savings Toward a Net Zero Energy Building, ASHRAE, 2011.

ANSI/ASHRAE Standard 90.1 (2019). Energy Standard for Buildings Except Low-Rise Residential Buildings.

Building Envelope Enclosure (January 2022). Measured Air Leakage Requirements in Context. Daniel Overby.

Exergy Guidebook for Building Professionals: How to use exergy to analyze and develop energy systems for the built environment. Sabine Jansen. Delft University of Technology, October 2014, Delft.

International Energy Agency. LowEx Communities – Optimized Performance of Energy Supply Systems with Exergy Principles (Annex 64).

Energy Design Resources. Advanced Variable Air Volume System Design Guide.

ASHRAE Journal, May 2014. Face Velocity Considerations In Air Handler Selection, Kent W. Peterson

EPA. The Inside Story: A Guide to Indoor Air Quality.

IEA. World Energy Outlook 2023.

Houssainy, Sammy, Landan Taylor, Paul Torcellini, and Sugirdhalakshmi Ramaraj. 2024. Decarbonizing Building Thermal Systems: A How-to Guide for Heat Pump Systems and Beyond. Golden, CO: National Renewable Energy Laboratory. BK-5500-87812. www.nrel.gov/docs/fy24osti/87812.pdf.

ASHRAE Journal, December 2021. Key Factors of Fan System Wire-to-Air Efficiency. B. Rajavel & Paul Bauch.

National Resources Canada, Air-Source heat Pump Sizing and Selection Guide, Version 1.0, 2020-12-21, Cat. No. M154-138/2020E-PDF (On-line) ISBN 978-0-660-36713-2

Building to Electrification Coalition, BC Commercial Building Electrification Guide, November 2024 B2E Commercial Building Electrification Guide | B2E

National Renewable Energy Laboratory (NREL), Decarbonizing Building Thermal Systems: A How-to Guide for Heat Pump Systems and Beyond, September 2024, NREL/BK-5500-87812 docs.nrel.gov/docs/fy24osti/87812.pdf

Author Biography

Mike Reimer is a principal at First Light Energy Solutions Ltd based in Vancouver, British Columbia. As a registered professional engineer, he has over 20 years of experience in HVAC, plumbing, and controls system design for both new construction and retrofits in number of building sectors. Mike is a voting member on the Standing Guideline Project Committee 36 (ASHRAE Guideline 36, higher-Performance Sequences of Operations for HVAC Systems), including participating in the Heat Pump Working Group. He also served on the ASHRAE Building Decarbonization Retrofit Guide Working Group. Additionally, he co-authored the Building to Electrification (B2E) Coalition's Commercial Building Electrification Guide.

Acknowledgments

I would like to acknowledge and extend my gratitude to Dawn-Marie Barreira, Eric Mazzi, Sarah Mayer, and Moises De Alba for their thoughtful review of this chapter's draft and for their constructive feedback. Their insights and recommendations greatly contributed to shaping the final version of this chapter.

Practice Questions

1. An electric-driven 10-ton roof top air conditioner has an EER of 9.5 (recall that 1 ton of cooling = 12,000 Btu/hr).
   (a). What is its COP?
   (b). What is its kW/ton rating?
   (c). What is its kW input at full capacity?
   (d). How many kWh are consumed to provide 500 ton-hr of cooling?
2. Moist air enters a duct at 10 °C, 80% relative humidity, and a volumetric flow rate of 150 m^3/min. The mixture is heated as it flows through the duct and exits at 30 °C. No moisture is added or removed, and the mixture

pressure remains approximately constant at 1 bar. Using a psychrometric chart, determine:

(a). The rate of heat transfer in kJ/hr.
(b). At the exit: relative humidity, dew point (○C), wet bulb temperature (○C), enthalpy (kJ/kg) and humidity ratio.

3. You are trying to select an air conditioner, but the three companies you are looking at provide different measures of performance. Company A's model has an EER of 9. Company B's model has a COP of 2.6. Company C's model is rated at 1.2 kW/ton. Convert all of the values to a common measure and state which company's model is the most efficient.
4. An HVAC system dehumidifies air from 60°F and 90% relative humidity to 20% relative humidity (only latent heat transfer occurs). The flow rate is 100 lbm/hr. Calculate the HVAC load in Btu/hr.

(a). 400 Btu/hr
(b). 500 Btu/hr
(c). 600 Btu/hr
(d). 700 Btu/hr
(e). 800 Btu/hr

5. The sensible heat factor for an HVAC process is 60%. The change in sensible heat ($\Delta h_{sensible}$) is 12 kJ/kg. The room airflow rate is 500 kg/hr. The EER for the HVAC unit is 11 kJ/Wh. Calculate the input electric power.

(a). 110,000 W
(b). 10,000 kJ/hr
(c). 909 W
(d). 1,364 W
(e). 12 kW

6. An HVAC system cools air from 50 to 30 Btu/lbm. The air flow is 5,000 cfm (cubic feet per minute). Calculate the tons of cooling delivered by the HVAC system.

(a). 0.625 tons
(b). 8.33 tons
(c). 450,000 tons
(d). 37.5 tons
(e). 47.5 tons

7. 100 kg per minute of air undergoes a cooling process from 30 °C and 30% relative humidity to 15 °C. No moisture is added or removed. Estimate the cooling load in kJ/min.
8. An airstream is cooled from 35 °C and 40% relative humidity, to 13 °C and 70% relative humidity. Estimate the sensible heat factor.
9. Estimate the sensible heat in Btu/hr is needed to heat 1,000 CFM of 35 °F air to 95 °F air?
10. An air conditioner removes 120,000 Btu/hr of heat and has an input power of 12 kW. What is the COP of the unit?
11. The heating load of a facility is due to a work force of 22 people including 6 overhead personnel, primarily sitting during the day; 4 maintenance personnel and supervisors; and 12 people doing heavy labor. Assume everyone works the same 8-hour day. Estimate the heating load due to the workforce in both kWh/day and Btu/day.

10

Heat Pumps

Dawn-Marie Barreira

Firstlight Energy Solutions, Canada

10.1 Introduction

While Chapter 9 provides an overview of HVAC concepts and systems, this chapter focuses specifically on heat pumps to meet the heating and cooling needs of a facility.

Heat pumps have become increasingly popular in the last decade due to their highly efficient heating and cooling capabilities. Heat pumps can achieve efficiencies exceeding 300% depending on the application, and therefore, offer a highly efficient alternative to combustion-based heating, such as boilers and furnaces, which are limited to efficiencies of 80%–95%. Not only are they highly energy efficient, but they also can provide significant reductions in operating GHG emissions through fuel switching by utilizing electricity rather than gas. Heat pumps can also be used to achieve GHG reductions in locations that utilize fossil fuels (such as gas) to generate electricity.

Heat pumps are available at a range of scales and are suitable for applications at residential, commercial, and industrial facilities. They are able to be integrated into existing heating and cooling systems, often referred to as a hybrid or dual fuel hybrid systems, to improve energy performance, and can also be used to replace existing systems to meet all of the heating and cooling requirements of a facility.

When considering implementing heat pumps, understanding the connection between heating and cooling is important. Historically, in most buildings, heating and cooling systems have been separated, with heating provided by a fossil-fuel-based system and cooling provided by an electric chiller. Heat pumps create an opportunity to integrate heating and cooling systems. While integrated systems can seem more complex, shifting the mindset on designing heating and cooling systems can result in reduced energy losses and highly efficient operation.

If designed well, heat pumps can be used to improve the thermal comfort of occupants by providing consistent temperatures, offering better zone control, and adding cooling where air conditioning does not exist. Adding the ability to efficiently cool a space improves the climate resiliency of buildings located in areas that did not require cooling in the past, which are increasingly experiencing extreme weather events such as heat domes and warmer summer temperatures (see Chapter 8).

Heat pumps also have some inherent challenges that need to be considered in design and application. Risk of refrigerant leakage must be managed and assessed in the life cycle analysis of GHG emissions for heat pumps. Maintenance downtime can be disruptive, and repairs can require specialized parts that may not be immediately available. Backup, temporary, or redundant heating and cooling systems may be recommended for applications that require uninterrupted service and resiliency.

Heat pump innovation is accelerating with significant advancements in technology developing each year. Cold climate heat pumps, for example, have seen significant increases in performance and continue to be a focus of the industry to meet the needs of the North American heat pump market. They will continue to play an important role in providing GHG emission reductions and supporting decarbonization and electrification efforts for both new and existing buildings.

The objective of this chapter is to provide an overview of how heat pumps work, common heat sources and sinks, performance and efficiency, heat pump technology and common applications, and considerations for integrating heat pumps into heating and cooling systems. This overview includes relevant information for residential, commercial, and industrial applications of heat pumps.

10.2 How Heat Pumps Work

10.2.1 Refrigeration cycle

A heat pump efficiently transfers heat from one source to another. Using this principle, heat pumps move thermal energy from a location that has available heat to a place where it is needed or remove heat from a location

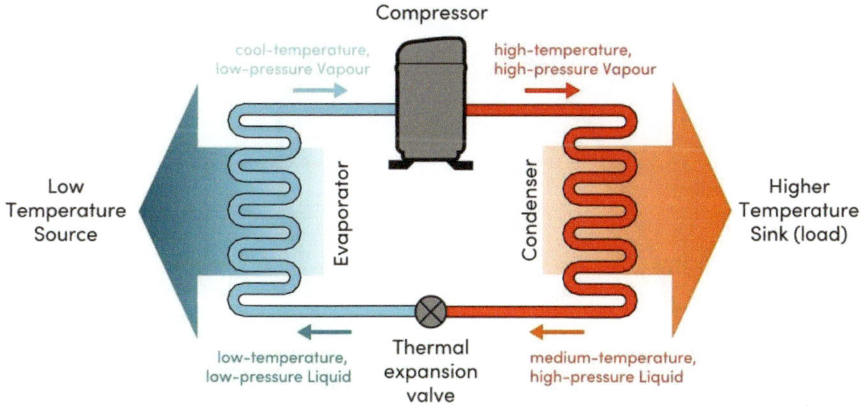

Figure 10.1 Vapor compression refrigeration cycle. (source: FirstLight Energy Solutions).

to a place where it can be rejected. It operates on the principle of extracting heat from a low-temperature source (such as outdoor air, ground, or water) and delivering higher temperature heat to a sink. As shown in Figure 10.1, the key components include an evaporator, compressor, condenser, and expansion valve.

Most heat pumps work on a vapor-compression refrigeration cycle. In heating mode, the evaporator absorbs heat from an external source, and the refrigerant vaporizes. The compressor then increases the pressure and temperature of the vaporized refrigerant, which is then condensed releasing heat. By reversing the direction of refrigerant flow, heat pumps can switch from heating to cooling. In cooling mode, the heat is absorbed by the refrigerant in the evaporator, and the compressor moves this heat to the condenser for release. Figure 10.2 shows the phase change process of the vapor compression cycle.

This cycle allows heat pumps to provide both heating and cooling functions, making them energy-efficient alternatives to traditional heating and air conditioning systems.

10.2.2 Heat pump components

Heat pumps have several key components that work together to transfer heat from one location to another, providing heating or cooling as needed to maintain desired indoor temperatures.

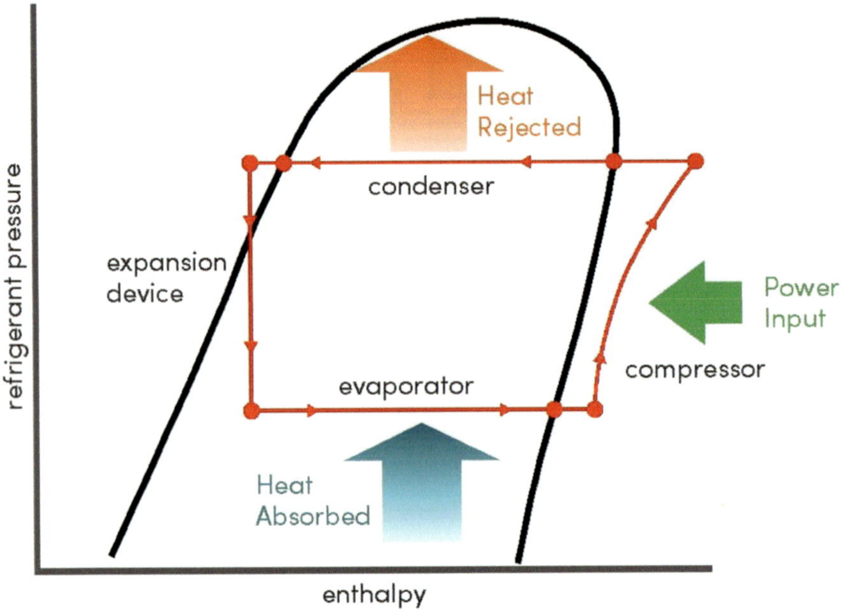

Figure 10.2 Phase change diagram for the vapor compression cycle. (source: FirstLight Energy Solutions).

The main components of a typical heat pump system are:

- **Compressor**. The compressor increases enthalpy (pressure and temperature) and circulates the refrigerant between the evaporator and condenser, facilitating the heat transfer process.
- **Refrigerant**. The refrigerant is circulated within the heat pump system and absorbs and releases heat as it changes state between liquid and gas.
- **Evaporator**. The evaporator absorbs heat. In heating mode, the evaporator is the heat exchanger that absorbs heat from the source, and in cooling mode, it is the heat exchanger that removes heat from the sink.
- **Condenser**. The condenser releases heat. In heating mode, the condenser is the heat exchanger that releases heat to the sink, and in cooling mode, the condenser is the heat exchanger that rejects heat to the source.
- **Expansion valve**. The expansion valve controls the flow of refrigerant into the evaporator, reducing its pressure and temperature.

- **Reversing valve.** The reversing valve is used to change the refrigerant flow direction, allowing the heat pump to switch between heating and cooling modes. Note that not all heat pumps contain reversing valves.
- **Defrost controls.** In cold climates, water vapor in the air freezes on the cold evaporator coil, which accumulates as frost or ice on the outdoor unit during heating mode. To prevent coil damage, defrost controls activate to melt the accumulated ice. Defrost cycles increase as the weather gets colder, using additional energy, which lowers the overall heat pump performance at colder temperatures.

The heat pump components and refrigerant direction in heating and cooling mode are shown in Figure 10.3.

10.2.3 Compressor types

Heat pumps can use different types of compressors, each with its own advantages and applications. The choice of compressor type depends on factors such as system size, efficiency requirements, noise considerations, and budget constraints. Four types of compressors are reciprocating, scroll, rotary, and centrifugal. Table 10.1 describes each of the four technologies.

The efficiency of each compressor varies by type, application, and compressor size. Technology is increasingly becoming more efficient, driven by innovation and minimum energy efficiency requirements. For typical and minimum efficiency requirements for heat pumps and heat recovery chillers, refer to ASHRAE Standard 90.1 Section 6: HVAC, Table 6.8.1-16 Heat-pump and Heat Recovery Chiller Packages – Minimum Efficiency Requirements.

10.2.4 Refrigerant considerations

As heat pumps gain traction in heating fuel switching applications, it is not without challenges. One of the major concerns with the growing heat pump market is the growth in refrigerants. Refrigerants typically have high global warming potential (GWP) and ozone depletion potential (ODP), causing concern for the impact of refrigerant releases on the atmosphere and environment. GWP and the impact of refrigerants are discussed in greater detail in Chapter 2. The chart in Figure 10.4 shows the range in GWP of typical refrigerants used in HVAC systems. As refrigerants with high GWP are phased out, lower and ultra-low GWP refrigerants are being introduced as replacements.

The Montreal Protocol was agreed upon in 1987 by many nations to phase out ozone-depleting substances from use and production. This included the phase out of many refrigerant types that were common in HVAC systems, such as CFCs and HCFCs. In 2016, an amendment to the Montreal Protocol was made, titled the Kigali Amendment, to include the phasing out of HFC refrigerants that have high GWP. Accordingly, several common refrigerants, such as R-410A, utilized in heat pumps and air conditioners, are being phased out, and newer refrigerants with lower GWPs are being introduced to the market.

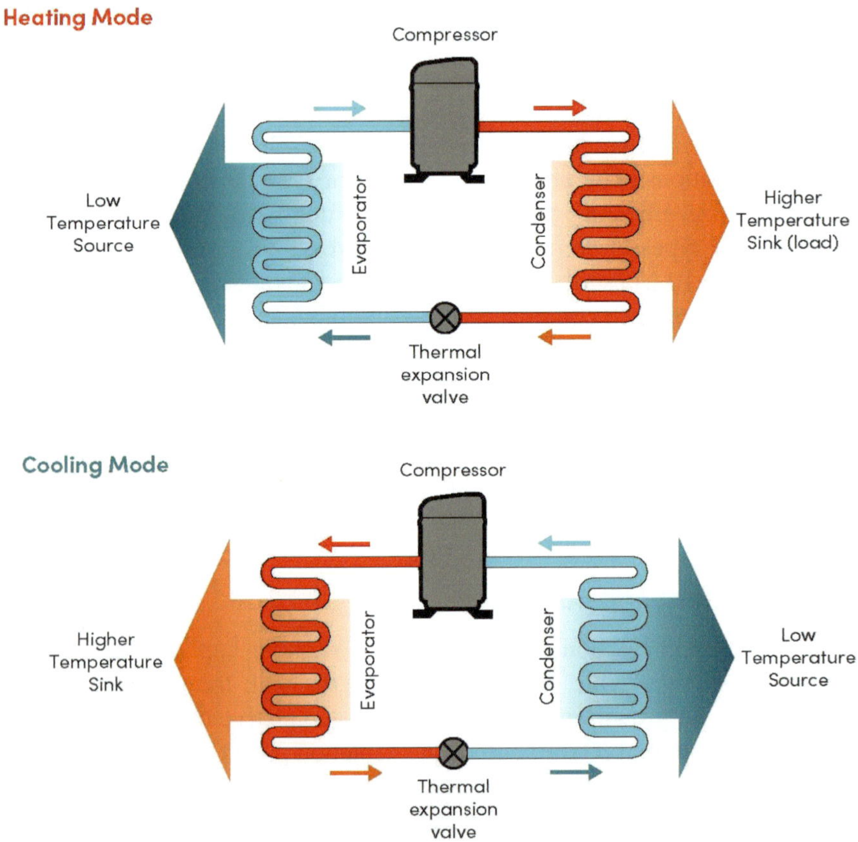

Figure 10.3 Heat pump components in heating and cooling mode. (source: FirstLight Energy Solutions).

Refrigerants are classified based on toxicity and flammability in ASHRAE Standard 34. Understanding the classification of the refrigerant is important in design to ensure that installations, particularly indoors in mechanical rooms, adhere to local building and safety code requirements.

The refrigerant market is rapidly evolving due to imminent legislation in the North American and global markets. Care should be taken when recommending and selecting heat pumps for design that the latest innovative, low-GWP refrigerants are selected to reduce the long-lasting climate impacts of refrigerant leaks.

No matter the refrigerant selected, equipment should be inspected annually by HVAC technicians as a part of a refrigerant leak management program to reduce the impact of refrigerant leaks on the environment. Refrigerant

Table 10.1 Compressor types.

Reciprocating 	Reciprocating compressors are the most common type of compressors used in industrial heat pump applications. They use a piston to compress the refrigerant and are reliable and relatively low cost. However, they can be noisy and have higher vibration levels compared to other compressor types.
Scroll 	Scroll compressors use spiral-shaped scrolls to compress the refrigerant. Scroll compressors are more energy-efficient and have fewer moving parts, reducing maintenance requirements. They are the most common in small heat pump applications.
Rotary (screw) 	Rotary or screw compressors use a rotating blade within a cylindrical chamber to compress the refrigerant. These compressors are compact, lightweight, and relatively quiet. They are common in larger commercial heat pump applications. Screw compressors are less efficient than scroll compressors but are more efficient than reciprocating compressors.
Centrifugal 	Centrifugal compressors use a rotating impeller accelerate the refrigerant. Centrifugal compressors are highly efficient and available in high capacities suitable for large-scale heating and cooling systems. They are typically more complex and expensive than other compressor types.

leakage is common and can exceed the GHG emissions of natural gas-fired heating systems over the lifetime of the equipment. Refer to ASHRAE Standard 228: Standard Method of Evaluating Net Zero Energy and Net Carbon Building Performance for additional information on embodied carbon related to refrigerants and refrigerant leakage.

10.2.5 Energy efficiency considerations

In addition to understanding the range in energy performance of different compressor types, the ability to operate at more than one capacity increases the energy efficiency of the compressor. Standard single-stage compressors cycle on and off frequently and only operate at full capacity. Multi-stage compressors include:

- Staged compressors allow heat pumps to operate at two capacities, low and high, to meet low and high loads. This reduces the amount of on and off cycles, increasing the efficiency and reducing wear. Multiple compressors can be installed in a bank to enable additional modulation control as compared to a standard single-stage compressor.
- Variable-speed compressors allow heat pumps to operate at multiple capacities to meet a range of thermal loads. These compressors run continuously at low speeds, reducing on and off cycles almost entirely, and consuming less power. These compressors are the most energy efficient, operating as low as 0.3 kW/ton and are able to respond quickly

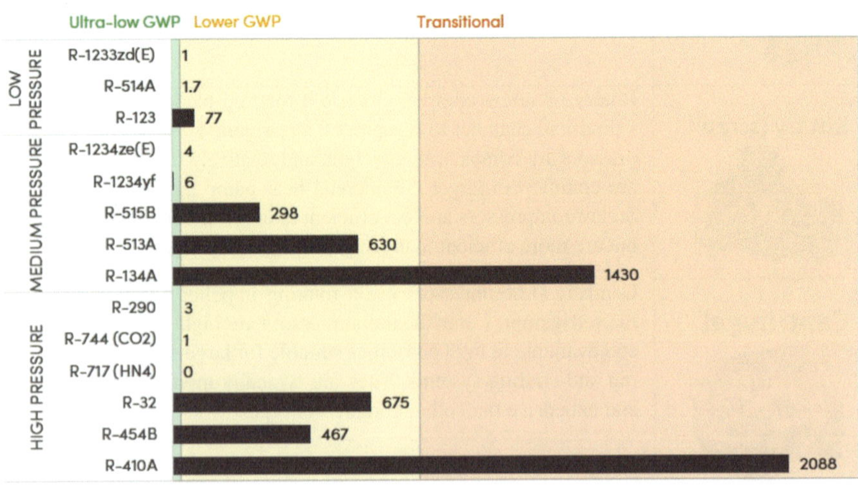

Figure 10.4 Typical refrigerants in heat pump systems and their GWP.

to fluctuations in outdoor air temperatures to meet indoor temperature requirements.

10.3 Heat Sources and Sinks

Heat pumps operate by transferring heat from a source (air, water, or ground) to a heat sink (desired location), utilizing a refrigeration cycle, as discussed in Section 10.2.1. The heat pump absorbs heat from the *heat source*, typically the outdoor air for air-source heat pumps (in heating mode) or the ground for ground-source heat pumps. The heat pump releases heat to the *heat sink*, typically the indoor space for heating or outdoor air for cooling for air source heat pumps.

Heat pumps can extract heat from various sources, which include:

In recommending or selecting a heat pump for a specific application, consider the quality, quantity, and availability of the heat source.

- Quality: Source temperature, temperature range, and fluctuations.
- Quantity: The quantity of heat available and whether the source has unlimited or limited potential.
- Availability: Frequency or timing of the availability of the source.

Heat pump selection and decision-making should be guided and informed by supporting information and data to determine the quality, quantity, and availability of the heat source. This is particularly key in existing building retrofits to ensure that reliable heating and cooling can be maintained. It is important to keep in mind that the system always needs to provide the worst-case scenario for peak design conditions (e.g., the hottest or coldest day of the year). If the quality, quantity, or availability of heat is insufficient, then a supplemental backup system must be considered.

The degree to which uncertainty is addressed through data analysis and feasibility studies should be balanced with the complexity and risks of the project. For example, the level of study required for adding a split heat pump system to a single-family home is significantly less than a sewer heat recovery

Thermal Energy Network (TEN) system that supplies heating and cooling to a community.

While outdoor air is the most common source and sink for a heat pump, many others exist, often within a building that is rich in thermal energy resources.

- **Exhaust heat recovery**. Recovering exhaust heat to use. This can be done by limiting free cooling and actively cooling the exhaust air to recover waste heat. Typical source temperature 20–25 °C.
- **Thermal storage**. Integrating thermal storage (see Chapter 17) into a centralized hydronic system for a building can create a consistent source.
- **Condenser loops**. Utilize low-grade waste heat in condenser water rather than heat rejection to the atmosphere using a cooling tower or fluid cooler. Typical source temperature 20–30 °C.
- **Wastewater/sewer heat recovery**. Capture heat from the building wastewater. Typical source temperature 15–20 °C.
- **Process heat**. Heat from process or refrigeration loads. Typical source temperature 20–40 °C.
- **Renewable resources**. Solar thermal or other renewable sources to provide a supplemental source of heating. Typical source temperature 20–60 °C.

Table 10.2 has been adapted from Table 9.1: Heat pump sources and sinks in Chapter 9: Applied Heat Pump and Heat Recovery from the 2020 ASHRAE Handbook – HVAC Systems and Equipment, and compares common heat sources and sinks, as well as their suitability and temperature ranges.

10.4 Heat Pump Performance

10.4.1 Heat pump efficiency key concepts

Chapter 9: HVAC Systems introduced the key concepts, terminology, and equations related to HVAC energy efficiency. Most importantly, the concepts that are relevant to heat pump efficiency and performance are:

- *Seasonal energy efficiency ratio (SEER)*, which is a measure of the cooling efficiency of a heat pump over an entire cooling season. The higher the SEER rating, the more efficient the heat pump is at cooling.

Table 10.2 Heat pump sources and sinks (adapted from 2020 ASHRAE Handbook HVAC Systems & Equipment, Chapter 9, Table 9.1).

		Suitability		Availability		Temperature	
Medium	**Examples**	**Heat source**	**Heat Sink**	**Location relative to need**	**Coincidence with need**	**Level**	**Variation**
Air							
Outdoor	Ambient air	Good	Good	Universal	Continuous	Variable	Extreme
Exhaust	Building ventilation	Excellent	Fair	Excellent	Excellent	Excellent	Low
Water							
Well	Ground-water well	Excellent	Excellent	Varies	Continuous	Excellent	Stable
Surface	Lakes, rivers, ocean	Excellent	Excellent	Limited	Continuous	Satisfactory	Depends on the source
Tap	Municipal water supply	Excellent	Excellent	Excellent	Continuous	Excellent	Low
Condensing	Cooling towers, refrigeration systems	Excellent	Poor to good	Varies	Varies	Favorable	Depends on source
Closed loop	Building water-loop heat pump systems	Good	Favorable	Excellent	As needed	As designed	As designed
Waste	Raw or treated sewage, gray water	Fair to excellent	Fair	Varies	Varies	Excellent	Low
Ground							
Ground	Buried or submerged fluid loops	Good	Fair	Varies	Continuous	Good	Low
Direct expansion	Refrigerant circulated in ground coils	Varies	Varies	Varies	Continuous	Varies	Low
Solar							
Direct or heated	Solar collectors and panels	Fair	Poor	Universal	Intermittent	Varies	Extreme
Industrial							
Process or exhaust heat	Distillation, molding, refining, washing, drying	Fair to excellent	Varies	Varies	Varies	Varies	Varies

- *Energy efficiency ratio (EER)*, which is another measure of cooling efficiency, but it represents the efficiency of the heat pump at a specific outdoor temperature rather than over an entire season. Higher EER ratings indicate better cooling efficiency at specific outdoor temperatures.

- *Heating seasonal performance factor (HSPF)*, which is a measure of the heating efficiency of a heat pump over an entire heating season. A higher HSPF rating indicates better heating efficiency.
- *Coefficient of performance (COP)*, which is a measure of the efficiency of a heat pump. It represents the ratio of energy (or power) output to energy (or power) input. A higher COP indicates a higher efficiency. The chart in Figure 10.5 provides a visual representation of the concept of COP and the relationship to the vapor compression cycle.

During simultaneous heating and cooling with a heat pump, the total COP (COP_{Total}) is the combined efficiency of the heat pump and is the combination of both heating and cooling COP. Heating COP (COP_H) is the efficiency in heating, and Cooling COP (COP_C) is the efficiency of cooling.

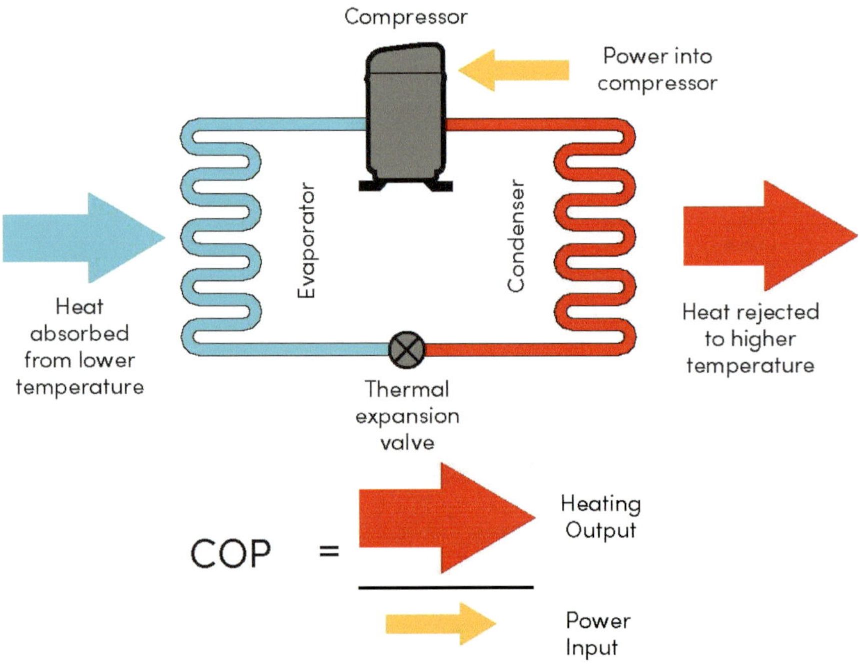

Figure 10.5 Refrigeration cycle and COP relationship for a heat pump in heating mode.

Example 10-1: Calculating COP

A $60,000 \text{Btu/hr}$ heat pump is used to heat a house at its full load while consuming 5 kW of electrical energy. What is the COP_H of the heat pump?

Solution:

Given:

Heat delivered to the house, $Q_H = 60,000 \text{Btu/hr}$

Electrical power input to the heat pump $= 5 \text{ kW} = 17,060 \text{Btu/hr} (1 \text{ kW} = 3,412 \text{BTU/h})$

Using the HP efficiency equations introduced in Chapter 9 to solve for COP_H :

$$COP_H = \frac{Q_H}{W} = \frac{60,000 \text{Bu/hr}}{17,060 \text{Btu/hr}} = 3.52$$

The COP_H of the heat pump is 3.52 , meaning that the heat delivered is 352% of the electrical energy used to move it from the outdoor evaporator to inside the house.

Example 10-2: Calculating Electrical Power Inputs

A heat recovery chiller is simultaneously providing heating and cooling to an office building. At its full load, it provides $400,000 \text{Btu/hr}$ of heat for heating and removes $300,000 \text{Btu/hr}$ of heat for cooling. The COP of heating is 4 and the COP of cooling is 3 . What is the full load electrical power input to the heat recovery chiller?

Solution:

Given:

Heat provided for heating, $Q_H = 400,000 \text{Btu/hr}$

Heat removed for cooling, $Q_C = 300,000 \text{Btu/hr}$

Using the HP efficiency equations introduced in Chapter 9 to solve for COP Cotal_t :

$$COP_{\text{Total}} = COP_H + COP_C = 4 + 3 = 7$$

Using the HP efficiency equations introduced in Chapter 9 to solve for the total electrical power input to the heat pump:

$$W = \frac{Q_H + Q_C}{COP_{\text{Total}}} = \frac{400,000 \text{Btu/hr} + 300,000 \text{Btu/hr}}{7} = \frac{100,000 \text{Btu/hr}}{3,412 \frac{\text{Btu/hr}}{\text{kW}}} = 29.3 \text{ kW}$$

The total electrical power input of the heat recovery heat pump is 29.3 kW .

10.4.2 Efficiency considerations

10.4.2.1 Temperature lift

Temperature Lift is the difference in temperature between the sink and the source. The temperature lift is inversely proportional to the heat pump efficiency (COP), meaning that the larger the lift required, the lower the efficiency of the heat pump:

$$\text{Temperature lift } (^\circ\text{C}) = \text{temperature sink} - \text{temperature source}.$$

The efficiency of heat pumps varies based on how hard the heat pump has to work (energy input) in order to meet the desired outcome (i.e., achieve the sink temperature).

For air-source heat pumps, the efficiency is impacted by outdoor air temperature, as shown in Figure 10.6. This can be particularly challenging in cold climates, and supplemental or backup heating may be required during extreme cold weather events.

Lift Example

If outdoor air temperature is –5 °C (23 °F) and the desired indoor temperature is 21°C (70 °F) then lift = 26 °C (47 °F). The heat pump will need to input more electrical power to meet setpoint than if the outdoor air temperature is 10 °C (50 °F) and the indoor temperature setpoint is 21 °C (70 °F), would be lift = 11°C (20 °F).

The same is true for the efficiency of water-source heat pumps, as shown in Figure 10.7.

It is important to note that when selecting heat pumps, there are limits on both the lift and the temperature output for each make and model. Selection criteria should consider the application requirements to ensure the heat pump is appropriate for the application. This is further discussed in Section 10.5 on heat pump applications.

Example 10-3: Calculating Power Input at Various Outdoor Air Temperatures.

A community centre would like to consider using an air-source heat pump to provide baseload heating for the hot water heating system. The ASHP needs to be sized to meet a heating load of $120,000\text{BTU}/\text{ hour}$. Heating water temperature must be 100°F to be seamlessly integrated into the existing system. Using the chart in Figure 10-6, how much electrical power will the heat pump consume at the following outdoor air temperatures of -10°C and 18°C.

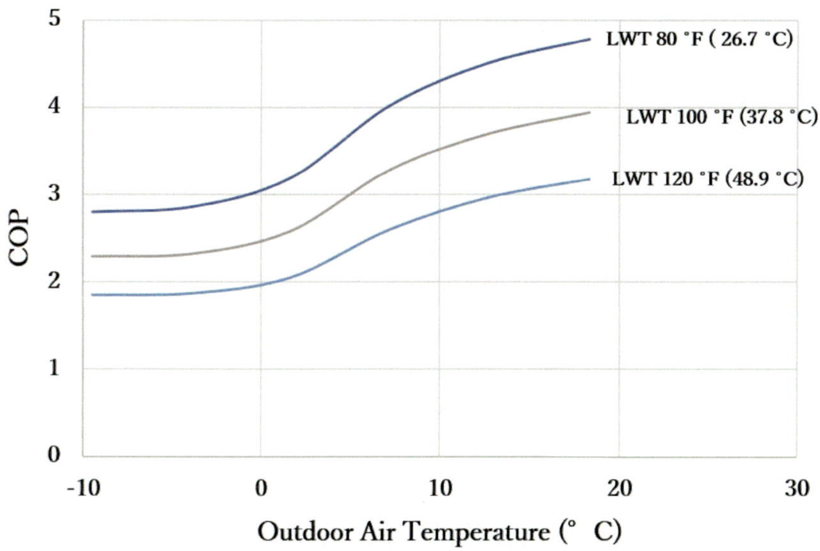

Figure 10.6 Heat pump performance (COP) of a typical ASHP comparing outdoor air temperature at several different leaving water temperatures (LWT).

Solution:

Given:

ASHP Heat required for heating, $Q_H = 120,000 BTU/hr$

From the chart in Figure 10.6,

$$COP_{H_r -10°C} = 2.3$$

$$COP_{H,18°C} = 3.75$$

Using the HP efficiency equations introduced in Chapter 9 to solve for the total electrical power input to the ASHP:

$$W_{-10°C} = \frac{Q_H}{COP_{H,-10°C}} = \frac{120,000 Btu/hr}{2.3}$$

$$= 52,174 Btu/hr \times \frac{1 \text{ kW}}{3,412 Btu/hr} = 15 \text{ kW}$$

$$W_{18°C} = \frac{Q_H}{COP_{H,18°C}} = \frac{120,000 Btu/hr}{3.75}$$

$$= 32,000 Btu/hr \times \frac{1 \text{ kW}}{3,412 Btu/h} = 9.4 \text{ kW}$$

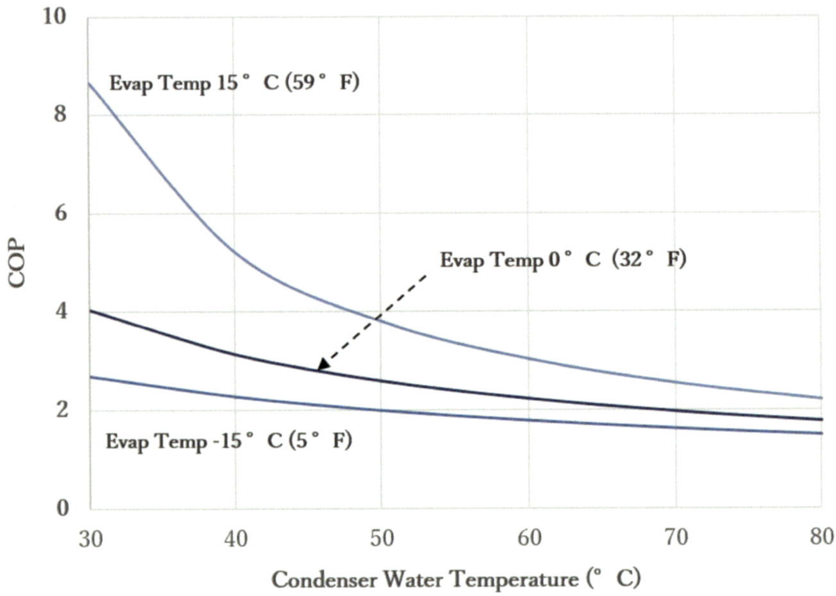

Figure 10.7 Heat pump performance (COP) of a typical water source heat pump, comparing evaporator water temperature to condenser water temperatures.

The total electrical power input of the ASHP at −10°C is 15 kW and at 18°C is 9.4 kW .

10.4.2.2 Size and load matching

Matching heat pump size to heating and cooling loads is important for optimizing efficiency. An oversized or undersized heat pump may not operate efficiently and can increase energy consumption and cause comfort issues. In addition, heat pump equipment can be very expensive, so carefully selecting the right equipment size to match the load has the potential to minimize the cost of the system.

Heat pumps, especially those used in space heating applications, do not operate at their maximum design output the majority of the time. Because they do not operate at full load all of the time, part load efficiency (COP) should be considered, as compressors can be less efficient at partial loads. Operating the compressors at part load can cause the compressors to cycle on and off, decreasing the efficiency and increasing the wear on mechanical parts and maintenance requirements. Multi-stage or VFD compressors can be considered to improve compressor efficiency at part loads.

It is important to ensure that the heat pump is sized appropriately for the heating and/or cooling loads for the application it serves. When modeling the electrical consumption of a heat pump based on the heating or cooling loads of a building, it is important to obtain the correct performance data for the heat pump model from the manufacturer, as derated part-load performance is not standard across manufacturers or equipment.

10.4.3 Supplemental heating

In applications where a supplemental or backup heating system is utilized for building heating (such as a heat recovery chiller with electric boiler back-up or an air-source heat pump with gas-furnace) there may be times when it is beneficial to switch to the full back-up heating device rather than continuing to run the heat pump, as it may not be cost effective to size a heat pump for the full heating load of the building. Additional heat pumps can also be used in a staged operating sequence to meet increased loads. Note that the supplemental heating system can be sized for either a partial or full heating load of the facility. The size of the supplemental system will depend on facility redundancy requirements, the size and performance of the heat pump, and the supply temperature requirements.

Supplemental heating can be provided by either gas or electric heating equipment; both can be considered hybrid heating systems. Dual fuel hybrid systems utilize gas-fired equipment as the supplemental heating source.

Determining the thermal balance point (or switchover) temperature of the heat pump is the point at which building heating load intersects the heat pump capacity, as illustrated in Figure 10.8. For an air source heat pump, since the heat pump capacity drops with outdoor air temperature, supplemental heating is needed to meet the heating demands when the heat pump is not sized to meet the full heating load.

When the outdoor air temperature exceeds the thermal balance point, the heat pump can fully meet the heating load of the facility. However, when the temperature falls below the balance point, supplemental heating will be necessary to satisfy the heating load of the building. In this range, the heat pump(s) and supplemental heating should operate together to optimize energy efficiency and minimize greenhouse gas emissions.

Switching over to a supplemental or backup heating system can also be driven by either GHG emissions or economics, before the shut-off temperature of the heat pump is reached. The balance point can be optimized on either GHG emission intensity or operating costs, as described below.

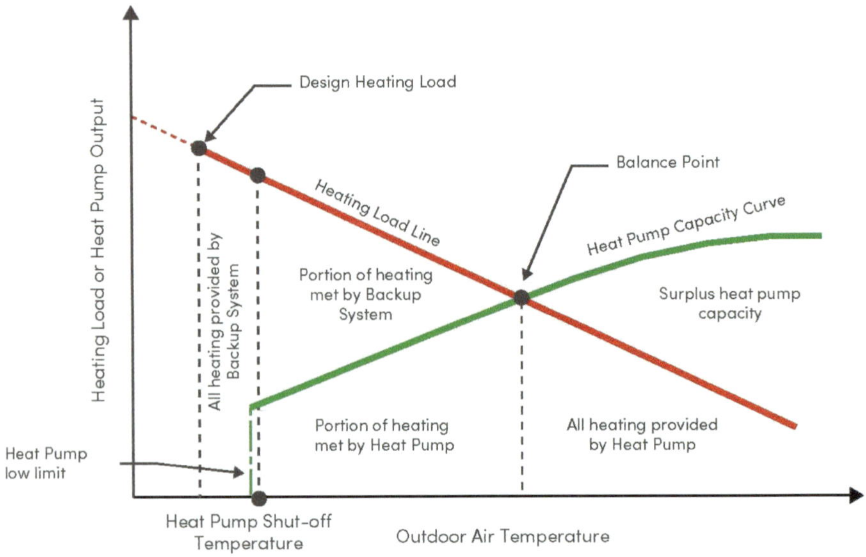

Figure 10.8 Example operational chart of a hybrid system with an outdoor air switchover temperature.

(source: FirstLight Energy Solutions).

GHG intensity balance point temperature

For heat pumps located in areas with electrical grids that have high GHG emission intensities, the GHG intensity balance point temperature (GHGi-BPT) can be estimated to determine when the heat pump COP is so low that operating the fossil-fuel-based heating source will result in lower GHG emissions. Heat pumps in this scenario need to have performance that exceeds COP_{min} to save GHG emissions.

$$\text{COP}_{\text{min}} = \frac{\text{GHG}_{\text{i, Electrical Grid}} \times \text{Efficiency}_{\text{backup}}}{\text{GHG}_{\text{i, backup fuel}}}.$$

Economic balance point temperature

Where operating costs are of primary concern, the economic balance point temperature (e-BPT) can be determined, where the cost of providing heating with the heat pump exceeds the cost of relying solely on the fossil-fuel-based heating source. Heat pumps in this scenario need to have performance that exceeds COP_{min} to save operating costs:

$$COP_{min} = \frac{\text{Cost of Electricity} \times \text{Efficiency}_{\text{backup}}}{\text{Cost of Backup Fuel}}.$$

The decision of whether to provide backup or supplemental heating depends on the local climate conditions, technical considerations, and financial constraints. It often involves an iterative process to find an optimal design solution that meets both GHG reduction targets and economic constraints.

Example 10-4: Calculating optimized COP on GHG emissions and economics in Seattle.

A commercial building located in Seattle, Washington has a dual fuel hybrid hydronic heating system that consists of an air-source heat pump and gas fired hot water boiler. The GHG Intensity of the electrical grid is $275\ \mathrm{kgCO_2e/MWh}$. The gas fired back-up boiler has an efficiency of 85%. The cost of electricity is $\$0.13/\mathrm{kWh}$ and the cost of gas is $\$16/\mathrm{GJ}$. Assume the emission factor for natural gas is $53\mathrm{kgCO_2e/GJ}$. Emissions factors for this example were obtained from the US Environmental Protection Agency 2024 Emissions Factor Hub.

a) What is the minimum COP that the heat pump can operate at if the dual fuel hybrid system is optimized for GHG Intensity?
b) What is the minimum COP that the heat pump can operate at if the dual fuel hybrid system is optimized for economics?

Solution:

Given:

$$\mathrm{GHG_{i,\,Electricity\ grid}} = 275\ \mathrm{kgCO_2e/MWh}$$
$$\mathrm{GHG_{i,\,gas}} = 53\ \mathrm{kgCO_2e/GJ}$$
$$\mathrm{Cost_{Electricity}} = \$0.13/\mathrm{kWh}$$
$$\mathrm{Cost_{gas}} = \$16/\mathrm{GJ}$$
$$\text{Efficiency}_{\text{Gas Boiler}} = 85\%$$

a) Solve for the minimum COP optimized on GHG intensity:

$$COP_{\min} = \frac{GHG_{i,\,\mathrm{Elec.}} \times \text{Efficiency}\ y_{\text{Gas boiler}}}{GHG_{i,\,\mathrm{Gas}}}$$
$$= \frac{275\ \mathrm{kg_{CO2e}/MWh} \times 0.85}{53\ \mathrm{kg_{CO2e}/GJ}} \times \frac{\mathrm{MWh}}{3.6\mathrm{GJ}} = 1.23$$

The minimum heat pump COP recommended for operation based on GHG intensity is 1.23 . The Owner should review the performance ratings for their ASHP and implement a strategy to only run the heat pump when the operating COP > 1.23 to reduce GHG emissions. Operating at a COP less than this value would result in an increase in GHG emissions vs. using the backup system.

b) Solve for the minimum COP optimized on economics:

$$COP_{\text{min}} = \frac{\text{Cost of Electricity } \times \text{ Efficiency } y_{\text{backup}}}{\text{Cost of Backup Fuel}}$$

$$= \frac{\$0.13/\text{kWh} \times 0.85}{\$16/GJ} \times \frac{1\text{kWh}}{0.0036\text{GJ}} = 1.92$$

Operating the heat pump at a COP less than 1.92 would result in an increase in energy cost vs. using the backup system. The Owner should review the performance ratings for their ASHP and implement a strategy to only run the heat pump when the operating COP > 1.92 to reduce operating costs.

Example 10-5: Calculating optimized COP on GHG emissions and economics in Denver.

a) A similar system to the previous example is in a building located in Denver, Colorado where the GHG intensity of the electrical grid is $513\ \text{kgCO}_2\text{e/MWh}$ (US EPA, 2024). What is the minimum COP that the heat pump can operate at if the dual fuel hybrid system is optimized for GHG Intensity?
b) The cost of electricity is $\$0.16/\text{kWh}$ and the cost of gas is $\$9/\text{GJ}$. What is the minimum COP that the heat pump can operate at if the dual fuel hybrid system is optimized for economics?

Solution:

Given:

$$\text{GHG}_{\text{i, Electricity grid}} = 513\ \text{kgCO}_2\text{e/MWh}$$
$$\text{GHG}_{\text{i, gas}} = 53\ \text{kgCO}_2\text{e/GJ}$$
$$\text{Cost}_{\text{Electricity}} = \$0.16/\text{kWh}$$
$$\text{Cost}_{\text{gas}} = \$9/\text{GJ}$$
$$\text{Efficiency}_{\text{ Gas Boiler}} = 85\%$$

a) Solve for the minimum COP optimized on GHG intensity:

$$COP_{\min} = \frac{GHG_{i,Elec.} \times Efficiency_{Gas\text{ boiler}}}{GHG_{i,Gas}}$$

$$= \frac{513\ \text{kg}_{\text{CO2e}}/\text{MWh} \times 0.85}{53\ \text{kg}_{\text{CO2e}}/\text{GJ}} \times \frac{1\text{MWh}}{3.6\text{GJ}} = 2.29$$

The minimum heat pump COP recommended for operation based on GHG intensity is 2.29 .

b) Solve for the minimum COP optimized on economics:

$$COP_{\min} = \frac{\text{Cost of Electricity} \times \text{Efficiency}\ y_{\text{backup}}}{\text{Cost of Backup Fuel}}$$

$$= \frac{\$0.16/\text{kWh} \times 0.85}{\$9/\text{GJ}} \times \frac{1\text{kWh}}{0.0036\text{GJ}} = 4.20$$

The minimum heat pump COP recommended for operation based on economics is 4.20 .

These two examples highlight that application of heat pumps can have very different environmental impact and value proposition based on the electrical grid GHG intensity and variable cost of electricity and natural gas in different regions.

10.4.3.1 Defrost cycles

Frost can accumulate on coils when the heat pump is operating in heating mode at temperatures near or below 0 °C (32 °F), and moisture in the air condenses on the coil, causing it to freeze. The frost limits the heat that can be transferred to the coil, decreasing the heat transfer efficiency of the coil. ASHPs use defrost cycles to remove frost on outdoor and fresh air heating coils, preventing equipment damage.

In a defrost cycle, the ASHP switches to cooling mode so that the hot refrigerant gas is sent to the outdoor coil (via the reversing valve) to melt the frost that has accumulated. Defrost cycles can be controlled by either a timed interval or on demand. The timed control enables defrost mode on a specified interval. Demand control requires a sensor to detect when frost has accumulated on the coil to enable defrost mode. This method is more efficient as it only enables defrost mode when required, which avoids unnecessary energy use.

10.4.3.2 Maintenance

Regular maintenance of heat pumps, including cleaning filters, coils, and ducts, is essential for maintaining the efficiency of the system, as well as minimizing the GHG emissions incurred by refrigerant leakage. Dirty or clogged components can restrict air flow or water flow and reduce efficiency. Additionally, periodic inspections by a qualified HVAC technician are recommended to identify and address any issues that may impact efficiency and avoid GHG emissions related to refrigerant leakage.

10.5 Heat Pump Technology Types and Applications

Choosing the appropriate heat pump source depends on factors such as regional climate, available space, and installation costs, with each source offering specific advantages and considerations.

Systems can use more than one technology, such as a heat recovery chiller and an air source heat pump, to provide an optimal solution for a building and allow for flexibility in "right sizing" to match the heating and cooling loads.

10.5.1 Air source heat pumps

Air source heat pumps (ASHP) extract heat from outdoor air as a heat source in winter (in heating mode) and reject heat to outdoor air as a heat sink in summer (in cooling mode). The efficiency of air-source heat pumps decreases in cold climates, as discussed in Section 10.4.2. Supplementary backup heating from gas-fired equipment may be required on the coldest days in some climate zones.

There are several types of ASHPs, each suited to specific applications. ASHPs can be split into two main types: air-to-air and air-to-water.

Air-to-air are the most common type of heat pumps; they transfer heat from the outdoor air to air-based systems. Examples of air-to-air ASHPs include:

- Packaged rooftop ASHPs contain both indoor and outdoor coils and components within the single unit. They can be used as a direct replacement for gas-fired packaged rooftop units. These units are commonly used in small and medium commercial buildings with decentralized HVAC systems. They can include supplemental back-up heating in the form of a gas or electric burner. Figure 10.9 shows an example of a packaged rooftop unit and air distribution system.

- Ductless split-system ASHPs utilize an outdoor condensing unit (which includes a compressor) connected to one indoor unit, referred to as a "mini-split," or multiple indoor units, referred to as a "multi-split." These systems are recommended for applications where there is no existing ductwork and can provide zone/ room level controls (NRCan, 2022). These units are commonly used in residential (both single-family homes and multi-unit residential buildings) and small commercial buildings with decentralized HVAC systems.

Air-to-water heat pumps (AWHP) transfer heat from the outdoor air to hydronic (water-based) central heating and cooling systems. Terminal units such as fan coils, radiators, or coils in air handling units are required to transfer and remove heat to and from the space. In retrofits, it is important to ensure temperature compatibility with the heat pump and the terminal units since many existing terminal units are designed to handle high-temperature heating water from a boiler (70–80 °C) and standard AWHPs typically operate most efficiently in the range of 45–60 °C. Examples of AWHPs include:

- 2-pipe AWHPs, which can operate in either heating or cooling mode, but not simultaneously. These units are found in small, medium, and large commercial buildings with centralized HVAC systems where simultaneous heating and cooling is not typically required.
- 4-pipe AWHPs are similar to 2-pipe AWHPs, except that they are able to provide simultaneous heating and cooling. They are highly efficient as they recover waste heat rejected from the cooling loads as a source delivering hot water for heating and chilled water for cooling at the same

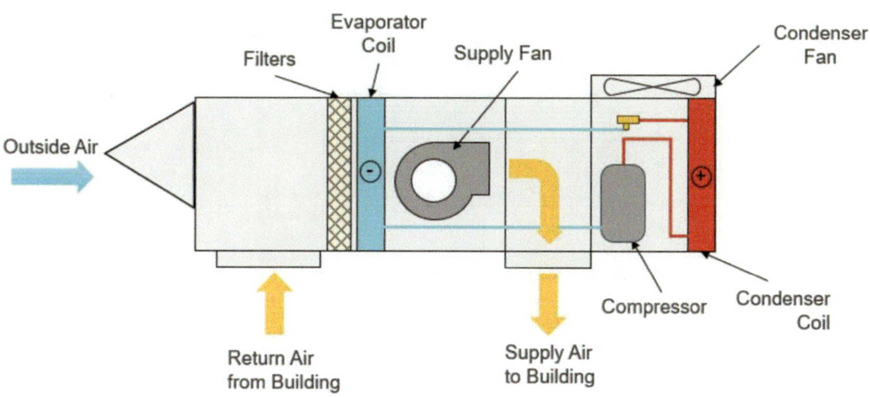

Figure 10.9 Packaged ASHP rooftop unit and air distribution system.

time. These units may be found in medium to large commercial buildings with centralized HVAC systems and are the preferred choice for applications with simultaneous heating and cooling loads. Figure 10.10 provides a diagram comparing operating modes of 2-pipe and 4-pipe AWHPs is an example of a packaged rooftop unit and air distribution system.

- Heat pump water heaters (HPWHs) are used in domestic hot water (DHW) heating applications. They can use indoor air as a source (in a parkade or mechanical room) to maximize efficiency by capturing waste heat in these areas. In addition, outside air source HPWHs are now available that use CO_2 as a refrigerant. HPWHs are 3 to 4 times more efficient than standard electric resistance water heaters (Caleffi, 2023). Often, HPWHs will have an electric resistance heater as backup. HPWHs can be used at a s mall scale in residential applications or at a larger scale in commercial buildings. For residential applications, the small HPWH is typically integrated into a storage tank. For commercial

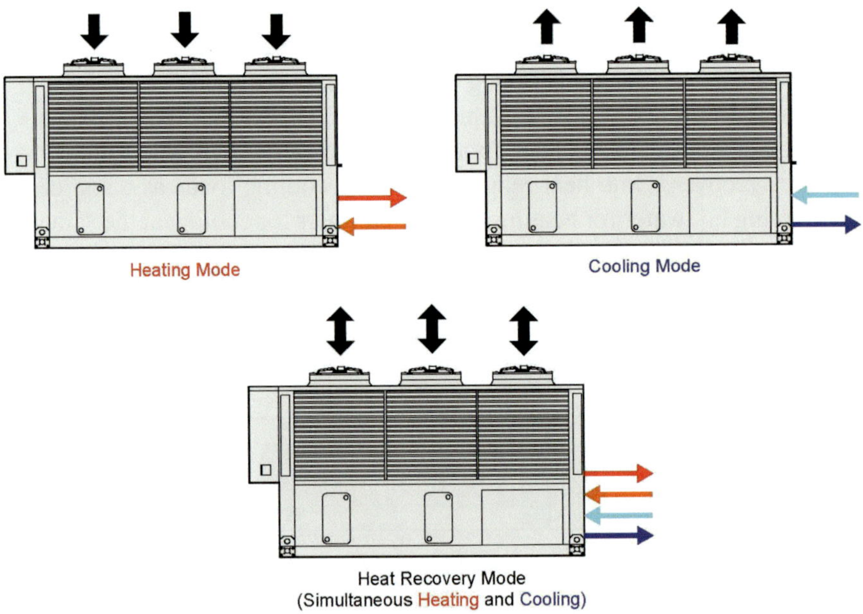

Figure 10.10 2-pipe ASHP compared to a 4-pipe ASHP. (source: FirstLight Energy Solutions).

applications, the HPWHs can provide heated water to one or more storage tanks, sized to accommodate the DHW needs of the facility. Figure 10.11 shows a typical domestic hot water HPWH.

10.5.2 Water source heat pumps

Water source heat pumps (WSHPs) extract and reject heat from and to a water source, typically to central hydronic heating and cooling systems. Water sources can include ocean, lakes, or wells, condenser water or chilled water loops, thermal energy storage within a building, or sewage wastewater. The efficiency of water source heat pumps varies based on the water source temperature and the desired output temperature (lift) as discussed in Section 10.4.2.

As with air-to-water heat pumps, terminal units are required to transfer and remove heat from the space. In retrofits, it is important to ensure

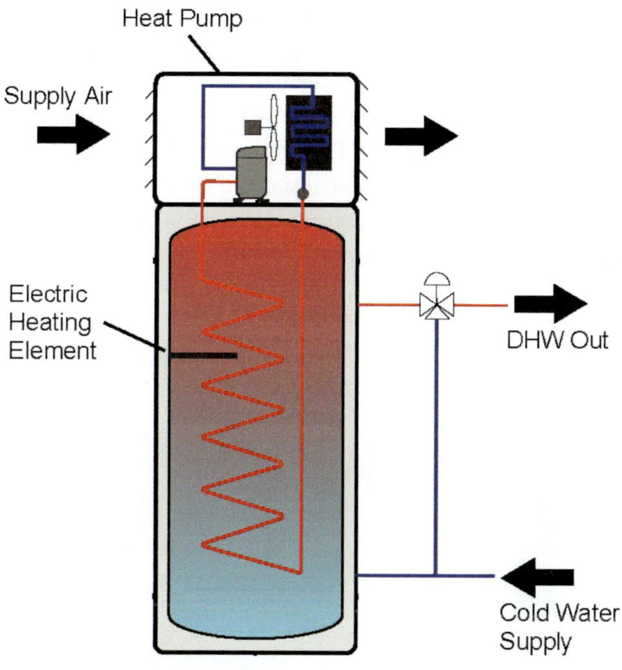

Figure 10.11 Domestic hot water heat pump water heater. (source: FirstLight Energy Solutions).

temperature compatibility with the heat pump and the terminal units. Examples of water source heat pumps include:

- Water-to-water heat pumps (WWHP) extract or reject heat using water sources like lakes, rivers, or wells. These are efficient but require proximity to a water source and are more suited to medium to large commercial buildings with centralized hydronic HVAC systems.
- Heat recovery chillers (HRCs) are a type of water-to-water heat pump that provides simultaneous heating and cooling. The main difference between a heat recovery chiller and a WWHP is that there is no reversing valve. They are highly efficient as they utilize the waste heat rejected from the cooling loads and lift the temperature to provide valuable heating water to building systems. These units are common in medium to large commercial buildings with centralized HVAC systems and are very effective for applications with simultaneous heating and cooling loads. Figure 10.12 shows an example of a heat recovery chiller supplying heating and cooling.
- Wastewater (sewer) heat recovery heat pumps extract heat from wastewater or sewer discharge from a building. They are best suited to applications that have large and consistent sewer loads, such as TEN systems, but can also be applied at a smaller scale in multi-unit residential or large commercial buildings. Figure 10.13 provides a simplified schematic of a sewer heat recovery heat pump installation.
- Small terminal water source heat pumps (WSHPs) can be connected to a central 2-pipe hydronic loop to provide heating and cooling to individual spaces or zones. The central loop is used as the heat source and the heat sink. This configuration is more common in multi-unit residential buildings and low-rise commercial buildings, where each unit has its

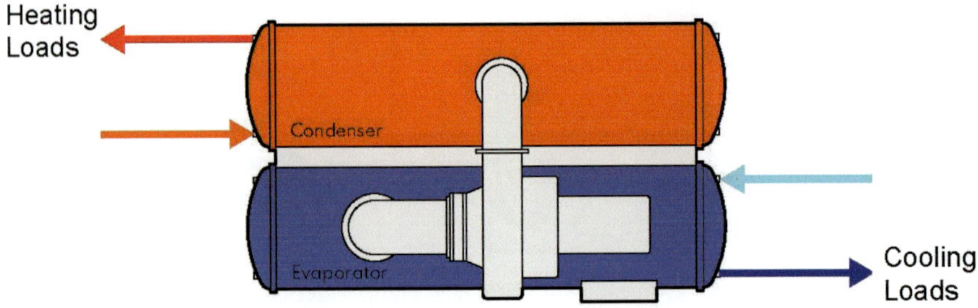

Figure 10.12 Heat recovery chiller.
(source: FirstLight Energy Solutions).

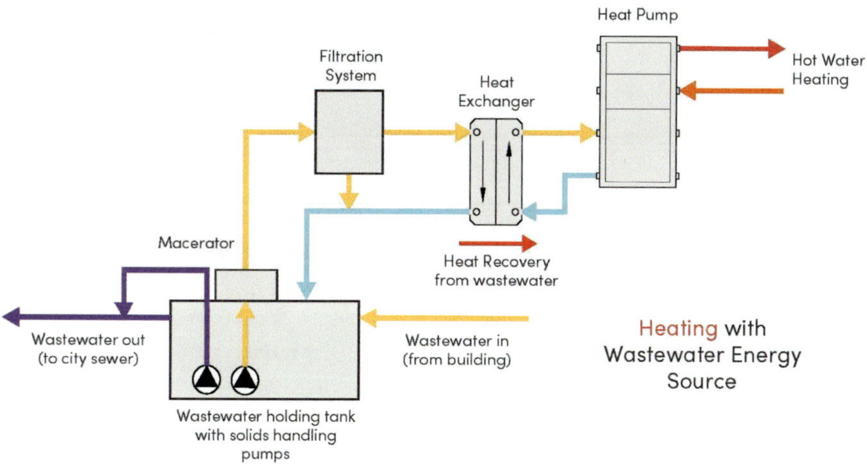

Figure 10.13 Sewer heat recovery heat pump schematic. (source: FirstLight Energy Solutions).

own heat pump to provide heating or cooling. Terminal WSHPs are very efficient, as they enable heat recovery in a building. For example, on a sunny winter afternoon, WSHPs can operate in cooling on the south side of a building where it is warmer with solar gains, rejecting heat into the central loop. The WSHPs on the north side can then pull heat from the loop. In this way, heat is transferred from one area of the building to another using minimal electricity.

10.5.3 Variable refrigerant flow systems

Similar to ASHPs, variable refrigerant flow (VRF) systems extract and reject heat from and to outside air using an outdoor unit. They can be connected to single or multiple indoor units such as fan coils, cassettes, or direct expansion (DX) coils (such as in an air handling unit or furnace). Where the systems differ is that VRF systems circulate refrigerant to each indoor unit based on the specific heating and/or cooling needs of each space. VRF systems require networks of refrigeration piping to be run throughout the building to all the terminal units. Figure 10.14 shows a schematic of the outdoor unit and indoor units connected in a VRF system. VRF systems enable simultaneous heating and cooling in different zones and can incorporate heat recovery.

VRF systems allow for zoning temperature control and are suitable in small and medium commercial as well as multi-unit residential buildings.

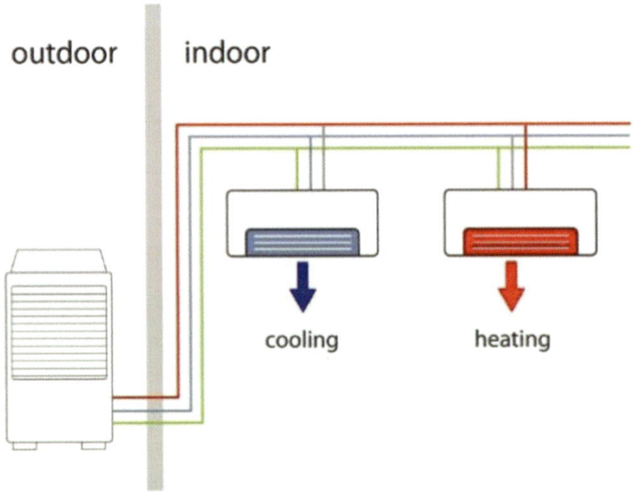

Figure 10.14 VRF system schematic.
(source: FirstLight Energy Solutions).

Large VRF systems require high volumes of refrigerant, which presents a significant risk if there is a leak anywhere in the system, as discussed in Section 10.2.3. As a result, VRF has average annual refrigerant leakage rates of 10% compared to 5% for AWHPs (ASHRAE Standard 228-2023).

A VRF system can have a hydronic heat exchanger (refrigerant to water) to combine a VRF system with a hydronic heating and cooling system. Figure 10.15 shows a schematic of a hybrid VRF system with a hydronic heat exchanger. This can be referred to as a Hybrid VRF system as it combines two system types (refrigerant and water). Hybrid VRF systems with heat recovery and water-based connections can be extremely efficient when simultaneous heating and cooling occur in the building.

10.5.4 Ground source heat pumps

Ground source heat pumps (GSHPs) extract heat from the ground (below the surface) as a heat source in heating mode and use the subsurface as a heat sink in cooling mode. Sources can include ground water or soil. GSHPs circulate a heat transfer fluid (typically a mixture of water and glycol) through either vertical boreholes or horizontal pipes buried in the ground. Both open

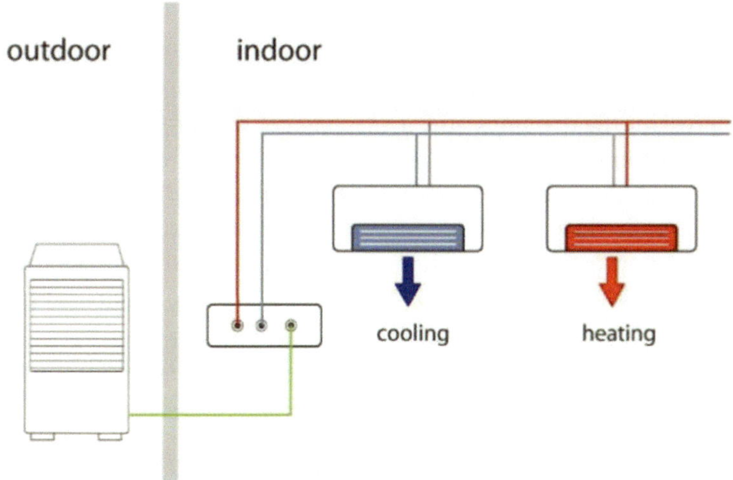

Figure 10.15 Hybrid VRF system schematic with hydronic heat exchanger. (source: FirstLight Energy Solutions).

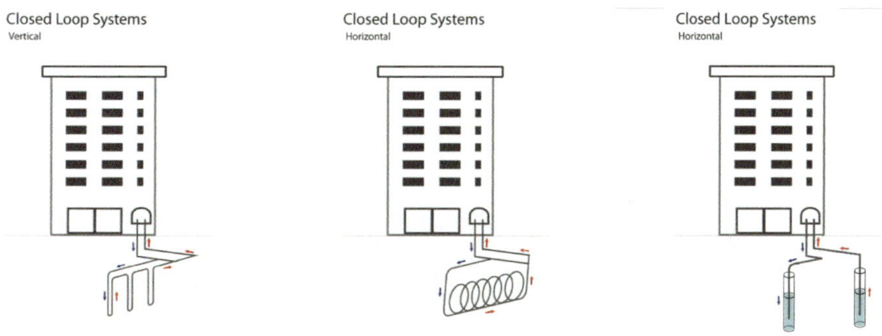

Figure 10.16 Geothermal systems.

and closed loop configurations can be considered. Figure 10.16 provides examples closed loop (vertical and horizontal) and open loop system.

GSHPs can be installed at a small scale in single family homes, or at a larger scale in multi-unit residential (MURBs) or commercial buildings. Due to the cost of drilling and the installation of ground piping infrastructure, GSHPs are more expensive than ASHPs, but they can achieve higher efficiencies in cold climate regions, as the ground temperature is warmer than the air during the winter.

10.5.5 Gas absorption heat pumps

Gas absorption heat pumps (GAHP) use an absorption refrigeration cycle rather than a vapor compression cycle (described in Section 10.2.1). While vapor compression heat pumps use a mechanical compressor to circulate refrigerant, absorption heat pumps use a combination of a refrigerant and an absorbent fluid (typically ammonia). GAHPs utilize outdoor air and waste heat from the gas fired generator as energy sources which are then transferred into a space via hydronic heating. Heat can also be rejected to outdoor air to provide space cooling. The key components of the gas absorption heat pump refrigeration cycle is shown in Figure 10.17.

While more efficient as compared to conventional gas-fired heating equipment, GAHPs are less efficient than electric heat pumps. Typical rated COP for GAHPs ranges from 1.2 to 1.4 in heating mode, and they can deliver heating water between 50 and 60 °C. In cooling mode, GAHP COP typically ranges from 0.6 to 0.8.

GAHPs are available in small-scale packaged units for residential and small commercial building applications. There are also large-scale GAHPs available that can be utilized in large commercial buildings and industrial applications. GAHPs may be favored in applications where increased efficiency is the goal, but electricity is much more expensive than gas, and/or where building electric service capacity is limited.

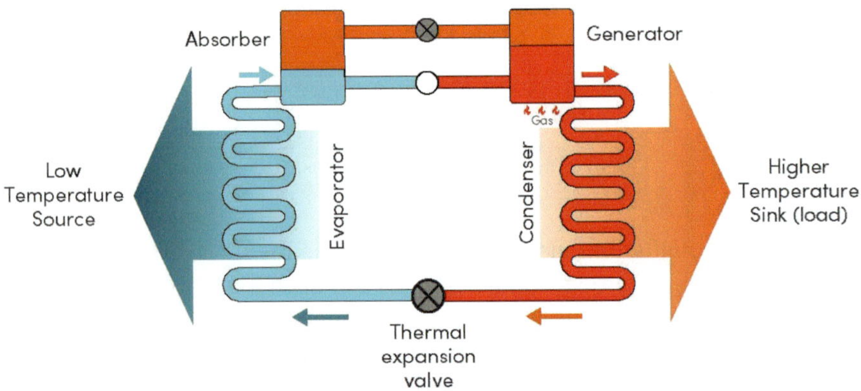

Figure 10.17 Gas absorption heat pump refrigeration cycle. (source: FirstLight Energy Solutions).

10.5.6 Gas engine heat pumps

Gas engine heat pumps (GEHP) use a gas-fired engine rather than a motor to drive the compressor. GEHPs use the vapor compression cycle (described in Section 10.2.1) to circulate refrigerant. GEHPs utilize outdoor air and can recover waste heat from the gas-fired engine cooling system or exhaust. Heat can also be rejected to outdoor air to provide space cooling.

While more efficient as compared to conventional gas-fired heating equipment, GEHPs are less efficient than electric heat pumps. Typical rated COP for GEHPs ranges from 1.2 to 2 in heating mode. In cooling mode, GEHP COP typically ranges from 1.2 to 1.6.

10.6 Integration Considerations

This section provides an overview of how heat pumps can be incorporated into a variety of building types, systems, and configurations. In most cases, switching to heat pumps will result in energy savings due to the improvement in efficiency. Although installing a heat pump does not always result in operating cost savings due to the difference in costs between gas and electricity.

Considerations provided are most applicable to retrofit applications, although similar considerations should be made for new construction during design.

Heat pumps play an important role in building decarbonization and can be implemented in a phased approach. It is crucial to establish a long-term plan for asset replacement in line with decarbonization goals and to stick to the plan. Heat pumps long-term assets such as boilers and other gas-fired equipment to support decarbonization efforts.

The following should be considered in most applications when integrating a heat pump:

- **Energy load analysis**. Conduct a thorough analysis of the heating and cooling load requirements for the specific facility. Consider building envelope characteristics such as insulation and window types, and occupancy patterns, which can impact the demand for heating and cooling. Ensure that the heat pump capacity aligns with the building's demands both in terms of peak demand and part-load conditions. Right sizing the heat pump for the heating and cooling loads that are expected the majority of the year is important to minimize system costs. Supplemental systems can be used to meet peak condition scenarios. Refer to Chapter 9, Section 9.5 for guidance on right-sizing for heating and cooling loads.

- **Energy efficiency**. Evaluate the energy efficiency of the heat pump, considering its design, heating, and cooling COP. Compare these metrics with the existing equipment to assess potential energy savings. Consider upgrades to building envelope components (insulation, air leakage reduction, windows) to reduce thermal loads required, and potentially reduce the capacity and capital cost of the heat pump system (refer to Chapter 3, Figure 3.1). In addition, implement best practices in building HVAC controls and improve lighting efficiency to reduce thermal loads. This can include using occupancy sensors, LED lighting, and implementing demand-controlled ventilation to further reduce demand on the heat pump system and downsize the capacity of the heat pump. For additional information on lighting, refer to Chapter 12. Building control strategies to reduce thermal loads and improve energy efficiency for buildings is discussed in Chapters 9 and 15.
- **System compatibility and integration**. Assess the compatibility of the existing building infrastructure with the heat pump system. Check if the new system can seamlessly integrate with the building's current controls, sensors, and building automation system (BAS). Ensure that the heat pump can be effectively integrated to optimize overall HVAC performance. Assess the compatibility of the existing ductwork with the air distribution requirements. Modifications may be necessary to ensure proper airflow and distribution for both heating and cooling modes.
- **Electrical infrastructure**. Verify the electrical requirements of the new heat pump and ensure that the existing electrical infrastructure, including panels and wiring, are compatible. Electrical upgrades may be required to accommodate the power demands of the new system if the existing system is gas-fired. Consultation with an electrical engineer is recommended to ensure that all applicable building codes are met.
- **Climate considerations**. Consider the local climate conditions, as discussed in Section 10.4.2; heat pumps can be impacted by cold temperature ranges or may have output limitations based on heating requirements. Cold climate specific heat pumps can operate at temperatures as low as −22 °F or −30 °C (Caleffi, 2022). As cold climate heat pump technology evolves, the efficiency at extremely low temperatures is improving. Back-up heating options must be evaluated and considered

for regions with extremely cold temperatures. Like all mechanical equipment, heat pumps can have maintenance failures that require special parts and downtime to repair. Resilient and redundant heating sources are recommended in cold climate regions.

- **Equipment weight**. Heat pumps can vary in size and weight and may exceed the load-bearing capacity of the roof or proposed installation location, which could lead to structural issues. Consult with structural engineers or building professionals to modify the structure at the location chosen to accommodate the additional weight and ensure that all applicable building codes are met.

10.6.1 Residential

Heat pumps can be integrated with existing HVAC and DHW systems in single-family or multi-unit residential homes and buildings (MURB). Adding heat pumps to homes also offers the ability to add cooling in situations where it does not already exist. As the climate changes, extreme weather events such as heat domes are making cooling resiliency an important decision-making factor in the residential sector (see Chapter 8).

When retrofitting a single-family home or MURB with heat pumps, an understanding of how the existing HVAC and DHW services are delivered should be established, as the appropriate solution(s) may vary based on the infrastructure that exists.

In most cases, heat pumps can be used to fully satisfy the heating and cooling needs of residential homes. In cold climates, additional backup heating sources such as electric resistance baseboard heaters or gas-fired furnaces may be required. It should be noted that cold climate heat pumps are currently available, with innovations in technology progressing rapidly every year. Figure 10.18 provides an overview of considerations when installing heat pumps in single-family homes or MURBs.

Partial retrofits can also be considered to improve the energy efficiency of a single-family home or MURB and reduce GHG emissions. These dual-fuel hybrid systems can offer fuel flexibility and can facilitate the energy transition from fossil fuels.

The following resources include additional information and guidance on retrofitting and selecting heat pumps for the residential sector (see References for the full citations):

- ASHRAE "Building Decarbonization Retrofits for Commercial and Multifamily Buildings"

Single Family Homes Considerations

- The existing HVAC and distribution system
- Peak heating and cooling loads for the home
- Back-up heating requirements
- Location for the Outdoor Unit
- Electrical capacity of the home and future additional loads (such as EV charging)
- Potential upgrades to building envelope to reduce the heat pump capacity and cost

MURB Considerations

- The existing HVAC and distribution system
- DHW production and distribution (integrated or segragated from heating system)
- Peak heating and cooling loads for the building
- Location for the outdoor unit and structural upgrade if required
- Back-up heating requirements
- Electrical capacity of the building and future additional loads (such as EV charging)
- Potential upgrades to building envelope to reduce the heat pump capacity and cost

Figure 10.18 Residential building heat pump considerations.

- NRCan "Energy Star Canada: Heating and Cooling with a Heat Pump"
- NRCan "Air-source Heat Pump Sizing and Selection Guide"
- Caleffi Hydronic Solution "Domestic Hot Water Heat Pumps"
- U.S. DOE "Energy Saver: Geothermal Heat Pumps"
- U.S. DOE "Energy Saver: Heat Pump Systems"
- U.S. DOE "Energy Saver: Air-source Heat Pumps"

10.6.2 Commercial

Heat pumps can be integrated with existing HVAC and DHW systems in commercial buildings. They can be applicable at all scales (small, medium, and large commercial), and the specific retrofit applications should be considered on a case-by-case basis, depending on the type of existing systems. As discussed in the HVAC overview in Chapter 9, applicable solutions will vary if the existing system is air-based or hydronic.

Understanding the heating and cooling loads of the building and how they fluctuate throughout the year is critical for right-sizing the heat pump appropriately. Chapter 9 discusses the methodologies for quantifying loads to use in equipment selection. Additional methodology specific to heat pumps is outlined in NRCan's "Air-source Heat Pump Sizing and Selection Guide."

It is also important to identify opportunities to reduce heating and cooling loads through, for example, controls optimization. Identify opportunities to recover waste heat whenever possible, for example, through passive and active exhaust heat recovery, by reducing free cooling, and by utilizing condenser water heat from existing chillers that would otherwise be rejected to the atmosphere using a cooling tower.

In older buildings, design heating water temperatures tend to be high (70–80 °C), with terminal units and coils designed for hot water boiler temperatures. As discussed in Section 10.4.2, operating heat pumps at high temperatures requires a higher lift, resulting in less efficient operation. In addition, air-to-water heat pump technologies may be limited to providing a maximum of 60 °C heating water. Testing can be done to determine the minimum temperature requirements and if space heating needs can be met with lower water temperatures, so backup boilers may still be required on the coldest days. If lower heating water temperatures do not meet space heating needs on the coldest days, backup electric or gas boilers may still be required.

High-temperature hot water heat pumps are available and can be considered in scenarios where the terminal units replacement cannot be justified or high-temperature heating water is required for a significant amount of the year. Although the overall efficiency may be lower, it can still be beneficial for energy efficiency, operating cost savings, and GHG emissions reduction.

Terminal unit replacement can be costly, and if considering replacing end-of-life terminal units, then ensure that they are compatible with low heating water temperatures. In air handling units, if a multi-row cooling coil already exists, the coil can often be used as a switchover coil to provide heating or cooling with the addition of control valves. In cases where the peak heating water temperature is a concern, an AWHP can be used for the majority of the year, without replacing terminal units, reverting to the backup heating system when necessary.

Commercial building applications also present an opportunity to incorporate thermal energy storage (TES). TES can be used to improve simultaneous heating and cooling loads. TES can also be used to reduce peak load demands and incorporate electric demand response abilities into a building. This becomes more important as the electrical grids reach new peak demand each year and utilities start to implement programs focused on peak demand management. Additional information on thermal energy storage is presented in Chapter 17.

Heat pumps can also be integrated with existing boilers and fossil-fuel-based systems to create dual-fuel hybrid systems. For many facilities, adding a heat recovery chiller to an existing hydronic heating system is an excellent way to achieve energy and GHG emission reductions. For additional information on boilers, refer to Chapter 20.

DHW heating systems are often oversized. When implementing a heat pump, this can cause issues due to both the cost and space required to match

Commercial Building Considerations

- The existing HVAC and distribution system
- Peak and simultaneous heating and cooling loads for the building
- Load reduction through controls optimization and/or building envelope improvements
- Design water temperatures of terminal units and coils
- Domestic hot water production and distribution (integrated or segregated from heating system)
- Space availability in mechanical rooms or roof
- Structural reinforcement if required for the new equipment
- Exhaust heat available for reclaim
- Cooling tower heat rejection loads available for reclaim
- Back-up heating requirements
- Electrical capacity of the building and future additional loads (such as EV charging)

Figure 10.19 Commercial heat pump considerations.

the existing equipment capacity. Measuring DHW consumption to establish base and peak loads so that HPWHs can be right-sized appropriately is key in all heat pump retrofits. Reducing the DHW loads as much as possible by implementing low-flow fixtures and shower heads is important to minimize the size of the HPWH.

Figure 10.19 provides an overview of considerations when installing heat pumps in commercial buildings.

It is important to remember that a one-size-fits-all approach does not work when integrating heat pumps with existing facilities, and this is especially true for commercial buildings.

The following resources include additional information and guidance on retrofitting and selecting heat pumps for the commercial sector:

- 2024 ASHRAE Handbook – HVAC Systems & Equipment
- ASHRAE "Building Decarbonization Retrofits for Commercial and Multifamily Buildings"
- NREL "Decarbonizing Building Thermal Systems: A How-to Guide for Heat Pump Systems and Beyond"
- NRCan "Air-source Heat Pump Sizing and Selection Guide"
- Caleffi Hydronic Solutions "Idronics: Air-to-Water Heat Pump Systems"
- Caleffi Hydronic Solutions "Idronics: Domestic Hot Water Heat Pumps"

10.6.3 Industrial

Heat pumps can be used in industrial applications and processes to optimize efficiency and reduce reliance on fossil-fuel-based heating. Industrial heat pumps can save more than 30% of process heating energy (Rightor et al., 2022). They can be used in production industries such as food and beverage, pulp and paper, wood products, chemical and petrochemical, steel, data centers, power and TEN production, sewage treatment, fabrics and textile, and automotive.

Industrial processes often require heat generation and subsequent cooling or refrigeration. This presents opportunities to recover and upgrade waste heat to, at least partially, meet the process or space heating requirements of the facility. If the waste heat available exceeds the needs of the process, or is not suitable to meet the needs, consideration can be made for surrounding buildings or communities to reduce wasted energy. A challenge for industrial heat pumps is that many industrial processes require temperatures in excess of 500 °C (930 °F). Commercially available technology is most commonly suitable for applications less than 100 °C (212 °F). Technology is increasingly becoming available for higher temperature applications, but is currently limited to around 150 °C (300 °F), although the market is evolving rapidly (WBCSD, 2022).

A few examples of industrial applications of heat pumps:

- Dairy process heating.
- Lumber drying in wood product manufacturing.
- Process water heating in pharmaceutical, pulp and paper, or textile manufacturing.

Industrial applications also include large-scale heat pumps that are commonly utilized in TEN systems. For additional information on TEN applications and industrial processes, refer to Chapters 19 and 20, respectively.

Figure 10.20 provides an overview of considerations when installing heat pumps in industrial facilities.

Each industrial application is specific and requires thorough study and analysis. The commercial availability of large-scale heat pumps suitable for high-temperature industrial applications has significantly increased in recent

Industrial Facility Considerations

- Heating and cooling loads for the process
- Space availability
- Waste heat available for reclaim
- Process temperature requirements
- Quality of waste heat available (low or high grade)
- Cooling tower heat rejection loads available for reclaim
- Upgrades to passive energy systems (piping and equipment insultation, reduced leakage) to reduce process heating and cooling loads
- Surrounding infrastructure such as industrial, commercial and residential buildings

Figure 10.20 Industrial facility heat pump considerations.

years, and the market will certainly continue to grow and evolve as the industry seeks innovative solutions to reduce GHG emissions and impact.

The following resources include additional information and guidance on retrofitting and selecting heat pumps for the industrial sector. (see References for the full citations):

- ACEEE "Industrial Heat Pumps" and "Industrial Heat Pumps: Electrifying Industry's Process Heat Supply"
- Industrial Heat Pump Alliance resources.
- Energy Innovation Policy and Technology LLC® "Decarbonizing Low-temperature Industrial Heat in the US"
- U.S. DOE "A Best Practices Steam Technical Brief: Industrial heat pumps for steam and fuel savings".

10.7 Summary

In this chapter, we have introduced heat pump technology in depth by providing an overview of how heat pumps work, including the refrigeration cycle, the key components, and compressor types. This includes energy efficiency considerations for equipment selection, a summary of the various types of air, water, and ground source heat pump products and their applications.

An overview of common heat sources and sinks was provided to understand the suitability of the heat medium for heat pump applications.

Key concepts in heat pump performance efficiency were introduced. Temperature lift provides an indication of how hard the heat pump has to work to meet the desired output and is an important concept to understand when considering the efficiency of a heat pump.

Various heat pump technology was discussed, and an overview of suitable applications for each was provided for the following heat pump technologies:

- Air source heat pumps
- Water source heat pumps
- Variable refrigerant flow systems
- Ground source heat pumps
- Gas absorption heat pumps
- Gas engine heat pumps

Lastly, considerations for integrating heat pumps into residential, commercial, and industrial facilities are provided. This includes general considerations applicable to all facilities, such as the heating and cooling loads of a facility, the potential for energy and operating cost savings, GHG emission reduction, compatibility and electrical and structural limitations of buildings, and implications on heat pump performance in the local climate.

References

ACEEE (2023). "Industrial Heat Pumps" https://www.aceee.org/industrial-heat-pumps

ASHRAE (2019) "ASHRAE 90.1-2019 Energy Standard for Buildings Except Low-Rise Residential Buildings".

ASHRAE (2020) "2020 ASHRAE Handbook – HVAC Systems & Equipment".

ASHRAE (2021) "2021 ASHRAE Handbook – Fundamentals".

ASHRAE (2023) "2023 ASHRAE Handbook – HVAC Applications".

ASHRAE (2023) "ASHRAE Standard 228-2023 Standard Method of Evaluating Zero Net Energy and Zero Net Carbon Building Performance".

ASHRAE (2025) "Building Decarbonization Retrofits for Commercial and Multifamily Buildings".

Natural Resources Canada (NRCan) (2020) "Air-source Heat Pump Sizing and Selection Guide".

Natural Resources Canada (NRCan) (2022) "Energy Star Canada: Heating and Cooling with a Heat Pump". https://natural-resources.canada.ca/energy-efficiency/energy-star-canada/about/energy-star-announcements/publications/heating-and-cooling-heat-pump/6817#b6

Caleffi Hydronic Solutions (2020) "Idronics: Air-to-Water Heat Pump Systems". N. 27 July 2020.

Caleffi Hydronic Solutions (2023) "Idronics: Domestic Hot Water Heat Pumps". N. 33 July 2023.

Houssainy, Sammy, Landan Taylor, Paul Torcellini, and Sugirdhalakshmi Ramaraj. 2024. Decarbonizing Building Thermal Systems: A How-to Guide for Heat Pump Systems and Beyond. Golden, CO: National Renewable Energy Laboratory. BK-5500-87812. www.nrel.gov/docs/fy24osti/87812.pdf.

IEA (2022). "The Future of Heat Pumps" IEA, Paris https://www.iea.org/reports/the-future-of-heat-pumps, Licence: CC BY 4.0

Industrial Heat Pump Alliance. https://industrialheatpumpalliance.org/

Rightor, E., P. Scheihing, A. Hoffmeister, and R. Papar. (2022) "Industrial Heat Pumps: Electrifying Industry's Process Heat Supply". Washington, DC: American Council for an EnergyEfficient Economy. aceee.org/research-report/ie2201.

Rissman, Jeffrey (2022). "Decarbonizing Low-Temperature Industrial Heat in the U.S." Energy Innovation Policy and Technology LLC®.

U.S. Department of Energy (2024) "Energy Saver: Air-source Heat Pumps". https://www.energy.gov/energysaver/air-source-heat-pumps

U.S. Department of Energy (2024) "Energy Saver: Geothermal Heat Pumps". https://www.energy.gov/energysaver/geothermal-heat-pumps

U.S. Department of Energy (2024) "Energy Saver: Heat Pump Systems". ht tps://www.energy.gov/energysaver/heat-pump-systems

U.S. Department of Energy (Energy Efficiency and Renewable Energy) (2023). "A Best Practices Steam Technical Brief: Industrial heat pumps for steam and fuel savings".

U.S. Environmental Protection Agency (U.S. EPA) (2024) "2024 GHG Emissions Factor Hub".

World Business Council for Sustainable Development (WBCSD) (2022). "Industrial Heat Pumps: It's time to go electric".

Author Biography

Dawn-Marie Barreira is a registered professional engineer and an Associate Principal at First Light Energy Solutions in Vancouver, British Columbia. She has worked with industrial and commercial facilities, developing and leading energy and GHG emission quantification, reduction planning strategies, and project implementation from within organizations and as a consultant.

Acknowledgments

I sincerely thank Eric Mazzi, Mike Reimer, and Scott Sinclair for their thoughtful review and constructive feedback on this chapter. Their insights were invaluable in refining the final version.

Practice Problems

Problem 1:

Vapor-compression (heat pump) cycles have four main processes: evaporation, condensing, compression, and throttling as illustrated in the diagram below.

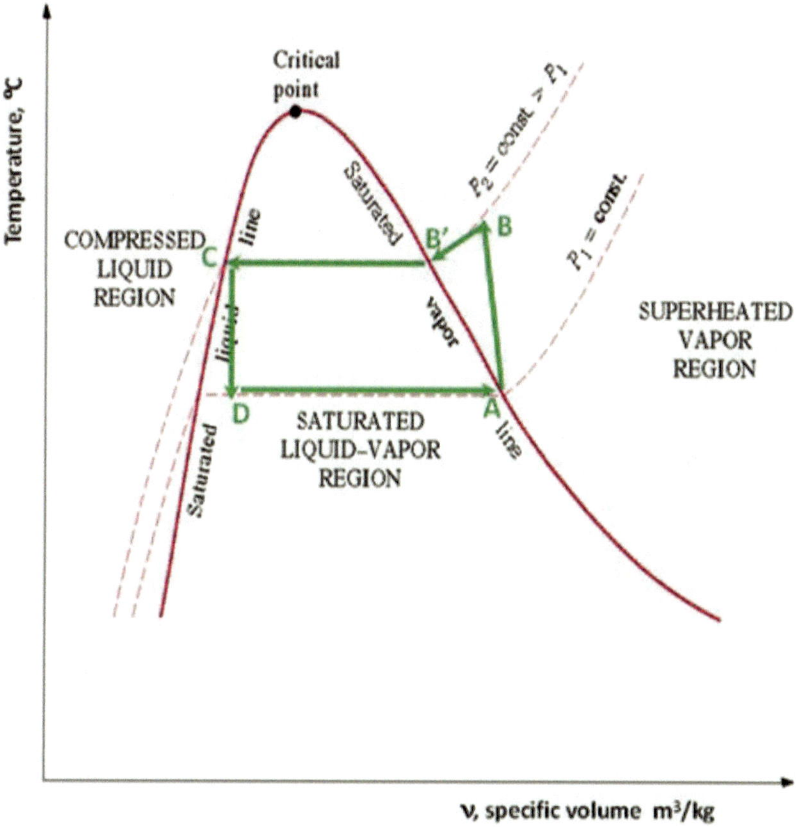

a) For the vapor compression system shown in the diagram, providing space cooling, state which of the four processes corresponds to each line in the diagram (fill in the blanks below)

A to B is ____________________________

B to C is ____________________________

C to D is ____________________________

D to A is ____________________________

b) Choose the correct option below: *higher* or *lower*

The outdoor temperature must be (higher or lower) than the temperature at point C.

The indoor temperature must be (higher or lower) than temperature at point A or point D.

Problem 2:

a) A 30 kW heat pump rooftop unit that provides heating and air conditioning has a cooling EER of 6.3 Btu/Wh. What is the COP_C of that unit? How many kW of input power does that unit draw when it operates at full load mechanical cooling?

b) The COP_H is 2.8. How much electrical energy is consumed at full heating load?

c) A rooftop air conditioner has an EER of 13.5 Btu/Wh. What is its kW_{input} per $kW_{thermal}$ rating? What is the kW_{input} per $ton_{thermal}$ rating?

d) How many kWh of electric input is used to provide 120 million kWh of air conditioning with a system having a COP of 3.0?

e) How many kWh electric input is used to provide 100 million kJ of heat removal (air conditioning) with a system having a COP of 5.0?

Problem 3:

A heat pump is rated to deliver 15,000 kJ/hr. The motor is rated at 120 volts and 14 amps (single phase), and the power factor is 0.95. For the rated conditions, calculate the EER of the unit (Btu/Wh).

Problem 4:

A heat recovery chiller is simultaneously providing heating and cooling to an office building. At its full load, it provides 850,000 Btu/hr of heat for heating and removes 650,000 Btu/hr of heat for cooling. The COP of heating is 4.2, and the COP of cooling is 3.2. What is the full load electrical power input to the heat recovery chiller?

Problem 5:

A library would like to consider using a VRF air-source heat pump system to provide heating to the building, to the existing gas-fired rooftop units with a thermal efficiency of 80%. Using the gas consumption (in thousand cubic feet) and ASHP technical data provided, estimate the electrical power consumption for the heat pump for the full year. Assume that the heat content of natural gas is 1 cubic foot = 1037 Btu.

Natural gas consumption and average monthly temperature at the library.

Month	**Average temperature (°F)**	**Gas consumption (MCF/month)**
January	41.0	506
February	42.8	542
March	46.4	344
April	51.8	137
May	57.2	59
June	62.6	76
July	68.0	58
August	68.0	55
September	62.6	71
October	53.6	184
November	46.4	455
December	39.2	690

ASHP technical Data

Outdoor air temperature (°F)	**COP_H**
14	2.02
23	2.39
32	2.76
41	3.13
50	3.50

Provide the solution by completing the table below.

Month	Average outdoor air temperature	Gas consumption (MCF)	Building energy load (MTBU)	Average COP	Heat Pump electrical consumption (kWh)
January	41.0	506			
February	42.8	542			
March	46.4	344			
April	51.8	137			
May	57.2	59			
June	62.6	76			
July	68.0	58			
August	68.0	55			
September	62.6	71			
October	53.6	184			
November	46.4	455			
December	39.2	690			

Problem 6:

A small domestic water heater at a community center is currently at the end of its life and requires replacement. The building maintenance team is considering replacing the water heater with a similar, gas-fired DHW heater (efficiency = 80%), an electric resistance DHW heater (efficiency = 98%), or a heat pump water heater (COP = 3.3). The annual DHW gas consumption is 650 GJ and costs $13.55 per GJ. The cost of electricity is $0.16/kWh. Compare the three options on energy, GHG emissions, and operating cost by completing the table below. Assume the emission factor for natural gas is 53 kg CO_2e/GJ and the emissions factor for the electricity grid is 375 kg CO_2e/MWh.

	Gas-fired DHW heater	Electric resistance DHW heater	Heat pump water heater
Energy consumption (kWh)			
GHG emissions (kg CO_2e)			
Operating cost ($)			

Problem 7:

A commercial building located in San Francisco, California, has a dual-fuel hybrid hydronic heating system that consists of an air-source heat pump and a gas-fired hot water boiler. The GHG Intensity of the electrical grid is 227 kg CO_2e/MWh. The gas-fired back-up boiler has an efficiency of 82%. The cost of electricity is $0.23/kWh, and the cost of gas is $19.16/GJ. Assume the emission factor for natural gas is 53 kgCO_2e/GJ. Emission factors for this example were obtained from the U.S. Environmental Protection Agency 2024 Emissions Factor Hub.

a) What is the minimum COP that the heat pump can operate at if the dual fuel hybrid system is optimized for GHG Intensity?

b) What is the minimum COP that the heat pump can operate at if the dual fuel hybrid system is optimized for economics?

c) If the facility were located in Detroit, Michigan, where the GHG intensity of the electrical grid is 555 kg CO_2e/MWh (U.S. EPA, 2024). What is the minimum COP that the heat pump can operate at if the dual fuel hybrid system is optimized for GHG Intensity?

d) The cost of electricity is $0.14/kWh and the cost of gas is $12.70/GJ. What is the minimum COP that the heat pump can operate at if the dual fuel hybrid system is optimized for economics?

Problem 8:

A natural gas boiler heating system is being converted to a natural gas absorption heat pump. The heat pump COP, boiler efficiency, and heating load vary based on outdoor air temperatures as provided in the table below. Estimate the annual fuel savings for this retrofit (GJ) and the annual GHG reduction (kg CO_2e). Assume the GHG emission factor from Chapter 2 (51.3 kgCO_2e/GJ HHV basis).

Outdoor air temperature ranges (°F)	**COP of natural gas heat pump**	**Baseline boiler efficiency (HHV basis)**	**Heating load GJ/hr**	**Hours**
59+	1.34	0.8500	620.99	2,405
54–59	1.29	0.8495	286.10	1,108
54–41	1.26	0.8473	851.83	3,299
41–28	1.21	0.8440	433.28	1,678
28–14	1.00	0.8405	65.84	255
14–(−4)	0.93	0.8365	3.87	15
−4 to −22	0.93	0.8320	0.00	0
< −22	0.93	0.8278	0.00	0
			2,261.9	8,760

Provide the solution by completing the table below.

Outdoor air temperature ranges (°F)	**Heating load GJ/hr**	**Hours**	**Baseline boiler fuel consumption. GJ**	**GAHP fuel consumption, GJ**	**Fuel savings, GJ**	**GHG emission savings, kg CO_2e**
59+	620.99	2,405				
54–59	286.10	1,108				
54–41	851.83	3,299				
41–28	433.28	1,678				
28–14	65.84	255				
14–(−4)	3.87	15				
−4 to −22	0.00	0				
< −22	0.00	0				
	2,261.9	8,760				

11

Building Envelope

Ali Muslim Syed

Applied Energy Consulting, Canada

Building envelope refers to the walls, roofs, floors, foundations, windows, and doors that enclose a building. The building envelope significantly impacts space heating, cooling, humidity control, and air conditioning. Depending on the weather, building envelope-related energy use can represent upwards of 60% of a building's annual purchased energy. Therefore, accurate estimation of thermal energy transmission from the building envelope is essential to ensure the right sizing of heating, ventilation, and air-conditioning (HVAC) equipment, lowering operational costs, mitigating greenhouse gas (GHG) emissions, and ensuring thermal comfort for the building occupants. As discussed in Chapter 2, investments in the improved building envelope can reduce the capital cost of energy efficiency or decarbonization equipment, reduce ongoing energy costs, and improve building resiliency.

This chapter introduces fundamental concepts of building envelope design related to heat, air, and moisture transmission. Common building envelope-related terminologies and heat transfer details across building envelopes are presented. The introduction to building infiltration, computational simulations, modeling, and energy efficiency considerations for building envelopes are also discussed. A discussion of building assemblies, material properties, and structural components as applied to the thermal envelope design is also provided.

11.1 Building Envelope Terminology

The following terminologies and definitions are commonly used to define the building envelope's heat and energy transfer characteristics.

Convection heat transfer coefficient (h). The convection heat transfer coefficient is the heat transfer per unit time and unit area induced by a unit temperature difference between the surface and the reference temperature in the surrounding environment, measured in W/m^2 °C or Btu/hr ft^2 °F.

Cooling degree days (CDD). CDD is the number of degrees in which a day's average temperature is above 18.3 °C or 65 °F (note that reference temperature can be adjusted in some cases).

Degree day (DD). A degree day is a measure of the comparison of outdoor temperatures for a location to a standard or base temperature (usually 18.3 °C or 65 °F). Degree Days = $\Delta T \times$ number of days, measured in °C days. Note that reference temperature can be adjusted in some cases.

Fenestration. Any glazed aperture in a building envelope. Fenestration includes glazing material, framing, and external and internal shading devices.

Heating degree days (HDD). HDD is the number of degrees Celsius a given day's mean temperature is below 18.3 °C or 65 °F (note that reference temperature can be adjusted in some cases).

Shading coefficient (SC). SC is the ratio or percentage of solar radiation incident on a glazing assembly that ends up inside a building as thermal energy, relative to the thermal energy gain for a 1-inch (3 mm) thick reference glass material. It can range from 0 to 1.

Sol-air temperature. Outside temperature, when in the absence of solar radiation, would give the same temperature distribution due to actual outdoor temperature plus incident solar radiation.

Solar heat gain coefficient (SHGC). SHGC is the percentage of solar radiation incident on a glazing assembly that ends up inside a building as thermal energy. It can range from 0 to 1.

Specific heat capacity. The heat required to change the temperature of a unit mass of material for a unit change in temperature, measured in J/kg °C or Btu/lbm °F.

Thermal conductance (C value). Thermal conductance is the reciprocal of thermal resistance, measured in W/m^2 °C or Btu/hr ft^2 °F.

Thermal conductivity (k). Thermal conductivity is the ability of a material to conduct heat, and it represents the quantity of thermal energy that flows per unit time through a unit area with a temperature gradient of 1° per unit distance, measured in W/m °C or Btu/hr ft °F.

Thermal emissivity (e). Thermal emissivity is the ratio of radiant flux emitted by a surface to that emitted by a black surface at the same temperature.

Thermal mass. Thermal mass is characterized by the mass and specific heat capacity of various building envelope components.

Thermal resistance (R-value). The R-value refers to the ability of insulation material to resist heat flow, measured in m^2 °C/W or hr ft^2 °F/Btu.

Thermal resistivity: Thermal resistivity is the reciprocal of thermal conductivity, measured in m °C/W or hr ft °F/Btu.

Thermal transmittance U or U-value. The overall heat transfer coefficient, or U-value, refers to how well heat is conducted over a series of resistant mediums. Its units are the W/m^2 °C or Btu/hr ft^2 °F.

Thermally heavy buildings: Buildings that have larger heat capacity and exhibit less change in energy consumption in response to changes in outside temperatures, especially for heating. This is due to their mass, specific heat capacity, and larger internal loads.

Thermally light buildings. Buildings that have a smaller heat capacity and exhibit relatively larger changes in energy consumption in response to changes in outside temperatures. Building envelopes are more important in thermally light buildings.

Visible transmittance (VT). VT is a fraction of the visible spectrum of sunlight transmitted through a window, door, or skylight glazing.

11.2 Fundamentals of Heat Transfer

Heat transfer or heat flow through the building envelope depends on its thermal resistance and the temperature difference across the envelope. Heat is thermal energy that flows due to a temperature difference, and always flows in the direction from hot to cold. There are three basic heat transfer modes: conduction, convection, and radiation.

The current chapter discusses the steady-state heat transfer modes to elaborate on the basic thermodynamic concepts. A detailed understanding of heat transfer requires transient simulations for accurate evaluation. Such analysis typically requires the use of heat transfer analysis software tools.

The three heat transfer modes are discussed below.

11.2.1 Conduction heat transfer

Conduction is the transfer of heat through solid materials such as walls, ceilings, roofs, window glass, floors, and doors from the hot side to the cooler side. Steady-state conduction is calculated through the following formula:

$$Q\left[W \text{ or } \frac{\text{Btu}}{\text{hr}}\right] = \frac{A\ (T_{\text{h}}-T_{\text{c}})}{\Sigma R},$$

where:

T_{h} = V hotter or higher temperature material (°C or °F)
T_{c} = colder or lower temperature material (°C or °F)
ΣR = sum of R values of individual layers of materials $\left(\frac{\text{m}^2\ °\text{C}}{\text{W}} \text{ or } \frac{\text{hr ft}^2\ °\text{F}}{\text{Btu}}\right)$
A = area (m^2 or ft^2).

If the R-value for a specific thickness is not known, conductivity (k value, W/m °C or Btu/hr ft °F) can be used for a given thickness (*L*) to find the overall R-value:

$$R = \frac{L}{k}.$$

In the case where conductance (C-value) is given, use:

$$C = \frac{1}{R}\left[\frac{W}{\text{m}^2\ °\text{C}} \quad \text{or} \quad \frac{\text{Btu}}{\text{hr ft}^2\ °\text{F}}\right].$$

Figure 11.1 shows a multi-layer construction assembly of three materials of different conductivity (k values), thicknesses (*L*), and temperature profiles across different layers.

To determine the conduction resistance of an assembly composed of multiple layers in series, the total thermal resistance (R_{total}) is simply calculated as the sum of the individual layers or components:

$$R_{\text{total}} = R_1 + R_2 + ... + R_{\text{N}},$$
$$i = 1, 2, 3, \ldots, \text{N}.$$

Steady-state thermal resistances (R-values) of building assemblies (walls, floors, windows, roof systems, etc.) are calculated from the thermal properties of the materials in the envelope assembly. ASHRAE Fundamentals (2025), Chapter 26, provides a comprehensive list of building assembly materials properties. Table 11.1 shows a sample of building materials' insulation properties.

11.2.2 Convection heat transfer

Convection heat transfer is associated with fluids (liquids or gases) in motion, generally through a pipe or duct or along a solid surface. Steady-state

Table 11.1 Insulation materials: design values (source: 2025 ASHRAE Handbook – Fundamentals, Guide to Energy Management: International Version – 8th ed).

	Material	Conductivity k W/m °C	Conductivity k Btu/hr ft °F	Thickness L mm	Thickness L inch	Thermal resistance R m^2 °C/W	Thermal resistance R hr ft^2 °F/Btu
Metals	Aluminum alloy 1,100	221	128				
	Bronze	29	17				
	Copper	393	227				
	Gold	298	172				
	Iron (cast)	48	28				
	Iron (wrought)	60	35				
	Nickel	60	35				
	Silver	424	245				
	Steel (mild)	45	26				
	Zinc, hot rolled	107	62				
Building boards	Asbestos cement	0.580	0.335				
	Gypsum or plasterboard	0.230	0.133	13	0.51	0.057	0.321
	Plywood	0.115	0.066	9.5	0.37	0.083	0.469
	Sheathing (impregnated or coated)	0.056	0.032	20	0.79	0.360	2.042
	Wood fiber – hardboard type	0.020	0.012				
Blanket and batt insulation	Mineral wool fibers (rock, slag, glass)	0.039	0.023				
	Wood fiber	0.036	0.021				
Board and slab insulation	Cellular glass	0.056	0.032				
	Corkboard	0.039	0.023				
	Glass fiber	0.036	0.021				
	Insulating roof deck	0.052	0.030	51	2.0	0.981	5.567

Table 11.1 *Continued.*

	Material	Conductivity k W/m °C	Conductivity k Btu/hr ft °F	Thickness L mm	Thickness L inch	Thermal resistance R m^2 °C/W	Thermal resistance R hr ft^2 °F/Btu
Board and slab insulation	Cellular glass	0.056	0.032				
	Corkboard	0.039	0.023				
	Glass fiber	0.036	0.021				
	Insulating roof deck	0.052	0.030	51	2.0	0.981	5.567
Loose-fill insulation	Mineral wool fibers (rock, slag, glass)	0.039	0.023				
	Vermiculite (expanded)	0.066	0.038				
Masonry: concrete materials	Cement mortar	0.720	0.416				
	Lightweight aggregates	0.245	0.142				
	Sand and gravel or stone	1.730	1.000				
	Aggregate stucco	0.720	0.416				
Masonry: brick, tile, block, and stone	Brick, common	0.720	0.416				
	Brick, face	1.300	0.751				
	Tile, hollow clay, 1 cell deep	0.520	0.300	102	4.0	0.196	1.113
Masonry: block, concrete, 3 oval cores	Sand and gravel aggregate	0.806	0.466	102	4.0	0.127	0.718
	Sand and gravel aggregate	0.780	0.451	153	6.0	0.196	1.113
	Cinder aggregate	0.520	0.300	102	4.0	0.196	1.113
	Cinder aggregate	0.505	0.292	153	6.0	0.303	1.720
	Stone, lime, or sand	1.800	1.040				
Plastering materials	Cement plaster, sand aggregate	0.720	0.416				
	Gypsum: lightweight Aggregate	0.230	0.133				
	gypsum: perlite Aggregate	0.220	0.127				
	gypsum: sand aggregate	0.810	0.468				

Table 11.1 *Continued.*

	Material	Conductivity k W/m °C	Conductivity k Btu/hr ft °F	Thickness L mm	Thickness L inch	Thermal resistance R m^2 °C/W	Thermal resistance R hr ft^2 °F/Btu
	Gypsum: sand aggregate metal lathe	0.830	0.480				
	Gypsum: vermiculite aggregate	0.250	0.144				
Roofing materials	Asphalt roll roofing	0.701	0.405	19	0.7	0.0271	0.154
	Built-up roofing	0.162	0.093	9.5	0.4	0.0588	0.334
Siding materials	Asbestos-cement	0.175	0.101	6.5	0.3	0.0371	0.211
	Asphalt insulating	0.510	0.295	13	0.5	0.0255	0.145
	Wood, bevel, 13 mm × 152 mm, lapped	0.091	0.053	13	0.5	0.1429	0.811
Wood	Maple, oak, and similar Hardwoods	0.160	0.092				
	fir, pine, and similar softwoods	0.116	0.067	20	0.8	0.1724	0.979

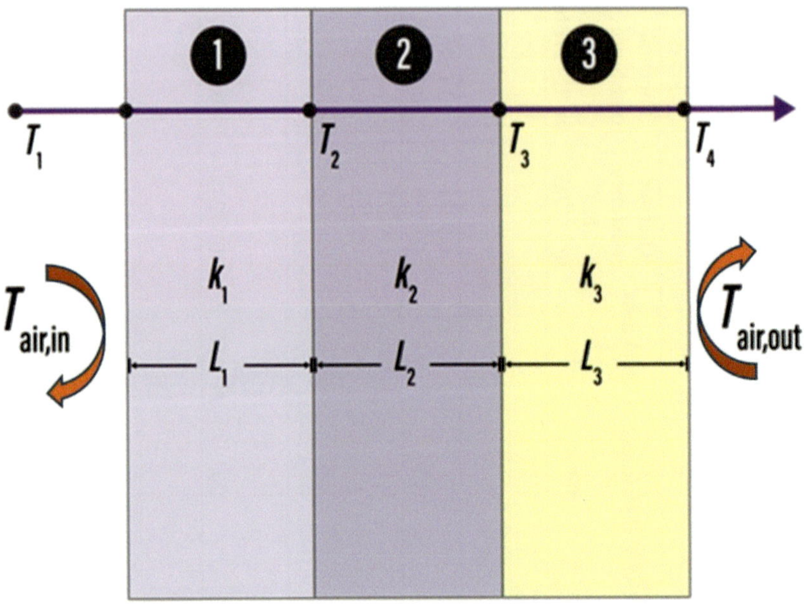

Figure 11.1 Heat conduction through a multi-layer construction.

convection heat transfer is expressed as:

$$Q\left[W \text{ or } \frac{\text{Btu}}{\text{hr}}\right] = \text{hA}\ (T_{\text{h}} - T_{\text{c}}),$$

where:

h = film coefficienct $\left[\frac{W}{\text{m}^2\ {}^\circ\text{C}} \text{ or } \frac{\text{Btu}}{\text{hr ft}^2\ {}^\circ\text{F}}\right]$

T_c = colder temperature, fluid or solid surface (°C or °F)

T_h = hotter temperature, fluid or solid surface (°C or °F)

A = area (m^2 or ft^2).

Refer to Figure 11.1 which shows a multi-layer construction assembly and the conduction heat transfer from the wall assembly to the surrounding films or layer of air.

For flat surfaces, the surface film coefficient is the amount of heat transferred from a surface to a gas (such as air) or from a gas to a surface per unit area of surface for each degree of temperature difference. Surface film coefficients for air are specified in terms of the surface resistance in the 2025 ASHRAE Handbook – Fundamentals, Chapter 26. In addition, some common values of convective resistance for surface films are given in Table 11.2.

Table 11.2 Convective resistance of air films.

	Direction of flow	R_{conv} m^2 °C/W	R_{conv} hr ft^2 °F/Btu
Still air (inside)			
Ceiling air film	Up	0.10	0.56
Floor air film	Down	0.17	0.96
Wall air film	Horizontal	0.13	0.74
Moving air (outside)			
25km/h (23 ft/s)	All	0.03	0.17
14km/h (13 ft/s)	All	0.04	0.23
7.0 km/h (6.4 m/s)	All	0.06	0.34

11.2.3 Radiation heat transfer

Radiation is thermal energy transfer by electromagnetic waves, so no transfer medium is necessary. The following formula gives the radiation power of a surface at temperature T:

$$Q\left[W \text{ or } \frac{\text{Btu}}{\text{hr}}\right] = \varepsilon A \sigma T^4$$

where:

ε = total emissivity (range $0 \leq \epsilon \leq 1$, where $\epsilon = 1.0$ for a black body)
$\sigma = 5.670 \times 10^{-8} \frac{\text{W}}{\text{m}^2\text{K}^4} = 0.1713 \times 10^{-8} \frac{\text{Btu}}{\text{hr ft}^2\text{R}^4}$ is the Stefan–Boltzmann constant
A = area (m^2 or ft^2)
T = absolute temperature (Kelvin, K or Rankine, R).

Note: calculation must use absolute temperature (do not use °C or °F).

11.2.4 Overall heat transfer calculation

Heat transfer through various mediums in series, for example, heat flow from the air on one side of a wall to the air on the other side can be analyzed using a thermal resistance network as illustrated in Figure 11.2 below. The total resistance for the system described above may be calculated from all the component resistances:

$$R_{\text{total}} = R_{\text{conv1}} + R_{\text{wall}} + R_{\text{conv2}}.$$

Heat transfer is inherently three-dimensional, and can occur through resistance in parallel. For example, the heat lost from the outer surface of a pipe can occur due to both the convective and radiative heat transfer mechanisms as illustrated in Figure 11.3.

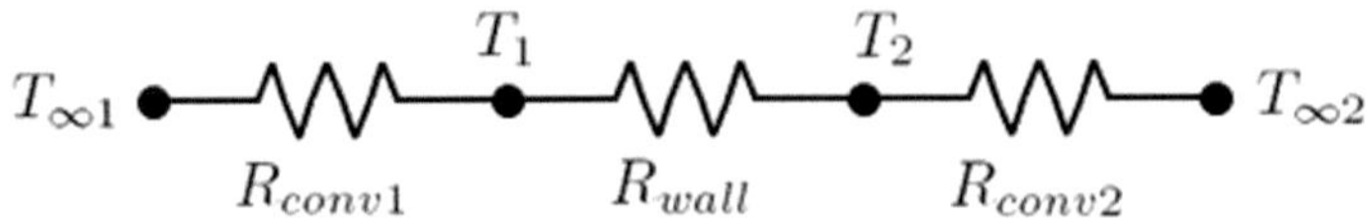

Figure 11.2 Thermal resistance in series.

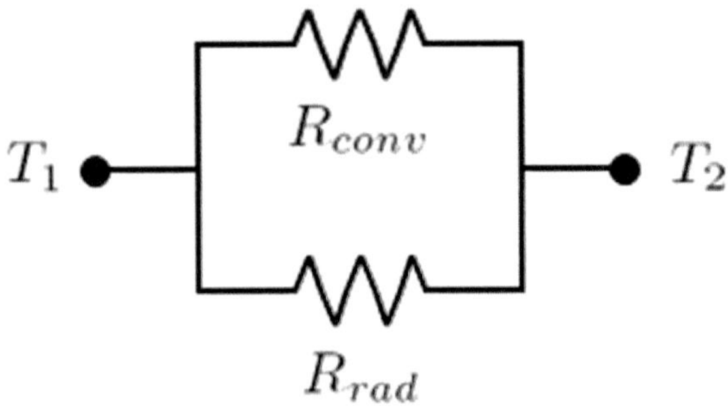

Figure 11.3 Thermal resistance in parallel.

The total thermal resistance for such an assembly is calculated by adding the inverse resistance of the two components, as follows:

$$\frac{1}{R_{\text{total}}} = \frac{1}{R_{\text{conv}}} + \frac{1}{R_{\text{rad}}}.$$

The total resistance for the system described above may be calculated from all the component resistances, as shown in Figure 11.4.

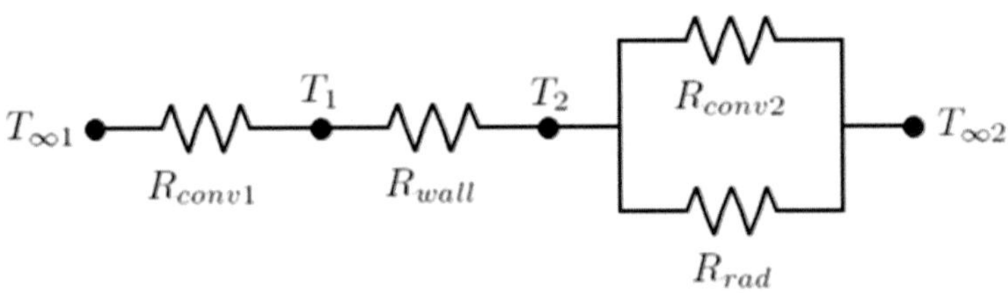

Figure 11.4 Total thermal resistance.

$$R_{\text{total}} = R_{\text{conv1}} + R_{\text{wall}} + \frac{R_{\text{conv2}}\, R_{\text{rad}}}{R_{\text{conv2}} + R_{\text{rad}}}.$$

Using the information provided above, the final heat transfer equations can be constructed as follows:

$$Q = \frac{1}{R_1 + R_2 + \cdots + R_{\text{N}}} \text{A}\ \Delta\text{T},$$

$$Q = \frac{1}{R_{\text{total}}} A\ \Delta T\ ,$$

$$Q_{\text{total}} = U\ A \Delta T,$$

where:

Q_{total} = rate of heat transfer for total surface area, W or Btu/hr
A = area of heat transfer surface, m^2 or ft^2
U = overall thermal conductance, $W/m^2\ °C$ or Btu/hr $ft^2\ °F$
UA = overall heat transfer coefficient, W/°C or Btu/hr °F.

For building applications, the UA value is sometimes called the "building load coefficient."

Example 11-1

A 1 cm thick (0.01 m) uninsulated mild steel tank storing a hot fluid, is shown in Figure 11.5. The fluid is heated to 90 °C, while ambient air is 20°. Estimate the heat loss from the uninsulated tank. Ignore the heat loss to the ground, and assume this cylindrical tank is large enough to be reasonably approximated as a flat surface. Assume the film resistance from the fluid to the inside wall of the tank to be zero. Lastly, assume a convective resistance of 0.088 $m^2\ °C/W$ from the tank wall and roof to the surrounding air.

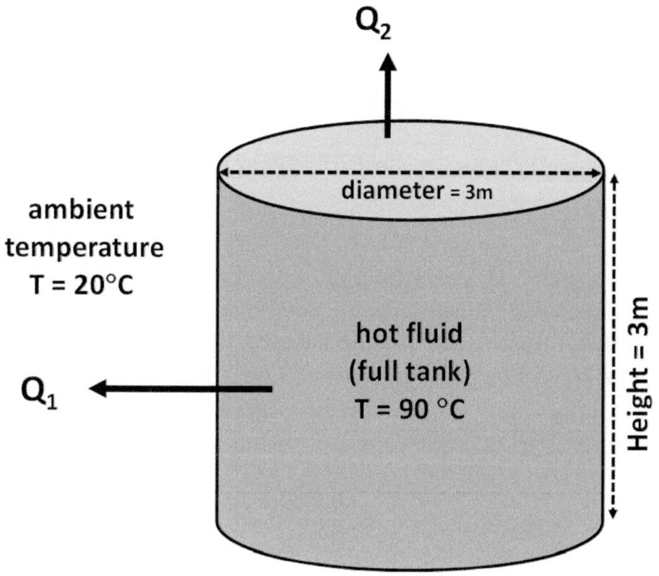

Figure 11.5 Tank storage of hot fluid.

The tank is 3 meters in diameter and 3 meters tall. Although this cylindrical tank is not truly a flat surface, it is large enough to be reasonably approximated as a flat surface. Smaller diameter cylinders, such as pipes, must be treated differently, as explained in the next section.

There are two primary sources of heat loss shown in Figure 11.5. Q_1 is the loss through the tank's walls, and Q_2 is the loss through the roof. Although the surface film coefficient will vary somewhat for vertical heat flow versus horizontal heat flow, we will ignore that here and assume that the $R_{\text{surface coefficient}}$ is equal for the roof and walls:

$$R_{\text{total}} = R_{\text{tank}} + R_{\text{surface coefficient}}.$$

The thermal resistance of the steel tank wall (obtained from Table 11.1) is quite small and is usually ignored, but we include it below for demonstration purposes. Note how small it is compared to the surface resistance:

$$R_{\text{tank}} = \frac{L}{k} = \frac{0.01\text{m}}{45\ \text{W/m}^2\ {}^\circ\text{C}} = 0.000222\frac{\text{m}^2\ {}^\circ\text{C}}{W}.$$

Assuming that the surface is dull metal, and the surface temperature is 50 °C, $R_{\text{surface coefficient}} = 0.088\ \text{m}^2/{}^\circ\text{C W}$:

$$U = \frac{1}{R_{\text{surface}} + R_{\text{tank}}} = \frac{1}{0.088222} = 11.34\frac{\text{W}}{\text{m}^2\ {}^\circ\text{C}}.$$

The total area is the area of the cylinder plus the area of the circular roof:

$$A = \text{area} = \pi\text{DH}+\pi\text{r}^2 = \pi(3\text{m})\ (3\text{m}) + \pi\ (1.5\text{m})^2 = 35.34\ \text{m}^2$$

$$\begin{aligned} Q_{\text{total}} &= \text{UA}\Delta\text{T} = \frac{11.34\ \text{W}}{\text{m}^2\ {}^\circ\text{C}}\left(35.34\ \text{m}^2\right)(90\ {}^\circ\text{C}{-}20\ {}^\circ\text{C}) \\ &= 28{,}045\ \text{W or } 28.05\ \text{kW}. \end{aligned}$$

To see the impact that insulation can have on these calculations, assume that aluminum-jacketed fiberglass insulation 2 cm (0.02 m) thick is added so that it covers the tank. Using Table 11.1, $k = 0.039$ W/m °C for fiberglass at a mean temperature of around 60 °C. Then, the new heat loss would be:

$$R_{\text{insulation}} = \frac{L}{k} = \frac{0.020}{0.039} = 0.5128$$

$$U = \frac{1}{R_{\text{tank}} + R_{\text{insulation}} + R_{\text{surface}}}$$
$$= \frac{1}{0.00022 + 0.5128 + 0.088} = \frac{1}{0.6010}$$
$$U = \frac{1}{0.6010} = 1.664$$
$$Q_{\text{total}} = \text{UA}\Delta T = \frac{1.664\ \text{W}}{\text{m}^2\ {}^\circ\text{C}}\left(35.34\ \text{m}^2\right)(90\ {}^\circ\text{C}{-}20\ {}^\circ\text{C})$$
$$= 4,117\ \text{W}, \text{Savings} = 28,053{-}4,117 = 23,930\ \text{W}, \%\ \text{Savings}$$
$$= \frac{23,936}{28,053} = 0.853 \text{ or } 85.3\%$$

Note that this is 85% saving for only 2 cm of insulation. For 4 cm of insulation, the heat loss would be:

$$Q_{\text{total}} = \text{UA}\Delta T = \frac{1}{0.00022 + 1.0256 + 0.088}(35.34)(90 - 20)$$
$$= 2,221\ \text{W}$$
$$\text{Savings} = 28,053 - 2,221 = 25,832\ \text{W}$$
$$\%\ \text{savings} = \frac{25,832}{28,053} = 0.921 \text{ or } 92.1\%$$

Example 11-1 demonstrates a characteristic of the cost-effectiveness of insulation. The first increment of insulation thickness yields by far the largest energy savings. Each additional increment of thickness increases the energy savings but rapidly decreases the cost-effectiveness. At some point, it becomes uneconomical to add any additional insulation. This will be discussed later in this chapter.

11.3 Infiltration

Infiltration is the uncontrolled air leakage through the building cracks and openings caused by wind pressure, pressure differences, temperature differences, and a difference between outside and indoor air densities. Air and moisture infiltration into a building can be a significant heating or cooling load source. In addition, uncontrolled moisture infiltration can potentially cause structural damage to a building over time.

Estimating air infiltration in the building envelope can be challenging due to multiple factors. However, two simplified approaches traditionally used are:

- Air change method
- Crack method

Air-change method. Air infiltration is estimated based on the assumed number of air changes per hour for a given building. Most of the air change rates are based on experience from measurement data on similar structures and the assessor's experience. Infiltration air change per hour (ACH) that a building experience depends highly on the building type and condition. Its age and general condition are highly subjective to the assessor's experience. The range of ACH can vary from 0.5 (very low) to 2.0 (very high).

The infiltration rate using the ACH method can be calculated using the following formula:

$$Q = \frac{(\mathrm{ACH})\,(V)}{C_{\mathrm{T}}}$$

where:

Q = infiltration rate in m^3/s or cfm
ACH = number of air changes per hour
V = gross space volume in m^3 or ft^3
C_T = constant of 3,600 for SI units or 60 for English units.

Crack method. Estimating air infiltration based on the characteristics of the windows, walls, doors, and pressure difference inside and outside the building. 2025 ASHRAE Handbook – Fundamentals, Chapter 16, Ventilation and Infiltration, states that the crack method is relatively more accurate among these two methods. The main problem in an accurate estimation of air infiltration is due to multiple structural components, prediction of pressure differences across surfaces, the impact of stack effect in taller buildings, lack of accurate information of how many times the doors open and close, etc.

The infiltration flow rate can be calculated using the following formula.

$$Q = A\,C(\Delta P)^n$$

where:

Q = infiltration rate in L/s or cfm for English units
A = effective leakage area of the cracks
C = flow coefficient in (L/s)/Pa^n or (cfm)/(inch of water)^n for English units

ΔP = pressure difference between outside and inside of a building in Pascals or inches of water)

n = exponent defining the nature of the flow in the crack, $0.4 < n < 1.0$, dimensionless.

ΔP used for the crack method itself depends on factors such as:

- Pressure difference due to wind
- Stack effect
- Relative pressurization within each zone of the building
- Presence of vestibules
- Operation of mechanical equipment (i.e., supply fans, exhaust fans, and vented combustion devices impact pressure difference across the building and the air change rate)
- Traffic through doors, etc.

Similarly, building features like internal partitions, stairwells, elevator shafts, smoke control zones, vents, chimneys, doors, etc. play a vital role in accurately estimating building pressure differences.

Example 11-2

Consider an automatic door with an area of 2 m^2 (21 ft^2) and n of 0.5. Assuming that for the building height and wind speed, the ΔP is 50 Pa (0.20 inch of water). Assuming 150 people per hour pass through the door, using Figure 11 Airflow Coefficient of Automatic Doors of 2025 ASHRAE Handbook – Fundamentals, Chapter 16, the C for doors without vestibules is approximately 170 L/(s m^2 Pa^n) or 540 cfm/[ft^2 (inch of water)^n]. Therefore, airflow through the automatic doors is:

$$Q = (2)\,(170)\ \ (50)^{0.5} = 2,404\,\frac{\mathrm{L}}{\mathrm{s}},\ \text{or}\ \ (21)\,(540)(0.2)^{0.5} = 5,071\ \text{cfm}.$$

For more details on determining the flow coefficient C for specific building geometry and operating conditions, 2025 ASHRAE Handbook – Fundamentals, Chapter 16, Ventilation and Infiltration, and Chapter 24, Airflow Around Buildings are recommended.

The 2025 ASHRAE Handbook – Fundamentals, Chapter 16 states that measuring the building infiltration air change rate is the only accurate way to determine the building infiltration air change rate. The most widely used method to measure infiltration air change rate is the trace method as dictated by ASTM Standard E741.

There are a few other standards worldwide that detail how to perform this test. Some of these include:

- The United States Army Corps of Engineers Air Leakage Test Protocol for Building Envelopes
- ISO 9972:2006 – Thermal performance of buildings – Determination of air permeability of buildings – Fan pressurization method
- ASTM E779 – Standard Test Method for Determining Air Leakage Rate by Fan Pressurization
- ASTM E1827 – Standard Test Methods for Determining Airtightness of Buildings Using an Orifice Blower Door
- ASTM E3158 – Standard Test Method for Measuring the Air Leakage Rate of a Large or Multizone Building
- ATTMA – Measuring Air Permeance of Building Envelopes (Dwellings)
- ATTMA – Measuring Air Permeance of Building Envelopes (Non-dwellings)

An accurate infiltration estimation for each project should be conducted based on the budget and accuracy required. The 2025 ASHRAE Handbook – Fundamentals, Chapter 16 provides some approximate infiltration rates for existing commercial buildings of 0.61 to 2.44 L/s/m^2 at 75 Pa (0.12 to 0.48 cfm/ft^2 at 0.30 inches of water). However, according to Emmerich and Persily (2005), existing commercial buildings in the U.S. have higher infiltration rates approximately 28.4 m^3/h m^2 (approximately 7.89 L/s/m^2 or 1.55 cfm/ft^2) at 75 Pa (0.30 inches of water).

Recent versions of ASHRAE 90.1 and ASHRAE 189.1 require continuous air barriers for commercial buildings. For example, ASHRAE 90.1 requires that the measured air leakage rate of the building envelope should not exceed 2.0 L/s per m^2 across a pressure difference of 75 Pa. Similarly, the maximum allowable leakage rates for different fenestration and door components, ASHRAE 90.1, 2022, prescribes the following maximum air leakage rates (refer to Section 5, ASHRAE Standard 90.1, 2022 for more details). Typical air leakage rates are shown in Table 11.3.

Table 11.3 Air leakage rates for fenestration and doors (source: ASHRAE 90.1, 2022).

Envelope component	Maximum leakage rate, L/s/m^2 @ 75 Pa	Maximum leakage rate, cfm/ft^2 @ 0.3-inch water
Swinging entrance doors, power-operated sliding doors, revolving doors	5.1	1.0
Curtainwall, storefront glazing	0.3	0.06
Skylights	1.5	0.3
Non-swinging overhead bay doors	6.6	1.3

For additional information on infiltration rates for old buildings, refer to Table 41, Chapter 6, Infiltration and Ventilation, Part I Load Estimation (Carrier, 1972).

11.4 Insulation Value of Windows

The overall window heat transfer consists of heat transfer paths through the center of the glass, the edge of the glass, and the window frame. Hence, it is imperative to estimate the overall U-value of fenestration that accounts for the combined effect of all heat transfer paths of a window. For a given fenestration, the center of glass and edge of glass U values are very different from the overall U value. ASHRAE Fundamentals (2025), Chapter 15 Fenestration provides a list of glazing U values for common types of window assemblies, with a sample shown in Table 11.4.

However, depending on the complexity of the window assembly construction, an accurate estimation of the effective U value of an assembly may require numerical simulation in software like THERM. THERM is a two-dimensional conduction heat-transfer analysis tool developed by the Lawrence Berkeley National Laboratory. THERM uses the finite-element method, which can model the complicated geometries of building products. THERM is available at https://windows.lbl.gov/software/therm.

11.5 Window Rating Method

The National Fenestration Rating Council provides window rating methods as a label. The council labels outline the energy performance ratings of a window in terms of U-value, solar heat gain coefficient, visual transmittance, and air leakage rate. Optimal window selection must be based on the energy performance, incorporating the design parameters and feedback from all concerned disciplines of the design team. Figure 11.6 provides an example of a window rating.

11.6 Cool Roof and Heat Island Reduction

Dark, non-reflective surfaces used for roofs absorb the sun's heat and radiate it, creating heat islands. According to the U.S. Environmental Protection Agency, urban areas can have temperatures of 1 to 3 °C warmer than surrounding suburban and undeveloped areas and as much as 12 °C warmer in

Table 11.4 U-value of various fenestration products.

						Glass + frame (National Fenestration Rating Council) U-value Btu/hr ft2 °F				
DOE-2 glass library entry	**Glass type code**	**# of pane**	**Gap thick-ness (inch)**	**Gap glass fill**	**Center glass U-value Btu / hr ft^2 °F**	**Aluminum no break**	**Aluminum with break**	**Aluminum clad wood**	**Wood vinyl**	**Fiberglass**
Double clear	2001	2	0.50	Air	0.49	0.65	0.58	0.51	0.51	0.49
Double clear	2002	2	0.50	Argon	0.46	0.62	0.55	0.49	0.49	0.46
Double clear	2003	2	0.25	Air	0.56	0.75	0.67	0.58	0.58	0.56
Double clear	2004	2	0.50	Air	0.48	0.64	0.57	0.50	0.50	0.48
Double clear	2005	2	0.50	Argon	0.45	0.61	0.54	0.48	0.48	0.45
Double low iron	2006	2	0.25	Air	0.57	0.76	0.68	0.59	0.59	0.57
Double low iron	2007	2	0.50	Air	0.49	0.65	0.58	0.51	0.51	0.49
Double low iron	2008	2	0.50	Argon	0.46	0.62	0.55	0.49	0.49	0.46
Double low iron	2009	2	0.25	Air	0.56	0.75	0.67	0.58	0.58	0.56
Double low iron	2010	2	0.50	Air	0.49	0.65	0.58	0.51	0.51	0.49
Double low iron	2011	2	0.50	Argon	0.45	0.61	0.54	0.48	0.48	0.45
Double tint bronze	2200	2	0.25	Air	0.57	0.76	0.68	0.59	0.59	0.57
Double tint bronze	2201	2	0.50	Air	0.49	0.65	0.58	0.51	0.51	0.49
Double tint bronze	2202	2	0.50	Argon	0.46	0.62	0.55	0.49	0.49	0.46
Double tint bronze	2203	2	0.25	Air	0.56	0.75	0.67	0.58	0.58	0.56
Double tint bronze	2204	2	0.50	Air	0.48	0.64	0.57	0.50	0.50	0.48
Double tint bronze	2205	2	0.50	Argon	0.45	0.61	0.54	0.48	0.48	0.45
Double tint green	2206	2	0.25	Air	0.57	0.76	0.68	0.59	0.59	0.57
Double tint green	2207	2	0.50	Air	0.49	0.65	0.58	0.51	0.51	0.49
Double tint green	2208	2	0.50	Argon	0.46	0.62	0.55	0.49	0.49	0.46
Double tint green	2209	2	0.25	Air	0.56	0.75	0.67	0.58	0.58	0.56
Double tint green	2210	2	0.50	Air	0.48	0.64	0.57	0.50	0.50	0.48

Table 11.4 *Continued.*

						Glass + frame (National Fenestration Rating Council) U-value Btu/hr ft2 °F				
DOE-2 glass library entry	**Glass type code**	**# of pane**	**Gap thick-ness (inch)**	**Gap glass fill**	**Center glass U-value Btu / hr ft^2 °F**	**Aluminum no break**	**Aluminum with break**	**Aluminum clad wood**	**Wood vinyl**	**Fiberglass**
Double tint green	2211	2	0.50	Argon	0.45	0.61	0.54	0.48	0.48	0.45
Double tint grey	2212	2	0.25	Air	0.57	0.76	0.68	0.59	0.59	0.57
Double tint grey	2213	2	0.50	Air	0.49	0.65	0.58	0.51	0.51	0.49
Double tint grey	2214	2	0.50	Argon	0.46	0.62	0.55	0.49	0.49	0.46
Double tint grey	2215	2	0.25	Air	0.56	0.75	0.67	0.58	0.58	0.56
Double tint grey	2216	2	0.50	Air	0.48	0.64	0.57	0.50	0.50	0.48
Double tint grey	2217	2	0.50	Argon	0.45	0.61	0.54	0.48	0.48	0.45
Double tint blue	2218	2	0.25	Air	0.56	0.75	0.67	0.58	0.58	0.56
Double tint blue	2219	2	0.50	Air	0.48	0.64	0.57	0.50	0.50	0.48
Double tint blue	2220	2	0.50	Argon	0.45	0.61	0.54	0.48	0.48	0.45

Figure 11.6 National Fenestration Rating Council label (source: www.nfrc.org).

the evenings. This phenomenon is called the heat island effect. Heat islands increase cooling loads in the summer, necessitating larger, more powerful air-conditioners that use more electricity, increasing cooling costs, producing more greenhouse gases, and generating pollution.

Some strategies help mitigate the heat island effect, such as installing solar panels on shading devices or using a vegetated roof to insulate a building and extend the roof's life. However, the most effective measure is installing a cool roof. A cool roof features a highly reflective outer surface that reduces the amount of heat conducted through the roof. A cool roof has high "solar reflectance" (i.e., ability to reflect sunlight on a scale of 0 to 1) and high "thermal emissivity" (i.e., ability to emit thermal radiation on a scale of 0 to 1), as shown in Figure 11.7.

On a sunny day, a cool roof typically lowers the roof temperature, thereby decreasing the amount of heat transferred into a building. Typical benefits of a cool roof include:

- Downsized air-conditioning equipment: A cool roof can reduce peak cooling demand by up to 40% in warm climates.
- Extended roof life: Cool roofs tend to last longer because they are less susceptible to thermal expansion and contraction.
- Reduced heat island effect.

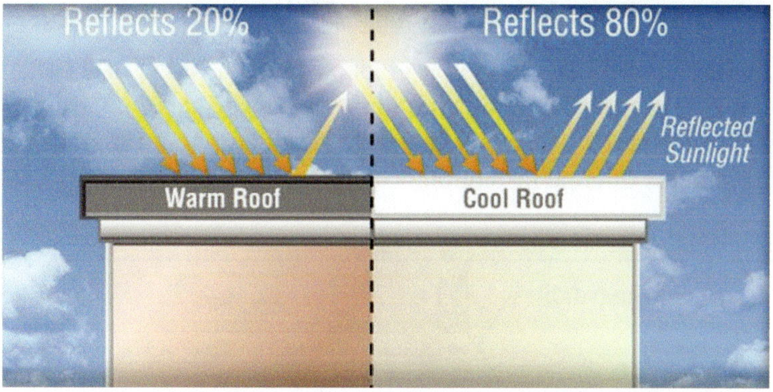

Figure 11.7 Cool roofs reflect sunlight (source: Lawrence Berkeley National Laboratory).

Table 11.5 Solar radiation properties of roofing materials (source: Lawrence Berkeley National Laboratory).

Material	Albedo (%)	Emittance (%)	Solar reflectance index (SRI)
White asphalt shingles	21	91	21
Black asphalt shingles	5	91	1
White granular-surface bitumen	26	92	28
Red clay tile	33	90	36
Red concrete tile	18	91	17
Unpainted concrete tile	25	90	25
White concrete tile	73	90	90
Galvanized steel unpainted	61	4	37
Aluminum	61	25	50
Siliconized white polyester over metal	59	85	69
Polyvinylidene fluoride (PVDF) white over metal	67	85	80
Black ethylene propylene diene ter-polymer (EPDM)	6	86	-1
Gray EPDM	23	87	21
White EPDM	69	87	84
T-EPDM	81	92	102
Chlorosulfanated polyethylene (CSPE) synthetic rubber	76	91	95

Cool roofs are scored based on the solar reflectance index (SRI) of a roofing material's ability to reject solar heat. SRI values of different roofing materials are given in Table 11.5.

11.7 Heat Loss Considerations for Slab on Grade

Most of the heat loss for the floors on grade is from the edge of a concrete floor slab. Hence, proper insulation around the slab's perimeter is essential in colder climates. Figure 11.8 shows heat loss factors for slab on grade for a range of placements of edge insulation.

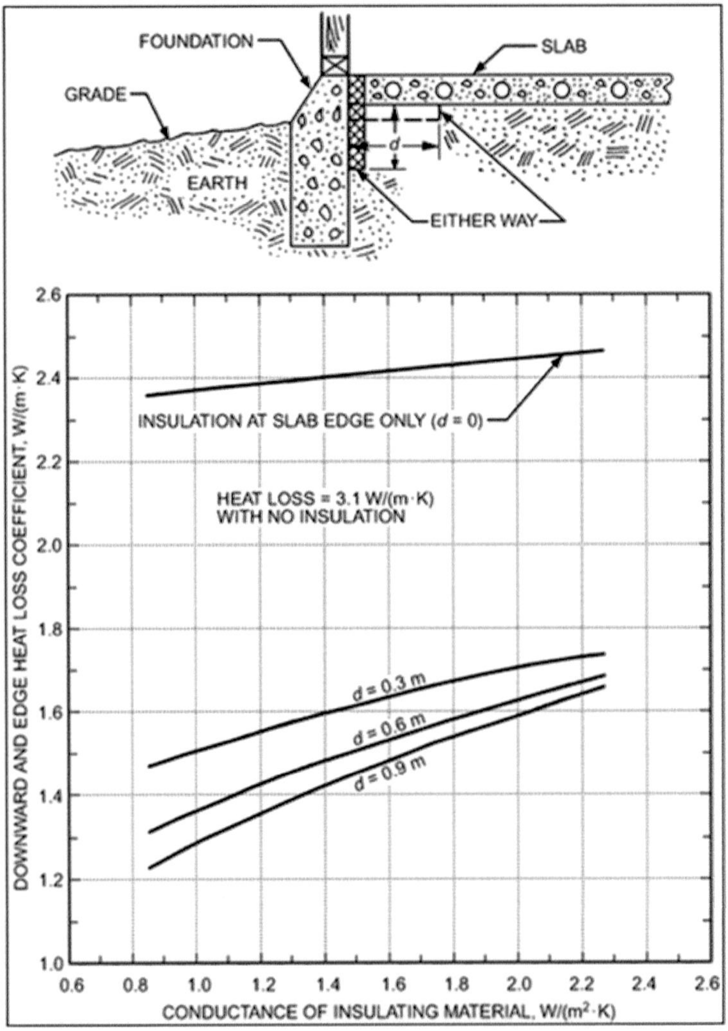

Figure 11.8 Downward and edgewise heat loss coefficient for concrete floor slabs on grade (source: ASHRAE Handbook, Systems, and Equipment, 2020).

ASHRAE Standard 90.1, 2022, provides the recommended slab edge insulation for slab-on-grade floors for myriad design configurations. Refer to ASHRAE 90.1 2022 for slab-on-grade floor insulation requirements.

11.8 Moisture or Hygrothermal Load Considerations for Envelope

In addition to the thermal (temperature-driven) loads, the building envelope is also subjected to moisture-related or hygrothermal loads. Therefore, the influence of the hygrothermal behavior of building assemblies must be considered for the building envelope design and the thermal loads. As recommended by (Künzel and Karagiozis, 2004), building envelope design (new and retrofits) should entail moisture control and mitigation measures. As per ASHRAE Fundamentals (2025), building envelope hygrothermal loads demonstrate diurnal and seasonal variations at the exterior surface and

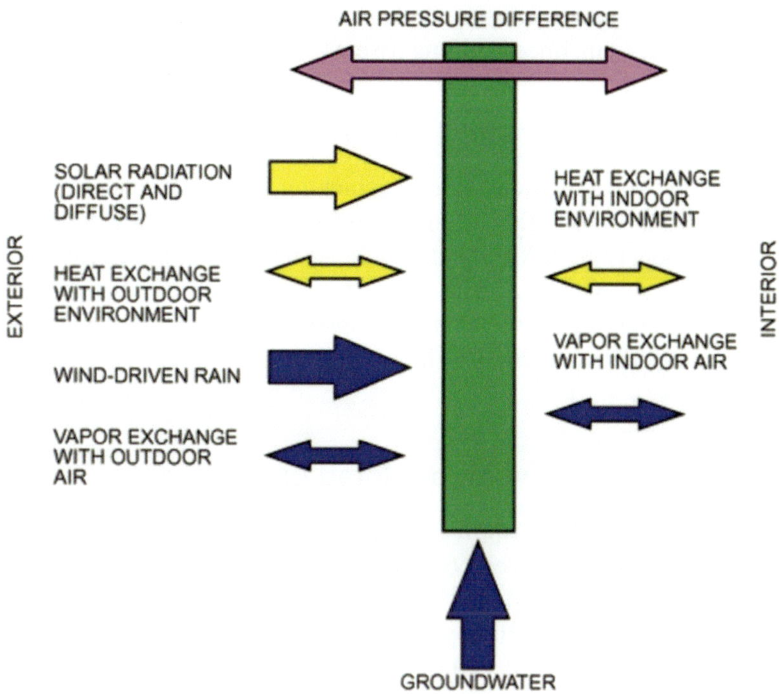

Figure 11.9 Moisture loads acting on building envelope (source: ASHRAE Handbook, Fundamentals, 2025).

mainly seasonal variations at the interior surface. Hence, it is incumbent to perform a precise analysis of the expected loads before designing a building envelope component. As shown in Figure 11.9, numerous factors determine the hygrothermal load on building envelope assembly, namely:

- Ambient temperature and humidity
- Indoor temperature and humidity
- Solar radiation
- Exterior condensation
- Long-wave radiant effects.
- Interior temperature differential.
- Wind-driven rain
- Construction moisture
- Ground and surface water
- Air pressure differentials

However, determining local conditions at the envelope's surface is complicated and requires specific experience. Computer simulations are sometimes necessary to assess the microclimate acting on differently oriented, overhang-protected, or inclined building assemblies.

11.9 Indoor Thermal Comfort and Building Envelope

Building envelope plays a crucial role in achieving indoor thermal comfort. Thermal comfort is a condition of mind expressing satisfaction with the thermal environment (ASHRAE Standard 55). As a part of building envelope design, it is important to consult with the design team to ensure that thermal comfort considerations are accounted for. Chapter 3 includes a discussion and some examples of understanding and improving thermal comfort for facility retrofits.

11.10 Estimation of Envelope-related Energy Use

Envelope-related energy use requires a detailed numerical analysis of transient heat transfer from the envelope over time (usually a season or a year), using a given time interval (usually hours or minutes). Several energy simulation software tools can be employed to do a detailed analysis of building energy use. ASHRAE Standard 140 (2020) specifies a standard test method for evaluating software's technical capabilities and applicability in calculating the thermal performance of buildings and their HVAC systems. A detailed comparison of energy simulation tools for building energy use (including those related to envelope) is presented in Syed (2020) and Syed (2021).

However, for an approximation of envelope-related energy use based on weather, a simplified approach can be used to estimate a building's energy use (generally accurate for most energy study requirements). This method is called the degree-day method. It should be noted that a degree-day method does not replace a rigorous ASHRAE Standard 140-compliant simulation.

Annual or seasonal energy loss through the envelope using the degree-day method can be estimated using:

$$E = 24\ \mathrm{U\ A\ DD} \left[\frac{\mathrm{Wh}}{\mathrm{yr}} \text{ or } \frac{\mathrm{Btu}}{\mathrm{yr}}\right],$$

where:

E = Energy, Wh/yr or Btu/yr

UA = Building load coefficient (product of U and Area), W/°C or Btu/ft^2 hr °F

DD = number of heating or cooling degree days, °C.day/yr or °F day/yr

Degree day is a measure of comparing outdoor temperatures for a location to a standard or base temperature. Degree days = $\Delta T \times$ number of days, measured in °C.days or °F.days.

The standard or base temperature for estimating degree days varies, though a typical base of 18.3 °C (65 °F) is commonly used. A common source of degree days calculation for locations worldwide is www.degreedays.net.

An alternative to using a fixed reference temperature of 18.3 °C (65 °F) is to use a facility-specific balance point temperature (BPT). BPT is the temperature where the building heat gains equal losses (ASHRAE Handbook Fundamentals, 2025, chapter 19). In principle, below the BPT, the facility requires mechanical heating, and above the BPT, it requires mechanical cooling. Chapter 3 includes an example of how to estimate the BPT graphically for a natural gas-heated building.

Example 11-3

Assume a building wall has an R-value of 0.47 m^2 °C/W (470 m^2 °C/kW). The temperature inside is kept at 20 °C in winter and 25 °C in summer. The plant operates 365 days/year, 24 hours/day. Assuming a heating plant efficiency of 0.80 and a cooling coefficient of performance (COP) of 2.5, estimate the cost of energy loss through the wall. Electricity costs $0.072/kWh and gas $10.00/GJ ($0.036/kWh). The building experiences 2,200 °C HDDs and 1,100 °C CDDs annually. The total wall area is 300 m^2:

$$\begin{aligned}
\text{Winter heating cost} &= \left(\frac{2,200\ ^\circ\text{C days}}{\text{year}}\right)\left(\frac{\text{kW}}{470\ ^\circ\text{C m}^2}\right)(300\ \text{m}^2)\\
&\quad \left(\frac{24\ \text{hr}}{\text{day}}\right)\left(\frac{1}{0.80}\right)\left(\frac{\$0.036}{\text{kWh}}\right)\\
&= \$\ 1,517/\text{year}\\
\text{Summer cooling cost} &= \left(\frac{1,100\ ^\circ\text{C days}}{\text{year}}\right)\left(\frac{\text{kW}}{470\ ^\circ\text{C m}^2}\right)(300\text{m}^2)\\
&\quad \left(\frac{24\ \text{hr}}{\text{day}}\right)\left(\frac{1}{2.5}\right)\left(\frac{\$0.072}{\text{kWh}}\right)\\
&= \$\ 485/\text{year}.\\
\text{Total energy cost} &= \$1,517 + \$485 = \$2,002/\text{year}
\end{aligned}$$

If the plant described above had a night setback program, then you would have to use degree hours calculated by manipulating the base temperature corresponding to the setback hours. For example, if the temperature were a setback to 13 °C, the heat loss would be calculated using heating degree hours determined from a 13 °C base for the respective hours of setback rather than the standard 18.3 °C base for all heating hours.

11.11 Insulation for Mechanical Systems

Thermal insulation reduces the energy consumption of HVAC systems and equipment. Different building codes prescribe a minimum insulation value for different HVAC components. A good starting point is ASHRAE Standards 90.1 and 90.2. In addition to energy efficiency, thermal insulation of mechanical components can provide the following benefits:

- Personnel protection against burns by controlling surface temperatures.
- Condensation control by keeping the surface temperature above the dew point of the surrounding air.
- Process control to minimize temperature fluctuations for a process.
- Freeze protection against cold temperature exposure of pipes or ducts.
- Noise attenuation by dampening mechanical vibration-related noise
- Fire safety by slowing the spread of flames.

11.12 Heat Transfer Calculations for Pipes

Calculations become more involved in determining the effects of insulating pipes because the heat flow is in a radial direction away from the pipe through the insulation to a larger surface area. The effect of dispersing the heat over a larger surface is to increase the insulation thickness. This effect is manifested in the calculations through a concept known as *equivalent thickness* d':

$$\mathrm{d}' = r_2 \ln \frac{r_2}{r_1},$$

where
d' = equivalent thickness
r_1 = outside radius of pipe
r_2 = outside radius of pipe plus insulation.

Table 11.6 shows the outside diameter of a pipe if the nominal size is given. Note that the radius equals half the diameter.

11.13 Energy Calculators for Equipment Insulation

The National Mechanical Insulation Committee (see https://insulation.org) has developed energy calculators for equipment to help designers and energy managers estimate the savings realized by installing insulation. This organization has calculators for vertical flat surfaces and for horizontal piping. PAROC Calculus is another tool to evaluate energy-efficient and economical insulation solutions for different HVAC, process industries, and marine applications (www.paroc.com/en/tools-documents/calculators/calculus).

Example 11-4
A 200 mm nominal size pipe has 50 mm of insulation. Estimate the equivalent thickness of insulation in mm:

$$\mathrm{d}' = r_2 \ln \frac{r_2}{r_1},$$

where $r_1 = 219.1$ mm/2 $= 109.55$ mm (from Table 11.6 for 200 mm standard pipe),
$r_2 = 109.55 + 50 = 159.55$ mm
$d' = 159.55 \ln \frac{159.55}{109.55} = 60.0$ mm or 0.060 m

The equivalent thickness can then be used in the normal equations to determine the effective R-value.

Table 11.6 Nominal pipe size (standard wall thickness) versus inside diameter.

Nominal pipe size (mm)	Inside diameter (mm)	Outside diameter (mm)	Outside surface area, (m^2/m)	Flow area (m^2)	Nominal pipe size (in)	Inside diameter (in)	Outside diameter (in)	Outside surface area, (ft^2/ft)	Flow area (in^2)
10	12.52	17.1	0.0538	0.0001231	3/8	0.493	0.675	0.1767	0.1909
20	20.93	26.7	0.0838	0.0003441	3/4	0.824	1.050	0.2749	0.5333
32	35.05	42.2	0.1325	0.0009649	1-1/4	1.380	1.660	0.4346	1.496
40	40.89	48.3	0.1516	0.001313	1-1/2	1.610	1.900	0.4974	2.036
50	52.50	60.3	0.1895	0.002165	2	2.067	2.375	0.6218	3.356
65	62.71	73.0	0.2294	0.003089	2-1/2	2.469	2.875	0.7527	4.788
80	77.93	88.9	0.2793	0.004770	3	3.068	3.500	0.9163	7.393
100	102.26	114.3	0.3591	0.008213	4	4.026	4.500	1.1781	12.73
150	154.05	168.3	0.5286	0.018639	6	6.065	6.625	1.7344	28.89
200	202.70	219.1	0.6882	0.032270	8	7.981	8.625	2.2580	50.03
250	254.50	273.0	0.8578	0.050870	10	10.02	10.75	2.8143	78.85
300	304.80	323.9	1.0174	0.072966	12	12.00	12.75	3.3379	113.10
350	336.60	355.7	1.1173	0.088985	14	13.25	14.00	3.6652	137.89
400	387.40	406.5	1.2769	0.117872	16	15.25	16.00	4.1888	182.65

Example 11-5

The same 200 mm pipe from the precious example is carrying fluid at 95 °C and insulated by 50 mm (0.050 m) of aluminum-jacketed fiberglass in a 20 °C ambient area. Estimate the heat loss per linear meter of pipe. Ignore the thermal resistance of the pipe wall, and assume the convective resistance of the fluid inside the pipe is zero. Assume the convective film thermal resistance to the surrounding air is 0.155 m^2 °C/W.

We will estimate the heat loss using the following: $Q = \frac{1}{R_{\text{total}}} A\Delta T$,

$$R_{\text{total}} = R_{\text{insulation}} + R_{\text{convection}} = R_{\text{insulation}} + 0.155.$$

From Table 11.1, the thermal conductivity for insulation is found to be $k = 0.039$ W/m °C:

$$R_{\text{insulation}} = \frac{d'}{k} = \frac{0.060}{0.039} = 1.538$$

The surface area for one linear meter of pipe is based on the circumference of the insulated pipe. The diameter of the insulated pipe = pipe diameter + 2 (insulation thickness) = 0.2191 m + 2 (0.050 m) = 0.3191 m. Thus, the area for each 1 m of insulated pipe is:

$$A = \pi(0.3191\text{m})(1\text{m}) = 1.0025 \text{ m}^2$$

Putting together the parameters above, the heat loss per linear meter of insulated pipe is:

$$\begin{aligned} Q &= \frac{1}{R_{\text{total}}} A\Delta T \\ &= \left(\frac{1}{1.538 + 0.155}\right)(1.0025)(95 - 20) \\ &= 44.41 \text{ W} \end{aligned}$$

Example 11-6

For the same pipe in the previous example, estimate the reduction in heat loss per linear meter of pipe. Ignore the thermal resistance of the pipe wall. Again assume the convective film thermal resistance inside the pipe is zero, and convective resistance to the surrounding air is 0.155 m^2 °C/W.

To determine the reduction in heat loss from insulating the pipe, we must calculate the heat loss for the uninsulated pipe. The heat loss for each one-meter length of uninsulated pipe is:

$$Q = \frac{1}{R_{\text{total}}} A\Delta T.$$

The area for each 1 m of uninsulated pipe is: $A = \pi$ (0.2191 m) (1 m) = 0.6883 m^2.

Therefore, the heat loss per linear meter of uninsulated pipe is:

$$Q = \left(\frac{1}{0.155}\right)(0.6883)(95 - 20)$$
$$= 333.06 \text{ W}.$$

The heat loss reduction per linear meter of pipe = 333.06 – 44.40 = 288.7 W or 0.2887 kW.

For energy management purposes, it is usually necessary to calculate the heat loss or gain over a year instead of for only an hour, as already shown. To do this, it is necessary to sum the hourly heat losses over the number of hours in a year.

Example 11-7

Suppose the pipe in the previous example carried hot fluid 24 hours/day for 365 days/year. Further, assume that the unit generating the hot fluid is 80% efficient using natural gas costing $10.0/GJ ($0.0360/kWh). Find the energy cost savings per linear meter of pipe from insulating it with 50 mm of aluminum jacketed fiberglass.

The annual fuel cost savings per linear meter of insulated pipe is:

$$\text{Fuel cost savings} = (0.2887 \text{ kW})\left(\frac{1}{0.80}\right)\left(\frac{\$\,0.036}{\text{kWh}}\right)\left(\frac{8760 \text{ hr}}{\text{yr}}\right)$$
$$= \$114 \text{ per year per linear meter of pipe.}$$

11.14 Thermal Bridging and Two-dimensional Assembly U-factor Calculation

Heat flow through building envelopes occurs in two and three dimensions across components such as a window frame, wall studs, roof, parapet, balconies, floor slabs, etc. When envelope insulation is interrupted, it results in thermal bridges, as shown in Figure 11.10. Hence, simplified one-dimensional heat flow cannot accurately calculate a multi-layered envelope's net or effective thermal resistance. When envelope assemblies are composed of dissimilar materials, there are the following methods to estimate the two-dimensional assembly U-factor as per ASHRAE Fundamentals 2025, Chapter 27, Heat, Air, and Moisture Control in Building Assemblies – examples:

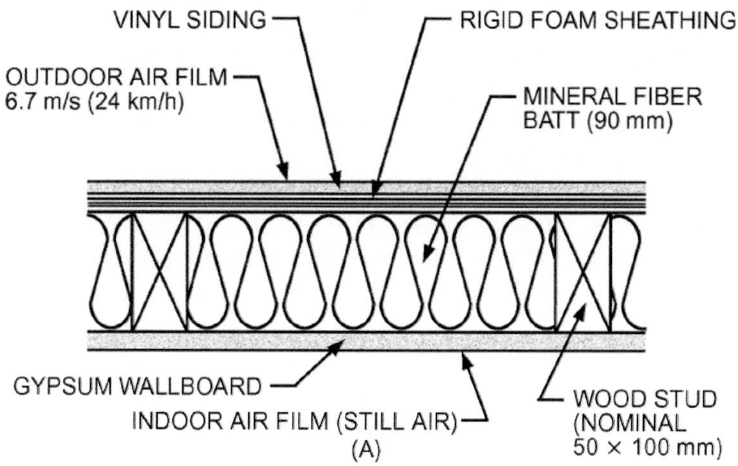

Figure 11.10 Multi-layer envelope and thermal bridging (source: 2025 ASHRAE Handbook – Fundamentals).

- Parallel-path method: Used when the thermal conductivity of the dissimilar materials is close in value.
- Isothermal-planes method: Used when materials with conductivities moderately differ from those of adjacent materials.
- Zone method and modified zone method: Used when adjacent assembly materials have a very high difference in conductivity.

The effective R-value calculation, accounting for thermal bridging for complex envelope assemblies, may require numerical simulation in software like THERM. However, for further reference, the framing factors for common construction types are discussed in BC Hydro, Building Envelope Thermal Bridging Guide, Version 1.5, 2020.

11.15 Categories of Insulation Materials

Insulation materials can be categorized into one of five major types:

(1) *Cellular insulations* are composed of small individual cells either interconnecting or sealed from each other to form a cellular structure. Glass, plastics, and rubber may comprise the base material, and various foaming agents are used.

(2) *Fibrous insulations* are composed of small diameter fibers that finely divide the air space. The fibers may be organic or inorganic and are

normally (but not always) held together by a binder. Typical inorganic fibers include glass, rock wool, slag wool, and alumina-silica.

(3) *Flake insulations* are composed of small particles or flakes that finely divide the air space. These flakes may or may not be bonded together. Vermiculite, or expanded mica, is flake insulation.

(4) *Granular insulations* are composed of small nodules that contain voids or hollow spaces. These materials are sometimes considered open-cell materials since gases can be transferred between individual spaces.

(5) *Reflective insulation* and treatments are added to surfaces to lower the long-wave emittance, thereby reducing the radiant heat transfer to or from the surface. Some reflective insulation systems contain multiple parallel thin sheets or foil spaced to minimize convective heat transfer. In addition, low-emittance jackets and facings are often used in combination with other insulation materials.

11.16 Common Insulating Materials

Some of the more popular types of materials with a discussion of some of their specific properties are given here. For further details, consult the 2025 ASHRAE Handbook – Fundamentals.

- **Mineral fiber-rock wool.** Mineral fiber insulation is made from molten rock. It is fairly impervious to heat and can be used in relatively high temperatures.
- **Fiberglass.** Probably the most popular type of insulation, fiberglass, can be obtained in blankets, batts, boards, and pipe covering. Although organic binders are frequently used, which limit temperature ranges somewhat, the cell structure is such that the limitations can sometimes be exceeded and still have acceptable results.
- **Foams.** Several types of foam insulation are available; some types have problems meeting fire hazard classifications but have very good k values. Others meet the fire hazard requirements but do not offer very good k values. Foams are particularly applicable to cold applications.
- **Calcium silicate.** A very popular type of insulation for high-temperature use, calcium silicate is spun from lime and silica. It is extremely durable and offers high thermal resistance.
- **Refractories-ceramic fiber.** Ceramic fibers, an alumina-silica product, are available in blankets or felt and could be used alone or added to existing fire bricks.

- **Refractories-fire brick.** Fire bricks are made for high-temperature applications. Made of refractory clay with organic binders, which are burned out during manufacture, they offer good thermal resistance and low storage of heat.
- **Others.** Other types of insulation include cellular glass, perlite, and diatomaceous earth. Each has advantages and disadvantages with which the energy manager must become familiar.

11.17 Concept of Super Insulation

Super Insulation is a crucial aspect of passive design. For passive design, the thermal transmittance levels are aimed at between 0.10 - 0.15 W/m^2K (0.017–0.026 Btu/hr ft^2 °F) and avoid thermal bridging and air leakage throughout the structure. For super insulation, using typical conventional insulating materials (mineral wool, polystyrene, cellulose), the thickness is needed in the range of 200–300 mm (8 to 12 inches). In addition, vacuum insulation materials have sometimes been used in the building industry.

With proper design, the super insulation, like it is used in passive houses, can show net positive cost payback. Additional factors that improve the economic viability of superinsulation can be:

- Energy costs are expected to mount in the future, increasing energy cost savings with super insulation.
- Increased thermal comfort, for example, no cold drafts, no molds, and a comfortable indoor climate.
- Reduced GHG emissions, helping fight climate change.

11.18 Building-integrated Photovoltaics

Building-integrated photovoltaics (BIPV) are solar power-generating products or systems that can be integrated into the building envelope and part of building components such as façades, roofs, or windows. BIPV serves the dual purpose of the building skin that simultaneously converts solar energy into electricity and provides building envelope functions such as:

- Weather protection (waterproofing, sun protection)
- Thermal insulation
- Noise protection
- Daylight illumination, and/or
- Safety

BIPV systems can be installed during the construction phase of a building or deployed during a retrofit of an existing building when one of the envelope components needs to be replaced. The built environment allows for many ways to integrate BIPV. In general, there are three main application areas for BIPV:

- Roofs (e.g., shingles, tiles, and skylights)
- Façades (e.g., cladding, curtain walls, and windows)
- Externally integrated systems (e.g., balcony railings and shading systems)

According to Natural Resources Canada, BIPV modules currently available on the market use either crystalline silicon-based (c-Si) solar cells or thin-film technologies such as amorphous-based silicon (a-Si), cadmium telluride (CdTe), and copper indium gallium selenide (CIGS). For skylight or curtain wall applications, for example, semi-transparency can be achieved with most technologies by spacing opaque cadmium telluride solar cells or making the thin film layer transparent. However, the module efficiency decreases with the increase of transparency as less sunlight is captured and converted into electricity by the photovoltaic layer. Refer to Chapter 18 for more details on solar PV systems.

11.19 Economic Insulation Thicknesses

The economic thickness of insulation minimizes the total life cycle cost. However, beyond the economic thickness, added insulation benefits little in energy savings and added labor and material costs. Hence, conducting economic analysis for building envelope design is very important.

As the thickness of insulation increases, the cost of material and installation increases. On the other hand, the cost of lost energy goes down but at a decreasing rate. In other words, the energy cost savings also go up, but at a slower rate than the cost of materials and installation. At some point, the total cost, which is the sum of the lost energy and material costs, reaches a minimum point. That amount of insulation is called economic thickness. This concept was illustrated in Example 11-1, where the additional insulation was shown to have a small improvement in savings.

As Figure 11.11 shows, the total cost curve is relatively flat in the immediate neighborhood of the economic thickness.

To determine this economic thickness, the energy manager needs to construct cash flow diagrams for the different alternative thicknesses and

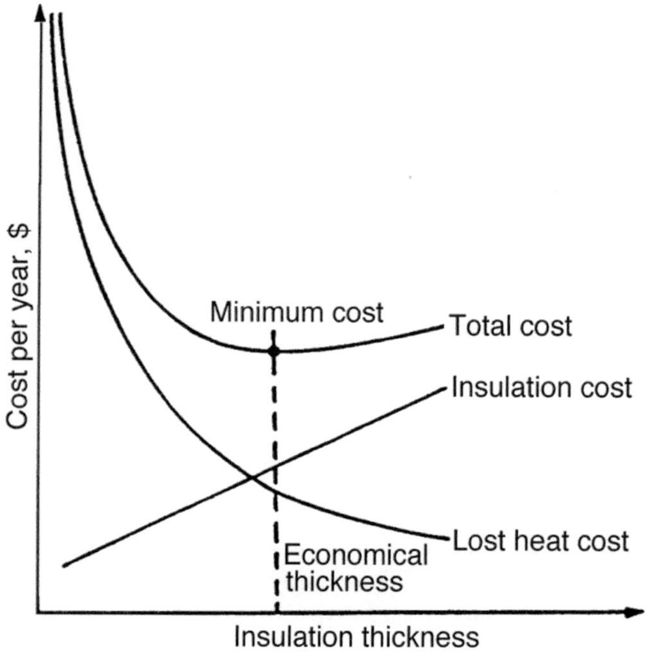

Figure 11.11 Illustration of the economic thickness of insulation.

calculate the annual equivalent cost for each increment. Since the cash flows include future fuel costs, careful handling of inflation is required (see Chapter 5).

11.20 Consideration for a Durable Envelope for Building Longevity

To ensure the durability of a building to perform its required functions, it is important to design against envelope failure. Building envelope failure is most commonly characterized as moisture-damaged cladding or roof components, as well as condensation on walls and windows. Drafts and uneven building temperatures, discoloration, and indoor air quality problems such as mold can result from building envelope failure.

To prevent premature deterioration of the building envelope, special attention should be given to vulnerable components and penetration of the building envelope. Chapter 3 includes some discussion and resources for building condition assessments, including the envelope.

Though the framework of design considerations related to building life and durability is beyond the scope of this book, building envelope failures are important in terms of the feasibility of energy upgrades. One common situation is found in low-income or energy-burdened households where building deficiencies can be substantial (energy equity issues are discussed in Chapters 1 and 3). In these cases, often energy-efficient insulation upgrades cannot be completed due to severe structural deficiencies or moisture problems. For any significant building envelope failures, it is recommended to ensure that the specialist is involved in the design process. Further details are available in CSA S478 (2019).

11.21 Considerations for Passive Solar Thermal Design of Envelope

Passive solar systems are design strategies or related technologies that use elements of the building structure, building glass, thermal mass, and building orientation to heat, cool, shade, and light buildings. Passive solar technologies can include direct gain heating, radiative cooling, natural ventilation and economizer cycles, natural lighting, light shelves, and shading systems. For example, a south-facing glass window can serve as a passive solar collector of heat for a building, and the interior of the building serves as a heat storage device. In addition, windows also provide significant amounts of natural lighting. As a result, a suitable passive solar design feature can potentially help reduce energy costs and improve facility resiliency (e.g., to maintain safe indoor temperatures longer in the event of loss of electric power).

Some considerations for the solar thermal design of a building as related to its development design can be:

- Using passive solar features like natural shades from the summer sun and/or windbreaks in the winter.
- Facility orientation should be chosen based on energy efficiency considerations, for example, minimizing sun exposure in the summer while using the sun for heating during the winter.
- Consider the use of earth-bermed structures. For example, use a large amount of backfill on northern and western walls to protect the facility from cold on the north side and heat on the west.
- Incorporate cool roof or site features to reduce the heat island effect.
- Utilizing overhangs or awnings on southern exposures to block the summer sun but allow the winter sun to enter. It is important to constitute

a multidisciplinary design team to incorporate site-specific and viable passive design features.

11.22 Considerations for Energy Efficiency Related to the Building Envelope

This section summarizes typical energy efficiency measures associated with the building envelope and considerations for energy management.

- Reduce building infiltration. In addition to unwanted outside air, infiltration can result in frozen pipes in extreme cases.
- Ensure that weatherstripping for the building entrance and overhead doors is in good condition.
- Older, operable windows with damaged frames can have high infiltration losses. Therefore, leakage should be managed.
- Conduct a building pressure test to quantify infiltration.
- Optimize building supply and return air plenum pressures to minimize infiltration.
- Focus on building exterior colors to ensure heat reflection instead of heat absorption for warmer climates.
- Ensure optimal wall and roof insulation during new construction, as post-construction insulation upgrades can be challenging and more expensive.
- Ensure that thermal breaks are provided for envelope design. Provide thermal breaks for window frames.
- Window design parameters should be selected using detailed energy simulation, quantifying the impact of glass, frame, and assembly U value, shading coefficient, and heat gain.
- Always select windows using the assembly U value, and not the center of the glass, edge of the glass, or frame U value only.
- High-performance glazing can help reduce or eliminate perimeter heating for certain climates, whereas equipment savings can help pay for highly efficient glazing.
- Use low-e coatings to help manage envelope cooling loads.
- Exteriors or interior shades should be employed based on the local climatic conditions.
- Envelope design should be an integrated multidisciplinary process to maximize passive solar gain and sunlight to reduce interior lighting.

- Building envelope optimization should be done using detailed building energy simulation modeling, accounting for all synergies with the mechanical, electrical, architectural, and site conditions.

11.23 Summary

Reducing heat loss or heat gain by adding insulation and reducing infiltration is often a cost-effective energy management opportunity. First, an energy manager must know the different types of insulation available and their respective advantages and disadvantages in specific applications. For example, different kinds of insulation are needed for buildings' walls and roofs, compared to insulation needed for high-temperature steam lines. Knowing how to perform heat loss-gain calculations is necessary to properly evaluate the effects of adding insulation and calculate the benefits of different insulation levels. Finally, understanding insulation economics is important for good energy management since the first increments of added insulation produce very cost-effective benefits. At the same time, there is a point of diminishing savings returns for increasing insulation levels.

References

ANSI/ASHRAE Standard 140-2017, Standard Method of Test For The Evaluation Of Building Energy Analysis Computer Programs

ANSI/ASHRAE/IES Standard 90.1-2022, Energy Efficiency Standard for Buildings Except for Low-Rise Residential Buildings

ANSI/ASHRAE/IES Standard 90.2-2018, Energy-Efficient Design of Low-Rise Residential Buildings

ASHRAE® Handbook - Fundamentals (SI Edition), 2025

BC Hydro, Building Envelope Thermal Bridging Guide, Version 1.5, 2020

Capehart, B.L., Kennedy, W.J., & Turner, W.C. (2016). Guide to Energy Management: International Version (8th ed.). River Publishers. https://doi.org/10.1201/9781003152002

Carrier System Design Manual Part 1: Load Estimating Paperback – January 1, 1972

CSA S478, 2nd Edition, April 2019 - Durability in buildings

Emmerich, S. and Persily, A. (2005), Airtightness of Commercial Buildings in the United States, 2005, Brussels, -1, [online], https://tsapps.nist.gov/publication/get_pdf.cfm?pub_id=860990 (Accessed April 14, 2023)

Roosa S., Doty S. and Turner W. (2020) Energy Management Handbook, Ninth Edition, River Publishers, 2020.

Gowri K., Winiarski D. and Jarnagin R. (2009). Infiltration modeling guidelines for commercial building energy analysis. Contract. 10.2172/968203.

National Energy Board. Canada's Energy Future 2018: Energy Supply and Demand Projections to 2040 DOI: https://doi.org/10.35002/wp48-nq03

Ng, L., Emmerich, S. and Persily, A. (2016), An Improved Method of Modeling Infiltration in Commercial Building Energy Models, Noting from paper Tech Note #1829 Pub.#914500, Atlanta, GA, [online], https://tsapps.nist.gov/publication/get_pdf.cfm?pub_id=915966 (Accessed June 29, 2021)

Standard 140-2020 – Method of Test for Evaluating Building Performance Simulation Software (ANSI Approved)

Syed A. (2021) EnergyPlus - Towards the Selection of Right Simulation Tool, Journal of Energy and Power Technology, August 2021

Syed A. (2020) Evolution of Building Thermal Simulation Methodologies, IJSEEP V3, Issue 6 on 20 November 2020

Bergman T., Lavine A., Incropera F., DeWitt D. (2017) Fundamentals of Heat and Mass Transfer, 8th Edition, ISBN: ES8-1-119-32042-5.

US Department of Energy, THERM, June, 2021 Available at: https://windows.lbl.gov/software/therm

Author Biography

Dr. Ali Muslim Syed is a professional engineer (P.Eng.) with extensive expertise in energy conservation, renewable energy systems, and sustainable urban infrastructure. He holds a Ph.D. in Environmental Design from the University of Calgary and a Master of Applied Science in Mechanical Engineering from Dalhousie University. He has provided energy management and sustainability services to clients in North America, Japan, and the Middle East. Dr. Syed holds several key certifications, including LEED Accredited Professional, Certified Measurement and Verification Professional, and Certified Energy Manager. His publication record in peer-reviewed journals highlights his engagement with research, particularly in low-energy, net-zero urban design, energy-sharing systems between buildings, and enhancing resilience to climate change. He is a speaker and trainer at conferences and industry events globally. Throughout his career, Dr. Syed has successfully managed large-scale projects across various sectors, emphasizing advanced energy auditing, lifecycle cost analysis, and commissioning strategies. His work is grounded in delivering practical, results-oriented solutions aimed at decarbonizing the built environment and enhancing energy efficiency to create a more sustainable future.

Practice Problems

1. The building envelope's overall thermal conductance, UA = 5,700 Btu/hr °F. The annual heating degree days = 2,700 °F day/yr. The consumption rate of natural gas is 470 MMBtu per year. Calculate the seasonal efficiency of the heating system.
2. The table below contains real temperature data from an airport weather station (except the temperatures have been rounded to the nearest integer for ease of calculation). Calculate the number of heating degree hours and number of HDDs for this 24-hour period.

Station name	Date (local standard time)	Air temperature (°F)	Heating degree hours (reference 65 °F)
Airport A	11-May-2017 01:00	46	
Airport A	11-May-2017 02:00	43	
Airport A	11-May-2017 03:00	43	
Airport A	11-May-2017 04:00	43	
Airport A	11-May-2017 05:00	43	
Airport A	11-May-2017 06:00	45	
Airport A	11-May-2017 07:00	50	
Airport A	11-May-2017 08:00	55	
Airport A	11-May-2017 09:00	61	
Airport A	11-May-2017 10:00	64	
Airport A	11-May-2017 11:00	66	
Airport A	11-May-2017 12:00	66	
Airport A	11-May-2017 13:00	66	
Airport A	11-May-2017 14:00	68	
Airport A	11-May-2017 15:00	70	
Airport A	11-May-2017 16:00	72	
Airport A	11-May-2017 17:00	73	

Station name	Date (local standard time)	Air temperature (°F)	Heating degree hours (reference 65 °F)
Airport A	11-May-2017 18:00	72	
Airport A	11-May-2017 19:00	64	
Airport A	11-May-2017 20:00	59	
Airport A	11-May-2017 21:00	59	
Airport A	11-May-2017 22:00	55	
Airport A	11-May-2017 23:00	52	
Airport A	12-May-2017 00:00	52	

3. A plant has 110 m of uninsulated hot water lines carrying water at 80 °C. The pipes are 100 mm in nominal diameter. It was decided to insulate with 50 mm calcium silicate snap-on insulation (k = 0.066 W/m °C) at \$600/m^2 installed cost. The boiler supplying the hot water consumes natural gas at \$9.5/GJ and is 80% efficient (HHV basis). Ambient air is 25 °C, and the lines are active at 8,760 hr/yr. Assume the convective film resistance inside the pipe is zero, and ignore the thermal resistance of the pipe wall. Assume the convective resistance to the surrounding air is 0.10 m^2 °C/W. Estimate the annual fuel savings in energy (GJ) and dollars. Estimate the net present value assuming a 12-year life and 6% discount rate.

4. A building consists of four walls that are each 8 feet high and 20 feet long. The wall is constructed of 4 inches of cork board, with 1 inch of plaster on the outside and 1/2 inch of gypsum board on the inside. Three of the walls each have one 6 × 4 foot, single-pane window with R = 0.7 hr ft^2 °F/Btu. The fourth wall has one 6 × 4-foot window and one 3 × 7-foot door made of one-inch thick softwood. The roof is constructed of 3/4 inch plywood with asphalt roll roofing over it. The inside temperature of the building is regulated to 78 °F by an air-conditioner operating with a thermostat. The air conditioner has a SEER of 8.0. The outside temperature is 95 °F for one hour.
Estimate the following:

   a. R-value of one of the walls with just a window.
   b. R-value of the wall with the window and the door.

c. Btu/hr that the air conditioner must remove to keep the inside temperature at 78 °F.
d. kWh of electric energy will be used in that 1-hour period by the air conditioner.

5. Repeat the previous problem, where the single-pane windows are replaced with triple-paned windows that have an R-value of 2.32 hr ft^2 °F/Btu.

6. A building consists of four walls that are each 8 feet high (2.438 m) and 20 feet long (6.096 m). Assume passive house level of insulation with $R = 40$ hr ft^2 °F/Btu (7.04 m^2 C/W) for walls, R 50 (8.81 m^2 °C/W) roof and a triple glazed window with R-value of 5 (0.881 m^2 °C/W). The inside temperature of the building is regulated to 78 °F (25.6 °C) by an air conditioner operating with a thermostat. The air-conditioner has a SEER of 8. The outside temperature is 95 °F (35.0 °C) for 1 hour. Estimate the following in SI units

   a. R-value (m^2 C/W) of one of the walls with just a window.
   b. R-value (m^2 C/W) of the wall with the window and the door.
   c. R-value (m^2 C/W) of the roof.
   d. Heat (kW) the air conditioner must remove to keep the inside temperature at 25.6 °C.
   e. kWh of electric energy will be used in that 1-hour period by the air conditioner.

12

Lighting

Cristian Suvagau

BC Hydro, Energy Management & Innovation, Canada

12.1 Introduction

The lighting system provides many opportunities for cost-effective energy savings with little or no inconvenience. In many cases, lighting can be improved, and operation costs can be reduced at the same time. Lighting improvements are excellent investments in most commercial businesses because lighting accounts for a large part of the energy bill – ranging from 30% to 70% of the total energy cost. With industrial facilities, lighting energy use represents only 5%–25% of the total energy, but it is usually cost-effective to address because lighting improvements are often easier to make than many process upgrades.

While there are significant energy-use and power-demand reductions available from lighting retrofits, the minimum lighting-level standards of the Illuminating Engineering Society (IES) should be followed to ensure worker productivity and safety. Inadequate lighting levels can decrease productivity, and they can also lead to a perception of poor indoor air quality. However, most lighting retrofits will increase productivity and are most likely to appeal to employees, clients, and guests, which can have a positive impact on the facility, business reputation, and/or especially sales.

Since everyone has experience with lighting, it can attract immediate employee attention and participation and can be used as a starting place for an energy management program. A lighting retrofit program can be a win–win proposition for the business owner and the employees as it can improve morale, safety, and productivity while reducing lifecycle costs.

This chapter provides a brief description of lighting systems, their characteristics, and retrofit options. Currently, most of the energy-efficient lighting upgrade projects use solid-state light (SSL) sources (i.e., LED) to replace the non-SSL sources (e.g., incandescent, fluorescent, high-intensity discharge) also referred to as legacy or conventional lamps.

12.2 Components of the Lighting System

A lighting system consists of light sources (lamps), ballasts (also drivers), and luminaires (fixtures). Each component will affect the performance, energy use, and annual operating cost of the lighting system. The use of lighting controls will help with energy management while bringing many non-energy benefits to occupants and facilities. This section discusses each of these components and provides the basic information on lighting technology needed to successfully implement lighting energy efficiency measures (EEMs). More information about lighting fundamentals is presented in Section 12.3.

12.2.1 Lamp characteristics

Lamps are rated several different ways, and each characteristic is a factor to consider in the lamp selection process. The basic ratings include luminous efficacy (lm/W), color temperature (Kelvins), color rendering index (CRI), cost ($), rated life (operation hours), and labor required for relamping. Lamps should carry recognizable name brands and should be purchased from a reputable vendor. Some off-brand lamps, particularly those from manufacturers not yet established, may have low light output and/or short lives. Table 12.1 presents general characteristics on many commonly used lamps of legacy/conventional (non-SSL) technologies. More data is available from lamp manufacturers and the IES Lighting Handbook [IES (2011)]. Information on SSL characteristics is presented in Section 12.2.5.

12.2.1.1 Luminous efficacy

The luminous efficacy of a lamp is an estimate of the light output (lumens) divided by the electrical power input (Watts) under test conditions. Lamps operating outside their design envelope may suffer reduced efficacy. For example, be cautious to use 25-Watt energy-saving T8 fluorescent lamps in environments with temperatures below 60 °F; at lower temperatures, they are prone to flickering, their light output is low, and they have short lives. Table 12.1 presents lamp data on many commonly used lamps. More data is available from lamp manufacturers and the IES Lighting Handbook [1]. Figure 12.1 shows the range of efficacy for different types and wattages of light sources. Higher wattage lamps can have higher efficacies. The largest

Table 12.1 Legacy light source characteristics [Woodruff, 2018, Table 13.4].

			High-intensity discharge (HID)			
	Incandescent, including tungsten halogen	**Fluorescent**	**Mercury vapor (self-ballasted)**	**Metal halide**	**High-pressure sodium (improved color)**	**Low-pressure sodium**
Wattages (lamp only)	15–1,500	15–219	40–1,000	175–1,000	70–1,000	35–180
Life* (hr)	750–12,000	7,500–24,000	16,000–15,000	1,500–15,000	24,000 (10,000)	18,000
Efficacy (lm/W) (lamp only)	15–25	55–100	50–60 (20–25)	80–100	75–140 (67–112)	Up to 180
Lumen maintenance	Fair to excellent	Fair to excellent	Very good (good)	Good	Excellent	Excellent
Color rendition	Excellent	Good to excellent	Poor to excellent	Very good	Fair (very good)	Poor
Light direction control	Very good to excellent	Fair	Very good	Very good	Very good	Fair
Source size	Compact	Extended	Compact	Compact	Compact	Extended
Relight time	Immediate	Immediate	3–10 min	10–20 min	Less than 1 min	Immediate
Comparative fixture cost	Low: simple fixtures	Moderate	Higher than incandescent and fluorescent	Generally higher than mercury	High	High
Comparative operating cost	High: short life and low efficiency	Lower than incandescent	Lower than incandescent	Lower than mercury	Lowest of HID types	Low
Auxiliary equipment	Not needed	Needed: medium cost	Needed: high cost	Needed: High cost	Needed: High cost	Needed: high cost

**Life and efficacy ratings subject to revision. Check manufacturers' data for latest information.*

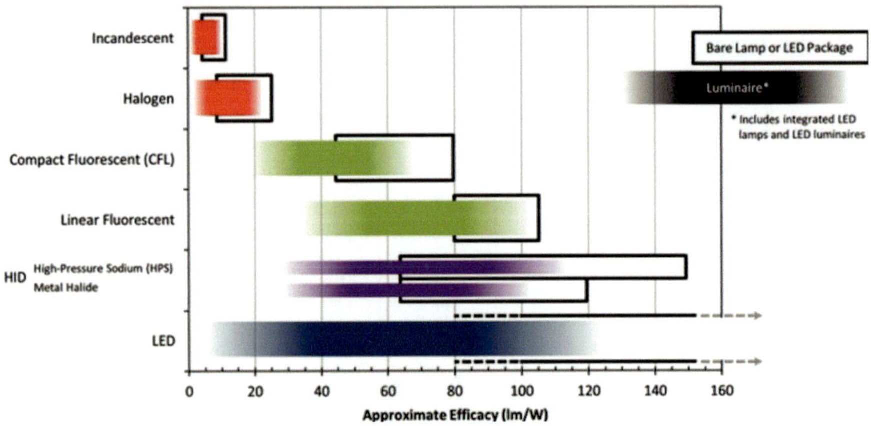

Figure 12.1 Range of efficacy for various light sources [DOE, 2013].

range is for LED lamps and luminaires, where the higher wattage lamps have a significantly higher efficacy.

12.2.1.2 Color temperature

The color temperature of a lamp describes the appearance of the light generated compared to the perceived color of a blackbody radiator at that temperature on the absolute temperature scale (i.e., Kelvin scale). Except for incandescent technologies, the industry is using the correlated color temperature (CCT) term. For example, while a warm-white fluorescent lamp rated at 3,000 Kelvins appears yellowish, a white fluorescent lamp rated at 4,100 Kelvins appears white, and a daylight fluorescent lamp rated at 6,300 Kelvins appears bluish and cool. Figure 12.2 shows the color temperatures of commonly used fluorescent lamps.

While hospitality applications usually require color temperatures between 2,800 and 3,500 Kelvins for a more relaxing mood, most of the commercial lighting applications use lights with CCTs from 3,000 to 4,100 Kelvins. Industrial and outdoor applications usually require cooler color temperatures between 4,100 and 5,000 Kelvins. The energy manager should be sensitive to lamp color when recommending lighting changes and should not recommend changing lamp color temperature unless nearly everyone is in favor of the proposed change. Understanding the color needs of the facility is important too. Some merchandise looks better at a particular color temperature. Example: meat in grocery stores looks more appealing under warm lamps that accentuate the red hues. The same meat in a meatpacking

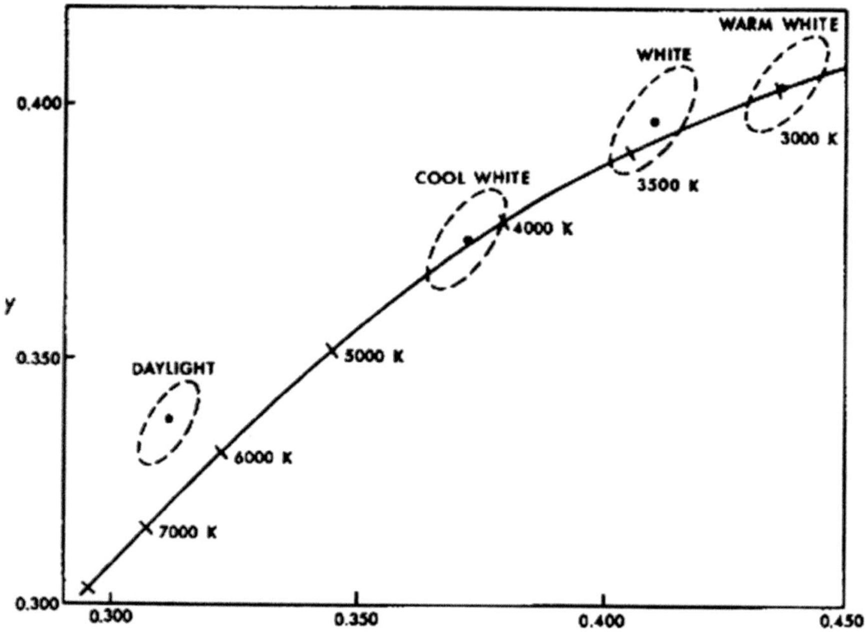

Figure 12.2 Fluorescent lamp color temperatures [IES, 2011, Figure 12.6– 12.17].

facility can be illuminated with lamps of a higher color temperature because visual appeal is not a factor.

12.2.1.3 Color rendering

CRI is a relative indication of how well colors can be distinguished under the light produced by a lamp with a particular color temperature. The index runs from 0 to 100, where a high CRI indicates good color rendering. Many commonly used lamps have poor CRIs (e.g., the CRI of a typical T12 cool white fluorescent lamp is 67).

Rare-earth elements in the phosphors of high-efficiency lamps (T8, T5) increase light conversion efficacy and color rendition. Light sources such as metal halide lamps have better color rendering abilities than high-pressure sodium lamps. The need for accurate color rendering depends on the particular task. Matching colors in a garment factory or photo laboratory will require a much higher CRI lamp than assembling large machine parts.

The color rendering abilities of commonly used lamps are described in Table 12.2.

Table 12.2 Color rendition of various lamp types.

	Incandescent, high-intensity discharge lamps					
	Filamenta	**Clear mercury**	**White mercury**	**Deluxe whiteb mercury**	**Multi-vapor**	**Lucaloxa**
Efficacy (lm/W)	Low	Medium	Medium	Medium	High	High
Lamp appearance effect on neutral surfaces	Yellowish white	Greenish blue-white	Greenish white	Purplish white	Greenish white	Yellowish
Effect on	Warm	Very cool,	Moderately	Warm,	Moderately	Warm,
"atmosphere"		greenish	cool, greenish	purplish	cool, greenish	yellowish
Colors	Red	Yellow	Yellow	Red	Yellow	Yellow
strengthened	Orange	Green	Green	Yellow	Green	Orange
	Yellow	Blue	Blue	Blue	Blue	Green
Colors grayed	Blue	Red, orange	Red, orange	Green	Red	Red, blue
Effect on complexions	Ruddiest	Greenish	Very pale	Ruddy	Grayed	Yellowish
Remarks	Good	Very poor	Moderate	Color	Color	Color
	color	color	color	acceptance	acceptance	acceptance
	rendering	rendering	rendering	similar to cool white fluorescent	similar to cool white fluorescent	approaches that of warm white fluorescent

	Fluorescent lamps						
	Cool[b] white	Deluxe[b] cool white	Warm[a] white	Deluxe[a] warm white	Daylight		Soft white/natural
Efficacy (lm /W)	High	Medium	High	Medium	White Medium-high	High	Medium
Lamp appearance effect on neutral surfaces	White	White	Yellowish white	Yellowish white	Bluish white	Pale yellowish white	Purplish white
Effect on "atmosphere"	Neutral to moderately cool	Neutral to moderately cool	Warm	Warm	Very cool	Moderately warm	Warm pinkish
Colors strengthened	Orange Yellow Blue	All nearly equal	Orange Yellow	Red Orange Yellow Green	Green Blue	Orange Yellow	Red Orange
Colors grayed	Red	None appreciably	Red, green blue	Blue	Red, orange	Red, green blue	Green, blue
Effect on complexions	Pale pink	Most natural	Sallow	Ruddy	Grayed	Pale	Ruddy pink
Remarks	Blends with natural daylight; good color acceptance	Best overall color rendition; simulates natural daylight	Blends with incandescent light; poor color acceptance	Good color rendition; simulates incandescent light	Usually replaceable with cool white	Usually replaceable with cool white or warm white	Tinted source

[a] *Greater preference at lower levels.*
[b] *Greater preference at higher levels.*
Source: General Electric Technical Pamphlet TP-I 19 [DOE (2013)]. Reprinted with permission of General Electric Co.

12.2.2 Lamp types

Lamps come in a variety of types and have a wide range of characteristics. Choosing the appropriate lamp type depends on the lighting task. Often, the least expensive lamp to buy is not the least expensive to operate. The energy manager should be familiar with the different options available to provide the desired lighting levels.

12.2.2.1 Incandescent lamps

Incandescent lamps render colors well, are inexpensive to purchase, and easily dimmed, small, and controllable, which is useful for product display. However, they have relatively short lifespans, low efficacy, and are susceptible to failure from heat and vibration. Incandescent lamps rated for long life or rough service have a correspondingly low efficacy. Some energy-saving lamps have a poorly supported filament and should not be used in environments with vibration or other mechanical stresses.

Most incandescent lamps tend to darken with age as tungsten is lost from the filament and deposited on the lamp walls. Figure 12.3 shows most of the commonly used lamps. Incandescent lamps have a color temperature between 2,400 and 3,100 Kelvins.

A-lamps. Also known as general service lamps, these lamps are low cost and commonly used in sizes of 20–1,500 Watts. They project light out in all directions. In old industrial plants, look for large A-lamps in pendant fixtures where the lamps are left on most of the time. These are good candidates for replacement with HID or LED lamps.

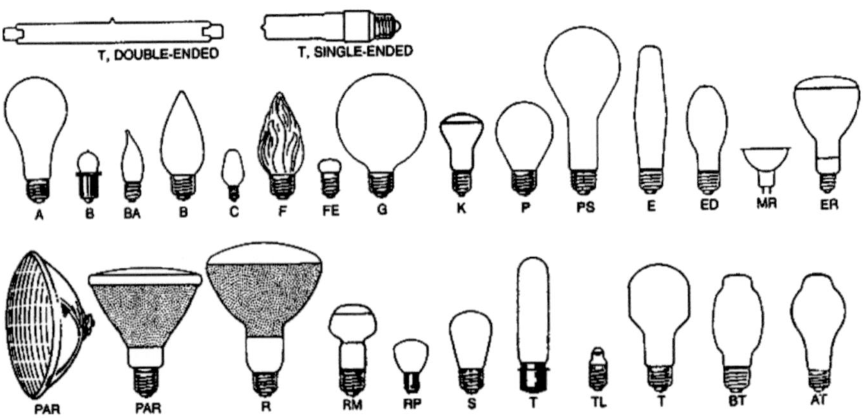

Figure 12.3 Incandescent bulb shapes and designations [IES, 2011, p. 8–5].

Reflector (R) lamps. These lamps are usually more expensive than A-lamps and offer better control of the direction in which light is cast due to a reflective paint on the lamp wall. They have a focal point at the back of the lamp, which results in the light from the lamp being dispersed broadly by the reflective surface of the lamp.

Ellipsoidal reflector (ER) lamps. These lamps cost about the same as R-lamps, but they are longer and have a focal point in front of the lamp. This location of the focal point results in the light being more concentrated as it leaves the lamp, and thus the beam is narrower than from an R-lamp.

Parabolic aluminized reflector (PAR) lamps. These lamps are more expensive than R and ER lamps but offer much better beam focus and control of dispersion. They consist of a parabolic reflector and a glass or plastic lens sealed permanently together as a unit. These lamps come in many combinations of size, beam spread, and wattage.

Quartz-halogen lamps. These lamps have a short life and low efficacy. They can be a good choice for areas, which need lighting on an irregular basis. These lamps should not be cleaned.

Halogen lamps. These lamps have a higher efficacy and cost than the lamps listed above and are available in many of the same lamp configurations.

12.2.2.2 Fluorescent lamps

Fluorescent lamps have high efficacy, long life, and low surface luminance. They are cool and are typically available in a variety of color temperatures from 2,800 to over 6,000 Kelvins. Figure 12.4 shows many of the commonly used fluorescent lamps.

Typical fluorescent lamps. Fluorescent lamps are available in standard, high output (HO), and very high output (VHO) configurations. The HO and VHO

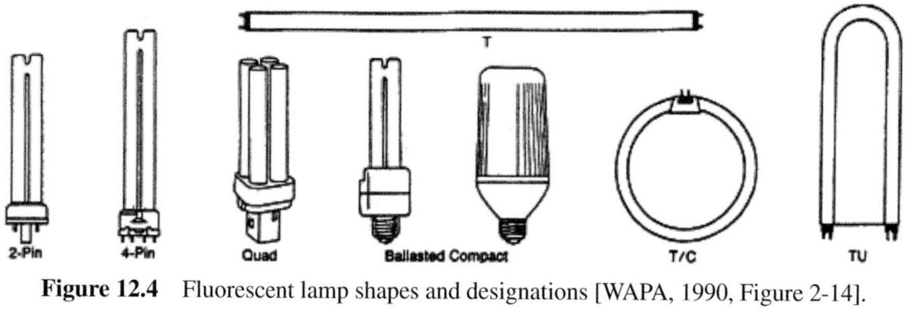

Figure 12.4 Fluorescent lamp shapes and designations [WAPA, 1990, Figure 2-14].

lamps are useful for low-temperature environments and areas where a lot of light is needed with minimal lamp space.

Energy-saving lamps. Energy-saving fluorescent lamps, which can replace the standard lamps, can reduce power demand and energy use by about 15%. They will also decrease light levels by about 3%–10%. These lamps can only be used with ballasts designed and rated for energy-saving lamps and should not be used in areas in which the temperature falls below 60 °F. However, amalgam mercury technology is now used in many modern fluorescent lamp types to allow a more stable light output, independent of ambient temperature or burning position. Table 12.3 shows some of the specifications that vary between many of the energy-saving fluorescent lamps available.

T-measures for lamps. The T-measure for a fluorescent lamp is the measure of the diameter of the lamp in eighths of an inch. Thus, a T12 lamp is twelve-eighths of an inch (or 1-1/2 inches) in diameter, while a T8 lamp is 1 inch in diameter and a T5 lamp is five-eighths of an inch

T8 lamps with electronic ballasts. T8 lamps produce an efficacy of up to 100 lm/W, the highest efficacy of any fluorescent lamp. They will not generally operate on standard ballasts rated for T12 lamps. They also use less of the toxic materials found in larger fluorescent lamps. T12 fixtures can be retrofitted with T8 lamps by using socket adapters and replacing the old ballasts with the T8 compatible ballasts. T8's (and in some cases T5 lamps) have become the standard retrofit from T12 technologies. T8 lamps have been highly diversified for longer life, higher output, or reduced wattage.

T5 lamps with electronic ballasts. T5 lamps currently exist in two designs, standard and high output (HO), and operate only on electronic ballasts. T5

Table 12.3 Energy saving fluorescent lamps [Woodruff, 2018, Figure 13-4].

Fluorescent options	Low-wattage	T8	Super-T8	T5	T5HO
Watts	32	25–30	32	14–28	54
Lumens	1,700–3,200	2,400–2,900	3,100+	2,900	4,400–5,000
Efficacy, lm/W	92–100	89–92	88+/90+*	104	93
Rated average life, hours	20,000	18,000–30,000	24,000+	20,000	20,000
CRI	75–85	82–85	81+	82–85	82–85

* *Programmed rapid start ballasts/instant start ballasts.*

lamps are a true energy-efficient alternative to T12's but lack the retrofit opportunity because of metric lengths (which do not fit in inch-sized luminaires), higher brightness, different lamp holder design, and little to no efficacy improvement of the lamp-ballast efficiency over T8's. However, the T5's are likely to be best used in new luminaires (designed around their smaller size, unique lengths, and thermal characteristics) typically used in indirect, slim profile, or high-bay applications.

Compact fluorescent lamps. These twin-tube (TT) and double twin-tube (DTT) lamps are designed to replace many frequently used incandescent bulbs. These lamps can be used to reduce energy use and power demand by over 70%. For example, a 13-Watt fluorescent TT lamp can be used to replace a 60-Watt incandescent lamp. The light produced is similar in appearance to that of an incandescent lamp (i.e., color temperature of 2,700 Kelvins for both). Since these lamps produce less heat, space cooling costs are also reduced. Many TT lamps/ballasts have a low power factor of about 0.5–0.6. A high-power factor ballast with low harmonic distortion should be specified if a large number of these lamps is used.

Compact fluorescent lamps can be installed as a screw-in or hardwired conversion kit and have a lifetime of at least10,000 hours. Frequent cycles of short operation hours will significantly reduce lamp life. One-piece screw-in compact fluorescent lamps have higher lifecycle costs than the hardwire models because the useful ballast life ends when the lamp burns out. Using the hardwire kits will eliminate the possibility of an uninformed maintenance worker throwing away a two-piece screw-in conversion kit when the lamp fails. Table D.1 in Appendix D lists the characteristics of some of the compact fluorescent lamps (legacy lighting).

12.2.2.3 High-intensity discharge lamps

These lamps are relatively expensive initially but offer low lifecycle costs due to long life and high efficacy. In general, the larger the HID lamp, the higher the efficacy. The HID lamp efficacy can be affected by the lamp position, and some of these lamps have a significant color shift and loss of efficacy near the end of their rated life. Figure 12.5 shows some of the commonly used HID lamps.

Mercury vapor lamps. Mercury vapor lamps were the first HID lamps. They can offer good color rendering and low-to-moderate efficacy. Self- ballasted mercury vapor lamps are a direct replacement for large incandescent lamps

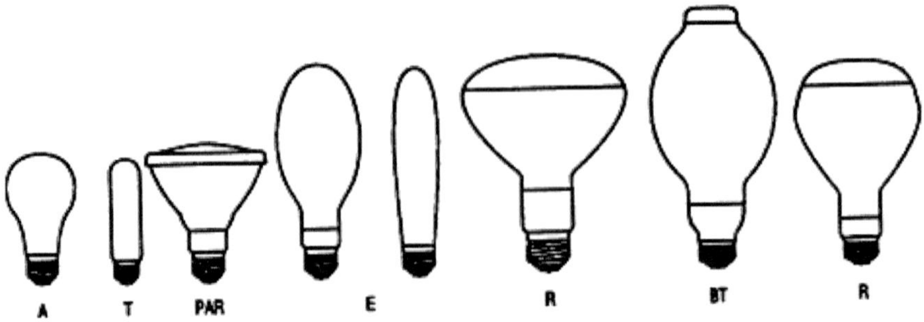

Figure 12.5 HID lamp shapes and designations [WAPA, 1990, Figure 2-16].

but have only 30%–50% of the efficacy of typical mercury vapor lamps. Mercury vapor lamps are currently being replaced with more efficient light sources such as metal halide, high-pressure sodium, and LEDs.

Metal halide lamps. Many of these lamps produce bright white light. Metal halide comes in a large variety of color temperatures from 3,000 Kelvins to over 20,000 Kelvins. They are used in applications that require good color rendering and high lighting levels such as sports facilities. They do not have as long a life as most other HID lamps. Some metal halide lamps can be installed in fixtures designed to operate mercury vapor lamps. Improvements in ballast (pulse-start) and lamp (ceramic) technologies have improved both energy efficiency, start/restriking time, and CRI and reduced the discoloration at the end of life for metal halide lamps.

High-pressure sodium (HPS) Lamps. HPS lamps offer high efficacy, long life, and a relatively small light source, which is easily controlled. The light produced is orange yellow (CCT of 2,200 Kelvins) and does not render colors well but is useful for tasks in which color rendering is not critical. These lamps were once considered usually the best choice for warehouses, factories, exterior floodlighting, and street lighting.

12.2.2.4 Low-pressure sodium (LPS) lamps

LPS lamps offer the highest efficacy of any light source (i.e., up to 180 lm/W). They have a long life but are fairly large compared to HID lamps. Their size requires a larger, more complex reflector design to efficiently utilize the light produced. The color of the light is nearly a monochromatic yellow under which very little color discrimination is possible. LPS lamps are common in European street lighting systems.

12.2.3 Ballasts

Light from discharge-type lamps (e.g., fluorescent, mercury vapor, metal halide, high-pressure sodium, and low-pressure sodium) is produced indirectly by a cathode exciting a gas in which an electrical arc forms which then emits light. In a fluorescent lamp, mercury vapor emits ultraviolet radiation, which strikes the sides of the lamp wall where phosphors convert it to visible light. A ballast is required to start and operate all discharge lamps.

Each lamp/ballast combination should have test results that state the ballast factor. The lumen rating for a lamp is based on a particular lamp/ballast combination with a ballast factor of 1.0. The ballast factor indicates the light output of a particular system relative to a standard test ballast on which the lamp lumen ratings (see more information in Section 12.3.1) are based. For example, if a four-foot F32T8 has a rated lamp light output of 3,000 lumens, and it is used with a ballast that has a ballast factor of 1.1 for the lamp/ballast combination, the system produces about 3,300 lumens.

The ballast label usually gives the electric current drawn for particular lamp types, states whether the ballast does not contain polychlorinated biphenyls (PCB), and shows a wiring diagram. Many older ballasts, especially those manufactured before 1980, utilize PCB oil that contains some of the most toxic chemicals known. Leaking PCB-laden ballasts should be handled as a hazardous waste material during disposal. Ballasts that do not contain PCB can currently be landfilled in most locations, but you should check local regulations regarding disposal legislation.

National standards require that all new ballasts have a minimum efficiency and contain no PCBs. Ballasts are usually labeled low-heat (LH), very low heat (VLH), super-low heat (SLH), energy-saving, or electronic. Not all electronic ballasts on the market will operate at higher efficiency than magnetic ballasts. Today's T-8 electronic ballasts also allow more configurations, such as ballast factor adjustments to provide the optimal amount of light for each system and each dollar invested. Use the ballast factor to determine the actual efficiency of the ballast. Many of these new ballasts have much longer operating lives than the ballasts they replace. Some low-efficiency replacement ballasts are still available but should not be used.

Unlike magnetic core-and-coil ballasts, some electronic ballasts can be dimmed, but they also can produce significant levels of harmonic distortion. When recommending electronic ballasts, specify ballasts with a total harmonic distortion (THD) less than 20%.

Strobing and flickering lamps strain the ballast. Strobing occurs as the ballast attempts to restart and operate the lamp(s). The problem is sometimes due to loose or corroded connections, but most often the lamp has come to the end of its useful life. As lamps age, their light output decreases and they become more difficult to start and operate. If checking the connections and lamps does not reveal the cause of the flickering, then ballast replacement may be necessary.

12.2.4 Luminaires (fixtures): lenses, diffusers, and reflectors

The luminaire is the complete lighting fixture. It consists of a housing socket, the light source (lamps), and the components that distribute the light, such as the lens, the diffuser, and the reflector.

The coefficient of utilization (CU) of a lighting fixture is the ratio of the light falling on the working surface to the light produced by the lamp in that fixture. Light is absorbed and converted to heat by surfaces, by lamp-to-lamp interactions, and by the lenses. Tables of common CU values can be found in the IES Lighting Handbook [IES, 2011]. The coefficient of utilization helps determine the number of luminaires and lamps required for a given application:

$$\begin{aligned}\text{Number of Luminaires} &= [(\text{Maintained Illumination in}\\ &\quad \text{Lux or Footcandles}) \times (\text{Area in Square Meters or Square Feet})] /\\ &\quad [(\text{Required Light Output/Fixture Lumens})\\ &\quad \times (\text{Coefficient of Utilization}) \times (\text{Light Loss Factor})]. \qquad (12.1)\end{aligned}$$

$$\begin{aligned}\text{Number of Lamps} &= (\text{Number of Luminaires})/\\ &\quad (\text{Number of Lamps in a Fixture}). \qquad (12.2)\end{aligned}$$

12.2.4.1 Types of fixtures

Some common types of lighting fixtures are listed below. Retrofit options are noted where appropriate.

Jars and globes. These fixtures typically have low CU values. Although they generally use incandescent lamps, compact fluorescent lamps can also be used in jar and globe fixtures.

Wall surface-mounted. These fixtures have moderate CU values.

Pendant. These fixtures utilize a variety of lamp types and fitted with globes, lenses, reflectors, or diffusers. These fixtures often hang from the ceilings of offices or classrooms in commercial applications or from medium or high-bay structures in production and warehouse areas. They can sometimes be lowered to increase the light levels at the work surfaces if the resulting light distribution is acceptable.

Track and recessed. These fixtures (sometimes referred to as downlights) may use baffle (a uniformed groove texture) trims when A-lamps or deeply recessed R-lamps are used. ERs are designed to cast more light out of recessed fixtures. If an adjustable-depth can-type fixture without a specular reflector looks bright inside, there is potential for using an ER lamp. However, a protruding ER lamp is not a good application.

Although there are some excellent downlights available, which have been designed for CFLs, in general a compact fluorescent screw-in lamp should not be used in a track or recessed fixture unless lighting levels can be reduced. A CFL casts nearly all its light out to the sides instead of downward where it is needed in a downlight.

Area lights. Many of these fixtures cast light in nearly all directions. Much of the light is lost to the sky and overhead trees, or it trespasses off the property. Cutoff luminaires that use reflectors to direct the light where it is needed can produce the same light levels for the area of interest with about a third of the power.

Streetlights. Some types such as top-mounted and cobra head lights cast light upward and away from the area of interest. Cutoff luminaires offer a good alternative here too. Figure 12.6 demonstrates the application of well-controlled lighting fixtures to reduce outdoor lighting costs of poorly applied fixtures.

Exit signs. Older signs typically use two 20-Watt or two 25-Watt incandescent lamps, which can be replaced by a 1−3 Watt colored (green or red) LED retrofit lamp. This will reduce operating costs by more than 90%. These LED exit lamps can last up to 90,000 hours when run continuously. New exit signs use mostly LEDs.

Floodlights. These specialized luminaires are used to illuminate area, parking, or facility yard applications. As these applications cast a large amount of lumen output, floodlights are attached to poles, structures, walls, and use mostly HID and LED sources.

12.2.4.2 Lenses, Diffusers, and Reflectors

Adding lenses or reflectors to a lighting fixture changes the light distribution pattern (i.e., photometrics). Replacing the lens or diffuser can be helpful in

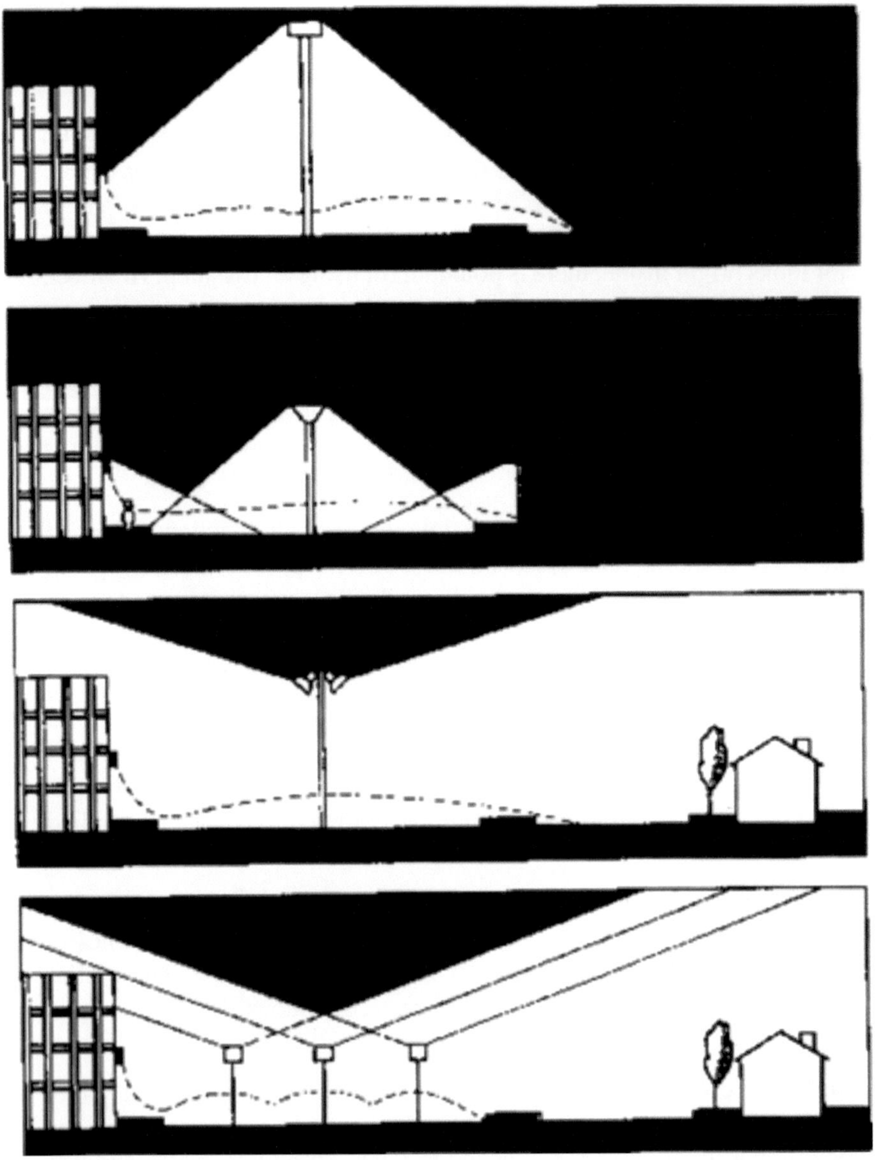

Figure 12.6 Effective outdoor lighting [National Lighting Bureau, 1987, Figure 11].

situations where the system was poorly designed, or the use of the space has changed. For example, many office areas that were lighted appropriately for traditional paperwork now have glare on video terminals because people are using computers instead of typewriters or pens. Polarized lenses, deep cell baffles, or recessed indirect diffusers can reduce the glare, but they also reduce the amount of light leaving the fixture.

Glare from the specular surfaces reduces visibility. This veiling glare can be reduced with careful fixture positioning and proper selection of diffusers. Glare on video display terminals can often be alleviated by adjusting the monitor's tilt or position. Figure 12.7 shows how glare can reach the eyes of a worker.

Most fixtures have built-in reflectors. A fixture with a baked white enamel finish has a reflectivity of about 0.88. The practice of removing half the lamps and adding a reflector to a fixture will reduce the amount of light leaving a fixture. Therefore, this is only practical when the area is overlighted. The performance of lenses, diffusers, and reflectors depends on environmental factors such as dirt accumulation, oxidation, glare, discoloration due to UV light exposure, and vandalism. Surface finishes react with airborne chemicals and lose some light transmission capability as they age. Choosing materials appropriate for the work environment and cleaning the fixtures regularly will

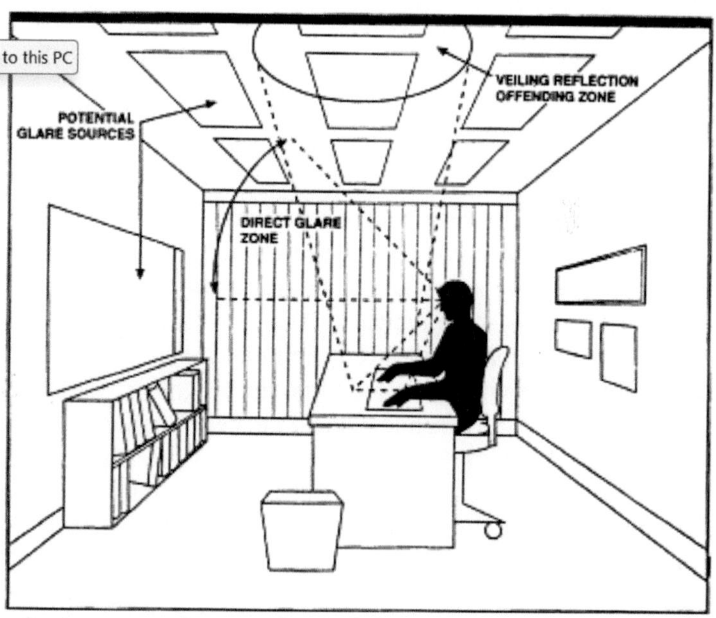

Veiling reflections commonly occur when the light source is directly above and in front of the viewer, in the "offending zone."

Figure 12.7 Veiling reflections [National Lighting Bureau, 1987, 1987, Figure 3].

Table 12.4 Comparison of acrylic and polycarbonate lenses [National Lighting Bureau, 1987, 1987, Table 4].

	Acrylic	High-impact acrylic	Polycarbonate
Light transmission	92%	90%	88%
Aging-light stability	10–15 yrs	10–15 yrs	3–4 yrs
Impact strength	1	10×	30× but degrades rapidly
Haze	Under 3%	Under 3%	Under 3% but degrades
Scratch resistance	Very good	Good	Good
Burning character (Underwriters Laboratory Class)	Slow	Slow	Self- extinguishing
Smoke generation	Slight	Slight	High
Resistance to heat	200 °F	180 °F	250 °F
Type smoke	Nontoxic	Nontoxic	Toxic
Relative cost	1×	2×	4×

prolong their useful lives. Outdoor lighting presents some special problems. Vandals often delight in shooting at exterior lighting fixtures. Polycarbonate lenses resist damage due to firearms and other projectiles. However, they also discolor from prolonged exposure to UV light produced by the lamp or the sun. Acrylic lenses typically last longer than even UV-inhibited polycarbonate lenses. Table 12.4 gives a comparison of the performance of acrylic and polycarbonate lenses over time.

12.2.5 Solid-state lighting

LEDs, or light-emitting diodes, are semiconductor devices that produce visible light when an electrical current is passed through them. LEDs are a type of SSL, as are organic light-emitting diodes (OLEDs) and light-emitting polymers. Figure 12.8 shows some typical LED lamps. LED lighting differs from incandescent and compact fluorescent lighting in several ways. LED lighting devices use light-emitting diodes to produce light very efficiently. A DC current passes through a semiconductor material, which generates light. The heat produced is absorbed into a heat sink. More LEDs create a module. Modules could be surface-mounted devices or chips on board for higher lumen output. LEDs require a driver to convert and transform the line voltage AC current to lower power DC.

Two semiconductor materials are commonly used, depending on the required color: AllnGaP (elements from the periodic table used for red and orange colors) and InGaN (similar but for green and blue). White light can be generated in a couple of ways by combining modules of red-green-blue

Figure 12.8 Typical LED lamps.

(RGB) LED clusters mounted together on a common base or by using a blue LED with a yellow-emitting phosphor. This creates a spectral wavelength distribution perceived as white. Colored LEDs are widely used as signal lights and indicator lights or for signage. RGB LEDs are mainly used in dynamic color-mixing applications.

12.2.5.1 Properties of LED lights

Directional light emission. One of the defining features of LEDs is that they emit light in a specific direction. Since directional lighting reduces the need for reflectors and diffusers that can trap light, well-designed LED fixtures can deliver light efficiently to the intended location. In contrast, fluorescent and "bulb" shaped incandescent lamps emit light in all directions. Much of the light produced by these lamps is lost within the fixture, reabsorbed by the lamp, or escapes from the fixture in a direction that is not useful for the intended application. For many fixture types, including recessed downlights, troffers, and undercabinet fixtures, it is not uncommon for 40%–50% of the total light output of fluorescent and incandescent lamps to be lost before it exits the fixture.

Lumen output. With LEDs, the metric sought is lumens rather than wattage. With continuous increasing energy efficiency, LED lamps and luminaires are improving the lumen output levels to successfully match all the conventional lighting equipment.

Useful life. The useful life of LED lighting products is defined differently than that of other light sources, such as incandescent or CFL. This is because LEDs typically do not "burn out" or fail. Instead, they experience lumen depreciation, where the amount of light produced decreases and light color appearance can shift over time. Instead of basing the useful life of an LED product on the time it takes for 50% of a large group of lamps to burn out (as is the case with traditional sources), LED product "lifetime" is set based on a prediction of when the light output decreases 30% (i.e., L70 from Figure 12.20). Good quality white LEDs in well-designed fixtures are expected to have a useful life of 30,000–50,000 hours or even longer, compared with typical incandescent lamp that lasts about 1,000 hours. A comparable CFL lasts 8,000–10,000 hours, and the best linear fluorescent lamps can last more than 30,000 hours.

Thermal management. A primary cause of lumen depreciation is heat generated at the LED junction. LEDs do not emit heat as infrared radiation like other light sources, so the heat must be removed from the device by conduction or convection. Thermal management is arguably the most important aspect of successful LED system design. This is usually done with a heat sink, which is a passive device that absorbs the heat produced and dissipates it into the surrounding environment, keeping LEDs from overheating and burning out. LED products use a variety of unique heat sink designs and configurations to manage heat, so they may look very different from each other.

Temperature response. The luminous output drops as temperature increases. While the maximum operating temperature for LEDs is normally 100 °C (212 °F), LEDs operate well in cold ambient temperatures.

Dimmability. As constant-current sources, LEDs are inherently dimmable. Their level of brightness is adjusted simply by controlling the current passing through the LED chips. Unlike conventional sources, dimming does not affect the efficacy or longevity of LEDs. In fact, dimming can extend the lifespan of LEDs by lowering their operating temperature. Also, the dimming range of LEDs is broader than that of compact fluorescent and high-intensity discharge lamps. They can turn down to less than 1% of full output, compared to 10%–30% of measured light output for compact fluorescents, and 30%–60% of lamp power for HID lamps. LED devices require a dimming (constant-current) driver in order to dim by using analog or digital control.

12.2.5.2 LED characteristics

LEDs offer the potential for cutting general lighting energy use nearly in half by 2035 [DOE, 2019a], saving energy dollars and carbon emissions in the process. Their unique characteristics – including compact size, long life and ease of maintenance, resistance to breakage and vibration, good performance in cold temperatures, lack of infrared or ultraviolet emissions, and instant-on performance – are beneficial in many lighting applications. The abilities to be dimmed and to provide color control are other benefits of LED lights.

Key aspects of lighting quality are color appearance (whether a white light appears more yellow/gold or bluer) and color rendering (the ability of the light source to render colors, compared to incandescent and daylight reference sources).

Color appearance. Color appearance is communicated using correlated color temperature (CCT) on the Kelvin (K) scale (see more in section 12.2.1.2). For most indoor lighting applications, warm-white (2,700–3,000 Kelvins) and in some cases neutral-white (3,500–4,000 Kelvins) light is appropriate. Cool-white LEDs with very high CCT (bluish in appearance) tend to offer higher efficacy at low cost but may not meet user expectations for color. An increasing number of high-efficacy LED products are available in warm-white or neutral-white colors, to the point where many have surpassed CFLs. Two light sources with identical CCTs can render object colors very differently due to the differences in spectra. While CCT provides an indication of whether a light source may appear yellowish or bluish in color, *D*uv is a supplementary metric that describes the distance of a light color point from the black body curve and can help identify sources with excessively greenish or pinkish hues. Tunable white LED lamps and luminaires can dynamically modify the CCT and the output of the emitted light (usually between 2,500 and 7,000 Kelvins) to match the daylight color dynamic throughout the day and entrain people's circadian cycle for better health and productivity.

Color rendition. The color rendering index (CRI) measures the ability of light sources to render colors, compared to either incandescent reference sources if warm in color, or daylight reference sources if cooler in color (see more in section 12.2.1.3). The leading high-efficiency LED manufacturers now claim a CRI of 80 or higher for phosphor-converted, warm-white devices. In general, a minimum CRI of 80 is recommended for interior lighting, with CRIs of 90 or higher indicating excellent color rendering. The CRI has been found to be inaccurate for RGB (red, green, blue) LED systems because it is poor at predicting the quality of the appearance of saturated color objects and does not correspond well to human perception of color quality. As a supplement to CRI, a lamp's R9 value describes how closely it renders a saturated red color sample, relative to the reference illuminant. CCT and CRI together only get you in the ballpark for selecting and matching lamp colors. Among numerous new color-rendering metrics proposed in recent years, TM-30 is the more prominent considering more sets of numbers (color fidelity, gamut area, and chroma shift visualization) for a better characterization. While waiting for the new metric wide-adoption, color rendering of LED products should be evaluated in person and in the intended application if possible.

Table 12.5 LED efficacy compared to conventional lighting technologies [DOE SSL Program, 2015].

Lighting product	Luminous efficacy, lm/W	Usable life, hours
LED A19 Lamp (Dimmable, Warm-White)	78	25,000
LED PAR38 Lamp (Warm-White)	70	28,000
LED T8 Tube (Neutral-White)	107	50,000
LED 6" Downlight (Warm-White)	64	40,000
LED Troffer 2ft × 4ft (Warm-White)	94	56,000
LED High/Low-Bay Fixture (Warm-White)	102	90,000
LED Street Light	96	50,000
OLED Luminaire	43	40,000
HID (High Watt) System	115	15,000
Linear Fluorescent System	108	25,000
HID (Low Watt) System	104	15,000
CFL A19 Replacement	70	12,000
Halogen A19	20	8,400
Incandescent A19	15	1,000

Efficiency. Two aspects of energy efficiency are important to consider: the efficacy of the LED device itself (source efficacy), and how well the device and fixture work together in providing the necessary lighting (luminaire efficacy). The efficacy seems also to depend on the color temperature, the cool white LEDs being more efficient than the warm white ones. How much electricity is used to provide the intended lighting service depends not only on the LED device but also on the lighting fixture design. Because they are sensitive to thermal and electrical conditions, LEDs must be carefully integrated into lighting fixtures. The efficiency of a poorly designed fixture that uses even the best LEDs will be only a fraction of what it would be if the fixture were well-designed, and the design can also affect lumen maintenance.

With continuous improvements in the fabrication of LED chips and the other device components, the energy performance of white LED products continues to improve rapidly. DOE's long-term research and development goal calls for cost-effective LED packages, producing 249 lumens per Watt by 2035 [DOE, 2019a]. Table 12.5 shows typical luminous efficacies for legacy and LED sources, including ballast losses as applicable.

Cost effectiveness. Costs of LED lighting products vary widely. Good quality LED products currently carry a significant cost premium compared to standard lighting technologies. However, costs are declining rapidly. For example, recent industry road mapping [DOE, 2019b] indicates prices for general service LED lamps have declined four times from approximately $12 per thousand lumens (kilolumens, or klm) in 2017 to approximately $4/klm in 2020. Overall LED costs are forecasted to further reduce 25% over the next 10 years. It is important to compare total lamp replacement, electricity, and maintenance costs over the expected life of the LED product.

12.3 Determining Lighting Needs

Lighting encompasses a wide range of visual tasks, operating conditions and economic considerations. The seeing task varies widely from small to large, dark to light, specular to diffuse, transparent to translucent, and flat or contoured surfaces. In addition, the tasks or viewer may be moving. For each of these conditions, the lighting must provide adequate visibility for the purpose – it could be reading in an office or assembling components into a finished product in industry.

Physical hazards may exist in the workplace and lighting must address safety as well as performance issues. The speed of operations may allow only a minimum inspection time, thus increasing the complexity of the visual task. The lighting must be designed to increase the speed and accuracy for the viewer.

12.3.1 Lighting metrics

The purpose of lighting is to enable us to see with optimum effectiveness and to create a pleasant environment. Lighting metrics are psychophysical units because they are based on the physical unit of power (Watt) and the way the psychological eye and brain respond to power. While the eye is where our visual system first creates a first impression (inverted image) of the visual environment, the optic nerve carries it to the brain, which reverts the image completing the visual process.

12.3.1.1 Luminous flux

This metric expresses the "total light output" of a light source (lamp or luminaire) and measure unit is the lumen (lm). Lumen output of lamps can be found in the lamp manufacturers' catalogues.

For example, in a recessed fluorescent fixture, a standard T8 fluorescent lamp provides an output of 2,800 lumens, while a tubular LED (TLED) replacement lamp (using the same fixture) only 2,100 lumens. However, the overall luminaire's (which could include several lamps) lumen output is rather similar as the fluorescent lamps have an omnidirectional light distribution (compared to the directional distribution of the TLED) and not all the generated light is leaving the fixture toward the visual task.

12.3.1.2 Luminous intensity

This metric expresses the intensity of a light source in a specific direction and is measured in candela (cd) or candlepower (cp). Candela and candlepower metrics are used primarily in conjunction with reflector lamps and luminaires and details can be found in the luminaire manufacturers' catalogues.

12.3.1.3 Luminance (brightness)

This metric expresses light directionally reflected from (or transmitted through) a surface and is the closest metric we have to express "brightness." It is expressed in units of intensity per unit area – namely, candelas per square meter (SI) or candelas per square foot (Imperial). High luminance can produce glare that can be reduced or avoided by using diffusers, shields, or fixture positioning (see more in Section 12.2.4.2).

12.3.1.4 Illuminance

This metric expresses the amount of light incident arriving at a surface. Illuminance depends upon lumens from the light source and is inversely proportional with the distance from the surface. It is expressed in lumens per unit area (typically square feet or square meters): foot-candles (Imperial) or Lux (SI). The conversion between these units is one foot-candle equals 10.764 Lux. For approximations, one can round off the ratio to 1:10 (i.e., 1 fc = 10 lx).

12.3.1.5 Light levels

A variety of techniques are available for estimating the lighting levels in a given space. The average illuminance method described in the IES Lighting Handbook [IES, 2011] incorporates the major variables affecting light utilization: amount of light produced by lamps, amount of light exiting the fixture, mounting height and spacing of fixture, fixture photometrics, lumen dirt depreciation, lamp lumen depreciation, ballast factor, and room surface finish characteristics. A worksheet to calculate lighting levels is also found in the IES Lighting Handbook [IES, 2011].

The IES has developed standards for appropriate lighting levels for typical applications. Lighting levels are generally expressed in terms of illuminance. Table 12.6 shows the illuminance category for several commercial and industrial applications.

Table 12.6 Lighting recommendations for specific tasks.

Area-activity	Illuminance category
Bakeries	D
Classrooms	D–E
Conference rooms	D
Drafting rooms	E–F
Hotel lobbies	C–D
Home kitchens	D–E
Inspection, simple	D
Inspection, difficult	F
Inspection, exacting	H
Machine shops	D–H
Material handling	C–D
Storage, inactive	B
Storage, rough, bulky items	C
Storage, small items	D

Once the appropriate illuminance category has been identified, Table 12.7 can be used to determine the range of illuminance values needed to achieve desirable lighting levels. In using these tables, the IES recommends that the lower values be used for occupants whose age is under 40 and/or where the room reflectance is greater than 70% and that the higher values be used for occupants more than 55 years old and/or where the room reflectance is less than 30%. For occupants between 40 and 55 years of age and where the reflectance is 30%–70% or where a young occupant is combined with low reflectance or an older person is in a high-reflectance environment, the intermediate values should be used. In addition, the need for speed and accuracy influences the amount of light needed, with higher speed and accuracy demanding more light.

Lighting levels can sometimes be reduced if there is sufficient contrast between the work and its surroundings. However, too much contrast between the work and the ambient environment will fatigue the eyes. For example, occupants of dark paneled offices who work with pencil on brightly illuminated white paper will often experience eye fatigue.

The illuminance values considered acceptable have changed through the years in response to emerging technology. Recommended light levels were relatively low in the incandescent lamp era due to restrictions resulting from heat production. Starting with the 1950s, more efficient light sources were developed, and energy costs plummeted, leading to a tripling of the recommended values [Pansky, 1985]. Rising energy and power demand costs beginning with the 1970s spurred the industry to reduce the recommended

Table 12.7 Illuminance categories and illuminance values.

		Ranges of illuminance		
Type of activity	**Illuminance category**	**Lux**	**Foot-candles**	**Reference work plane**
Public spaces with dark surroundings	A	20–30–50	2–3–5	
Simple orientation for short temporary visits	B	50–75–100	5–7.5–10	General lighting throughout spaces
Working spaces where visual tasks are only occasionally performed	C	100–150–200	10–15–20	
Performance of visual tasks of high contrast or large size	D	200–300–500	20–30–50	
Performance of visual tasks of medium contrast or small size	E	500–750–1,000	50–75–100	Illuminance on task
Performance of visual tasks of low contrast or very small size	F	1,000–1,500–2,000	100–150–200	
Performance of visual tasks of low contrast and very small	G	2,000–3,000–5,000	200–300–500	
size over a prolonged period				
Performance of very prolonged and exacting visual tasks	H	5,000–7,500–10,000	500–750–1,000	
Performance of very special visual tasks of extremely low contrast and small size	I	10,000–15,000–20,000	1,000–1,500–2,000	Illuminance on task, obtained by a combination of general and local (supplementary lighting)

lighting levels to the minimum necessary to provide adequate illumination. This trend has stabilized now with the advent of LEDs that can balance much better energy efficiency and adequate light-level needs.

12.4 Maintaining the Lighting System

In addition to a proper choice of light sources, ballasts, and luminaires, the effectiveness (consistent level of lighting) and energy efficiency of a lighting system depend on maintenance policies. Maintenance includes both cleaning and relamping.

12.4.1 Luminaire maintenance

The performance of lenses, diffusers, and reflectors depends on environmental factors such as dirt accumulation, oxidation, vandalism, and degradation due to ultraviolet (UV) light exposure. Typical fluorescent lamp performance under various temperature conditions is shown in Figure 12.9.

Lamps, fixtures, reflectors, lenses, and diffusers collect dust and insects. Dust accumulation on lighting fixtures and on surfaces adjacent to lighting fixtures reduces light utilization by up to 40% and increases heat production. Periodic cleaning of the fixtures will maintain higher and more uniform light

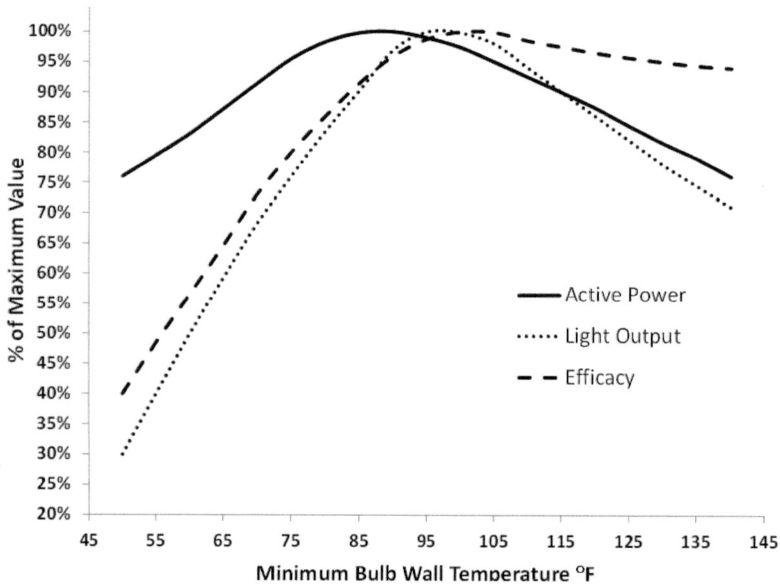

Figure 12.9 Lamp performance vs. temperature [adapted from IES, 2011].

levels. All lamps should be cool before cleaning. Gloves should be worn when cleaning any mirror-like reflective part of a luminaire. Quartz lamps should not be cleaned.

Outdoor lighting has some special maintenance problems. Poorly sealed gaskets allow insects to clog the lenses of outdoor lighting fixtures. Dead insects can completely block out light. Overgrown vegetation can also reduce lighting levels from outside fixtures. Small trees planted under or near lighting fixtures can quickly grow to block the light source. Regular trimming of shrubs and trees will help to fully utilize these light sources.

12.4.2 Establishing the lighting system maintenance schedule

Establishing a good maintenance schedule for a lighting system takes three steps. First, you must determine the maintenance characteristics of the

Table 12.8 Maintenance categories for luminaire types.

Maintenance category	Top enclosure	Bottom enclosure
I	None	None
II	None Transparent with 15% or more uplight through apertures Translucent with 15% or more uplight through apertures Opaque with 15% or more uplight through apertures	None Louvers or baffles
III	Transparent with less than 15% uplight through apertures Translucent with less than 15% uplight through apertures Opaque with less than 15% uplight through apertures	None
IV	Transparent unapertured Translucent unapertured Opaque unapertured	None Louvers
V	Transparent unapertured Translucent unapertured Opaque unapertured	Transparent unapertured Translucent unapertured
VI	None Transparent unapertured Translucent unapertured Opaque unapertured	Transparent unapertured Translucent unapertured Opaque unapertured

luminaires in your facility. Table 12.8 lists the maintenance categories for a variety of luminaires.

The next step is to determine what dirt conditions the luminaires are likely to experience. Table 12.9 shows dirt conditions for representative areas. Once you know both the maintenance category from Table 12.7 and the appropriate dirt conditions for the facility from Table 12.8, then use the graphs in Figure 12.10 to set the luminaire maintenance schedule. These graphs show the effect that dirt accumulation has upon lighting levels over a period of months.

These graphs are for average conditions. Actual conditions may warrant a more frequent cleaning schedule than the graphs indicate. Using open fixtures and diffusers often reduces the potential for light loss since there are fewer surfaces for settling, and particles and dust can be more easily removed from these fixtures. However, an environment with potentially explosive dust (such as flour, corn, coal, etc.) must have sealed explosion-proof fixtures. Typical lamp lumen depreciation, luminaire dirt depreciation, and room surface dirt depreciation curves for fluorescent lamps are shown in Figure 12.11.

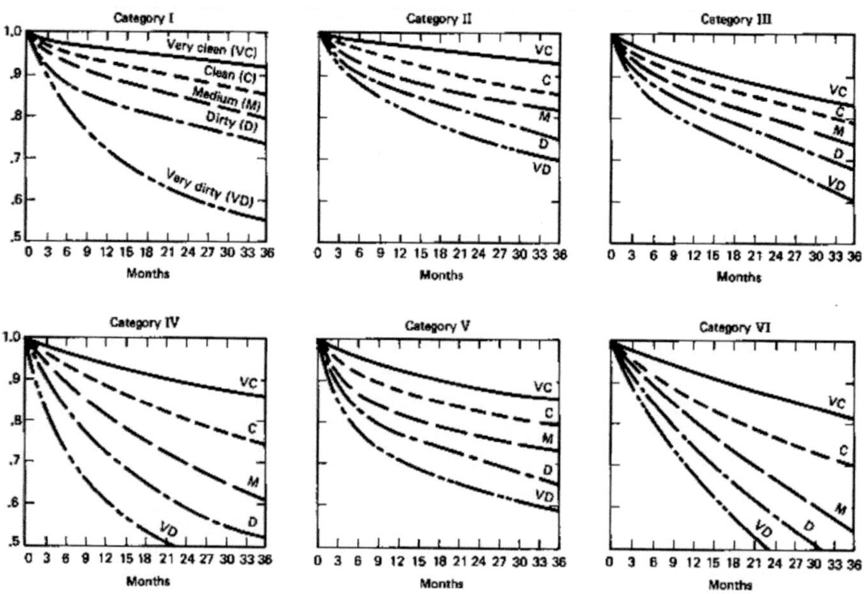

Figure 12.10 Luminaire dirt depreciation factors (fraction of full light output is the scale on the y-axis) [IES, 2011].

12.4.3 Relamping strategies

The usual strategy for replacing lamps in many facilities is to wait until a lamp burns out and then replace it (called spot relamping). This relamping strategy is not necessarily the best one for a facility to follow because it does not consider such factors as labor costs or lumen depreciation. It is often more economical to replace all the fluorescent and HID lamps in a facility at one time (called group relamping). Spot relamping is more labor intensive and results in less efficient lighting than group relamping. However, spot relamping can be more practical for lamps with a short life such as incandescent lamps.

The reason that spot relamping results in a loss of lighting efficiency is that the amount of light that comes from a lamp declines with the age of the lamp. Performance curves, such as those in Figure 12.12, show how the light output is reduced as a function of ordinary usage [Roberts, 2014]. Note that the life of a lamp is measured in hours of use rather than installed

Table 12.9 Degrees of dirt conditions.

	Very clean	Clean	Medium	Dirty	Very dirty
Generated dirt	None	Very little	Noticeable but not heavy	Accumulates rapidly	Constant accumulation
Ambient dirt	None (or none enters area)	Some (almost none enters)	Some enters area	Large amount enters area	Almost none excluded
Removal or filtration	Excellent	Better than average	Poorer than average	Only fans or blowers if any	None
Adhesion	None	Slight	Enough to be visible after some months	High-probably due to oil, humidity, or static	High
Examples	High-grade offices, not near production; laboratories; clean rooms	Offices in older buildings or near production; light assembly; inspection	Mill offices; paper processing; light machining	Heat treating; high-speed printing; rubber processing	Similar to "Dirty" but luminaires within immediate area of contamination

Source: Courtesy of the Illuminating Engineering Society of North America.

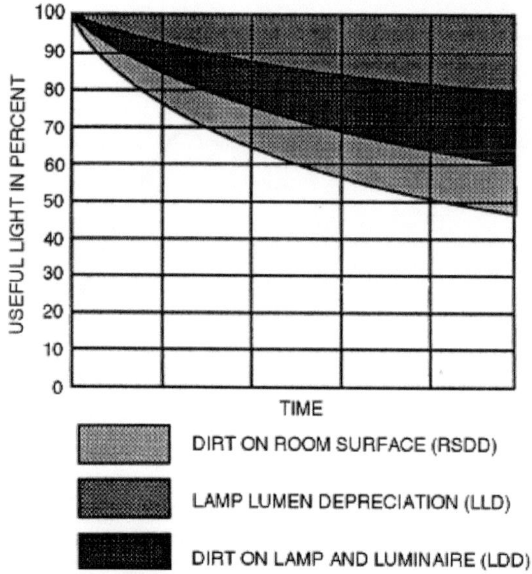

Figure 12.11 Lumen maintenance curve for fluorescent lighting systems [WAPA, 1990, Figure 12.20].

hours. In addition to degraded performance of individual lights, the total lighting system performance is decreased as individual lamps burn out. Typical mortality curves [GE, 1969a; GE, 1969b; Roberts, 2014], showing the percentage of lamps in service as a function of time, are given in Figures 12.13, 12.14, 12.15, and 12.16.

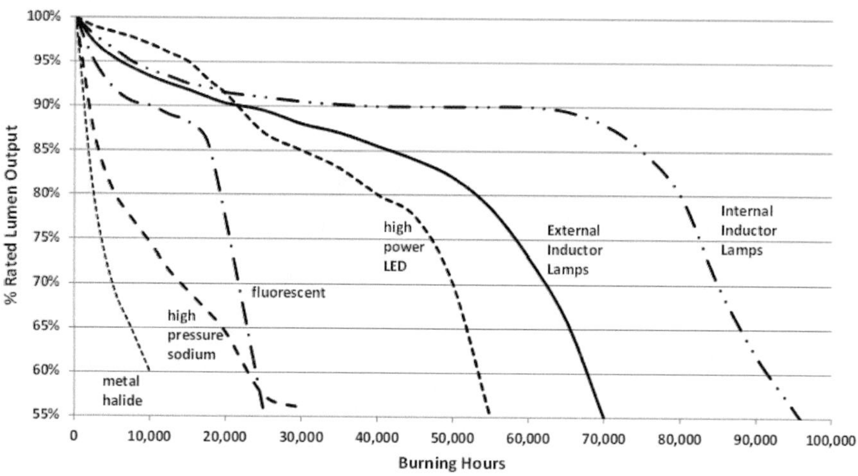

Figure 12.12 Typical lumen maintenance curves for various light sources (adapted from Roberts, 2014).

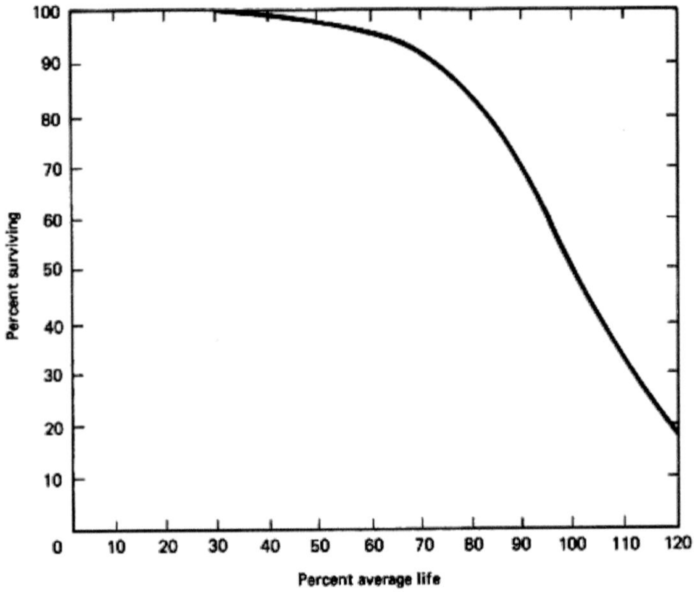

Figure 12.13 Typical mortality curve for fluorescent lamps (GE, 1969; courtesy of General Electric).

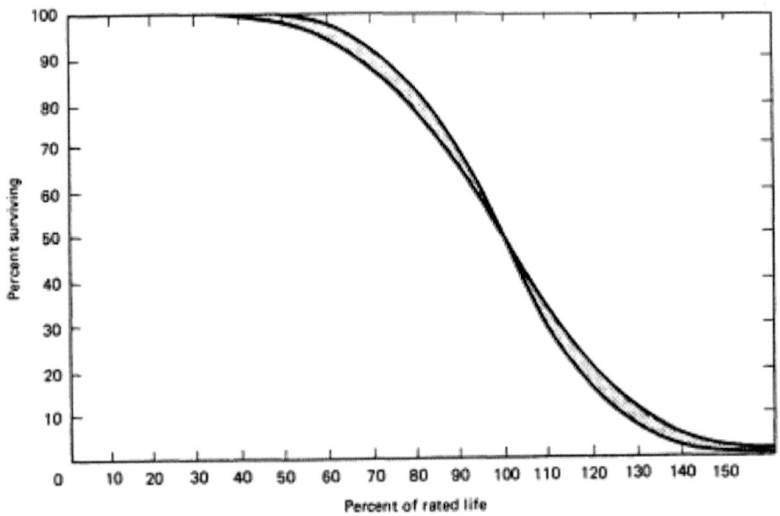

Figure 12.14 Typical mortality curve for filament lamps (GE, 1969; courtesy of General Electric).

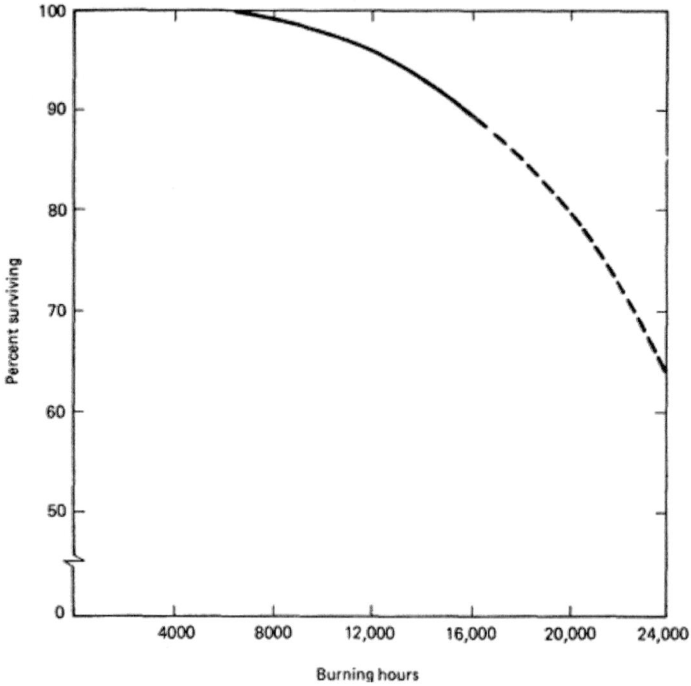

Figure 12.15 Typical mortality curve for HID (mercury vapor) lamps (GE, 1969; courtesy of General Electric).

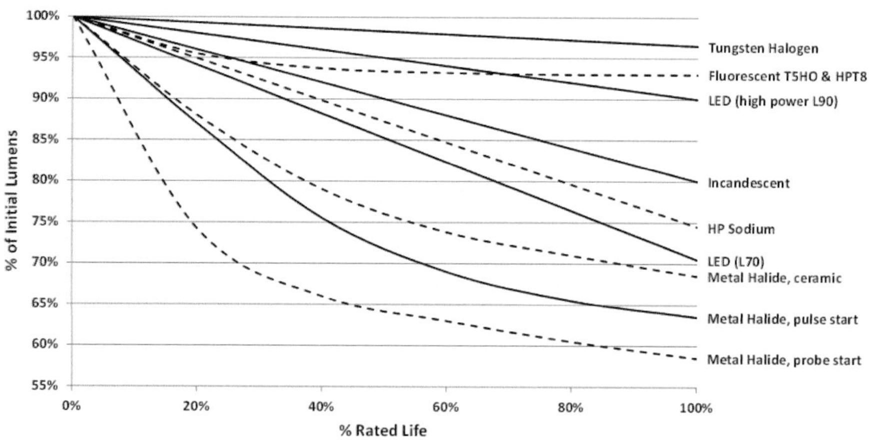

Figure 12.16 Mortality curve for various lighting sources (IES RP-36-15).

As these figures show, lamp mortality at 85%–100% of rated life is about 25 times that of lamps aged 0%–70% of rated life. Therefore, fixtures should be group relamped when lamps are between 70% and 80% of the rated lamp life. Figure 12.17 shows an example of how ambient temperature influences the life of the LED sources (a streetlight luminaire in this example).

The average useful fluorescent lamp life is shown in Figure 12.18. Since most four-foot fluorescent lamps are rated for 20,000 hours, they should be replaced after 16,000 hours of operation. Lamp life ratings indicate the point at which half of the lamps are likely to have failed. Fluorescent lamp lifetimes are rated at 3 hours per start, and HID lamps are rated at 10 hours per start. Operating the lamps for shorter periods will reduce lamp life.

The benefits of group relamping are as follows.

1. **Reduce labor costs.** Spot relamping can require up to 30 minutes to move furnishings or equipment, set up, replace the lamp, and put equipment away (e.g., ladder, lamps, tools). In group relamping, a fixture can usually be relamped and cleaned in 5 minutes. The loss-of-use cost of replacing a few lamps before they have burned out is generally less than the increased cost of the labor to individually replace burned out lamps.

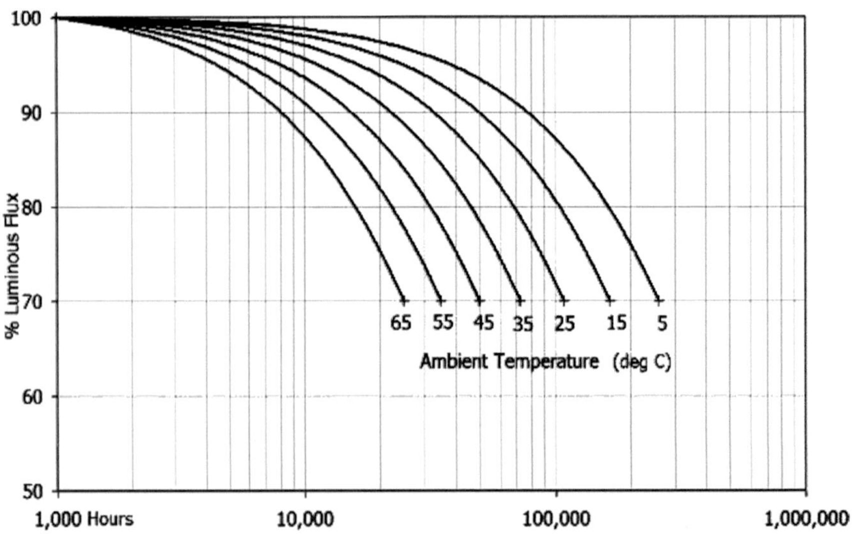

Figure 12.17 Example of mortality curves for LEDs (DOE, 2008, Figure 5).

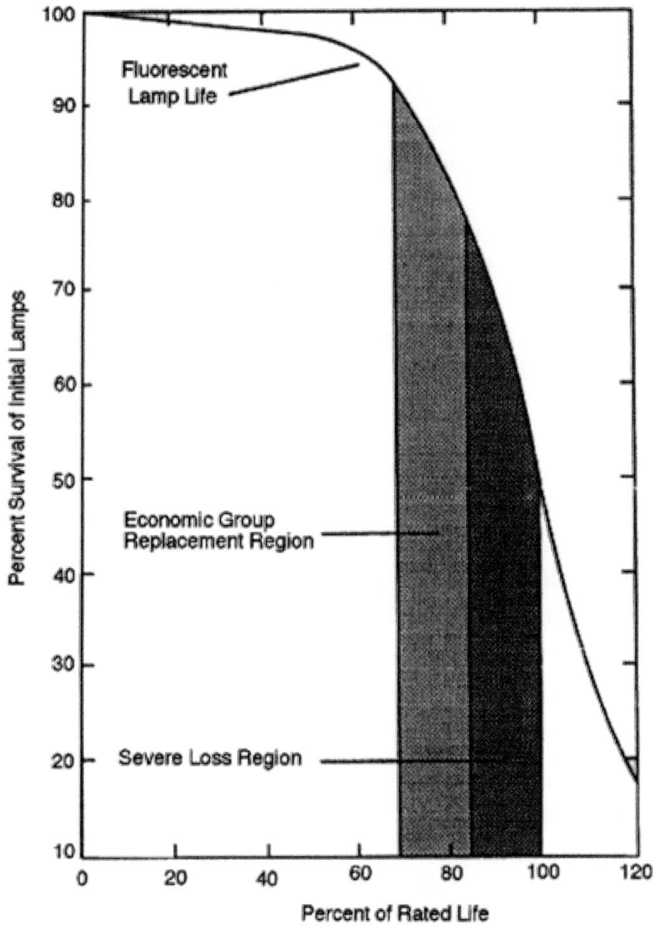

Figure 12.18 Mortality curve for fluorescent lighting systems [WAPA, 1990, Figure 2-22].

2. **Reduce lamp costs.** Purchasing a large number of lamps at one time allows for high volume discounts. Fewer purchases result in less time spent ordering, receiving, and stocking lamps.
3. **Allow lamp maintenance to be scheduled.** Relamping can be scheduled for slow periods for the maintenance staff. Scheduled relamping also allows a regular schedule to be set for regular inspection and cleaning of lamps and fixtures.
4. **Maintain higher and more uniform lighting levels.** The light output of a lamp decreases with age. Group relamping ensures that all lamps have a high light output. Fewer lamp failures, less flickering, and reduced

swirling and color shift produce a safer, more comfortable work environment. If the lighting system was over-designed to allow for loss of light levels, group relamping may allow for some delamping.

5. **Reduce inventory needs.** Fewer lamps must be stored in inventory since fewer spot failures occur in group relamped fixtures when the relamping interval is set correctly.
6. **Ensure the correct lamp use.** Spot relamping often results in the installation of a variety of lamp types with inconsistent light output levels, lifespans, and colors. This can occur when the inventory of spare lamps runs out and the person who purchases a quick replacement does not know the correct lamp type, cannot locate the correct type, or does not realize that the cheapest lamp may not be the best value. Group relamping provides an opportunity to install the newest energy-efficient lamps.
7. **Extend ballast life.** Ballasts have to work harder to start and operate strobing lamps that are near the end of their life.
8. **Reduce interruptions in work area.** Group relamping prevents most of the unplanned lamp replacements.

General Electric [GE, 1969a] has provided the following cost formulas for determining relamping costs:

Spot replacement costs:

$$C = L + S. \tag{12.3}$$

Group relamping costs:

$$C = \frac{L + G}{I} \tag{12.4}$$

where: C = total replacement cost per lamp L = net price per lamp

S = spot replacement labor cost per lamp G = group replacement labor cost per lamp

I = group relamping interval (% of rated lamp life)

Example 12.1

An office building contains a number of small (400 ft^2) rooms, each of which has four two-lamp fluorescent fixtures. Every time a maintenance person changes lamps, they must bring a ladder into the room and clear away furniture. It takes the person 15 minutes to replace one lamp. It takes 25 minutes to replace all the lamps in a room and clean the luminaires if all the work is done at one time. The lamps cost $8.50 each, and labor costs are

\$50/hour. The lamps are used about 2,000 hours/ year. The average lamp life is 20,000 hours. Determine whether group relamping with $I = 0.8$ is preferable to spot relamping for this building.

Solution

Find the cost for spot relamping and group relamping using equations above and select the method with lowest cost.

$$\begin{aligned} L &= \$8.50 \text{ per lamp} \\ S &= (\text{time to replace lamp in hours}) \times (\text{cost of labor per hour}) \\ &= 15/60 \times \$50 = \$12.50 \\ G &= (\text{time to replace lamp in hours}) \times (\text{cost of labor per hour}) \\ &= (25/8)/60 \times \$50 = \$2.60. \\ I &= 0.8 \end{aligned}$$

and thus:

$$\begin{aligned} C_{\text{spot}} &= 8.50 + 12.50 \\ &= \underline{\$21.00/\text{lamp}} \\ C_{\text{group}} &= (8.50 + 2.60)/0.80 \\ &= \underline{\$13.88/\text{lamp}} \end{aligned}$$

Therefore, the decision should be made to use group relamping. The reduced labor cost more than offsets the cost of the lost lamp life. In fact, the savings from group relamping is almost twice the full cost of each lamp.

12.5 The Lighting Survey

To perform an objective evaluation of the lighting system, the energy auditor must gather the following data: lighting needs and objectives, hours and days that lighting is required, lighting levels, type of lamps, age of lamps, age of the lighting installation, ambient environment of lighting fixtures (e.g., dust exposure, air temperature, etc.), room surface characteristics, type of ballasts, condition of fixtures, and relamping practices. Building plans are not useful unless the facility was built as planned and few modifications have been made.

Forms are useful for recording the specific lighting data needed. Chapter 3 includes a sample lighting data collection form. Figure 12.19 shows a sample data collection form for recording the lighting system condition, while Figure 12.20 illustrates a sample form for recording lighting needs.

Location ______________________ **Light type (HPS, FL, etc.)** ______________

Lamps ______________

		Watts			Condition of system*	
Lighting Level (Lux)	**No.**	**Each**	**Total**	**% burning**	**Luminaires**	**Reflective surfaces**

Figure 12.19 Lighting system condition form.

Location	*Hours when light is needed*	*Importance of color rendition*	*Task lighting possible?*	*Light levels required (fc)*

Figure 12.20 Lighting needs form.

Two basic surveys should be conducted to look for savings opportunities: one to see how the facility operates while in production and another to determine the lighting practices when the facility is dormant or shut down for the night.

12.5.1 Interviews with facility users

Interviews with the facility occupants, such as managers and workers, help the energy auditor to evaluate the relamping and maintenance practices, determine problems with the lighting system, ascertain the employees' satisfaction level, find out when light is needed, and uncover the potential for cost savings. Depending on the size and complexity of the facility, in-person interviews may not be sufficient and online surveys should be considered. Chapter 3 provides a sample response from an energy audit online survey, and a sample audit survey form is provided in the appendix. The first question for an interview or survey should be: Are you satisfied with your lighting? Major retrofits such as fixture replacement and color changes affect everyone in the work environment, so the opinions of a representative sample of all users for a facility should be obtained.

12.5.2 Measuring light levels

A light meter is needed to measure illuminance levels. An inexpensive digital light meter is practical and rugged for screening lighting levels and determining relative values. A self-calibrating digital light meter (photometer) is very useful to get fast, accurate, and repeatable measurements. Using a light meter, the auditor should record light intensity readings for each area and task of the facility.

Taking notes on the types of tasks performed in each area will help the auditor select alternative lighting technologies that might be more energy-efficient. Other items to note are the areas that may be infrequently used and are potential candidates for controlling the lighting with occupancy sensors, or the areas where daylighting may be feasible.

12.6 Regulatory and Safety Issues

The lighting industry is encountering increasing safety and environmental concerns. Some of the materials used in lighting fixtures are, or may soon be, labeled hazardous for disposal.

For example, older ballasts and capacitors may contain polychlorinated biphenyl (PCB) oil, which should be sent to a facility certified for handling hazardous wastes. Federal laws in both Canada and US require proper disposal of leaking ballasts containing PCB oil. Fluorescent lamps contain mercury vapor (Hg), antimony (Sb), cadmium (Cd), and other toxic chemicals. T5 and T8 lamps use less phosphor material and mercury gas than the conventional T12 lamps, reducing disposal costs and environmental impacts.

12.6.1 Lighting safety issues

Traditional lamps are fragile and break easily. Broken fluorescent lamps are difficult to transport and recycle. Areas subject to vibration or other mechanical stress should be illuminated with durable lamps or with fixtures that have adequate containment for broken lamps within the fixture housing. Delamping inexpensive fluorescent lighting fixtures can also be hazardous if the lamp pins come in contact with the fixture housing.

Insulation placed on top of lighting fixtures recessed in the ceiling may pose a fire hazard unless the fixture is rated for insulation contact. The insulation can increase the operating temperature of the lamps and ballasts of fluorescent or HID fixtures, which will reduce the lifespan of all the system components.

HID lamps have arc tubes that operate at high temperatures. Pieces of hot arc tube can fall from the fixture if the lamp wall is fractured. Some manufacturers recommend using lenses or fixture housings capable of containing incendiary materials.

12.6.2 Energy legislation

Regulatory restrictions on conventional light sources result in changes to both the performance and price of the affected conventional sources, and force market trends to more efficient technologies. These include both standards prescribed via congressional action (e.g., Federal Energy Policy Act – FEPA 1992 [US Congress, 1992] or the General Service Incandescent Lamp Standards established in Energy Independence and Security Act – EISA 2007) as well as energy efficiency standards that are promulgated by DOE (e.g., the Fluorescent Lamp Efficacy Standard of 2015).

These restrictions will require facility managers to take a hard look at the emerging high-efficiency lighting technologies when many of the lamps currently in use become difficult or impossible to find.

In addition, the 2005 Energy Policy Act created tax incentives for installing energy-efficient equipment. Significant tax rebates are available

Table 12.10 U.S. Minimum Energy Efficiency Regulations.

Product	**Effective date**	**Description**	**Market impact**
General service lamps	EISA 2007 - took effect between 2012 and 2014	The maximum wattage standards require a 25% efficiency increase for all general service lamps	Eliminates 40, 60, 75, and 100 W medium screw-based incandescent light bulbs
Medium screw-based CFLs	Energy Policy Act 2005 CFLs manufactured on or after January 1, 2006 must meet new requirements	All CFLs used must meet selected specifications in the 2001 ENERGY STAR® Standard, including: minimum initial efficacy levels, minimum lumen maintenance at 1,000hrs and 40% of rated life, rapid cycle stress test, and minimum lamp life of 6,000hrs	Market transition toward more efficient lamps
lamps	DOE regulation applies to lamps manufactured on or after July 14, 2012	Prescribes minimum efficacy standards for covered products in the 40 – 205 W range, determined by lamp spectrum, lamp diameter, and rated voltage. Certain small diameter, elliptical reflector, and bulged reflector incandescent reflector lamps (IRLs) are excluded.	halogen infrared technologies
Fluorescent lamps	2009 DOE amendments to EPACT 1992 effective in 2012 2015 DOE amendments effective January 26, 2018	New efficacy requirements for 4-foot medium bipin, 2 -foot U-shaped, 8 -foot slimline, 8-foot high output, 4-foot miniature bipin standard output, and 4-foot miniature bipin high output by specific CCT ranges	Eliminates T12 (few exemptions) and low efficacy T8 lamps from the market
Fluorescent ballasts	After T12 ballast regulation set for 2009, new DOE regulation applies to covered fluorescent ballasts manufactured on or after November 14, 2014	Prescribes minimum ballast luminous efficiency standards that will effectively shift the fluorescent market from T12 ballasts to T8 and T5 electronic ballast systems. Exemption for dimming, sign and residential ballasts	Because covered magnetic ballasts are unlikely to meet the standards, it removes T12 systems from the marketplace over time
Mercury vapor ballasts	Energy Policy Act 2005 - can no longer be manufactured or imported on or after January 1, 2008	Bans the sale of mercury lamp ballasts essentially banning the sale of new mercury vapor fixtures	MV ballasts are no longer available for purchase in Canada and the United States

Table 12.10 *Continued*

Product	Effective date	Description	Market impact
Metal halide fixtures	EISA 2007 - minimum efficiency standards to HID fixtures manufactured on or after January 1, 2009 2014 DOE amendments - effective February 10, 2017	Sets minimum efficiency requirements for MH fixtures between 50 and 1,000 W, determined by wattage and voltage Prohibits probe start ballasts in MH fixtures greater than 500 W and less than or equal to 1, 000 W	Increasing the efficacy and price of metal halide fixtures accordingly
Illuminated EXIT signs	Energy Policy Act 2005 - effective January 1, 2006 - can no longer be used in new installations.	Must meet ENERGY STARÂő efficiency levels (input power <= 5 Watts)	
Pedestrian crosswalk modules		Essentially forces the use of LEDs	
Traffic signal heads			
Torchiere fixtures	Energy Policy Act 2005effective January 1,2006	Torchieres shall not be capable of operating with lamps that total more than 190 Watts	
Transformers	Energy Policy Act 2005 - effective January 1,2007	Low voltage dry-type transformers must meet NEMA TP-1 standards	

in US if the lighting project saves energy. More information about tax incentives (and utility incentives) can be found at www.dsireusa.org and www.lightingtaxdeduction.org. Canada's Energy Efficiency Act of 1992 provides for the making and enforcement of regulations concerning efficiency standards for energy-using products, as well as the labelling of energy-using products and the collection of data on energy use. Legislation is kept updated through amendments.

Government regulations (changes/impacts specific to lighting are documented in Table 12.10) have mostly eliminated the use of general service incandescent lamps, T12 fluorescent lamps, and electromagnetic ballast technologies while tightening the specifications of HIDs. This also requires that LED lighting achieve higher efficacy levels and lower price points before the market starts to shift. In general, the US and Canada continuously harmonize their efficiency policies to ensure market consistency.

12.7 Identifying Potential Energy-Efficient Measures

Lighting is used primarily for illumination of workplaces and living spaces, for safety, and for decoration. In each of these uses, the same three questions can be asked: (1) How much light is needed? (2) How must the light be controlled? (3) How can lighting be provided most efficiently? When examining an existing system of lighting, the answers to these questions can be used to decrease lighting cost, improve lighting quality, and improve lighting efficiency.

Lighting improvements provide cost savings in a number of ways: reduced energy use and power demand; reduced heat production; lower lifecycle lamp costs; reduced need for maintenance; and increase in safety and productivity. In some instances, improving lighting quality increases worker productivity, and can result in greater profitability for a facility since the benefits of even a small change in worker productivity can vastly outweigh the relamping and maintenance costs. Figure 12.21 presents the results of a study of productivity in an office building when the lighting levels were first cut in half and then restored.

In examining the lighting system, the energy auditor should ask three questions: first, whether the light is needed; second, whether the correct amount of light is being used; and third, what the most cost-effective lighting technology to use to supply the correct quantity and quality of light is. To locate the energy management opportunities, the auditor should specifically:

- Identify and characterize the visual tasks and determine the contrast of the work to the surrounding surfaces.
- Look for the potential to use daylighting and task-specific lighting to displace high ambient, artificial lighting levels.
- Determine the appropriate lighting levels and the quality of light needed.
- Select alternative lighting systems to meet the needs and analyze the cost-effectiveness of each alternative.
- Select the best alternative to implement.

The remainder of this section provides some specific recommendations for areas that can result in cost-effective improvements in lighting systems. The advice of a qualified lighting consultant should be solicited before undertaking any major lighting retrofit to ensure the task is completed with the best available technology and the lowest lifecycle costs.

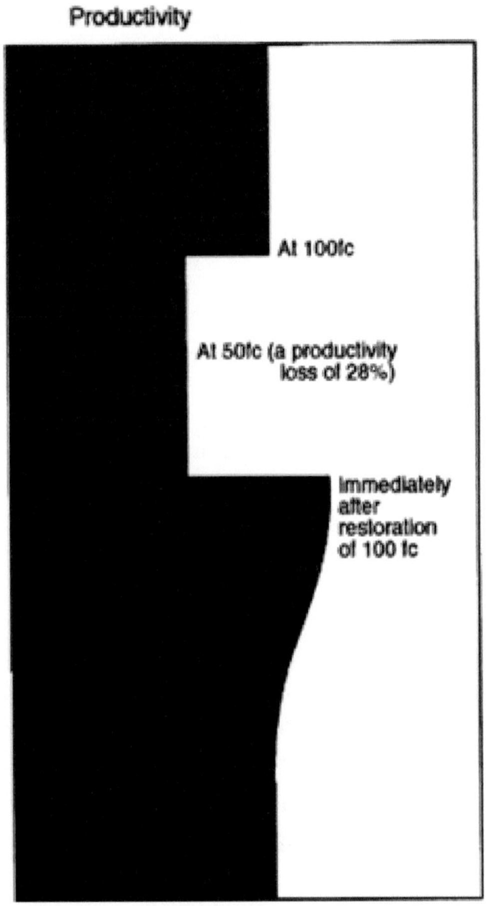

Figure 12.21 Productivity loss from light reduction Woodruff (2018).

12.7.1 Delamping

Major savings can be obtained by removing some of the lamps that are producing excessive levels of illumination. The lighting levels in many facilities have been over-designed to allow for poor maintenance and relamping practices. The first place to check for excessive light levels is in corridors and at workstations. The range of foot-candles for each task was given in Tables 12.4 and 12.5. Other places where lighting should be examined carefully include warehouses and storage areas.

There is often a good potential for delamping fixtures when planned group relamping is practiced. For example, a fluorescent light fixture with four new

F40T12, 40-Watt lamps may provide over twice the actual illumination specified by IES standards for a particular task. In such a case, half of the lamps can often be removed while still providing the area with sufficient light levels and light distribution patterns. Hallways with four-lamp fixtures frequently offer a good opportunity for removing half the lamps while maintaining adequate light levels.

If two lamps are removed from a four-lamp fixture, it is usually better to remove either the inboard or outboard set. Which set depends on the fixture design? Measuring the light levels after removing each set of lamps in turn will reveal which set should be removed. The best light distribution is typically achieved when the lamps are centered in the fixture. There are low-cost kits available for repositioning lamps. Although some reflector salespeople advertise their product as capable of maintaining the same fixture light output with half the lamps, much of the increase in light levels is due to centering the lamps in the fixture.

The ballast for rapid-start lamps will continue to consume some power even when the lamps are removed. Disconnecting the ballast and capping the power leads will eliminate this power draw and provide a readily accessible replacement ballast already mounted in the fixture. The ballasts in instant-start fixtures automatically disconnect when the lamps are removed. Leaving burned-out lamps in instant-start fixtures reduces ballast life.

Example 12.2

The packing and shipping area in a plastic jar production facility is lighted with 50 fluorescent fixtures that each have four F40T12 40-Watt lamps. Assume that each light fixture has two ballasts, that each ballast serves two lamps, and that ballast consumption is an additional 15% of the lamp power. It operates two 8-hour shifts per day for 250 days a year. Light-level measurements show that the average illumination is about 110 foot-candles. Is delamping warranted, and if so, how much can be saved if electricity costs 8 cents per kWh? Assume category D from Table 12.5 applies.

Solution

The IES illumination level standard for packing and shipping – which are material handling tasks – is 50 foot-candles as an upper limit for tasks involving large items. Thus, half of the lamps can be removed if the resulting light distribution pattern is still acceptable. Each light fixture has two ballasts,

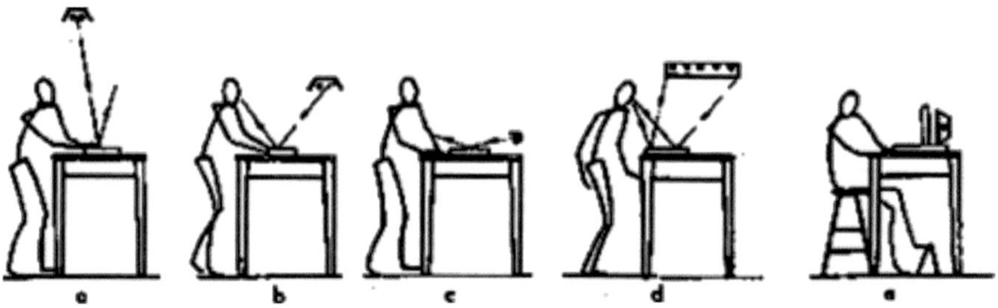

Figure 12.22 Supplementary luminaires [IES, 2011]. Examples of placement of supplementary luminaires: (a) Luminaire located to prevent veiling reflections and reflected glare; reflected light does not coincide with angle of view. (b) Reflected light coincides with angle of view. (c) Low-angle lighting to emphasize surface irregularities. (d) Large-area surface source and pattern are reflected toward the eye. (e) Transillumination from diffuse source.

and each ballast serves two lamps. The two lamps and ballast removed will save 80 Watts for the two lamps, and an additional 15%, or 12 Watts, for the ballast. The total energy cost savings from this delamping can be calculated as follows:

$$\begin{aligned}\text{Cost savings} &= (92 \text{ Watts/fixture}) \times (50 \text{ fixtures}) \times (16 \text{ hours / day}) \\ &\quad \times (250 \text{ days / year}) \times (1\text{kWh}/1{,}000) \times (\$.08/\text{kWh}) \\ &= \$1{,}472/\text{ year}\end{aligned}$$

12.7.2 Task lighting

Ambient lighting levels can be reduced when adequate task lighting is supplied for the work. Ambient lighting levels of 25 foot-candles (269 Lux) or less are frequently sufficient if the individual work areas have sufficient light from task-dedicated lighting fixtures. Figure 12.22 demonstrates the placement of supplementary luminaires.

12.7.3 Relamping

Replacing existing lamps, ballasts, and luminaires with newer, more energy-efficient models offers the potential for significant savings on lighting system costs in many facilities. Lamp replacement is often a simple procedure if the new lamp works with the existing ballast and fixture. For example, replacing existing F40T12, 40-Watt lamps with F40T12 34-Watt lamps in appropriate areas offers an easy and cost-effective lighting system improvement. Replacing existing F40T12 40-Watt lamps with 25 to 32-Watt T8 lamps requires a change of ballasts and lamp sockets in addition to the lamp change. This may still be very cost-effective, but it costs more for equipment and labor, and must be analyzed to see if it is truly cost-effective for a particular facility.

As another example, a 2' × 4' lighting troffer (fixture) using four F40T12, 34-Watt energy-saving lamps can be retrofitted with three F25T8 lamps and an electronic ballast. These alternatives will maintain or increase light levels and reduce energy use by about 35%.

There are many potential lamp substitutions possible, and it is important to know what kinds of substitutions are reasonable. Table 12.11 shows some of the LED lamp and luminaire replacements to legacy technology lamps, which can produce lower lifecycle costs (see more in section 12.7.7). Many other lamps are also suitable for replacing commonly used lamps.

Example 12.3: Upgrade T12 Lighting System to T8

A hospital had 415, 4' 40WT12 fluorescent fixtures, which operate 24 hours/day, year-round. The lamps and ballasts were replaced with 4' 25WT8ES lamps and electronic ballasts, which saved about 30% of the energy, and provided higher quality light. Although the T8 lamps cost a little more (resulting in additional lamp replacement costs), the energy savings quickly recovered the expense. The 25WT8 system produces about 25% less lumen output. But since the original design was over lit, it actually increases the lighting quality. In addition, because the T8 system produces less heat, air conditioning requirements during summer months will be reduced. Conversely, heating requirements during winter months will be increased. The air conditioner COP = 2.6 and the A/C runs at full load 2,000 hours/year. The heating efficiency is 80% and operates at full load 1,500 hours per year. Fuel costs $4.00/MCF (1 MCF = 1,000 ft^3. Assume 1 MCF = 0.003413 kWh).

Data and assumptions:

- Number of fixtures . . . . 415 fixtures
- Number of lamps per fixture . . . . lamps/fix
- Annual operating hours . . . . 8,760 hours
- kWh cost $0.050/kWh
- kWcost $120/kW-year
- Retrofit cost per fixture $40/fixture

Existing fixture data:

- Input Watts per 2 lamp T12 fixture . . . . Watts/fix
- Lamp life . . . . 20,000 hours

Table 12.11 Generic LED replacements for legacy lamps.[1]

Legacy lamp	**Total input (incl. ballast[2]) wattage**	**LED replacement[3]**	**Average[4] generic LED wattage**	**Demand savings**
Incandescent				
Incandescent/halogen < 40 W	25	LED lamp < 450 lm	4	84%
Incandescent/halogen A lamp = 40 W	40	LED lamp ≥ 450 lm < 600 lm	6	85%
Incandescent/halogen A lamp = 60 W	60	LED lamp ≥ 600 lm < 1,000 lm	10	83%
Halogen reflector < 60 W	40	LED reflector lamp ≤ 600 lm	9	78%
		LED hardwired ≤ 600 lm	9	78%
Halogen reflector ≥ 60 W and ≤ 100 W	75	LED hardwired > 600 lm	12	84%
		LED reflector lamp > 600 lm and ≤ 1,500 lm	15	80%
Fluorescent				
4' T12-40W fluorescent lamp	40	Tubular LED lamp[5] Type A)[6] < 1,800 lm and <2,000 lm	20	50%
4' T8-32W fluorescent lamp	32	Tubular LED lamp (Type A) < 1,500 lm and < 2,500 lm	17	47%
4' T5HO-54W fluorescent lamp	54	Tubular LED lamp (Type A) < 3500 lm and < 4,000 lm	32	41%
2' 4-lamp T12 magnetic ballast[6]	100	LED (2 × 2) troffer luminaire or retrofit kit > 2000 lm and ≤ 40 W[7]	30	70%
4' 2-lamp T12 magnetic ballast	74	LED (2x×2) troffer luminaire or retrofit kit > 2,000 lm and ≤ 40 W	35	53%

Table 12.11 *Continued.*

Legacy lamp	Total input (incl. ballast[2]) wattage	LED replacement[3]	Average[4] generic LED wattage	Demand savings
		LED (1 × 4) troffer luminaire or retrofit kit >2,000 lm and ≤40 W	35	53%
		LED (4') low output strip luminaire or retrofit kit >1,500 lm and < = 35W	34	58%
4' 3 lamp T12 magnetic ballast	115	LED (2 × 4) low/medium output troffer luminaire or retrofit kit > 3,000 lm and ≤50 W	48	58%
		LED (4') medium output strip luminaire or retrofit kit >3,000 lm and < = 55W	48	58%
4' 4 lamp T12 magnetic ballast	148	LED (2 × 4) high output troffer luminaire or retrofit kit >3,000 lm and ≤70 W	60	59%
		LED (4') high output strip luminaire or retrofit kit >5,000 lm and < =90 W	65	56%
8' 2 lamp T12 magnetic ballast	126	LED (8') medium output strip luminaire or retrofit kit >3,000 lm and < =70 W	65	48%
8' 1 lamp T12 HO magnetic ballast	121	LED (8') medium output strip luminaire or retrofit kit >3,000 lm and < =70 W	65	46%
8' 2 lamp T12 HO magnetic ballast	237	LED (8') high output strip luminaire or retrofit kit >5,000 lm and < =110 W	100	58%
4' 2 lamp T8 electronic ballast	59	LED (4') low output strip luminaire or retrofit kit >1,500 lm and < =35 W	31	47%
4' 3 lamp T8 electronic ballast	86	LED (2 × 4) low/medium output troffer luminaire or retrofit kit > 3, 000 lm and ≤ 50 W	44	49%
		LED (4') medium output strip luminaire or retrofit kit > 3, 000 lm and <= 55 W	44	49%

Table 12.11 *Continued.*

Legacy lamp	**Total input (incl. ballast[2]) wattage**	**LED replacement[3]**	**Average[4] generic LED wattage**	**Demand savings**
4' 4 lamp T8 electronic ballast	112	LED (2 × 4) high output troffer luminaire or retrofit kit > 3, 000 lm and ≤ 70 W	57	49%
		LED (4') high output strip luminaire or retrofit kit >5,000 lm and <= 90 W	62	45%
8' 2 lamp T8 electronic ballast	112	LED (8') medium output strip luminaire or retrofit kit >3,000 lm and >=70 W	62	43%
8' 1 lamp T8HO electronic ballast	114	LED (8') medium output strip luminaire or retrofit kit >3,000 lm and >=70 W	65	43%
8' 2 lamp T8HO electronic ballast	172	LED (8') high output strip luminaire or retrofit kit >5,000 lm and >=110 W	91	47%
HID (MH/HPS)				
50 W MH/HPS	62	LED luminaire >500 lm and ≤1,500	12	80%
		lm Mogul screw-base LED[5,8] >500 lm and <=1,500 lm	17	73%
70 W/75W MH	97	LED luminaire >1,500 lm and ≤3,000 lm	23	76%
		Mogul screw-base LED >1,500 lm and ≤ 3,000 lm	25	74%
70 W HPS	92	LED luminaire > 1,500 lm and ≤ 3, 000 lm	27	71%
		Mogul screw-base LED > 1,500 lm and ≤ 3, 000 lm	30	67%
100 W MH/HPS	130	LED luminaire > 3, 000 lm and ≤ 5, 000 lm	31	76%
		Mogul screw-base LED > 3,000 lm and ≤ 5, 000 lm	40	69%
150 W MH/HPS	188	LED luminaire > 5, 000 lm and ≤ 8, 000 lm	56	70%
		Mogul screw-base LED > and ≤ 8, 000 lm	76	60%

Table 12.11 *Continued.*

Legacy lamp	**Total input (incl. ballast[2]) wattage**	**LED replacement[3]**	**Average[4] generic LED wattage**	**Demand savings**
175 W MH/HPS	210	LED luminaire $> 8,000$ lm and $\leq 10,000$ lm	64	70%
		Mogul screw-base LED > 8,000 lm and $\leq$ 10, 000 lm	93	56%
200 W HPS	240	LED luminaire > 10,000 lm and $\leq 12,000$ lm	72	70%
		Mogul screw-base LED > 10,000 lm and $\leq$ 12, 000 lm	96	60%
250 W MH/HPS	295	LED luminaire $> 12,000$ lm and $\leq 16,000$ lm Mogul screw-base LED > 12,000 lm and $\leq$ 16, 000 lm	91 110	69% 63%
350/360 W MH	400	LED luminaire $> 16,000$ lm and $\leq 19,000$ lm	110	73%
		Mogul screw-base LED $> 16,000$ lm and $\leq$ 19, 000 lm	160	60%
400 W MH/HPS	455	LED luminaire $> 19,000$ lm and $\leq 21,000$ lm	150	67%
		Mogul screw-base LED $> 19,000$ lm and $\leq$ 24, 000 lm	201	56%
750 W MH (including pulse start)	825	LED luminaire $> 24,000$ lm and $\leq 60,000$ lm	307	63%
		Mogul screw-base LED $> 24,000$ lm and $\leq$ 60, 000 lm	350	57%

Table 12.11 *Continued.*

Legacy lamp	**Total input (incl. ballast[2]) wattage**	**LED replacement[3]**	**Average[4] generic LED wattage**	**Demand savings**
1,000 W MH/ HPS	1,100	LED luminaire $> 60,000$ lm and $\leq 90,000$ lm	520	53%

Notes:

1. This table includes information from multiple sources based on author's experience.
2. Wattage include ballast losses, where applicable, assuming use of a standard ballast. Actual ballast losses to be experienced depend on the specific type of ballast involved and operating conditions that affect its performance.
3. This table does not indicate all possible LED replacement options to lamps, lamps and ballasts, or complete fixtures. All numbers reported in the table are approximations, and in certain cases assumptions are made about the types of fixtures and other conditions involved. Consult manufacturers for accurate data relative to direct replacements possible for a given installation or other restrictions that may apply.
4. Values for LED lamp and luminaire replacements are given as average between similar range products across the North American market as there is no standardization for lumen or wattage as in the case of legacy light sources.
5. LED fluorescent and HID lamp replacements (within existing luminaires) tend to provide lesser lumens than the complete LED luminaires that are designed for LED technology with superior efficiency.
6. Values in this table are given for LED lamp replacements for linear fluorescent lamps of Type A (work on existing ballast). Type B (self-ballasted) or Type C (dedicated LED driver) are available from many manufacturers and average wattage values may be different.
7. For LED luminaires replacing fluorescent luminaires with multiple lamps, range values are both expressed in lumen and wattage to reflect products with superior efficiency for average office-type illumination levels. These respective values are inspired from the Design Lights Consortium's (DLC) Quality Product List (v. 5.0 and v. 5.1) and 2021 lighting prescriptive incentive program of BC Hydro.
8. Values in this table are given for LED lamp replacements for HID lamps of Type A (work on existing ballast). Type B (self-ballasted) are available from many manufacturers and average wattage values may be different.

- Lamp cost $1.50/lamp

Proposed fixture data:

- Input Watts per 2 T8 25W fixture ....46 Watts/fix
- Lamp life ....30,000 hours
- Lamp $3.00/lamp
- T8 ballast $10
- Total number of ballasts needed .... 415 ballasts
- Total number of lamps needed .... 830 lamps

Solution

kW savings

$$\begin{aligned}&= (\#\text{fixtures})[(\text{Existing input Watts/fixture}) \\ &\qquad - (\text{Proposed input Watts/fixture})] \\ &= (415)[(86\ \text{Watts/T12 fixture}) - (46\ \text{Watts/T8 fixture})] \\ &= 16.6\ \text{kW}\end{aligned}$$

kWh savings

$$\begin{aligned}&= (\text{kW savings})(\text{Annual Operating Hours}) \\ &= (16.6\ \text{kW})(8,760\ \text{hours/year}) \\ &= 145,416\ \text{kWh/year}\end{aligned}$$

Air conditioning savings due to interactive effects

$$\begin{aligned}&= (\text{kW savings})(\text{Air Conditioning Hours/year})(1/\text{Air Conditioner COP}) \\ &= (16.6\ \text{kW})(2,000\text{hours})(1/2.6) \\ &= 12,769\ \text{kWh/year}\end{aligned}$$

Additional gas cost due to interactive effects

$$\begin{aligned}&= (\text{kW savings})(\text{Heating Hours/year})\ (0.003413\text{MCF}/\text{kWh}) \\ &\qquad (1/\text{Heating Efficiency})(\text{ Gas Cost }) \\ &= (16.6\ \text{kW})(1,500\ \text{hours}\ /\text{year})(0.003413\text{MCF}/\text{kWh}) \\ &\qquad (1/0.8)(\$4.00/\text{MCF}) \\ &= \$425/\ \text{year}\end{aligned}$$

Additional lamp replacement cost

$$= [(\#\text{fixtures})(\#\text{lamps/fixture})]\ [((\text{annual operational hours/proposed lamp life})\ (\text{proposed lamp cost})) - (\text{annual hours operation/present lamp life})\ (\text{existing lamp cost})]$$
$$= [(415\text{ fixtures})(2\text{ lamps/fixture})]\ [(8,760\text{ hours/30,000 hours})(\$3.00/\text{T8 lamp}) - (8,760\text{ hours/ 20,000 hours})(\$1.50/\text{T12 lamp})]$$
$$= \$726/\text{year}$$

Total annual dollar savings

$$= (\text{kW savings})(\text{kW charge}) + [(\text{kWh savings}) + (\text{(Air Conditioning savings})](\text{kWh cost}) - (\text{Additional gas cost}) - (\text{lamp replacement cost})$$
$$= (16.6\text{ kW})(\$120/\text{kW year}) + [(145,416\text{kWh}) + (12,769\text{kWh})](\$0.05/\text{kWh}) - (\$425/\text{ year}) - (\$726/\text{ year})$$
$$= \$8,750/\text{ year}$$

Implementation cost

$$= (\#\text{ fixtures})(\text{Retrofit cost per fixture})$$
$$= (415\text{ fixtures})(\$40/\text{ fixture})$$
$$= \$16,600$$

Simple payback

$$= (\text{Implementation Cost})/(\text{Total Annual Dollar Savings})$$
$$= (\$16,600)/(\$8,750/\text{ year})$$
$$= 1.9\text{ years}$$

Example 12.4: Upgrade T8 Lighting to LED

Calculate the annual savings from replacing 32-Watt F32T8/office lamps with 17W TLED T8lamps in one hundred (100) two-lamp fixtures that are operated continuously. Assume the following:

The F32T8 lamps cost $2.00 each and last for 20,000 hours. The TLED T8 lamps cost $7.00 each and last for 50,000 hours. Electric energy costs

\$0.050 per kWh. The demand charge is \$5.50 per kW per month. The ballasts are not changed, and the facility is not air conditioned.

Solution

Annual energy savings (ES):

$$\begin{aligned}\text{ES} =& (\text{\# of fixtures}) \times (\text{\# of lamps/fixture}) \times (\text{wattage of low-efficiency lamps} - \text{wattage of high-efficiency lamps}) \times (\text{annual operating hours})\\ =& (100 \text{ fixtures}) \times (2 \text{ lamps/fixture}) \times (32 - 17) \text{ Watts/lamp} \times (5,000 \text{ hours/yr})\\ =& 200 \text{ lamps} \times 15 \text{ Watts/lamp} \times 1 \text{ kW}/1,000 \text{ Watts} \times 5,000 \text{ hr/yr}\\ =& 15,000 \text{ kWh/yr}\end{aligned}$$

Energy cost savings (ECS):

$$\begin{aligned}\text{ECS} &= \text{ES} \times (\text{cost of electricity})\\ &= 15,000 \text{kWh/yr} \times \$0.05/\text{kWh}\\ &= \$750/\text{yr}\end{aligned}$$

Demand reduction (DR):

$$\begin{aligned}\text{DR} &= (\text{\# of lamps}) \times (\text{wattage reduction})\\ &= (200 \text{ lamps}) \times (15 \text{ Watts}) \times 1 \text{ kilowatt}/1,000 \text{ Watts}\\ &= 3 \text{ kW}\end{aligned}$$

Annual demand cost savings (ADCS):

$$\begin{aligned}\text{ADCS} &= \text{DR} \times \text{Demand Charge}\\ &= 3 \text{ kW} \times \$5.50/\text{kW/month} \times 12 \text{ months/yr}\\ &= \$198/\text{yr}\end{aligned}$$

Annual lamp cost savings (ALCS):

The office lights only have a lifetime of 20,000 hours, where the replacement lamps have a lifetime of 50,000 hours. Each of these costs must be annualized to determine the actual cost savings. The total number of lamp hours used in one year is found from multiplying the number of lamps times the hours of use in one year.

Total annual use $= 200 \times 5{,}000$ hr $= 1{,}000{,}000$ lamp hours

32-Watt lamp cost:

To compute the number of office lamps needed for one year, divide the total annual lamp hours needed by the life of one office lamps.

Number of lamps needed $= 1{,}000{,}000/20{,}000 = 50$ lamps
The annual cost is: 50 lamps/yr $\times$ \$2.00/lamp $=$ \$100

17-Watt TLED lamp cost:

Number of lamps needed $= 1{,}000{,}000/50{,}000 = 20$ lamps
The annual cost is: 20 lamps/yr $\times$ \$7.00/lamp $=$ \$140
ALCS $=$ \$100 $-$ \$140 $= -$\$40 per year

The total annual savings from this relamping EEM is the sum of all the savings calculated above.

Total annual cost savings:

$$= \$750 + \$198 - \$40 = \$908$$

As illustrated in the example above, note that changes in lighting energy consumption usually imposes interactive effects with HVAC systems. For air-conditioned facilities, the operating costs associated with lower wattage or more efficient lamps will be reduced because they produce less heat. The cooling system will consume less energy due to the lower amount of heat produced from lighting. For space heating, the interactive effects may result in increased consumption of the heating energy to compensate for the reduced heating provided by the lighting system.

Lighting options should be compared to the existing system in a recently relamped and cleaned condition. A reasonable estimate of the lighting levels in the relamped system can be made from measuring lighting levels and noting the age/operation hours of the lamps. Some lamp sale representatives demonstrate how their lamp is superior to the lamps in use by relamping a fixture for comparison with adjacent fixtures. The results can be misleading due to light losses caused by lamp lumen depreciation (i.e., old lamps) and lumen dirt depreciation (i.e., the relamped fixture is usually wiped clean during lamp installation). Make sure both fixtures are clean and contain new lamps before comparing lamp alternatives.

12.7.4 Ballasts

Ballasts are an important part of a lighting system, and each ballast uses from 5% to 20% of the power of the lamp it is associated with. Furthermore, the ballast draws some power even if the lamp has been removed. Therefore, when a lamp is removed from a fixture, the ballast should usually be removed too. The ballast can be stored for future use, saving additional replacement costs. Ballasts should also be replaced if they overheat or smoke.

When older coil and core ballasts in a lighting fixture fail, replace with an electronic ballast. New, electronic ballasts are much more efficient than the older magnetic ones and offer desirable features such as dimming capabilities. T8 and T5 fluorescent lamps are specified with electronic ballasts. LED technologies use drivers that can be analog or digital, and in most of the cases support dimmability.

12.7.5 Lighting control technologies

12.7.5.1 Control strategies

Areas that are seldom occupied do not need constant light, yet lights are frequently left on in such places. A good lighting design includes a good controls design. The goal of an effective control system is to support the lighting application goals, which often translates to:

1. Efficient energy management.

- *Lighting energy* – controls can affect the energy (kWh) by reducing the amount of power drawn by the lighting system during operation and/or the number of operating hours, thereby reducing utility energy charges and thereby increase profitability and competitiveness for customers.
- *Lighting demand* – controls can reduce the amount of power (Watts) drawn by the lighting system, reducing utility demand charges, particularly during peak demand periods, when demand charges are highest.

2. Providing a productive visual environment.

- *The right amount of light* – lighting controls provide the ability for users to adjust light levels based on changing needs or individual preference, either through dimming or through bi- or multi-level switching. Access to controls could also increase worker satisfaction by providing personal control of their lighting systems in office, work benches, and other environments.

Switches. Many types of switches are available for controlling lighting. Switches should be installed in the areas in which the fixtures are controlled.

Rewiring to reduce the number of fixtures controlled by a single switch increases the ability of occupants to control the amount of lighting that is used. While HID lamps take up to 15 minutes to regain full light output after restarting and fluorescent lamps and ballasts are affected by frequent switching, the LED lamps and luminaires start instantaneous and are not sensitive to switching frequency.

Photocells. Automatically operate the exterior lighting between dusk and dawn. Photocell input can also be used as a basis for controlling interior lighting. Some energy management control systems can use photocell input data for automatically adjusting indoor lighting levels to maintain a constant value when dimmable ballasts are used.

Daylight harvesting. Automatically operate interior lighting to utilize daylight to supplement artificial lighting. Lamps are automatically switched or dimmed as required, maintaining the illumination of the space within design parameters, avoiding the situation of overlighting, or underlighting due to lumen maintenance depreciation of lamps.

Dimmers. Dimmers are good for areas that require low ambient lighting levels most of the time with an occasional need for bright lighting. Dimmers must match the load being dimmed and are typically incompatible with different lamp types/sources. Solid-state dimmers operate by reducing the voltage supplied to the lamps. This reduces energy use and extends lamp life. While fluorescent and HID lamps cannot be dimmed without dimmable ballasts, most LED lamps and luminaires have dimmable drivers.

3. Providing light where it is needed.

An effective control system ensures that the lighting system operates (and consumes energy that costs the owner money) only when it is needed. Determining when the lighting system should be operating depends on how the space is occupied:

- *For predictable occupancy* – a time-*based* strategy can be considered. For example, a switching system can be scheduled to automatically shut off the lights by area/ floor.

Timers. Timers can be used to control indoor lighting in smaller and less used areas (e.g., mechanical/ electrical rooms) or outdoor lighting. Adjustments should be made to simple timers about four times per year to prevent unnecessary operation of equipment. Timers can be used in conjunction with photosensors to reduce lighting costs if the lighting can be turned off before dawn.

- *For non-predictable occupancy* – a threshold-event-based strategy can be considered. For example, occupancy sensors can be used to automatically turn on and shut off lights in areas depending on people presence.

Occupancy sensors. Occupancy sensors can also be used to reduce unnecessary lighting use. Infrared sensors are directional and useful for active areas. Ultrasonic and microwave sensors are non-directional. The sensor's coverage of the area must be complete, or nuisance cut- offs will occur, and the occupants will remove the sensors.

Example 12.5

The prisoner holding cells in a courthouse utilize two fixtures with two F32T8 lamps each to illuminate an 8' × 12' area. The cells are usually occupied a maximum of 45 minutes per day, but they are illuminated about 12 hours per day, 5 days each week, and 50 weeks per year. How much can you save by installing occupancy sensors in these cells? Calculate the simple payback period and ROI if the sensors cost $70 installed and have a 10-year life. Electricity costs 8 cents per kWh.

Solution

Assume that the occupancy sensor has a delay built into its operation, and that the lights will operate 1 hour a day, 5 days a week, and 50 weeks a year. The total input of the T8 luminaire is 59 W. Thus, the total savings from the occupancy sensor is found from:

$$\begin{aligned} \text{Cost Savings} &= (\text{2 fixtures/ cell}) \times (\text{59 Watts/fixture}) \times (1\,\mathrm{kW}/1{,}000\,\mathrm{Wh}) \\ &\quad \times (\text{11 hours / day}) \times (\text{5 days / week}) \\ &\quad \times (\text{50 weeks / year}) \times (\$.08/\mathrm{kWh}) \\ &= \$25.96/\text{ year per cell} \\ \text{SPP} &= \text{Installed cost/annual savings} \\ &= \$70/\$25.96/\text{ year} \\ &= 2.7\text{ years} \\ \text{ROI} &= \text{Solution for } i \text{ to:} \\ &\quad \$25.96(\mathrm{P/A}, i, 10) = \$70 \\ \text{ROI} &= 56.9\%. \end{aligned}$$

12.7.5.2 Networked lighting controls

In recent decades, technological development has increasingly automated the control functions and allowed integration of devices into larger, more flexible systems. The advent and low-cost availability of solid-state digital components and the development of the internet has allowed lighting controls to advance from simple analog line voltage dimming and on/off switching to digital lighting control systems with greater capabilities (e.g., zoning, scene control, individual luminaire addressability, continuous dimming/load trimming, device and energy monitoring, remote diagnostic, demand response, etc.).

Networked lighting control (NLC) systems consist of individually/group addressable luminaires and control devices connected (wired or wireless) within an intelligent network using software. Wireless lighting architectures offer increased system flexibility and lower implementation costs, being better suited for lighting upgrades.

Many fixtures now offer integrated occupancy sensors (also photocells for daylight measurement and dimming). These luminaire-level lighting control fixtures offer a new level of automation for the energy management program. Another opportunity is to give "power to the people" by integrating lighting and environmental comfort controls to the occupant via a web interface.

NLC systems are typically used in conjunction with the installation or upgrade of lighting systems with LED lighting. At a very high-level, the potential energy savings from the addition of NLC system can therefore increase on average by about 49% from just the LED hardware conversion [DLC, 2020].

Lighting systems provide the physical infrastructure to incorporate non-lighting sensors to facilitate the deployment of hardware required, and to support dispersed communication technologies. NLC systems become a "de facto" framework for connectivity, providing non-energy benefits such as improved health and productivity (see controls, Chapter 15 for more information).

12.7.6 High efficiency fluorescent high-bay luminaires

Improvements in fluorescent lighting have resulted in T5 and T8 lamps that are superior in many ways to HID luminaires in medium and high-bay applications in retail, manufacturing, warehousing and recreational, etc. The numerous benefits of high-performance lamps (e.g., instant start, lower wattage and lumen depreciation rates, high CRI, reduced glare) have created a new trend in the lighting industry: the replacement of HID luminaires with high-efficiency fluorescent lighting. For example, one can now replace a 400 W MH luminaire with a four-lamp F54-T5HO luminaire, saving about 50% on energy, while producing about 20% less maintained light output. The modular construction of these fluorescent luminaires allows a redesign of the application to better match or improve the required horizontal and vertical illuminance values of the high-bay applications, and is more efficient than using the existing HID wattage options. Having multiple lamps and ballasts also allows step-dimming, and the instant start characteristic of fluorescent technology allows for effective use of occupancy sensors (i.e., warehousing) to boost the demand and energy savings.

12.7.7 LED replacements

12.7.7.1 LED replacement lamps

The common factor for all LED lamp replacements is that you need to match the light output (lumens) and not the wattage scale of the existing conventional lamps. While LED lights were only used in low output equipment (e.g., exit signs) before 2010, currently LED lighting equipment can match most of all output levels produced by conventional light sources of various wattages (see Table 12.11).

For incandescent lamps. Both omni-directional (A-type) and directional (reflector) LED replacement lamps constitute a growth area in the lighting industry and technologies and designs are rapidly advancing. In many applications with high maintenance costs (e.g., airport runways) and other signaling/marking applications), LEDs are reducing lamp replacement and labor costs. Manufacturers frequently update their product offerings, and new manufacturers continue to enter the field. High efficacy values combined with very long life (see Figure 12.12) make the LED lamps the "de facto" replacements for all incandescent lamp technologies. Currently, there are LED replacements for almost all the various types of incandescent and halogen lamps with equivalent luminous (or beam-candle) output and mimicking 2,700–3,000 K CCTs for low to high wattage lamps. Some LED bulb solutions may look like familiar light bulbs and some may not, but LEDs can better match the performance of traditional light bulbs. While the average LED lamp replacements have a CRI in low 80s, there are some notable exceptions with CRIs in mid-90s.

For fluorescent lamps. LED lamps marketed to replace linear fluorescent products are an emerging product category with the potential to deliver significant energy and maintenance cost savings. Linear LEDs replace T12, T8, and T5 lamps in almost all lengths and CCTs. While the CRIs are similar with the fluorescent technology, their efficacy and expected useful life are much higher (see Figure 12.12). Primary types of linear and TLED have emerged on the market:

- UL Type A: TLED lamp with internal driver that is designed to operate on a linear fluorescent lamp electronic ballast (mostly instant or program start). These lamps are the easier retrofit on the market, and they do not require re-certification. When replacing, check the ballasts for compatibility (as indicated by the TLED manufacturer). Replace first the defective and aged ballasts before relamping with TLEDs.
- UL Type B: TLED lamp with internal driver that must be connected directly to line voltage for power. This option is able to provide dimming but is more expensive as it involves electrical work to disconnect, remove, or bypass the existing ballast and it requires luminaire re-certification.
- UL Type C: TLED lamp with external driver that is designed to replace both the linear fluorescent lamp and fluorescent lamp ballast. This is the most expensive option as it requires that both lamp and ballast be replaced with LED technology but provides dimmability and the longest longevity. It also requires re-certification.

In addition, some products can operate under multiple scenarios such as with a fluorescent ballast and also when the ballast is replaced with a compatible LED driver. These hybrid products, also called dual-mode products, are currently available in Types AB and AC. Smart TLEDs also exist in the market, allowing wireless controllability an integration with dimmers and control sensors.

Since the TLEDs provide light using only a half of the tube's circumference, look at the lumens per Watt delivered by the lamp and luminaire together, not just the lamp itself. Consider LED tubes coated for diffused light to avoid glare and choose medium-to-wide distributions (larger than 120 degrees). TLEDs are best in over lit and/or low temperature environments.

For HID lamps. While the number of HID lamps is small compared to other types of lamps, HID lamps have a large impact on energy use as annual HID energy use is second only to linear fluorescent lamps [DOE, 2019b]. Screw-in LED lamps exist now to replace HID wattages from 50 to 400 W or more with better life, efficacy, CCT, and CRI characteristics. For high-bay and high-power HID applications, it is better to use LED luminaires. Mogul screw-based (most of the HID lamps have E39 large, mogul sockets) LED replacement lamps can have omnidirectional or directional lighting distribution. While most LED replacements for HID lamps have an internal driver and thus require rewiring of the existing luminaire to bypass the ballast (UL Type B), some have external drivers and can use the existing ballast (UL Type A). LED mogul lamps are suited for both indoor and outdoor applications and can be mounted enclosed (always check with the manufacturer first) or open, but may offer lower levels of illumination, and their overall size could be different from the HID lamps. Good practice is to test them first with the existing luminaires.

Note of caution: Where screw-in LED replacement lamps are used in open luminaires within dirty/dusty spaces, the dirt/dust will be attracted to the lamp's heat and might quickly accumulate on its surfaces. This will affect the lumen output and might create an insulating layer that helps to retain heat within the lamp. Both conditions will affect the lamp's performance and lifespan. Such is the case in industrial sites like sawmills.

Example 12.6

A school with 100, 2-lamp, 32 W, T8 troffers had the 32 W lamps replaced with 18 W tubular LED type A, plug-in replacements. Lights were on 4,000 hours a year, and electricity cost $0.10/kWh. How many kWh were saved, and what was the payback if the new lamps were $12 each?

Solution

Savings calculation:

$$\text{kWh/yr} = 200(32 - 18\ \text{W})(4,000\ \text{h/yr})/1,000 = 11,200$$
$$\$/\text{yr} = (11,200\text{kWh/yr})(\$0.10/\text{kWh}) = \$1,120/\text{yr}$$
$$\text{New lamps cost } 200 \times \$12 = \$2,400$$
$$\text{Simple Payback } = \$2,400/\$1,120/\text{yr} = 2.14 \text{ years}$$

But the new lamps last much longer than fluorescent ones, so payback is less than two years.

12.7.7.2 LED luminaires and retrofit kits

Although the range of energy efficacy for LED luminaires and LED lamps is similar, the potential for larger form factors in dedicated LED luminaires allows for more lumen output. Hence, LED luminaires and retrofit kits are better replacement products for legacy lighting technologies as they are designed to maximize the light distribution coming from the LED chips while ensuring their efficient thermal management.

Retrofit LED kits tend to be less expensive than new LED luminaires, and they are largely used for recessed ceiling mounted, incandescent, CFL, or fluorescent applications. Always check with the authorities having jurisdiction if you need re-certification using retrofit kits.

LED luminaires are a better replacement than retrofit kits for conventional lighting installations since they are designed for the LED technology, ensure dimmability, and control connectivity, thus providing significant energy savings and visual comfort. Many LED luminaires also use specialized, nano-optical technology advancements for diffusers and refractors to make the light go where it is needed in order to suit the application. LED luminaires have superior efficiency than their conventional technology counterparts (it is not uncommon for 40%–50% of the total light output of fluorescent and incandescent lamps to be lost before it exits the fixture). While most LED light fixtures have LEDs built in as a permanent light source, some offer replaceable LED modules.

Lighting applications with higher light output usage (likely previously using HO fluorescents or HIDs) use predominately LED luminaire replacements rather than lamps or retrofit kits. LED luminaires are very resistant to vibration and built for rough use. While some of these luminaires keep a similar HID luminaire form factor, others use innovative form factors efficiently using modularity, scalability, and adjustability to adapt to the application. Many of the LED luminaires can incorporate control modules and sensors (occupancy and daylight harvesting), making it easier to configure networked lighting control systems for additional energy savings and operational cost benefits.

LED technology allows for manufacturing smaller light fixtures so that designers can put lighting fixtures closer to the task where the light is needed. This breakthrough changes the design because instead of the conventional approach to "blast more light through an efficient fixture, lens and then through the air," we can have a target lit from a close range (or even "back-lit") with much less light. Over the next decade, thin film "light panels" will be hitting the market with new applications for creating "soft ambient light" even from walls!

Beyond being low-Wattage, close-range light sources, LEDs can offer multicolored light without a lot of heat or maintenance. For a retailer, they can light a window display for products with different colors, creating different moods and attracting different buyers. The same principle applies for signs – more flexibility.

LED penetration (installed stock) to the lighting stock in US and Canada is aggressively increasing. Currently, on average, about 35% across all application subsectors is expected to grow to near full saturation by 2035 [DOE, 2019b].

Example 12.7

A commercial store is presently lit with 28 two T8-32 W fluorescent lamp fixtures at 59 Watts per fixture including ballast. What are the annual savings of replacing the existing lighting system with 18 LED luminaires (44 Watts per fixture)? Assume 6,000 hours of operation per year, an energy cost of $0.12 per kWh, and a demand cost of $8.50 per kW-month.

Solution

$$\begin{aligned}
\Delta \mathrm{kW} &= [(28 \text{ fixtures })(0.059 \text{ kW/ fixture })] \\
&\quad - [(20 \text{ fixtures })(0.044 \text{ kW/ fixture })] \\
&= 0.772 \text{ kW} \\
\text{Demand \$ savings } &= (0.772 \text{ kW})(\$8.50/\text{kW} - \text{mo})(12\text{mo/yr}) \\
&= \$78.74/\text{yr} \\
\text{Energy \$ savings } &= (0.772 \text{ kW})(6,000\text{hrs/yr}))(\$0.12/\text{kWh}) \\
&= \$555.84/\text{yr} \\
\text{Total \$ Savings } &= (\$79 + \$556)/\text{yr} = 635/\text{yr}.
\end{aligned}$$

12.7.8 Other lighting EEMs

12.7.8.1 Exterior lighting

Exterior lighting is another area in which lighting energy is often wasted. In motels, for example, peripheral lighting is often left on both day and night. Such waste can be easily corrected with a timer or with a light switch turned on and off by a "dusk-to-dawn" photocell. Each of the perimeter and outside lights should be carefully considered to see when it should be on, how much light is needed for the intended function, and whether more efficient lighting sources would work as well as those now being used.

12.7.8.2 Daylighting

Windows and skylights are often used to add light in a given area. One problem is that windows admit radiant heat as well as light, and it may be more expensive to remove the heat than to supply the light. In that case, the windows should be treated with exterior-mounted solar screens, louvers, or a reflective film with a low shading coefficient and a high percentage transmission of visible light.

12.7.8.3 Environmental factors

An area can appear to be dark if the walls, floors, or ceilings are painted (or otherwise covered) in dark colors or are greasy or dirty. Using light colors for paint and flooring or cleaning these surfaces more often can make the existing light more effective and thereby save money.

12.7.9 Selecting lights for a new facility

12.7.9.1 Modern lighting design

Technology is advancing at an unprecedented rate, and these changes affect how organizations are structured, their flexibility, marketing, operations, etc. We are living in a digital world with more computer screens, technology, and active lighting systems. Because humans are mostly visual communicators, lighting is evolving to fulfill the needs of our rapidly changing society. For these reasons, it is recommended to consult with a lighting designer prior to a retrofit/redesign. There are many variables, and the impact on human productivity can completely obliterate or enhance the financials of a project.

New solid-state (LED) lighting technologies do offer many advantages over the conventional light sources. But the design should consider how to better use their novel form factors to provide a better visual environment that is both energy-efficient and aesthetically pleasant. For example, a linear luminaire can be successfully used in direct/indirect suspended mounting selectively using one distribution (up/down) or both to optimize the lighting for audio-visual presentations or for general illumination in a classroom or office.

In an increasingly connected world, new lighting design should incorporate advanced, networked control abilities to allow for occupants' comfort, effective energy management, and optimization of operation and maintenance costs.

12.7.9.2 Cost effectiveness

Any time a new facility is built or an existing facility is expanded, there is a significant opportunity to save energy costs and improve indoor environmental quality by selecting and installing cost-effective, energy-efficient lighting systems at the time of construction. It is almost always less expensive to install correct equipment the first time than to retrofit existing equipment. Greater first costs may produce significantly lower operating costs and provide cost-effective savings for the facility. Unfortunately, many design decisions are made based on first cost rather than lifecycle costs, which include operation and maintenance costs. Utilizing lifecycle costing can assure the lowest lighting costs throughout the life of the lighting system.

Example 12.8

Gator Plastics Company is experiencing such a growth in demand for their products that they are planning on adding a new production room. Their industrial engineer is responsible for selecting the lighting system to be installed. The engineer has identified two alternative lighting systems. Alternative 1 uses 40 fluorescent light fixtures with four 32-Watt T8 lamps in each fixture with a total input of 98 Watts per luminaire. This system costs $2,000 to purchase and $2,000 to install. Alternative 2 uses 40 fixtures with three 25-Watt T8 lamps, together with one three-lamp electronic low-ballast-factor ballast with a total input of 86 Watts per luminaire. This system costs $2,300 to purchase and $2,000 to install. If each lighting system lasts six years, the lights are used 2,000 hours per year, electricity costs $0.08 per kWh, and the company's investment rate is 10%, which alternative should the engineer select? Calculate the three standard economic evaluation measures for each alternative – SPP, ROI, and benefit–cost ratio.

Solution

Alternative 1: The operating cost is found as follows:

$$\begin{aligned}\text{Annual cost} &= (40 \text{ fixtures}) \times (98 \text{ Watts/fixture}) \times (2,000 \text{ hours}) \\ &\quad \times (1\text{kWh}/1,000 \text{ Wh}) \times (\$0.08/\text{kWh}) \\ &= \$627.20\end{aligned}$$

Alternative 2:

$$\begin{aligned}\text{Annual cost} &= (40 \text{ fixtures}) \times (86 \text{ Watts/fixture}) \times (2,000 \text{ hours}) \\ &\quad \times (1\text{kWh}/1,000 \text{ Wh}) \times (\$0.08/\text{kWh}) \\ &= \$550.40\end{aligned}$$

Economic evaluation: Alternative 2 costs $300 more than Alternative 1, but it saves ($627.20 – $550.40) = $76.80 per year in operating costs. To determine if this additional cost is a good investment, we need to calculate the three standard economic performance measures.

$$\begin{aligned} \text{SPP} &= \text{Initial cost/Annual cost savings} \\ &= \$300/\$76.80.40 = \underline{3.9} \text{ years} . \end{aligned}$$

The ROI is found by solving:

$$\begin{aligned} (\$76.80)(\text{P}/\text{A}, i, 6) &= \$300 \\ \text{ROI} &= 12.6\% \end{aligned}$$

Using the methods from Chapter 5 (Economics) to find the present worth of the annual savings gives:

$$\begin{aligned} \text{Benefit} \quad - \quad \text{cost ratio} &= \$76.80 \times (\text{P}/\text{A}, 10\%, 6)/\$300 \\ &= \$76.80 \times 4.355/\$300 = \underline{1.15}. \end{aligned}$$

Thus, for many companies, the added cost of Alternative 2 would be considered a good investment.

12.8 Lighting Checklist

Lighting and maintenance costs can be reduced with a concurrent improvement in worker productivity, safety, and comfort. Table D.2 [adapted from IES, 2011] in the appendix presents a checklist of energy-saving guidelines for lighting developed by IES.

12.9 Summary

The lighting system in a facility is an important area to examine and to improve in terms of energy efficiency, indoor environmental quality, health and safety, and productivity. This chapter has introduced the basics of lighting systems, described the components of the system, and provided suggestions for ways to improve systems. Lighting technology is changing at a rapid pace, and new equipment and systems are being developed and marketed aggressively. Major energy savings opportunities exist in most existing lighting systems, and additional cost-effective savings are often possible even in relatively new systems since technology is continually improving.

References

DLC (2020) Energy Savings from Networked Lighting Control (NLC) Systems with and without LLLC. Design Lights Consortium, https://designli

ghts.org/wp-content/uploads/2021/01/Energy-Savings-From-Networked-Lighting-Controls-with-and-without-LLLC_FINAL_09242020.pdf

DOE (2008) Demonstration Assessment of Light Emitting Diode (LED) Street Lighting. 2008. DOE & PG&E report. www.energy.gov/sites/prod/files/2014/05/f15/emerging_tech_report_led_streetlighting.pdf

DOE (2013) Energy Efficiency of LEDs Fact Sheet, DOE Building Technologies Program. www1.eere.energy.gov/buildings/publications/pdfs/ssl/led_energy_efficiency.pdf

DOE (2015) DOE SSL Program R&D Plan, prepared by Bardsley Consulting, SB Consulting, SSLS, Inc., LED Lighting Advisors, and Navigant Consulting, Inc., May 2015.

DOE (2019a) Solid-State Lighting R&D Opportunities. DOE www.energy.gov/sites/prod/files/2019/01/f58/ssl_rd-opportunities_jan2019.pdf

DOE (2019b) Energy Savings Forecast of Solid-state Lighting in General Illumination Applications. www.energy.gov/sites/default/files/2020/02/f72/2019_ssl-energy-savings-forecast.pdf

GE (1969a) Lighting Maintenance. General Electric Technical Pamphlets TP-119 & TP-105, General Electric Co., Cleveland OH.

GE (1969b) Technical Pamphlet TP-105. General Electric Company, Cleveland, OH.

IES (2011) The Lighting Handbook, Tenth Edition. Illuminating Engineering Society.

IES/NALMCO RP-36-15 (2015) Recommended Practice for Lighting Maintenance. Illuminating Engineering Society

National Lighting Bureau (1987) Profiting from Lighting Modernization. Pamphlet.

Roberts M. (2014) Environmental Aspects of Magnetic Induction Lamps. Indulux Technologies Inc.

Pansky, Stanley H. 1985. Lighting Standards, Tracing the development of the lighting standard from 1939 to the Present, Lighting Design & Application, Illuminating Engineering Society.

U.S. Congress (1992) Energy Policy Act of 1992., P.L. 102-486 (H.R. 776).

WAPA (1990) Electric Utility Guide to Marketing Efficient Lighting. Western Area Power Administration (WAPA) and American Public Power Administration.

Woodroof, Eric A., Walerczyk, S., Hauber, F. (2018). Lighting, Chapter 13, in Energy Management Handbook, Ninth Edition, Fairmont Press/ River Publishers.

Bibliography

- CIBSE (2018) The SSL Lighting Handbook. www.cibse.org
- DiLouie, C. (2020) Lighting Controls Handbook. Taylor & Francis.

Author Biography

Dr. Cristian Suvagau has been practicing and teaching architectural lighting design and energy efficiency in Europe and North America for over 35 years. He is an energy management engineer with BC Hydro, in BC, Canada. Cristian is a Professional Engineer and holds a Ph.D. degree from TUC Bucharest, Romania. Cristian is a member of the National Board of Directors of the Illuminating Engineering Society (IES), Chair of the C419 Committee on Performance of Lighting Equipment of the Canadian Standards Association (CSA), and President of the Canadian National Committee of the International Commission on Illumination (CNC/CIE). Cristian also participates actively with standard-developing associations like CSA, IES, and CIE to develop performance standards for lighting and control systems. An adjunct professor with the New York Institute of Technology – Vancouver, Cristian is teaching lighting classes and is often presenting at industry seminars, while authoring numerous technical lighting articles for publications around the world.

Acknowledgments

I would like to thank Brian Friedman for performing a review of an early version of this chapter.

Practice Questions

12.1

When performing an energy survey, you find 12 two-lamp F40T12 security lighting fixtures turned on during daylight hours (averaging 12 hours/day). The lamps draw 40 Watts each, the ballasts draw 12 Watts each, and the lights are currently left on 24 hours per day. Assume one photocell can control all 12 fixtures.

Energy cost:	\$0.10/kWh
Peak power cost:	\$12/kW
Lamps:	\$15/lamp
Photocell (installed):	\$200/cell

How much can you save by installing a photocell? What is the payback period of this investment?

12.2

You count 120 four-lamp F40T12 troffers that contain 34-Watt lamps and two ballasts (156.4 W per fixture including ballast). How much can you save by installing:

a) 4-Tubular LED-T12 lamps (Type-B c/w ballast bypass) at \$50/fixture. 80 W per fixture including ballast.
b) 3-F32T8 lamps and an electronic ballast at \$70/fixture. 91.2 W per fixture including ballast.

Assume the same energy costs as in Problem 12.1
What is the simple payback period (SPP) and what is the return on investment for each alternative assuming equipment life of seven years?

12.3

During an audit, you count 25 exit signs with two 20-Watt incandescent lamps each. The 20-Watt incandescent lamps have a 2,500-hour lifespan and cost \$15 each. The 2-Watt LEDs have a 50,000-hour lifespan and cost \$25 each. Assume the same energy costs \$0.12 per kWh and \$14 per kW peak demand. Estimate annual utility cost savings and simple payback period by replacing the two 20-Watt bulbs with two 2-Watt green LED screw-in replacement kits. Include utility bill savings and lamp replacement cost savings.

12.4

During a lighting survey, you discover 36 150 high-pressure sodium cobra-head streetlights (190 W including ballast) operating 4,300 hours per year on photocells. There is no demand charge, and energy costs \$0.13 per kWh. Estimate annual cost savings by replacing these fixtures with 70-Watt LED cutoff luminaires. Assume the lamp life and lamp cost are the same for both technologies.

12.5

You find a factory floor that is illuminated by 84 250-Watt high-pressure sodium downlights (300 W per fixture including ballast). This facility operates two shifts per day for a total of 18 hours, five days per week, and 52 weeks per year.
Assume that the lights are contributing to the facility's peak demand, and the rates are $0.10/kWh and $12/kW peak demand.
Estimate the savings from retrofitting an industrial facility with 95-Watt LED downlights.
What will happen to the lighting levels?

12.6

An exterior loading dock in Chicago uses F32T8 32-Watt lamps in enclosed fixtures (29.5 W per fixture including ballast factor). They are considering a move to use 25-Watt lamps (23 W per fixture including ballast factor). Assume electricity rates are $0.13 per kWh and $12.00 per kW peak. What is your advice?

12.7

A light industrial facility uses 100 A19 lamps (100 W each), which need to be replaced. The two options are replace-in-kind with 100-Watt A19 incandescent, or install 14-Watt LED. A19 lamps cost $5.00 and last 750 hours. The LEDs cost $15.00 each, and last 12,000 hours. The electricity time-of-use rate schedule is:

- Mon–Fri 2–7pm $0.15/kWh
- All other hours $0.10/kWh

The facility operates two shifts during weekdays, from 7 am to 11 pm, Monday through Friday, for 50 weeks per year.
Estimate the present value cost for both the A19 incandescent and the LEDs for three years. Assume cash flows occur at the end of each year and use a discount rate of 8%. Report the NPV after three years, and IRR.

12.8

A retail shop uses a 1,000-Watt HPS floodlight (1,100 W per fixture including ballast) on the corner of the building to illuminate the parking lot. Some of this light shines out into the roadway. Assume the light does not contribute to the shop's peak load, and operates 360 hours per month. The facility has an inclining block rate schedule as follows:

- \$0.080 per kWh first 1,800 kWh monthly
- 0.210 per kWh over 1,800 kWh monthly

Estimate the annual electricity cost savings by installing a 62,000 lumen LED floodlight (553 W per fixture including ballast). Assume the electricity consumption for all facility loads, not counting the floodlight, is constant at 1,600 kWh per month. What problems can you anticipate from the light "trespass" effect? How would you recommend improving the lighting?

12.9

A commercial pool uses four 300-Watt quartz-halogen floodlights. The lights do contribute to the facility's peak load, and the electric rates are \$12.50 per kW demand charge, and \$0.11 per kWh energy.
Estimate the energy and power cost savings from using two 250-Watt pulse-start metal halide floodlights (300 Watts including ballast)? What will happen to the lighting levels? Assume 4,380 operating hours annually.

12.10

You notice that the exterior lighting around a manufacturing plant is frequently left on during the day. You are told that this is due to safety-related issues. Timers or failed photocells would not provide lighting during dark overcast days. Propose a solution.

12.11

A manufacturing facility uses F96T12HO lamps to illuminate the production area. Lamps are replaced as they burn out. These fixtures are about 15 years old and seem to have a high rate of lamp and ballast failure. Propose a new lighting solution for these problems.

12.12

Consider a lighting system where the original (baseline) lighting is provided by nine 100-Watt incandescent lights, operated 24 hours/day, 7 days/week, and 365 days/yr. The efficacy of the incandescent lighting is 10 lm/W.

A. For the incandescent lighting systems:

- What is the energy consumption (kWh/yr)?
- What are the lumens and lumen-hours per year provided by incandescent lighting?

B. The same lumens of lighting are provided by conversion to T-8 fluorescent lighting at 70 lm/W.

- What is the electricity savings in kWh/yr attributable to the T-8 installation? (kWh/yr saved)
- Adding to the T-8 installation, a timer system is installed, switching off lights from 10 pm to 7 am (37.5% savings). What are the incremental (additional) electricity savings due to the timer? (kWh/yr saved)
- What is the total electricity saved? (kWh/yr saved)

C. Now re-order the installation sequence:

- A timer system is installed first on the incandescent system, switching off lights from 10 pm to 7 am (37.5% savings). What are the incremental (additional) electricity savings due to the timer? (kWh/yr saved)
- With the timer system installed, the same T-8 lighting system is installed. What are the electricity savings in kWh/yr attributable to the T-8 installation? (kWh/yr saved)
- What is the total electricity saved? (kWh/yr saved)

13

Facility Electrical Systems

Eric Mazzi

Mazzi Consulting, Canada

13.1 Introduction

Electrical power is generally the most costly source of energy to operate homes, commercial buildings, and industrial facilities. It is important to understand how electrical systems work, to define basic terms (e.g., voltage, current, power), to understand system designs (e.g., single-phase and three-phase), and to be familiar with other electrical phenomena (e.g., power factor, power quality). This chapter introduces the basics for electrical systems and describes how to calculate electrical loads and electrical savings in facilities.

13.2 Basic Electrical Systems in Buildings and Facilities

Electricity is the flow of electrical charge through a conductive material, such as power cords or distribution wiring, and powers equipment such as lighting, cooling systems, data centers, motors, heaters, and a myriad of other end-use equipment.

There are two types of electric current: AC (alternating current) and DC (direct current).

- AC current is the standard for electric systems in facilities.
- DC current exists in equipment such as batteries, solar cells, fuel cells, or from power supplies inside equipment such as computers.

In AC systems, the voltage changes directions many times per second, moving first positive and then negative. In North America, the design frequency is 60 Hertz (Hz). Many other countries in the world use 50 Hz electric power.

AC power is most often used because it has advantages for electric utilities since they can step up the generated voltage to transmit and distribute electricity efficiently to buildings and facilities. Using higher voltage reduces the transmission and distribution losses from the resistance in the power lines. Using transformers, the AC can also easily be changed to different voltages for different uses in facilities.

When the flow of the electricity is in one direction only, it is direct current (DC). What happens in direct current is that negative charge flows toward positive charge, creating a direct electrical current. DC power maintains a constant voltage and flows in only one direction. There are DC motors, battery powered inverters, and electronic equipment that run directly on DC. DC power also has some advantages in terms of variable speed controls.

13.3 Electrical System Definitions

Now, let us define some basic terms that are commonly used for electrical systems. To use an analogy, electricity flowing through a circuit is similar to water running through a hose or pipe. When you open a water faucet, the pressure in the water line forces the water to flow at some rate into the hose. Friction reduces the force and rate of the water flow before it exits the hose. When electricity flows through a wire from a source such as a generator, voltage is like the water pressure to provide force to move the electric energy, current is a measure of the flow, and electric resistance in cables and equipment are like friction in water systems.

Generators are devices for producing electrical energy by moving a coil of wire (usually driven by a turbine) in a magnetic field. This is usually located at a utility central station power plant, but it could be located at a facility to provide distributed generation.

Current is a flow of electric charge in a closed circuit. It is measured in Amperes, or Amps.

Voltage is the force causing current to move in a closed circuit. It is measured in Volts.

Hertz is the unit of frequency of our AC power systems.

Resistance is a measure of a material's ability to oppose the flow of electrical energy in a closed circuit. It is measured in Ohms. All electrical conductors have some small resistance per unit length of wire.

Conductors are materials through which electrical current moves freely, such as wires, and have low values of resistance. Larger wire sizes of the same material, such as copper, have less resistance smaller than wire sizes.

Insulators are materials that block the flow of electricity.

Resistors resist, but do not totally block, the flow of electricity. They are used to control the flow of current. Current can move either way through a resistor.

Capacitors store up electrical energy as potential energy while current is flowing into them, and then release the energy later. Capacitors are usually made by rolling up two strips of metal foil separated by a piece of insulating material. Capacitors are often used for power factor correction.

Inductors store up electrical energy as magnetic energy, and release it later. Inductors are made by wrapping wire around an iron core. Inductors, or inductive effects, exist in fluorescent ballasts, LED lighting, variable frequency drives (VFDs), transformers, and induction motors. Power factor correction capacitors are often connected to inductive loads to compensate for the inductive effect.

Ground is the reference point in a circuit from which voltage is measured. Sometimes this point is connected to the actual ground (Earth), either through a grounded electrical circuit, water pipe, or some other method.

Inverter is a device that converts DC power to AC power. Inverters are used in solar PV and battery systems.

Rectifier is a device that converts AC power to DC power, such as to supply electricity to a DC motor or to charge a battery.

13.4 Voltages in AC Power Systems

In general, utilities that supply power in the North America (and most other 60 Hz countries) are required to provide power to the incoming point of a facility in multiples of 120 Volts. Thus, incoming equipment, such as circuit breaker panels, are rated in multiples of 120 Volts. The common voltages are 120, 240, 480, and 600 Volts. Higher voltage levels are 1,200, 2,400, 4,800, and 13,200 Volts. In addition, utilities are obligated by their regional governing authorities (usually called Public Utility Commissions, or Public Service Commissions) to regulate the voltage within a fairly narrow range, such as plus or minus 5% or 10%.

13.5 Phases and Frequencies in AC Power Systems

The number of phases in an AC power system is given by the number of different sine waves that make up the power. The frequency of the power system is given by the frequency of the sine wave or waves.

13.5.1 Single-phase AC electrical systems

Single phase is a power system having only one sine wave, and the frequency of that sine wave is 60 Hz in North America (see Figure 13.1). For example, in most homes or small buildings where only single-phase AC systems supply the electric power, the voltage is 120/240 volts. It is brought to the building with four wires, one being a ground. One is a neutral wire, and the other two (called "hot" wires) have voltages of 120 volts with respect to the neutral wire. The voltage difference between the two "hot" wires is 240 volts.

13.5.2 Three-phase AC electrical systems

Three-phase is a power system having three sine waves, and the frequency of each sine wave is 60 Hz in North America. The three sine waves are separated in time with a phase angle of 120° difference (see Figure 13.2). These systems are commonly used in commercial and industrial applications due to their higher power capacity and increased efficiency compared to single-phase systems. In North America, typical three-phase voltage levels are 208, 480, and 600 volts. Three-phase systems can be connected in two configurations: delta and wye (also called star). In a wye configuration, one end of each of the three phases is connected to a common neutral point, while in a delta configuration, the phases are connected in a loop without a neutral point.

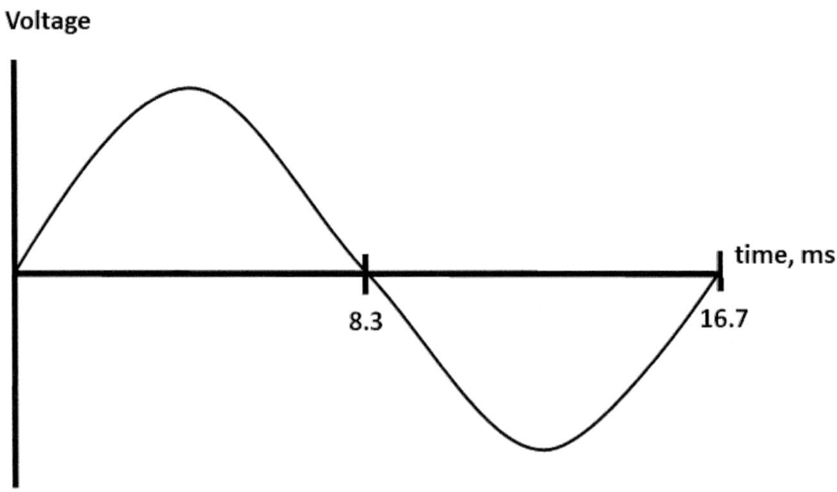

Figure 13.1 Single-phase electric voltage (60 Hz).

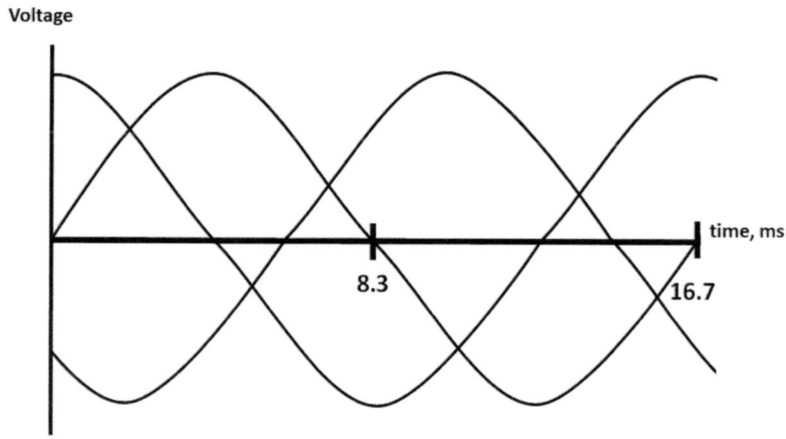

Figure 13.2 Three-phase electric voltage (60 Hz).

13.6 Basic Relationships of Voltage, Current, and Resistance

In DC circuits, and in AC circuits with only resistors in them, there is a fairly simple relationship between V (voltage), I (current), and R (resistance). In AC power systems, examples of loads that are pure resistive include incandescent lights, resistance space heaters, resistance water heaters, and electric resistance ovens. Loads like LED lights, computer equipment, and induction motors all have inductor effects as well as resistive effects.

13.7 Ohm's Law

For DC circuits, or for single-phase AC circuits with only resistors, we have the following relationship called Ohm's law:

$$V = I \times R \text{ or Volts} = \text{Amps} \times \text{Ohms}$$
$$\text{or } I = V/R$$
$$\text{or } R = V/I$$

Example 13-1:

A 100-Ohm resistor is connected to a source of 120 volts. What is the current through the resistor?

Solution:

$$I = V/R = 120\ \mathrm{V}/100\ \mathrm{ohms} = 1.2\ \mathrm{amperes}$$

13.8 Electrical Power for DC and Pure Resistive AC Loads

For DC circuits, or for single-phase AC circuits with only resistors, there is a relatively simple expression for power. Current (I), voltage (V), and resistance (R) are also related to electrical power (P_e) (measured in Watts), as follows:

$$\mathrm{Watts} = \mathrm{Volts} \times \mathrm{Amps},\ \mathrm{or}$$
$$P_e = V \times A = V \times I$$

Example 13-2:

In Example 13-1, how much power is the resistor dissipating (i.e., converting to heat)?

Solution:

$$P_e = V \times I = (120\ \mathrm{volts}) \times (1.2\ \mathrm{amps}) = 144\ \mathrm{Watts\ or}\ 1.44\ \mathrm{kW}$$

13.9 Ohm's Law for Power for DC and Pure Resistive AC Loads

In DC circuits, and in single-phase AC circuits with only resistors in them, if we substitute the voltage from Ohm's law ($V = I \times R$) into the expression for power, the result is:

$$P_e = VI = I \times R \times I = I^2 R$$

This relationship is called Ohm's law for power, and it tells us that for a fixed value of resistance, the power dissipated in that resistor is proportional to the square of the current through the resistor. Thus, if the current through a resistor is doubled, the dissipated power will go up by a factor of 4. When the main purpose of the power is to produce work like motor shaft power, the losses in the system due to resistances are often referred to as the "I^2R losses."

Example 13-3:

Rework Example 13-1 to find the power dissipated by the resistor using the $P = I^2R$ formula directly.

Solution:

$$P_e = I^2R = (1.2)^2 \times 100 = 1.44 \times 100 = 144 \text{ Watts or } 1.44 \text{ kW}$$

When dealing with more complex loads, such as inductive or capacitive loads in AC circuits, the calculations become more complex, and you will need to consider the power factor, reactive power, and apparent power in your calculations.

13.10 Power Factor in General AC Circuits and Systems

Power factor is a parameter which quantifies the ratio of real power to total or apparent power (terms defined below). It is important for both technical and financial reasons. In terms of system performance, lower power factor is associated with greater current for a given work output for a piece of equipment. Higher current can cause a number of performance problems such as overheating equipment, reduced capacity, and more frequent failures. Financially, most customers have a power factor charge in one form or another, as described in Chapter 4.

To understand power factor, it is necessary to understand different types of electric current and power. The current required by induction motors, transformers, fluorescent lights, induction heating furnaces, resistance welders, etc., is made up of three types of power:

1. **Real power (working power or active power):** This is the power that is converted by the equipment into useful work, such as turning a lathe, making a weld, or pumping water. The unit of measurement of the real power (P_e) produced from working current is the kilowatt (kW).
2. **Reactive power:** This is the power that is required to produce the magnetic flux necessary for the operation of induction devices. Without magnetizing current, energy could not flow through the core of a transformer or across the air gap of an induction motor. The unit of measurement of the reactive power (P_q) is the kilovoltamperes reactive (kVAr).

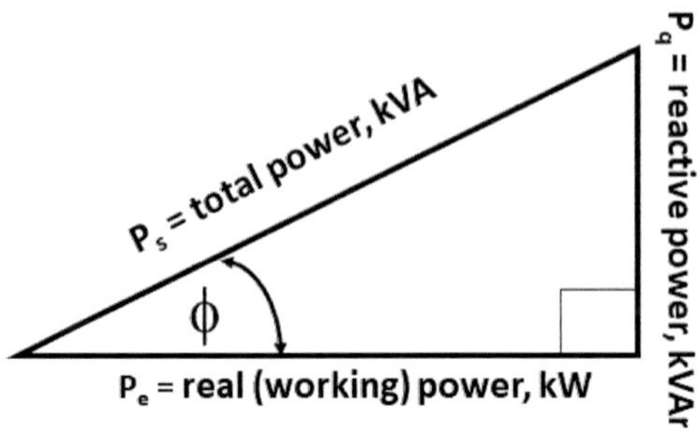

Figure 13.3 Power triangle.

3. **Apparent power (total power):** This is the power that is the product of voltage and amps metered in a circuit. It is made up of the vector sum of the reactive power and the real power. The unit of measurement of apparent power (P_s) associated is the kilovoltampere (kVA).

Power factor (PF) is the ratio of real power being used in a circuit, expressed in Watts or kilowatts, to the apparent power drawn from the power line, expressed in voltamperes or kilovolt-amperes: $\mathrm{PF} = P_e/P_s$.

The relationship of P_e (kW), P_q (kVAR), and P_s (kVA) in an electrical system can be illustrated by a right triangle as illustrated in Figure 13.3. The angle between the kW and kVA legs is known as the *phase angle*. The cosine of this angle (ϕ) is the power factor: $\mathrm{PF} = \cos(\phi)$. Based on the Pythagorean theorem from geometry: ${P_s}^2 = {P_e}^2 + {P_q}^2$.

For inductive loads, current lags the voltage in phase and this is called a lagging power factor. For capacitive loads, current leads the voltage and this is described as a leading power factor. The power factor value can range from greater than zero to 1.0. For resistive loads, current and voltage are in phase, which means the power factor equals exactly one ($\mathrm{PF} = 1.0$). Whenever power factor is discussed, it is typically assumed to be lagging unless stated otherwise.

For a DC power system, there is no reactive power, and thus no power triangle. In addition, for DC power systems, the power factor is always 1.0, or 100%.

13.11 Power Factor Correction Capacitors

For induction motors and other AC inductive equipment, the most cost-effective way to provide reactive power is usually to provide capacitors for the load, instead of having to buy the reactive power from the electric utility.

Capacitors can usually be put on the motors and other equipment to increase their power factors to 90% or higher.

Table 13.1 can be used to find the capacitance needed (kVAr) to improve the power factor of a given electrical power load in a building or a facility. The change in kVAr (ΔP_q) is found by multiplying the table factor by the real power load in kW. The example below illustrates this method.

$$\Delta P_q = \text{Real power} \times \text{Table Factor} \tag{13.1}$$

Example 13-4:

A facility has a power factor of 0.72, and a real power load of 800 kW. How many kVAr of capacitance is needed to add to the facility load to have a new power factor of 0.92?

Table 13.1 Power factor correction table.

target PF → initial PF ↓	0.80	0.81	0.82	0.83	0.84	0.85	0.86	0.87	0.88	0.89	0.90	0.91	0.92	0.93	0.94	0.95	0.96	0.97	0.98	0.99
0.50	0.982	1.008	1.034	1.060	1.086	1.112	1.139	1.165	1.192	1.220	1.248	1.276	1.306	1.337	1.369	1.403	1.440	1.481	1.529	1.590
0.51	0.937	0.963	0.989	1.015	1.041	1.067	1.093	1.120	1.147	1.174	1.202	1.231	1.261	1.291	1.324	1.358	1.395	1.436	1.484	1.544
0.52	0.893	0.919	0.945	0.971	0.997	1.023	1.049	1.076	1.103	1.130	1.158	1.187	1.217	1.247	1.280	1.314	1.351	1.392	1.440	1.500
0.53	0.850	0.876	0.902	0.928	0.954	0.980	1.007	1.033	1.060	1.088	1.116	1.144	1.174	1.205	1.237	1.271	1.308	1.349	1.397	1.458
0.54	0.809	0.835	0.861	0.887	0.913	0.939	0.965	0.992	1.019	1.046	1.074	1.103	1.133	1.163	1.196	1.230	1.267	1.308	1.356	1.416
0.55	0.768	0.794	0.820	0.846	0.873	0.899	0.925	0.952	0.979	1.006	1.034	1.063	1.092	1.123	1.156	1.190	1.227	1.268	1.315	1.376
0.56	0.729	0.755	0.781	0.807	0.834	0.860	0.886	0.913	0.940	0.967	0.995	1.024	1.053	1.084	1.116	1.151	1.188	1.229	1.276	1.337
0.57	0.691	0.717	0.743	0.769	0.796	0.822	0.848	0.875	0.902	0.929	0.957	0.986	1.015	1.046	1.079	1.113	1.150	1.191	1.238	1.299
0.58	0.655	0.681	0.707	0.733	0.759	0.785	0.811	0.838	0.865	0.892	0.920	0.949	0.979	1.009	1.042	1.076	1.113	1.154	1.201	1.262
0.59	0.618	0.644	0.670	0.696	0.723	0.749	0.775	0.802	0.829	0.856	0.884	0.913	0.942	0.973	1.006	1.040	1.077	1.118	1.165	1.226
0.60	0.583	0.609	0.635	0.661	0.687	0.714	0.740	0.767	0.794	0.821	0.849	0.878	0.907	0.938	0.970	1.005	1.042	1.083	1.130	1.191
0.61	0.549	0.575	0.601	0.627	0.653	0.679	0.706	0.732	0.759	0.787	0.815	0.843	0.873	0.904	0.936	0.970	1.007	1.048	1.096	1.157
0.62	0.515	0.541	0.567	0.593	0.620	0.646	0.672	0.699	0.726	0.753	0.781	0.810	0.839	0.870	0.903	0.937	0.974	1.015	1.062	1.123
0.63	0.483	0.509	0.535	0.561	0.587	0.613	0.639	0.666	0.693	0.720	0.748	0.777	0.807	0.837	0.870	0.904	0.941	0.982	1.030	1.090
0.64	0.451	0.477	0.503	0.529	0.555	0.581	0.607	0.634	0.661	0.688	0.716	0.745	0.775	0.805	0.838	0.872	0.909	0.950	0.998	1.058
0.65	0.419	0.445	0.471	0.497	0.523	0.549	0.576	0.602	0.629	0.657	0.685	0.714	0.743	0.774	0.806	0.840	0.877	0.919	0.966	1.027
0.66	0.388	0.414	0.440	0.466	0.492	0.519	0.545	0.572	0.599	0.626	0.654	0.683	0.712	0.743	0.775	0.810	0.847	0.888	0.935	0.996
0.67	0.358	0.384	0.410	0.436	0.462	0.488	0.515	0.541	0.568	0.596	0.624	0.652	0.682	0.713	0.745	0.779	0.816	0.857	0.905	0.966
0.68	0.328	0.354	0.380	0.406	0.432	0.459	0.485	0.512	0.539	0.566	0.594	0.623	0.652	0.683	0.715	0.750	0.787	0.828	0.875	0.936
0.69	0.299	0.325	0.351	0.377	0.403	0.429	0.456	0.482	0.509	0.537	0.565	0.593	0.623	0.654	0.686	0.720	0.757	0.798	0.846	0.907
0.70	0.270	0.296	0.322	0.348	0.374	0.400	0.427	0.453	0.480	0.508	0.536	0.565	0.594	0.625	0.657	0.692	0.729	0.770	0.817	0.878
0.71	0.242	0.268	0.294	0.320	0.346	0.372	0.398	0.425	0.452	0.480	0.508	0.536	0.566	0.597	0.629	0.663	0.700	0.741	0.789	0.849
0.72	0.214	0.240	0.266	0.292	0.318	0.344	0.370	0.397	0.424	0.452	0.480	0.508	0.538	0.569	0.601	0.635	0.672	0.713	0.761	0.821
0.73	0.186	0.212	0.238	0.264	0.290	0.316	0.343	0.370	0.396	0.424	0.452	0.481	0.510	0.541	0.573	0.608	0.645	0.686	0.733	0.794
0.74	0.159	0.185	0.211	0.237	0.263	0.289	0.316	0.342	0.369	0.397	0.425	0.453	0.483	0.514	0.546	0.580	0.617	0.658	0.706	0.766
0.75	0.132	0.158	0.184	0.210	0.236	0.262	0.289	0.315	0.342	0.370	0.398	0.426	0.456	0.487	0.519	0.553	0.590	0.631	0.679	0.739
0.76	0.105	0.131	0.157	0.183	0.209	0.235	0.262	0.288	0.315	0.343	0.371	0.400	0.429	0.460	0.492	0.526	0.563	0.605	0.652	0.713
0.77	0.079	0.105	0.131	0.157	0.183	0.209	0.235	0.262	0.289	0.316	0.344	0.373	0.403	0.433	0.466	0.500	0.537	0.578	0.626	0.686
0.78	0.052	0.078	0.104	0.130	0.156	0.183	0.209	0.236	0.263	0.290	0.318	0.347	0.376	0.407	0.439	0.474	0.511	0.552	0.599	0.660
0.79	0.026	0.052	0.078	0.104	0.130	0.156	0.183	0.209	0.236	0.264	0.292	0.320	0.350	0.381	0.413	0.447	0.484	0.525	0.573	0.634
0.80		0.026	0.052	0.078	0.104	0.130	0.157	0.183	0.210	0.238	0.266	0.294	0.324	0.355	0.387	0.421	0.458	0.499	0.547	0.608
0.81			0.026	0.052	0.078	0.104	0.131	0.157	0.184	0.212	0.240	0.268	0.298	0.329	0.361	0.395	0.432	0.473	0.521	0.581
0.82				0.026	0.052	0.078	0.105	0.131	0.158	0.186	0.214	0.242	0.272	0.303	0.335	0.369	0.406	0.447	0.495	0.556
0.83					0.026	0.052	0.079	0.105	0.132	0.160	0.188	0.216	0.246	0.277	0.309	0.343	0.380	0.421	0.469	0.530
0.84						0.026	0.053	0.079	0.106	0.134	0.162	0.190	0.220	0.251	0.283	0.317	0.354	0.395	0.443	0.503
0.85							0.026	0.053	0.080	0.107	0.135	0.164	0.194	0.225	0.257	0.291	0.328	0.369	0.417	0.477
0.86								0.027	0.054	0.081	0.109	0.138	0.167	0.198	0.230	0.265	0.302	0.343	0.390	0.451
0.87									0.027	0.054	0.082	0.111	0.141	0.172	0.204	0.238	0.275	0.316	0.364	0.424
0.88										0.027	0.055	0.084	0.114	0.145	0.177	0.211	0.248	0.289	0.337	0.397
0.89											0.028	0.057	0.086	0.117	0.149	0.184	0.221	0.262	0.309	0.370
0.90												0.029	0.058	0.089	0.121	0.156	0.193	0.234	0.281	0.342

Solution:

Go to Table 13.1 and go to the row for the original power factor of 0.72. Now, go across that row to the column with a top marking of the new power factor, 0.92. Where that row and column intersect, read the number in the table, which is 0.538. Then,

$$\Delta \mathrm{P_q} = 800 \times 0.538 = 430.4 \text{ kVAr}$$

13.12 Power in Single-Phase AC Systems

Many electric loads in facilities are single-phase loads. As mentioned earlier, most residential and small commercial buildings only have single-phase AC 120/240-Volts power available. Devices like full house air conditioners, electric hot water heaters, and electric clothes dryers are almost always operated on 240 Volts AC, single phase.

For single-phase AC systems, the equation for total power is:

$$P_s = I_L \times V_L/1,000 \text{ in units of kVA}$$

where I_L is line current (amps) and V_L is line voltage (Volts).

Because $P_e = \mathrm{PF} \times P_s$, the equation for real power is:

$$P_e = \mathrm{PF} \times I_L \times V_L/1,000 \text{ in units of kW}$$

Example 13-5:

Find the real power drawn by a 240-Volts AC, single-phase electric resistance water heater, if it is drawing 20 amps of current.

Solution:

Since the water heater is an electric resistance device, its power factor is 100%. Thus, the real power drawn is:

$$P_e = V \times I \times \mathrm{PF} = (240)(20)(1.0) = 4,800 \text{ Watts} = 4.8 \text{ kW}$$

13.13 Power in Three-Phase AC Systems

Almost all electric motors over 1 hp, as well as all large pieces of electrical equipment in a facility, run on three-phase AC power.

The equation for the power drawn by a general three-phase load is:

$$P_s = (3)^{\frac{1}{2}} \times I_L \times V_L/1,000 \text{ in units of kVA}$$
$$P_e = \text{PF} \times (3)^{\frac{1}{2}} \times I_L \times V_L/1,000 \text{ in units of kW}$$

where I_L is line current and V_L is line voltage.

It is important to note that the equations for three-phase total and real power shown above are valid for both wye and delta designs (see the next section below).

Example 13-6:

A three-phase AC induction motor draws 40 kW of real power, 50 kVA total power, and operates with an average voltage of 2,300 Volts. Estimate (a) the reactive power, (b) the power factor, and (c) the input motor current.

Solution:

(a) Draw a power triangle and use the Pythagorean theorem to solve for P_q.

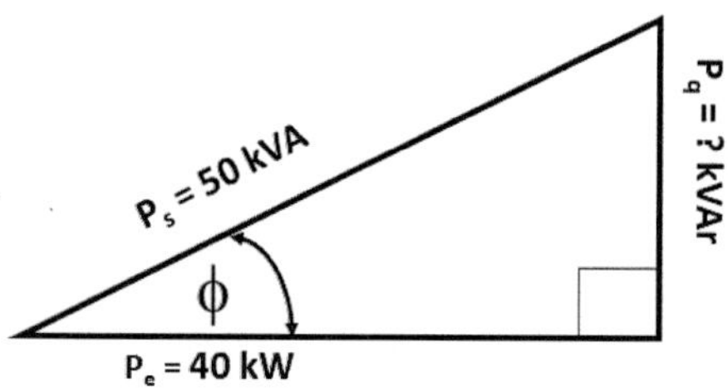

$$P_q^2 = P_s^2 - P_e^2$$
$$P_q^2 = (50)^2 - (40)^2 = 900$$
$$P_q = 30 \text{ kVAr}$$

(b) PF $= P_e/P_s = 40/50 = 0.80$ or 80%.

(c) $P_s = (3)^{\frac{1}{2}} \times I_L \times V_L/1{,}000$ for three-phase AC systems, where I_L is line current and V_L is line voltage:

$I_L = 1{,}000 \times P_s \,/\, [(3)^{\frac{1}{2}} \times V_L]$
$I_L = 1{,}000 \times 50 \text{ kVA} \,/\, [(3)^{\frac{1}{2}} \times 2{,}300 \text{ V}]$
$I_L = 12.6$ amps

13.14 Three-Phase Wye and Delta Systems

A common electrical design used in commercial and industrial applications is called a five-wire grounded wye system (sometimes described as "star" system). A three-phase wye system is shown in Figure 13.4. A low-voltage version of the wye design, used for small buildings, would have 120 Volts (single phase) or 208 Volts (three phase). A higher voltage version, for larger buildings and facilities, is a 277 Volts (single-phase) and 480 Volts (three-phase). In a five-wire grounded wye system, the high-voltage three phase is 1.732 (the square root of 3) times higher than the low-voltage single phase.

For wye designs, the individual phase and line-to-line voltage, current, and power equations are (Chapman, 2012):

$$I_L = I_{\text{phase}}$$
$$V_L = (3)^{\frac{1}{2}} \times V_{\text{phase}}$$
$$P_e = \text{PF} \times (3)^{\frac{1}{2}} \times I_L \times V_L/1{,}000$$

Grounded wye systems are generally considered to be safer and more flexible than the older ungrounded systems. The flexibility comes from the

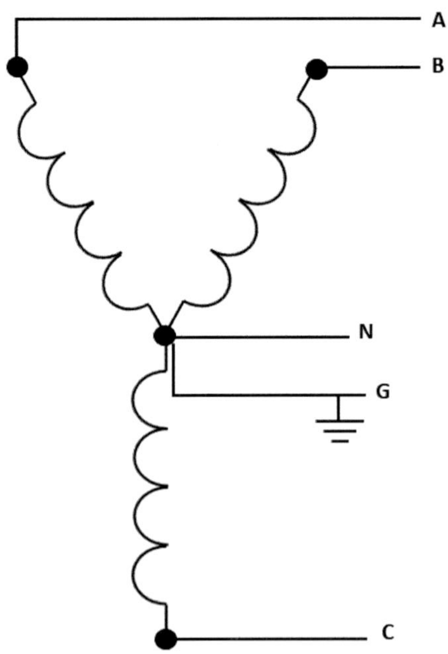

Figure 13.4 Three-phase grounded three-phase wye system.

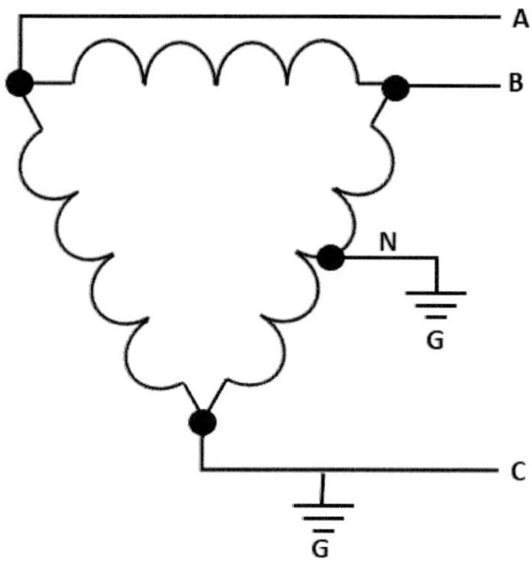

Figure 13.5 Three-phase grounded three-phase delta system.

ability to handle single-phase plug loads that operate at 120 Volts, or lighting circuits that operate at 120 or 277 Volts, from the same system that feeds the three-phase circuits for motors, equipment for heating, air conditioning, elevators, and industrial machinery.

Another common electrical design is the delta design system, illustrated in Figure 13.5.

For delta designs, the individual phase and line-to-line voltage, current, and power equations are shown below (Chapman, 2012). For a 480-volt delta system, the phase and line-to-line voltages are both equal to 480 Volts.

$$I_L = (3)^{\frac{1}{2}} I_{\text{phase}}$$
$$V_L = V_{\text{phase}}$$
$$P_e = \text{PF} \times (3)^{\frac{1}{2}} \times I_L \times V_L/1{,}000$$

It is important to note that the equation for power consumption is the same for both wye and delta three-phase designs.

13.15 Power Quality

AC power was introduced at the beginning of this chapter, and illustrated in Figures 13.1 and 13.2 as perfect sine waves. However, real-power systems have imperfections, which are described as power quality. For example, actual current and voltage measurements will exhibit waveform distortions (i.e., not smooth sine waves as shown in Figure 13.2), the magnitude of the voltage for each phase in a polyphase system may not be equal, and the supply of power may not be absolutely constant over time.

Harmonics are electromagnetic signals that are multiples of the fundamental frequency (e.g., multiples of 60 or 50 Hz). There are a wide range of indicators of power quality problems such as equipment derating, poor control, overheating, reduced efficiency, and increased equipment failures. Power quality problems are also potentially safety issues. Ultimately these problems can degrade the services delivered by energy systems such as poor thermal comfort in buildings, manufacturing defects, or unplanned production outages.

A definition of power quality is "...power quality is generally meant to express the quality of voltage and/or the quality of current and can be defined as: the measure, analysis, and improvement of the bus voltage to maintain a sinusoidal waveform at rated voltage and frequency. This definition includes all momentary and steady-state phenomena" (Masoum and Fuchs, 2015). A framework for classifying power quality problems is illustrated in Figure 13.6 (adapted from Masoum and Fuchs, 2015). The cause of power

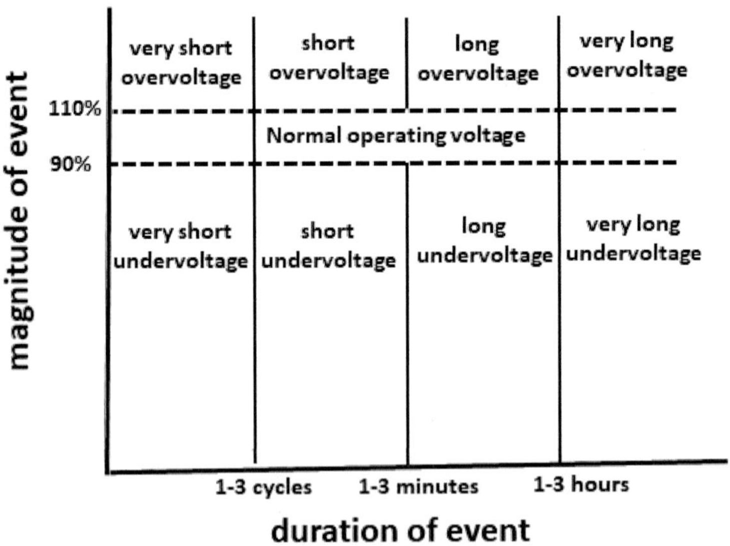

Figure 13.6 Classification of power quality events by magnitude and duration.

quality problems can be the utility supply or the facility distribution and loads, or both.

The main phenomena that are the underlying causes of electromagnetic and power quality disturbances are outlined in Table 13.2. This table is provided to illustrate the wide variety of phenomena involved in power quality problems. Detailed descriptions of the causes, diagnostics, and solutions to these phenomena can be found in many resources (Masoum and Fuchs, 2015; Krarti, 2017; Belu, 2019; and ASHRAE, 2023).

Harmonics are one of the common phenomena causing power quality problems. Symptoms of harmonic problems potentially include noisy or vibrating transformers, tripping circuit breakers, and overheating equipment. Sources of harmonics in power systems include variable frequency drives (VFDs), LED lighting, electric vehicle (EV) chargers, arcing devices, computers, and data centers.

A common problem resulting from harmonics is to derate motors, as measured by the harmonic voltage factor. The National Electrical Manufacturers

Table 13.2 Phenomena causing power quality disturbances (Masoum and Fuchs, 2015).

Conducted low-frequency phenomena
Harmonics, inter-harmonics
Signaling voltage
Voltage fluctuations
Voltage dips
Voltage imbalance
Radiated low-frequency phenomena
Magnetic fields
Electric fields
Conducted high-frequency phenomena
Induced continuous wave voltages or currents
Unidirectional transients
Oscillatory transients
Radiated high-frequency phenomena
Magnetic fields
Electric fields
Electromagnetic field
Steady-state waves
Transients
Electrostatic discharge phenomena
Nuclear electromagnetic pulse

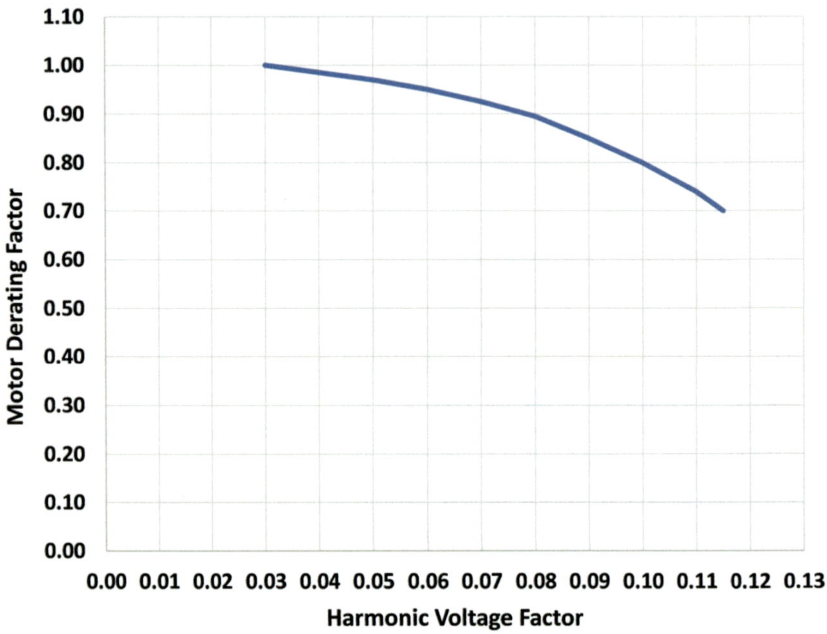

Figure 13.7 Harmonic voltage factor effect on rated output of polyphase induction motors up to 500 hp (based on data in NEMA, 2016).

Association (NEMA, 2016) defines this factor as:

$$\text{Harmonic voltage factor} = \sum \left[V_n^2/n\right]^{\frac{1}{2}}$$

where:

V_n = voltage of each harmonic

n = odd order (i.e., multiplier of the fundamental frequency) of harmonic, starting with the 5th harmonic and higher, and not counting harmonics divisible by 3.

Figure 13.7 provides a chart of derating factor for inductive motors based on the harmonic voltage factor.

Example 13-7:

For a 75-hp induction motor, an electrician measures harmonic voltages as follows:

5th harmonic 0.108 V
7th harmonic 0.074 V

11th harmonic 0.061 V

13th harmonic 0.048 V

Calculate the harmonic voltage factor, estimate the motor derating factor, and the derated full load output of the motor.

Solution:

$$\text{Harmonic voltage factor} = \left[(0.108)^2/5 + (0.074)^2/7 + (0.061)^2/11 + (0.061)^2/13\right]^{\frac{1}{2}} = 0.0603.$$

From Figure 13.7, the motor derating factor is 0.95.

The derated output of the motor = (0.95) × (75 hp) = 71.3 hp.

Voltage imbalance for three-phase motors is another common power quality problem. Voltage imbalance for a three-phase motor is defined as:

$$\text{Voltage imbalance} = \frac{\text{(maximum deviation from average voltage)}}{\text{(average voltage)}}.$$

The impact of voltage imbalance on motor rating for induction motors is provided in Figure 13.8.

Example 13-8:

For a 75-hp induction motor, an electrician measures the line voltage supplied to each phase of a three-phase motor. Phases A, B, and C have measured voltages of 476, 471, and 456 V, respectively. Calculate the voltage imbalance, estimate the motor derating factor, and the derated full load output of the motor.

Solution:

$$\text{Average voltage} = (476 + 471 + 456) = 467.7 \text{ V}$$

Phase	Voltage	Phase deviation	Phase deviation
A	476 V	8.3 V	1.8%
B	471 V	3.3 V	0.7%
C	456 V	−11.7 V	−2.5%

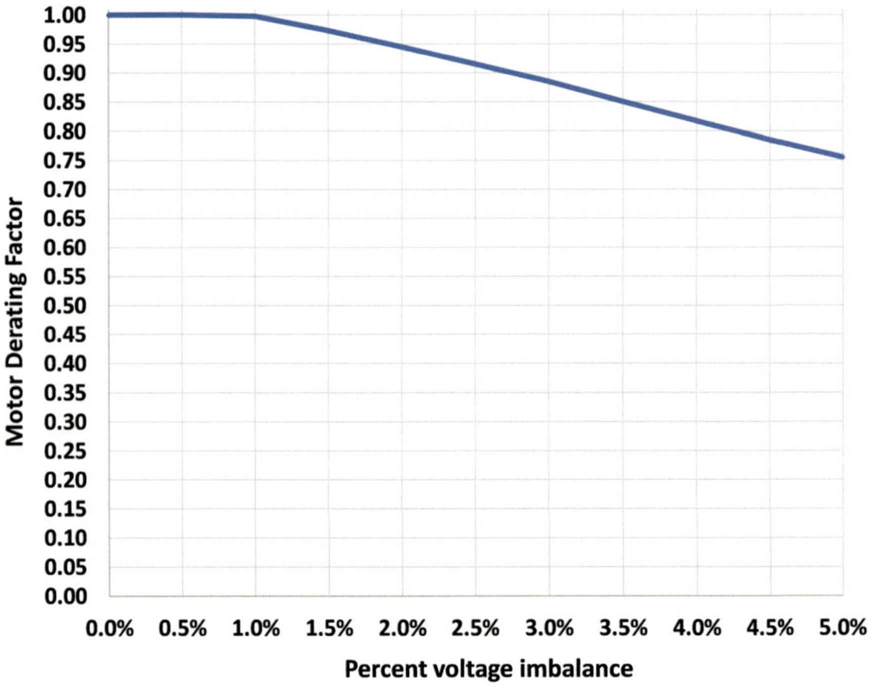

Figure 13.8 Voltage imbalance effect on rated output of polyphase induction motors up to 500 hp (based on data in NEMA, 2016).

The maximum percentage deviation is phase C at 2.5% (use the absolute value). From Figure 13.8, the motor derating factor is 0.92. The estimated output of the derated motor is: (0.92) × (75 hp) = 69.0 hp.

If there are indicators of power quality problems in facilities, the first step is to engage a qualified specialist electrician or engineer to perform testing and diagnosis. IEEE 519 is one of the main standards that provide guidance for power quality parameters for facilities. There are a number of solutions to improve power quality including devices such as harmonic filters, special rectifiers, voltage compensators, special transformers, and uninterruptible power supply (Krarti, 2017). These devices can help mitigate the effects of power quality issues such as harmonic distortion, voltage imbalance, and other disturbances, improving the overall reliability and efficiency of the electrical system. It is important to note that the selection and implementation of these devices should be performed by trained professionals to ensure their proper operation and effectiveness.

Table 13.3 Common electric power parameters for EV charging systems.

Charging system	Typical power supply voltage and phase	Typical output power	Typical charging time for empty BEVs[1]
Level 1	120 V single-phase	AC 1.4–2.0 kW	40–50 hours
Level 2	208–240 V single-phase	AC 7–10 kW	4–10 hours
Level 3 (DC fast charging)	208–600 V three-phase	DC 22–350 kW	20–60 minutes

[1] www.transportation.gov/rural/ev/toolk it/ev-basics/charging-speeds. Charging times for PHEV vehicles (which have smaller batteries) are about one-fifth the time for Level 1 and Level 2 chargers.

13.16 Electric Vehicle (EV) Charging

Facilities today often are (or eventually will be) equipped with electric vehicle supply equipment (EVSE). EVSE (also referred to as charging stations) can comprise a significant load for single family homes, multiunit residential buildings, and commercial facilities. EVSE comes in three common designs, as summarized in Table 13.3 (CEC, 2018; Pollution Probe, 2020).

Passenger vehicles (e.g., for employees or fleet vehicles) come in two basic designs: battery electric vehicles (BEV), which run on 100% electricity, and plug-in hybrid vehicles (PHEV), which can run on an internal combustion engine or battery. BEVs have relatively large batteries and longer driving range on a single charge, such as 200–500 km (120–300 miles). PHEVs have smaller batteries and a short range in electric mode such as 30–80 km (20–50 miles). Level 1 chargers are for home applications to slowly charge batteries, and these are mostly suited for PHEVs. Level 2 chargers are generally considered essential for BEVs. Both Level 1 and Level 2 charges deliver AC power to the vehicle onboard rectifier, which converts the power to DC. An example of a Level 2 charger installed in a parkade is shown in Figure 13.9. Level 3 chargers deliver DC power direct to the car battery.

There are several issues that facility energy managers need to consider with EVSE systems in their facilities. As shown in Table 13.3, EV chargers can add significant peak loads to facility electric systems. Installation of EV

Figure 13.9 Level 2 charger with two ports in a parking garage (source: https://images.nrel.gov).

chargers could require upgrades to utility service connections or in-facility equipment such as transformers. The peak power demands (in kW or kVA) can represent a substantial cost for utility bills, in addition to the energy (kWh) charges. If the facility has time-variant pricing such as time of use rates (see Chapter 4), then managing the timing charging activities (e.g., through EV charger software and controls) may be critical to manage costs and to keep loads within the rated capacity of the facility electric infrastructure. Additionally, EV chargers often contribute to harmonics and reduced power factor in facility electric systems. EV chargers may have dedicated meters used for billing or third-party payment systems (Tiedemann and Sulyma, 2023).

13.17 Summary

This chapter provides an introduction to electricity basics for facilities. Power systems can be of AC or DC design. Single- or three-phase AC power are by far the most common designs, operating at frequencies of either 50 or 60 Hz. The basic equations to measure and calculate power consumption for these systems are provided. Electric loads can be inductive (e.g., induction motors), resistive (e.g., incandescent lamps), or capacitive (e.g., power factor correction devices). The components of electric power are total (kVA units), real (kW units), and reactive power (kVAr units). Real power is defined as the power that provides useful work such as movement of materials on a conveyer or pushing air through an HVAC system. Power factor is the ratio of real power to total power. Lower power factor results in higher current to deliver the same real power. In turn, higher current leads to inefficiencies and increased equipment failures due to increased I^2R losses. Installation of capacitors is a common solution to maintain power factor within targeted limits (e.g., 0.95 or higher). The basics of power quality are introduced. There are many types and causes of power quality problems, although the two most common types are harmonics and voltage imbalance. Harmonics are low-magnitude voltages or currents operating at multiples of the fundamental frequency (i.e., multiples of 50 or 60 Hz). Voltage imbalance is where voltages supplied in a polyphase system are unequal. Indicators and consequences of poor power quality include reduced energy efficiency, poor control, overheating equipment, and more frequent failures. The basics of EVSE in facilities are provided, including required voltage and phase, and peak electric loads. Charging systems are classified as Level 1 (AC slow charge, useful for PHEV), Level 2 (AC faster charge for BEVs), and Level 3 (DC fast chargers). Metering and management of EVSE systems is increasingly becoming common for facility and energy managers.

References

ASHRAE (2023) "ASHRAE Handbook – HVAC Applications" Chapter 57 "Electrical Considerations."

Belu R. (2019) "Industrial Power Systems with Distributed and Embedded Generation" The Institution of Engineering and Technology. London, UK.

CEC (2018) "Electric Vehicle Charger Selection Guide" funded by the California Energy Commission. Written by authors from 5 organizations. https://afdc.energy.gov/files/u/publication/EV_Charger_Selection_Guide_2018-01-112.pdf

IEEE 519-2014 "IEEE Recommended Practice and Requirements for Harmonic Control in Electric Power Systems" Institute of Electrical and Electronics Engineers. www.ieee.org

Krarti (2017) "Energy-Efficient Electrical Systems for Buildings" CRC Press.

Masoum M. and Fuchs E. (2015) "Power Quality in Power Systems and Electrical Machines" 2nd edition. Elsevier.

NEMA (2016) "Motors and Generators" ANSI/NEMA MG 1-2016. National Electrical Manufacturers Association. www.nema.org

Pollution Probe and Delphi Group (2020) "Guide to Electric Vehicle Charging in Multi-Unit Residential Buildings" https://www.nrcan.gc.ca/sites/nrcan/files/energy/pdf/Revised_Guide_to_EV_Charging_in_MURBs_ENG_ACC.pdf

Tiedemann K and Sulyma I (2023) "Literature Review of EV Charging Station Rates and EV Technology Incentive Programs" CEATI Report No. T211700-7074. www.ceati.com

Author Biography

Eric Mazzi is a consultant and educator focusing on technology and policy for demand-side energy systems. He specializes in performing independent technical analysis, M&V, policy studies, and educational services. He advocates and contributes to efforts to reduce energy consumption and achieve net-zero GHG emissions, while maintaining or improving energy services and advancing energy equity. His experience includes air quality consulting, utility maintenance, and five years as full-time engineering faculty at the University of British Columbia. He volunteers as a board member for the U.S. Northwest Energy Coalition, and the International Performance M&V Protocol committee. Eric teaches part-time, delivering university and professional certification courses. He is an adjunct professor of energy management at the New York Institute of Technology – Vancouver. He has B.Sc. and M.Sc. degrees in mechanical engineering, and a Ph.D. degree in environmental policy. He is a registered engineer in California and British Columbia.

Acknowledgments

I would like to thank chapter reviewer Zaid Al-Chalabi for his thoughtful review and commentary for this chapter. I would like to acknowledge the contributions of the authors of the 8th edition of this book, as some parts of this chapter are similar to the 8th edition, especially the introduction to electrical system basics.

Practice Questions

1. A three-phase motor is connected to a centrifugal fan. The motor nameplate, full load current is 40 amps. The fan operates at 45% efficiency and 1,790 rpm. Motor line voltage is measured at 2,300 V and line current is measured at 20 amps. Calculate the apparent power input to the motor.
2. A single-phase load of 5 kW operates at power factor of 0.6 lagging. It is proposed to improve this power factor to 0.95 lagging by connecting a capacitor across the load. Calculate the kVAr rating of the capacitor required.
3. The rating of a single-phase electric resistance water heater is 2,000 Watts, at 230 Volt. Calculate:

   (a) Rated current
   (b) Resistance of the water heater in Ohms
   (c) Actual power drawn when the measured supply voltage is 210 Volts

4. The term K in the following equation in a three-phase AC supply system is? Reactive power = kVAR = $\sqrt{3} \times V \times I \times K$

   (a) $K = \cos(\phi)$
   (b) $K = \sin(\phi)$
   (c) $K = 1 - \sin(\phi)$
   (d) $K = 1 - \cos(\phi)$

5. The approximate kVAR compensation required for improving the power factor of a 100-kW load from 0.95 leading power factor to unity power factor is:

   (a) 328 kVAR
   (b) 750 kVAR
   (c) 1,000 kVAR
   (d) None of the above

6. An induction motor load is 400 kVA (P_{e1}) with a power factor of 0.75 (PF_1). The induction motor is connected in parallel with a synchronous motor. The combined load is 450 kW (P_{et}) with a power factor of 0.90 (PF_t). Estimate: (a) the kVAr of the synchronous motor (P_{q2}), (b) the kVA for the synchronous motor (P_{s2}), and (c) the power factor of the synchronous motor (PF_2). Hint: $P_{et} = P_{e1} + P_{e2}$ and $P_{qt} = P_{q1} + P_{q2}$ and $P_{st} = [{P_{et}}^2 + {P_{qt}}^2]^{1/2}$.

7. Describe how harmonic distortion of current versus distortion of voltage may be different in terms of the symptoms or operating impacts. Note: You will need to review additional sources such as those listed in the References of this chapter.
8. List some of the key design and operation issues that should be considered if retrofitting electric vehicle supply equipment system to an existing multi-family residential building. Note: You will need to review additional sources such as those listed in the References of this chapter.

14

Electric Motors and Drives

Eric Mazzi

Mazzi Consulting, Canada

14.1 Introduction

Electric motors are the single largest end use of electricity globally, with estimates ranging from about one-third to one-half of total electricity generated. In the U.S. and Canada, the estimates vary, but most sources state that motors consume roughly 25% of electricity in residential buildings, 40% commercial buildings, and 70% in industry. Motors can vary in size from 1/100th to 100,000 hp. Three-phase induction motors typically range from 1 to 10,000 hp. Motors are found on almost every piece of equipment used to perform a process in industrial operations. HVAC systems are composed of equipment to move air, water, flue gas, and refrigerants with fans, pumps, and compressors, all driven by electric motors. Electric motors provide efficient, reliable, and long-lasting service, and most require comparatively little maintenance. Despite these advantages, however, motors can be costly to operate if they are not properly selected and maintained. The annual cost of operating a motor can often be 5–10 times the original purchase price of the motor.

The purpose of this chapter is to introduce the basics of electric motors, mainly focusing on AC induction motors. Motor variables such as load factor, efficiency, and power factor are described, which are important to accurately measure or estimate motor electrical loads. Motor controls are introduced, with an emphasis on variable speed drives to save energy.

14.2 AC Induction Motors

Electric motors are devices that convert electric power into mechanical power. Most AC motors are induction motors, which are simpler, lighter, and cheaper than most alternatives. Estimates for the U.S. are that roughly 90% of motors in facilities are AC induction motors (Krarti, 2017).

Motors are rated by the amount of mechanical shaft power they deliver at full load, which is printed on motor nameplates. In English units, motors are rated in horsepower (1 hp = 0.7457 kW). In SI units, motors are rated in kW. The power delivered by a motor is the product of speed and torque as follows:

SI units:

$P_{\text{motor, out}} = (\omega \times T_{\text{motor, out}})/9{,}550$

where:

$P_{\text{motor, out}}$ = motor output power in kW
ω = motor speed in RPM
$T_{\text{motor, out}}$ = torque in Nm (Newton meters).

English units:

$P_{\text{motor, out}} = (\omega \times T_{\text{motor, out}})/5{,}245$

where:

$P_{\text{motor, out}}$ = motor output power in hp
ω = motor speed in RPM
$T_{\text{motor, out}}$ = torque in ft lbf (foot pounds-force).

AC induction motors range in size from fractional hp to thousands of hp. Motors over 500 hp (or over 600 V) are generally considered custom manufactured. There are several operating parameters of AC induction motors that are of critical importance in energy management, including speed, electricity consumption, load factor, energy efficiency, and power factor.

Another common variable is open vs. enclosed design. Open drip proof is a common, low-cost motor design where air can ventilate from inside the motor, but also the internal components such as windings and bearings are exposed to the surrounding environment. Totally enclosed fan-cooled motors are a common design for applications, such as outdoor operation, to protect the internal components from moisture and particles. Fan-cooled motors incorporate an external fan to keep the motor cool.

14.2.1 Synchronous speed of induction motors

Synchronous speed (ω_{sync}) is the no-load speed of an AC induction motor. Without any load, the speed of the induction motor is keyed to the electrical power supply frequency. Thus, the synchronous or no-load speeds of AC induction motors are always multiples of 60 for 60 Hz systems, and 50 for 50 Hz. The no-load or synchronous speed of an induction motor is also a function of its mechanical construction, and how many poles it was built with. A pole pair is a matching electromagnet on the stator and the rotor of the motor with a north pole and a south pole. The equation for the speed of an induction motor is related to the number of poles and the frequency of the applied voltage in Hz.

$$\omega_{\text{sync}} = 2 \times f \times 60/n$$

where:

ω_{sync} = motor synchronous speed in RPM
n = the number of poles
f = electric power frequency in Hz (i.e., 60 or 50 Hz).

Common synchronous speeds for 60 Hz electric systems are 3,600, 1,800, 1,200, 900, and 600 RPM. For 50 Hz systems, common synchronous speeds are 3,000, 1,500, 1,000, 750, and 500 RPM.

Example 14-1

Calculate the synchronous speed for a motor with 6 pole motor, for both 60 Hz and 50 Hz electric power.

Solution

For six poles ($n = 6$):

- At 60 Hz, $\omega_{\text{sync}} = 2 \times 60 \times 60/6 = 1{,}200$ RPM
- At 50 Hz, $\omega_{\text{sync}} = 2 \times 50 \times 60/6 = 1{,}000$ RPM

Full load speed ($\omega_{\text{nameplate}}$) is the speed of the motor when operating at the full load output. Full load speed is stamped on the nameplate, and is typically 2%–3% lower than the no-load or synchronous speed.

Slip is the term for the difference between the synchronous speed of a motor and the actual speed of the motor at any operating condition. The full load slip is the largest value, and it is the difference between the synchronous speed and the nameplate (full load) speed. Percentage slip is defined as follows:

$$\% \text{ slip} = 100\% \times (\omega_{\text{sync}} - \omega_{\text{meas}})/\omega_{\text{sync}}.$$

14.2.2 Electric power consumption for motors

Motor power consumption is calculated using the basic electric power equations introduced in Chapter 13, using either the single-phase equation or three-phase equation (the design phase will be stamped on the nameplate). Almost all AC induction motors over 1 hp use three-phase power. Electric power consumed by motors is calculated as follows:

$$P_{\text{e, motor}} = \text{PF} \times I_L \times V_L/1{,}000 \text{ for one-phase AC motor}$$
$$P_{\text{e, motor}} = \text{PF} \times (3)^{1/2} \times I_L \times V_L/1{,}000 \text{ for three-phase AC motors.}$$

14.2.3 Load factor for AC induction motors

Motor load factor (MLF) is defined as the output power of the motor at a given operating condition, divided by the nameplate (full load) output power. MLF is often described using different terms such as "motor loading" or just "load factor." Note that MLF is not directly a function of the input electric power, as it is a ratio of two output (mechanical) power values:

MLF = motor load factor = operating output or load/rated output or load.

A motor is a load-driven device, which means that the motor provides the amount of power required by the load (assuming sufficient motor capacity). If the load on a 20 hp motor is a fan requiring 10 hp of shaft power to drive it, the load factor on the motor is 10 hp/20 hp, or 50%. The input electric power to the motor will only be what is needed to drive the actual load, and most often it will not be the full rated load of the motor.

Typical MLF values on an annual basis are in the range 40%–60% for HVAC motors. In industrial applications, equipment is often oversized, including the motors. Estimates are that industrial motors are typically loaded 60%–70% (McCoy and Douglass, 2014). When energy studies are performed on motor-driven equipment, analysts should attempt to obtain measurements to confirm actual motor loading to ensure accuracy in estimating energy savings.

Motor loading is a function of the ratio of the actual (or operating slip) to the full load design slip. The equation for this relationship is shown below. What this means is that measurement of the actual rotating speed of a motor, combined with nameplate design information, is a method of measuring actual motor loading. It should be noted that accurate measurement of speed at low slip combined make it challenging to obtain accurate results using this method.

$$\mathrm{MLF} = (\omega_{\mathrm{sync}} - \omega_{\mathrm{meas}})/(\omega_{\mathrm{sync}} - \omega_{\mathrm{nameplate}}).$$

Another measurement that can be used to estimate actual MLF is motor current:

$$\mathrm{MLF} = (I_L)/(I_{\mathrm{nameplate}})/[(V_L)/(V_{\mathrm{nameplate}})].$$

With this formula, it is often reasonable to assume that the actual voltage equals nameplate voltage, and this formula simplifies to: MLF = $(I_L)/(I_{\mathrm{nameplate}})$. Using the ratio of metered current to nameplate current (i.e., full load current) is generally only valid if the MLF is 50% or greater (McCoy and Douglass, 2014). This is because motor current is typically linear with motor load only in the range of 50%–110% of motor load for induction motors. In cases where there is significant magnetizing current (see Chapter 13), using measured motor current to estimate MLF can produce large errors. Nonetheless, if it can be confirmed that motor loading is greater than 50%, using motor current is typically the most accurate method to estimate actual MLF.

A third method for estimating actual motor load is to use the measured input power as a ratio to the input power at nameplate (full load) operation (McCoy and Douglass, 2014):

$$(P_{\text{motor, in}})/(P_{\text{motor,in @full load}}).$$

Example 14-2

The speed of an induction motor is measured to be 3,564 RPM with a tachometer. The motor operates at 60 Hz power supply and the nameplate is shown in Figure 14.1. Based on this nameplate, (a) estimate the actual output power of the motor in horsepower, and (b) state the number of poles in the motor design.

Solution

(a) From the nameplate, the motor is 60 Hz, 10 hp (7.5 kW), and the full load speed is 3,520 RPM. The synchronous speed is not stated on the nameplate, which is common. Since the full load speed is usually about 2%–3% lower than the synchronous speed, and the synchronous

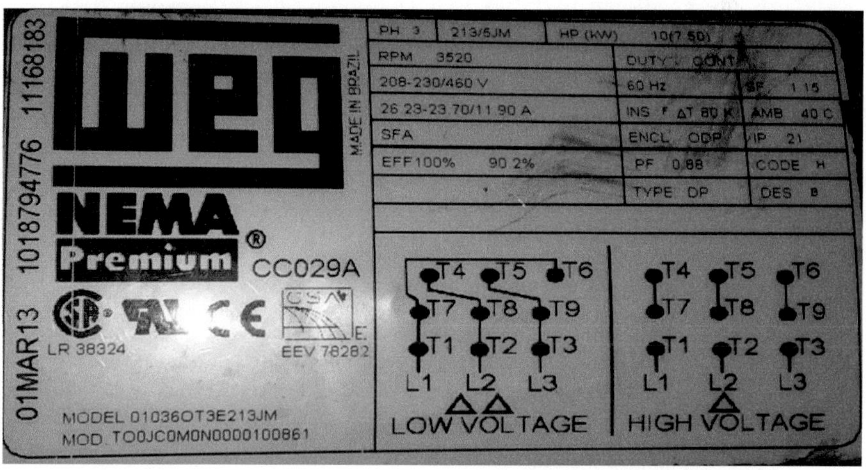

Figure 14.1 Sample nameplate for a 10 hp, three-phase AC induction motor.

speed is a multiple of the 60 Hz frequency, the synchronous speed for this motor is known to be 3,600 RPM. Using the equation above, the estimated motor load is:

MLF $=$ (3,600 – 3,564)/(3,600 – 3,520) $=$ 36/80 $=$ 0.45 or 45%

$P_{\text{motor, out}} = 0.45$ (10 hp) $=$ 4.5 hp.

(b) The number of poles is known based on the synchronous speed and frequency:

$n = 2 \times f \times 60/\omega_{\text{sync}}$ $=$ (2) (60) (60)/1,800 $=$ 4. Thus, the motor has four poles.

14.2.4 Motor efficiency

Motor efficiency (η_m) defined as the ratio of the output (mechanical) power divided by the input electric power. The efficiency of the motor at full load conditions is typically stamped on the nameplate (and sometimes 50% and 75% motor loading as well). It is important to note that efficiency is not constant for all operating conditions of a motor (see examples in Figures 14.2 and 14.5). The equation for motor efficiency is:

$$\eta_m = P_{\text{motor, out}}/P_{\text{e,motor}}.$$

Motor standards in North America are set by National Electrical Manufacturers Association (NEMA, www.nema.org). Internationally, the International Electrotechnical Commission (IEC) standards are typically used. The terminology for motor efficiency classes and the correlation between these two standards are as shown in Table 14.1 (Almeida, 2014; Ferreira, 2018). Currently, new motors are often selected in the premium (IE3) and super-premium (IE4) categories, but the ultra-premium (IE5) category is offered by some manufacturers. Ultra-premium motors are typically not induction type, and are usually permanent magnet, switched reluctance, or synchronous designs. Over recent decades, these standards have evolved with the application of more advanced motor designs and materials to reduce motor losses, roughly 10% for each standard level.

Table 14.1 Motor efficiency standards.

NEMA	IEC
Standard efficiency	IE1
High efficiency	IE2
Premium efficiency	IE3
Super-premium efficiency	IE4
Ultra-premium efficiency[1]	IE5

Table 14.2 Nominal efficiencies of NEMA Premium AC induction motors (600 V or less).

Nameplate Hp	3,600 RPM	1,800 RPM	1,200 RPM	900 RPM
Open design				
1	77.0%	85.5%	82.5%	75.5%
2	85.5%	86.5%	87.5%	86.5%
5	86.5%	89.5%	89.5%	88.5%
7.5	88.5%	91.0%	90.2%	89.5%
10	89.5%	91.7%	91.7%	90.2%
20	91.0%	93.0%	92.4%	91.0%
25	91.7%	93.6%	93.0%	91.0%
50	93.0%	94.5%	94.1%	92.4%
75	93.6%	95.0%	94.5%	94.1%
100	93.6%	95.4%	95.0%	94.1%
200	95.0%	95.8%	94.5%	94.1%
300	95.4%	95.8%	95.8%	
400	95.8%	95.8%		
500	96.2%	96.2%		
Enclosed design				
1	77.0%	85.5%	82.5%	75.5%
2	85.5%	86.5%	88.5%	84.0%
5	88.5%	89.5%	89.5%	86.5%
7.5	89.5%	91.7%	91.0%	86.5%
10	90.2%	91.7%	91.0%	89.5%
20	91.0%	93.0%	91.7%	90.2%
25	91.7%	93.6%	93.0%	90.2%
50	93.0%	94.5%	94.1%	92.4%
75	93.6%	95.4%	94.5%	93.6%
100	94.1%	95.4%	95.0%	93.6%
200	95.4%	96.2%	95.8%	94.5%
300	95.8%	96.2%	95.8%	
400	95.8%	96.2%		
500	95.8%	96.2%		

A sample of motor efficiency standards as specified by NEMA is provided in Table 14.2 (NEMA, 2016). This table lists the nominal efficiency for a sample of motors to meet NEMA Premium requirements for AC induction motors rated for 600 V or less. Nominal efficiency is the average efficiency obtained by testing a representative group of motors.

14.2.5 Motor power factor

Power factor (PF) was introduced in Chapter 13. Motor power factor is simply the ratio of real electric power consumed by the motor divided by total (or apparent) power consumed by the motor. For a motor, the power factor is calculated using the motor power consumption equations introduced above.

PF $= 1{,}000 \times P_{e,\text{ motor}}/(I_L \times V_L)$ for one-phase AC motor
PF $= 1{,}000 \times P_{e,\text{ motor}}/[(3)^{1/2} \times I_L \times V_L]$ for three-phase AC motors.

It is important to note that power factor is not constant for all operating conditions of a motor. Like motor efficiency, power factor varies across the operating range of motor loading. See Figure 14.2 for an example of variation of power factor with motor loading.

Example 14-3

For the motor nameplate in Figure 14.1, calculate the full load PF for the motor operating at 208 Volts and 26.23 amps (full load amps from the nameplate), and compare to the nameplate PF.

Solution

Input real power: $P_e = 10$ hp (0.7457 kW/hp)/0.902 = 8.267 kW
Input total power: $P_s = (3)^{1/2}$ (208) (26.23)/1,000 = 9.449 kVA
PF = 8.267/9.449 = 0.875, which is close to the nameplate PF of 0.88.

14.2.6 Motor specification and nameplate data

Every motor has a nameplate that lists critical information about the motor design. A sample motor nameplate is provided in Figure 14.1. A sample motor specification sheet is provided in Figure 14.2. Some of the key design data for this specific motor include:[2]

- Motor power is the design output power at full load = 500 hp (or 373 kW)
- Number of poles = 4
- Synchronous speed (not shown in the specification) = $(60)(60)/2 =$ 1,800 RPM
- Full load speed = 1,784 RPM
- Full load current (amps at design load and voltage) = 63.6 amps
- Design voltage = 4,000 Volts
- Efficiency at rated output power = 95.0% (note that efficiency does not change much as MLF drops to 50%).

---

[2] This is an actual motor specification and is an example of how both SI and English units are often used for the same piece of equipment. For example, the motor is rated in hp, with torque in kg m.

AC INDUCTION MOTOR DATA SHEET

Model No.or RFQ No		Item No.		Rev. No.	[ ]
Project Name	Motor - 500HP	Project No.		Quantity	1 set

GENERAL SPECIFICATION

Frame Size		5011
Type		HLE5-5011-48E
Enclosure(Protection)		TEFC (IP 55)
Method of Cooling		IC411
Rated Frequency		60 Hz
Number of Phases		3
Insulation Class		■ F ☐ B ☐ H
Temp. Rise at full load (by resistance method)	at 1.0 S.F	80 deg.C
Motor Location		☐ Indoor ■ Outdoor
Altitude		Less than 1000 meter
Relative Humidity		Less than 90 %
Ambient Temp.		40 deg.C (Max.)
Duty Type		Continuous (S1)
Service Factor		1.15 on SWP / SF 1.0 on VFD
Mounting		■ B3 ☐ B5 ☐ V1 ☐
Bearing	Type	Anti-friction
	DE/N-DE	6320C3 / 6319C3
	Lubricant	Polyrex EM Grease
External Thrust		Not applicable
Coupling Method		■ Direct ☐ V-Belt
Shaft Extension		■ Single ☐ Double
Terminal Box	Main	■ Steel ☐ Cast Iron
	Aux.	■ Yes ☐ No
Rotation(Viewed from DE)		☐ CW ☐ CCW ☐ BI-Direction
Application		--
Area classification		Non-Hazardous
Type of Ex-Protection		N/A
Applicable Standard		IEC, IEEE, NEMA,CSA

ACCESSORIES

1. Winding Temp. Detector(Pt 100 Ω) : 2ea/Phase
2. Bearing Temp. Detector(Pt 100 Ω) : 1ea/Bearing
3. Space Heater : 1Phase, 120V, 180W

PERFORMANCE DATA

Rated Output		500 HP
Number of Poles		4
Rotor Type		Squirrel Cage
Starting Method*		DOL or VFD (12~60Hz)
Rated Voltage		4000 V
Current	Full Load	63.6 A
	Locked-rotor**	650 %
Efficiency	50% Load	94.2 %
	75% Load	94.9 %
	100% Load	95.0 %
Power Factor(p.u)	50% Load	0.78
	75% Load	0.85
	100% Load	0.88
Speed at Full Load		1784 r.p.m
Torque	Full Load	203.6 kg·m
	Locked-rotor**	90 %
	Breakdown**	220 %
Moment of Inertia (J)	Load	-- kg·m²
	Motor	8.5 kg·m²
Sound Pressure Level (No-load & mean value at 1m from motor)		85 dB(A)
Vibration		3.8mm/sec(peak)
Permissible number of consecutive starts	Cold	2 times
	Hot	1 time
Paint	Munsell No.	7.5BG6/1.5

SUBMITTAL DRAWING

Outline Dimension Drawing	\	Motor Weight(Approx.)
B3		2360 kg

REMARK

1. The motor shall be manufactured base on multi-level VFD inverter.
VFD Duty winding, Variable Torque, 5:1 Speed Range

Figure 14.2 Sample AC induction motor specification.

- Power factor at rated output power = 0.88 or 88.0% (note that PF drops significantly as MLF drops to 50%, and typically drops even further below 50% loading).
- Service factor = 1.15 (i.e., the motor can deliver 1.15 times 500 hp = 575 hp for short time periods under design operating conditions).
- Full load torque = 203.6 kg m (multiply this by 9.81 to get torque in Newton meters, i.e., 1,998 Nm torque).

14.3 Motors Other than AC Induction

While AC induction motors are by far the most common, there are other types of motors that are used for specific applications.

- DC motors have favorable characteristics in terms of speed and torque for some applications, and work well in variable speed operation (DOE, 2014; Drury and Hughes, 2019). In particular, DC motors are well suited for high torque at low-speed applications. If the power supply for a system is DC, then DC motors have an advantage. With AC power supplies, DC motors require rectifiers, which adds to the cost.
- AC synchronous motors operate at a constant speed, which is the synchronous speed of the motor (Krarti, 2017; Chapman, 2012). Synchronous motors also have a power factor of one, or sometimes are designed to provide capacitive load (i.e., with a leading power factor – see Chapter 13). Large, customized industrial motors (e.g., in the range of 5,000–50,000 hp or larger) are often synchronous motors, such as used for natural gas compressors.
- Switched reluctance is a special motor design where the rotor is made of shaped iron to generate a magnetic field (McCoy and Douglass, 2014; Hydro Quebec, 2014). These motors operate with reluctance torque, which is torque induced in iron material in the presence of an external magnetic field. Speed and torque are precisely controlled with an electronic power converter and controller, and the motor requires the use of a variable frequency drive (VFD). These motors have advantages such as: high power for the motor size, high torque at low speeds, large turndown ratio, and ability to run as both a motor and a generator (e.g., for electric power regeneration).
- AC permanent magnet motors (sometimes described as brushless) operate as the name implies, with the use of permanent magnets instead of induced magnetic fields. These motors have a flatter efficiency curve across the motor loading range, and typically are much more efficient at low motor loads (ASHRAE, 2020; Hydro Quebec, 2014). Permanent magnet motors, like switched reluctance, require a VFD to operate, which sometimes is built into the motor assembly.
- AC synchronous reluctance motors operate with torque induced as described above for a switched reluctance motor. These motors also operate at synchronous speeds, which is one of the advantages. These motors are typically more efficient than AC induction motors across the motor load range (Chapman, 2012; Hydro Quebec, 2014). An example

of an application that benefits from this motor design is for cooling towers, to provide efficient variable speed control in a small unit under warm and moist conditions.

It should be noted that there are more specialized motor designs than summarized above (Drury and Hughes, 2019). Essentially all of these alternative motor designs will have higher initial costs as compared to AC induction motors. However, if selected and designed effectively, these specialized motor designs can provide energy savings and improved reliability to make the investment worthwhile.

14.4 Energy Efficiency Improvements for Motor-driven Systems

To identify and quantify energy saving opportunities for motor-driven systems, it is important to consider the whole system from electricity supply to the end-use device as illustrated in Figure 14.3.

Starting on the left side of the diagram, electric power is supplied, then converted to mechanical power by motors, and then converted to fluid power in common end-use devices such as pumps, fans, and compressors (or in non-fluid end use systems, power is converted to mechanical work such as material conveyers). Each component in the system shown in Figure 14.3 has conversion efficiencies, which present potential opportunities for savings:

- The power supply could include transformers, VFDs, and other equipment (e.g., cables, breakers, switchgear) with efficiency designated as η_p. If the motor is a DC motor, the power supply could include a rectifier.
- The motor converts input electric power to mechanical power with efficiency designated as η_m. Motor efficiency is discussed earlier in this chapter.
- A coupling or drive connects the motor to the shaft of the end-use equipment. This could be a simple mechanical coupling, gear box, hydraulic coupling, magnetic or eddy current coupling, or belt and sheave (pulley) drive, with efficiency designated as η_c.

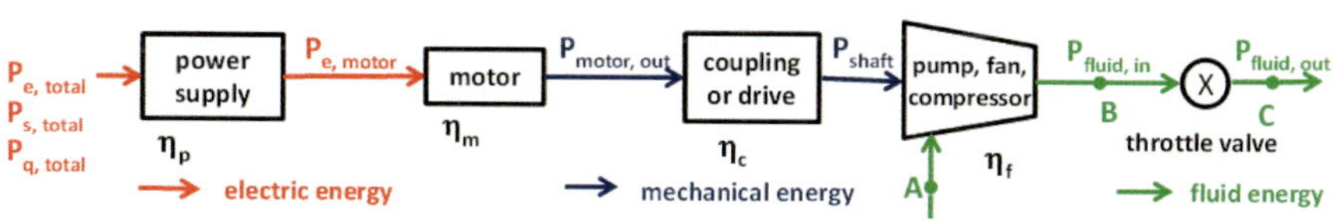

Figure 14.3 Electric motor driven systems.

- The pump, fan, or compressor converts shaft power to fluid power to raise the pressure of the fluid, such as water or air, with efficiency designated as η_f. Chapter 21 includes more detailed information on energy savings for pumps, fans, and compressors.

Installing new, more efficient motors is one simple measure to consider. Motor replacement on a retrofit basis is often challenging to meet payback, NPV, or other economic requirements. However, for new system designs and end of life replacement, it is often economical to invest in the most efficient motor that is commercially available. One factor that can increase retrofit costs is due to the mounting configuration. If the new motors are a different size and frame than the existing, then it increases the installation costs.

Example 14-4

A 20 hp enclosed motor with a load factor of 70% and an efficiency of 89% will be replaced by a new 20 hp enclosed motor operating at the same load factor. The new motor has an efficiency of 91% at 70% loading. Estimate the real power and annual energy savings, assuming 5,000 hours per year operation. Estimate annual cost savings assuming energy is \$0.10 per kWh and the demand charge is \$14 per kW monthly.

Solution

Baseline motor:

$P_{\text{e, motor}} = P_{\text{e, motor,out}}/\eta_m = (0.70)\ 20\ \text{hp}\ (0.7457\ \text{kW/hp})/0.89 =$ 11.730 kW.

New motor:

$P_{\text{e, motor}} = P_{\text{e, motor,out}}/\eta_m = (0.70)\ 20\ \text{hp}\ (0.7457\ \text{kW/hp})/0.91 =$ 11.472 kW

Power savings = (11.730 – 11.472) = 0.258 kW

Energy savings = 0.258 kW (5,000 hr/yr) = 1,291 kWh/yr

Energy cost savings = 1,290 kWh/yr (\$0.10/kWh) = \$129/yr

Demand cost savings = 12 months (\$14/kW-month) (0.258 kW) = \$43/yr

Total utility bill savings = \$172/yr or \$1,720 over 10 years (undiscounted).

This example illustrates that the energy cost savings for a modest improvement in motor efficiency (2 percentage point improvement in this case) may not justify much investment. However, if a motor needs replacement, then the incremental cost may still be well worth it to pay for the higher efficiency motor.

14.5 Rewinding vs. Replacement of Electric Motors

If a motor has failed, it could be replaced in-kind or replaced with a higher efficiency version (e.g., super-premium or ultra-premium, if available). Alternately, a failed motor could be sent to a shop to be repaired and rewound. This is a common practice, and motors may be rewound several times to extend the motor life for decades. The initial cost of rewinding a motor is often about 60% less than the cost of purchasing a new motor. Some sources claim that motor rewinding commonly results in efficiency losses of 1−2 percentage points (i.e., similar efficiency difference as in Example 14-4). However, it is also reported that rewinding can maintain efficiency and improve motor reliability if best practices are followed (Ferreira, 2018). Motor industry test data shows that rewinding can limit efficiency losses to well under 1 percentage point (EASA, 2019). One analysis of lifecycle costs of rewinding 1,800 RPM motors found that replacement was more economical for motors smaller than 20 hp, and that rewinding was more economical for motors ranging 20−75 hp (McCoy and Douglass, 2014). Overall, there is no clear, single approach to motor rewinding that works for most circumstances. Like any energy or maintenance management decision, the best practice is to obtain relevant data (e.g., motor rewind cost, rewind efficiency data for similar motors, cost of electricity) and analyze the options for the specific application.

14.6 Speed Reduction for Motor-driven Systems

Many applications that are currently controlled to operate at constant speed are well suited for variable speed operation. For example, in many pump and fan applications, flow is controlled by using restrictive devices, such as throttle valves or dampers, or bypass methods. Although these flow control methods have advantages, speed control is often a more efficient and cost-effective option for many systems (see Chapter 21). Similarly, in many material handling systems, variable speed operation can increase system efficiency and improve system reliability. For example, in older conveyor systems, lines are controlled by energizing and de-energizing a series of motors. These frequent starts and stops result in higher operating temperature and cause more frequent equipment failures. Using variable speed drives can smooth out line motion for more efficient and reliable operation. Table 14.3 provides a summary of three different load types, common applications, and the potential for energy savings with variable speed operation.

Table 14.3 Potential for speed reduction to save energy based on load type.

Load type	Common applications	Potential for energy savings with variable speed
Variable torque	Centrifugal fans and blowers Centrifugal pumps Centrifugal compressors	Reduced speed reduces electricity consumption following power law equation ($n = 1.8–2.8$)
Constant torque	Air compressors Mixers Conveyers Positive displacement pumps Lifting hoist	Reduced speed reduces electricity consumption in direct proportion to speed reduction
Constant shaft power	Machine tools Lathes Drilling and milling Punch presses	No electricity savings at reduced speeds

The benefits of motor speed control potentially include lower system energy costs, improved system reliability, fewer maintenance requirements, and more effective process control. Referring to Figure 14.3, in general, there are multiple options to vary the speed of end-use equipment including:

- **Electric power supply**. Vary the voltage or frequency, but frequency control is much more efficient (Chapman, 2012).
- **Motor**. Motor speed can be changed by electrically changing the number of motor poles (multi-speed motors), or adding resistance in specific motor designs (Chapman, 2012). These methods are technically viable, but rarely used today.
- **Coupling**. Speed control with the coupling design that connects the motor shaft and end-use equipment shaft. This can be achieved with belts and sheaves or gearboxes (to change from one fixed speed to another), mechanical variators, fluid couplings, magnetic couplings, and eddy current drives (ASHRAE, 2020; Saidur, 2012).

Although there are multiple methods to vary speeds of motor-driven systems, the focus in this chapter is on VFDs.

14.7 Using Variable Frequency Drives (VFDs)

VFDs have substantial advantages in comparison to other speed control options (Drury and Hughes, 2019). They are highly efficient, reliable, and flexible, and the VFD can be bypassed to perform maintenance without having to take the motor out of service. VFDs can be incorporated into closed-loop control systems, and their speed adjustment ratios are often more precise than those of other speed control systems. If they are used in place of mechanical drive options, VFDs can also improve system reliability by removing potential failure modes and requiring less maintenance because they have fewer mechanical components.

VFDs work by converting fixed frequency AC input power to DC power to vary voltage and frequency, and then back to AC power to supply the motor. To accomplish this, VFDs have rectifiers, controls, inverters, and sometimes active cooling systems. As such, VFDs consume power and can introduce significant harmonics and power factor reduction. Figure 14.4 provides a sample specification for an industrial VFD illustrating efficiency, power factor, and harmonic effects. In analyzing VFD conversions for energy savings, it is important to include VFD energy losses. Figures 14.5 and 14.6 provide the results of laboratory tests of a 100 hp VFD and motor system, providing an example of how motor and VFD efficiency vary with speed. In general, the energy efficiency of VFDs can range from 2% to 15% depending on many factors. Typical values assumed in energy studies range from 3% to 7%, but the best practice is to use the manufacturer specifications where available.

VFD Performance:

☐ Required HP (kW): **2970 hp**

☐ Constant Torque ☑ Variable Torque

☑ Required Speed Range: Min: 1440 RPM Max: 1890 RPM

☐ Minimum Acceleration Torque: ______

☐ VFD Service Factor: 1.00

☐ Minimum Load Inertia: ______

◊ Efficiency (%Speed)

%Load	25	50	75	100
25	90	94	96	97
50		95	96.5	97
75			96	97
100				96.5

◊ Power Factor (%Speed)

%Load	25	50	75	100
25	95	95	97	98
50		98	98	98
75			98	98
100				98

◊ THD Current (%) (%Speed)

%Load	25	50	75	100
25				
50				
75				
100				1.6

◊ Overload Capacity

◊ 150% for ______ minutes out of ______ minutes

◊ 200% for ______ seconds out of ______ minutes

Figure 14.4 Sample specification for a 2,970 hp industrial VFD.

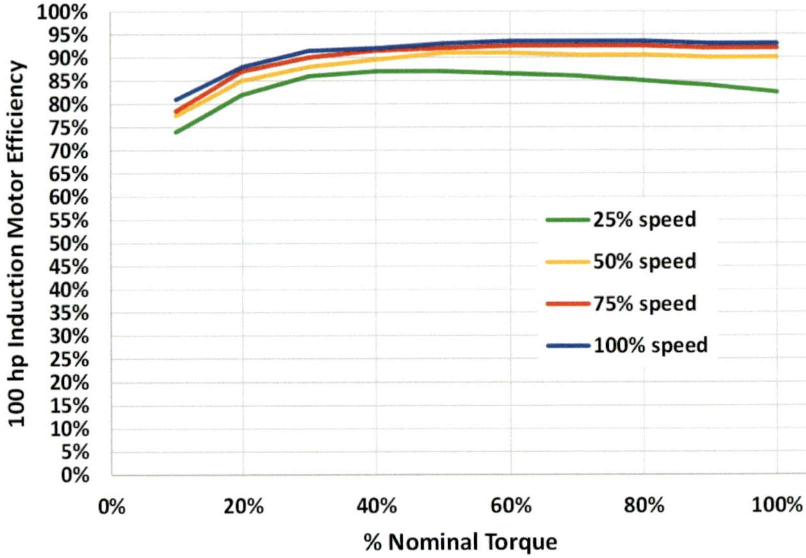

Figure 14.5 Measured motor efficiency for 100 hp induction motor at different motor speeds (Hydro Quebec, 2009).

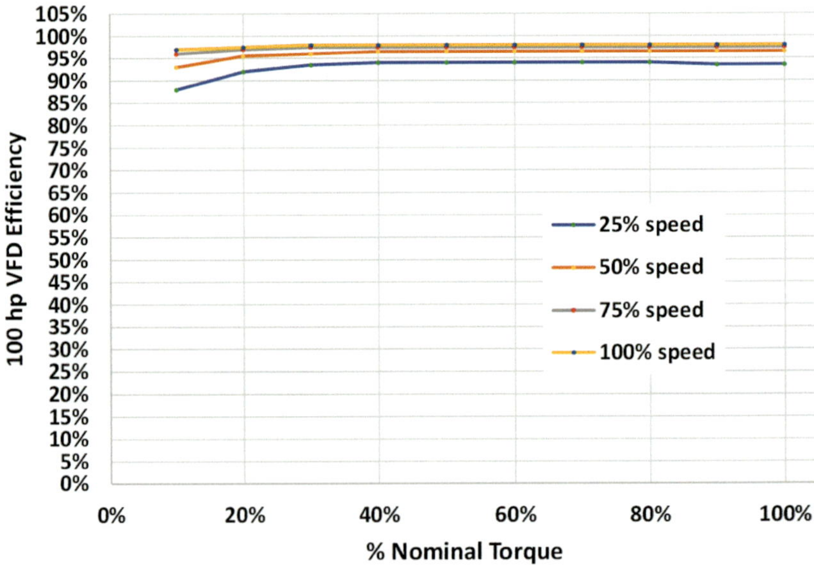

Figure 14.6 Measured VFD efficiency for 100 hp induction motor at different motor speeds (Hydro Quebec, 2009).

Example 14-5

An industrial plant has three cyclone fans used for ventilating stack gases. The three fans were replaced with one larger fan and motor having a VFD. The cost of the new fan, motor, and installation of new ductwork was \$18,250. The cost of the VFD was \$20,000. The electrical consumption from the new system dropped 500,000 kWh/yr from that of the old fan system. If the facility paid \$0.061/kWh, calculate the simple payback period (SPP) for this measure.

Solution

The electrical cost savings from the new system is found as

Annual savings = (500,000 kWh/yr) (\$0.061/kWh) = \$30,500

The total cost of implementation is the sum of the costs for the new fan, motor, ductwork, installation, and the VFD:

Total implementation cost $= \$18,250 + \$20,000 = \$38,250$

SPP = \$38,250/\$30,500/yr = 1.25 years.

14.8 Variable Speed Operation for Centrifugal Fans and Pumps

A common energy savings measure involves using VFDs for centrifugal fans on air or gas systems, and centrifugal pumps for water or other liquid systems. For centrifugal systems, there are equations that relate speed (ω) with flow ($\dot{V}$), pressure (p), and shaft power (P). These equations are:

Flow and speed: $\frac{\dot{V}_1}{\dot{V}_2} = \frac{1}{2}$

Pressure and speed: $\frac{p_1}{p_2} = \left(\frac{1}{2}\right)^2$

Mechanical power and speed: $\frac{P_{\text{shaft}1}}{P_{\text{shaft}2}} = \left(\frac{1}{2}\right)^n$.

These equations are commonly referred to as "affinity laws." However, these equations are only valid under certain conditions. For applications where there is negligible gravity and pressure forces (i.e., low pressure requirement), i.e., only kinetic energy and friction losses, then the power equation exponent is $n = 3$. However, for real-world systems, it is rarely the case that gravity and pressure forces are negligible (see Chapter 23 for a brief overview of fluid mechanics). To be conservative, some sources recommend using $n = 2$ for energy efficiency analyses (Doty, 2020; Krarti, 2017).

Another challenge with the power equation above is that shaft power is rarely measured in field projects.[3] Typically, it is electric power that is measured and electric energy savings that is of primary interest. For most practical applications, the following equation is more useful:

$$P_{e2,\ total} = P_{e1,\ total}\ (\omega_2/\omega_1)^n.$$

In the common situation where static head and gravity forces are significant, common practice is to use $n = 1.8\text{–}2.8$. For high-pressure applications, $n = 1.8$ may apply, while for low-pressure applications, $n = 2.8$ may apply. As a general guideline, using $n = 2.4$ for centrifugal fan and pump applications is reasonable. To be more conservative in making energy saving estimates, use n as low as 1.8. If systems have low static pressures and minimal changes in elevation, then consider using n as high as 2.8.

Example 14-6

A VFD-driven centrifugal pump with a 5 hp motor is used on a chilled water line to supply the largest flow rate needed on the hottest day of the year, and runs fully loaded at 1,760 RPM. On a mild day, only 60% of the maximum flow is needed. The motor efficiency is constant at 92% in the range of 60%–100% loading. Estimate: (a) the new speed of the motor at 60% flow, (b) the ratio of pressure head at 60% flow to pressure head at 100% flow, (c) electric power consumption at reduced speed assuming $n = 3$, (d) electric power consumption at reduced speed assuming $n = 2.1$, and (e) compare electric power savings estimated using these two different values for n.

Solution

(a) $\omega_2 = \omega_1 \frac{\dot{V}_2}{\dot{V}_1} = 1{,}760\ (0.60) = 1{,}056$ RPM

(b) $\frac{p_2}{p_1} = \left(\frac{2}{1}\right)^2 = (0.60)^2 = 0.36$

The pressure head at lower speed is estimated to be 0.36 times the initial value.

(c) Initial electric power consumption:

$P_{e1} = (5\ \text{hp})\ (0.7457\ \text{kW/hp})/0.92 = 4.053$ kW.

Estimated final power consumption assuming $n = 3$:

$P_{e2} = P_{e1}\left(\frac{2}{1}\right)^3 = (4.053\ \text{kW})\ (0.6)^3 = 0.875$ kW.

(d) Estimated final power consumption assuming $n = 2.1$:

---

[3] Examples where shaft power was measured are provided in two motor and VFD lab studies cited elsewhere in this chapter. See Hydro Quebec, 2009 and Hydro Quebec, 2014 in the references.

$$P_{e2} = P_{e1}\left(\tfrac{2}{1}\right)^{2.1} = (4.053 \text{ kW})\,(0.6)^{2.1} = 1.386 \text{ kW}.$$

(e) Electric power savings assuming $n = 3$: $P_{e1} - P_{e2} = 4.053 - 0.875 =$ 3.18 kW savings.

(f) Electric power savings assuming $n = 2.1$: $P_{e1} - P_{e2} = 4.053 - 1.386 =$ 2.67 kW savings.

14.9 Tools for Motor Systems Management

The U.S. Department of Energy's MEASUR tool (www.energy.gov/eere/amo/measur) is open source and designed to assist industry in identifying and analyzing motor efficiency opportunities and to better manage electric motor systems. With MEASUR, users can access cost and performance data (current, efficiency, and power factor) from thousands of industrial electric motors and perform several tasks to help with motor management. These tasks include performing a comparative benefits analysis for replacing existing in-service motors with possible alternatives, motor maintenance support, and calculating the lifecycle costs for potential motor efficiency improvement projects. MEASUR includes functions for analyzing technical performance and cost for different motor drives, and for rewinding.

14.10 Summary

Electric motors convert electric power to mechanical power, and consume roughly one-third to one-half of global electricity generation. As such, effective design and maintenance of motors is essential for any energy management initiative. Key parameters for motor performance are operating speed, slip, motor load factor, energy efficiency, and power factor. Estimates are that roughly 90% of motor power in the U.S. and Canada use AC induction motors. Induction motors are generally reliable and cost-effective, but the effective selection and use of induction motors requires careful attention to the motor load range (process conditions), efficient controls, and effective maintenance including rewinding practices. Alternative motor designs include AC synchronous motors, switched reluctance motors, DC motors, AC synchronous reluctance motors, and permanent magnet motors. Although these alternative designs have higher initial costs, in many applications, the benefits of efficiency, reliability, and process control make them worthwhile. Variable speed operation of motor-driven equipment is effective for many applications including conveyance systems and, especially, centrifugal pumps and fans. Speed control can be accomplished in several ways, but VFDs are the most common and best method in most applications. When integrating VFDs into motor and process controls, it is important to consider the full load range of operation, and account for VFD efficiency, power factor, and power quality. There are many resources available to help analyze, design, and operate motor and drive systems. U.S. DOE's MEASUR tool is highlighted as one of the most useful tools.

References

Almeida A. (2014) "Technical and Economical Considerations on Super High-Efficiency Three-Phase Motors" IEEE Transactions on Industry Applications, Vol. 50, No. 2.

ASHRAE (2020) "ASHRAE Handbook – HVAC Systems and Equipment" Chapter 45 "Motors, Motor Controls, and Variable Frequency Drives."

DOE (2014) "Improving Motor and Drive System Performance: A Sourcebook for Industry"

Doty, S. (2020) "Commercial Energy Auditing Reference Handbook" (4th edition). River Publishers.

U.S. Department of Energy Office of Advanced Manufacturing.

Drury B. and Hughes A. (2019) "Electric Motors and Drives" Newnes.

EASA (2019) "The Effect of Repair / Rewinding on Premium Efficiency/IE3 Motors" Electrical Apparatus Service Association (EASA) and the Association of Electrical and Mechanical Trades (AEMT). https://easa.com

Ferreira F., Silva A., Aguiar V., Pontes R., Quispe E., de Almeida A. (2018) "Overview of Retrofitting Options in Induction Motors to Improve their Efficiency and Reliability" 2018 IEEE International Conference on Environment and Electrical Engineering.

Hydro Quebec (2009) "Benchmark Study to Establish a Test Protocol for Determining the Efficiency ff Variable Frequency Drives (VFDs)" CEATI Report T071700-7028. www.ceati.com

Hydro Quebec (2014) "Efficiency Testing of Switched Reluctance, Synchronous Reluctance and AC Permanent Magnet Motors" CEATI Report T101700-7053. www.ceati.com

Krarti (2017) "Energy-Efficient Electrical Systems for Buildings" CRC Press.

McCoy G. and Douglass J. (2014) "Premium Efficiency Motor Selection and Application Guide – A Handbook for Industry" U.S. DOE.

NEMA (2016) "Motors and Generators" ANSI/NEMA MG 1-2016. National Electrical Manufacturers Association. www.nema.org

Paes R., Toner B., Bankay G., and Agashe G. (2021) "Energy Optimization for Adjustable-Speed Drive Applications" IEEE Industry Applications Magazine. September/October.

Saidur R., Mekhilef S., Ali M.B., and Mohammed H.A. (2012) "Applications of variable speed drive (VSD) in electrical motors energy savings" Renewable and Sustainable Energy Reviews vol 16.

Author Biography

Eric Mazzi is a consultant and educator focusing on technology and policy for demand-side energy systems. He specializes in performing independent technical analysis, M&V, policy studies, and educational services. He advocates and contributes to efforts to reduce energy consumption and achieve net-zero GHG emissions, while maintaining or improving energy services and advancing energy equity. His experience includes air quality consulting, utility maintenance, and five years as full-time engineering faculty at the University of British Columbia. He volunteers as a board member for the U.S. Northwest Energy Coalition, and the International Performance M&V Protocol committee. Eric teaches part-time, delivering university and professional certification courses. He is an adjunct professor of Energy Management at the New York Institute of Technology – Vancouver. He has B.Sc. and M.Sc. degrees in mechanical engineering, and a Ph.D. degree in environmental policy. He is a registered engineer in California and British Columbia.

Acknowledgments

I would like to thank the chapter reviewer Markus Zeller for his careful review of a draft of this chapter.

Practice Questions

1. A fan is driven by a 400 hp (2,300 V, three-phase) electric motor and operates with a power factor of 0.87. Average motor amperage drawn is 50 A. The fan operates 6,500 hours each year. Electricity costs $0.085 per kWh. The fan can be upgraded at a capital cost of $28,000 to save 20% of baseline energy consumed.

(a) Estimate the annual energy savings (kWh/yr).
(b) Estimate the simple payback period (years) based on energy (kWh) savings.
(c) Estimate the net present value of the project over a 10-year lifetime using an 8% interest rate.

2. An electric motor runs at a measured speed of 590 RPM when driving a load requiring a torque of 200 Nm. If the motor input power is measured at 15 kW, calculate the motor efficiency.
3. An electric motor has the following nameplate information:
   - 4,500 hp
   - 4,000 V
   - 584 full load amps
   - 890 RPM
   - 3-phase, 60 Hz
   - 8 poles

Laboratory test data for the motor are provided in the table below:

MLF	Measured efficiency, %	Measured power factor	Measured line current, amps	Measured slip, %	(a) Motor speed, RPM	(b) Motor output power kW	(c) Motor input power, kW	(d) Motor input power, kVA	(e) Motor input power, kVA
0.25	0.952	0.603	210.6	0.15					
0.50	0.969	0.794	314.6	0.31					
0.75	0.972	0.849	439.9	0.47					
1.00	0.970	0.862	579	0.65					
1.15	0.968	0.859	669.5	0.78					

Complete the table above by estimating the following for each motor loading:

(a) Motor speed (RPM) based on the % slip.
(b) Motor output power (kW) based on rated output and MLF.
(c) Motor input real power (kW) based on motor output power and motor efficiency.
(d) Motor input real power (kVA) based on motor input real power and power factor.
(e) Motor input real power (kVA) using the formula: $P_s = (3)^{1/2} \times I_L \times V_L/1,000$ assuming nameplate voltage.

Using the test data and the completed table:

(f) Make a chart of motor RPM vs. MLF. Comment on the linearity or lack thereof. How well does this test data set support the method of using measured motor speed to estimate actual, operating MLF?

(g) Make a chart of motor RPM vs. line current. Comment on the linearity or lack thereof. How well does this test data set support the method of using measured line current to estimate actual, operating MLF?

(h) Using the motor speed calculated for part (a), estimate the MLF. Compare this estimated MLF to the MLF in the first column of the table. Assuming the MLF in the first column is accurate, estimate the error in the calculated MLF.

4. Consider a common pump system that consumes 100 kW of electric power input to the power supply transformer. Refer to Figure 14.3. Assume the following efficiencies:

   - VFD efficiency = 97.0%
   - Motor efficiency = 91.0%
   - Coupling efficiency = 98.0%
   - Pump efficiency = 72%

   Given the efficiencies provided, estimate

   (a) P_{fluid} = the fluid power delivered by the pump.
   (b) The overall system efficiency.

5. For the motor nameplate shown below:

   (a) What is the synchronous speed in RPM?
   (b) Estimate the input real power at full load based on nameplate amps, voltage, and power factor.

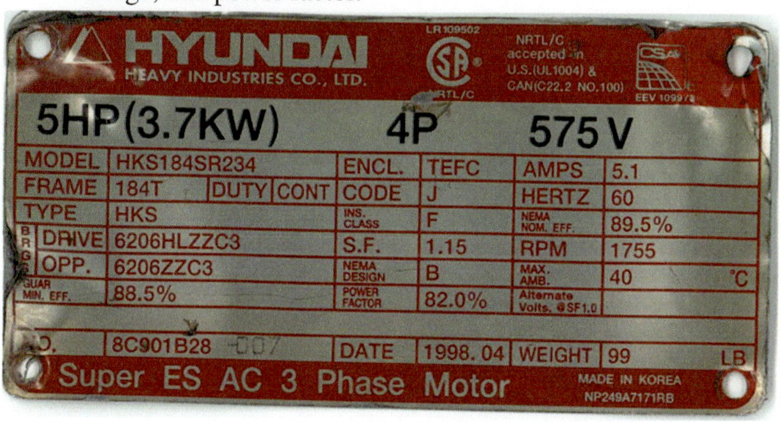

(c) Estimate the input real power at full load based on the rated output power and nameplate NEMA efficiency. How does this value compare to the result in part (b)?

6. The shaft power for a damper-controlled, centrifugal fan is measured to be 50.0 kW operating at 3,580 RPM. After retrofit with variable speed drive controls, the fan is able to deliver the pressure and flow required by the process at 2,506 RPM. Assuming the power law exponent for speed change for this centrifugal equipment is 2.6, estimate the expected shaft power savings (in kW) from this reduction in fan speed.
7. Summarize the key elements and activities involved in effective management program for electric motors and drives in an industrial plant. Search for and use the U.S. DOE's "Improving Motor and Drive System Performance" publication, www.greenmotors.org, and other sources if you choose. Summarize in one page or less.

15

Introduction to Control Systems for Energy Management

Eric Mazzi

Mazzi Consulting, Canada

15.1 Introduction

Equipment can be controlled to save energy, reduce energy and O&M costs, reduce environmental impact, and maximize profits while still delivering the desired services such as indoor environment (temperature, illumination, humidity, safety) or the quality and quantity of industrial product output. Energy system controls can be as simple as manually turning off a switch. However, optimizing systems to meet performance requirements while saving energy and O&M costs typically involves automated controls ranging from simple clocks to sophisticated digital systems. A good rule is that each control should be: (1) as simple and reliable as possible, while also, ideally, (2) connected digitally to extend its usefulness and value. This chapter gives an overview of energy system controls.

Every piece of energy-consuming equipment has some form of control system. Most controls do one or some combination of three simple actions: turn something on or off, increase or decrease a manipulated variable, or set a schedule for operation. Lights have on/off switches and some have simple timers and dimmer controls. Motors have on/off switches and some have variable speed controls. Air conditioners have thermostats and fan switches, and some have night setback controls or timers. Large air conditioning systems have extensive controls consisting of several thermostats, valve controls, pump controls, motor speed controls, and possibly scheduling controls to optimize the operation of all system components. Large heating systems have modulating controls on the boilers and adjustable speed drives on pumps and variable air volume fans. For HVAC, controls manage the three S's: setpoints, sequences, and schedules.

These controls are necessary for the proper operation of the equipment and for the comfort and basic safety of occupants and operators. Controls enable equipment and systems to be operated in a manner that reduces the power and energy requirements of equipment that consume energy, optimizes the production of energy such as solar PV, and/or manages energy storage.

15.2 Control System Elements and Processes

Fundamentally, process control technology manipulates variables in order to achieve desired process variable (also called controlled variable) outcomes in response to process disturbances. This relationship is illustrated in Figure 15.1 (Hughes, 2015).

- Manipulated variables are changes in devices that effect changes to the process with the aim to achieve desired process variables. Examples of manipulated variables are motor speed, heating element power, and valve position.
- Process variables are the variables to be controlled as an outcome of the process equipment. Examples include temperature, moisture content, pressure, or liquid level. Process variables are sometimes described as the controlled variable (Honeywell, 1997).
- Disturbances are changes in process conditions such as change in outdoor air temperature, change in occupancy for a building, or change in feedstock composition for an industrial process.

In order to achieve its basic function, controls systems typically have four elements as illustrated in Figure 15.2 (Hughes, 2015):

- Process. Equipment and systems built to perform a specific function. For example, cooling systems are designed to maintain safe and comfortable indoor air temperature. Temperature is an example of a process variable.

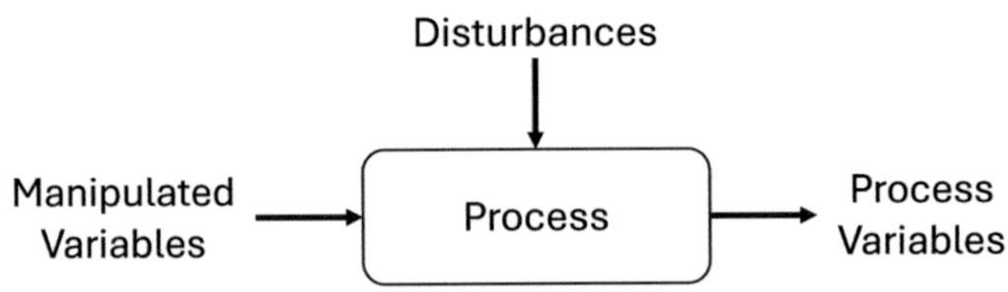

Figure 15.1 Diagram illustrating the basic function of control systems (Hughes, 2015).

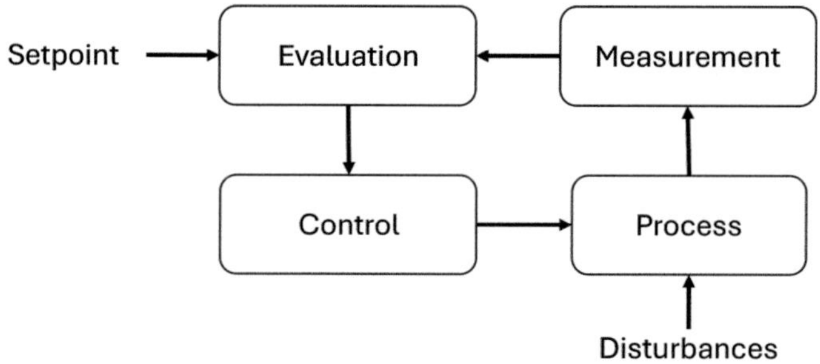

Figure 15.2 Four elements of controls systems.

- Measurement. Using a sensor to quantify the process variable and convert it into an analog or digital signal. For example, an immersion temperature sensor used for hydronic heating water.
- Evaluation. Measured value of the process variable is compared to the target value or setpoint. Then the amount of corrective action is determined by a controller device.
- Control. The process whereby system components exert a direct influence on the process by receiving an input and utilizing input–output algorithms to output a signal to a final control element. For example, a control valve, damper, or electric motor.

Control systems use control loops to achieve their purpose. A control loop has a sensor providing an input (e.g., tank level sensor), a control unit that processes the input signal to produce an output signal (e.g., a PID controller), and a final control element or device (e.g., actuated valve). Control loops can be open or closed as illustrated in Figure 15.3 (Honeywell, 1997). Open control loops would not have the feedback components shown with red dashed lines in the figure. Open control loops assume a fixed relationship between the manipulated variable and the controlled variable such that feedback is not critical. An example of an open loop control is a cooling system that sets the chilled water valve condition based on a program curve vs. outdoor air temperature, and thus would not adapt to disturbances such as changes in internal heat gains. A closed loop space cooling system would use temperature sensors to provide feedback to the controller, employing one of the common control modes.

Control loops can be operated in three ways: manual, automatic, or cascade. With manual controls, an operator directly adjusts the controller output.

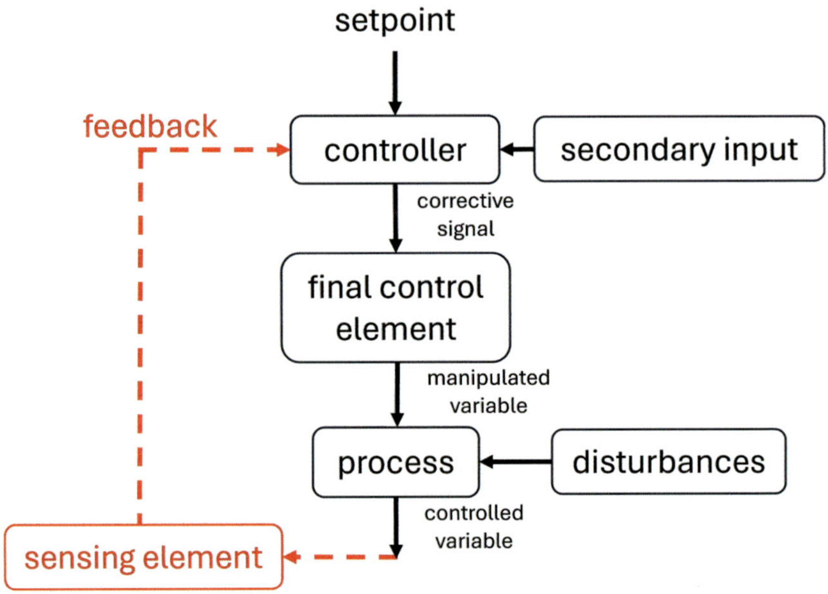

Figure 15.3 Diagram showing the difference between open and closed control loops. Open control loops would not have the components shown in red dashed lines.

Automatic control is when the controller output is calculated according to programming to achieve the process variable setpoint. Automatic control is advantageous in that the control system responds to process disturbances and setpoint changes without operator action. With cascade control loops, two or more controllers are arranged so that the primary (master) controller output becomes the setpoint for a secondary (subordinate) controller. Each controller measures a different variable in the same process, forming two nested feedback loops. Compared to basic automatic loops, cascaded control loops can respond faster to process disturbances and improve stability.

In terms of control system hardware, the common components are (Hughes, 2015; ASHRAE Handbook Fundamentals 2021, Chapter 7; Honeywell, 1997):

- Sensors. Measure process variables. Common sensors measure temperature, pressure, level, flow, and analytical variables (e.g., pH, oxygen). ASHRAE Handbook Fundamentals 2021, Chapter 38, provides a useful summary of measurement and instruments.
- Transmitters. Communicate data between components such as a sensor and the input to a controller. Some common communication signal

designs are 3–15 psig (pneumatic), 4–20 mA (electric), and 0–10 VDC (electric).

Example 15-1

A transmitter is connected to a pressure sensor which is used to send a signal to a controller. The instrument is designed for 50–500 kPa input and 4–20 mA output. If the instrument is linear and accurately calibrated, estimate the expected output signal at 203 kPa input pressure.

Input percentage of full scale = (203 – 50)/(500 – 50) = 0.340 or 34.0%

Expected output = 4.0 mA + [(0.340) × (20 – 4)] = **9.44 mA**.

- Actuators. They receive signals from controllers to physically change a final device such as a valve or damper.
- Final control devices. These devices physically manipulate the process; common devices are a valve (e.g., to throttle pump discharge flow), VFD (e.g., to change motor speed), or damper (e.g., to throttle inlet flow to a fan). The specific relationship between the control device position and controlled variable is important and can be specified in the design and selection of the device. Figure 15.4 provides an example for a flow control valve where the valve can be quick opening, linear, or equal percentage (ASHRAE Handbook Fundamentals 2021, Chapter 7).
- Controllers. They are devices that sense changes in the controlled variable (typically from a sensor and transmitter) and induce corrective action in the form of an output signal to a final control device. Controllers can be designed to operate using different control modes as described in the next section.
- Networks. They control equipment communications using networks such as Ethernet, Modbus, or BACnet^{TM} to exchange data between system components.
- Interfaces. Graphical displays of various types are human–machine interfaces to provide O&M staff the ability to monitor performance data, intervene manually to change control settings, or view alarms.

With control system devices such as sensors, transmitters, and valves, *it is critical to ensure that these devices are calibrated to operate within acceptable limits*. It is tempting to see data on the graphical display such as for a damper position that, for example, indicates the damper is 65.87% open and assume that is the case. However, experienced field staff find that often a visual inspection of the actual damper position could reveal it to be far from the indicated position – e.g., even stuck fully open or fully closed.

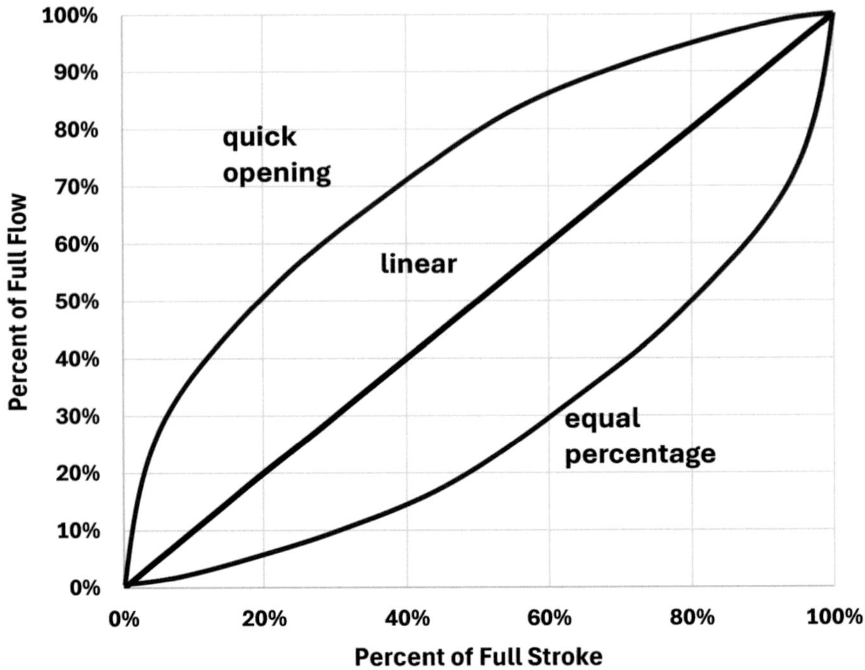

Figure 15.4 Example of a characteristic curve for a flow control valve.

It is the same for critical sensors such as flow meters for liquids and gases, which often are found to be inaccurate. *Fundamental to the success of control upgrades is to ensure that all critical devices are calibrated and maintained to acceptable standards and accuracy*. Otherwise, the data displayed in control system dashboards or downloaded to data files may be far from reality.

Control system components can be classified into two types in terms of how data is generated and shared: analog or digital. Analog devices operate continuously with, theoretically, an infinite number of values between a minimum and maximum value. Examples of analog inputs to a controller are from temperature and humidity sensors. Examples of analog outputs from a controller are electric power frequency (Hz) and damper actuator position. Digital components operate with data in binary form. Examples of digital inputs to a controller are tank liquid level low/high sensors and motor on/off sensors. Examples of digital controller outputs are lights on/off and valve open/closed. Figure 15.5 depicts the difference between an analog and digital control signal for a control device (Honeywell, 1997).

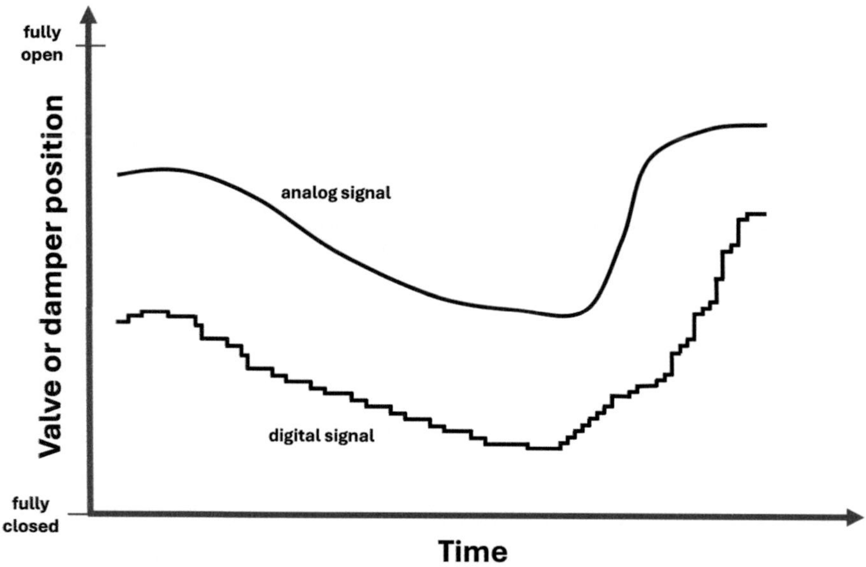

Figure 15.5 Comparison of an analog signal vs. digital signal.

15.3 Control Modes and Strategies

In this section, we provide an overview of basic control modes and strategies. Then the following section describes the main technologies used to implement the various control modes and strategies. Almost all of the basic control modes can be implemented using the different technologies in the next section. For example, on/off or proportional control can be implemented with pneumatic, electric, electronic, or direct digital controls (DDC).

Two-position or on/off control is where the final control element occupies only one of two positions, except for a brief time when the control element is moving from one position to the other. Figure 15.6 provides an example of how two-position control works. If this was for a heating system, the controlled variable would be indoor air temperature. The control differential is the difference between the temperature when the heat turns on and off. Because of inertia and overshoot, the operating temperature briefly rises after heat is turned off and drops after heat is turned on. The operating differential is the difference between the highest and lowest operating temperatures.

Modulating control mode is where the controller output can vary across the full range as illustrated in Figure 15.7 (Hughes, 2015). The figure

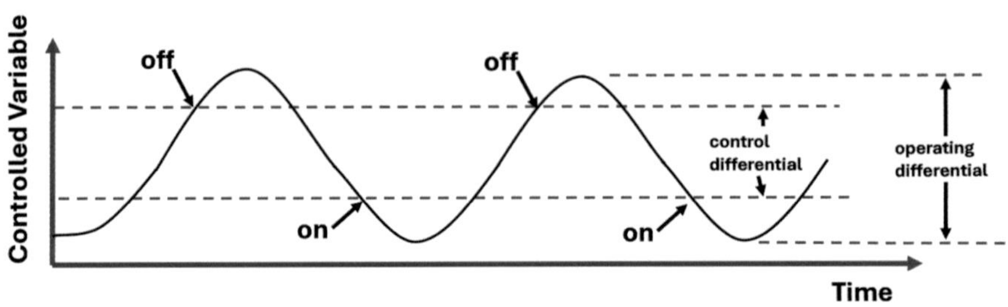

Figure 15.6 Controlled variable vs. time for a two-position controller.

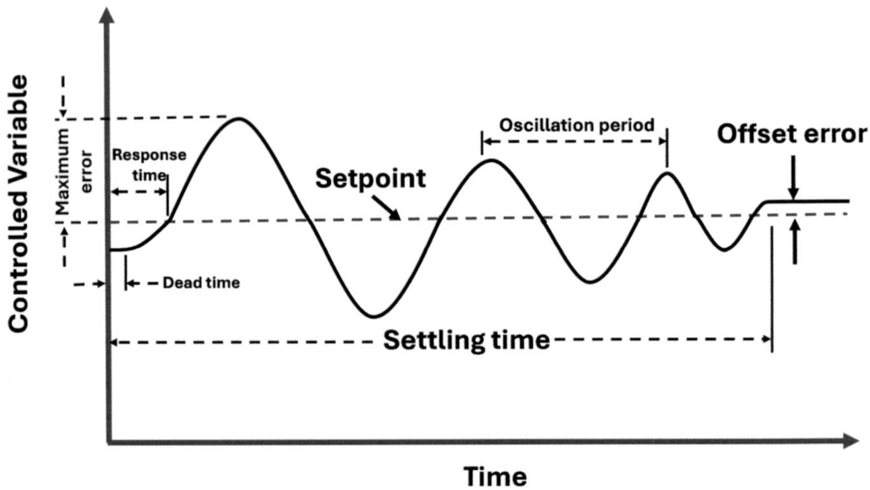

Figure 15.7 Controlled variable vs. time for a modulating controller.

illustrates some important concepts and definitions for modulating controls (Honeywell, 1997; Hughes, 2015):

- Dead time is the time from when the controller output changes to when the controlled (or process) variable begins to change.
- Response time is the time from when the controller output changes to when the controlled variable achieves the setpoint.
- Settling time is the time from when the controller output changes to when the controlled variable achieves steady state.
- Oscillation period is the time between peaks in the controlled variable.
- Error, in general, is the difference between the process variable and setpoint at any time. Offset error is the difference between the controlled variable and setpoint once the process achieves steady state.

- The maximum error is the largest difference between the controlled variable and the setpoint.

Another important concept in control systems is deadband. Deadband is the range of the controlled variable in which no corrective action is taken by the controller, and no energy is used (Honeywell, 1997). A common example is a thermostat controlling both heating and cooling. Deadband would exist, for example, if the thermostat was programmed to call for neither heating nor cooling between the temperature range of 16 and 20°C (61 and 68°F).

The main modulating control modes include proportional, integral, derivative, and adaptive control (ASHRAE Handbook Fundamentals 2021, Chapter 7; Honeywell, 1997).

- *Proportional control* is a control mode where the controlled device is set in proportion to the instantaneous value for error.
- *Proportional plus integral control* mode is where the proportional control signal is added to a term that represents the sum (which is mathematically an integral) of error over time. Adding the integral term has the effect of minimizing, if not eliminating, the offset.
- *Proportional plus integral plus derivative (PID) control* mode is where a third term is added representing the rate of change of error (which is mathematically the derivative). Adding the derivative term has the effect of reducing overshoot and reducing the time to reach a stable control condition.

Figures 15.8 and 15.9 illustrate how PID controls work (Hughes, 2015). Figure 15.8 is a diagram that describes the mathematics[1] for each function of a PID controller. Figure 15.9 provides an illustration of how the controlled variable varies over time, comparing proportional only vs. proportional plus integral vs. PID. Proportional only control tends to require more time to stabilize and exhibits relatively large offset. Adding integral control works to minimize offset. Then adding derivative tends to reduce overshoot and achieve a minimized offset in less time.

PID controllers are widely used in energy system processes. However, the full functionality of PID controllers is not always needed. Table 15.1 provides a general guideline for how PID control modes are usually used (Hughes, 2015). This table describes how control modes are typically programmed for common control loops for flow, level, temperature, pressure, and analytical

---

[1] How integrals and derivatives are defined is beyond the scope of this book; see any calculus textbook.

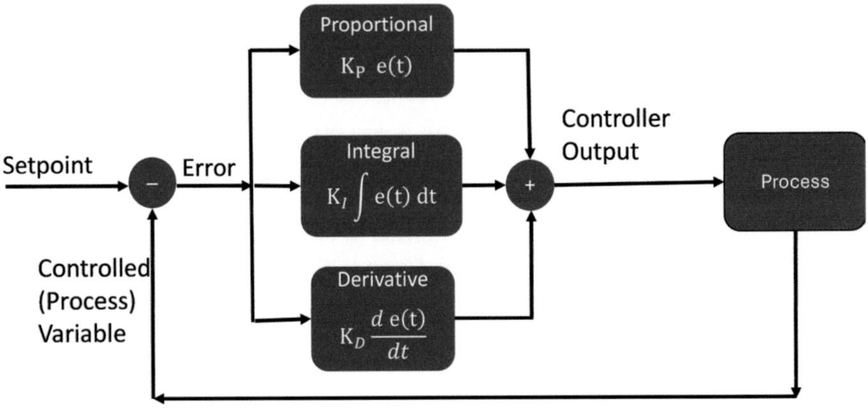

$$\text{Error at time, } t = e(t) = \text{Setpoint} - \text{Process Variable}$$

$$\textbf{Controller Output} = \mathbf{K_P\,e(t) + K_I \int e(t)\,dt + K_D \frac{d\,e(t)}{dt}}$$

Figure 15.8 Diagram illustrating how PID controllers work.

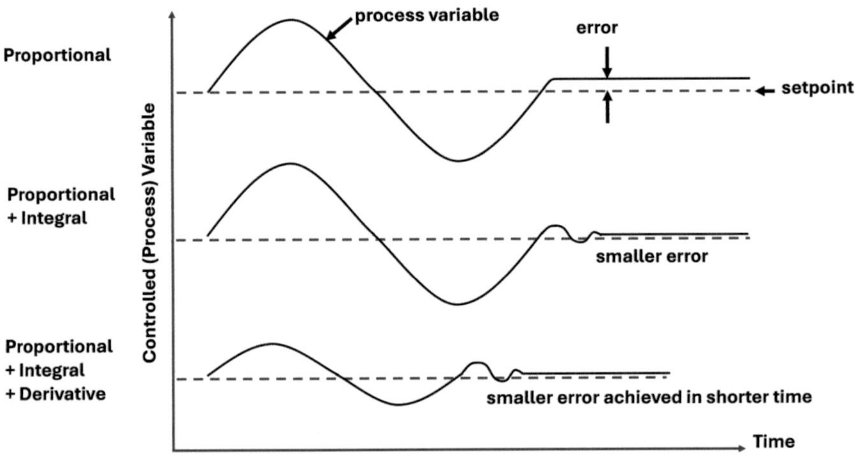

Figure 15.9 Illustrative charts of controlled variable vs. time comparing proportional, integral, and PID control modes.

variables. Analytical variables are process variables such as pH and the concentration of moisture or chemicals in process fluids (e.g., boiler water treatment as one example).

Floating control is a control mode that works as a hybrid combination of on/off and modulating control (Williams, 2025, Chapter 11). With floating

Table 15.1 Guideline for the application of PID modes in control system designs

	Proportional	Integral	Derivative
Flow	Always	Usually	Never
Level	Always	Usually	Rarely
Temperature	Always	Usually	Usually
Pressure	Always	Usually	Sometimes
Analytical (e.g., pH, chemical composition)	Always	Usually	Sometimes

control, the controller has a two-position output, but the control device can "float" across a continuous operating range between the on and off positions of the controller. In between the on and off positions, the process variable is not directly controlled as the value floats. For float control to work, the actuator must react slowly and the sensor must have a fast response.

Adaptive control is an advanced control mode that uses feedback rules to dynamically adjust the controller response (ASHRAE Handbook Fundamentals 2021, Chapter 7). With adaptive controls, the tuning parameters (e.g., the proportional or integral constants for a PID controller) change based on the characteristics of the process. Adaptive controls may use neural networks, which are computational models designed to work similar to biological neural networks. Neural networks use machine learning,[2] which is where computer systems learn and adapt without following specific instructions (i.e., "self-learning").

15.4 Control System Technologies

15.4.1 Manual controls

The earliest controls were manual, requiring a person to do something to control a piece of equipment, like turning a valve to control water flow, or using a switch to turn lights or equipment off when not needed. In the late 1800s, when heating systems advanced from individual fireplaces to central heat generation, the state of the art was to signal via a manually rung bell from the room where it was cold down to the boiler room, where people would hear the bell and stoke the boiler to provide more heat. This was remote control, albeit in a hands-on form.

Today, even control adjustments done manually can lead to dramatic savings. For example, lights are often left on at night, but turning them

[2] www.sagepub.com "Introduction to Artificial Intelligence."

off whenever possible would save energy (see Chapter 7 for methods to increase energy-saving behaviors). Exhaust and makeup air fans are another opportunity for savings through manual control. These fans are often located at the top of a high ceiling, and they are frequently left on unnecessarily because it is easy to overlook that they are running. Turning off exhaust fans, even manually, saves energy in two ways: first, electricity is no longer required to run the fan motor; and second, conditioned air is no longer being exhausted. Consider the following example:

Example 15-2

A fan with a 5-hp (3.73 kW) motor is exhausting air from a welding area, 24 hours/day, 7 days/week, at a rate of 10,000 ft3/min (4,720 L/s). But the plant only runs two shifts, from 8:00 am to 12:00 midnight, 5 days/week. Estimate the cost savings for turning the fan off at night and on weekends. Estimate the avoided GHG (CO2e) in metric tons (1 ton = 1,000 kg) assuming a natural gas emission factor[3] of 26.1 lbm CO2e/MMBtu (50.2 kg CO2e/GJ) and electricity GHG intensity (see Chapter 2) of 854 lbCO2e/MWh (0.388 kg CO2e/kWh).

Assumptions: The space is not cooled and is heated to 65 °F (18.3 °C) by a gas furnace that is 80% efficient. The efficiency of the motor is 84% and the motor is fully loaded. Gas costs \$5.00/MMBtu (\$4.74/GJ) and electricity costs \$0.08/kWh. There will be no demand savings since peaking does not occur at night. The outside temperature averages 30 °F (−1.1 °C) for the hours the fan can be shut off.

Solution

For electricity and heating, the annual hours for savings are as follows:

$$\left(\frac{5\text{ days}}{\text{week}}\times\frac{8\text{ hr}}{\text{day}}+\frac{2\text{ days}}{\text{week}}\times\frac{24\text{ hr}}{\text{day}}\right)\times\frac{52\text{ weeks}}{\text{yr}}=4,576\text{ hr}.$$

Electricity savings:

The electric energy savings from turning the motor off during nights and weekends is found by multiplying the motor load in kW times the number of hours saved (number of hours the motor is not running). This energy savings in kWh is then multiplied by the energy cost to get the dollar cost savings.

[3] U.S. EPA Greenhouse Gas Equivalencies Calculator, natural gas is 0.0053 metric tons CO2e per therm, www.epa.gov/energy/greenhouse-gas-equivalencies-calculator-calculations-and-references

$$\text{Electricity savings} = 5\text{ hp} \times \frac{0.7457\text{ kW}}{\text{hp}} \times \frac{1}{0.84} \times \frac{4,576\text{ hr}}{\text{yr}}$$
$$= 20,311\text{ kWh}.$$
$$\text{Annual electricity cost savings} = 20,311\text{kWh}\times\$0.08/\text{kWh} = \textit{\$1,625}.$$

Heating savings is calculated similar to electricity savings based on sensible heating of air from 30 to 65 oF, as described in Chapter 9 and the formula sheet in Appendix A. Use air density = 0.075/ft^3, specific heat = 0.24 Btu/lbm °F, and 80% boiler efficiency.

$$\text{Heat savings} = \frac{10,000\text{ft}^3}{\text{min}} \times \frac{60\text{ min}}{\text{hr}} \times \frac{4,576\text{hr}}{\text{yr}} \times \frac{0.075\text{lbm}}{\text{ft}^3} \times \frac{1}{0.80} \times \frac{0.24\text{Btu}}{\text{lbm}^\circ\text{F}} \times (65^\circ\text{F} - 30^\circ\text{F})$$
$$= 2,162 \times 10^6\text{Btu or } 2,162\text{MMBtu/yr fuel savings.}$$

Annual heating cost savings = 2,162 MMBtu × \$5.00/MMBtu = \$10,810.
Total annual cost savings = \$12,435.
To convert this savings to carbon emissions reduced in metric tons (1 ton = 1,00 kg):

carbon reduced from electricity savings:
20,311 kWh × 0.000388 tons CO2e/kWh = 7.88 tons CO2e

carbon reduction from using less fuel:
2,162 MMBtu × 1.055 GJ/MMBtu × 0.0532 tons CO2e/GJ = 121.3 tons CO2e

Total annual CO2e reduction = 129 tons.

15.4.2 Pneumatic controls

In the early and mid-twentieth century, pneumatic controls, where air pressure changes send signals to equipment, were widely used. Pneumatics were used in plumbing and HVAC as well as many industrial processes to control water flow, air flow, cooling and heating refrigerant, and more complicated sequences of operations. Figure 15.10 provides an example of a pneumatic control system for a hydronic heating valve (Honeywell, 1997). In these systems, all components are pneumatic, such as relays, thermostats, actuators,

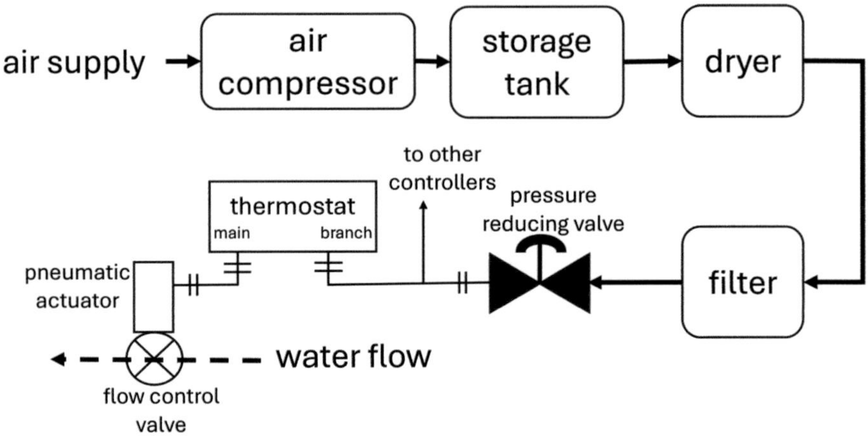

Figure 15.10 Example of a pneumatically controlled hydronic heating valve with thermostat.

and sensors (e.g., pneumatic sensors were designed for 3–15 psi output). Figure 15.11 illustrates the relationship between throttling range, setpoint, and branchline pressure for this type of system. Pneumatic controls are considered safe, due to the absence of electric spark risk, often require low maintenance, and are reliable in harsh environments. However, these controls are also less precise than modern controls, are difficult to change settings, and there are relatively few maintenance staff trained and experienced in pneumatics. There are still some specialized systems that utilize pneumatics, but as electronics evolved, most building and process controls migrated to electronic controls – first simple programmable controllers, and later, DDCs and computer-based systems.

15.4.3 Electric and electronic controls

Electric and electronic controls use electricity as the energy source to power controls and communicate between devices. Electricity has key advantages as compared to purely mechanical or pneumatic systems (Honeywell, 1997; ASHRAE Handbook Fundamentals 2021, Chapter 7):

- Electricity is widely available and it is relatively easy and inexpensive to install power and communication cables.
- Control system components can be designed to be located long distances apart without incurring significant added costs or loss of performance.

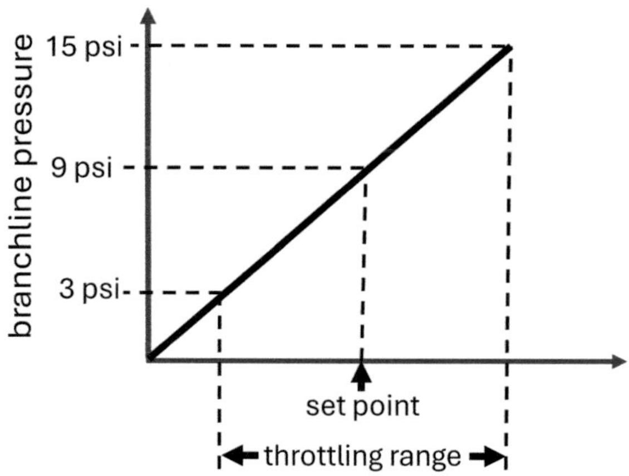

Figure 15.11 Example of relationship between throttling range, setpoint, and branchline pressure for a pneumatically controlled thermostat.

- Signals from sensors/transmitters can be used to control one or several devices. For example, several actuators can be controlled by one controller.

One potential drawback of using electricity is the potential for spark and fire risk (e.g., sawmills have traditionally used pneumatics for this reason). However, modern electric technology has evolved to reduce risks so that electricity-based control systems can be used even in environments where potentially combustible material exists.

Purely electric control equipment includes valve or damper actuators, process controllers, relays, motor starters, etc. Control signals can be designed to use low voltage (e.g., 10 V DC) or line voltage (e.g., 120 V AC).

Electronic controls are also powered by low voltage electricity, and are solid state systems that use semiconductor components such as diodes and transistors. Electronic control systems are typically designed as follows:

- Controller: low voltage and solid state
- Inputs: 0–1 V DC, 0–10 V DC, 4–20 mA, resistance element
- Outputs: 2–10 V DC or 4–20 mA
- Control mode: two-position, PID

Advantages of electronic controls include (Honeywell, 1997):

- Sensors, actuators, and final control devices can be located at long distances from the controllers.

- Controllers can work with a wide variety of inputs.
- Electronic control systems allow for programming of complex control and override schemes.
- Using universal communication systems (e.g., BACnet^{TM}) can interface into many types of actuators.
- Electronic controls enable effective use of human–machine interface displays.

15.4.4 Direct digital controls

DDC systems are those that use data in digital form to perform controller functions. DDC control systems may have both analog and digital inputs and outputs in various parts of the system, but the fundamental control algorithms (software) are digital. With these systems, operators may enter setpoints, set process limit variables, or enter PID constants, but the controller output is determined by the control algorithms. DDC controls may be designed as stand-alone systems, operating one or multiple control loops, or they may be integrated into industrial process automation or building automation systems (BAS). Individual controllers can perform as central controllers for small buildings (e.g., under 50,000 ft^2 or about 5,000 m^2) with up to about a dozen control points and a highly regular pattern of use.

An example of the use of DDCs is to program them for demand shedding. A controller can sense that the electrical demand is approaching a targeted maximum level, and the unit then shuts off equipment to keep the demand from passing that critical level. As discussed in Chapter 4, peak demand and/or peak time period (under a TOU rate) electricity charges can be a large part of an electrical bill, so the cost savings can be significant.

Another example of the use of DDCs is for combustion fuel and air flows for heaters and boilers. A controller that senses CO_2, O_2, or perhaps CO levels in the exhaust can continuously adjust the combustion air intake to yield optimum thermal efficiency while maintaining acceptable levels of pollution and stable combustion.

15.4.5 Building automation systems

Computerized BASs have typically been deployed in buildings over 100,000 ft^2 (about 10,000 m^2) with more than 100 control points. BASs that are equipped with multiple DDC controllers enable sharing of information using established communication protocols between not only HVAC systems but also other building service systems (e.g., security, fire protection, maintenance). These are powerful tools that provide the capability to implement advanced energy-saving applications. ASHRAE Guideline 13 and ASHRAE Standard 135 provide more detailed guidance for networking and interoperability for these systems.

The combination of BAS and DDC technologies are often utilized as an energy management control system. The use of BASs has been demonstrated to result in significant energy savings, typically stated to be at least 8% of baseline consumption for both electricity and heating fuels (O'Grady, 2021). Based on the proven benefits of BASs, it is widely recognized that the lack of BASs for smaller buildings is a pressing need. Smaller commercial and institutional buildings (say, less than 50,000 ft^2 or 5,000 m^2) seldom are equipped with a BAS. One industry white paper reports that only 13% of commercial buildings 50,000 ft^2 and smaller in the U.S. are equipped with a BAS.[4]

Since 2010, BASs and energy management systems have been moving away from running on their own wired or wireless digital network and toward being 100% digitally connected to the building's information technology data network. BASs and energy management systems are sometimes referred to as Energy Management Information Systems (EMIS), though the term EMIS can refer to a subset of functionality specific to tracking energy. Since the technology is changing very rapidly and there are many vendors in the field, each introducing new equipment and control apps, a potential user should consult several vendors and be well prepared to discuss the facility's current and anticipated future needs.

15.4.6 Control system design strategies

We conclude this section on control system technologies by summarizing common design strategies. In addition to the different control modes and available technologies, there are three common control system design strategies that can be chosen. Using HVAC as an example, Table 15.2 describes these three common strategies (https://armstrongfluidtechnology.com): stand alone, DDC, or BAS. Each strategy is defined as well as descriptions of key attributes of the strategy including scope, programming, flexibility, cost, and best applications. These same strategies can be applied to non-HVAC systems, such as automation of industrial processes.

[4] www.nexuslabs.online/untapped-87 report cites CBECS 2018 survey data.

Table 15.2 Comparison of HVAC control strategies

	Stand-alone control	**Direct digital control (DDC)**	**Building (or process) automation systems (BAS)**
Definition	Individual devices or controllers that operate locally without network connection	Microprocessor controllers handling multiple inputs and outputs with digital logic	Centralized system that integrates multiple DDC controllers and subsystems into one platform
Scope	Single piece of equipment (e.g., thermostat on a rooftop unit)	Equipment-level systems (e.g., air handling units, boilers)	Entire building or campus, integrating multiple systems such as HVAC, security, energy, etc.
Programming	Minimal or none; typically factory set	Customizable logic using software; supports use of advanced sequences	High-level supervisory program, scheduling and optimization
Flexibility	Low: fixed functionality	Medium: flexible sequences for specific equipment	High: scalable; can integrate across disciplines and sites
Cost	Lowest initial cost	Moderate (controller and software)	Highest initial cost, but savings in energy and O&M costs
Best application	Small, simple equipment with limited needs	Mid-size systems needing reliable control and some networking	Large or multiple facilities requiring integration, monitoring, and optimization

15.5 Control Applications for Energy Efficiency

This section provides general guidance on the approaches to identify and develop the scope for controls-based EEMs, followed by some case studies.

15.5.1 Process controls strategies for energy efficiency

The general strategy for using process controls to achieve energy efficiency (and decarbonization) improvements is to operate process (controlled) variables precisely within the desired range to meet service requirements in quantity and quality (see the discussion of energy services in Chapter 1). The goal is to avoid under-delivering or over-delivering service requirements. For example, with ventilation air, the goal is to deliver the minimum required flow in the ideal temperature range (i.e., not over-cooling in summer or overheating in winter). Service requirements are generally established based on:

(a) applicable codes (e.g., ASHRAE 62.1 for ventilation) and (b) facility user preferences. For every application of controls to save energy, it is critical to understand both code requirements and facility-specific user preferences for satisfactory service. *A common error for retrofit projects is to assume that past or present setpoints in existing or baseline control systems suffice for meeting service requirements.* In general, this is not necessarily true. The baseline level of service may have been falling short of facility user needs and/or exceeding required service levels. Any deficiencies in existing or baseline systems, whether under- or over-delivering services, will not be known unless one collects baseline data. Conducting surveys (see the description of surveys for energy audits in Chapter 3) or using RCx procedures for establishing owner project requirements (see Chapter 16) are available methods for establishing service requirements for control system upgrades. Tracking work orders related to equipment performance problems (for both buildings and industries) or occupant complaints for HVAC can also be useful.

With this general strategy in mind, the following points provide guidance on how to improve the energy efficiency of HVAC systems using controls (ASHRAE Handbook HVAC Applications 2023, Chapter 48; Doty, 2020).

- Operate equipment only when needed. For HVAC, start systems as late as possible to achieve minimum comfort conditions at the required time. At night for unoccupied spaces, setback winter temperatures to 60°F (16°C).
- Sequence heating and cooling to avoid simultaneous heating and cooling, unless required for humidity control or other reasons.
- Avoid simultaneous humidification and dehumidification.
- Provide only the heating and cooling actually needed. Use controls to reset hot or chilled water (or air).
- Use variable speed fans and pumps to provide the minimum pressure required for all end uses (avoid over-pressurizing).
- Provide heating or cooling from the most efficient source. Depending on the system design, this could involve heat pumps, backup boilers, outdoor air, or recovered heat.
- For domestic hot water, provide the temperature needed for service requirements (e.g., to limit bacteria growth), while avoiding overheating which can be a scalding hazard.
- For fluid systems, recirculation can be a source of unwanted heat loss (for heating systems) or heat gain (for cooling), as well as consuming power for pumps and fans. In general, reduce recirculation of air, process

gases, and liquids to the minimum practical flow rate and scheduled hours.

Additional specific guidance for implementing controls for HVAC can be found in ASHRAE Guideline 36-2024 and ASHRAE Handbook HVAC Applications 2023, Chapter 65 "Occupant-Centric Sensing and Controls."

Table 15.3 provides a list of examples of EEMs that utilize process controls to save energy (Doty, 2020; Williams, 2025; ASHRAE Handbook HVAC Applications 2023, Chapters 43 and 48). Examples are provided for common building systems as well as a few industrial examples. Then in the following sections, some case studies are provided.

Table 15.3 Examples of energy efficiency measures using process controls

Control EEM	Typical baseline	EEM description
Space heating hot water reset	Hydronic heating system produces constant temperature hot water	Reduce or vary hot water temperature as a function of outdoor air temperature. Alternatively, use polling of hot water end uses so that at least one end use control valve is at least 90% open.
Boiler or heater oxygen trim controls	Fixed air/fuel ratio controls without O_2 monitoring	Meter boiler or heater exhaust O_2 and use as a feedback signal to fuel/air controller to reduce O_2 to the practical lower limit for stable combustion and operate within pollutant limits (e.g., CO).
CO_2 demand control ventilation	Constant ventilation rate 8,760 hours annually	Install CO_2 sensors at all critical control locations and use as feedback signal for ventilation fan speed. Has applications for residential, commercial, and industrial. Always ensure ventilation meets safety and reliability needs.
Chilled water reset	Constant chilled water supply temperature	Vary chilled water supply temperature based on return water temperature and/or end use chilled water valve positions so that the coil with the largest cooling load maintains at least 90% open.
HVAC variable volume fan optimization	Constant air supply pressure, far exceeding minimum pressure required	With variable speed controls, reduce supply air pressure so that at least one end use (terminal unit) damper is at least 90% open.
Timed tenant override	Tenant override of setback HVAC temperatures have no set time limit	When tenants override HVAC temperatures during setback time periods (e.g., weekends), set time limit to 2 hours to return to setback conditions.

Table 15.3 *Continued.*

Control EEM	Typical baseline	EEM description
Heating or cooling optimal start/stop	Heating or cooling systems startup earlier and/or shutdown later than needed to meet comfort conditions	Schedule heating or cooling systems to start up (or ramp up) as late as possible, and/or to shut down as early as possible (e.g., shut down 1/2 hour before end of occupancy). Validate with occupant data (e.g., surveys) that comfort conditions are maintained.
Programmable lighting ballast	All lighting for a given zone, floor, or room is according to one schedule, even though needs are more granular	Provide illumination when and where needed using "addressable" lighting ballast that can be controlled individually, to customize scheduling by room, area, or fixture.
Window or door sensors interlock with thermostat	Space heating or cooling remains on even if windows and outside doors are open	Install window and door sensors interlocked with heating or cooling thermostats such that mechanical heating/cooling shuts down after time delay (e.g., over 15 minutes of open window). Ensure adequate ventilation is always provided.
Compressed air systems with multiple compressor controllers[5]	Compressed air systems with multiple compressors and each controlled individually	Network controls use the on-board compressor controls' microprocessors linked together to stop/start, load/unload, and modulate production to achieve end use flows and pressures at minimum electricity consumption.
Industrial furnace pressure controls[6]	Industrial furnaces operate with negative pressure (i.e., $<$ ambient) causing significant infiltration	Pressure gauge in furnace connected to pressure controller to maintain a slightly positive pressure in the furnace chamber to avoid infiltration. Furnace flows are regulated by varying supply or exhaust fan speeds or damper settings. Ensure equipment safety under positive pressure conditions.
Industrial boiler automatic blowdown	Constant flow, manually set boiler water blowdown	Conductivity or dissolved solids monitoring and feedback to blowdown control valve and controller.
Electric vehicle (EV) load scheduling controls	EV charging occurs without consideration of electricity price schedule or demand response incentives	EV load scheduling controllers set schedules for vehicle charging based on time-of-use rates schedules or peak pricing incentives (see Chapter 4). Shift some or all EV charging to off-peak hours.
Demand-based heating and cooling plant operation	Heating or cooling plant is enabled based on building schedule and outside air temperature threshold	Enable and disable heating and cooling plants based on zone(s) demand for heating and cooling to prevent systems from running at low/no load.

Table 15.3 *Continued.*

Control EEM	Typical baseline	EEM description
Differential pressure setpoint reset	Pump VFDs control to fixed differential pressure	Reset differential pressure setpoint based on zone demand to maintain one coil control valve at 90% open. (Note: Needs careful coordination if implementing with demand-based temperature setpoint reset.)
Pump staging based on flow	Lead/lag pump staging is based on VFD pump speed	Stage pumps based on flow to keep the pumps operating near their best efficiency point to reduce pump power.
Occupied standby setbacks	Heating and cooling setpoints are maintained at occupied setpoints throughout scheduled occupancy	Where zones include occupancy sensors; implement mild setback/setup of temperatures when the space has been unoccupied for more than 5 minutes to push these zones into deadband mode to minimize airflow and reduce reheating at the zone level.
Separate space heating and cooling temperature setpoints	Space with single setpoint (e.g., 22°C or 72°F) with tight control. Zone can oscillate between heating and cooling	Provide separate heating and cooling setpoints with a deadband between them where no active heating or cooling is commanded. Provide a defined deadband to prevent control hunting, reduce simultaneous heating and cooling, and reduce thermal energy.
VAV dual maximum airflow logic	Supply airflow is maximum (100%) at peak cooling and at minimum (typical 30%) during deadband and heating	Use different heating and cooling airflow maximums and a lower minimum airflow than conventional sequences, saving fan and reheat energy and improving comfort by reducing drafts during deadband and heating mode.
Supply air temperature reset (VAV)	Supply air temperature is fixed at a constant setpoint to meeting peak cooling requirements	Reset supply air temperature as a function of outdoor air temperature and zone demand to avoid excess cooling of supply air at off-peak conditions; reduces reheat energy.
Stage variable speed centrifugal chillers based on best efficiency point	Lag chiller is staged on when the lead chiller is at or near full output	Centrifugal chiller efficiency varies significantly with lift. Often, variable speed centrifugal chillers have a best efficiency point when running at part load conditions and when lift is lower. Staging based on best efficiency point reduces compressor power.
Stage condensing boilers based on best efficiency point	Lag boiler is staged on when the lead boiler is at or near full output	Modulating condensing boilers are typically more efficient at low load since the heat transfer area to thermal output is higher, increasing flue gas condensation. Staging boilers are lower (but stable) loads maximize boiler efficiency.

Table 15.3 *Continued.*

Control EEM	Typical baseline	EEM description
Agricultural pumping with VFD for demand response	Fixed irrigation schedule without consideration of electricity price schedule or demand response incentives	Using VFD controls, reduce or pause water pumping during peak periods (e.g., based on time-of-use rates or peak pricing incentives; see Chapter 4) and increase water pumping during off-peak.

15.5.2 Case study: Control upgrades for a mixed use facility

A large, mixed use facility (over 2 million square feet) was built in the 1960s with some upgrades installed in 2020. An energy audit identified four control measures with the results shown in Table 15.4. The audit focused on thermal savings for natural gas, but also revealed significant savings in electricity.

Table 15.4 Controls-based EEM case study for a large mixed use facility

EEM	How it saves energy	Estimated energy savings
Install differential pressure sensor and VFDs to control secondary heating pumps	Baseline is constant speed hydronic pumps when enabled. Pressure sensor used as indicator of demand to operate pumps at less than 100% speed. Reduces heat losses in piping network and pump motor power.	1,100 GJ/yr (1,040 MMBtu/yr) natural gas 60,000 kWh/yr electricity
Recommission BAS (replace or repair sensors and actuators; reprogram schedules and update sequences). See also Chapter 16	Optimize HVAC schedules Optimize temperature setpoints Reduce or eliminate idle losses	3,750 GJ/yr (3,560 MMBtu/yr) natural gas 32,000 kWh/yr electricity
Install DHW recirculation demand controller	Baseline is continuous DHW recirculation flow, which results in water heating to maintain temperatures. With demand controller, DHW recirculation pumps run only when needed based on hot water fixture demand sensors. Reduces heat loss and pump power.	740 GJ/yr (700 MMBtu/yr) natural gas 6,200 kWh/yr electricity
Demand control ventilation on air handling units	Baseline is ventilation based on set schedules with fixed minimum damper opening, and no CO_2 metering. VFD with CO_2 or occupancy sensors to enable heating to match demands, reducing fuel consumption and electric fan power.	925 GJ/yr (880 MMBtu/yr) natural gas 29,000 kWhyr electricity

The facility baseline natural gas consumption was about 28,500 GJ (27,000 MMBtu), and so the total identified gas savings represents 23% of baseline consumption – all with control upgrades.

15.5.3 Case study: Industrial dryer controls

This case study involves an industrial drying process[7] as illustrated in Figure 15.12. A wet solid material enters from the left onto a porous drying belt. Burners are fired to achieve a target discharge temperature, belt speed is varied, and fans are run at full speed as the baseline operation. The controlled variable is the solid material moisture content as it exits the dryer to achieve a setpoint of 7% or less. The EEM involves a change in control strategy to use the underbed fan speed as the main manipulated variable. The retrofit involved replacing the discharge moisture meter to improve its accuracy, adding underbelt humidity sensors, careful optimization testing under both winter and summer conditions, and reprogramming of existing controls. The new control scheme reduced the amount of waste heat as well as reduced the fan electric power consumption based on reduced speeds. Estimated energy savings are as follows:

- 38,000 GJ/yr (36,000 MMBtu/yr) natural gas
- 1,700,000 kWh/yr electricity

15.5.4 Case study: Automatic boiler blowdown

A facility with two, natural gas-fired 400 boiler horsepower saturated steam boilers operates at 125 psig or 860 kPa (each boiler rated output is 13.39 MMBtu/hr or 14.13 GJ/hr). It is estimated that retrofitting the boilers with an automated boiler blowdown system will reduce blowdown flow from 2% to 1% of total flow as compared to baseline controls using manually set, constant flow blowdown. The system will use a conductivity sensor to feedback to a PID controller as illustrated in Figure 15.12. The estimated annual fuel savings is 391 GJ/yr (371 MMBtu/yr).

Example 15-3

Using an automated boiler blowdown controller, an industrial boiler is assessed to reduce blowdown from 8% to 6% of boiler water flow.[8] The boiler

---

[7] ASHRAE Handbook HVAC Applications (2023), chapter 31 provides a useful introduction to industrial drying systems.

[8] This example is from: www.energy.gov/sites/prod/files/2014/05/f16/steam9_blowdown.pdf

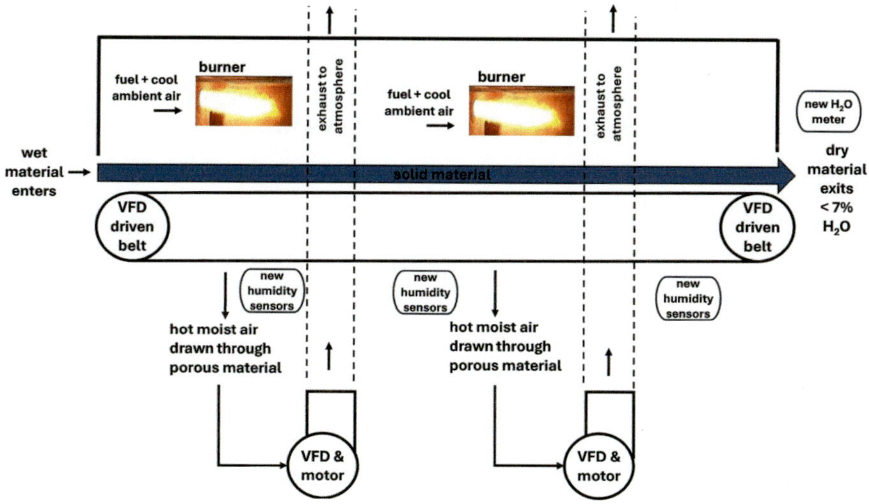

Figure 15.12 Controls-based EEM for an industrial drying process.

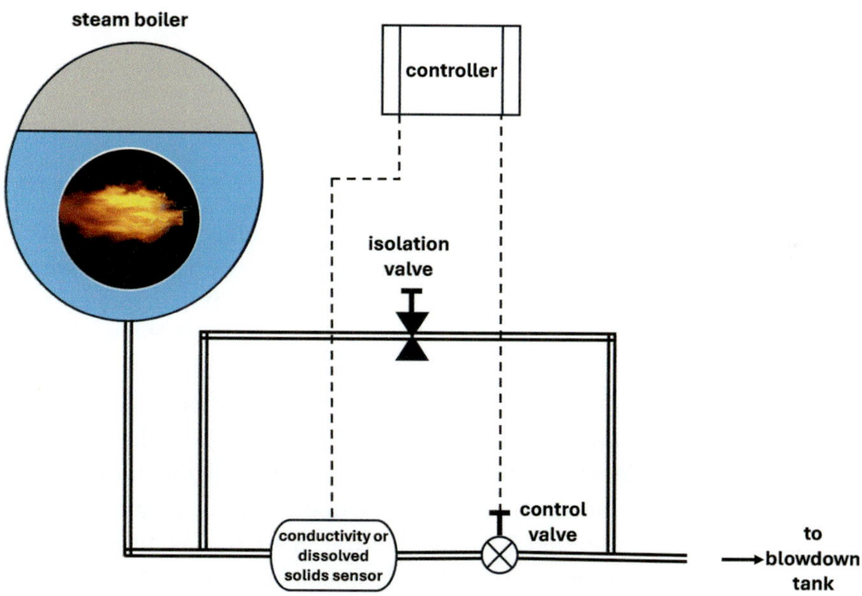

Figure 15.13 Controls-based EEM case study for an industrial boiler.

operates at 100,000 lb/hr (45,000 kg/hr) saturated steam at 150 psig (1,000 kPa) pressure (366°F or 186°C saturation temperature) and 8,650 hr/yr.

Makeup water is 60°F (15.6°C) and the boiler operates at 80% efficiency (HHV basis). Fuel cost is \$8 per MMBtu (7.60 per GJ). Water, sewage disposal, and chemical treatment costs are \$0.004 per gallon (\$1.06 per m^3). Estimate the total cost savings by reducing the blowdown.

Solution

Baseline boiler feedwater: 100,000 lb/hr/(1 – 0.08) = 108,696 lb/hr

EEM boiler feedwater flow: 100,000 lb/hr/(1 – 0.06) = 106,383 lb/hr

Makeup water savings: 108,696 – 106,383 = 2,313 lb/hr

Enthalpy of feedwater is 338.5 Btu/lbm and enthalpy of water at 60°F is 28.1 Btu/lbm.

Enthalpy added to makeup water = 338.5 – 28.1 = 310.4 Btu/lbm

Annual fuel savings = (2,313 lb/hr × 8,650 hr/yr × 310.4 Btu/lbm)/0.80) × MM/10^6

= 7,763 MMBtu or 8,190 GJ

Annual fuel cost savings = 7,763 MMBtu × \$8/MMBtu = \$62,104

Annual water and chemical savings = 2,313 lb/hr × 8,650 hr/yr × \$0.004/gal × gal/8.34 lbm

= \$9,596

Total annual cost savings = 62,104 + \$9,596 = *\$71,700*

15.6 Conclusion

In this chapter, we have examined the basics of how control systems can be used to improve energy efficiency and thus support decarbonization for buildings and industrial facilities. Control systems operate by manipulating variables that control processes in response to disturbances, with the goal of achieving desired values for process (controlled) variables. Control systems are composed of control loops that can be three types: manual, automatic, or cascading. A design choice for control system components is the use of analog vs. digital data. Common control modes are introduced and illustrated with schematics and diagrams including: two-position control, proportional, integral, derivative, floating, and adaptive controls. Different control technologies have been used historically including manual controls, pneumatic, electrical, electronic, and DDC. PID controls are the workhorse of modern control systems, although the use of adaptive controls (e.g., machine learning or artificial intelligence) in combination with PID hardware is increasing. The basic strategy involved in designing energy-efficient control system upgrades, whether for new facility designs or retrofits, is to design systems to accurately meet service requirements in quantity and quality. To do so, the objective is to operate process (controlled) variables precisely within the desired range. The goal is to avoid under- or over-delivering relative to the ideal service requirements. Many control-based EEMs are identified and three case studies based on actual projects are provided.

References

ASHRAE Handbook Fundamentals (2021), chapter 7 “Fundamentals of Control”

ASHRAE Handbook Fundamentals 2021, Chapter 38 “Measurement and Instruments”

ASHRAE Handbook HVAC Applications (2023), chapter 43 “Supervisory Control Strategies and Optimization”

ASHRAE Handbook HVAC Applications (2023), chapter 48 “Design and Application of Controls”

ASHRAE Handbook HVAC Applications (2023), chapter 65 “Occupant-Centric Sensing and Controls”

ASHRAE Handbook HVAC Applications (2023), chapter 41 “Computer Applications”

ASHRAE Handbook HVAC Applications (2023), chapter 31 “Industrial Drying Systems”

ASHRAE Guideline 13-2024 Specifying Building Automation Systems

ASHRAE Guideline 36-2024 “High-Performance Sequences of Operation for HVAC Systems”

ASHRAE Standard 135-2016, “BACnet^{TM} – A Data Communication Protocol for Building Automation and Control Networks”

Doty (2020) “Commercial Energy Auditing Reference Handbook” Chapter 7 “Automatic Controls Strategies”. River Publishers.

Honeywell Corporation “red book” (1997) www.av8rdas.com/uploads/1/0/3/2/103277290/gray_manual.pdf

Hughes T. (2015) “Measurement and Control Basics” 5th edition. International Society of Automation.

O’Grady T., Chong H., & Morrison G. (2021) “A systematic review and meta-analysis of building automation systems” Building and Environment 195, 107770.

Williams (2025) “Energy Management Handbook” Chapter 11

Author Biography

Eric Mazzi is a consultant and educator focusing on technology and policy for demand-side energy systems. He specializes in performing independent technical analysis, M&V, policy studies, and educational services. He advocates and contributes to efforts to reduce energy consumption and achieve net-zero GHG emissions, while maintaining or improving energy services and advancing energy equity. His experience includes air quality consulting, utility maintenance, and five years as full-time engineering faculty at the University of British Columbia. He volunteers as a board member for the U.S. Northwest Energy Coalition, and the International Performance M&V Protocol committee. Eric teaches part-time, delivering university and professional certification courses. He is an adjunct professor of Energy Management at the New York Institute of Technology – Vancouver. He has B.Sc. and M.Sc. degrees in mechanical engineering, and a PhD in environmental policy. He is a registered engineer in California and British Columbia.

Acknowledgments

Contributions to this chapter were made by Peter Scanlon, Mike Reimer, and Sheila Glessman. All contributors are gratefully acknowledged.

Practice Problems

1. A multi-unit residential building has natural gas-fired hydronic baseboard heating that is controlled in each suite by a conventional wall-mounted thermostat. Heating fuel is paid for through association fees, so there is no direct or obvious incentive for individual condominium owners to limit the use of the heating system. Energy auditors observed that a high percentage of occupants simultaneously open windows during winter while running the heating system. Describe a retrofit solution that could be implemented to both save heating energy while maintaining or improving occupant service needs. Focus on identifying a low capital cost solution – i.e., assume replacing the entire building heating system is not an option.
2. Describe how floating control works. Review some of the literature listed in the references, and identify a common application where floating control is beneficial. Hint: ASHRAE Handbook, Williams (2025), and Honeywell (1997) sources may be helpful.

3. What is the main reason or benefit to use integral action in combination with proportional control vs. proportional control alone?
4. Describe the type of process that can obtain the most benefit from the use of PID control as compared to two-position control.
5. For the control loop shown below, state whether it is an open or closed loop. In a few sentences, describe how the control loops works to achieve the setpoint value of the controlled variable.

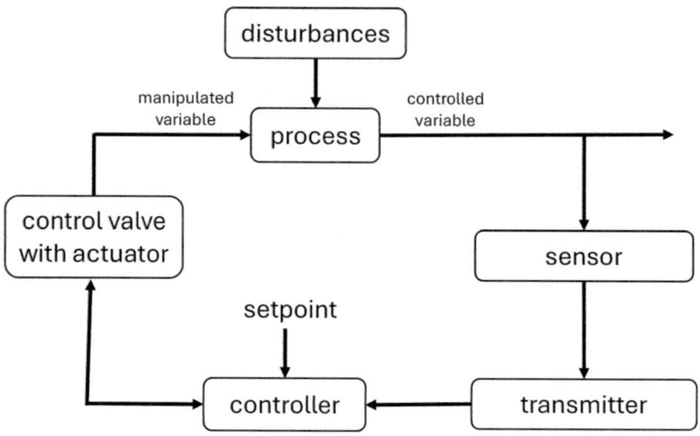

6. A sensor outputs a 4–20 mA signal to a transmitter, and delivers a 0–10 V DC signal to the controller. If the transmitter is linear and properly calibrated and a 6.0-V signal is input to the controller, what is the expected value of the sensor output?
7. Explain what is meant by dead time for a process.
8. Discuss some of the advantages of wireless devices and networks. Hint: ASHRAE Handbook, Williams (2025) and Honeywell (1997), and Hughes (2015) sources may be helpful.
9. An industrial compressor operates in a natural gas fracking operation and is designed to maintain a constant pressure of 1,400 psig (9,650 kPa) input to the gas treatment system. List at least three process disturbances that might occur, necessitating corrective action from the compressor controller.
10. List at least 10 functions or actions that a building automation system (BAS) can perform to save energy and/or improve building performance and occupant satisfaction. Hint: ASHRAE Handbook HVAC Applications 2023, Chapter 48, and other sources in the References section will be helpful.

16

Commissioning for New and Existing Buildings

Craig Hatch

CFMS Alberta Ltd., Canada

16.1 Introduction

Commissioning (Cx) is a process. The language used to describe commissioning varies from one source to another, but its purpose is always just to provide a diverse set of technical review and verification services based on experience, knowledge, skills, and abilities to owners and project teams. Commissioning, throughout project design, construction, and operation, results in better buildings that work effectively and efficiently.

The American Society of Heating, Refrigerating and Air-Conditioning Engineers (ASHRAE) Standard 202-2018, "The Commissioning Process for Buildings and Systems," and ASHRAE Guideline 0-2019 – "The Commissioning Process," define commissioning as:

> "A quality-focused process for enhancing the delivery of a project. The process focuses upon verifying and documenting that all the commissioned systems and assemblies are planned, designed, installed, tested, operated, and maintained to meet the Owner's Project Requirements." (ANSI/ASHRAE/IES, 2018)

The Building Commissioning Association's National Conference on Building Commissioning defines total building commissioning as:

> "A systematic process of assuring by verification and documentation, from the Design Phase to a minimum of one-year after construction, that all facility systems perform interactively in

accordance with the design documentation and intent, and in accordance with the owner's operational needs, including preparation of operation personnel." (Stum and Bjornskov, 2017)

The latter definition includes training of operations personnel as part of the commissioning process, which is key to maintaining a consistently high level of building performance.

There are five types of commissioning:

- New building commissioning, often abbreviated as "Cx," that occurs during design and construction of new facilities.
- Existing building commissioning (EBCx) (ANSI/ASHRAE/IES, 2018) also known as Retro-commissioning and Re-commissioning (collectively, RCx) that concern themselves with the operations of existing buildings.
- Ongoing commissioning (OCx).
- Monitor-based commissioning (MBCx)
- Building envelope (enclosure) commissioning (BECx)

Various definitions of commissioning exist and most, simply stated, describe Cx as a quality assurance process to verify that the owner's goals and design intent are met in new construction. EBCx aims for more efficient operations of existing buildings and can be thought of as a building tune-up that restores a facility to its original operating condition to meet today's operational needs as closely as possible. It can also include a phase that modernizes a facility with low-cost and no-cost energy efficiency measures (EEMs).

The concept of commissioning can be applied to any building system or subsystem but for purposes of this chapter we will limit ourselves to addressing "energy-using" systems of a building, which is in concert with LEED's definitions and applications of Cx and EBCx. Chiefly, this means heating, ventilating, and air conditioning (HVAC), lighting and lighting controls, domestic hot water (DHW), and renewable energy (e.g., solar or wind) systems.

To say we live in unprecedented times is an understatement. Political, social, scientific, and environmental factors (see Chapter 2) seem to be in flux to a degree unseen in modern times. While events can be unsettling, opportunities for a better future abound. This is especially true in the built environment world. Here are two indicators that point to making this a great time to commit to reducing energy consumption and costs by increasing building systems efficiency:

Return to office: The lasting impact of the global pandemic on how day-to-day business is conducted is still being defined, but there is no question changes will be made. According to a report from the International Energy Agency (IEA, 2020), the ongoing crisis presents an opportunity to shift to a more efficient energy system with multiple benefits. This is an opportune time for companies to take stock of their current resources and requirements. What does "return to office" mean? Will it be a return to a pre-COVID status quo? Do the efficiencies of remote working translate to a new way of operating going forward? If your organization has or is considering a hybrid approach to office attendance, this could be the ideal time to evaluate your building system needs. Systems that have been sitting idle or under-used may warrant careful inspection before being brought back fully online.

Greater environmental awareness: Addressing employee health and safety concerns is also a great time to improve environmental, social, and governance goals of your organization. Making your building more energy efficient can create sustainability benefits beyond the bottom line. Committing to a program of energy reduction and efficiency can pay dividends in positive public awareness, an improved environment, and better work performance. A qualified expert in LEED, Certified Energy Manager^{TM} (CEM), and other certifications can help you reach goals of net-zero carbon emissions.

16.2 New Construction Building Commissioning (NCCx)

16.2.1 Why do we need NCCx?

The New Construction Building Commissioning Best Practice (Stum and Bjornskov, 2017) is intended to cover a general new construction commissioning process that is applicable to most building systems and assemblies. It is process driven, with allowable flexibility for unique conditions, and where a practice seems unclear by its varied application, more detail is given. For example, the document will give a general statement about the best practice of functional testing but will not delve into the particulars of the best practice functional testing of any specific piece of equipment or assembly. The document describes the recommended process of commissioning but does not offer guidance on the systems that should be commissioned or the specific rigor that should be applied since those are project specific. The only notable exception is the specific addressing of building automation system (BAS) and controls because they have been and continue to be the focus of most, if not all, building commissioning projects. Details of most of the elements and means and methods for their application can be found in the documents referred to in the online BCxA Knowledge Center (www.bcxa.org/resources). In many places, the best practices indicate who is expected to be responsible for the completion of a given task. Where absent, the responsible party may be any qualified member of the Cx team.

A case study survey of a mixed-use building of office and retail in Seoul S. Korea: After conducting HVAC commissioning from January to December in 2014, deficiencies found from testing and balancing and functional testing was corrected, and actual annual energy consumption of the building was monitored, and these data were used for dynamic simulation to verify performance. After carrying out the HVAC commissioning for one year, the actual annual energy consumption of the building was recorded, analyzed, and compared against the simulation results for verification. After finding a performance gap in an HVAC system in a case study building and correcting these findings, it was simulated with a dynamic simulation software tool. The energy consumption was reduced by 2.1% for fan control, 1.4% for condenser water pumps, 0.4% for primary chilled water pumps, and 1.4% for the secondary chilled water pump. Therefore, the energy reduction of the case study building by HVAC commissioning was 5.3% of total energy consumption. A key point to emphasize is that commissioning is the process

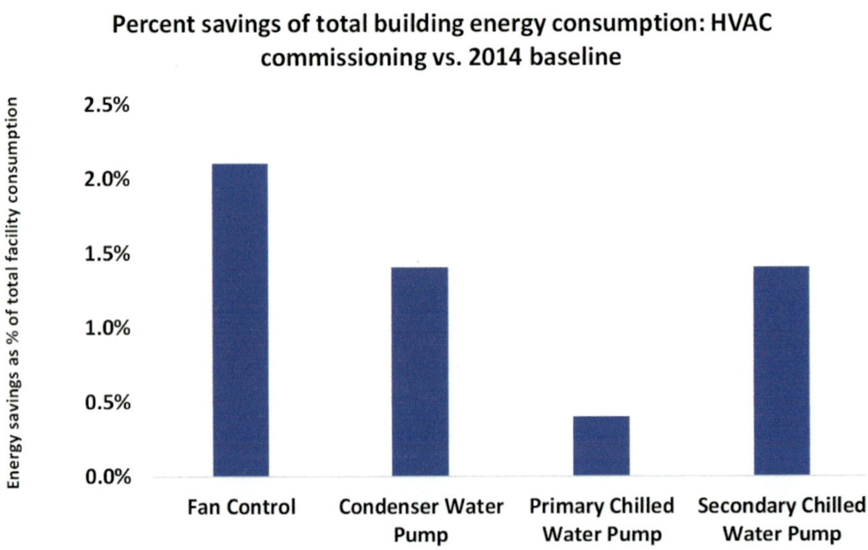

Figure 16.1 Energy savings after HVAC commissioning (Kim and Altan, 2018).

of tuning systems to better meet the owner's project requirements (OPR) (this can be thought of as the quantity and quality of services introduced in Chapter 1), while at the same time reducing energy consumption and energy cost.

16.2.2 What does NCCx cost?

One of the most important objectives of owners and project teams is to associate cost with value. Figure 16.2 describes the process of constructing new buildings, and illustrates that the opportunity to influence costs is highest in the early phases, and is very low once the building construction is completed. The process of commissioning is intended to maximize a facility's asset value: delivering the expected performance to owners and occupants (comfort, safety, security, reliability, etc.) at a "reasonable" cost. It must be said at the outset, however, that commonly used methods for generalizing the cost of commissioning can only be considered "back of the envelope" estimates with large variations in resulting budgets by providers because of how providers apply the same nominal scope of commissioning differently.

Management of the process of commissioning is a service, usually defined in construction terms as "soft cost" related to the CxP's level of involvement in a project and the scope and complexity, and amount of time and phases from predesign through occupancy they are engaged in. Every project and all owners' expectations are different.

Commissioning costs are typically understood to be the price of delivering the management of the process of commissioning through technical consulting services; however, the cost of the time of the CxP does not include all the commissioning costs. There is some debate as to what should be included in the cost of commissioning when determining the first cost for an owner in both new and existing buildings. There are also costs to the construction manager, contractor, architect/engineer, and owner for their participation in the commissioning process.

A 2015 Building Commissioning Association (BCxA) survey found that the average cost of new construction commissioning was 1.5% (Stum and Bjornskov, 2017). But when the highest and lowest "outliers" were taken out of the mix (which represents small, simple project, or complex facilities with energy efficiency directives from predesign through occupancy), the overall average dropped to 1%. The U.S. General Services Administration, Federal Energy Management Program, and other Department of Energy data generally collaborate that, on average, as of 2016, new construction commissioning costs between 0.5% and 1.5% of the total project cost. The caveat is that all projects are different, and the current cost of commissioning depends on

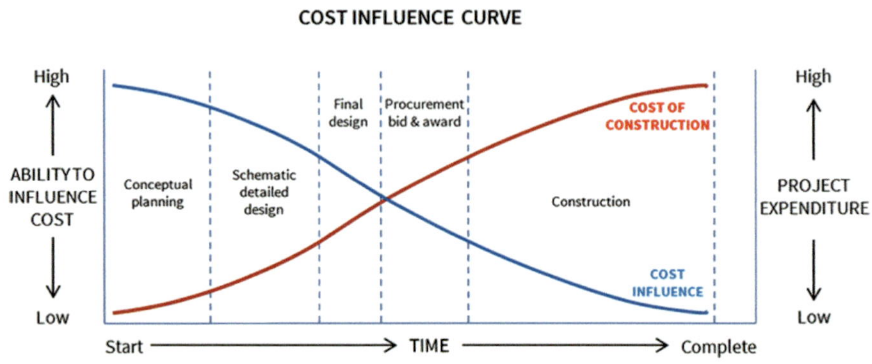

Figure 16.2 Value of commissioning for new construction.

the type, size, location, and complexity of the building and its systems and assemblies.

Costs are typically estimated in one of these ways:

- **Dollars per square foot**: Depending on the building type and commissioning scope, or dollars per ft^2 depending on the energy use intensity (EUI), or energy used per ft^2 per year. In general, the EUI is lower for large and/or multistory facilities and higher for small and/or single-story facilities. Table 16.1 provides cost/benefit metrics.

Table 16.1 NCCx commissioning cost/benefits metrics (Crowe, 2020).

Cost/benefit metric	NCCx		
	Median	**25th–75th Percentile range**	**Sample size (projects)**
Energy savings cost	13%	9%–30%	7
per m^2 (cost per ft^2)	$11.08 ($1.03)	$4.71–$23.76 ($0.53–$2.21)	140
Simple payback	4.2 years	1.5–10.8 years	36

- **Percentage of construction cost**: This may or may not account for CxP involvement prior to the construction phase. Sometimes the construction costs referenced are only the mechanical and electrical.
- **Percentage of total project delivery cost (design fees and construction fees)**.

16.2.3 Commissioning team

The goal of commissioning – to deliver a building that functions in accordance with the OPR – is shared by the entire project team. Therefore, the ideal commissioning team will typically include the designers, suppliers, builders, and future operators of all the systems to be commissioned, although the formal Cx team might vary with each project. In fact, some owners and commissioning providers (CxPs) have a narrow view of who is really on the team and might list only those who actively take part in certain tasks like testing, leaving the design team, and facility group off the list. This can compromise commissioning effectiveness on all but the simplest of projects. Others include multiple owner representatives, developers, user groups, tenants, and vendors in addition to the typical CxP, owner representative, contractor, and design team. Though many individuals may be impacted by the commissioning process or participate in some way, a risk with a large commissioning team is that many team members might not be interested in the process except where it specifically impacts them. Regardless of the specific team makeup, the level of interface also varies by project phase. For example, members during design may not be as active during construction.

The following core group typically provides input into the Cx process and is impacted by it throughout most phases:

- Owner (or representative, i.e., project manager)
- Developer (when providing the building for an owner)
- Design team (architect and engineering firms(s), especially mechanical and electrical)
- Contractor (and relevant subcontractors)
- CxP
- Facility operations staff

16.2.4 NCCx phases

16.2.4.1 Pre-design phase

Often, little attention is given to commissioning during predesign, resulting in budget decisions and attitudes that may preclude the incorporation of an optimal commissioning program in later phases. Not incorporating a CxP or at least giving serious thought to commissioning during predesign will also likely result in less thorough engagement by the owner in developing the OPR and could miss the advantage of improved communication between owner and designers. The earlier in the process commissioning is started, the magnitude of commissioning benefits increases and the cost for implementing recommendations coming out of commissioning decreases.

Commissioning provides project benefits even in the predesign phase by laying the groundwork and objectives for the project, defining the scope and plan for commissioning, bringing a CxP on board, and beginning the essential team-building process.

The overall objective in the predesign phase relative to commissioning is to develop and set in place a program for an enhanced and successful delivery of the envisioned project, including:

- Bring a CxP on board and identify the commissioning team.
- Develop the OPR.
- Define the initial commissioning plan.
- Verify that the design intent is consistent with the OPR.
- Integrate commissioning into the overall project delivery process.

The commissioning team during predesign consists of the party who will bring on board the CxP and all the available stakeholders and contributors to the OPR:

- Owner/owner's representative
- Building occupant/tenant representatives (when available)
- CxP
- Design team personnel
- Operations and maintenance (O&M) personnel (when available)
- Construction team personnel (when available)

The following list identifies the activities of the predesign phase of a project. The activities and tasks are shown in the order they would normally be executed:

- Identify Cx leadership and team
- Order of activities in predesign
- Develop the OPR for the project
- Develop the initial commissioning plan
- Define the commissioning budget
- Compile the predesign phase commissioning documentation

16.2.4.2 Design phase

After the predesign phase, the next step is to apply the quality process of commissioning to the design itself. During the design phase, the commissioning process confirms that design documentations (plans, specifications, basis of design (BOD), etc.) are consistent with each other, include the commissioning requirements, and meet the OPR.

The primary objective of commissioning during design is to facilitate the design documents meeting the OPR.

The BOD is the documentation of the primary thought process and assumptions behind design decisions that were made to meet the OPR. The scope of the BOD varies with the project. Sometimes the BOD includes most building systems and assemblies. In other projects, it only includes what is in the OPR, which may include only those systems and assemblies that are commissioned.

This overall objective is supported by the following goals:

1. Communicating the commissioning requirements to other project team members.
2. Building engagement and cooperation among the project team members.
3. Verifying, through review, that the design documentation is consistent with the OPR and BOD.
4. Ensuring that commissioning requirements are included in the construction documents.

These goals can best be met through the systematic and disciplined process of commissioning. The commissioning team during predesign consists of the party who will bring on board the CxP and all the available stakeholders and contributors to the OPR:

- Owner/owner's representative
- Building occupant/tenant representatives (when available)
- CxP
- Design team personnel
- Operations and maintenance (O&M) personnel (when available)
- Construction team personnel (when available)

16.2.4.3 Construction phase

Commissioning during construction is where the bulk of the commissioning effort is applied. What is more, the effort has a high likelihood of being successful with the process established during design. Construction phase commissioning is also a well-coordinated quality assurance process that includes installation, start-up, functional testing, documentation, and training. During the construction phase, the commissioning team works to ensure that equipment, systems, and assemblies are property installed, integrated, and operating in a manner that complies with the OPR and the contract documents.

Commissioning during construction includes several objectives, with the primary desired outcome to simply have all systems and assemblies working properly at turnover. Supporting this outcome are the following objectives:

- Submitting specified systems.
- Installing submitted systems and assemblies properly.
- Systems and assemblies functioning and performing properly at turnover.
- Properly training O&M personnel.
- Leaving needed O&M documentation with staff.

During construction, the number of parties involved in the commissioning process increases significantly. In addition to the previous players, the construction team is added. Specifically listed, the team now includes:

- Owner's representative
- Construction management representative (when available)
- CxP and any sub-consultants
- Design team (architect and engineering design consultants of commissioned systems or assemblies)
- Contractors and subcontractors of commissioned systems and assemblies (including vendors and suppliers of larger equipment)
- Building occupant or user group representatives
- Personnel responsible for the building's operation and maintenance (O&M)

On small projects, the key active players might be limited to the owner, general contractor, mechanical, controls, and balancing subcontractors. On larger projects, the architect and mechanical and electrical designers would be added along with the tenant representative and O&M representative(s) and vendors of large equipment and other specialty designers and subcontractors, such as for building enclosures, process mechanical, fire alarm, and others.

16.2.4.4 Occupancy phase

The occupancy and operation phase normally begins at substantial completion, when the building is turned over to the owner, though some construction phase activities may still be in process. In the occupancy and operations phase, all uncompleted activities from the construction phase are finished (project closeout activities) as well as the long-term process for ensuring building performance over time being developed and put into place (ongoing maintenance and performance activities). The ongoing commissioning activities ideally start during this period (covered under ongoing commissioning

(OCx)). During this period, the operations staff take over the operation of all systems from the contractor and begin their maintenance program.

Objectives of the occupancy phase, independent of OCx, include:

- Facilitate the continued engagement of the Cx team and verify the completion of outstanding Cx issues.
- Complete any seasonal/deferred functional testing, O&M staff training, and occupant orientation.
- Complete systems and commissioning documentation.
- Ensure an adequate maintenance program is in place.
- Optimize system performance.
- Survey occupants, formally check in with operations staff, and assess issues.

During occupancy, the primary members of the commissioning team continue from the construction phase, including:

- Owner/owner's representative
- O&M personnel
- CxP
- Contractors
- Design team

The owner, O&M personnel, and the CxP are the primary players during this phase. The contractor is also still involved, and the design team may be involved when design issues arise.

16.3 Existing Building Commissioning (EBCx)

16.3.1 Why do we need existing building commissioning?

The scope of EBCx includes existing buildings, systems, and assemblies that may or may not have been previously commissioned. The scope may be limited to specific systems and assemblies, like energy using equipment, or broadened to include other systems such as those that affect comfort such as temperature control or air quality impacting systems, life safety, and other mission critical or redundancy equipment in healthcare, industrial process and similar facilities, building envelope elements such as roofs, windows and doors, curtain walls and other building envelope assemblies, or even aesthetics such as landscaping and irrigation.

The purpose of EBCx is most often to address facility conditions such as:

- A change in building use, ownership, or functional requirements.

- The building uses more energy than expected.
- The building has problems that impact the occupants or functional program.
- The systems and assemblies are in poor condition, or don't perform properly.
- An addition or alteration to the building.

EBCx addresses these conditions by following a specifically designed, complete, and methodical process. Key process elements include the following, which have a positive impact on building asset management and energy efficiency:

- Comparison of the facility's current functional requirements with the original basis of design.
- Comparison of field verified current performance with the original basis of design and opportunities to optimize performance.
- Implementing measures to address performance deficiencies and take advantage of performance optimization opportunities.
- Identify and leverage new systems, equipment, and technology that may not have existed when the building was constructed to enhance and optimize performance.

The EBCx objectives are to make the building systems perform interactively to meet the current facility requirements (CFRs) and owner project objectives and provide the tools to support ongoing commissioning (OCx), performance tracking, and improvement of system performance over time.

As a best practice, it is advisable for most owners to hire an independent third party to provide commissioning services and act directly as the owner's advocate.

16.3.2 What does EBCx cost?

The cost of the existing building commissioning may begin with an estimate based on anticipated project cost. However, due to diverse reasons for – and forensic issues related to – commissioning an existing building, costs cannot be as clearly predicted as for new construction commissioning. Cost is dependent on scope, such as:

- The scope is limited to HVAC and lighting or is expanded to include other systems and assemblies.
- The scope includes only low-cost or no-cost measures.

Table 16.2 EBCx key cost/benefits metric (all data) (Crowe, 2020).

Cost/benefit metric	EBCx		
	Median	25th–75th Percentile range	Sample size (projects)
Energy savings cost	6.4%	3.4%–12.4%	446
per m^2 (cost per ft^2)	\$2.84 (\$0.26)	\$1.60–\$6.03 (\$0.15–\$0.56)	985
Simple payback	1.7 years	0.8–3.5 years	656

- The scope includes the costliest components, equipment replacements, and/or rebalancing.

Additionally, the rigor of energy savings estimates, the extent of monitoring, benchmarking, tracking, measurement and verification (M&V), and energy management information system (EMIS) software associated with the project can dramatically impact commissioning costs.

16.3.3 Costs and risk factors in EBCx

All risks have associated costs. The success of the EBCx process will be determined by how well one can anticipate and manage the risks and costs associated with the facility.

Because they are live, functioning places, existing buildings come with challenges for conducting commissioning activities. Risk cannot be avoided, but it can be mitigated. And, by definition, all risks have associated costs. The potential liability placed on the commissioning team is not insignificant. The success of the EBCx process will be determined by how well one can anticipate and manage the risks and costs associated with the facility.

New construction begins with an essentially blank slate, sequenced activities, and a relatively reliable schedule from planning through delivery. EBCx scheduling, on the other hand, is dependent on transitional activities involving existing problem identification, the facility's current versus planned function, and working around occupants' space, business needs, safety, and security – not to mention unexpected findings along the way.

Adapting the EBCx activities to avoid or at least minimize interruption of current operations can cause delays or unexpected schedule changes that should be considered during the planning process. The following are some typical EBCx risk management scenarios.

- Equipment accessibility: Questions for assessing (and managing) impacts on occupants:
    - Will there be an impact on occupants' productivity?

- There is a risk of inadvertent injury if a ceiling tile falls?
- Is it reasonable to cover office equipment and work surfaces?
- How much will interrupt operation of these systems negatively affect normal occupancy?
- Will indoor air quality or life safety issues occur that affect occupants?
- Will it be necessary to start and stop systems or run them in various operational modes for periods of time?

- Security risks: Questions for accessing (and managing) impacts on building security:
  - Is there access control or surveillance equipment in place to record areas where the CxP will have access during site work or testing?
  - Is it possible to make the access control provisions for commissioning staff the same as for building staff?
  - If the facility is a high security risk, can security personnel escort CxP's around the building during commissioning activities?
  - Can investigation or implementation work be phased out or staged to reduce impact?
  - Environmental hazards: Are there asbestos, mold, lead, or other hazards?

16.3.4 Who is involved in EBCx?

The decision to commission an existing building typically begins with an internal decision by the owner, facility manager, and other stakeholders affected by the potential alterations and/or replacements, such as occupants and building operations staff. All these parties are part of the EBCx team. Then, as part of the planning phase, or at least prior to the initial assessment phase, the CxP is assigned from current staff, or a third-party consultant is brought on board as the CxP. The CxP leads the EBCx process. Later, during the investigation phase, vendors involved in implementing measures will be part of the team.

16.3.5 EBCx process

This section provides details of the fundamental EBCx process and is divided into five primary phases. Some guidelines merge the planning and initial assessment phases, but we separate the planning phase and represents the task the owner typically does prior to hiring the CxP.

16.3.5.1 Planning Phase

The objectives of the planning phase are to select the building, document the owner's operational needs for the building, and engage a CxP.

This phase consists of the up-front work by the owner in preparation for bringing on a CxP and the actual investigative work of the EBCx process. The tasks of the planning phase in general chronological order include:

- Screen the building for EBCx viability
- Perform deferred maintenance
- Define the owner's EBCx goals and objectives
- Define the roles and responsibilities
- Define the CFR
- Define scope of work, schedule, RFP/RFQ, and engage a CxP

16.3.6 Initial assessment phase

Following the planning phase, the initial assessment phase begins. The objectives of this phase are to expand the planning process to gain a preliminary understanding of how and why the building systems and equipment are currently operated and maintained, how much potential there might be for improvement, and where to focus the investigation phase efforts, culminating in an initial EBCx plan.

To begin this phase, the building should be screened for viability and the CFR developed or enhanced, if not done during planning. The assessment entails reviewing building documentation, interviewing the facility operating and managing staff, and making initial site inspections of all primary equipment and assemblies in the EBCx scope. If available, the CxP conducts a preliminary review of control sequences, set-points, control parameters, and schedules. Actual equipment testing, measurements, trending, and diagnostics are done during the investigation phase, not in the initial assessment phase. The CxP identifies problems and improvements solely by acquiring information readily available, such as interviewing facility staff and conducting a preliminary review of the control systems and other written documentation. It is important that the initial assessment considers equipment life expectancy, current condition, control system capability, and physical limitations of the facility to provide a pragmatic expectation and focus for the project.

The tasks of the initial assessment phase in order include:

- Screen building for viability (if not already done)

- Define the CFR (if not already done)
- Review building documentation
- Develop an initial EBCx plan and M&V requirements (if applicable)
- Analyze energy data (monthly billings or 15-minute interval data)
- Conduct a kick-off meeting
- Perform an initial site walk-through and BAS review
- Set up baseline monitoring (when needed)
- Perform initial facility staff interviews (manager and operators)
- Perform building occupant interviews
- Document findings, including:
  - Master list of findings
  - Potential findings list
  - Tracking checklist
- Meet with the owner to confirm focus for investigation phase
- Determine means and methods for diagnostic monitoring, trending, and functional testing
- Update EBCx plan and scope

16.3.7 Investigation phase

The objectives of the investigation phase are to evaluate systems and assemblies in detail and either confirm proper operation or identify and document measures for improvement. Proper operation is based on the owner's current operational needs and requirements defined by the CFR and good practice.

The investigation phase consists of a detailed examination comparing actual building conditions and system performance with the CFR, such as only ensuring that the facility is as energy efficient as possible. On a technical level, the investigation phase reveals existing conditions in the building, evaluates the need for change, and identifies potential improvements that will align the facility better with the CFR. The findings are recorded in the master list of findings or facility improvement measures (FIMs) and the costs and benefits of recommendations are estimated and recommendations prioritized. The owner evaluates the FIMs and decides which will be implemented and when. Some issues discovered during the investigation phase can be corrected with minimal time and effort, do not require additional evaluation, and may be corrected immediately at low or no cost.

The investigation phase includes the following tasks given in order:

- Conduct a kick-off meeting
- Investigation phase planning and schedule update meeting

- Periodic Cx issues meetings and coordination
- Check calibrations of BAS trended points
- Set up trends and monitoring equipment and retrieve and analyze data
- Conduct additional detailed documentation review
- Conduct detailed site survey, including condition assessment
- Finalize building personnel and occupant interviews
- Conduct and evaluate functional testing
- Gather sufficient data for savings calculations and detailed implementation recommendations
- Perform energy savings and cost calculations
- Set up energy performance monitoring for baseline and ongoing tracking
- Update and prioritize master list of findings and recommendations
- Review with owner the individual findings and the evaluation of whether the CFR objectives are being met
- Finalize M&V plans (when applicable)
- Perform pre-implementation measurements and benchmarking (when applicable)
- Complete scope of work for implementation and handoff phases

16.3.8 Implementation phase

The primary objectives of the implementation phase are to implement FIM recommendations selected from the master list of findings and to verify that the changes and additions to systems and assemblies were installed, set up, and operating correctly.

Implementation activities are defined in the implementation plan created at the end of the investigation phase and added to the EBCx plan. Most EEMs and FIMs will be made immediately upon the start of this phase, but others may be planned for later (months or even years) to accommodate seasonal issues and budget cycles or because they need to be conducted after another EEM and FIM or event has occurred, such as equipment replacement.

In the implementation phase, the CxP may play a secondary role, because the facility staff contract for the measure installation or perform it themselves. In such projects, CxP is primarily waiting for the completion of installations to return to verify the proper performance. However, CxP can provide a value-added service by verifying the progress and that there are not any questions about the recommendations. Proactively engaging the CxP can add both perspective and engineering expertise to mitigate weakening measures based on lack of information.

The tasks of the implementation phase are provided below in order and include these steps:

- Complete additional needed investigation and engineering
- Assist in obtaining quotes from vendors (answer questions by vendors and facility staff)
- Assist in implementing EEMs and FIMs (answer questions by vendors and facility staff)
- Verify proper performance of EEMs and FIMs
- Adjust energy savings estimates and measure costs
- Set M&V monitoring and metering (when in scope)
- Prepare summary report of implementation phase

16.3.9 Handoff phase

The primary objective of the handoff phase is to ensure that the facility staff and occupants have the understanding and documentation necessary to properly operate and maintain the changes and improvements made during the EBCx process, so that the performance improvements persist over time.

The handoff phase involves the CxP engaging more closely with operations staff in a way following the technical efforts of investigation and implementation. The CxP's "people skills" are important for teaching facility staff and encouraging and motivating them to take the new features and control routines seriously. Most EEMs and FIMs can be easily changed, overridden, or disabled by operations staff. Not infrequently, this occurs when conditions change, and technical decisions need to be made by staff to maintain comfort and mitigate occupant complaints or deal with equipment that does not seem to be working properly with the new EEMs and FIMs installed. If the staff do not understand the "why's" of the new measures or control routines and how they work, they are more apt to disable them to manage other, more pressing problems. The CxP's role is to assist the facility staff in gaining understanding so that it does not happen.

At the end of the investigation phase, the CxP would have developed a scope of work for the handoff phase and added it to the EBCx plan. The primary tasks of handoff are to:

- Train and orient facility staff on the implemented measures
- Update or rewrite the sequences of operation
- Update existing or compile a new systems manual
- Update preventative maintenance procedures

- Implement performance tracking
- Develop ongoing commissioning recommendations
- Write a report of the handoff phase

There are some optional activities that can provide value to the project but are not always absolutely needed. These include steps to:

- Create digital (computer-aided design) online schematic drawings of major systems
- Develop an occupant user's guide
- Develop an OCx plan and assist in implementation
- Conduct M&V activities
- Hold a lesson learned meeting

16.3.10 Why do EBCx?

The EBCx process is implemented to ensure that the current systems are operating and performing to the CFR and is focused on improving energy efficiency to reduce operating costs. The EBCx process for each facility is based on the CFR, which details the owner's needs and goals for the facility. The decision to commission existing buildings may be based on one or many factors:

- The desire to increase energy savings and address operating, performance, and comfort concerns
- Fix existing systems, replace, or update aging equipment
- Take advantage of utility incentive programs
- Upgrade asset value
- Respond to institutional policies or requirements for governing bodies
- Track performance over time

The benefits of EBCx can be significant in terms of short- and long-term return on investment, energy savings, and building asset value. Many existing buildings have never undergone a commissioning process, even during new construction. Even when buildings have been commissioned, facility requirements often change over time. Operational efficiencies of facility systems tend to degrade. Without updating documentation of performance and comparing to expectations, buildings may perform well below their potential, use more energy than necessary, and cost more to operate than warranted. EBCx responds to the owner's desire to maintain or improve building performance, solve comfort and operational problems, and reduce operating costs. It is important for the owner to understand that custom tailoring of the EBCx type, approach, or scope will be necessary to most efficiently achieve each mix of specific objectives selected and that a building's appropriateness and the EBCx cost, and duration will vary depending on the mix. Much more than with new construction commissioning, the project goals and objectives must be matched with the most appropriate type and approach of commissioning applied to effectively meet the goals.

The specific objectives for EBCx vary from project to project, but some common objectives include the owner's desire to:

- Verify that a facility and its systems meet defined CFR
- Reduce operational costs by saving energy and reducing demand
- Identify and resolve building system operation, control, and maintenance problems
- Reduce or eliminate occupant complaints and increase tenant satisfaction
- Improve indoor environmental comfort and quality and reduce associated liability
- Document systems operation
- Identify and recommend enhancements to maintenance programs
- Identify O&M personnel training needs and provide training
- Minimize operations risks in process or mission-critical facilities
- Increase asset value for improved return on investment (ROI) or sale
- Extend equipment life
- Ensure ongoing performance of previous improvements
- Assist in achieving upgrades or improvements required to meet a non-residential rating system such as LEED for existing buildings, Green Globes, ENERGY STAR, or regulatory measures such as the California Energy Commission's Existing Building standards (Title 24)
- Participate in utility programs for energy savings and financial incentives.

16.4 Monitoring-Based Commissioning (MBCx)

16.4.1 Why do we need monitoring-based commissioning?

MBCx is characterized by using facility data as the primary indicator of building operation. The key elements of MBCx are: (1) collection of building data, typically for energy-metering and energy-consuming systems; (2) data analysis through automated fault detection and diagnostics (FDD) with follow-up review; and (3) energy performance tracking.

MBCx leverages the use of an EMIS as the main tool to facilitate the collection, aggregation, and analysis of data, as shown in Figure 16.3. EMIS is a broad term applied to a rapidly evolving family of tools that monitor, analyze, and control building energy use and system performance. Most frequently, an EMIS is stand-alone software, but it may be an extension to a BAS. Reference to EMIS in this section is the functionality that they provide and not necessarily to specific software.

A distinction can be made between energy meter-level EMIS, which uses metering data only, and system-level EMIS, which involves the granularity of energy monitoring and functional monitoring of individual systems.

Steps in the MBCx process leverage functions commonly available in both meter-level EMIS and system-level EMIS, including FDD and energy performance tracking.

Examples of meter-level EMIS include:

- Benchmarking and monthly utility bill analysis
- Monitoring, targeting, and reporting (MT&R, see Chapter 1)
- Advanced EMIS

Examples of system-level EMIS include:

- BAS
- FDD
- Automated system optimization

The implementation of MBCx can include installation of stand-alone software, module(s) added to a facility's BAS, or using native functions of the BAS. Using continuous monitoring, collection, and automated analysis of data, aspects of MBCx can approach true real-time commissioning of the facility. While the technology-enabled methods and techniques can support significant elements of the OCx process, the software mainly determines the performance of a commissioned building.

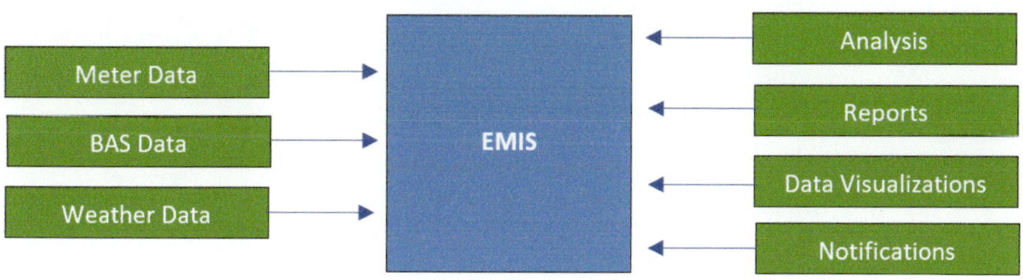

Figure 16.3 Typical function of an EMIS.

Figure 16.4 illustrates the uses of EMIS for an MBCx program. MBCx is used to maintain and improve building performance over time by identifying anomalies in the data that could signify a problem, providing a platform for follow-on investigation, verifying corrective actions, and tracking performance against metrics set forth in the CFR. The scope of data requirements, analysis, and metrics are unique to each building.

What makes MBCx unique is the technology that enables constant monitoring of a facility. Along with a range of other functions, a BAS will perform FDD to highlight systems and equipment that are not performing optimally.

These tools allow energy managers to record energy consumption and identify root causes over time (such as ambient conditions or other common

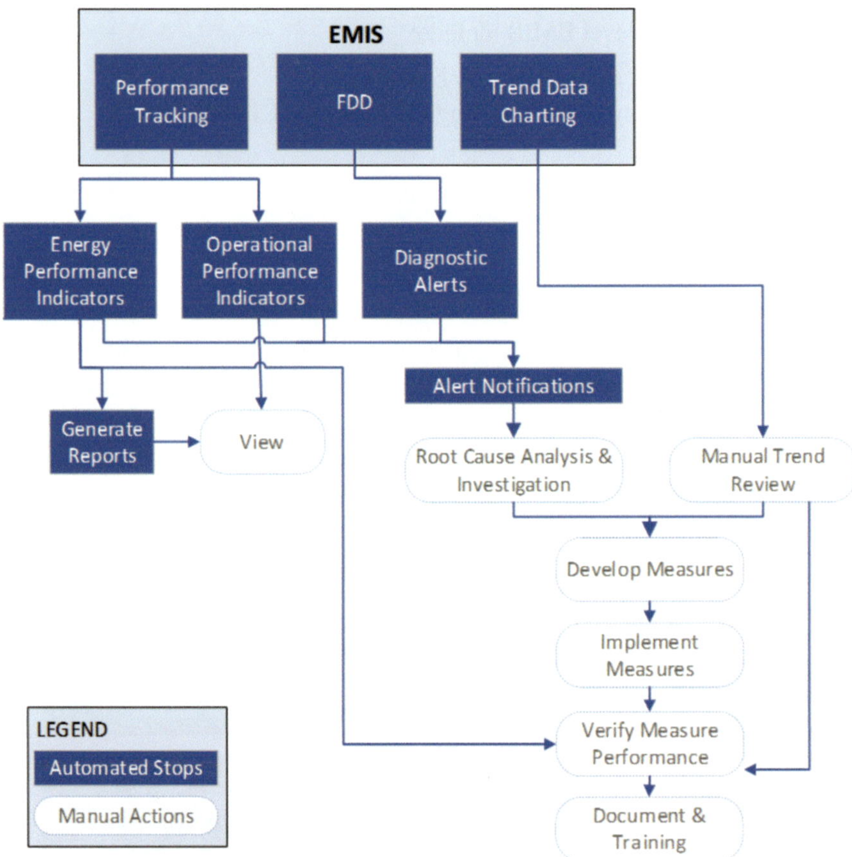

Figure 16.4 Use of EMIS functions in an MBCx program (Stum, Bjornskov, 2017).

causes in commercial buildings). They also enable users to create a predictive model relating building energy consumption and use the model to:

- Identify periods of best performance
- Quantify savings
- Set targets for reduced consumption
- Compare actual consumption to best practice
- Determine when energy consumption is higher than expected
- Initiate actions to correct variances

Due to the highly involved process of using a BAS, vendors will often offer training on creating facility-based benchmarks and managing the alerts that the system provides.

16.4.2 MBCx costs and cost savings

MBCx, also known as continuous optimization, persistent commissioning, or ongoing commissioning, is like EBCx in that it is a process applied to existing buildings to keep them performing at optimal levels. But where EBCx requires personnel to periodically tour the facility to identify potential measures, MBCx uses sensors and software in addition to standard EBCx practices to provide a real-time account of systems within the building. This method results in more persistent energy savings than conventional commissioning. The potential for cost savings is illustrated in Figure 16.5.

EMIS tools are used in the MBCx process to organize, present, visualize, and analyze the data. With campaign data from over 400 million ft^2 of installed space, the results achieved by owners which implemented EMIS, along with associated technology costs. The study's EMIS users that reported savings achieved the median cost savings of \$0.19/ft^2 and 7% annually, with savings shown to increase over time. For 35 portfolio owners, the median base cost to install an EMIS was \$0.03/ft^2, with an annual recurring software cost of \$0.02/ft^2 and an estimated annual labor cost of \$0.03/ft^2. There are two types of EMIS systems: energy information systems and FDD systems. It was found that FDD systems have both higher savings and higher costs (Kramer, 2020).

Twenty-seven participants submitted energy data for all or a subset of their buildings (in total, 687 buildings with 94 million ft^2). The number of buildings reported by each participant ranged from 1 to 335. Two participants reported savings results determined from interval data analysis tools. Two participants estimated savings using engineering calculations. The energy savings from the other 23 participants were calculated using monthly bill analysis (Kramer, 2020).

Figure 16.6 shows the energy and cost savings results for each participant since the installation of their EMIS. The participant energy savings ranged from −6% to 28%, and the median was 7%. The median cost savings was $0.19/ft^2. In total, these 27 participants are saving 790 billion Btu/year and $18 million/year, comparing the most recent year for which data are available to the baseline year before the EMIS installation.

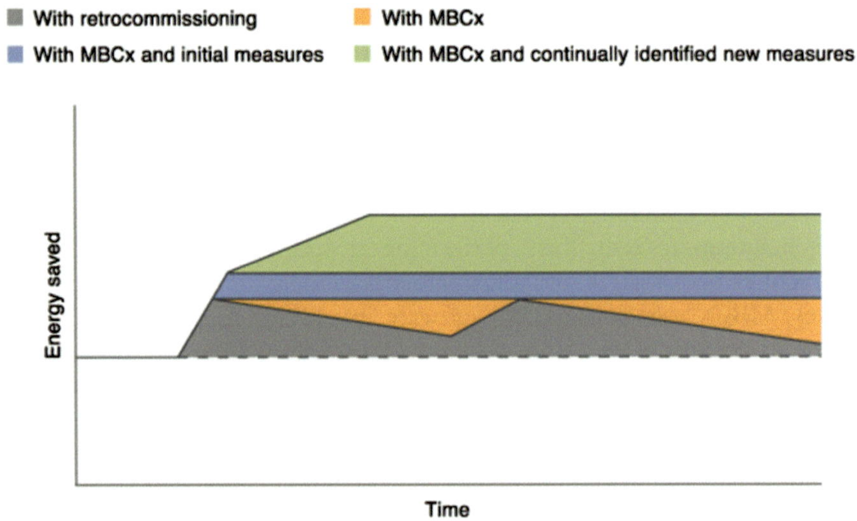

Figure 16.5 MBCx can save more energy than retro-commissioning (Mills and Mathew, 2014).

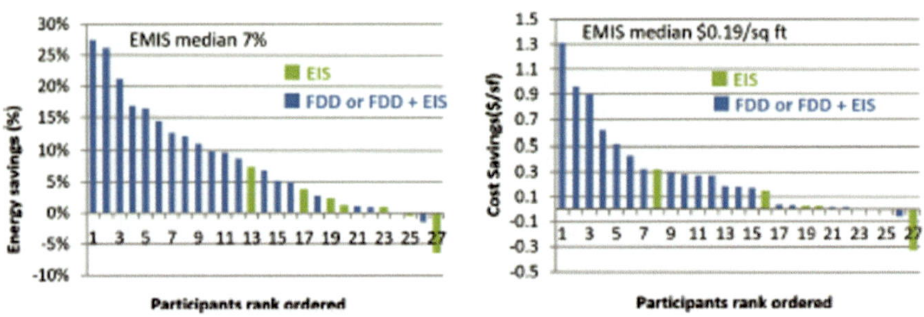

Figure 16.6 Participant energy savings (left) and cost savings (right) for participants since EMIS installation (Kramer, 2020).

16.4.3 MBCx phases

There are four phases to MBCx programs:

1. Data collection
2. FDD
3. Optimization
4. Energy performance tracking

Data collection: Monitored data typically includes data from the BAS, energy metering systems, and weather data. It may also include other data sources relevant to tracking the performance of the facility, such as production data for a manufacturing facility or access control data for a visitors' center. How well the data represents building systems, and the resolution and quality of that data will ultimately determine the efficacy of the MBCx to impact building performance.

FDD: FDD is a data-based method that uses algorithms to detect equipment-level or system-level faults and diagnoses their causes. It drastically reduces the time required for analysis by standard methods by applying rule-based algorithms to the data such that the analysis is automated and can be applied to very large datasets on a real-time basis. Most applications will use a combination of fault detection and FDD. Fault detection alone would notify a building operator of an issue, but it would require that operator to investigate the data further to determine possible causes. To be categorized as FDD, potential causes of the fault must be identified in the tool. The platform may also prioritize faults based on cost or criticality. FDD is an advanced technique that can be applied as part of MBCx, with the unique elements of performing detailed analysis of component, equipment, and system operating data.

FDD can be performed by stand-alone software with standard or custom rule sets, and some BAS systems contain customizable FDD modules native to their system. Alternatively, some owners have achieved success by programming FDD into the existing BAS, but the process is often limited by the available processing resources and network bandwidth.

Optimization in the context of MBCx refers to the steps taken to improve building performance over time beyond the baseline of how the building was originally intended to operate. Optimization activities go beyond correcting situations in which systems are not operating as they should; they take advantage of opportunities to save energy, reduce maintenance costs, extend the lifecycle of building equipment, or improve operational performance. For example, if a static pressure set-point was inadvertently overridden to the maximum of its reset range, releasing this override, and confirming a proper reset based on variable air volume damper positions would be maintaining operation over time. However, the one step further of fine-tuning the limits and behavior of the control loops for the static pressure would be improving operation or optimization.

Optimization is included as a core activity in MBCx because the abundance of data can provide insights for opportunities for optimization. The additional savings from optimization also helps justify the program for a more favorable business case.

Measures for optimization could include simple adjustments of control set-points to modified sequences of operation. In any case, improving upon the design basis almost always requires engineering review to ensure any modifications are appropriate for the equipment and will maintain compliance with relevant codes and guidelines.

Energy performance tracking. Tracking of energy performance indicators allows the CxP to identify when systems deviate from a known baseline to prompt corrective action. Baseline values for each criterion are set when the building is known to be working as intended, typically following a Cx or EBCx process. Energy performance criteria can be system- or load-type specific or could also be set for whole-building energy use. For example, chiller plant kW per ton is a system-level criterion, while hole-building electric demand kW is a meter-level criterion. Baseline criteria can also be normalized for external load factors, such as weather and occupancy, such that reasonable comparison can be made to the adjusted baseline value.

Energy performance tracking can also encompass energy management activities, such as tracking whole-building energy and cost metrics for individual or groups of buildings.

In contrast to FDD, notification when an energy performance indicator has deviated from an expected baseline does not directly provide any feedback as to the potential causes in sufficient detail to prompt maintenance action. However, the notifications can be essential to identify anomalies for further investigation that may otherwise go undetected if only reviewing system operation. Facility managers can interact with their energy data and analysis with the use of EMIS. Functions of EMIS often include visualizations of data, an energy dashboard, and an interface to perform common energy analysis or comparisons.

16.5 Ongoing Commissioning (OCx)

16.5.1 Why do we need OCx?

OCx is a process that is intended to maintain and continuously improve buildings performance over time. The goal of OCx is to achieve the following benefits:

- Maintain the CFR throughout the life of the facility.
- Ensure that systems are operating in accordance with the current design intent and associated energy performance.
- Continuously improve building performance by implementing monitoring and optimization.
- Reduce ongoing operational costs.
- Identify system deficiencies prior to impact on building occupants.
- Improve system reliability.

The focus of OCx is to maintain performance over time. Any buildings with significant or widespread operational problems, equipment deficiencies, or high energy use likely require the application of EBCx prior to transitioning to OCx. In such cases, a hybrid approach may be employed where the tools and process for OCx are implemented while an EBCx process is applied to identify and address corrections requiring capital investment.

For a building that was recently commissioned, either through a new construction or existing building process, a transfer to OCx ensures the persistence of performance after the project is complete and turned over to the owner. While NCCx or EBCx is set up in phases and applied within a project with a clear end. OCx is a continuous process to maintain and even improve performance over time. This concept is shown in Figure 16.7. Degradation in performance can be reversed through the EBCx process on a periodic basis – typically, no more frequent than every five years. OCx uses similar steps to NCCx and EBCx, but OCx is continuously applied to issues and systems to avoid degradation occurring at all.

The primary goal of the OCx process is verifying and documenting that a building operates according to its functional requirements, meaning the OPR or CFR can be maintained, resulting in the indoor building environment being maintained within the required parameters and the systems operating efficiently and effectively.

Even though previous NCCx or EBCx activities included requirements for documentation and training for facility staff to aid in the successful

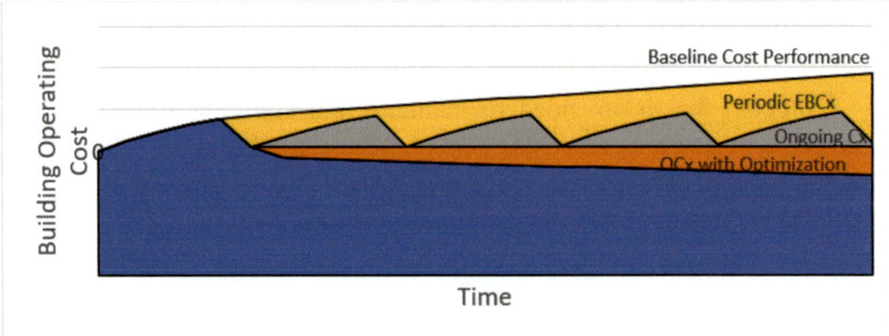

Figure 16.7 Building operating cost over time – a comparison of EBCx and OCx.

turnover to facility operations, building performance will still degrade over time. However, even with the best training and available documentation, it is all too common for building performance to degrade over time. The most common causes of performance degradation include:

- Renovations and the building HVAC system.
- Repurposing of space.
- Failure of mechanical or control components.
- Inadequate preventative maintenance.
- Control sensor accuracy drift.
- Operator error.
- Operator overrides.
- Oversizing.

16.5.2 Justifying investment in an OCx program/building a business case

Before deciding to implement an OCx program, it is important to set expectations for the overall program to obtain buy-in to start the program and for its continued support from decision-makers. Performing a cost–benefit analysis in the planning stage is essential to evaluate the total costs, estimate the benefits, and provide metrics for the expected costs and benefits.

The costs of an OCx program should include first costs associated with implementing an OCx program and costs to support the program over time. From the owner's perspective, it is important to consider internal and external costs, including budget allowance for implementing measures.

16.5.3 Costs of an ongoing commissioning program

OCx costs include the following main categories:

- Program initialization (including EBCx if not recently Cx'd)
- EMIS implementation: licensing, setup, and maintenance (MBCx)
- EMIS third-party monitoring and reporting subscription services (MBCx)
- Staff training
- OCx activities by CxP and owner staff
- Measure implementation (labor/materials)

First costs associated with initialization of an OCx program would include the cost of the owner OCx champion and the cost of an in-house OCx specialist or consultant as third-party CxP, as well as any needed initial EBCx to bring the building up to a level of good operation prior to implementing OCx.

- The OCx process resulted in energy savings across a range of 5%–20%, with a weighted average of 8%.
- Simple payback of OCx projects were like those reported for EBCx projects, in the one- to five-year range, with a median of 1.5 years (ASTM, 2012).

16.5.4 Commissioning team

Forming an OCx team is challenging as it requires additional skills and expertise than Cx or EBCx and must integrate to the maintenance and operation of the facility. Unlike new construction projects in which all players are in well-understood and predefined roles, the structure and procedures of a maintenance organization vary widely. OCx must successfully integrate into facility operations to understand and maintain conditions, to complete maintenance and repair items identified in OCx, and to coordinate the implementation of measures recommended in the OCx process, whether by maintenance staff or third-party service providers.

A building owner or facility manager wishing to implement OCx will at some point need to consider whether the OCx program can be supported by in-house staff or will use a third-party consultant. In making these decisions, some key factors to consider include:

- Is the building stock relatively centralized, such as a traditional college campus, or is it spread across a large region, such as a community

college system, large research university, or a commercial property owner or retail business chain?
- Are there capable staff currently available or would the staff need to be hired?
- How will the issues and problems that get identified by the OCx program be addressed?
- Will repairs and corrections be made by in-house maintenance personnel or contractors, or possibly both?

Model #1: Owner as a CxP. An employee of the owner, typically housed in a facility management, maintenance, or energy management department, serves as the CxP and is the primary user of the EMIS and leads the testing, analysis, and development of measures. This individual is typically a licensed engineer or an energy manager, who understands system operation, has a minimum working knowledge of BAS, and can apply engineering judgment to determine appropriate corrective actions.

Model #2: Consultant as a CxP. An independent consultant is the CxP and leads the OCx process. In this scenario, there still must be a designated owner's representative who champions OCx and will facilitate and coordinate activities across the stakeholders.

16.5.5 Phases of OCx

OCx comprises the sequential phases as shown in Figure 16.8.

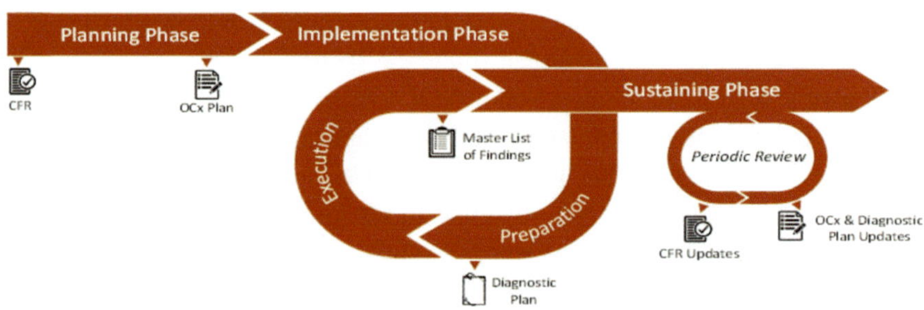

Figure 16.8 Sequential phases of OCx programs (BCxA Ongoing Cx Best Practice, 2020).

16.5.5.1 Planning phase

When developing a plan for OCx (ideally within the NCCx or EBCx process unless OCx is developed as a stand-alone process), the objective of the planning phase is to confirm and clearly document the owner's project goals and objectives; evaluate the applicability of OCx to the project; scope, design, and plan for the applicable technology tool(s) to be implemented to support the project; and document the plan for OCx.

The planning phase includes these activities:

- Engaging a CxP
- Defining and documenting owner's OCx goals and objectives
- Conducting a preliminary cost–benefit analysis
- Identifying and forming a multi-disciplinary team
- Performing condition assessment, documenting the preliminary master list of findings
- Reviewing the CFR
- Conducting a gap analysis for existing technology infrastructure
- Determining technology use and design
- Developing and formalizing OCx plan including
    - Project management plan
    - Change management plan

16.5.5.2 Screen candidate buildings and systems

While all buildings may experience some benefit from OCx, the owner is encouraged to apply a rigorous screening and selection process to identify the best candidates for OCx. The best candidates will maximize the value of the OCx program by ensuring the potential benefits justify the investment.

In general, buildings can be poor candidates for OCx if the systems cannot accept changes and modifications to improve performance. The feedback of information and analysis in OCx must result in actions to fix or improve the building or system.

Alternatively, if the buildings have widespread operational issues, excessively high energy use, or recent changes in occupancy, the building likely requires a complete EBCx process prior to performing OCx. In other words, the facility should be brought back to its original or current functional performance requirements prior to establishing an OCx program.

While most buildings can benefit from the commissioning process, limited resources typically require that the most advantageous buildings be addressed first. The typical assessment process consists of:

- Building physical condition
- Mechanical system preliminary evaluation
- Mechanical system condition

- Building control systems
- Building energy usage
- Assessing opportunity

16.5.5.3 Screening criteria for OCx

The screening criteria for OCx are provided in Table 16.3.

Table 16.3 Screening criteria to determine suitability for an OCx program.

Screening criteria – motivation for Ocx
Recent Cx, EBCx, or Energy Project
Energy intensive
Critical facility
Screening criteria – opportunity
Energy intensive use
Use of energy savings control strategies: systems with VFD, occupancy controls, setbacks, resets
Screening criteria – things to avoid
Widespread failures
Planned capital upgrades/renovations
Deferred maintenance
Barriers and/or excessive labor required to test
Many service calls, complaints
Poor indoor air quality, environmental control
Analog controls
Legacy, unsupported BAS with limited integration capabilities
Limitations on system changes (physical or organizational/legal)

16.5.5.4 Implementation phase

The objective of the implementation phase is to develop a diagnostic plan, apply the tasks described therein, and identify specific EEMs and FIMs. The implementation phase has been divided into two sequenced sub-phases – preparation, followed by execution. During the implementation phase, preparation and execution sub-phases may each take several iterations until the desired outcome is achieved (prior to moving on to the sustaining phase).

During both sub-phases, the CxP continues to engage the OCx team, especially personnel involved in O&M. The findings, recommendations, and subsequent analysis are presented to the owner for further prioritization and selection of measures for implementation. In the execution sub-phase, the CxP applies the documented metrics and diagnostic tasks and compiles the master list of findings.

The implementation phase includes activities as listed below.

Preparation:

- Engage OCx team selected during planning
- Create a diagnostic plan
- Align data monitoring with diagnostic plan
- Implement diagnostic plan
- Create or update facility guide

Execution:

- Coordinate preventive maintenance
- Conduct technical/technology application training
- Update and prioritize master list of findings
- Establish communication/coordination methods and procedures
- Take corrective actions
- Update OCx plan

16.5.5.5 Sustaining phase

The primary goal of a successful OCx project is the ongoing, sustained building performance for the long term, and ideally for the life of the building. As a process complementary to NCCx and EBCx, OCx provides the building owner with assurance that the investment in building systems and the optimization of those systems continues to operate in an optimal manner. As such, the sustaining phase is the critical means of moving from implementation of the OCx approach to a business-as-usual means of managing building performance.

In the sustaining phase, systems and improvements made are maintained over time, and the processes established for OCx are scaled down within the Cx team and operations staff to those required to continue optimal building operation. Additionally, as with any dynamic system, maintenance procedures must be established to support the ongoing value and validity of data, analysis, and fault detection methods identified in the diagnostic plan.

An OCx project advances to the sustaining phase when the findings arising during implementation have been acted upon, and the rate of identifying new findings has been reduced significantly, therefore meeting the owner's goals and objectives. The building has then reached a reasonable level of stability and performance in accordance with the CFR. In addition, the diagnostic plan and corrective actions are now fully integrated into the owner's workflow where OCx actions are business-as-usual for the O&M team. Technology infrastructure is fully implemented, with the technology maintenance staff making system adjustments as applicable to maintain the technology system integrity, sensor calibrations, and updates as necessary with CFR and facility changes.

The sustaining phase includes these activities:

- Implement operating and maintenance procedures
- Transition to stable operations
- Monitor and report results

16.6 Building Envelope Commissioning (BECx)

16.6.1 Why do we need BECx?

Building envelope commissioning (BECx), also known as building enclosure commissioning, is a systematic, technically driven process to document the performance of the building enclosure as outlined in the OPR. The specifics and execution of BECx vary widely from project to project, as there are differences in owner missions, project types, building enclosure assemblies, building uses, climates, occupant desire, architect visions, financial constraints, delivery methods, and so on. The process may be as simple as completing a series of predefined questionnaires and checklists or as complex as rigorous quality assurance process and performance verification for mission-critical facilities with high levels of contamination control with multiple redundancies.

The American Society for Testing and Material (ASTM) defines BECx as "a process that begins with the establishment of the OPR and endeavors to ensure that the exterior enclosure and those elements intended to provide environmental separation within a building or structure meet or exceed the expectations of the owner as defined in the OPR" (ASTM, 2012).

The goal of BECx is to define the specific performance requirements of the project at hand (including schedule and budget), clearly communicate them to the project team, and then work to ensure the project meets these criteria throughout the construction process.

This section further defines the process of BECx, identifies the potential value to the project team, and supports the efforts of facility staff to set expectations of the BECx provider (BECxP) and full project team to achieve the desired performance of the project.

16.6.2 Why commission the building enclosure?

The Cx process is often viewed in the light of mechanical, electrical, and plumbing systems or discrete and often repetitive pieces of equipment or equipment types that function together to create an operating system to achieve an outcome. They include fans, belts, motors, pumps, coils control, and so on, where each piece has specific operating criteria. Generally, these components are accessible after occupancy for routine maintenance, service, and replacement. The building enclosure, on the other hand, requires a similar commissioning process. However, the enclosure is composed of multiple materials and assemblies with less-defined boundaries – often with one assembly being provided by a variety of manufacturers and installers. It must perform in all possible weather conditions, and this performance can only be confirmed when the building enclosure assembly is complete and a load (air, water, sound, etc.) is applied.

Additionally, once the project is complete, the layers of the enclosure are interstitial (buried) within the construction assembly and are not accessible, but they are expected to remain intact and unchanged for the life of the building, with few exceptions. Thus, the most reliable means to achieve performance targets during construction is to have an expert with technical knowledge of the design and installation process of the project-specific building enclosure systems integrated into the design process who visually observes the installation and testing of the work. This crucial step in confirming completion is intended to:

- Mitigate contractual risk
- Improve O&M
- Prevent indoor environmental quality problems
- Meet regulatory or program requirements
- Support energy performance
- Investigate/validate air tightness
- Enhance security and resiliency

16.6.3 What does BECx cost?

By utilizing the building enclosure commissioning process, delays in construction, long-term costs, and building service interruptions can be avoided. Owners can realize lifetime savings because they are not prematurely addressing typical maintenance issues such as resealing a facility due to moisture penetration, or replacing an area where air infiltration was caused from a breach in the building enclosure. These issues are anticipated and addressed during design and are validated in the construction phase with the quality assurance process of mock-up testing, field observations, and in-field performance testing of installed elements.

Building enclosure commissioning services have been documented to provide cost savings many times the cost and effort of recladding a building. Utilizing the enclosure commissioning process, qualified BECx providers help ensure that the mission of the institution remains the primary focus in the building – giving owners and operators more security that their investment is performing to the optimum level of performance for the long term.

According to Evan Mills, PhD, a researcher at Lawrence Berkeley National Laboratory, building enclosure commissioning, or BECx, should be viewed as "the single-most cost-effective strategy for reducing energy, costs, and greenhouse gas emissions in buildings today" (Sullivan, 2013).

Mills studied the benefits of BECx, noting that commissioning only costs about \$1.16/ft^2 for new construction and \$0.30/ ft^2 for existing buildings on average, with a payback period of as little as 14 months. Savings associated with using BECx from both maintenance and energy savings average about 16% for existing buildings and 13% for new construction. The main benefit is that whole-building energy savings are guaranteed, thanks to the pivotal role of the enclosure in determining efficiency performance.

"Further enhancing the value of commissioning, its non-energy benefits surpass those of most other energy-management practices," including major first-cost savings through right-sizing of HVAC equipment, Mills has stated. "When accounting for these benefits, the net median commissioning project cost was reduced by 49% on average, while in many cases they exceeded the direct value of the energy savings" (Sullivan, 2013).

16.6.4 BECx process

Building enclosure commissioning is a more recent development and less mature than say new construction commissioning. There are two documents generally referenced as the "standards". The first is ASTM E2813-12, "The Standard Practice for Enclosure Commissioning (E2813)." The second is the National Institute of Building Sciences (NIBS) Guideline 3-2012 (NIBS, 2012), which is built upon ASHRAE Guideline 0-2005, "The commissioning Process", more recently, ASHRAE Standard 202-2013, "Commissioning Process for Buildings and Systems." In an effort to clarify the two documents, in 2012 NIBS and ASTM developed a memorandum of understanding to combine the documents into the ASTM ANSI-certified process to develop enforceable mandatory language for the standard practice. This move has allowed NIBS to focus on the development of a training program to those standards.

The BECx process, as documented by NIBS, closely aligns with the recommended timing of new construction Cx processes beginning with predesign and OPR and BOD development through design, procurement, construction, functional performance verification, occupancy, and operations.

A notable difference between the BECx process and the typical new construction commissioning process is when functional testing is accomplished. Some testing on electrical equipment occurs in early construction, with the rest in mid or late construction; mechanical testing is almost all done late in construction. However, BECx testing occurs in all phases. On mock-ups, it occurs at the very beginning of construction, and testing installed assemblies occurs as the enclosure progresses, with some testing of final air tightness and roof integrity done late in construction.

Another of the primary differences in the BECx process vs. the new construction commissioning process for mechanical and electrical systems is the oversight of the commissioning provider during installation. Whereas the contractors are somewhat on their own during installation for mechanical and electrical systems during the construction phase, BECx is highly dependent upon observations during installation and early functional performance verification, often in preconstruction mock-up and just after the first assembly is installed. This has become particularly important with the rise in high-performance framed structures in which the layering of building enclosure systems makes discovering a noncompliant issue early a matter of great importance. The more that performance verification is delayed into the construction process, the greater the cost risks. These cost risks include: demolition of work to access a needed repair, rework of an application, and inability to observe or test properly and identification of noncompliance assemblies. As the costs to rework or access components become larger, the momentum to follow through with the planned testing may fade.

A dynamic program starts by setting a strong "foundation" for the project prior to design and during the development of the OPR. Like a building's foundation, the OPR establishes a base and sets the tone for a project, giving it a basic structure by which all parties, from architect to subcontractors, must abide. It is during the OPR process that the BECx provider defines the way a building enclosure will be commissioned for the owner with a commissioning plan that fits within the boundaries of the BOD.

The process begins with collaboration and project integration by engaging the project architect in review of their proposed building enclosure components and materials. The building enclosure CxP begins by reviewing the OPR developed by the team and the architect and refines basic selections of systems that include the cladding type and roof assembly.

The BECx process includes all the typical general procedures and deliverables as for mechanical and electrical systems, including:

- OPR and BOD
- Cx plan (typically part of the main Cx plan)
- Design reviews
- Specifications
- Submittal reviews
- Construction checklists
- Field observations
- Functional performance testing
- Issue logs and reports
- System manual contributions
- Documentation verification
- Training verification
- Reports
- Warranty period evaluations
- Preventative maintenance program review

16.7 Commissioning Certifications

There are several organizations that offer commissioning providers options to seek certification. Some organizations have accreditation with The American National Standards Institute (ANSI), which is a private non-profit organization that oversees the development of voluntary consensus standards for products, services, processes, systems, and personnel in the United States.

Certified Building Commissioning Professional – CBCP

The Association of Energy Engineers (AEE) offers commissioning providers the CBCP designation. The CBCP certificate is recognized by the U.S. Department of Energy (DOE).

Building Systems Commissioning Certification

The National Environmental Balancing Bureau (NEBB) offers commissioning providers the Building Systems Commissioning Certification designation.

Building Commissioning Professional Certification (BCXP)

The American Society for Heating, Refrigeration and Air Conditioning (ASHRAE) offers commissioning providers the BCXP Certification designation. This certification is ANSI accredited.

Certified Commissioning Authority (CxA)

The Associated Air Balance Council (AABC) Commissioning Group (ACG) offers commissioning providers the CxA certification designation. This certification is ANSI accredited and recognized by the DOE.

Certified Commissioning Professional (CCP)

The Building Commissioning Association's (BCA) Building Commissioning Certification Board (BCCB) offers commissioning providers the CCP certification designation. Additionally, commissioning firms may also be certified as Certified Commissioning Firms (CCF). These certifications are ANSI accredited.

Certified Building Enclosure Commissioning Provider (CBECxP)

The International Institute of Building Enclosure Consultants (IIBEC) offers building enclosure commissioning providers the CBECxP certification designation. This certification is ANSI accredited.

Building Enclosure Commissioning Process Provider (BECxP) and Accredited Commissioning Authority + Building Enclosure (CxA+BE)

University of Wisconsin-Madison's College of Engineering offers building enclosure commissioning providers the BECxP and CxA+BE certification designations.

16.8 Summary

Commissioning is the process of assuring that a building is designed and constructed to meet the owner's intent and that it operates and continues to operate at peak efficiency. There are five types of commissioning and in this chapter, we explained the differences between: new construction, existing building, monitoring-based, ongoing, and building enclosure commissioning systems. Once the differences and phases are understood, the energy manager will be able to recognize what commissioning process will be useful in conducting an energy management process for existing buildings.

References

ANSI/ASHRAE/IES (2018) "Commissioning Process for Buildings and Systems" Standard 202-2018.

ASTM (2012) "ASTM E2813-12 Standard Practice for Building Enclosure Commissioning."

Crowe E., Mills E., Poeling T., Curtin C., Bjornskov D., Fischer L., and Granderson J. (2020) "Building Commissioning Costs and Savings Across Three Decades and 1,500 North American Buildings." Energy and Buildings 227, Issue C.

Granderson J., Crowe E., Curtin C., and Mills E. (2018) "Value of Commissioning 2018 Market Survey" Published by the Building Commissioning Association. Building Commission Association www.bcxa.org.

IEA (2020) "Energy Efficiency 2020" International Energy Agency. www.iea.org/reports/energy-efficiency-2020.

Kim Y.K. H. Altan H. (2018) "Analysis of Energy Reduction through HVAC Commissioning: A Case Study of a LEED Certified New Building" 2018 5th International Conference on Renewable Energy: Generation and Applications.

Kramer H., Lin G., Curtin C., Crowe E. and Granderson, J. (2020). "Building analytics and monitoring-based commissioning: industry practice, costs, and savings" Energy Efficiency vol 13, p. 537.

Mills E. and Mathew P. (2014) "Monitoring-based Commissioning: Benchmarking Analysis of 24 University Buildings in California" Energy Engineering vol 111, issue 4.

NIBS (2012) "NIBS Guideline 3-2012 Building Enclosure Commissioning Process BECx" National Institute of Building Sciences www.nibs.org.

Stum K., and Bjornskov D. (2017) "The building commissioning handbook" (3rd edition). Building Commissioning Association and APPA. www.bcxa.org

Sullivan C.C. (2013) "Calculating the ROI of building enclosure commissioning" Building Design + Construction www.bdcnetwork.com/calculating-roi-building-enclosure-commissioning.

Acknowledgments

The author would like to thank Karl Strum and Diana Bjornskov who authored the Third Edition of the Building Commissioning Handbook, written and published by The Building Commissioning Association and APPA (2017).

Author Biography

Craig Hatch is a Certified Commissioning Professional from Calgary, Alberta, Canada. He has been providing new construction and existing building commissioning and building automation systems design services since January 2001. Craig has been involved with the Building Commissioning Association since 2007 and is the current Communications Director for the International Board of Directors.

Practice Problems

1. Under what conditions would OCx not be advisable to transition directly into OCx, and what would need to be done prior to launching OCx?
2. The goal of EBCx is to make building systems perform interactively to meet the _______ and focused on ________ to reduce ________.
3. What documents should be used to develop functional performance tests?
   - A. Owner's Project Requirements, project specifications, industry guidelines, and approved submittals
   - B. Manufacturer's listing of inspections and testing used on factory testing and acceptance
   - C. Test procedures used on the previous project for similar equipment
   - D. The commissioning plan on the last project completed for this client
4. Describe fault detection and diagnosis, state which of the five types of commissioning FDD is used, and briefly discuss how it fits in with commissioning.
5. Functional performance testing of the building systems has been successfully completed. What is the next step in the commissioning process?
   - A. Submit the final commissioning report
   - B. Perform owner training
   - C. Test inter-relationships between systems and subsystems
   - D. Recommend beneficial occupancy

17

Energy Storage

Sheila Glesmann

Pando Alliance, USA

17.1 Introduction to Energy Storage

Energy storage is a promising approach for effectively managing electrical and thermal loads by decoupling energy generation and use. Today, a variety of both grid-scale and Behind-the-Meter (BTM) Commercial and Industrial (C&I) applications rely on energy storage technologies, and the options are quickly expanding. Storage can benefit the end user directly and/or provide grid services in some circumstances. Storage is a rapidly developing field, and this chapter will touch on the most common technologies, from residential- to grid-scale, emphasizing thermal and battery storage, which are increasingly used in BTM applications.

Any energy storage approach works by capturing energy in a physical storage device or medium, which is referred to as the charging step, and releasing that energy at a specific desired discharge rate for the use case. Energy storage enables power generation and utilization to be asynchronous; for example, solar energy captured during daylight can be utilized at night. Similarly, thermal storage can be used to produce chilled water or ice using grid power when electricity rates are low and use the ice or water to enhance cooling during daytime when electricity rates are higher. These two cases exemplify diurnal energy storage, a common application for energy managers. Other cases can be of shorter term, for example as equipment startup support, or longer term, as in energy availability to manage grid outages or extreme weather. Energy storage is increasingly part of energy management in both new-build and retrofit cases, and basic evaluation techniques are needed to assess the options.

The physical range of methods for energy storage includes thermal (sensible energy and/or latent energy with phase-change materials, or thermochemical storage; thermal storage can apply to cooling or heating), electrical (using batteries or capacitors), or mechanical (kinetic or potential energy using flywheels, pumped hydropower, compressed air storage, or gravity-based mechanical systems). Hydrogen is an emerging tool for energy storage. Electric vehicles' bidirectional integration, aggregation of domestic hot water heaters, and a few other specialty applications are also emerging.

This chapter describes the technologies briefly, with the deepest discussion in battery technology for the storage of electricity. Electricity stored in batteries can either be discharged BTM within a building/campus directly or exported to the grid. In either case, the battery output power is DC and an inverter is required to produce AC power. Thermal storage is widely deployed and is another prime BTM technology.

A few important concepts for any energy storage technology selection are the scale and lifetime of the technology, the energy capacity and density, the power rating (for electric systems), the duration of discharge at rated power, the number of useful cycles for a given application, and the factors that degrade performance or life. These elements need to match the use case.

In this chapter, terminology is defined, followed by a discussion of the commercial deployment of energy storage. An overview of technologies currently in use or being developed is next, with an emphasis on rapidly developing battery storage. The reader should be able to use the chapter guidance and resources to learn about the range of energy storage approaches and how they are applied. Resources are also provided for further reading and cost modeling in this rapidly developing area.

17.1.1 Benefits and value stacking

Energy storage can drive several different benefits, creating value streams that go beyond utility bill savings. Their usefulness or value is specific to the project under consideration. Often some "value stacking," such as quantifying the value of avoiding hours of disruption in operations, *beyond* direct rate savings is needed to justify the investment in energy storage. Policies and incentives are important inputs for analysis of any project. Some benefits are quantifiable directly through cost savings or revenue generation, while others are non-monetary benefits that are valued by some customers or jurisdictions, depending on their policies (for example, reducing carbon footprint).

Advantages of energy storage can include:

- *Load shifting or levelizing/peak load reduction*, resulting in time-of-use energy charge reductions, known at the utility level as energy arbitrage, and/or peak demand charge reductions:
    - Equipment cost reductions can also result. For example, if a heating system includes a boiler and hot water storage, the storage can

enable the boiler to be sized at the peak demand averaged over a longer time scale, such as a day to a week, instead of an hour. This can significantly reduce the boiler size.

- *Load firming*, by making energy available on demand instead of when generated, which can enable variable renewable energy to be available for use on-demand. Electricity markets place a value on firm capacity.
- *Improved efficiency or energy use* resulting in:
    - *lower costs* and
    - *reductions in greenhouse gas and other emissions.*
- Electrical modulation to *maintain frequency or improve power quality.*
- Improved *resiliency*, or recovery from adverse events, including black start in some cases (see Chapter 8).

The design of an energy storage system as well as its operation, e.g., charging methods, are very case-specific. System economics and environmental benefits can be defined by how the system is energized. For example, excess nuclear power generated in order to keep a nuclear plant at steady load can be stored in a pumped hydro reservoir. Renewable energy such as solar and wind, for which there is no fuel cost, can be the charging source for batteries and enable firm power supply as well as cost predictability.

Value stacking of these benefits can justify investment in energy storage. Any given project must be designed to capture the value streams that are available for the location and use case. To encourage deployment and improve electric reliability and resiliency, many jurisdictions at the federal and/or state/province levels offer energy storage incentives (a carrot approach) or requirements (a stick approach). These are then implemented by utilities as incentives to their customers when applicable. Figure 17.1 exemplifies mechanisms available to create value in the categories of *revenue, cost savings, resiliency*, and *environmental.* Quantification of each depends on the customer's goals and priorities, utility billing structures available, and rules and incentives in place at the time and location of a project. The range of categories for C&I BTM installations that any new project could consider is shown in the figure, and each of these mechanisms may or may not have any dollar value associated with it for a given project.

Grid revenue:

Utility programs can enable energy storage to participate in demand response or other grid benefits. Some C&I projects can bid into the wholesale market, opening up more revenue opportunities such as frequency regulation, reserves, or resource adequacy.

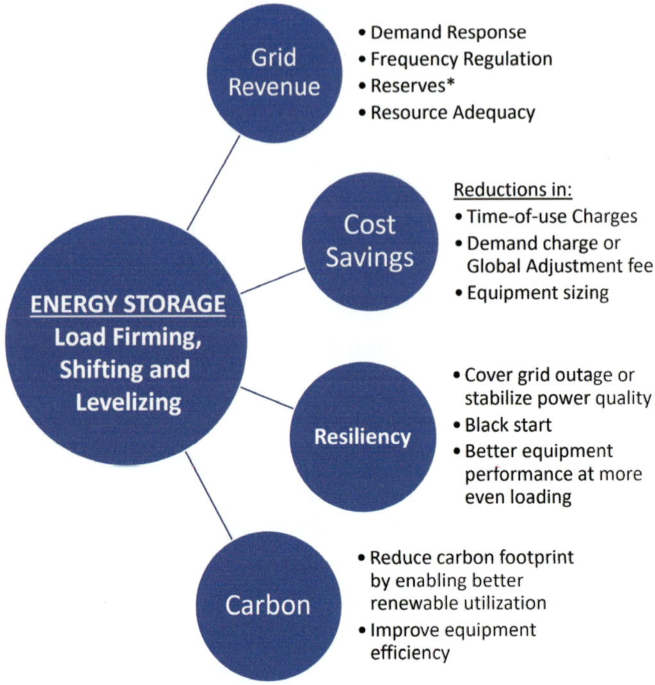

*Reserves can be spinning reserves, available immediately in the event of a power failure,
Or non-spinning reserves, available soon after a failure.

Figure 17.1 Energy storage benefits.

Cost savings:

Time-variant rates can drive energy cost savings when electricity consumption is reduced during higher-rate periods. In addition, demand charges can be reduced by replacing part of peak load with storage discharge. This can apply on a diurnal, seasonal, or annual basis. Costs can become more leveled and predictable if peak loads can be reduced. Pricing may add a premium at discrete high-demand periods, as in Ontario's Global Adjustment Fee policy. This incentivizes large users to reduce their consumption. The impact of reducing peak load during just five hours a year is disproportionally beneficial, since these charges contribute up to 40%–60% of a Class A (>5 MW) user's electricity bill (Ontario Reg. 429/04: Adjustments Under Section 25.33 of the Act, 2014).

Resiliency:

Grid disruptions can be managed by energy storage if the system is designed to do so. For example, some batteries can provide black start capabilities for distributed equipment. Either a loss of power or inconsistent power quality can be addressed in an energy storage system design. Another benefit of increased resiliency derives from reduced cycling of equipment, leading to lower maintenance and more reliable operation.

Environmental/GHG emissions:
Two main benefits can accrue on the environmental side. One is the increased utilization of renewable energy by having storage (according to Lawrence Berkeley National Lab, in 2019, 70% of all BTM energy storage systems were paired with solar generation (Standalone Storage vs. Solar-Plus-Storage | EnergySage, 2021)). The second is improved equipment operational efficiencies resulting from conditions that the timing of storage enables. For example, chillers have higher efficiencies at cooler condenser water temperatures. So, generating chilled water overnight can both reduce daytime electrical use and make the chiller more efficient. Improved energy efficiency reduces greenhouse gas and other pollution when part of the supplied energy is from fossil fuels.

A way that the value of an individual project is often shown is in a stacked column chart that depicts potential $/kW/yr or internal rate of return (IRR) contribution for each item and in total. Figure 17.2 shows this concept for similar cases with a couple of differences. In the first case, utility demand response and local incentive payments are available. Project IRR is much improved over the second case, without these two benefits. Also, if resiliency or environmental benefits can be quantified, those can add value to achieve an acceptable project return, depending on customer goals. A data center,

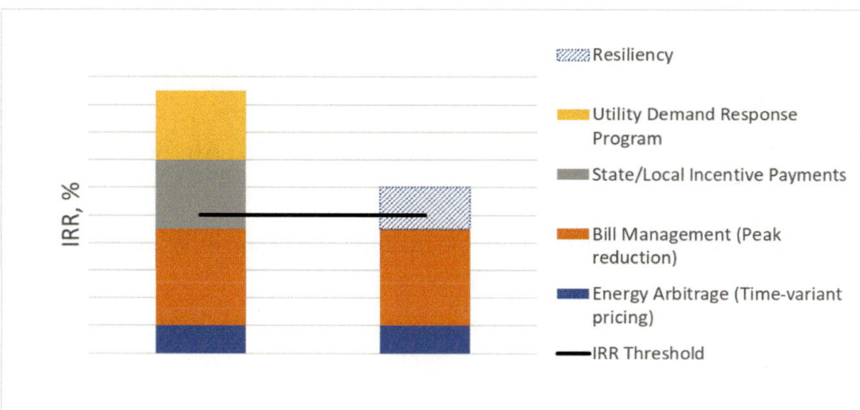

Figure 17.2 Value stacking concept.

for example, may value resiliency very highly. Downtime can result in direct revenue hits as well as damage to their reputation with customers. So, the second example of value stacking may be justified although direct dollar-based IRR is under the target threshold.

17.1.2 Terminology and common design considerations (Bowen, Thomas et al.)

Some introductory terminology is needed to understand energy storage, regardless of the method of storage. Storage can contain heat or cooling, electricity, chemical energy, or potential energy. Certain terminology is universal, although definitions are somewhat inconsistent among reference sources. Each term below is defined as used in this book.

Most storage systems degrade over time, reducing capacity. An initial design decision for any system is the minimum capacity needed ("target" capacity) and how much margin to include. Once significant deterioration occurs, replacement is needed.

17.1.2.1 General terms

Cycle life. The number of charge/discharge cycles a system can provide while retaining $\geq$ target capacity. Cycle life is different from calendar life (age) or shelf life, which is the length of time until a system experiences significant loss of capacity when not actively cycling.

Charge rate. How fast the system charges in watts (electricity) or Btu/hr (thermal). For batteries, charge controllers manage charge rate.

Discharge rate. How fast the system discharges in watts or Btu/hr. This can be over a range of rates and the limits are defined by the system. For batteries, Discharge Rate is also known as C-rate (see below).

Energy capacity. Maximum usable energy stored in Wh, Ah, or Btu (for thermal storage).

> **Energy density.** Energy capacity on a volumetric basis, in Wh/L or Btu/ft^3. It helps bracket storage dimensions.
>
> **Specific energy.** Energy capacity on a weight basis, in Wh/kg or Btu/lbm. It helps bracket storage weight.

Rated power capacity. Instantaneous discharge rate in watts, amps, or Btu/hr at a reference time period, such as 1 second, 1 hour, or 20 hours.

Power density. Power at rated discharge per volume in W/L or Btu/hr-ft^3. It helps bracket storage dimensions.

Specific power. Power produced on a weight basis, in W/kg or Btu/hr-lbm. It helps bracket storage weight.

Round-trip efficiency. Ratio of usable energy discharged from the storage system to energy to charge the storage system. For example, a 85% round-trip efficiency means that 85% of the energy used to charge a system is available for use (85% is a good assumption, per NREL, 2023a). This does not necessarily account for idling losses, which is self-discharge over time when a system has a constant *SOC* (see below). Losses are inevitable in any system, and these range by technology and installation. Note that battery suppliers may quote DC–DC efficiency, whereas AC–AC efficiency better reflects the full system's power losses in the case of batteries, to account for inverter losses.

State of charge (SOC). Percentage of the charge available at a given point in time, theoretically from 0% to 100% for mechanical systems. For batteries, there is a tighter range than for mechanical systems. Batteries vary and may be limited to 10%–90% SOC is the available energy divided by Energy Capacity, expressed as a percentage.

Time-variant pricing. This idea is introduced in the chapter on billing rates and is key to economic analysis of energy storage. For many commercial and industrial customers, off-peak rates can be used to charge energy storage. Various rate structures may apply, such as time-of-use rates or market rates (see Chapter 4).

17.1.2.2 Battery-specific terms

Cell. The basic building block of a battery system. For example, a typical 18650 cell is cylindrical with a diameter of 18 mm and length of 65 mm. It is slightly larger than a standard "AA" battery and commonly used for lithium-ion cells. The voltage and current design values for a cell vary with battery chemistry and materials. Two other leading configurations of battery cells are prismatic and pouch cells (both rectangular packets), which carry the benefit of tighter packing than cylinders.

C-rate. Discharging a battery more quickly (higher rate) shortens the duration that power is available. Expressed in 1/time, so discharge at 1C is the amount of power that will be available for 1 hour. 0.05C corresponds to the 20-hour discharge rate (as may be specified for lead-acid batteries), and 5C corresponds to the 12-minute discharge rate. Batteries specify a C-rate that corresponds to their anticipated duration of discharge. They also may be discharged over a range of C-rates, which can impact capacity and *cycle life*.

Depth of discharge (DoD). The percentage of battery capacity that has been discharged compared to a fully charged state. In general, DoD impacts the *cycle life* of batteries, i.e., deeper DoDs lead to deterioration of battery capacity sooner (in fewer cycles).

Rated capacity. Amount of charge that can be withdrawn from a battery at specific rate, e.g., a lithium-ion battery may be rated for 10 Ah at 0.5C (2-hour duration). It could also operate at 20 Ah at 1C (1-hour duration). Capacity and cycle life can be impacted by operating at different C-rates than specified.

State of health (SOH). The number of remaining cycles in a battery's life – useful for replacement planning.

17.2 Global Energy Storage Technologies in Use

Batteries are the fastest-growing segment of energy storage globally. Both utility-scale and commercial and residential BTM installations are rapidly increasing, particularly in conjunction with renewable power generation.

While batteries are the fastest-growing, by far the largest deployed energy storage is pumped hydro. In 2023 this grid-scale technology represents about two thirds of global energy storage, with 181 GW installed. Batteries are the next-largest with 87 GW installed, while all other storage technologies total 4 GW.

Publicly-available data and projections for energy storage deployment, with some breakouts by type of storage, are accessible from the International Energy Agency (IEA (2024) *International Energy Agency, 2024*), the US Department of Energy's Global Energy Storage Database (GESDB) (National Technology & Engineering Sciences of Sandia, LLC 2021), the U.S. Energy Information Administration (EIA - U.S. Battery Storage Market Trends, 2025), and the American Clean Power Association's US Energy Storage Monitor (ACP, 2025), among other sources.

Figure 17.3 shows the current dominance of pumped hydro, as well as the rapid growth of batteries anticipated to 2030. In the figure, 2030 STEPS represents the IEA's Stated Policies Scenario, while 2030 NZE represents IEA's Net Zero Emissions by 2050 Scenario. Growth is anticipated in BTM batteries from 33 GW in 2023 to 177 GW under the 2030 STEPS case, a

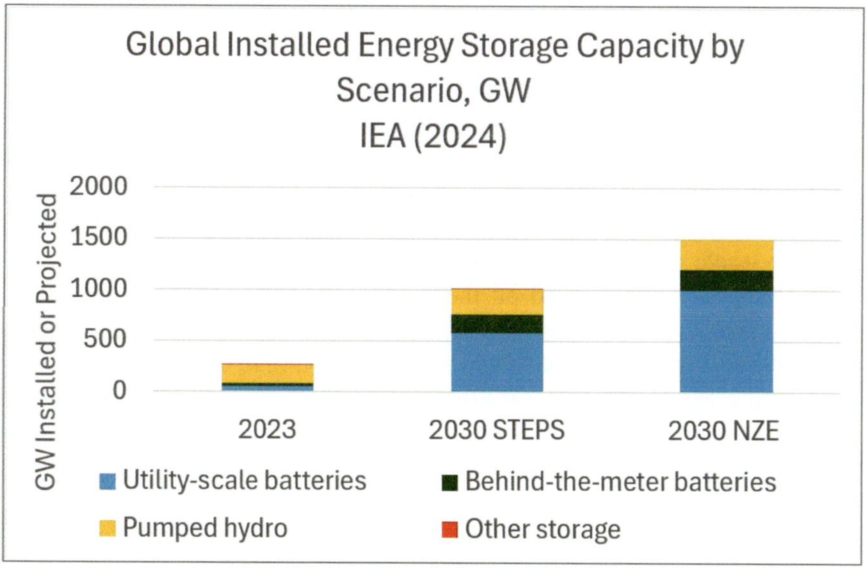

Figure 17.3 Global Installed Energy Storage.

five-fold increase globally. Utility-scale battery deployments are also rapidly growing, which drives technology development that will benefit smaller, BTM installations. "Other storage" in Figure 17.3 includes compressed air energy storage, flywheel and thermal storage. Hydrogen electrolyzers are not included. These methods collectively grow from 4 GW to 5 GW under the 2030 STEPS case, a 25% increase.

In the US, installed BTM battery capacity was about 2.3 GW in 2023, with rapid growth each year. The residential sector is driving over 60 MW/quarter of growth in 2025, while commercial is driving over 20 MW/quarter. The other key storage technologies described in the sections below have less installed capacity, and vary in their growth rates. An approximate breakdown for combined US and Canada installations from about 2022 is 2.2 GW batteries, <1 GW Compressed Air Energy Storage (CAES), <1 GW thermal storage, and 0.1 GW flywheel storage.

Some storage technologies are only practical at utility or grid scale. Below is a brief description of the dominant energy storage technologies and their typical applications. A summary of key energy storage metrics is shown in Table 17.3, following the technology descriptions.

17.2.1 Mechanical methods (more commonly used on grid side)

Pumped hydro. Pumped hydro utilizes the gravitational potential energy of water between higher- and lower-level reservoirs. This is the most common energy storage technology and has been in use since the early 1900s. Pumping power often utilizes the off-peak excess power of a base-load nuclear power plant. Efficiencies are up to 70%–85% and the system is highly reversible. These utility-scale systems can provide significant load (tens to hundreds of MW). Pumped hydro is used on the grid by utilities, for short-term or seasonal reserves. Globally some new projects are underway today, at large scale. The biggest growth region is China (Energy Storage Grand Challenge Energy Storage Market Report, 2020). Pumped hydro requires large-scale and specific geography, i.e., availability of adequate sites with elevations for pumping water. The scale, as well as length of time to obtain permits and complete construction (5–10 years), limits the applicability to grid-scale systems. Pumped hydro uses generators that can operate in reverse and pump water uphill to charge the system, and release it through the generators to produce power in the discharge mode. They have long lifetimes (30–100 years) and high power rating (up to 3,000 MW).

Flywheels. Using rotational kinetic energy, flywheels are reversible systems that charge and discharge power through applying varying torque in conjunction with a power electronic converter. Efficiency can be up to 95% but drops significantly with discharge rate. They have flexible power charge/discharge rates and quick response time. Discharge is typically seconds in duration, to provide power immediately and for a brief period to cover disruptions in power supply, until other backup generation comes up. Also, flywheels can smooth and manage power quality to protect equipment and systems, as in critical hospital or data center applications. Technical advances in materials and design are broadening the uses for flywheels and extending their useful discharge duration. While they can operate for many cycles with low maintenance, they are limited in energy capacity and tend to self-discharge over time.

Compressed air energy storage. A compressor pressurizes air into a reservoir, where it is stored. Energy is recovered by first heating the air and then expanding it through an electrical generator. This reversible process can obtain efficiencies up to 70% if heat is conserved. Compressed air energy storage can utilize an underground cavern and fulfill utility functions such as black start, renewable utilization, and diurnal peak load management. CAES has been in operation since 1978, when the first salt-cavern system was started in Germany, in conjunction with a nuclear plant that charges it when demand is low. Similar to pumped hydro, this technology is only effective at large scale where geography allows, and has a long life of 20–40 years.

Gravity-based mechanical systems. A rather new concept wherein physical weights in the form of railcars or blocks are raised and then released to utilize their potential energy, generating electricity. Efficiencies can be up to 80%–90% for these utility-scale electro-mechanical systems, and they can use renewable energy for the charging cycle.

17.2.2 Electro-chemical methods (used both grid-side and BTM)

Batteries. Currently the fastest-growing segment of global energy storage, secondary (rechargeable) batteries can be used for grid-scale and BTM storage. Two battery types are discussed here: cell batteries, which are well-established, and flow batteries, which are emerging technologies. Within these types, there is a range of chemistries, scale, and applications.

Cell batteries use closed systems that produce electrical energy through reduction–oxidation (redox) chemical reactions within each cell. There are three basic components to a battery cell: an anode (the negative terminal during battery discharge, where the oxidation takes place), a cathode (positively charged terminal), and an electrolyte that produces ions. The electrolyte or a separator also provides a barrier so that electrons do not cross through the battery directly but allows ions to migrate from the anode to the cathode through a circuit, generating DC electricity.

Rechargeable or secondary batteries are fundamental devices that convert reversibly between electrical and chemical energy using redox reactions. Efficiencies can be up to 95%. Once the reactants are depleted by discharging, the process can be reversed by applying electricity to recharge the battery, restoring the chemical species in the anode and cathode electrodes. The rechargeability of the systems gradually degrades over time and with use, and so are limited in both calendar life and in cycle life. Cells are combined in series (to increase the voltage) or in parallel (to increase the current).

Pb-acid cell batteries are the original rechargeable batteries, first developed in 1860. They are based on reactions between lead, water, and sulfuric acid. The electrolyte solution is primarily sulfuric acid when fully charged, and a dilute sulfuric acid when discharged. Additional subtypes used today include gel electrolytes and sealed batteries. The installed base of lead-acid batteries is large, and materials are usually recycled. Most familiarly, vehicle startup batteries are lead-acid. These batteries have a high current output and are designed for shallow discharge. Other applications such as vehicle power (e.g., in forklifts), small-scale off grid power, and uninterruptible power supplies may use lead-acid deep discharge batteries. These are designed for lower current output over a longer period of time (low C-rate), and the number of cycles they last is an inverse function to the DoD. Optimal cycles versus DoD can often be achieved at about 50% discharge per cycle.

Deep discharge batteries are the ones of interest for energy storage applications. Corrosion and sulfation of any lead-acid battery eventually leads to declines in performance, requiring replacement. Lead-acid batteries have less energy density than newer lithium-chemistry batteries, but in some cases may have advantages in cost, reliability, lifecycle emissions, or environmental footprint since they can be almost fully recycled. Further reading on lead-acid batteries for energy storage use is available; see May et al., 2018. Figure 17.4 shows the impacts that DoD, SOC, temperature, and storage time have on performance characteristics for a specific battery provided by Rolls Battery. In addition, current estimates are that batteries will deteriorate to 70% of initial energy capacity in 10 years.

Example 17-1

A 12-V, 12-Ah lead-acid deep discharge Rolls Battery is recycled and replaced when the capacity reaches 7.2 Ah. In operations, it is discharged to a minimum SOC of 20% before recharging to 100%. The storage temperature is controlled at 25°C (77°F). Refer to Figure 17.4.

1. About how many cycles will this battery provide before being disposed/recycled? How much would this change if the minimum SOC was 50%?
    a. Capacity of 7.2/12 Ah = 60%. Reading the cycle life at 60% capacity, 650–700 cycles correlate to an 80% DoD (= 20% minimum SOC). This would improve to ~1,200 cycles if the minimum SOC was 50%.
2. What capacity will the battery actually provide at the operating temperature initially? What residual capacity will it provide after one year of storage?
    a. Initially, the capacity will be 12 Ah (100%), and after one year it will drop to 75% or 0.75×12 Ah = 9 Ah.

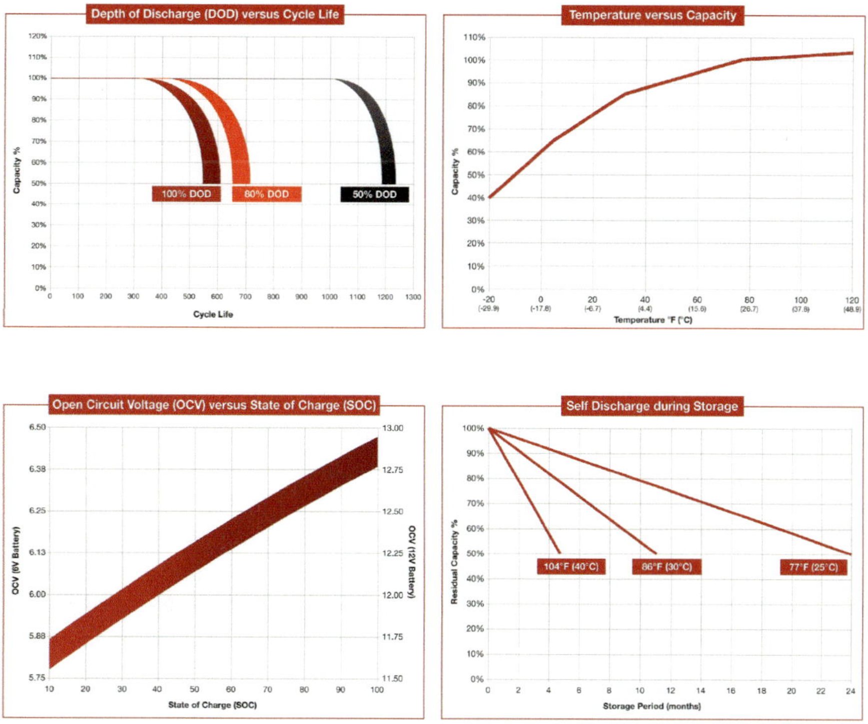

Figure 17.4 Example of lead-acid battery performance characteristics. From: http://support.rollsbattery.com/en/support/solutions/articles/4346-agm-discharge-characteristics

3. What is the open circuit voltage range in operations when discharged to 50% SOC?
   a. The open circuit voltage remains above 12 V, in the range of 12.15–12.3 V.
4. If the battery is used at freezing temperature, what is its capacity early in its life?
   a. About 85% capacity or 0.85×12 Ah $=$ 10.2 Ah.

The concepts demonstrated by the chart and example above are key to understanding battery life more broadly. Batteries inherently deteriorate over time and use, and this needs to be taken into account in design and selection.

- Self-discharge increases with temperature
- Capacity reduces as temperature drops

- Cycle life is impaired by higher DoD
- Voltage drops with SOC
- Longer shelf life reduces capacity
- Higher temperature decreases calendar life

Lithium-ion cell batteries are a more recent development. Lithium-ion chemistries in closed cell batteries are beginning to dominate new installations in energy storage. Although battery technologies are available in many chemistries, about 90% of stationary-scale energy storage batteries are lithium-ion.

Rechargeable lithium batteries were briefly commercialized in the 1980s, but safety issues prevented their further use. In 1991, Sony commercialized a modified version of lithium batteries called lithium-ion battery (LIB), where the metallic lithium was replaced by a graphite-based anode, thereby improving the safety of these batteries. Energy density, power density, and the overall safety of LIBs have been continuously improving since the early 1990s. In the past decade, LIBs have had an accelerated development, partly because of their adoption for use in electric cars. They have higher energy densities compared with lead-acid batteries, resulting in smaller size and weight. Costs have come down such that analysis is required to make an economic choice between these technologies.

Manufacturers are making LIBs that meet specifications and voltages to substitute for lead-acid batteries. There are a number of LIB varieties with a wide range of specifications in terms of energy, power, safety, and cost. LIBs are generally characterized by the choice of their cathode material chemistry, e.g., lithium cobalt oxide, lithium nickel manganese cobalt oxide, lithium iron phosphate (LFP), etc. The LFP chemistry, for instance, offers cost advantages and higher safety margins than other cathode chemistries. This is at the expense of lower energy density; thus, they are more suitable for stationary than mobile applications. LFP is dominant as of 2021 in new stationary installations, according to NREL, 2023b.

Similar to lead-acid batteries, lithium-ion batteries' performance also depend on temperatures, DoD, SOC, and shelf storage time. The specifics for a given product will vary, but in general LIBs are higher in energy density, and have higher cycle life, than lead-acid.

A *battery energy storage system (BESS)* consists of much more than just the electro-chemical cell itself. As batteries are scaled up to larger sizes, the controls, installation, and safety processes get more complex. The cell is a building block, and many cells are combined to achieve energy storage at scale.

An electric vehicle's main battery, for example, may contain thousands of 18650 cells, each rated nominally at 3.6 V and 2.8 Ah for an individual cell energy capacity of 10 Wh. Groups of cells are connected in parallel into modules, increasing the current while maintaining voltage. These modules are connected in series to increase the voltage available. Inside a rigid housing, modules are connected in series and parallel as determined by current and voltage requirements. A battery management system monitors, coordinates, and protects the cells' voltage, currents, temperatures, demand, and cycling. The battery management system has a balancing function to maintain consistency among cells during operation, also optimizing and preventing out-of-range conditions. Combined with modules and a ventilation system, this creates a battery pack. For large stationary energy storage installations, modules are assembled by the tens into racks, which can be then installed by the tens into a container. To generate AC power from battery cells, an inverter, or power conditioning system, is required. The full system of battery cells, battery management system, inverter, and overall control system is referred to as a BESS. For large systems, fire protection and safety requirements can be extensive (New York State Battery Energy Storage System Guidebook, 2020).

The following example illustrates the buildup of cells into functional energy storage for a large residence or small commercial building. The two products dominating residential market share as of 2022 are LG Chem RESU10H @1.9 hours' design duration and Tesla PowerWall @ 2.7 hours' design duration.

Example 17-2

For a Tesla Powerwall capacity of 13.5 kWh and battery voltage of 50 V, assume individual cells are of 4.2 V and 3 Ah. What number and configuration of cells could produce this voltage and capacity?

Voltage (V) $\times$ Current (A) $=$ Power (W)

13,500 Wh / 50 V $=$ 270 Ah

To achieve 50 V with 4.2 V cells, 50/4.2 $=$ 11.9. 12 cells must be in series

To achieve 270 Ah with 3 Ah cells, 270/3 $=$ 90 cells must be in parallel

90 $\times$ 12 $=$ 1,080 cells are needed. Each cell has a capacity of 4.2 V $\times$ 3 Ah $=$ 12.6 Wh

BESS scale. Typically, grid-scale BESS are on the order of 5–100 MW in size, with some smaller systems down to 1 MW in distribution applications. These are offered at 1 to 8-hour duration. BESS for C&I use range from about 2 kW to 2 MW, reflecting the rated discharge of the system. Durations are up to several hours, depending on the discharge rate. If the full power of the system (e.g., 1 MW) is needed, reasonable designs today result in practical limitations of 2–4 hours' duration at full rating. But research in this field is progressing rapidly with emphasis on storage capacities, duration, cycle life improvement, and cost. System designs accommodate longer usage hours if the drawdown rate is less than full power rating, for example, at 0.5C. Batteries are specified by both power (kW) and energy (kWh), and it is important to clearly understand the basis of any cost quote provided.

Cost components of a battery system. As background to costs and cost quotes, an understanding of BESS capacity is needed. There are several losses and design margins included in any system, which are well-described by Energy Storage Grand Challenge Energy Storage Market Report, 2020. As they point

Tools for Modeling and Specifying BESS

Guidance and resources from the U.S. Department of Energy are described in "Energy Storage Valuation: A Review of Use Cases and Modeling Tools," publication No. DOE/OE-0029, at www.osti.gov/servlets/purl/1873889/ REopt is an important tool highlighted in the above review. The current version of REopt can be found at reopt.nrel.gov. This tool can screen for applicability of solar, wind, battery storage, or chilled water storage for a particular site. Remember to carefully review all assumptions in the resulting report to ensure they are correct for the use case.

For battery-based storage, here are some key elements when specifying BESS requirements:

Power and energy:

- Nameplate rating for (duration in hours). The system must maintain this capability over the expected lifetime out, there is no "common lexicon for discussing the end use energy capacity of a storage facility" – the implication being that it is essential that the user accurately specifies the system and knows what is included and excluded in a given cost quote. The user needs to know the available delivered energy and the manufacturer needs to design for adequate additional capacity to cover, for example, unusable capacity below minimum SOC, design margin, cell imbalance effects, and losses.

NREL maintains an annual technology baseline (ATB) for utility-scale, commercial-scale, and residential-scale battery storage systems that includes capital and operating cost evaluations (May et al., 2018) (NREL, 2023b). Sources differ in how much of the cost basis is attributed to energy (kWh) vs. power (kW). These two sources however, come to a similar conclusion on total installed capital estimates for a 600-kW, 4-hour duration LIB system in 2022. An example breakdown and total is shown in Table 17.1, based on the 2023 ATB.

- 50% and if shallow discharge, 70% of maximum output power, for (duration in hours)

Efficiency:

- Minimum 85%âĹŠ90% AC round trip for Li-Ion and lead acid technologies
- Minimum 70% AC round trip for flow batteries

Ambient temperature range

Monitoring requirements (include voltage, current, power, and PF).

Data acquisition system shall have (30/60/90) days on site data storage and capability to be remotely accessed and data downloaded.

Sizing the energy and power capacity of the battery required is the biggest cost driver. Costs in general have trended downwards because of rapid developments in battery chemistry and cell level technology, design improvements, system integrations, and economies of scale. They are also highly use-case specific. Recent experience in 2024 shows that a behind-the-meter 50 kWh battery system supporting a solar installation can be procured for about

Table 17.1 Estimated mid-level costs for residential- and commercial-scale BESS in 2023 (NREL, 2023a).

	Commercial 600 kW, 4-hour BESS	Residential 5 kW, 4-hour BESS
Battery energy component, $/kWh	241	547
Battery power component, $/kW	893	625
Battery capital cost component, $	$203k	$7,575
Total installed capital cost (CapEx)	$1.32 M	$21,640
Annual O&M @ 2.5% of CapEx	$33k	$541
Assumed round trip efficiency	85%	85%

$500/kWh, all-in. To underscore the pricing trends, electric vehicle battery costs decreased by a factor of five from 2010 to 2016 and are projected to reduce by half again by 2025 (Bowen, Thomas et al 2018). Costs are steadily reducing for battery-based stationary systems, and new technologies are enabling longer storage durations. BloombergNEF reported late in 2024 that they expected prices to continue dropping, with LIB battery packs reaching $115/kWh in 2024 ("Lithium-Ion Battery Pack Prices See Largest Drop Since, 2017, Falling to $115 per Kilowatt-Hour: BloombergNEF" 2024). However, battery packs are less than half of the BESS system cost. In addition to battery pack costs, inverter, power conditioning, fire safety, overall controls, O&M, and decommissioning costs should also be considered in project planning. Updated pricing from vendors and clarification of scope are essential to the initial phase of evaluating energy storage cost and feasibility.

Flow batteries. These systems differ from cell batteries in that the liquid electrolytes are stored separately and brought together at controlled rates. This process is more reversible than cell batteries since analytes can be refreshed, and energy can be fully discharged without deterioration. Improvements to cathode and anode materials and design are progressing with the goal to increase the cycle life of these key components. Efficiencies are in the range of 60%–85%.

Redox flow batteries are now being demonstrated at grid-level MW scales. Their energy densities are lower than those of other battery alternatives; however, this is not as a critical parameter in stationary installations as it is in other sectors such as transportation. Example flow battery chemistries include zinc bromine or vanadium redox, and several other successful chemistries have been identified. They have become more cost-competitive for grid-scale applications. Packaged systems of the kW to one MW scale are available and may be suitable for BTM uses.

Table 17.2 Comparison among batteries, supercapacitors, and capacitors (Uke et.al., 2021).

	Batteries	**Supercapacitors**	**Capacitors**
Specific energy, Wh/kg	Highest 25−50 (lead acid) to 100−300 (Li-ion)	Low <1−20	Lowest <0.5
Specific power, W/kg	Low <10−500	High Up to 10,000	Very high 1,000–500,000
Cycle life (dependent on use case)	Up to 10,000	Up to millions	10,000– >100,000

Capacitors and supercapacitors. Capacitors and supercapacitors together fill a role of higher power density with rapid charge and discharge times as well as frequent cycling. Compared to batteries, they have very high cycle life and fast response, often less than 1 second. However, their specific energy is much lower than that of batteries. Table 17.2 compares some key characteristics of these technologies. The applications for capacitors include smoothing rapidly changing power loads. They store electrical energy in an electrical field using conductors and a dielectric or electrolyte materials and use a power electronic converter to charge and discharge. Capacitors are vital to circuit management, for example, by enabling alternating current but blocking direct current, or correcting power factor. They are often used in conjunction with battery systems to bridge gaps in specific power requirements. The high power density and low energy density of supercapacitors enables them to bridge between batteries and capacitors, delivering short (seconds to minutes), rapid bursts of high power, although at high cost (Rashid Khan & Latif Ahmad, 2025).

Supercapacitor efficiency has the potential to be the highest of any technology discussed. There is no theoretical limit to cycle life for capacitors, and so with proper operation and maintenance, they can last millions of cycles.

17.2.3 Thermal storage methods (used both grid-side and BTM) (Roosa 2018)

In commercial and industrial cases, the largest single consumer of energy in a facility can be the cooling and heating systems. The ability to shift the timing of this demand can have a large impact on peak load. As described above, load shifting and peak load reduction are tools to enable energy bill savings through energy arbitrage and demand charge management. Through demand response programs, this can lead to revenue generation. Thermal energy storage systems generate heat or cold during lower-demand times and store it for use during peak times. This can be in several forms and over a wide range of scales, as described below.

Hot or chilled water storage. The most common BTM method of thermal energy storage is pre-cooling water and storing it in tanks for later use in a chiller-based cooling system. The water can be used directly in the cooling system or mixed with chiller water output. It reduces the peak load of the chiller, reducing required equipment size. The design reduces or can even eliminate on-peak chiller operation. Tanks are best stored underground or well insulated. Efficiency drops with exposure to sunlight or ambient conditions and longer storage. Adding storage to a chiller system is a straightforward retrofit that requires the least energy of the cooling storage options.

Facilities or district heating systems that have boilers can also store hot water in tanks to levelize thermal load, thereby reducing boiler sizing and improving the reliability of heating when demand peaks.

Hot or chilled water storage represents straightforward ways to store energy. The basic formula for change in sensible energy storage in a solid or liquid medium (see Appendix formula sheet) is:

$$\dot{Q} = \dot{V} * \rho^{*} C_p{}^{*} \Delta T.$$

To calculate the sensible energy that can be stored in a given volume, use the formula:

$$C = mc_p \Delta T,$$

where C = heat content in Btu or kWh, m = mass of storage medium in lbm or kg, c_p = 1 Btu/lbm-F or 4.2 kJ/kg °C (for water), and ΔT is the temperature difference between the stored water and the return water, in °F or °C. This formula is used in the examples in Section 17.3.

Phase-change materials. These processes use latent heat to store energy. Certain salts, for example organic–inorganic eutectic urea CH3COONa * (H2O)3 (T_m = 30 °C), have relatively low melting points and are eutectic or easily (uniformly) melted salts. They can be built into the closed side of a tank-enclosed heat exchanger design whereby a salt solution is solidified by chiller output water on the shell side. When needed to enhance cooling, the shell side is run with warmer building outlet water and cooled as the salt material melts. The example urea solution contains about 80 Btu/lbm of cooling energy, reducing the volume required for storage significantly compared with a chilled water system. The Encyclopedia of Energy Storage (Sevault et.al., 2022) has a chapter on the use of phase-change materials in buildings with more details and a calculation example.

Ice storage. Storing ice rather than water reduces the required volume of storage by 75%–90%. Since ice contains 144 Btu/lbm of cooling energy, it is more efficient than water at 1 Btu/lbm-°F. However, more energy is required to create ice, with evaporator temperatures around 10–25 °F versus the mid-40s °F for chillers. These systems are available in packaged units that can retrofit to existing chilled water systems. They have higher storage density than water or many other phase change materials, but are less energy efficient.

Molten salt storage. These systems store sensible heat at grid scale in conjunction with renewable power generation. Solar energy can be used to heat salts with a high heat capacity over a wide range of temperatures. For example, sodium nitrate and potassium nitrate have melting points of 300–500 °C and volumetric heat capacities of 1,670–3,770 kJ/m^3 °C. An example salt blend used in solar plants has a liquid temperature range of 149–538 °C, enabling high energy densities to be stored at large scale in the form of sensible heat (Dodaro, John 2015). The heated molten salt is stored until needed and then can generate power by producing steam through a heat exchanger. The steam is used as the motive force in a turbine-generator. The heat can alternately be used directly for process support. The cost of these systems may be prohibitive in many cases, but the potential for large scale and coupling with generation technologies such as nuclear or renewable power is an advantage. Commercial molten salt systems in conjunction with concentrated solar power provide about 15–1,800 MWh of thermal storage for later use (National Technology & Engineering Sciences of Sandia, LLC 2021).

Both phase-change material methods, whether using salts or ice, require about 30% higher auxiliary power than water-based systems.

River Publisher's *Energy Management Handbook* (Roosa 2018) is a good resource for example problems and comparative economics of various cooling storage systems.

Thermochemical (Cabeza et al. 2021). This grid-scale developing method uses a reversible chemical reaction or sorption between two materials to store energy. The reaction products are stored until needed, which may be seasonal storage or diurnal. They are characterized by high heat storage capacity (higher than latent or sensible heat systems above), very low thermal losses during storage, and selection of an appropriate working pair of materials, such as silica gel or zeolite with water. An example is the dissociation of ammonia (NH_3) at a solar receiver (Silva-Pérez, 2017). The mixture of nitrogen and hydrogen is then stored in a tank until discharge through a steam generator, producing power and NH_3. Newer research on materials such as metal carbonates is identifying higher energy density materials that may provide a future pathway to broader use. Economic viability of these systems requires further technological improvements.

17.2.4 Developing and combined technologies

Hydrogen. Hydrogen as energy storage is getting increasing attention, research, and testing. Significant challenges remain for the cost-effective production and storage of hydrogen. The most economical way to produce hydrogen is through reforming of natural gas (methane reforming), which produces CO_2 as a byproduct. Most hydrogen today is produced from the fossil fuels: natural gas, coal, or oil. Water electrolysis is another way some

Table 17.3 Energy storage metric ranges for various technologies.

Method	Discharge duration magnitude	Max cycles or lifetime	Energy density (wH/L)	Round trip efficiency
Batteries:				
Lithium ion battery	Hours	1,000–10,000 Cycles*	200–400	~85%
Flow battery	Hours	12,000–14,000 Cycles	2–6	60%–85%
Lead acid battery	Hours	6–40 YEARS Hundreds of cycles	50–80	80%–90%
Mechanical systems:				
Pumped hydro	< 1 day	30–100 Years	0.2–2	60%–85%
Compressed air	Hours to a day or two	20–40 Years	2–6	40%–70%
Flywheel	Seconds to minutes	20,000–100,000 Cycles	20–80	70%–95%
Thermal energy storage:				
Sensible heat only	Hours	Determined by tank, piping, and pump life	10–50 kWh/ton	50%–90%
Phase-change materials ice molten salt	Hours	30 Years	~100 kWh/m^3 70–210 wH/L	75%–90%
Thermo-chemical storage			Up to 250 kWh/ton	75%–95%
Hydrogen	Not commercial; potential for broad range	5–30 Years	600 (at 200 bar)	25%–45%

Sources: Environmental and Energy Study Institute, 2019, IRENA, 2013, and National Technology & Engineering Sciences of Sandia, LLC 2021.

*Cycles are very dependent on DoD, *T*, design, etc. Lower DoD enables higher cycles.

hydrogen is produced. This has the benefit of having no carbon footprint if renewable energy is used, but it is very energy-intensive. Storing hydrogen brings inefficiencies as well. Hydrogen in gaseous form has very low volumetric energy density (3 kWh/m^3), while it has a high energy density by weight (33 kWh/kg). Typically, hydrogen is stored under high compression, with pressures from 350–700 bar. This translates to a loss of 10%–15% of the energy in compression energy. Storage of hydrogen in solids, like carbonaceous materials or metal hydrides, is an alternative methods of storing hydrogen. This solid-state-based approach leads to very high volumetric energy densities but at the expense of reduced gravimetric energy density and overall requires further research and development (Samantaray et.al., 2021). To store liquid hydrogen, a temperature of −253 °C must be reached, and attaining this is also a significant source of energy loss. After storage, hydrogen is utilized through a fuel cell, with efficiencies of about 40%–60%.

The overall energy conversion, from electricity to production of hydrogen by electrolysis and then conversion to electricity by fuel cells, is about 25% (Rufer, 2017). Although this is a low energy efficiency, the advantages of a carbon-free and transportable fuel and storage medium are driving further development. Hydrogen can potentially be at scale for seasonal storage, providing the opportunity to reduce curtailment of excess load by renewables, thus utilizing more renewable generation.

Bidirectional EV charging. Some EV battery charging systems are now being designed for two-way flow of electricity, enabling vehicle batteries to be used as a BTM or grid resource. This area is under rapid development and will likely be implemented over the next decade as EV deployment grows. Major vehicle manufacturers such as BMW are evaluating the use cases, controls system underpinnings, and pilot-scale implementation to enable commercial operation (Ostermann et.al., 2020). Peak shaving, for example, is being evaluated for both residential and commercial BTM applications. This is likely to be a key area for future energy management project evaluation and optimization, especially for fleet vehicles. Bidirectional charging can lead to improved resiliency as well as increased renewable utilization. A recent study showed that most vehicle batteries are often available, illustrating the potential scale of the unused energy resource available in EVs:

> "Statistically, private used cars in Germany are parked at home 20:15 hours per day, which equals 84% of the time in which they could be used as storages." (Ostermann et.al., 2020)

Pre-heating or pre-cooling. At a smart neighborhood project, researchers demonstrated that grid-connected controls anticipating energy usage patterns could effectively reduce peak electricity use by hot water heaters and HVAC systems by over 15%. This was accomplished through a combination of pre-heating hot water during low-demand times, pre-cooling conditioned spaces, and responding to grid events (Buckberry et al., 2020).

Hot water heater aggregation. Increasingly, energy aggregators are combining multiple commercial or residential hot water heaters into a larger-scale demand response or other grid revenue resource. Intelligent controls are essential to this function, as each water heater must be tied into the system and controls anticipate usage needs.

Batteries with solar. Most energy storage today is in combination with solar power generation. Tools are available online to examine use cases and these are helpful to identify site-specific conditions and incentive programs. NREL's ATB and PV Watts calculator are both available to scope and cost these systems. The use of PV Watts is presented in Chapter 18.

Passive phase change materials integration. Distinct from active systems that are driven by additional power input to circulate materials through heat exchangers, passive materials in building walls or structures are earlier in their development. These passive systems are described in Chapter 11 on building envelope design.

17.3 Example Problems of Battery and Thermal Energy Storage

17.3.1 Hospital battery example: battery energy storage system design

In a simplified case of a small hospital, typical hourly loads during one day of the six-month cooling season are shown in Table 17.4. The potential benefit of BTM battery storage is to reduce energy charges by load shifting to off-peak hours. We apply the Fort Collins municipal utility rates presented in Chapter 4, Table 4-4, for this case. Since the rates apply Monday–Friday, each weekday, the battery will charge during off-peak hours and discharge during peak. We will first calculate appropriate battery sizing, and then the energy charge savings that the designed battery would provide.

Battery sizing will depend on the hospital's goals and constraints (such as financing ability or power supply limits). We will focus on the case where

Table 17.4 Small hospital design load profile.

Hour	Load, kW	Summer M-F $/kWh (100 days)	Non-Summer M-F $/kWh (150 days)
0:00	110	0.0754	0.0754
1:00	105	0.0754	0.0754
2:00	105	0.0754	0.0754
3:00	105	0.0754	0.0754
4:00	108	0.0754	0.0754
5:00	105	0.0754	0.0754
6:00	106	0.0754	0.0754
7:00	99	0.0754	0.0754
8:00	103	0.0754	0.0754
9:00	116	0.0754	0.0754
10:00	134	0.0754	0.0754
11:00	143	0.0754	0.0754
12:00	153	0.0754	0.0754
13:00	158	0.0754	0.0754
14:00	160	0.2715	0.0754
15:00	163	0.2715	0.0754
16:00	164	0.2715	0.0754
17:00	162	0.2715	0.2503
18:00	169	0.2715	0.2503
19:00	166	0.2715	0.2503
20:00	159	0.0754	0.2503
21:00	151	0.0754	0.2503
22:00	147	0.0754	0.0754
23:00	144	0.0754	0.0754
Average Load	135		
Total	3,235 kWh	$437/day	$385/day

100% of peak load is shifted to non-peak, as this is the maximum useful battery size.

The battery output will be sized to cover the peak load period for each of summer and non-summer rate seasons. For example, summer peak energy load is calculated by adding each hourly load from 14:00 to 19:00 in Table 17.4, for a total of 985 kWh. The peak discharge power is also a key design point. This is seen at 18:00 to be 169 kW. Non-summer conditions are examined as well, to ensure the battery size covers both seasons:

	Summer	Non-summer
Energy used during peak pricing, kWh	985	807
Peak power discharge, kW	169	169
Battery duration	= 985/169 = 5.8 hours	5 hours

Based on this analysis, the battery specification for 100% load shifting from peak is 170 kW and 6-hour duration. Shorter duration battery configurations can also work, such as a 4-hour battery with power output of 250 kW. This latter battery would be discharged at a lower C-rate, still achieving 6 hours of energy supply. The sizing selection would also be based on obtaining quotes and specifications from suppliers to compare availability and costs, as well as any grid power supply constraints. Next, we calculate the rate savings.

The energy to charge the system is greater than the discharge because of losses. A round-trip efficiency of 85% is a reasonable assumption, so:

985 kWh/0.85 = 1,159 kWh needed to charge the BESS in summer

807 kWh/0.95 = 949 kWh needed to charge the BESS in non-summer

The cost of charging is based on these energy demands and uses the rates in Chapter 4, Table 4-4.

1,159 kWh × \$0.2715/kWh = \$87.36 to charge in summer

The energy cost that is offset when the battery is discharging in summer is:

985 kWh × \$0.2715/kWh = \$267.39/day

With about 100 days of summer peak rates, this is a savings of (\$267.39−\$87.36) × 100 = \$18,003.03.

Table 17.5 shows the results for both seasons:

Table 17.5 Summer and non-summer energy costs and savings.

	Summer	Non-summer
Energy to charge system, kWh	1,159	949
Cost to charge system, \$	\$87.36	\$71.58
Energy offset in discharge, kWh	985	807
Avoided cost, \$	\$267.39	\$201.96
Number of days at rate	100	150
Net Gain, \$/yr	\$18,000.03	\$19,558.06

The seasons are additive, so the annual energy charge reduction is \$37,561.09/yr.

We next examine the facility power supply impact, to see whether the kW demand at the hospital will increase because of the BESS.

Quantifying the inefficiency in the BESS charging: 1,185 kWh – 985 kWh = 174 kWh increase in use.

Adding this to the base facility energy use of 3,235 kWh results in 3,409 kWh. With over 18 hours of charging, the total levelized load is 3,409 kWh/18 hours = 189 kW. This is an increase of 20 kW over the baseline power consumption that peaks at 169 kW. Power supply needs to be examined to ensure this can be supported.

Figure 17.5 shows the daily pattern of facility use, battery charging and battery discharging, as well as the grid power supply. During peak load, the battery is discharging to meet facility usage and the grid draw is zero. During off-peak load, the battery is charging at a rate to target levelized grid load each hour. If no grid power constraints exist, the charging pattern could vary, resulting in much higher grid draw at times, and a more traditional charging pattern that charges the battery quickly at first, and then more slowly as it becomes full.

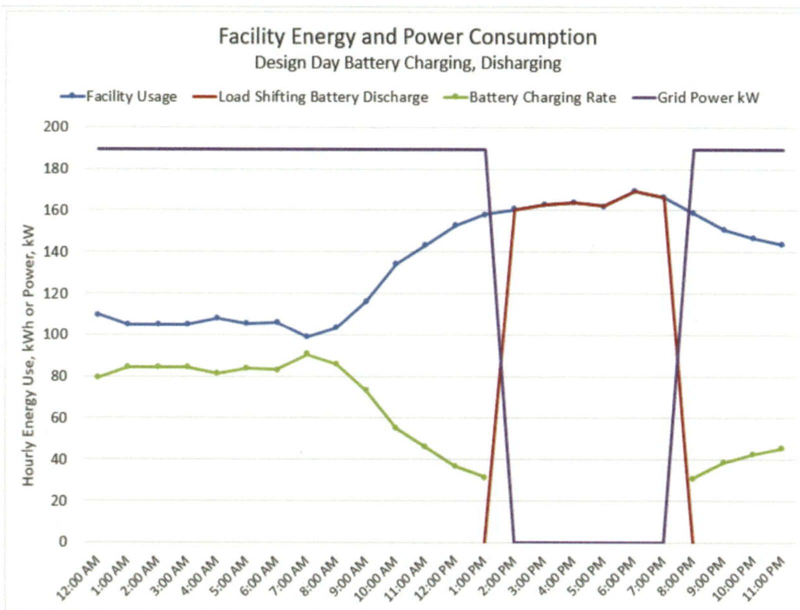

Figure 17.5 Facility power and energy usage with 100% load-shift BESS in place.

17.3.2 Commercial building chiller example

A commercial building is evaluating adding thermal storage to reduce the required size of a chiller. The load schedule is shown in Table 17.6. It is also shown graphically in Figure 17.6.

The default chiller sizing is 800 tons to cover the building's peak load. For this example, we will ignore losses and design factors of safety. We will size the chiller to shift load off of peak electricity pricing times. First, we size the chiller energy load, and then calculate the mass and volume of chilled water or ice needed.

Load shift case. Assume the peak electricity rates are from 1 PM to 7 PM, a total of 6 hours. To shift chiller load to zero for 6 hours, the entire chiller daily load needs to be produced in 18 hours. The chiller size to handle this shifted load is

$$13,150 \text{ ton-hr}/18 \text{ hours} = 731 \text{ ton}.$$

Next, we size the storage needed to support the load shift. Based on Table 17.6, the load from 1–6 PM is 800 ton and from 6–7 PM is 750 ton, for a total energy demand of 800 ton $\times$ 5 hours $+$ 750 ton $\times$ 1 hour $=$ 4,750 ton-hr. The storage size is based on this energy.

For chilled water storage, assume the storage is at 6 °C (42 °F) and return water is at 14 °C (56 °F), for a $\Delta T = 8$ °C. Referring to the Appendix formula sheet provides the conversion factors needed below.

Using $C = mc_p\Delta T$, where c_p for water $=$ 4.2 kJ/kg °C and $\Delta T = 8$ °C, we first convert the energy from ton-hr to kJ, and then solve for m.

Table 17.6 Daily cooling system load demand, tons.

Load demand, tons	Time of day	Number of hours	Total ton-hr
250	10 PM to 3 AM	5	1,250
400	3 AM to 7 AM 9 PM to 10 PM	5	2,000
500	7 AM to 8 AM	1	500
600	8 AM to 11 AM 8 PM to 9 PM	4	2,400
750	11 AM to 1 PM 6 PM to 8 PM	4	3,000
800	1 PM to 6 PM	5	4,000
Total		24	13,150

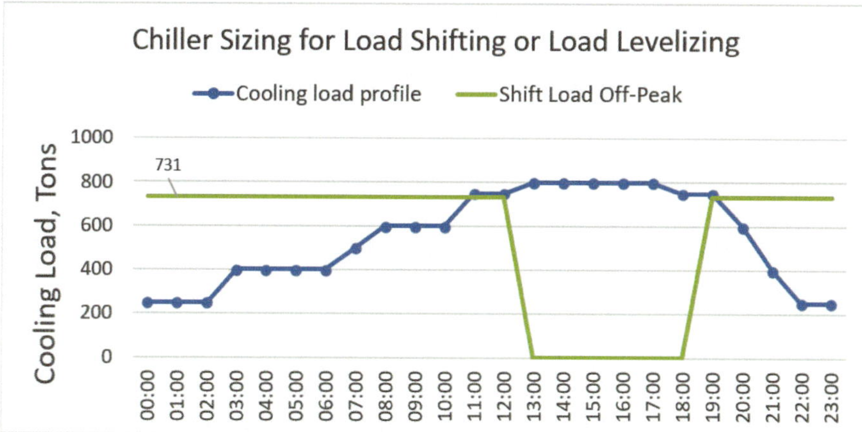

Figure 17.6 Chiller sizing for load shift case.

Stored energy load is 4,750 ton-hr. To convert to kJ, we have the following: 4,750 ton-hr $\times$ 3.51685 kJ/s/ton $\times$ 3,600 s/hr $=$ 60,138,135 kJ. Now the formula is

$$60{,}138{,}135\ \text{kJ} = m \times 4.2\ \ \text{kJ/kg°C} \times 8°\text{C}$$

So:

$$m = 1{,}789{,}825\ \text{kg water.}$$

Since the density of water is 1 kg/L, and a cubic meter is 1,000 L, this translates to 1,790 m^3 of chilled water storage.

Next we will size the ice storage for comparison. The stored energy load is unchanged from the chilled water calculation, so the formula for ice (latent heat only):

$$60{,}138{,}135 = m \times \text{heat of fusion.}$$

For ice, heat of fusion $=$ 335 kJ/kg. So: m $=$ 60,138,135 kJ/335 kJ/kg $=$ 179,517 kg.

And the density of ice is 0.919 kg/L. So: 179,517 kg/0.919 kg/L/1,000 L/m^3 $=$ 195 m^3.

From this comparison, we have seen that storing chilled water requires about nine times the volume of using ice. Also that the chiller size can be smaller (731 ton rather than 800 ton, in this example) if the chiller can be run at full load when building demand is low, as long as that energy can be stored.

17.4 Summary

Energy storage is increasingly used in many BTM applications to manage load, benefitting energy and equipment costs, environmental footprint, and/or resiliency, as well as in some cases, providing revenue. These benefits need to be weighed against the cost of storage for the given use case. There are a range of technologies for energy storage, with the most common BTM methods being thermal storage and batteries. On average, the largest energy use for residential sites, and about 40% for commercial sites, is heating and cooling. Chiller size can be reduced by adding thermal storage. Most typically, chilled water (or ice) is generated and stored when rates are lower overnight, and then utilized during the day when cooling loads are high, reducing peak load. Alternatively, hot water storage enhances heating capacity and reduces boiler size by reducing peak load. Batteries can be charged using grid power, but are increasingly paired with solar generation, improving utilization of variable renewable energy. Battery pack costs are dropping rapidly, with innovation being driven by electric vehicle demand as well as goals to improve greenhouse gas footprint. It is critical to obtain current market pricing and understand available incentives to best advise customers in this rapidly developing field.

Author Biography

Sheila Glesmann has extensive experience in project development, industrial operations, and commercializing emissions control technology. She has managed projects from weeks to years in duration, and from under $5k to over $350 million in scope. Her largest project to date has been the development and operation of the Red River Plant, an activated carbon producer in Coushatta, Louisiana. Her Maryland-based consulting practice is focused on industrial emissions control and management, as well as behind-the-meter energy management. She is a registered Professional Engineer in California and Maryland and a Certified Energy Manager. Her website is sincenergy.com.

Acknowledgments

I am grateful to Arman Bonakdarpour for his peer review of this chapter, which benefited greatly. Also I would like to thank Eric Mazzi for his excellent review, and for inviting me to join this noteworthy project to update the GTEM.

Practice Problems

1a. List an example of a potential benefit of energy storage in each category shown below.

- Environmental
- Cost
- Resiliency

1b. How would you evaluate value stacking in a given case?

2a. Describe the directional impact of each of the following on battery capacity:

- Extreme cold
- Hot temperatures
- DoD
- High cycle count
- Storing battery in a charged state for a long period

2b. List five items you would consider and include when specifying a BESS.

3. Greenhouse BESS Problem:

Quebec Hydro is offering a special rate for greenhouse heating and lighting. A greenhouse in Quebec qualifies for a reduced electric rate if it opts in to eliminate the extra heat and light electricity load attributable to crops during critical peak pricing (CPP) periods, which will be up to 5 hours in length. If it does not opt in, the greenhouse will be charged the CPP rate of $0.55/kWh for all electricity used during CPP periods, and the normal rate at other times.

Using the following operating assumptions, size and calculate the potential benefit and payback period of a battery energy storage system:

Reference (non-crop) load, kW Daily kWh	50
Crop power load (attributable to heat and light for crops), kW	150
# of 5-hour events per year	20
Annual crop energy load, kWh	900,000
Rates, $/kWh:	
CPP	0.55
Crop load with opt-in	0.06
Normal	0.10

a. What savings result from the reduced rate on an annual basis?
b. What size battery energy storage system (BESS) and approximate costs would be required to enable zero grid crop load during CPP periods?

Use Table 17.1 "Commercial" column for cost estimate.

c. Calculate the simple payback period for the BESS investment.

4. For the example chiller problem in the chapter text, show the case of levelized load that is constant over the 24-hour period. Use the same assumptions as in the text example.

a. Size the chiller
b. Calculate chilled water storage size in kg and m^3
c. Calculate ice storage size in kg and m^3

5. Greenhouse thermal storage problem.

A commercial greenhouse facility is procuring a propane-fired boiler and wants to incorporate thermal hot water storage to levelize the demand and reduce the boiler sizing. Design an option for boiler and storage sizing for this facility. Thermal hot water storage is very similar to chilled water storage in terms of the energy capacity of the water and the use of only sensible heat.

The thermal hot water tank will levelize the load such that peak hours can be covered if demand is higher than the boiler size.

Figure 17.7 shows the hourly thermal demand from the greenhouse over one year. Also shown is the outdoor temperature. This supports the following data:

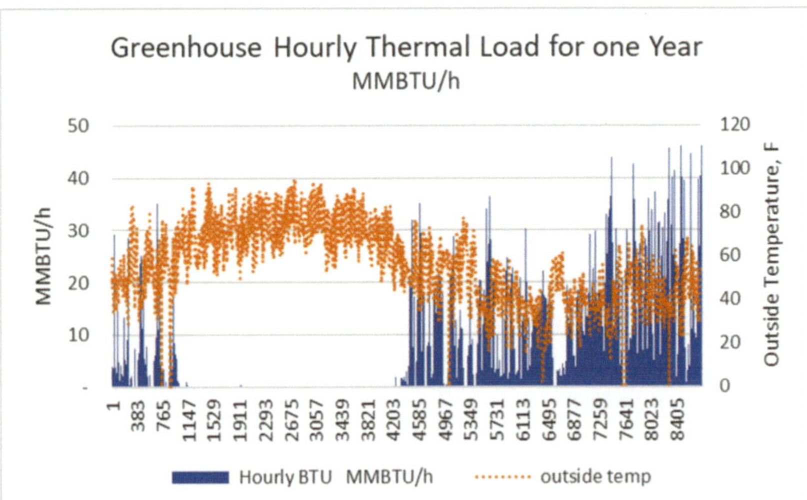

Figure 17.7 Commercial greenhouse actual thermal load demand. (source: Sinc Energy)

Peak design hour	40 MMBtu
Peak day	500 MMBtu
Peak week	2,500 MMBtu

Assume 80% storage efficiency and a design supply water temperature of 180 °F and return water temperature of 110 °F.

a. Calculate the thermal load of the facility in kWh and the implied boiler size in kW.

	MMBtu	kWh	Implied boiler size, kW
Peak hour			
Peak day			
Peak week			

b. Size boiler at 110% of peak day in kWh/day, kW and MMBtu/hr. How does this kW boiler size compare to the size required if the design had to address peak hourly demand?
c. Add storage to accommodate peak hourly load for 20 hours (remember to account for 80% efficiency factor here)

kWh	MMBtu

d. Calculate mass of water needed (lbm) and storage volume (gallons) using $C = mc_p\Delta T$.
e. Is a 7-MW boiler and 750 thousand gallon storage tank a reasonable design recommendation for this facility? Why or why not?

References

ACP. (2025, Q2). U.S. Energy Storage Monitor. American Clean Power. https://cleanpower.org/resources/u-s-energy-storage-monitor/

Bowen, Thomas et al. 2018. "2018 NREL Grid Scale Battery Storage 74426." NREL.

Buckberry, Heather, Jennifer Burke, Michael Starke, Helia Zandi, Jeffrey Munk, Supriya Chinthavali, Christopher Winstead, et al. 2020. "Smart

Technologies Enable Homes to be Efficient and Interactive with the Grid." ORNL/TM-2020/1507, 1615193. https://doi.org/10.2172/1615193.

Cabeza, Luisa F., Ingrid Martorell, Laia Miró, A. Inés Fernández, Camila Barreneche, Luisa F. Cabeza, A. Inés Fernández, and Camila Barreneche. 2021. "1 - Introduction to Thermal Energy Storage Systems." In Advances in Thermal Energy Storage Systems (Second Edition), edited by Luisa F. Cabeza, 1–33. Woodhead Publishing Series in Energy. Woodhead Publishing. https://doi.org/10.1016/B978-0-12-819885-8.00001-2.

Dodaro, John. 2015. "Molten Salt Storage." 2015. http://large.stanford.edu/courses/2015/ph240/dodaro2/.

EIA. 2020. "Utility-Scale Battery Storage Costs Decreased Nearly 70% between 2015 and 2018 - Today in Energy - U.S. Energy Information Administration (EIA)." October 23, 2020. https://www.eia.gov/todayinenergy/detail.php?id=45596.

EIA - U.S. Battery Storage Market Trends. (2025). https://www.eia.gov/analysis/studies/electricity/batterystorage/

"Energy Storage Grand Challenge Energy Storage Market Report." 2020, 65.

Environmental and Energy Study Institute. 2019. "Fact Sheet | Energy Storage (2019) | White Papers | EESI." February 22, 2019. https://www.eesi.org/papers/view/energy-storage-2019.

EPRI. 2020. "UNDERSTANDING USABLE ENERGY IN BATTERY ENERGY STORAGE SYSTEMS." December 2020. https://www.epri.com/research/programs/053125/results/3002019753.

IEA (2024), Global installed energy storage capacity by scenario, 2023 and 2030, IEA, Paris https://www.iea.org/data-and-statistics/charts/global-installed-energy-storage-capacity-by-scenario-2023-and-2030, Licence: CC BY 4.0

IRENA. 2013. "Thermal Energy Storage." January 1, 2013. https://www.irena.org/publications/2013/Jan/Thermal-energy-storage.

Lithium-Ion Battery Pack Prices See Largest Drop Since 2017, Falling to $115 per Kilowatt-Hour: BloombergNEF." 2024. BloombergNEF (blog). December 10, 2024.

May, Geoffrey, Alistair Davidson, and Boris Monahov. 2018. "Lead Batteries for Utility Energy Storage: A Review | Elsevier Enhanced Reader." Journal of Energy Storage 15 (2018):145–57. https://doi.org/10.1016/j.est.2017.11.008.

National Technology & Engineering Sciences of Sandia, LLC. 2021. "DOE Global Energy Storage Database." 2021. https://sandia.gov/ess-ssl/gesdb/public/index.html.

"New York State Battery Energy Storage System Guidebook." 2020. NYSERDA. December 2020. https://www.nyserda.ny.gov/All%20Programs/Programs/Clean%20Energy%20Siting/Battery%20Energy%20Storage%20Guidebook.

NREL. 2023a. "2023 ATB Commercial Battery Storage." Tableau Software. 2023. https://public.tableau.com/views/2023CMTS/TechSummary?:embed=y&:toolbar=no&Technology=Commercial%20Battery%20Storage&Core%20Metric%20Parameter=CAPEX,Fixed%20O%26M,Variable%20O%26M,OCC&:embed=y&:showVizHome=n&:bootstrapWhenNotified=y&:apiID=handler0.

NREL. 2023b. "2023 Cost Components." Tableau Software. 2023. https://public.tableau.com/views/2023CostComponents/CostComponents?:embed=y&:toolbar=no&Technology=Commercial%20Scale%20Battery%20Storage&:embed=y&:showVizHome=n&:bootstrapWhenNotified=y&:apiID=handler1.

"Ontario Reg. 429/04: Adjustments Under Section 25.33 of the Act." 2014. Ontario.Ca. July 24, 2014. https://www.ontario.ca/laws/view.

Ostermann, Adrian, Mathias Mueller, and Sebastian Faller. 2020. "Bidirectional Charging Management – Developing a Measurement Concept for Pilot Operation in Germany." In NEIS 2020; Conference on Sustainable Energy Supply and Energy Storage Systems, 1–6.

Ramasamy, Vignesh, Jarett Zuboy, Michael Woodhouse, Eric O'Shaughnessy, David Feldman, Jal Desai, Andy Walker, Robert Margolis, and Paul Basore. 2023. "U.S. Solar Photovoltaic System and Energy Storage Cost Benchmarks, With Minimum Sustainable Price Analysis: Q1 2023." Renewable Energy.

Rashid Khan, Humaira, and Abdul Latif Ahmad. 2025. "Supercapacitors: Overcoming Current Limitations and Charting the Course for next-Generation Energy Storage." Journal of Industrial and Engineering Chemistry 141 (January):46–66. https://doi.org/10.1016/j.jiec.2024.07.014.

Roosa, Stephen A. 2018. Energy Management Handbook. Ninth. River Publishers.

Rufer, Alfred. 2017. Energy Storage Systems and Components: Systems and Components. Edition 21.8.2017. Boca Raton: CRC Press.

Samantaray, Sai Smruti, Seth T. Putnam, and Nicholas P. Stadie. 2021. "Volumetrics of Hydrogen Storage by Physical Adsorption." Inorganics 9 (6): 45. https://doi.org/10.3390/inorganics9060045.

Sevault, Alexis, Fride Vullum-Bruer, and Olaf Lehn Tranås. 2022. "Active PCM-Based Thermal Energy Storage in Buildings." In Encyclopedia of Energy Storage, edited by Luisa F. Cabeza, 453–69. Oxford: Elsevier. https://doi.org/10.1016/B978-0-12-819723-3.00008-1.

Silva-Pérez, M. A. 2017. "17 - Solar Power Towers Using Supercritical CO2 and Supercritical Steam Cycles, and Decoupled Combined Cycles." In Advances in Concentrating Solar Thermal Research and Technology, edited by Manuel J. Blanco and Lourdes Ramirez Santigosa, 383–402. Woodhead Publishing Series in Energy. Woodhead Publishing. https://doi.org/10.1016/B978-0-08-100516-3.00017-4.

"Standalone Storage vs. Solar-Plus-Storage | EnergySage." 2021. 2021. https://www.energysage.com/energy-storage/should-you-get-storage/standalone-vs-solar-plus-storage/.

Uke, Santosh, Satish Mardikar, Ashwani Kumar, Yogesh Kumar, and Meenal Gupta. 2021. "A Review of π-Conjugated Polymer-Based Nanocomposites for Metal-Ion Batteries and Supercapacitors." Royal Society Open Science 8 (October). https://doi.org/10.1098/rsos.210567.

18

Solar Energy for Buildings

Arman Bonakdarpour

University of British Columbia, Canada

18.1 Introduction to Solar Energy

The global demand for renewable energy technologies has intensified, driven by the imperative to reduce greenhouse gas emissions (see Chapter 2) and the constraints of fossil-fuel-based energy sources. Within the realm of renewable options, solar photovoltaic technology is currently undergoing a golden age, propelled by technological advancements, resulting cost reductions, and governmental incentives. Solar energy can be harnessed through various approaches, and we will provide a brief overview of these methods. This chapter focuses primarily on the behind-the-meter solar photovoltaic technology and its use for buildings. An overview of solar thermal is included.

Solar photovoltaics (PV): Invented at Bell Labs in the 1950s, solar photovoltaic (PV) cells enable the direct conversion of solar radiation into electricity. The scalability of electricity production through solar PV technology, ranging from Watts to Gigawatts, has led to its substantial growth worldwide. Solar PV cells have the capability to harness both direct (unobstructed) and diffuse radiation (scattered, such as on cloudy days). This implies that even in regions with less-than-ideal conditions, such as northern Europe with a relatively high percentage of cloudy climates, solar PV technology can still generate significant levels of solar power. Solar photovoltaic (PV) technology can be deployed at small to medium scales ($\approx 1-1{,}000$ kW_{dc}) in a behind-the-meter configuration for building applications. This includes installations on rooftops of residential, commercial, or industrial buildings, aiming to minimize or eliminate the reliance on external power sources. Solar photovoltaic (PV) installations can be implemented at significantly larger scales ($\approx$ 10–1,000 MW_{dc}) in a front-of-the-meter configuration, such as solar farms for utilities, and integrated into the electric utility distribution and transmission infrastructure. When combined with complementary technologies like advanced batteries, solar PV offers intriguing possibilities in energy arbitrage, contributing to a notable decrease in power demand charges, especially in the presence of advanced electricity rate structures (see Chapter 4). Additionally, the reduction of soft costs and government incentives has played a crucial role in fostering market penetration and fostering the accelerated growth of solar PV systems. At present, the installed capacity of solar PV-based technology in the U.S. is about 100 GW_{dc}.

Concentrated solar power (CSP): Solar electricity can also be produced by using mirrors to concentrate solar radiation and yield high temperatures (750–1,200 °F, or 400–650 °C) to produce steam and generate electricity through a thermal power plant. This application of solar energy requires concentrators such as parabolic trough mirrors, solar power towers, or heliostats and is only effective at larger scales (10–300 MW_{dc}) for the utility level. The concentrators require primarily direct sunlight (unlike solar PV). This requirement, therefore, restricts the geographical regions where CSP plants can be constructed. For example, southwest regions of the U.S. are ideal for building CSP plants. The use of molten salts, which can store heat through the day, allows for electricity production at night as well. CSP can also be integrated with conventional thermal plants, e.g., a natural gas power plant, for a more effective use. At present, the installed capacity of CSP in the U.S. is about 1.5 GW_{ac}. Although CSP has had continuous growth, particularly in countries like the U.S., China, Israel, and Spain, its growth lags that of PV technology. More on CSP approach can be found elsewhere (Heller, 2017).

Solar thermal collectors: Solar radiation can also be used to pre-heat water or air for their use in buildings, thereby reducing heating costs. Both technologies are generally well-developed and can be integrated with buildings. For water pre-heating, there are several different types of collectors (panel pool heater, flat plate, and evacuated tubes) that allow for different levels of water pre-heating. These systems require other mechanical components such as pumps, valves, sensors, control units, and storage tanks. Some regions like China and Europe still see large interest in these systems, but competition with solar PV has made its penetration more difficult. An excellent, but somewhat more advanced, source on solar thermal systems is that of Duffie and Beckman (2023). German Solar Energy Society also provides a good handbook on solar thermal energy (German Solar Energy Society, 2010).

Passive solar approaches: Solar heat can also be harvested without the use of any active (power consuming) components and entirely in a passive mode. Examples of passive solar approaches include advanced windows, Trombe walls, and building strategies and designs, which can maximize the use of solar energy for space heating. These topics are covered in Chapter 11 (Building Envelope).

18.2 Global and the U.S. Trends in Solar PV Technology

In 2022, solar photovoltaic (PV) achieved a record-breaking global growth capacity, boasting the largest growth rate among all renewable energy sources at 26%, as reported by the International Energy Agency (IEA 2024). IEA also reports that this growth rate matches the expected rated for the Net Zero Emissions by 2050 but requires continued development in the supply chain and policy support (IEA 2024).

In the U.S., solar PV has grown with an annual average rate of 24% over the last decade, reaching over 140 GW_{dc} in 2023, sufficient to power over 27 million homes in the U.S. Figure 18.1 shows the growth of installed solar PV capacity by market and total CSP growth in the U.S. for the period of 2007–2021.

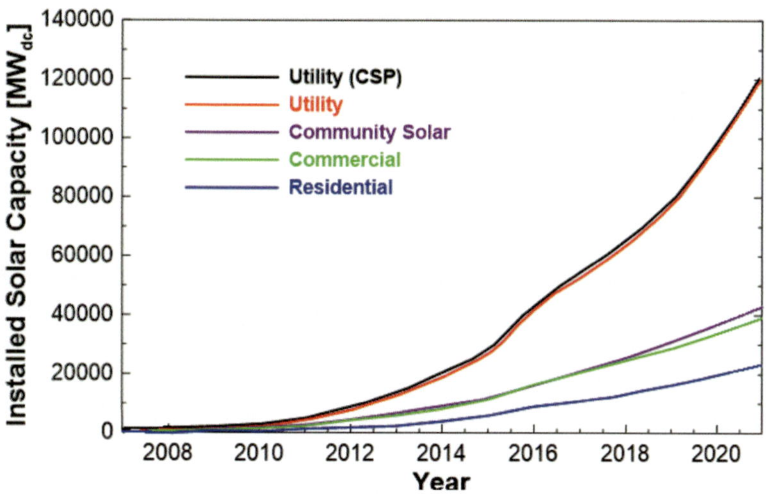

Figure 18.1 Growth of cumulative installed solar PV capacity by market and total CSP growth in the U.S. for the period of 2007–2021. Adapted from (Solar Energy Industries Association 2024).

With this growth, more than 263,000 Americans employed, and the industry has generated more than $35 billion of private investment.

18.3 Solar Energy Related Information Resources

Numerous resources and websites provide up-to-date information on solar technology, policies, costs, installers, and training. For a more in-depth exploration of these topics, we recommend that interested readers explore these resources. This section highlights a few of these sources, and additionally, more specific resources and databases will be introduced later in this chapter.

National Renewable Energy Laboratory: Originally founded as the Solar Institute in the late 1970s and located in Golden, Colorado, NREL is one of the main U.S. National Labs sponsored by the U.S. Department of Energy (DOE, 2024). NREL's core focus is research and development in several key renewable energy technologies, including solar, wind, biomass, and geothermal. Solar activities at NREL include R&D in photovoltaics, concentrated solar power, grid system integration, market research, and analysis (NREL 2024a). In the area of photovoltaics research, NREL conducts both novel research and measurements and characterization of available PV cells. NREL provides, free of charge, the System Advisor Model, a software package for the techno-economic assessment of renewable energy solutions with a strong emphasis on solar energy, encompassing various financial options such as behind-the-meter, power purchase agreements, and technologies like PV, thermal, batteries, etc. There is a significant amount of online guidance and support for the System Advisor Model, and it is well-maintained with regular updates. NREL also provides a simplified but effective PV assessment tool called PVWatts® Calculator, which we will use later in this chapter (NREL 2024b).

North American Board of Certified Energy Practitioners®: Since 2003, this board has been providing widely recognized certifications for professionals in the field of renewable energy (NABCEP, 2024). These certificates are well-suited for both skilled professionals and those new to the fields of photovoltaics, solar heating, and small wind technologies. This organization offers three general Associate certificates and eight Board Certification programs in these related areas. Board Certifications in solar include design, installation, commissioning and maintenance, technical sales, and inspection. Various local schools and organizations offer courses that provide the required training for these certificates.

Solar Energy International: This is a well-established organization that has been providing training for the global solar energy workforce for the past 30 years. The training covers publications, hands-on experience, safety, and technical skills for individuals and small-scale projects to utilities wishing to operate large-scale solar farms (Solar Energy International 2013).

Solar Reviews: Recognized as one of 5,000 fastest-growing companies in 2020 (in the U.S.), Solar Reviews, founded in 2012, provides accurate and current solar-related information for homeowners. These include reviews of up to 30,000 solar installer reviews, technology reviews and descriptions, and an advanced online solar calculator for accurate savings estimations (Solar Reviews 2024a).

18.4 Solar and Insolation and Collector's Orientation

Solar irradiance and insolation: Intensity of solar radiation, measured in the units of W/m^2, is referred to as *irradiance*. Irradiance is an instantaneous quantity and changes throughout out the day. *Insolation*, (also known as irradiation, solar exposure, solar insolation, or insolation), however, is the amount of solar energy received in kWh/m^2 for a given period. Insolation is also commonly expressed in terms of *peak sun hour*, which is the amount of time in hours with an equivalent irradiance of 1,000 W/m^2. For example, if the total insolation for a specific location and time of the year is 5.5 $kWh/m^2/day$, the corresponding sun hours is 5.5 hours. This, however, does not mean that the Sun is necessarily radiating at a constant rate of $1{,}000/m^2$, throughout the day (or 5.5 hours); rather it means that the total radiation over the length of the day averages to 5.5 hours of one sun (1 sun = 1,000 W/m^2) radiation. Figure 18.2 is an illustration of this concept.

Solar radiation can be estimated using mathematical models. However, the most straightforward and accurate way of obtaining solar radiation is through direct measurements. Many weather stations (e.g., airports) have solar radiation measurement equipment and record the radiation throughout the year. These data are widely available through the internet and can be accessed for many places around the globe. One common data format is the *typical meteorological year* (TMY). TMY data are averaged over years, and for each month, a typical meteorological year is selected, effectively averaging the effects of clouds, for instance. The overall result is that solar radiation data available may not necessarily be accurate for a particular day, but the annual average values are very close. In solar projects, there is typically a focus on yearly production.

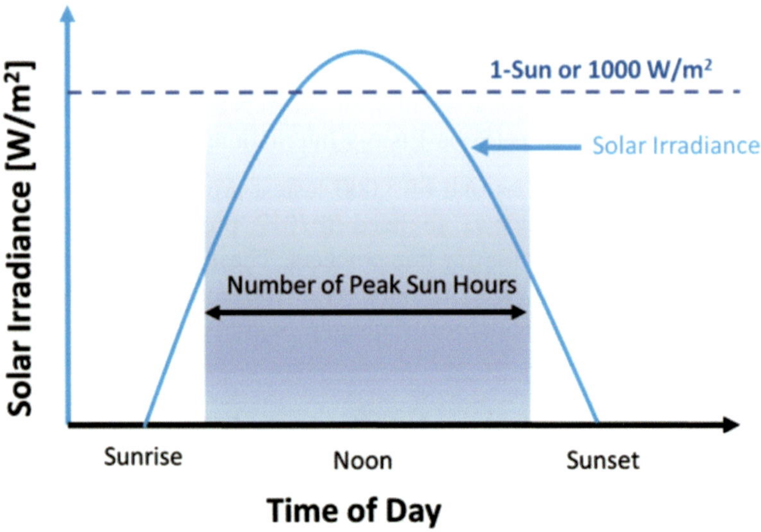

Figure 18.2 The area under the curve, which is the total radiation received, is equal to number of peak sun hours × 1,000 W/m^2.

Solar coordinates and the Sun's trajectory: Solar irradiance is a function of the Sun's angular coordinates (altitude and azimuth) in the sky, which constantly change during the day and vary throughout the year. The altitude measures the elevation of the Sun in degrees above the horizon and ranges from 0° to 90°, while azimuth is a measure of the angular distance of the Sun from the northern point along the horizon and ranges from 0° to 360°. Figure 18.3 shows the solar coordinates. The Sun's daily trajectory in the sky begins with sunrise above the horizon (near the East), passes through the sky, reaches a maximum altitude at solar noon at the southern point (or northern for the southern hemisphere), and then ends at sunset (near the west). In the northern hemisphere, this trajectory takes the longest path during the summer solstice (June 21) and shortest path during the winter solstice (December 21). During the equinoxes (March 21 and September 23), the Sun takes an average trajectory and takes exactly 12 hours to complete (Figure 18.4). We will use these simple concepts in Box A to evaluate optimum configuration(s) for a fixed solar collector.

Orientation of solar collectors: To harvest the maximum irradiance throughout the day, the collector would naturally have to track the Sun continuously. Exact tracking of the Sun is possible with the aid of motors that provide

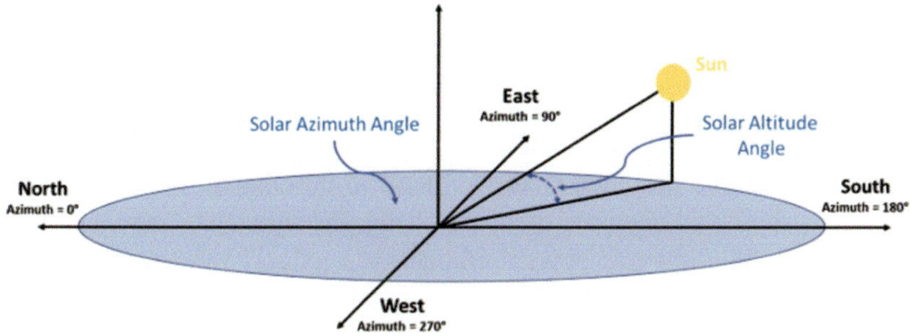

Figure 18.3 Azimuth and altitude angles describe the coordinates of the Sun in sky. Note: Here azimuth is shown with respect to north = 0. Alternative definitions of azimuth use south as the reference zero point.

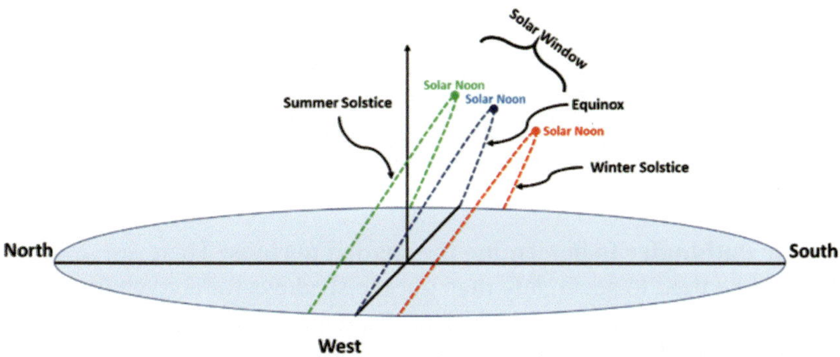

Figure 18.4 The solar window covers the region of the sky where the Sun's pathway falls during a year. The Sun will not be found outside of this region for this given (mid-northern) latitude.

two-axis tracking. However, this approach is quite costly and is usually used when concentrators, like parabolic dishes, are employed. Single-axis tracking is more common than two-axis tracking, including for some large-scale solar thermal plants that use parabolic trough concentrators. Residential or commercial building rooftop or ground-based solar PV panels are nearly always installed with no tracking mechanism. If a proper racking system is used, then there may be the possibility of changing the angle of the tilt at regular intervals to receive higher insolation. However, rooftop racking systems are less frequently used. For rooftop applications, the prevalent method involves installing modules on rails secured into roof structures, parallel to the roof. The relevant parameter in this case is the roof pitch (rise over run) or the angle of the roof. If solar modules are mounted on the ground or a flat rooftop, racking is utilized, and the angle can be adjusted. We will investigate the impact of geographical location (i.e., latitude), roof pitch, and angle tilt (when racking is used) in Box A.

Shading: Obstructions that cast shadows need to be considered, and a site survey would provide an accurate measure of the obstacles that cast shadows on the modules. Shading has several impacts on solar PV systems. First, a shaded region would naturally receive less insolation, resulting in a proportional loss of produced electricity. Therefore, it is imperative that, in the early stage of site assessment, a proper determination of possible shading is performed. There are several tools that allow one to determine the annual shading in monthly percentages for a desired location. We will discuss these methods here and see a practical application in Box B. Shading could also have a detrimental impact on the PVs and power produced. Because of the semiconducting nature of the PV cells, current will not be able to flow through a PV module properly when it is partially shaded, and this could lead to overheating and damage. To overcome this issue, solar PV modules use internal bypass diodes that allow current to flow in these situations and avoid potential hazards.

Using Solar Pathfinder to determine the shaded periods: There are several tools that can be used to accurately map out the shadows for a given location and incorporate into the design calculations. Solmetric, for example, provides a handheld digital device called SunEye (www.solmetric.com). Equipped with fisheye lens, digital compass, GPS, and tilt sensor, SunEye allows the user to perform onsite survey of the desired location and assess the shade. This device enables the collection of several data points in quick succession. Even for a small residential rooftop, at least four data points are taken to generate adequate shading information. The shade data can then be imported into the user's software and used to correct the solar insolation.

Another tool that could be used is a manual device for tracing shade patterns by Solar Pathfinder (www.solarpathfinder.com). The Solar Pathfinder™ uses a polished and transparent plastic dome to provide an overall view of the desired site. All objects in the view, such as buildings, trees, etc., show reflections on the surface of the dome. Afterward, a pen is employed to trace the shadow boundaries on the Sun chart diagrams, which are placed beneath the dome. These diagrams can be observed through the dome or photographed for subsequent analysis (Figure 18.5). The Solar Pathfinder can be used at any time of the day (when the Sun is shining) or

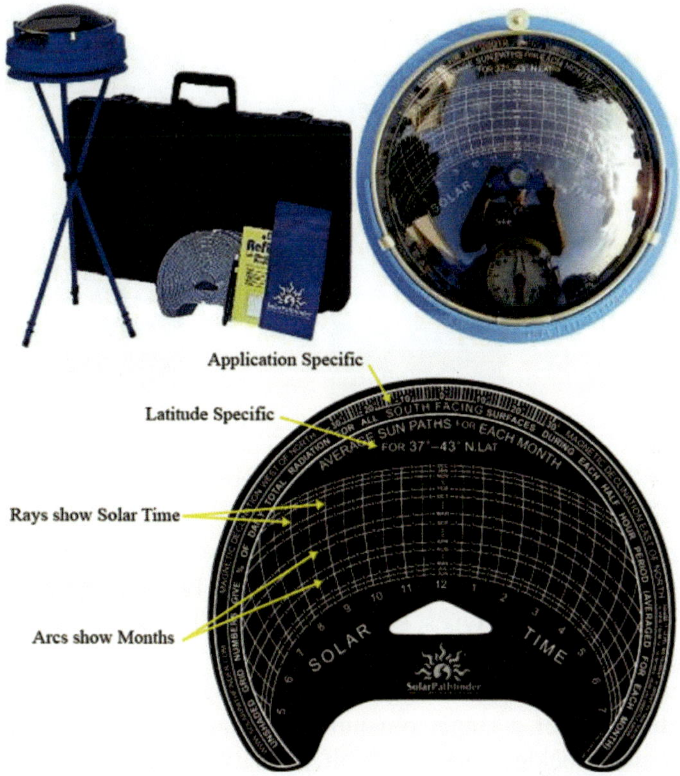

Figure 18.5 The Solar Pathfinder kit showing the dome, sun chart, and the case. Close view of the dome showing the compass (for proper alignment), level bubble, and the reflection of the objects on to the dome. An appropriate solar chart corresponding to the latitude (or sufficiently close) is inserted under the dome where the boundary of shadows can be traced (or a photo can be taken) and a solar path diagram sheet good for latitude 37° to 43° North. Adapted from (Solar Pathfinder 2024).

year, irrespective of the Sun's position in the sky because it uses reflections of the objects, which are independent of the Sun's position in the sky.

18.5 Photovoltaics Solar Cells and Electricity Production

18.5.1 PV system components

Photovoltaics: The first solar cells for converting solar radiation to electricity were invented in the early 1950s when p-n junction devices and semiconductors were first introduced. At the heart of a solar cell, there is a p-n

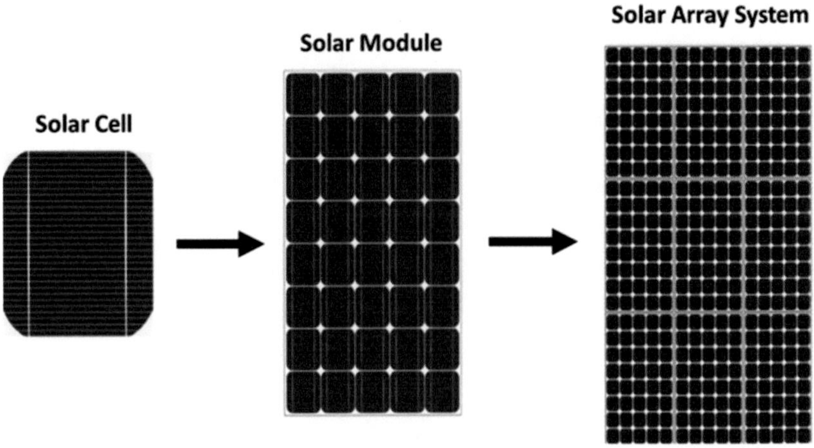

Figure 18.6 Solar cell, module (with 40 cells), and a solar array with nine modules.

junction semiconductor layer that allows for the absorption of photons and the generation of an electric current. The most widely used semiconductor in solar cells is Si, but other types, such as CdTe, CIGS,[1] and organic-based PV cells, have also been developed. The semiconducting properties of a p-n junction can be tuned to absorb specific wavelengths, but the maximum theoretical efficiency of a single p-n junction solar device is 33% and is referred to as the Shockley–Queisser limit. To harvest photons over a lager range of wavelengths, different p-n junction semiconductors can be stacked together in tandem to increase the overall efficiency, but this approach is costly.

To enhance power output, solar photovoltaic cells are connected in series, forming a collective unit known as a solar module or solar panel. A solar module is the smallest unit usually acquired and employed in the design of a larger solar array. Manufacturers provide specifications at the module level. Subsequently, based on the needed voltage and current levels, modules are combined in series and/or parallel configurations to create a solar array (Figure 18.6).

Solar photovoltaic cells: Solar cells are semiconducting devices that consist of p-type (hole-doped) and n-type (electron-doped) materials. They can absorb photons to generate what are known as electrons and holes, along with a voltage of around 0.5 V. When we place a solar cell under radiation

[1] CIGS cells are manufactured from copper, indium, gallium, and selenide.

and connect it to a load, an electric current passes through the load, delivering power. Figure 18.7 (top) shows the composition of a solar cell and a solar cell circuit delivering current. To have power, we need both voltage and current ($P = I \times V$); so it is instructive to consider the $I-V$ (current–voltage) curve of a solar cell. Figure 18.7 (bottom) shows a typical $I-V$ curve of a solar module, the corresponding power ($P-V$) curve, and the important points on these curves. The maximum power point (MPP) corresponds to the point on the $I-V$ curves where the maximum power is delivered.

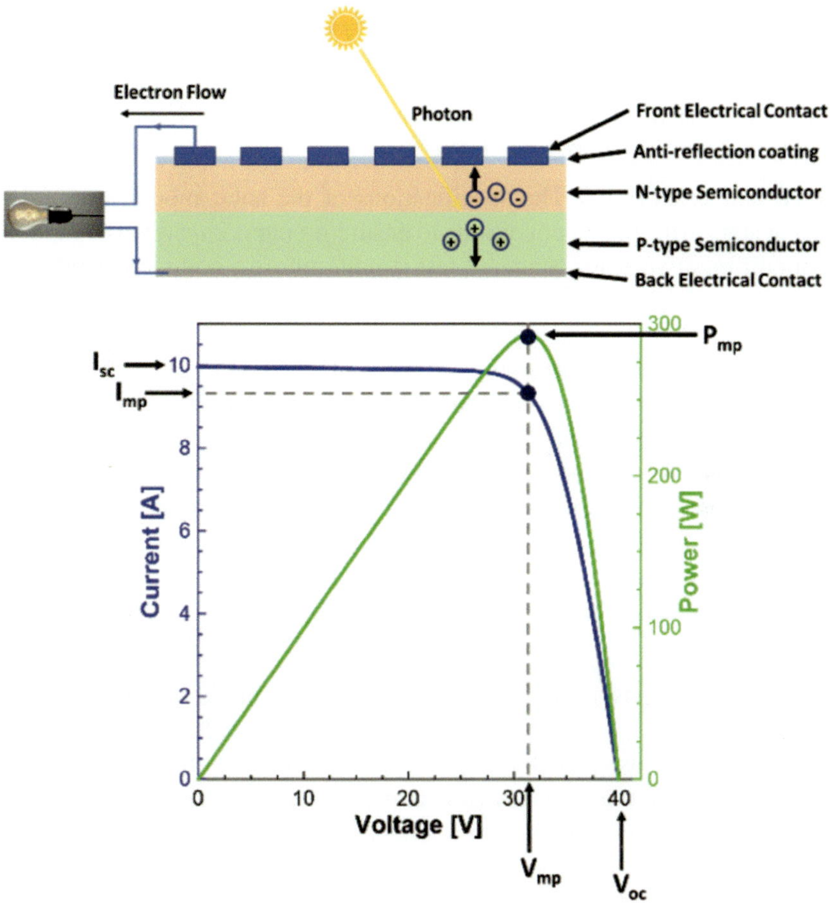

Figure 18.7 Top: p-n semiconductors, electron–hole separation and generation of electricity in a solar PV cell. Bottom: Schematic diagram of a solar PV module's $I-V$ and power curves with specific voltage/current/power points indicated.

- *Short circuit current (I_{SC}): The current that is drawn when both leads of the PV cell (or module) are shorted. No power is delivered and voltage is forced to zero.*
- *Open circuit voltage (V_{OC}): The leads of the cell (or module) are open and not connected to a load. Current is zero and cell voltage is at its maximum.*
- *Maximum power point (P_{mp}): The point on the $I-V$ curve where the load is drawing maximum power.*
- *Current at maximum power point (I_{mp}): The current that corresponds to the maximum power point.*
- *Voltage at maximum power point (V_{mp}): The voltage that corresponds to the maximum power point.*

The $I-V$ curve itself is a function of operating conditions (e.g., temperature and irradiation) as we will see later.

Standard test conditions: The specifications of the solar modules are all reported at standardized test conditions to ensure proper comparison of modules from different vendors. Standard test conditions are a common standard in testing PV modules, which include: 1,000 W/m^2 of radiation, a cell temperature of 25 °C (77 °F), and an air mass index of 1. Note that the temperature here refers to the cell temperature, not the ambient temperature. The cell temperature is always greater than the ambient temperature because the cell heats up under solar radiation. The specified power (and voltage, current) of the module is referred to as the rated power (and rated voltage, current) at standard test conditions. Solar module specifications also provide additional parameters that allow one to estimate the power output at non-standard test conditions. These conditions are set in a lab and are used to characterize all modules, so the $I-V$ curves are expressed under a standardized set of conditions.

Solar modules: To boost power output, individual cells are linked in series to form a solar module. A solar PV module (or PV panel) serves as the standard unit in solar PV systems, available for purchase and assembly to configure the desired PV system. A module is a series connection of many cells. Figure 18.8 shows a commercial PV module from LONGi Group (LR4-72HPH-440M) and its respective $I-V$ curves. This module consists of 144 cells in series, which provide a V_{mp} of 41.1 V_{DC}, an I_{mp} of 10.71 A_{DC}, and a rated power of 440 W_{DC}.

A typical solar array consists of parallel and series combinations of modules to provide the desired power and the respective voltage and current

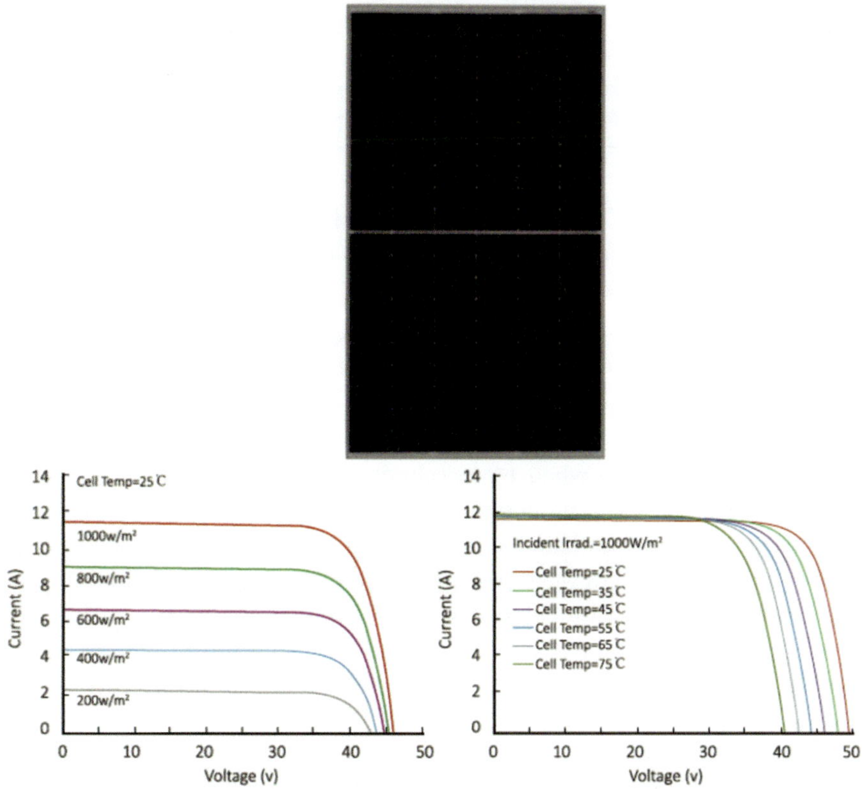

Figure 18.8 Top: LR4-72HPH-440M PV module from LONGi. Bottom left: The $I-V$ curves of the LR4-72HPH-440M modules at different solar radiations from 300 to 1,000 W_{DC}/m^2 at 25 °C. The I_{sc} is directly dependent on the radiation level; so doubling the radiation leads to a doubling of the I_{sc}, but input radiation has a rather negligible effect on the module voltage. Bottom right: The impact of temperature on the $I-V$ curves is also clear. The $I-V$ curve at 1,000 W_{DC}/m^2 at 75 °C has a I_{sc}, which is very close to that of 1,000 W_{DC}/m^2 at 25 °C, but its voltage (V_{mp} and V_{oc}) is reduced considerably ($\approx$ 10 V_{DC} in the V_{oc}). Adapted from (LONGi 2024).

levels. A series combination of modules is referred to as a string. To increase the system voltage, PV modules are added in series and to increase the current, strings are connected in parallel. The voltage and current of the overall array must meet certain electrical codes, and, therefore, the number of modules and their series/parallel configuration must be determined correctly. Residential and commercial codes set the maximum voltage to 600 and 1,000/1,500 V. The choice of an inverter could also impose further voltage

Table 18.1 Module, string, and array voltage/current/power.

	Module	**String**	**Array**
V_{mp} (V_{DC})	41.1	411	411
I_{mp} (A_{DC})	10.71	10.71	21.41
MPP (W_{DC})	440	4,400	8,800

limitations on the strings. **Table 18.1** shows the MPP-rated power of module, string, and the resulting for a parallel combination of two strings each having 10 modules (LR4-72HPH-440M) in series.

Thermal effects: Temperature significantly influences the voltage, current, and power output of solar cells. As temperature rises, the voltage of a solar cell decreases, while its current experiences a slight increase. Consequently, the overall effect leads to a drop in power output when temperatures rise (Figure 18.8). Conversely, when temperatures fall, the opposite occurs – module voltage increases. The coefficients of voltage (and power) loss in solar cells depend on the semiconductor type used in cell construction and are expressed as a percentage of rated power/voltage (%/°C) or directly in terms of power/voltage (W_{DC}/°C or %V_{DC}/°C) for every degree above (or below) the standard testing condition, which is 25 °C (77 °F). When PV modules are exposed to the Sun, they could easily be 15–30 °C (59–86 °F) above the ambient temperature. For sizing reasons, it is, therefore, important to know the maximum and minimum of the rated voltage (power) that solar module may produce for specific site. Applications of the thermal coefficients, when expressed in terms of percentage of rated power/voltage (%/°C), require knowledge of the deviations from 25 °C (i.e., standard test conditions), hence, the cell temperature under the desired operating conditions. An additional parameter, called normal operating cell temperature (NOCT), allows us to estimate the module temperature. NOCT for a module is determined in test labs under a constant radiation of 800 W_{DC}/m^2 with an ambient temperature of 20 °C (68 °F). When this parameter is known, e.g., from the product specification sheet, the module temperature can be calculated according to:

$$T_{\text{cell}} = T_{\text{ambient}} + \frac{\text{NOCT} - 20}{0.80} \times S, \tag{18.1}$$

where $T_{ambient}$ is the temperature of the module environment and S is the input radiation in kW/m^2.

Example 18.1, Thermal effects: We intend to use LR4-72HPH-440M module in a location where the highest and lowest temperatures are 42 and –18 °C

Table 18.2 LONGi PV Module (LR4-72HPH-440M) specifications.

Nominal efficiency	**20.2%**	**Max power current (I_{mp})**	**11.46 A**	**P_{mp} thermal coefficient**	**−0.350 %/°C**
Max power point (P_{mp})	440 W	Open circuit voltage (V_{oc})	48.9 V	V_{oc} thermal coefficient	−0.270 %/°C
Max power voltage (V_{mp})	41.1 V	Short circuit current (I_{sc})	11.46 A	I_{sc} thermal coefficient	0.048 %/°C

and the corresponding radiation levels are 750 and 600 W/m^2, respectively. Using the values provided in Table 18.2, we will calculate the module open circuit voltage and V_{mp} values for these extreme cases. The NOCT value of this module is 45 °C.

Example 18.1 Solution: We first need to calculate the module temperature at these conditions and then using the appropriate thermal coefficients, we will calculate V_{oc} and V_{mp}.

a) For $T_{ambient} = 42$ °C, the cell temperature would be:

$$\begin{aligned} T_{\text{cell}} &= 42 + \frac{45 - 20}{0.80} \times 0.750 = 65.4\ ^\circ\text{C} \\ \Delta T &= 65.4 - 25 = 40.4\ ^\circ\text{C} \\ \Delta V &= 40.4\ ^\circ\text{C} \times (-0.00270/\ ^\circ\text{C}) \times 48.9\ \text{V} = -5.34\ \text{V} \\ \text{Module } V_{\text{oc}} &= 48.9 - 5.34 = 43.6\ \text{V} \end{aligned}$$

The product specification sheet does not provide the thermal coefficient of V_{mp}. In this case, it is sufficient to use the P_{mp} coefficient, which is given as −0.35%/°C:
$\Delta V = 40.4$ °C × (−0.0035 %/°C) × 41.1 V = −5.81 V
Module $V_{mp} = 41.1 - 5.81 = 35.3$ V

b) Similarly, for $T_{ambient} = -18$ °C, the cell temperature would be:

$$\begin{aligned} T_{\text{cell}} &= -18 + \frac{45 - 20}{0.80} \times 0.750 = 5.44\ ^\circ\text{C} \\ \Delta T &= 5.44 - 25 = -19.6\ ^\circ\text{C} \\ \Delta V &= -19.6\ ^\circ\text{C} \times (-0.0027/\ ^\circ\text{C}) \times 48.9\ \text{V} = 2.58\ \text{V} \\ \text{Module } V_{\text{oc}} &= 48.9 + 2.58 = 51.5\ \text{V} \end{aligned}$$

For V_{mp}, we have:
$\Delta V = -19.6$ °C × (−0.0035/°C) × 41.1 V = 2.82 V
Module $V_{mp} = 41.1 + 2.82 = 43.9$ V

The extreme voltage values are important considerations for the design to ensure that electrical code and inverter's required input can be met.

Derate factors: When deploying PV modules, several factors contribute to the reduction of output AC power. Soiling or the accumulation of dust on PV modules, for instance, can result in reduced power output. In areas where soiling is a concern, periodic washing and cleaning of the modules can help reduce this source of loss. In rainy locations, soiling becomes a smaller issue. Deviations from the nameplate power rating due to manufacturing faults are another possible source of power loss. When there are slight differences in PV modules, either due to manufacturing or time-dependent deteriorations, we have a module mismatch loss, meaning the total power would not be a simple sum of the two different powers but slightly less than the total sum. Annual degradation, ohmic losses in the wiring, system availability (i.e., when the grid may be down for repair or due to accidents), snow coverage, and inverter losses are also relevant and important loss factors. Naturally, the exact breakdown of these derate factors is system- and location-dependent, but in general, we are interested in an overall derate factor that is a multiplicative product of these individual derate factors. Current PV systems have overall derate factors of $80\%-85\%$.

Example18.2 Overall derate factor and output AC power: The rated power of a PV module is 440 W_{DC}. What is the overall derate factor and the produced AC power if we have the following losses: soling $= 2\%$, module mismatch $= 2\%$, degradation $= 1\%$, wiring $= 4\%$, nameplate rating $= 1\%$, snow $= 2\%$, availability $= 2\%$, inverter $= 3\%$?

Example 18.2 Solution: The overall derate factor and the AC generation from the module are:

$$\text{Derate factor} = 0.98 \times 0.98 \times 0.99 \times 0.96 \times 0.99 \times 0.98 \times 0.98 \times 0.97 = 0.842 \text{ or } 84.2\%$$

$$P_{AC} = P_{DC} \times \text{Derate Factor} = 440 \text{ W}_{DC} \times 0.842 = 370 \text{ W}_{AC}.$$

String inverters: An inverter is a critical part of any solar PV system, unless all the loads are DC. The basic function of inverters, specifically designed for solar PV applications, is to convert the DC voltage of PV modules to AC using power electronics and diodes. Inverters have gradually become more sophisticated and are able to perform several other important functions besides DC/AC conversion. There are several different types of inverters for solar applications. For solar PV systems that are grid-tied systems, high-quality inverters – those that produce high-quality AC sine waves – must be used because when excess power is produced, electricity will be exported to the grid. For stand-alone systems, lower-quality inverters may be used, particularly if loads are simple (e.g., lights and heaters), but more delicate loads like motors would still operate better with high-quality inverters. Grid-tied inverters have an automatic mechanism to disable the PV system when grid power is interrupted. This safety feature, known as anti-islanding, helps prevent potential accidents, especially when electricians may be working on the grid. For inverters used in multi-mode PV systems, where batteries are integrated along with grid connection, an additional feature called a transfer switch is mandatory. The transfer switch enables the system to connect the critical loads to the battery bank and switch back to the grid/PV when power reconnects. Inverters also have specific internal power electronics and a built-in digital control system that ensures the PV modules operate near their MPP point. This functionality is referred to as *maximum power point tracking*. An example of a solar PV inverter is the *Fronius 5.0-1*, which we will examine in more detail in Box D. Inverters have a typical operating voltage range that needs to be provided by the PV modules. Beyond this range, the efficiency of the inverter would reduce drastically, and below a threshold voltage, the inverter will not be able to function. Proper MPP tracking functionality also requires a certain voltage range from the PVs and is generally smaller than the overall voltage range of inverters (Figure 18.9). Technical specifications of the *Fronius 5.0-1* inverter are also provided in Figure 18.9.

Microinverters: These devices are much smaller than string inverters and allow DC–AC conversion at the module level; otherwise, they provide all the functionalities of string inverters. Some vendors offer microinverters that can be used with several modules. Microcontrollers allow for more control and can perform at the MPP for each individual module. Also, the output of each module can be tracked and monitored using the corresponding app on a cell phone or computer. This means the output of each module is AC and, in general, simplifies the wiring. Microinverters have generally been more expensive than string inverters, but they are gaining more popularity. Additionally, there are vendors who offer combined PV module and microinverter in one package to have a module with AC output. The output of microinverters is connected in parallel, and string sizing is not required.

DC optimizers: Like microinverters, DC optimizers are small units that can be combined with each module and ensure that the module is working near its MPP. However, their output remains DC. Therefore, a string inverter is still required.

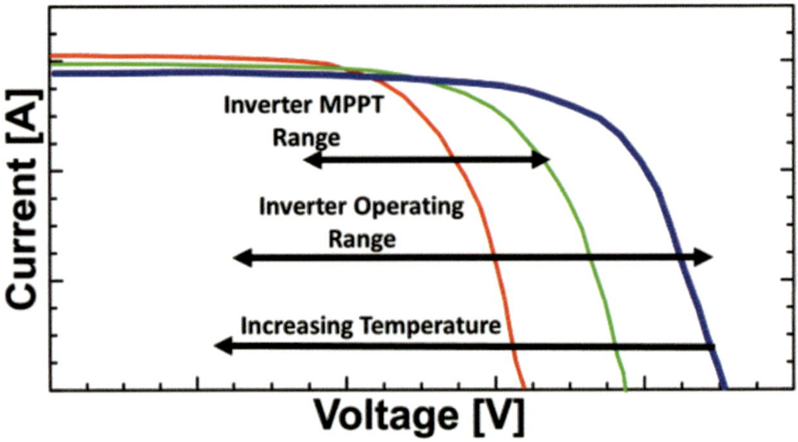

INPUT DATA	Fronius 5.0-1
Recommended PV power	4.25–5.75 kW
Nominal input current	13.8 A
Max useable input current	23.4 A
Operating voltage range	260–600 V
MPP tracking voltage range	230–500 V
OUTPUT DATA	
Nominal power output	5,000 W
AC output voltage	208/240/277 V
GENERAL DATA	
Max efficiency	96.2%

Figure 18.9 Top: A schematic illustration of inverter's general and MPP tracking operating ranges. Bottom: Operational specifications of the Fronius 5.0-1 inverter.

Balance of system components: A PV system also requires smaller components collectively referred to as the balance of the system. The electrical aspects of PV systems must adhere to the country's specific code, for example, the National Electrical Code in the U.S. Code Article 690 specifically deals with solar photovoltaic systems, and it also refers to several other National Electric Code articles for proper electrical installation. Details of electrical installations include the proper choice of conductors for wiring (type, sizing) for adequate ampacity under different temperatures, the choice of overcurrent devices like fuses or circuit breakers, proper grounding (which typically includes grounding on both the DC and AC sides of a PV system), and connectors for connecting the modules, and other items like splices, plugs, and receptacles (if needed), disconnect switches, and lightning protection devices. In systems where batteries are used, proper battery storage and venting (if lead-acid batteries are used) and charge controllers are required, which must also follow applicable codes.

Combiner boxes are used to combine two or more strings. These special types of junction boxes allow two or more PV strings to be combined in a box and provide a single output for the inverter. A typical combiner box includes terminal blocks, fuses, and fuse holders. Diodes are also common elements that may be used for various functions in solar PV circuits. Blocking diodes may be inserted in the circuit where they are needed to prevent undesired reverse flow of current, for example, from one string to another or from unwanted discharge of batteries. The PV modules themselves have internal diodes that allow them to mitigate shading to some extent; however, additional diodes, called bypass diodes, may also be used, which allow current to pass shaded modules in a string. Other important components include disconnect switches that must be included in various locations in a PV circuit, for example, on the DC side and AC side of the circuit. The purpose of these switches is to disconnect the system on either side at the time of repair and maintenance. If batteries are used, a disconnect switch on the battery side is also needed. Some inverters do have disconnect capabilities, therefore eliminating the need for dedicated disconnect switch(s). Besides national electrical codes, there may be other more local-based codes that installers would need to follow and implement.

Mechanical installation: There are a variety of possibilities for the installation of PV modules. Tracking systems with pole mounts, ground-based installation, and racking, which allow for a specific tilt angle and may also allow the user to change the tilt angle when needed, are all possible design options. However, the most common installations of solar modules are fixed ground-mount systems (for utility-scale solar farms) and fixed rooftops, i.e., using the fixed angle of the roof for building applications. In the case of flat rooftops, a racking system will be used with the desired tilt angle.

On tilted rooftops, the most common installation is to use standoff mounts, which allow a few inches of gap between the PV modules and the surface of the roof. These mounts consist of rails, brackets, clips, and flashing and are directly mounted, using lag screws, into the roof structures like the roof rafters (Figure 18.10). This method allows for air circulation and natural cooling, rainwater runoffs, and simplifies any required repair or maintenance. Local building codes must be followed for the mechanical installation.

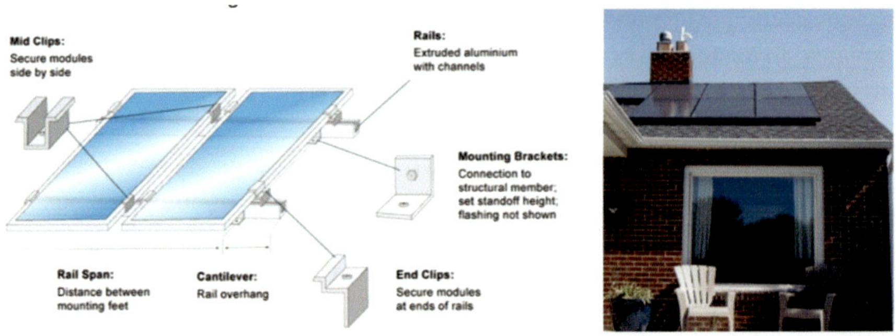

Figure 18.10 Left: Typical mechanical components (rails, clips, and cantilevers) used for rooftop PV installation. Adapted from Solar Energy4 International (2013). Right: Shingle-based rooftop solar PV system.

Batteries and charge controllers: Batteries and charge controllers for energy storage systems are vital components for off-grid systems and need to be sized appropriately for integration with the PV system. For off-grid systems, a careful assessment of critical loads is the first step in the PV solar design. This requires users to choose their loads frugally and aim for high-efficiency appliances to start. The design of the system then generally begins with battery sizing and the degree of autonomy, i.e., the period when PV power may not be available, followed by sizing the PV modules.

For the grid-tied system, where grid reliability is good, batteries may not be necessary, but can certainly be used to provide backup power for critical loads in the event of a power outage. In these systems, when the grid is down, the inverter would disconnect the PV from the grid (for safety reasons) and transfer the loads to the battery bank until the grid becomes available again. Batteries are, however, becoming more popular and are particularly advantageous in jurisdictions where time-variant rate structures are used (see Chapter 4). In these locations, the cost of electricity varies depending on the time of day, season, or other parameters, and it could be economical to charge the batteries when house demand is low (typically in the morning hours) and discharge the batteries when rates are higher (e.g., during a peak season or peak time of day). This mode of operation is referred to as energy arbitrage. Lead-acid batteries have been the most widely used type of batteries, but more recently Li-ion batteries are becoming more competitive because of their price reduction. Li-ion batteries have higher energy densities and are maintenance-free. With the battery bank, a charge controller device needs to be used as well to monitor charging and discharging of the batteries and ensure, for instance, that overcharging or overdischarging would not occur.

18.5.2 PV system configuration

PV system configurations can be generally classified in three categories: (1) off-grid, (2) grid-tied, and (3) multi-mode systems. We discuss these approaches to different extents in this chapter.

Off-grid: These systems are entirely independent of the grid. Examples could include remote areas with no grid connections and recreational cabins. The design of off-grid systems has to be performed very carefully, and loads should be chosen frugally. Batteries are critical in these systems. It is also suggested to have a secondary source of power (e.g., a diesel generator) for times when there is an insufficient amount of sunlight, and batteries have been discharged. The design of off-grid systems typically takes these steps: (1) choosing the loads, (2) sizing the battery, and (3) sizing the photovoltaic modules.

Grid-tied: Grid-tied systems are the most common configuration of solar modules in buildings. In this configuration, when sunlight is available, the PV modules provide the required power for the load, and any excess will be exported to the grid (Figure 18.11). When modules cannot provide sufficient power, the grid will provide the required electricity. The majority of utilities offer programs such as net metering (see Chapter 4), which allows the user to buy from the grid and sell back (Figure 18.12). In the past, utilities typically

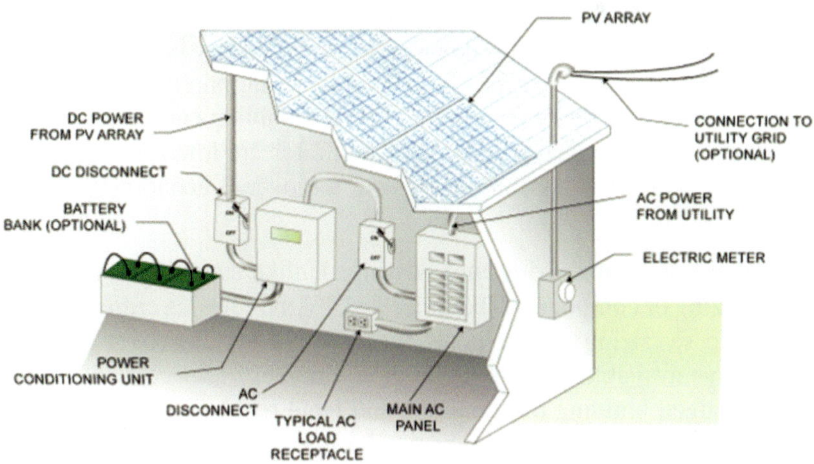

Figure 18.11 Grid-connected PV systems are the most common approaches in building PV systems. Battery storage capability is also becoming more popular as batteries become more affordable. Adapted from (Dunlop 2012).

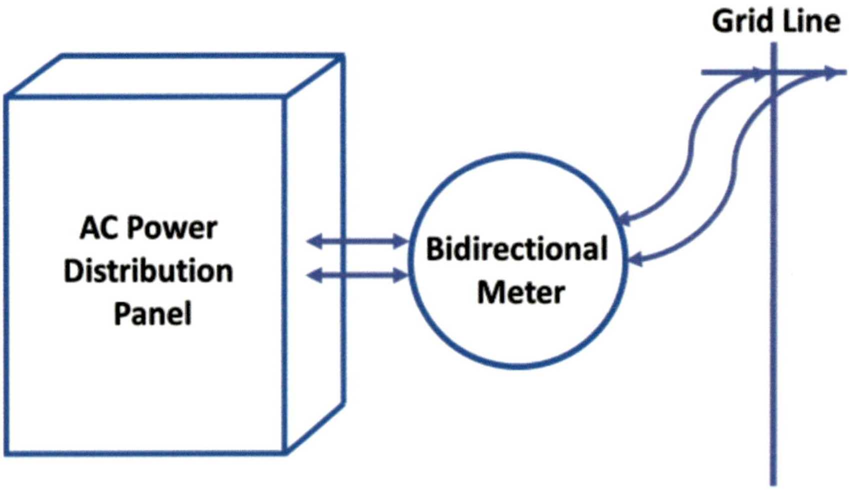

Figure 18.12 A bidirectional meter allows for net metering to be operational. Energy produced and exported to grid is logged and can be used as energy credit toward the next billing period when net metering is offered by the utility.

provided a higher buy-back rate to encourage solar module installations. Nevertheless, with the declining costs of installations, buy-back rates have also decreased in recent years.

Multi-mode: Multi-mode systems are grid-tied and include batteries, providing power even when the grid is down and/or solar energy is not available. The addition of batteries, however, provides interesting opportunities when time-variant electricity rate structures are present, enabling energy arbitrage. This allows the owner to charge batteries when rates are low and discharge batteries when rates are high, thus improving the overall economics of the system.

Simple grid-tied systems sizing and costing: Unlike stand-alone applications where sizing of components (batteries and PV modules) is critical, there are no specific set sizing requirements for grid-tied systems because grid power is always available. The sizing is then a matter of space availability (roof area), budget, building owner goals, and local regulations. Installation of PV modules on parts of the roof where solar insolation is low (e.g., north-facing roof segments and/or heavily shaded regions) is not recommended due to low-capacity factors, leading to low-value propositions. Budget is an especially important factor and may very well be the key decision-making factor to set an upper bound on the size of the system. Finally, the client may have a specific goal in mind (e.g., meeting 100% or 50% of the load with solar PVs). It may often be practical to size a few different systems and analyze the productivity and cost to help in the decision-making process. We will work out an example for the city of Seattle where a client would like to meet 50% of her load using PV solar modules. Relevant and recent technical and cost information is used with appropriate references for a practical sizing and costing scenario (see Boxes A–E).

Maintenance: Maintenance of solar PV systems is generally low and may mostly depend on the location's climate and the type of installation. Typical maintenance may include visual inspections of the PV modules and other components for signs of physical damage, possible overheating, and damage that may be caused by moisture. If nearby trees and vegetation have grown, periodic trimming and cleaning may be necessary. Accumulation of soiling (dust in the air, bird droppings), particularly in areas where rainfall levels are low, would be required. If batteries are used, they may have to be inspected for possible electrolyte leakage, corrosion, and may require terminal cleaning. Flooded lead-acid batteries need periodic addition of deionized water to maintain their electrolyte levels.

18.5.3 PV system costing and financing

Capital costs of installed PV systems: Hardware and soft costs of solar PV systems have been continuously dropping in the past two decades. Cost reduction is not limited only to the PV module, although PV modules have represented a significant portion of the cost reduction. According to a recent article in Solar Reviews, the average cost of solar PV system in the U.S. is in the range of \$2.60–\$3.35 per Watt of solar, which, for a typical size of 6 kW_{DC} a median solar PV system in the U.S., amounts to \$15,600–\$21,100. These prices reflect an 80% reduction compared to those of 2011 (Solar Reviews 2024b). For more detailed cost analysis and figures, we refer the interested reader to *Tracking the Sun*, a biannual and an authoritative source of PV cost trends from Lawrence Berkeley National Laboratory (LBL 2024). *Tracking the Sun* has detailed trends (e.g., system size, type of building, soft/hard costs) for well over one million grid-connected systems in the U.S.

Installed price declines reflect both hardware and soft cost reductions. The "Tracking the Sun" report categorizes the overall installed PV system costs

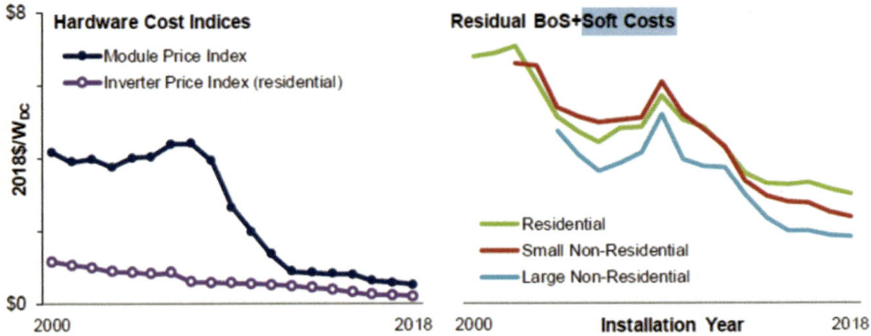

Figure 18.13 Left: Reduction of hardware (PV and inverter) costs from 2000 to 2018. Right: Reduction residual balance of system and soft costs from 2000 to 2018. The residuals are the difference of the installed price for each sector minus the hardware costs for the corresponding sector. Adapted from (LBL 2024).

as (i) hardware costs, which are primarily the PV modules and inverters, and (ii) balance of system and soft costs. Balance of system includes items such as racking, wiring, switches, combiner boxes, and soft costs such as design, installation, labor, permit, inspection, and installer margins. The hardware costs, primarily PV modules, have shown significant reductions over the past two decades. According to "Tracking the Sun," since 2014, about 64% of residential PV systems' cost reduction has resulted from hardware cost reduction, and the remaining 36% is due to balance of system and soft costs. Figure 18.13 shows the cost trends over the past two decades. PV cost reduction has been a consequence of technology improvement, volume manufacturing, and higher demand. Regarding soft costs, initiatives from government programs like the U.S. DOE's Solar Energy Technology Office actively aim to research and mitigate soft costs. These efforts involve identifying cost barriers, offering training materials for the skilled workforce, promoting market transparency, addressing social aspects, and fostering the adoption of solar technology, among other measures (DOE, 2024).

There are several online calculators that, using local information for the installers, electricity costs, and satellite imagery for calculating shades, provide cost estimates, savings, and payback periods. Solar Reviews has the world's first AI-capable solar calculator, and EnergySage provides total system costs of a number of installers for a given zip code location, thus allowing interested clients to choose among possible installers (EnergySage, 2024a).

Operation and maintenance costs of installed PV systems: An NREL working group has been tracking and examining the cost of operation and maintenance (O&M) for different implementations of solar PV (i.e., residential, commercial, utility, etc.) for several years. These O&M costs include not only the items discussed earlier in Section 5.2 but also insurance, property tax, component replacement, PV, and inverter replacements. Of the replacement costs, PV and components are generally low; furthermore, PVs generally have a long-time power warranty of about 25 years. Manufacturer warranties cover defects and mechanical wear and typically last around 10 years. The inverter replacement is more significant and can be either considered as a one-time cost (e.g., around year 12) or an annualized cost.

Incentives: Governmental-based incentives have been instrumental in the advancement and market penetration of solar PV. Depending on the country and local jurisdictions, a variety of options may be available. Although some countries have reduced their incentives in the past 10 years due to the overall reduction of PV costs, the incentives still play a significant role in the wider acceptance and adoption of this technology. In the U.S., the Database of State Incentives for Renewables & Efficiency® provides all the state (and federal) level incentives for renewables and efficiency upgrades (DESIRE, 2024). In recent years, the U.S. federal government increased the solar tax credit from 26% to 30% and has extended this credit until 2032. This allows the customer to reduce their income tax by up to 30% of the overall PV system cost, and the credit can be carried forward if not fully utilized during the first year. Essentially, the total capital cost can be reduced by up to 30% when the client has income tax obligations. Local regions may offer additional incentives, such as reduced sales tax on PVs or special discounts for locally produced PVs. Also, some electric utilities offer incentives to offset the initial capital cost.

A widespread method for energy trading between residential clients and utilities is the net metering program, commonly provided by many utilities in North America. With net metering, surplus power generation is sent back to the grid, accumulating kWh credits for future electricity consumption (see Chapter 4). If, by the year's end, the net electricity export is positive, the utility compensates the customer based on market prices. When net metering programs were introduced (2000–2010), the buy-back rates were generally much higher than the retail rate.

Box A. Assessing the solar resource using PVWatts® Calculator.

Impact of geographical location on average daily insolation: To see the solar insolation for a few selected cities in North America, we use NREL's online PVWatts® Calculator (https://pvwatts.nrel.gov/index.php). The results presented here were obtained using the Website Version 8.2.1 of PVWatts® API version 8.0. To start the simulation, enter the name of the city and press *Go*. Use the default resource data (NREL International) and press *Go to System*, Info. Most common roof pitches fall in the range of 4/12 (= 18.4°) to 9/12 (= 36.9°). Here we take an array type = fixed (roof mount), a roof pitch of 7/12 (tilt = 30.3°), and an orientation due south (azimuth = 180°). If you click on the ⓘ next to the *Tilt (deg) box*, you can read more about the tilt and pitch/angle values. We will leave all the other input parameters to default values because we are only interested in the insolation data for now. Press *Go to PVWatts® results* and inspect the Solar Radiation (average daily kWh/m^2 for each month). The average daily kWh/m^2 over a year is provided on the bottom row of the table.

Solar insolation (kWh/m^2/day)						
Month	Edmonton (53.5 °N)	Seattle (47.6 °N)	Boston (42.4 °N)	Omaha (41.3 °N)	Los Angeles (34.1 °N)	Phoenix (33.5 °N)
January	1.61	1.84	3.65	3.62	4.97	5.44
February	2.93	2.69	4.34	4.61	5.39	6.04
March	4.39	3.53	5.29	5.15	6.10	6.88
April	5.77	5.13	5.45	5.46	6.78	7.65
May	6.13	5.89	5.96	5.65	6.30	7.64
June	6.04	5.78	5.66	6.62	6.90	7.65
July	6.26	6.22	6.26	6.64	7.14	6.87
August	5.93	6.36	5.87	6.02	7.45	6.92
September	4.89	4.78	5.47	5.93	6.97	7.03
October	3.12	3.10	4.32	4.99	6.13	6.49
November	2.05	1.79	3.28	3.96	5.53	5.94
December	1.67	1.53	2.65	3.12	4.56	5.36
Average daily	**4.23**	**4.05**	**4.85**	**5.15**	**6.19**	**6.66**

Observations: Although Edmonton (Canada) is further north than Seattle, it has a slightly higher average insolation. This results from higher insolation during the months of February to May when more sunny days help with the insolation. Seattle, however, benefits from higher insolation during summer months when the rainy season is over. A comparison of Boston and Omaha is also interesting, as both have similar latitudes. Omaha, due to more sunny days and better climate, receives more insolation for every single month of the year. The same observation applies to Los Angeles and Phoenix, as they share nearly identical latitudes. Phoenix with more clear skies has higher insolation for all months except July/August. Note that the overall merit of a solar project PV project depends on other factors besides the solar resource. These include local installation and commissioning costs, electricity rates and incentives, available roof space, roof material and pitch/orientation, shading, and the client's energy consumption. In numerous regions of Europe, where insolation levels are lower compared to various parts of the world, including the U.S., the growth of solar PV on rooftops has been noteworthy over the past two decades.

Impact of roof pitch and orientation on average daily insolation: With PVWatts® Calculator, it is very straightforward to look at solar resource on part of rooftop with a given orientation direction (azimuth) and pitch. For example, for the segment of the roof that is facing east and has a pitch of 4:12, we enter 90° for the azimuth and 18.4° for the tilt angle. We set the array type = fixed (roof mount). We now examine the impact of azimuth and tilt for some common roof pitch angles for a house in the city of Seattle.

Average daily insolation ($kWh/m^2/day$) for Seattle (47.6 °N)					
Azimuth/Tilt →	0°	4:12 (18.4°)	7:12 (30.3°)	12:12 (45°)	90°
↓ North (0°)	3.47	2.78	2.30	1.72	0.82
East (90°)	3.47	3.33	3.18	2.93	1.90
South (180°)	3.47	3.93	4.05	4.02	2.65
West (270°)	3.47	3.44	3.34	3.13	2.07

Observations: When the tilt angle = 0°, the solar collector receives equal insolation for every azimuth angle by symmetry. For the other tilt angles (or roof pitches), the insolation received by the collector decreases in the order of south, west, east, and finally north-facing roof segments. In the northern hemisphere, the south-facing roof is the best portion of the roof. West- and east-facing roof segments are nearly the same, and the north-facing roof receives the least amount of insolation. We would, therefore, populate the south-facing roof first before adding panels to the west- and east-facing roofs. These are all in the absence of any possible shading. If obstacles (e.g., buildings or trees) are present, which may cast shadows in certain periods of the year, their effect needs to be considered, and the amount of solar insolation for the specified time needs to be reduced accordingly.

Impact of tilt angle on average daily insolation: Herein we can explore the impact of tilt angle on the insolation levels. We examine three tilt angles: 0, 47.6° (= latitude), 32.6° (latitude – 15), and 62.6° (latitude + 15) and 90°. We keep the azimuth = 180 (facing south) and set the array type = fixed (roof mount.

Average daily insolation (kWh/m²/day) for Seattle (47.6 °N)													
Tilt	**Jan**	**Feb**	**Mar**	**Apr**	**May**	**Jun**	**Jul**	**Aug**	**Sept**	**Oct**	**Nov**	**Dec**	**Annual**
0	1.02	1.80	2.78	4.55	5.76	5.90	6.21	5.74	3.77	2.15	1.09	0.83	3.47
32.6°	1.88	2.73	3.56	5.13	5.85	5.72	6.16	6.35	4.81	3.15	1.83	1.57	4.06
47.6°	2.12	2.93	3.61	4.94	5.40	5.18	5.61	6.05	4.85	3.32	2.02	1.78	3.98
62.6°	2.22	2.96	3.46	4.48	4.65	4.37	4.77	5.41	4.60	3.31	2.08	1.89	3.68
90°	2.07	2.57	2.70	3.05	2.78	2.47	2.72	3.51	3.48	2.77	1.89	1.81	2.65

Observations: In some situations, it is possible to set the tilt angle (e.g., on a flat rooftop and using racking infrastructure). We can note that the choice of tilt = *latitude* angle provides the most uniform insolation over a period of one year, whereas *latitude + 15°* and *latitude – 15°* angles are biased toward higher productions in the winter and summer months, respectively. It is then up to the client's needs and the specific load/applications to choose the more optimum tilt angle. By choosing finer angles in PVWatts, the total insolation can be maximized if one desires.

Box B. Determination of shaded regions.

Using tools such as those described in Section 4 of this chapter, shade patterns for a house in the city of Seattle can be represented by a solar chart, as shown in the example figure below. Each horizontal line corresponds to a specific month, and the vertical lines represent hours of the day, with divisions every 30 minutes. Portions of the Sun's path that fall inside the red lines indicate times (hours and months) when the Sun would be blocked by obstructions, and PV panels would experience shadow. The numbers written on each sun path show the percentage of the horizontal arc area. For example, in the month of December, there is shadow from 8:00 AM to 11:30 AM. The numbers on the Sun's path (1, 2, 3, 5, ...) designate the percentage of the Sun's radiation for a given day. We will now determine the percentage of time during which the Sun will be obstructed for the month of December. There will be shadow in the morning hours, and these are clearly labeled by red circles:

December Shade % = 3% + 5% + 6% + 7% + 8% + 9% = 38%

Therefore, only 62% of the Sun's radiation will be available at this site for the month of December. (Note: The line crosses between 11:00 AM and 11:30 AM, but we take a conservative approach and assume that this entire time, corresponding to 9%, is all obstructed).

Other months that experience shading are Jan (38%), Nov (38%), Dec (38%), Feb (18%) and Oct (18%). We will use these shading data in Box C.

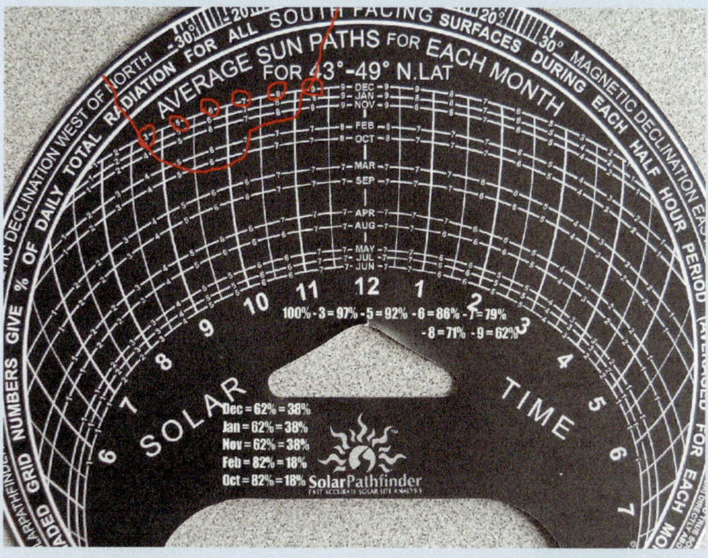

Solar chart shows the periods of time where solar radiation will be blocked because of the obstacles present. (Source = www.solarpathfinder.com/PF?id=icZJmYac)

Box C. Electricity production from a 5-kW solar rooftop in Seattle using PVWatts® Calculator.

Combining Solar Pathfinder and PVWatts results: We will conduct a comprehensive analysis of solar production for a rooftop in the city of Seattle. We assume that shading at our location can be described by the measurements in the figure in Box B. In PVWatts, enter the city as Seattle and proceed to *System Info*. For the system size, enter 1 kW for now; later, we will scale this as desired. PVWatts has three basic PV modules with fixed efficiencies and thermal derate factors: (1) standard (efficiency = 19%), (2) premium (efficiency = 21%), and thin film (efficiency = 18%). Currently (in 2023), typical efficiencies of commonly used solar modules are in the range of 20%–24%. For a more accurate calculation, proper efficiency and thermal factors need to be used. For our simple calculation, it suffices to use the premium type here. For the tilt angle, enter 30.3° corresponding to a roof pitch of 7:12 and choose fixed (roof mounting) for the array type. For the system losses, enter 15%, which is a typical value. This corresponds to an overall derate factor of 0.85. Alternatively, one may click on ⓘ to see (or enter) the individual loss factors that give rise to the overall derate factor. Azimuth will be 180°, i.e., facing toward south.

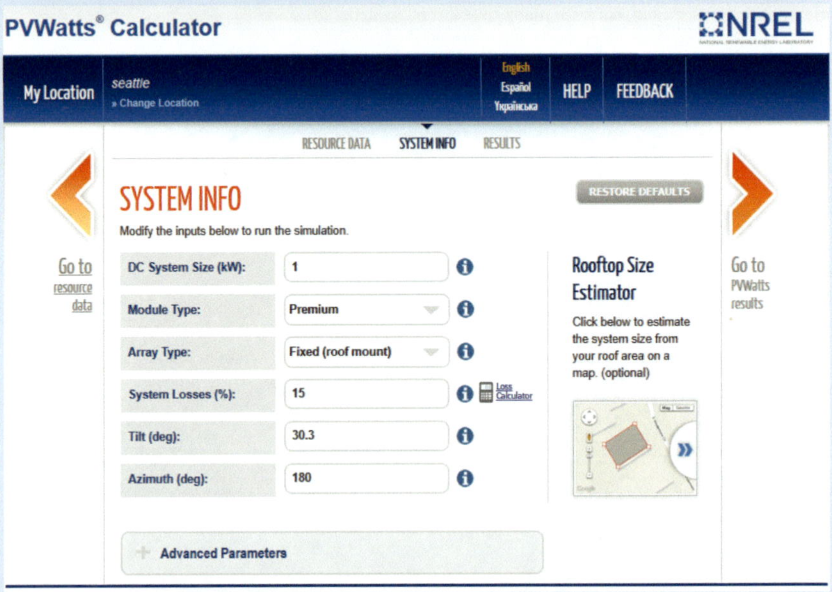

Screenshot of PVWatts showing the system information for a 1-kW PV system.

Next press *Go to PVWatts Results* to see the solar radiation and the AC energy (kWh) generated.

Month	Solar Radiation (kWh / m^2 / day)	AC Energy (kWh)
January	1.84	45
February	2.69	59
March	3.53	84
April	5.13	115
May	5.89	133
June	5.78	125
July	6.22	138
August	6.36	142
September	4.78	105
October	3.10	73
November	1.79	42
December	1.53	38
Annual	4.05	1,099

Screenshot of PVWatts showing the system information for a 1-kW PV system.

These values do not consider any possible shading. We now use the monthly shade percentages Nov/Dec/Jan/Feb from Box B. For example, in February, there is 38% shading, so the available solar insolation and the AC kWh produced will be multiplied by 0.62. The PVWatts Calculator reports 59 kWh for the month of February. After correction for shading, we have 0.62 × 59 = 36.6 kWh. The table below shows the monthly and annual values after correction for shading. Next, we will use these kWh values to size a PV system and do a simple solar array design with commercially available PV module and DC/AC inverter. In our location, a 1 kW_{DC} solar PV system would produce about 1,028 kWh_{AC} of electricity in one year.

Average daily insolation ($kWh/m^2/day$) for Seattle (47.6 °N)				
Month	**Solar radiation ($kWh/m^2/day$)**	**AC energy (kWh_{AC})**	**Available Sun %**	**AC energy (kWh_{AC}) – corrected**
Jan	1.84	45	62	28
Feb	2.69	59	82	48
Mar	3.53	84	100	84
Apr	5.13	115	100	115
May	5.89	133	100	133
Jun	5.78	125	100	125
Jul	6.22	138	100	138
Aug	6.36	142	100	142
Sept	4.78	105	100	105
Oct	3.10	73	82	60
Nov	1.79	42	62	26
Dec	1.53	38	62	24
Annual	**4.05**	**1,099**		**1,028**

Capacity factor: We will calculate the DC capacity factor of this rooftop PV system for no shade and with shade.

No shade: Capacity factor = 1,099 $kWh_{AC}/(1\ kW_{DC} \times 365\ d \times 24\ h/d) = 12.54\%$

With shade: Capacity factor = 1,028 $kWh_{AC}/(1\ kW_{DC} \times 365\ d \times 24\ h/d) = 11.73\%$

Higher levels of insolation, better tilt angles (or even tracking the Sun), improved derate factors, and reduced shade all help with increasing the capacity factor. A higher capacity factor, even by 0.5%, could lead to noticeable impact on the amount of energy produced and ultimately the value of the system.

Box D. A simple PV system design for Seattle.

Sizing and designing a solar array for the Seattle house: Assume our client desires to produce 50% of his/her annual electricity consumption, which is 9,500 kWh_{AC}, by solar PV. We saw that a 1-kW_{DC} system in Seattle with the shade described in Box B, produces 1,028 kWh_{AC}/year. Therefore, to meet 50% of the load, we need to scale our size accordingly (round up the annual consumption to 10,000 kWh_{AC}):

$10{,}000\ kWh_{AC} \times 0.5 \div 1{,}028\ kWh_{AC}/kW_{DC} = 4.86\ kW_{DC}$

Thus, a 5-kW_{DC} system would be able to meet the client's goal of producing 50% of the average load by PV modules.

System component PV module: We will use the LR4-72HPH-440M PV module from LONGi, which we encountered earlier. This module has a rated power of 440 W_{DC}. The number of required modules is:

$5{,}000\ W_{DC}/440\ W_{DC} = 11.36 \rightarrow 12$ modules

Therefore, to be more precise, our system size would be $12 \times 440\ W_{DC} = 5.28\ kW_{DC}$.

For the inverter, we will pick Fronius 5.0-1, which we have also seen earlier. This inverter is appropriate for a DC input power of 4.25–5.75 kW_{DC}. The table below displays the specifications of our selected module and inverter. It is worth noting that the market offers a vast array of modules and inverters. The selection of equipment may hinge on various factors crucial for designers and installers, such as availability, cost, ease of use, warranty, and more.

PV specifications (standard test conditions) of PV module and DC/AC inverter			
LR4-72HPH-440M PV module		**Fronius 5.0-1 inverter**	
Rated power	440 W	Recommended PV power	4.25–5.75 kW_{DC}
Open circuit voltage	48.9 V	Nominal input current	13.8 A
Rated current I_{mp}	10.71 A	Max usable input current	23.4 A
Maximum power voltage V_{mp}	41.1 V	MPP tracking voltage range	230–500
OCV temperature coefficient	−0.270 %/°C	DC startup	260 V
Power temperature coefficient	−0.350 %/°C	Max input voltage	600 V
		Nominal output power	5,000 W
		AC output voltage	208/240/277 V

Array configuration: We now need to determine the series/parallel configuration of our 12 modules. For this, we will need to consider the extreme cases where thermal effects would change the output voltage (and therefore power) of the PV modules. We first consider the V_{oc} at the coldest temperature and make sure that it would be less than 600 V maximum input level of our inverter. The 600 V is also an upper voltage limit for residential buildings. The lowest and highest temperatures ever recorded in Seattle were −18 and +42 °C, respectively. Although these are extreme and represent very short periods, we will use these for thermal derating calculations. Thus, we use the V_{oc} and V_{mp} values calculated earlier in Example 18.1.

Thus, the number PV modules can go in series is 600 V_{DC} ÷ 51.5 V_{DC} = 11.65 modules. Accordingly, we can use up to 11 LR4-72HPH-440M PV modules and still be within the working limits for the coldest day of the year. Since we only need 12 modules, and because we will need equal-sized strings, two strings with six modules would be sufficient. We now need to check the MPP tracking range of the inverter, to make sure there is sufficient voltage for the warmest days of the year. We use the V_{mp} for this calculation:

6 × 43.9 V_{DC} = 263.4 V_{DC}

This voltage is in the range of MPP tracking and sufficient as DC startup. The rated current of our module is 10.71 A_{DC}; therefore, two strings in parallel provide about 21.42 A_{DC}, which is also acceptable by our Fronius inverter.

We then conclude that two strings of LR4-72HPH-440M PV modules with one Fronius 5.0-1 inverter would be able to deliver 50% of our client's electricity consumption.

Box E. Financial analysis of the solar PV rooftop in Seattle.

Simple costing and incentives: We will carry out a simple cost and financial analysis of our 5.28 kW_{DC} system to assess the financial viability of the proposed PV system. The purpose of this exercise is to demonstrate a basic solar financial example. In general equipment costs, energy pricing and financial parameters like general inflation rates and escalation of energy prices are quite variable and may require a more detailed analysis for better accuracy. The main incentive in Seattle area is the U.S. federal investment tax credit, which in 2024 amounts to 30% of the capital cost. In addition, solar PV systems are also except from sales tax, which amounts to 6.5%. For solar PV systems up to 100 kW, Seattle City Light, the public utility, offers a net metering program that provides kWh credits for the excess energy produced. These credits are valued at the retail rate and can be used to offset future energy bills (Seattle City Light, 2024a).

According to EnergySage.com, the average cost of a 4–5 kW_{DC} system in the state of Washington is about \$2.92/W_{DC} in 2024 (EnergySage, 2024b). The total capital cost of system is, therefore, 5,280 W_{DC} × \$2.92/W_{DC} = \$15,417 and includes equipment cost, installation, permit, and commissioning. We assume the lifetime of the system is 25 years. The string inverters have a typical lifetime of about 12 years; so we will budget for a replacement in the year 12 at an estimated cost of \$1,500. This cost could be reduced if microinverters with longer lifetime are used or alternatively if a 25-year warranty can be purchased for the inverter at a reduced cost.

Payback period: To calculate the payback period, we calculate the energy savings per year and then divide the total costs by the savings to obtain the payback period in years. We ignore the discounting in this example but include a 3.5% increase in energy prices per year. We assume the project is entirely financed by downpayment (no loan). The average cost of electricity in Seattle is about \$0.13/kWh in 2024 (Seattle City Light, 2024b). This increase in electricity costs is based on the historical rise in electricity pricing. Residential and small-scale solar projects are quite often assessed by this approach.

Capital cost = 5,280 W_{DC} × \$2.92/W_{DC} = \$15,417

Capital cost after federal tax credit = \$15,417 – (0.30 × \$15,417) = \$10,792

Inverter replacement = \$1,500

Total costs = \$10,792 + \$1,500 = \$12,292

Total benefits = 5.28 kW_{DC} × 1,028 kWh_{AC}/kW_{DC}/yr × \$0.13/kWh_{AC} × $\Sigma\ (1 + 0.035)^n$ = \$28,433 (n = 25 years)

Or an equivalent annual cost savings of \$28,433/25 yr = \$1,137/yr

Payback period = \$12,292/\$1,137/yr ≈ 11 years

Net present value: A more rigorous metric for the economic viability of energy projects can be obtained by the net present value method. By this approach, we discount the future costs and benefits to the present time and then calculate NPV as NPV = Σ Benefits – Σ Costs. A positive NPV then indicates financial viability. These types of calculations are best performed in a spreadsheet where costs and benefits are tabulated over the lifetime with inflation as appropriate and then converted to present time using the appropriate discount factor. Here, we will calculate the NPV using a historical inflation rate of 2.5% and assume the client has a personal discount rate of 3%. The discount rate refers to an alternative investment that the client may have. For more details on financial calculations, please refer to Chapter 5. The client would like to finance the project with a 15% cash downpayment and is able to finance the remaining 85% with a low interest energy improvement loan with 2% interest with a 10-year loan period. We assume there will be no salvage value after 25 years. Although solar PV systems typically do not require any O&M, we allocate about 0.01 \$/W of system (about \$50 per year) for O&M, which may include cleaning. Loan payments are not affected by inflation, whereas inverter replacement and O&M costs are impacted. Energy rate increases according to the escalation rate of 3.5%.

Cash downpayment = 0.15 × \$10,792 = \$1,619

Annual loan payment (fixed) = CRF × 0.85 × \$10,792 = \$1,021

Where capital recovery factor (CRF) = $0.02/(1 - (1 + 0.02)^{-10}) = 0.1113$

Inverter replacement = \$1,500 × 1.025^{12} = \$2,017

The inverter costs have been inflated at a rate of 2.5% per year. Here 1.025^n is the inflation factor for the nth year.

The costs for every year need to be discounted to the present value using the appropriate discount factor, for example, discounting third year's costs (loan + O&M) to present the time we have:

$(\$1{,}021 + 56.9) \times (1 + 0.03)^{-3} = \$986$

The electricity values generated (i.e., savings) for year *n* can be calculated as:

$5.28\ kW_{DC} \times 1{,}028\ kWh_{AC}/kW_{DC}/yr \times \$0.13/kWh_{AC} \times (1.035)^n = \$706 \times (1.035)^n$, where $(1.035)^n$ accounts for the increased value of savings in year *n* due to escalation of electricity pricing.

As an example, the third-year benefits discounted to present time are:

$\$782 \times (1.035)^3 \times (1 + 0.3)^{-3} = \$715.$

The NPV can then be calculated as: Σ Benefits – Σ Costs = \$18,790 – \$14,217 = \$4,572. The NPV is positive and indicates that the project is financially viable.

18.6 Solar Thermal

Introduction

Solar energy can be also used to provide direct heating. There are many examples, ranging from passive housing strategies, effective solar windows, and solar walls to solar cookers, and pre-heating of building water and air. A good reference on these topics is Solar Energy by Andy Walker (Walker 2013). In this section, we will briefly review the solar thermal collectors used in pre-heating of water.

Solar thermal collectors:

The basic unit of a solar thermal system, or the equivalent of solar PV for thermal energy, is called a solar thermal collector. There are three types of collectors that provide different levels of heat collection and vary in their design and cost. The principle of operation in all these collectors is to circulate a fluid, commonly water or a mixture of water and antifreeze, through the collector exposed to solar radiation, allowing it to absorb the heat energy. The purpose of the collector is to concentrate the heat and raise the fluid temperature. Figure 18.14 shows the three types of solar collectors.

Unglazed collectors: These collectors are the simplest and least expensive types of thermal collectors. The most common collectors of this type consist of sheets of plastics like polypropylene with special coatings to enhance durability, extruded with tube structures inside that allow the flow of liquid through them. The color is black to enhance the absorptivity of sunlight, which reaches up to about 95%. These collectors are ideal for

Figure 18.14 Left: Unglazed EPDM (ethylene propylene diene monomer) rubber collectors for heating pool water. Middle: Flat-plate collectors mounted roof for water heating. Right: A series of evacuated tubes. Adapted from (Walker 2012).

low-temperature applications such as heating swimming pools. Water is circulated through these collectors and can be heated to about 10 °C (50 °F) above the ambient temperature. Unglazed collectors are flexible with no rigid frames and can be conveniently rolled on a large surface like a roof and be easily removed. A typical system would include collectors, a pump, filter, strainer, check valves, flow sensors, and valves. A conventional pool heater may still be required to provide additional heating when needed or when solar irradiation is low. As a simple guide, the surface area of the collector would need to be in the range of 50%–100% of the pool area and could cost about $3,000.

Glazed flat-plate collectors: These types of collectors, also referred to as mid-temperature collectors, allow for higher levels of heat collection and can heat water up to a temperature of about 50 °C (122 °F). The design of flat-plate collectors attempts to minimize heat loss by placing a set of copper tubes, where water absorbs the heat, inside a well-insulated box. The copper tubes are attached to a heat-collecting absorber plate, typically made from copper. The box is covered by a glass sheet with anti-reflection coatings, allowing radiation to enter the box and be collected by the absorber plate. The bottom and sides of the box are well insulated (e.g., fiberglass insulation) to minimize any heat loss (Figure 18.15). The total cost of a water heater system for an average house in the U.S. is typically in the range of $2,000–$6,000, depending on the system size and complexity.

Evacuated-tube collectors: These collectors allow for the highest amount of heat collection and temperatures greater than 50 °C (122 °F) can be reached. The heat absorbers and copper tubes in this design are contained inside glass tube, which are evacuated to significantly reduce heat loss by either conduction or convection. Because of this significant reduction of heat loss,

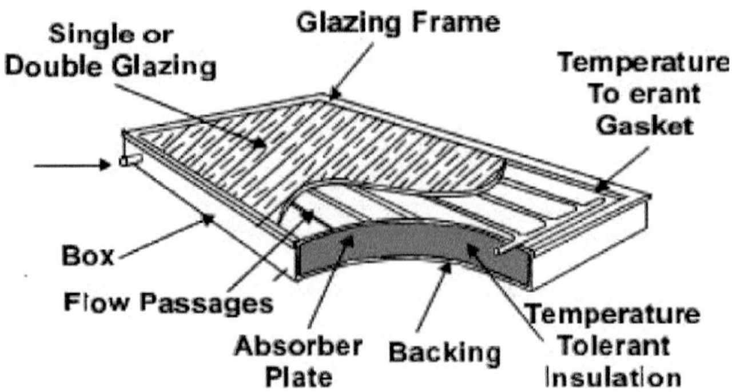

Figure 18.15 Components of a glazed and insulated flat-plate collector. Adapted from (Walker 2012).

evacuated tubes work well even in cold temperatures or cloudy conditions. Evacuated tubes could cost up to 50% more than flat-plate collectors.

Heat removal and efficiency: We will consider a simple analysis of heat gains and efficiencies of typical solar thermal collectors. In a solar collector, heat gained by the fluid is the difference between solar energy absorbed by the collectors and the amount of heat lost from the collector to the environment. The heat transferred or removed by the fluid is also proportional to the fluid mass flow rate and difference in the temperature of the collector and that of the ambient. We can then express the rate of useful heat from the collector (in W) by the following equation:

$$\begin{aligned}\dot{Q}_{\text{useful heat}} &= \dot{Q}_{\text{absorbed solar}} - \dot{Q}_{\text{lost to ambient}} \\ &= \alpha\tau AI - AU\left(T_{\text{collector}} - T_{\text{ambient}}\right), \end{aligned} \tag{18.2}$$

where α, τ, A, and I are the absorptivity and transmissivity coefficients, and A and I are the area (in m^2) and solar irradiation (in W/m^2). U is the heat loss coefficient (in W/m^2 °C). $T_{collector}$ and T_{ambient} correspond to temperature (in °C) of the collector and the ambient, respectively.

$$\dot{Q}_{\text{useful heat}} = \dot{m}c_p\left(T_{\text{fluid out}} - T_{\text{fluid in}}\right). \tag{18.3}$$

The rate of useful heat can also be expressed in terms of the mass flow rate ($\dot{m}$), fluid het capacity (c_p), and the temperature difference of output and input fluid ($T_{fluid\ out} - T_{fluid\ in}$). It should be noted that temperatures of input and output fluid and the ambient are easily accessible and well defined, but the $T_{collector}$ is not an easily defined or accessible parameter because of the thermal gradients across the collector. Thus, it would be of great help to avoid $T_{collector}$ in the analysis. The efficiency of the collector is the ratio of rate of useful heat removed to the input solar power which is *AI*. Using this definition and eqn (18.2), we can then express the collector's efficiency as:

$$\eta_{\text{collector}} = \frac{\dot{Q}_{\text{useful heat}}}{AI} = \alpha\tau - U\frac{T_{\text{collector}} - T_{\text{ambient}}}{I}. \tag{18.4}$$

Instead of using the collector temperature, we use the fluid-in temperature by introducing the heat removal factor or F_R. This leads to the well-known Hottel–Whillier–Bliss equation of water collector's efficiency:

$$\eta_{\text{collector}} = F_R\alpha\tau - F_RU\frac{T_{\text{fluid in}} - T_{\text{ambient}}}{I}, \tag{18.5}$$

where F_R is called the collector heat removal factor and depends on the geometry and construction of the collector and is specified by the manufacturer. Since the values of $\alpha\tau$ change depending on the solar angle, the F_R factor can be multiplied by an incident angle modifier parameter $K_{\alpha\tau}$ to capture this dependency; however, typical test data are provided for $K_{\alpha\tau}$ value of 1. A plot of η vs. $T_{fluid\ in} - T_{\text{ambient}}$ then results in a straight line whose slope and intercept depend on the type of the collector. When $T_{fluid\ in}$ is closest to the T_{ambient}, collectors show maximum efficiency in the removal of heat and

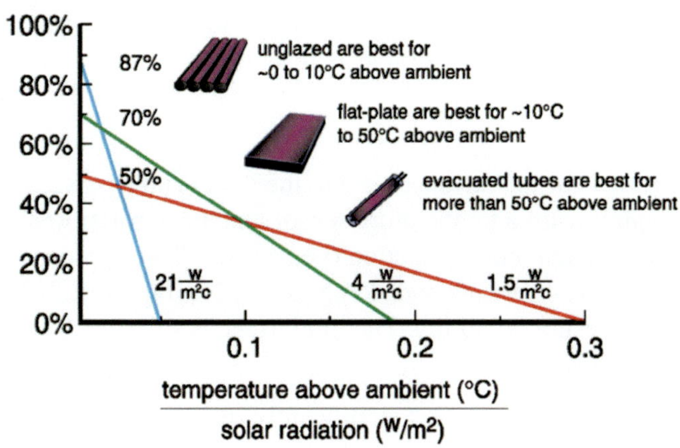

Figure 18.16 Efficiency of solar radiation to heat for unglazed, flat-plate, and evacuated tube collectors plotted vs. the difference between the fluid temperature and that of the ambient normalized by the solar radiation in W/m^2. Adapted from (Walker 2012).

as the difference in $T_{fluid} - T_{\text{ambient}}$ increases, i.e., fluid becomes hotter, efficiency drops. If the fluid stops flowing, efficiency could drop to near 0% corresponding to a situation where all incoming radiation is absorbed and not removed. This situation could lead to extreme temperatures within the collector and possible accidents and destruction of the collector. Figure 18.16 shows the efficiency curves of various heat collectors. This allows users to decide on the type of collector depending on the required heating temperature.

Solar thermal systems:

Most of the solar water collectors are used to pre-heat water to a certain temperature, and additional heating may be required for the desired setpoint to be reached. There are several design approaches of water solar heater systems with respect to system complexity, flexibility, and cost. In general, systems are either direct, where the water used is heated directly in the collector, or indirect where a working fluid (e.g., a mixture of water and antifreeze) would collect the heat and pass it to the water using a heat exchanger. We will briefly discuss these approaches here. Figure 18.17 shows a few possibilities.

Direct and passive: One of the simplest approaches consists of a collector and a tank mounted on a higher elevation. Under the thermosyphoning property of hot water, thermal convection would lead to hotter water filling the tank; therefore, an electric pump is not required (i.e., passive). An additional tank (auxiliary tank) may also be used in the system. The heated water will be directly used without the use of any secondary loops in the system (i.e., direct). These systems are quite simple with low capital and O&M costs, but they have low tolerance to freezing and corrosion and must be deployed carefully.

Direct and active: A further advancement to the direct and passive approach is addition of a pump with a temperature sensor and a controller to allow for a better and more efficient heat collection (i.e., active). This approach allows the user to place the pre-heat tank and storage at any location (e.g., basement) and provides more design flexibility. The fluid usage is still direct; therefore, freeze protection is low.

Indirect and closed-loop: This is a significantly improved design that separates the heat collection in the thermal collectors from the actual water used. The primary circuit uses water and antifreeze that circulate through the collectors and collect the heat from solar radiation (i.e., indirect). A heat

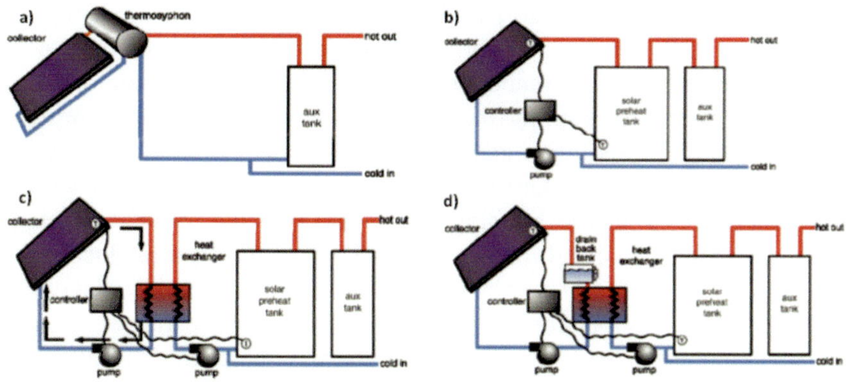

Figure 18.17 (a) Passive and direct system with thermosyphon, (b) active, direct, and open-loop system, (c) active, indirect, and closed-loop system, and (d) active, indirect, closed-loop with drainback system. Adapted from (Walker 2012).

exchanger is then used to transfer the collected heat to water. This system has excellent freeze protection, but it comes at a higher cost.

Indirect and closed-loop with drainback tank: A variation of the previous design is to add a drainback tank that allows the user to drain the solar collector fluid when the system is not being used for a period. Draining the collectors avoids any accidental loss of expensive fluid in case temperature rises and safety valves would discharge (and waste) the fluid.

18.7 Concluding Remarks

Solar PV will continue to be a key renewable energy technology for both large-scale utilities and smaller individual commercial and residential buildings. Improvements in technology, volume manufacturing, and wider customer acceptance – accelerated by governmental incentives – have been key factors in the global reduction of costs. According to NREL, the U.S. residential PV system costs (using 2022 USD) have dropped from \$8.70/W_{DC} in 2010 to \$3.16/W_{DC} in 2022. Most of these reductions have been associated with hardware; so, currently, soft costs are becoming more dominant in the cost breakdown. Technology continues to develop, and the use of hardware such as AC-based PV modules, bifacial modules (where PV cells exist on both sides of the panel), and microinverters is expected to increase in the coming years. Recent advances and cost reduction in batteries provide interesting opportunities for energy arbitrage and, with the growing demand in electric vehicles, offer attractive prospects for distributed electricity storage. Certain jurisdictions are mandating increased use and adoption of PV modules. It is worth mentioning that the installation of solar PV during the construction of a building is generally more efficient than retrofits, and NREL estimates that PV system costs are reduced by about $\$0.65/W_{DC}$ (approximately 25%) in new buildings. It is also possible to have certain provisions, such as a dedicated conduit from the potential solar array location to the mechanical/electrical room, in anticipation of future solar installation.

Solar thermal collectors continue to be popular for pre-heating water in buildings. A combined solar thermal and PV system is interesting, as it allows for the generation of both electricity and heat. One new direction involves utilizing the waste energy of PV panels, which have a rather low efficiency, in conjunction with solar thermal collectors; however, further work is needed to improve the overall technology.

Author Biography

Arman Bonakdarpour is an Associate Professor of Energy Management at the New York Institute of Technology (NYIT) in Vancouver, where he teaches courses on sustainable energy technologies, solar energy, batteries/hydrogen, and energy modeling. In the summer of 2024, Arman was invited by the U.S. Department of Energy to lead a team of students in the National U.S. Collegiate Solar District Cup. His team advanced to the final round of the Solar-Plus-Storage Competition, ultimately securing third place in their division during the second and final phase. In 2025, he received NYIT's Presidential Award for Excellence in Student Engagement in Research, Scholarship, or Creative Activities. Arman is also an Adjunct Professor at the University of British Columbia (UBC) and has been leading research projects in the Department of Chemical and Biological Engineering, focusing on sustainable electrochemical energy and water technologies. His work spans batteries, fuel cells, electrolysis, water treatment, and CO_2 conversion, and has resulted in over 85 peer-reviewed publications, 12 patents, and the training of more than 50 undergraduate and graduate research students. He is a co-founder and technical advisor at Mangrove Lithium, which is currently building North America's first commercial electrochemical plant for the production of battery-grade lithium hydroxide (LiOH) and lithium carbonate (Li_2CO_3).

Acknowledgments

I would like to thank Brodie Yyelland for his thoughtful review of a draft of this chapter.

References

DESIRE (2024) "Database of State Incentives for Renewables & Efficiency?" [Online]. Available: www.dsireusa.org.[Accessed:06-Jan-2024].

"Soft Costs - Solar Energy Technologies Office." [Online]. Available: www.energy.gov/eere/solar/soft-costs.[Accessed:06-Jan-2024]. U.S. Department of Energy Solar Energy Technologies Office

Duffie J. and Beckman W. (2013) "Solar Engineering of Thermal Processes" 4th ed. Wiley.

Dunlop J. P. (2012) "Photovoltaic Systems" 3rd ed. National Joint Apprenticeship and Training Committee for the Electrical Industry.

EnergySage (2024a) [Online]. Available: www.energysage.com.[Accessed: 06-Jan-2024].

EnergySage (2024b) "Solar Energy in Washington State" [Online]. Available: www.energysage.com/local-data/solar-panel-cost/wa[Accessed:06-Jan-2024].

German Solar Energy Society (2010) "Planning and Installing Solar Thermal Systems: A Guide for Installers, Architects and Engineers" 2nd ed. Routledge.

Heller, P. Ed. (2017) "The Performance of Concentrated Solar Power (CSP) Systems: Analysis, Measurement and Assessment" 1st ed. Woodhead Publishing.

IEA (2024) "Solar PV" [Online]. Available: www.iea.org/energy-system/renewables/solar-pv.[Accessed:06-Jan-2024].

LBL (2024) "Tracking the Sun 2023 edition" [Online]. Available: https://emp.lbl.gov/tracking-the-sun.[Accessed:06-Jan-2024].

LONGi Module (2024) [Online]. Available: https://natec.com/wp-content/uploads/2020/04/Datasheet-Longi-Solar-Mono-Silver-Frame-LR4-72HPH-425-455M-Hi-Mo4.pdf.[Accessed:06-Jan-2024].

NABCEP (North American Board of Certified Energy Practitioners?) (2024) [Online]. Available: www.nabcep.org.[Accessed:06-Jan-2024].

NREL (2024a) "Solar Research" [Online]. Available: www.nrel.gov/solar.[Accessed:06-Jan-2024].

NREL (2024b) "PVWatts Calculator." [Online]. Available: https://pvwatts.nrel.gov/index.php.[Accessed:06-Jan-2024].

Seattle City Light (2024a) "Solar Power" [Online]. Available: www.seattle.gov/city-light/residential-services/home-energy-solutions/solar-power#citylightsolarprograms.[Accessed:06-Jan-2024].

Seattle City Light (2024b) "Residential Rates" [Online]. Available: www.seattle.gov/city-light/residential-services/billing-information/rates.[Accessed: 06-Jan-2024].

Solar Energy Industries Association (2024) "Solar Industry Research Data" [Online]. Available: www.seia.org/solar-industry-research-data.[Accessed:06-Jan-2024].

Solar Energy International (2013) "Solar Electric Handbook: Photovoltaic Fundamentals and Applications" 2nd ed.

Solar Energy International (2024) [Online]. Available: www.solarenergy.org.[Accessed:06-Jan-2024].

Solar Pathfinder (2024) [Online]. Available: www.solarpathfinder.com/PF?id=icZJmYac.[Accessed:06-Jan-2024].

Solar Reviews (2024a) [Online]. Available: www.solarreviews.com.[Accessed:06-Jan-2024].

Solar Reviews (2024b) "How much do solar panels cost in 2024?" [Online]. Available: www.solarreviews.com/solar-panel-cost.[Accessed:06-Jan-2024].

Walker A. (2013) "Solar Energy - Technologies and Project Delivery for Buildings" Wiley.

Walker A. (2012) "Solar Water Heating" www.nrel.gov/docs/fy13osti/56706.pdf.NREL/PR-7A40-56706.7thAnnualNorthAmericanPassiveHouseConference,September27-30,2012DenverCO.

Practice Problems

18.1 PV module orientations: Discuss the various options for setting the tilt angle of solar PV modules and the optimizations that can be made. Should we consider mounting solar modules on the north-facing portions of the roof in the northern hemisphere?

18.2 PV modules, thermal effect, and system size: We have a residential PV system consisting of two parallel strings, each containing 10 modules in series. Module specifications are: Rated power = 320 W, V_{oc} = 56 V, Temperature coefficient of $V_{oc} = -0.27\%/°C$, I_{sc} = 6.1 A, and NOCT = 45 °C. Average daily insolation for the year is 4.9 hour/day. Assume an overall derate factor of 0.8. What is the size of the system and the expected annual generation of electricity? Can we have more than 10 modules, if in the coldest day of year, the ambient temperature reaches −10 °C and the Sun's radiation is 0.7 kW/m^2? Can we add a third string to our system if our DC−AC inverter has a maximum allowable DC current input of 20 A? What is the maximum possible rated power of our system?

18.3 Electricity generation and costing: Our annual energy consumption is 8,000 kWh/yr. If the average daily solar insolation is 5.8 kWh/m^2/day, and assuming an overall derating factor of 0.8, how much energy does a PV system with a rated DC power of 6 kW produce? What is the excess energy? Assume our system is grid connected with net metering in place. The cost of purchased electricity (from the grid) is $0.10/kWh and the selling cost is $0.11/kWh. What are the amounts of annual savings and annual profit generated? If the cot per wattage is $2.7/W, what is the total cost of our system and how long would the simple payback period be? (Note: To simplify, we have neglected any O&M and time value of money here).

18.4 Efficiency of solar thermal collectors: We have a solar collector with these specifications: $\alpha = 0.96$, $\tau = 0.97$, U =10 W/m^2 °C, and FR = 0.96. Solar radiation is I = 750 W/m^2 and the ambient temperature is 21 °C. Calculate the efficiency of the collector when the $T_{fluid\ in}$ = 42 °C and the temperature of the input fluid when the efficiency is zero.

18.5 Solar thermal collectors and cost savings: A family uses solar collectors to reduce their water heating costs for at least six months of the year on average. When heating is not sufficient, an electric heater supplies the additional required heat. During these six months, the collectors heat the water up to an average of 53 °C (127 °F) from a cold temperature of 18 °C (64 °F). The family uses about 15 tons of water per month. The cost of electricity is $0.12/kWh. Calculate the cost savings resulting from this solar water heating system.

19

Thermal Energy Networks

Jeff Giffin

Corix, USA

19.1 Introduction

19.2 Definition and scope

Thermal energy networks (TENs) are infrastructure systems designed to distribute thermal energy (heating and cooling) across multiple buildings or facilities. These networks use centralized or decentralized energy sources and typically employ water in a closed loop as the energy carrier. TENs can operate at various temperatures and are engineered to optimize energy efficiency, reduce greenhouse gas emissions, and enhance system resilience.

TENs share characteristics with other utility systems, such as electricity, water, sewer, and gas, in that they utilize public right-of-ways under streets and sidewalks to convey thermal energy from/to buildings and benefit from economies of scale, diversification of customer loads, and energy sources.

TENs are particularly relevant in dense urban environments, educational, corporate, medical or research campuses, and industrial parks where aggregated thermal demand allows for efficient energy sharing and thermal distribution. However, there are examples around the world with TENs ranging in size from just a few building connections to neighborhoods, cities, and even regions.

A note about nomenclature: In this chapter, we have chosen to use the term thermal energy networks (TENs) instead of district energy, community energy, or neighborhood energy. While these terms are effectively interchangeable, "TENs" more accurately describe the energy infrastructure in use and is a modern term associated with higher efficiency, the option for two-way thermal energy exchange, and lower carbon emissions.

19.3 Global importance

According to the World Economic Forum (2023),[1] heating alone accounts for over 40% of global energy-related carbon emissions (∼16 GtCO2e/year). Producing heat with low carbon systems is thus of critical importance to achieving net-zero emissions by 2050. TENs are scalable, flexible, and enable integration of waste heat sources and zero-carbon energy technologies not feasible at individual building levels, and are viewed as a key tool in decarbonizing industrial and residential heating loads.

19.4 Historical Evolution of TENs[2]

19.4.1 Ancient foundations

Centralized distribution of thermal energy is a hallmark of advanced civilizations, dating back thousands of years. In ancient Rome, aqueducts and hypocaust systems circulated hot water and air to heat public baths and buildings. In Chaudes-Aigues, France, geothermal water was first used in the 14th century to heat homes.

19.4.2 Industrial steam age (First Wave)

Germany's first TEN was established in Frankfurt in 1896 to supply heat to a large hospital, and the New York Steam Company began operations in Manhattan in 1882.[3] Most heating systems at the time used steam for heating and coal as a fuel source[4] and centralized steam boilers offered significant economies of scale, efficiency, labor, and cost savings. In 1903, Frederiksberg (an enclave in Copenhagen) Denmark[5] pioneered the use of waste incineration to heat buildings.

---

[1] https://www.weforum.org/stories/2023/03/heat-heres-why-its-the-elephant-in-the-room-for-decarbonization/

[2] https://dbdh.org/district-heating-history/

[3] Now owned by Con Edison, the New York Steam System is one of the largest district heating systems in the world with thousands of customers served.

[4] https://archive.curbed.com/2017/11/30/16716472/old-house-fireplae-coal-stove-history-heating#:~:text=Coal%20became%20mainstream%20in%20post,That%20translated%20into%20skinnier%20chimneys.

[5] https://wtert.org/wp-content/uploads/2024/01/100-years-of-waste-incineration-in-denmark.pdf

19.4.3 Post-war expansion (Second Wave)

During and after WWII, these highly fuel-efficient heating networks gained traction in urban areas due to the fuel scarcities resulting from diversion to the war effort, embargos, and reduced trade. Following the war, governments invested in TENs infrastructure to support urban growth and energy security.

19.4.4 Nordic leadership (Third Wave)

In response to the 1970s oil shocks, engineered supply shortages related to geopolitical conflicts, the Nordic countries (Sweden, Denmark, Norway, and Finland) began transitioning to centralized heating systems powered by local energy sources including bioenergy and waste heat. Today ~60% of buildings in Sweden are connected to thermal energy networks largely fueled by waste heat and renewable sources. This helped Sweden decouple energy use from carbon emission and achieve one of the lowest carbon emissions per capita of any industrialized nation, as shown in Figure 19.1.

19.4.5 North American innovation (Fourth Wave)

Around the Millenium, cities like Toronto, St. Paul, and Vancouver developed innovative energy systems such as the Deep Lake Water Cooling in Toronto and Sewage Heat Recovery in Vancouver.

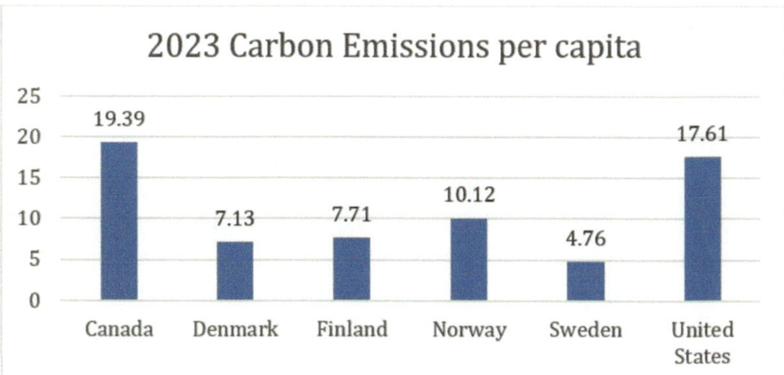

Figure 19.1 Per capita carbon emissions (EDGAR).

19.4.6 Sector integration and equity (Fifth Wave)

We are now in the era of electrified heating with significant interactive effects between electrical and thermal grids. Recent developments focus on urban TENs to decarbonize heating while promoting resilience, climate equity, and job creation.

19.5 Drivers of TENs Adoption

19.5.1 Economic drivers

TENs provide shared energy infrastructure, professional oversite, and economies of scale to reduce capital and operational expenditures while enabling the use of lower cost energy sources like waste heat or biomass. Additionally, bulk energy procurement helps lower variable energy rates to customers.

19.5.2 Environmental drivers

TENs enable the integration of renewable energy sources, waste heat recovery, thermal energy storage, and building-to-building energy sharing, resulting in high net fuel efficiency and significantly lower carbon emissions with enhanced resiliency (see Chapter 8 for more on resiliency).

19.5.3 Technical drivers

TENs offer high reliability, simplified building HVAC systems, and supports electrification and grid stability.

19.6 System Types and Generations

The evolution of TENs is provided in Table 19.1, categorized as different generations of system designs over time (Ostergaard and Svendsen, 2017; Lund, 2014; Buffa et al, 2019). The earliest systems were steam only for heating. Over time, system designs trended toward the use of water as a liquid with progressively lower temperatures and the option to provide cooling as well as heating. Another trend is that the carbon emission intensity of heating systems has generally reduced with each generation. Table 19.1 summarizes the main technical differences and key pros and cons of each generation. TENs are more feasible and require lower construction costs when designed for new construction. However, experience has shown that retrofitting of buildings to connect to TENs is possible.

Table 19.1 Evolution of thermal energy networks.

Temperatures	Generation	Description	Pros	Cons	Potential energy sources	CO2e
→ → → Decreasing water temperature → → →	First Generation	Steam systems >212 °F (100 °C)	Reliable, mature technology, high-energy density can provide heating, DHW, sterilization, and humidification	Very high thermal losses, water consumption Safety concerns due to high temperatures and pressures	Combustion of fossil or renewable fuels and electric boilers	→ → → Decreasing Carbon Emissions → → →
	Second Generation	High-temperature water >212 °F (100 °C)	Reliable, mature technology provides heating and DHW with some flexibility to modulate system supply temperatures	High thermal losses Safety concerns due to high temperatures and pressures	Combustion of fossil or renewable fuels and electric boilers	
	Third Generation	Medium-temperature water <212 °F (100 °C)	Meets all new and existing building hydronic heating temperatures	Medium thermal losses Some safety concerns due to temperatures and pressures	Combustion of fossil or renewable fuels, electric boilers, industrial heat pumps	
	Fourth Generation	Low-temperature water <158 °F (70 °C)	Meets new and some existing building hydronic temperatures including DHW with storage. Low thermal losses, suitable with certain types of plastic piping	Does not meet the temperature requirements of many older hydronic buildings	Combustion of fossil or renewable fuels, electric boilers, commercial heat pumps, waste heat, geoexchange	
	Fifth Generation	Ultra low-temperature water <125 °F (52 °C)	Ultra low distribution losses, suitable with a wide range piping materials	Not compatible with existing buildings. Requires new buildings designed for low temperature heating water with on-demand DHW or DHW boosters	Combustion of fossil or renewable fuels, electric boilers, commercial heat pumps, waste heat, geoexchange	
	Sixth Generation	Ambient or condenser water loops <100 °F (38 °C)	Negligible distribution losses, Low cost, uninsulated piping. (typically HDPE), bi-directional energy sharing	Requires heat pumps in every building or suite resulting in more equipment and higher maintenance costs. High reliance on electrical grid and limited ability to shed load	Combustion of fossil or renewable fuels, electric boilers, commercial heat pumps, waste heat, geoexchange	

19.7 Key Components of TENs

19.7.1 Energy centers

Most TENs have one or more energy centers where the heating and cooling generation equipment is located. An energy center typically houses gas boilers and electric chillers but can also incorporate more advanced thermal generation equipment such as heat pumps (see Chapter 10), combined heat and power engines (see Chapter 20), gas and steam turbines, geoexchange, bioenergy, waste heat recovery (see Chapter 20), and thermal energy storage (see Chapter 17). By centralizing the energy generation of many different customers, TENs are able to leverage industrial equipment and professional oversite to deliver highly reliable, efficient, and resilient energy generation.

19.7.2 Distribution piping system

The distribution piping system connects the energy center to customer buildings. Modern TENs use a variety of piping materials and insulation techniques to distribute thermal energy in closed loops (supply and return piping) with minimal water and thermal distribution losses. Isolation valves are installed in strategic locations so that repairs or expansions can be made without affecting all customers on the network. Like water, gas, and electric grids, distribution lines can also be connected together to form rings or loops (see Figure 19.2), which provide additional resiliency by allowing connected buildings to be fed from multiple directions.

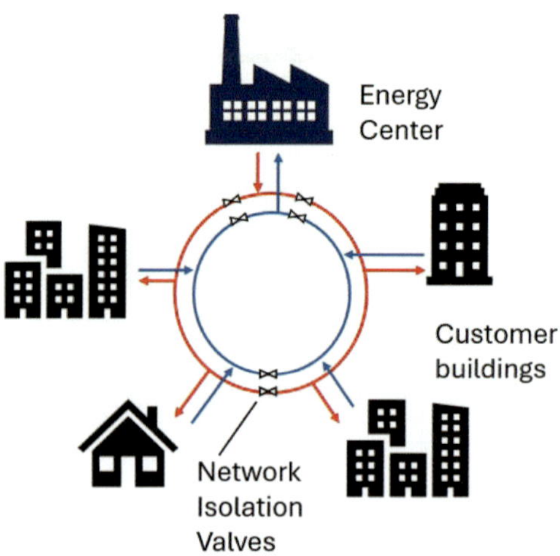

Figure 19.2 Example of a two-pipe ring topology TENs with isolation valve.

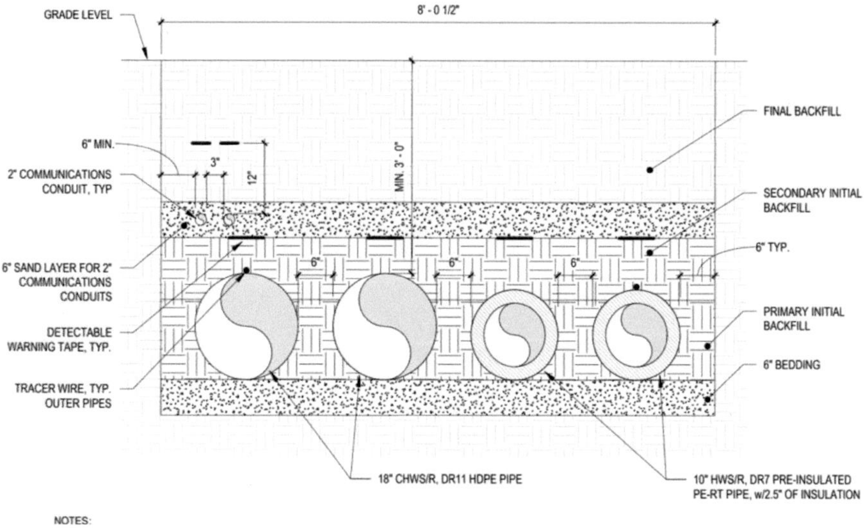

Figure 19.3 [6]

In Figure 19.3 CHWS/R is chilled water supply/return. HWS/R is heating water supply/return.

Some Sixth Generation systems (typically geoexchange systems) are taking the ring system further and use a single pipe solution, where the ring acts as supply and return with each building connected in series.

For higher temperature applications pre-insulated, all welded steel piping is generally used, and, with good water treatment, this proven approach can result in extremely long-lived (70+ years) piping systems. Seamless Schedule 40 piping is most commonly used, while EN 253 piping (a European piping standard) is also used throughout North America and around the world. High-density polyethylene (HDPE) piping is commonly used on lower temperature applications like chilled water, condenser water, and geoexchange. Fabrication of HDPE piping systems employs heat and compression to fuse piping ends together, creating a permanent bond. Polyethylene raised temperature resistance piping is a newer type of HDPE that is suitable for heated water applications up to 180 °F (82 ťC) and is becoming more popular due to its ease of use. Pre-insulated cross-linked polyethylene piping is also popular on smaller sized diameter branch lines and service connections. Figure 19.3 is a diagram for a four-pipe system utilizing HDPE and polyethylene raised temperature resistance piping.

19.7.3 Energy transfer stations

Heat exchangers are typically used to hydraulically separate buildings from TENs as illustrated in Figures 19.4 and 19.5. These energy transfer stations allow each system to operate independently with differing pressures, flows,

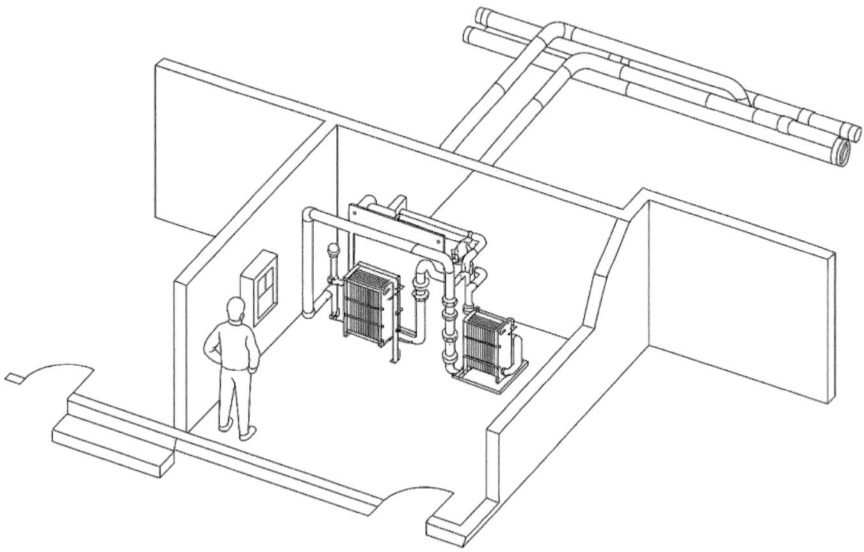

Figure 19.4 Typical heating and DHW energy transfer station layout.

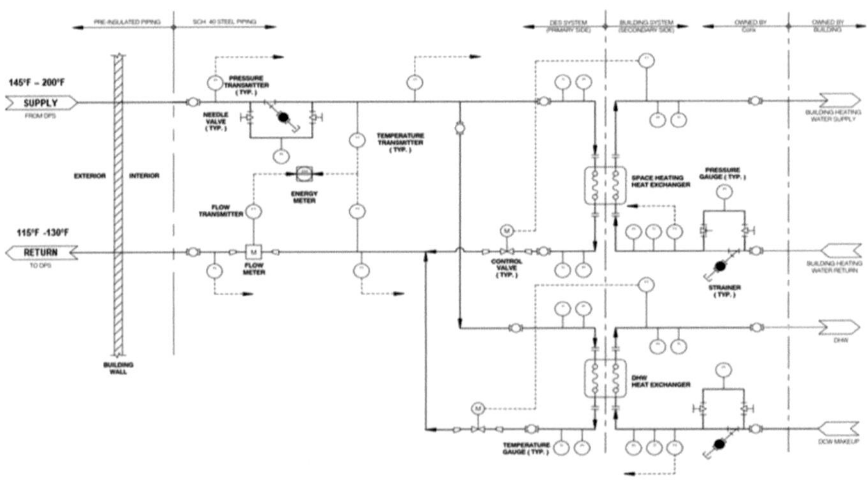

Figure 19.5 Simplified schematic of a heating and DHW energy transfer station.

and temperatures and prevent contamination or leaks from one system into the other.

Plate and frame or brazed plate heat exchangers are typically used in closed-loop applications (see ASHRAE Handbook – HVAC Systems and Equipment 2024, Chapter 47). They are cost effective, provide good heat transfer efficiency, and require a very small footprint. For domestic hot water systems, double walled plate and frame heat exchangers with a visible air gap are recommended to ensure process heating water never mixes or contaminates the potable water system.

Heat exchangers also provide a convenient demarcation point and ownership boundary for operations and maintenance as well as customer billing.

19.7.4 Thermal energy metering

Accurate metering of thermal energy is critical in today's high-performance buildings. In addition to utility billing, these meters provide important data for energy use benchmarking and monitoring-based commissioning. Unfortunately, all too often, low cost and unreliable energy meters are installed in buildings, which provide misleading information. Below are some general guidelines for choosing thermal energy meters.

Magnetic and ultrasonic flow meters with a +/−0.2% accuracy are generally recommended as these flow sensors have relatively low pressure drops, good rangeability, and accuracy while requiring very little maintenance. For fiscal meters used to bill customers, a sealed energy calculator with matched temperature sensors that meet international standards (OIML R75, EN1434 or CSA C900) for thermal energy metering is generally required. However, some regions like Canada have additional specific requires for fiscal meters that should be considered.

19.7.5 Connected buildings

Thermal energy networks work best when connected to buildings that utilize centralized hydronic (water) systems for heating, cooling, and DHW. Examples include fan-coils, perimeter and in-floor radiation systems, air handling units with and without reheat coils, water source heat pumps, and water source VRF systems.

Electric baseboards, air source heat pumps, gas-fired roof top units (RTUs), and packaged terminal air conditioners are incompatible with TENs.

In new building construction, it is generally recommended to design heating water systems for the lowest temperature as practical. Not only does

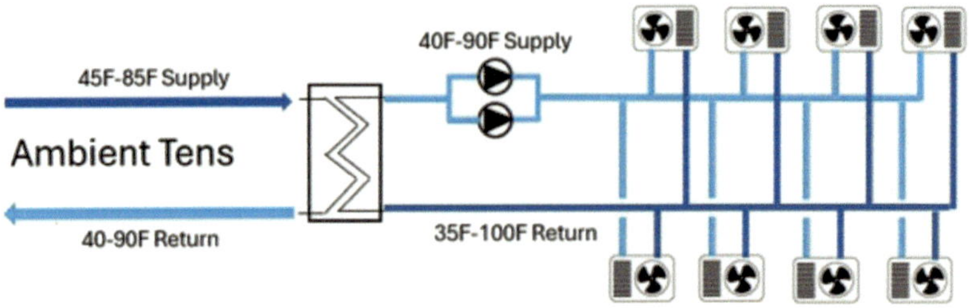

Figure 19.6 Ambient TENs with heat pump fan coils in each suite.

it help reduce distribution losses, but it also helps enable the integration of other waste heat sources. This often means sizing terminal unit heating coils for a maximum of 140 °F or less for the incoming water temperature. Using cooling coils as switch-over for heating applications is a common way to insure ultra-low temperature heating without added costs.

Buildings that are connected to ambient or condenser water TENs require heat pumps in each building or suite to elevate the temperature to useful levels. This can be done centrally using four-pipe heating and cooling distribution system or decentralized using a two-pipe condenser water system with distributed water source heat pump fan coils in each suite. Figure 19.6 is an example of a two-pipe ambient TENs.

19.8 TENs Design Principles

19.8.1 N − 1 redundancy

TENs are known for reliability and aim to achieve the golden standard of five-nines. This means that customers can count on the TENs to be available to deliver energy services 99.999% of the time. One of the ways that TENs achieve this is through redundancy of equipment. N − 1 refers to having enough thermal capacity to meet the system's peak demand loads with the largest single piece of equipment taken out of services, i.e., boiler, chiller, pumps, etc. In practice, this means installing smaller, similarly sized pieces of equipment vs. larger unique pieces of equipment.

19.8.2 Diversification of loads

Just like other utilities (electrical, gas, water, and sewer), TENs benefit from diversification of loads, which helps to reduce the overall peaks and troughs of the system and allows for "right sizing" of central plant equipment and distribution piping systems. A good example of this is what happens when you combine residential and commercial buildings into a single thermal energy network. Residential buildings are occupied 24/7 and tend to be relatively constant in their needs for heating, DHW, and cooling. Commercial spaces are typically occupied less than 50% of the time and tend to be more peaky with large morning warm-ups and generally require more year-round cooling.

Combining different building types into a thermal energy network can have a number benefits including:

Peak demand diversification. It is common that diversified loads, i.e., the peak thermal demands experienced at the energy center, are 60%–80% of the connected buildings design loads. This diversified demand is primarily attributed to the diversity of customers. However, thermal inertia in the distribution piping system also plays a key role by smoothing out the demand profiles.

Simultaneous heating and cooling. Residential buildings have year-round needs for heating due to DHW loads and can provide an excellent heat sink for recycled waste heat from commercial buildings or other cooling loads. For larger networks, the simultaneous heating and cooling overlap can range from as low as 20% of annual heating loads to as high as 88%. (See University of Stanford in the case studies section of this chapter[7].)

19.8.3 Distribution piping design

Good hydraulic design is critical to ensure efficient and reliable flow of thermal energy through a piping network and requires making trade-offs between costs, efficiency, pipe sizes, delta T's, and future load growth. It is recommended to employ the services of an experienced hydraulic modeler. Below are some key design constraints to keep in mind.

Delta Ts (ΔTs). The larger the ΔTs (difference between the supply water and return water temperature), the more energy can be transferred through the same sized pipe. This is very important for TENs, which may have many miles of distribution piping installed and a need to maximize value. Hot water TENs are commonly designed for a 60–80 °F ΔT (33–44 °C) at peak conditions but operate at lower supply water temperatures and ΔTs during

[7] https://sesi.stanford.edu/

off-peak times. To avoid freezing conditions or the need for glycol, chilled water systems tend to be limited to smaller ΔTs (15–20 °F or 8–11 °C) resulting in much larger pipe sizes (~2× diameter) of hot water pipes for same amount of energy transferred.

Water velocity. Like ΔTs, higher velocities also increase energy transferred through a given sized pipe. Recommended velocity limits for TENs are 7 ft/s (2.1 m/s) for smaller diameter pipes ($\leq$ 8 inch or 200 mm) and 11.5 ft/s (3.5 m/s) for larger diameters ($>$ 8 inch or 200 mm).

Pressure drop. Pressure drop is also a consideration and typically limited to 2.5 ft head loss per 100 feet (250 Pa/meter) to maintain pump efficiency and system balance.

Example 19-1

Estimate the maximum thermal energy (MMBtu/hr and GJ/hr) that can be delivered through a closed-loop TEN with a 12-inch (1.0 ft) internal diameter pipe, assuming 11.5 ft/s maximum velocity and maximum ΔT = 70 řF for heating. Hint: use the sensible heat formula from Appendix A: Q = GPM × 500.5 × ΔT in Btu/hr.

First, calculate the flow rate in GPM (see conservation of mass formula in Appendix A):

Flow = velocity × area = 11.5 ft/s × (π/4) × (1.0 ft)^2 = 9.028 ft^3/s

9.028 ft^3/s × (7.48 gallon/ft^3) × (60 s/min) = 4,052 GPM

Then the maximum heat delivery is:

Q = 4,052 × 500.5 × 70 = *141,962,000 Btu/hr or 142 MMBtu/hr (150 GJ/hr)*

Supply water temperature reset. During peak demand periods, it is common for TENs to increase heating and decrease cooling supply temperatures. This ensures that when it is needed the most, the design day ΔTs are met and the distribution system is maximized while ensuring energy efficiency during off-peak times. This can be achieved through simple outdoor air temperature resets (see Chapter 15 for more information on reset controls) or more direct measurements such as energy transfer station valve position feedback or system flow rates.

Thermal insulation of piping systems. All hot water and many chilled water TENs use insulation on the piping system to prevent thermal losses. Factory installed, pre-insulated piping (see Figure 19.7) is the most common because it is easy to install and provides a long lasting thermal barrier. A well-insulated modern TENs should have very low thermal distribution losses, i.e.,

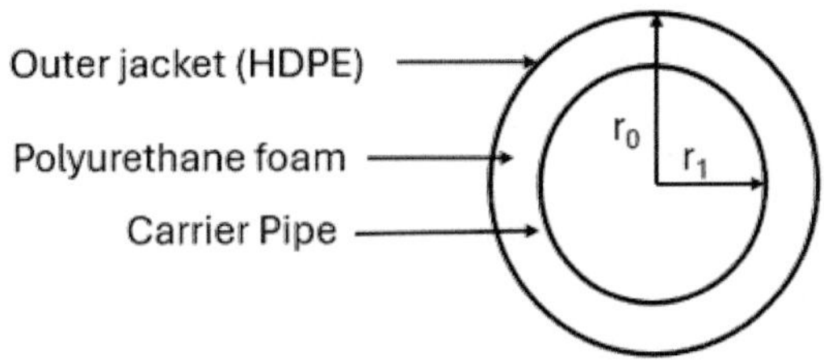

Figure 19.7 Pre-insulated distribution piping.

<5% of annual system heating loads met. You can estimate the thermal losses by using the manufacturer-provided thermal conductivity (k) values per unit length, measured in Btu/(hr-°F ft2-ft) and the following formulae:

$Q = 2\,\pi\,L\,(T_i - T_o)/[(\ln(r_o/r_i)/k\,]$

Q = heat transfer from cylinder or pipe (W, Btu/hr)

k = thermal conductivity of piping material (W/mK or W/m^{o}C, Btu/(hr ^{o}F ft^2 ft))

L = length of cylinder or pipe (m, ft)

π = pi = 3.1416

T_o = temperature outside pipe or cylinder (K or ^{o}C, ^{o}F)

T_i = temperature inside pipe or cylinder (K or ^{o}C, ^{o}F)

ln = the natural logarithm

r_o = cylinder or pipe outside radius (m, ft)

r_i = cylinder or pipe inside radius (m, ft)

Example 19-2

Estimate the distribution losses from 1,000 feet of 12" nominal steel pipe (12.75"OD, or 0.531 ft radius) with 2"of closed cell polyurethane insulation (16.75" total outer diameter, or 0.698 ft radius) operating at 158 °F. Assume an average ground temperature of 50 °F and a total thermal conductivity k-valve of the piping material (pipe + insulation + outer jacket) of 0.013 Btu/(hr-°F ft2-ft).

Also see Chapter 11 for additional examples.

Solution

Use the following formula and assume the resistance from pipe wall and outer jacket is included in the *k*-factor noted above.

$Q = 2\,\pi\,L\,(T_i - T_o)/[(\ln(r_o/r_i)/k]$

$Q = 2 \times 3.1416 \times 1{,}000 \times (158\ °F - 50\ °F)/[(\ln(0.698/0.531)/0.013]$

$Q(158\ °F)$ = *32,243 Btu/hr*

32,243 Btu/hr × kW-hr/3,412 Btu = *9.4 kW*

Ambient and condenser water loops. Usually operate between 40 and 90 °F (4–32 °C) and do not require insulated piping because the ground and the loop temperatures are close to the same. In fact, some ambient loops actually benefit from the uninsulated piping by creating a geoexchange effect where the ground actually provides beneficial cooling in the summer and heating the winter months.

Satellite Energy Centers. When pipes get over 24" (~600 mm) in diameter, the costs and installation can become very expensive and impractical. A good way to avoid pipes this large is to include satellite energy centers in the design. This allows for right-sized distribution piping systems, which preserve flexibility to accommodate future loads while also improving resiliency.

19.8.4 Load duration curves

Load duration curves are powerful tools for visualizing demand patterns and "right sizing" equipment or renewable energy systems. Figure 19.8 is an example of a thermal energy load duration curve from an existing university campus heating system. The university is currently using natural gas and has a goal of reducing the carbon footprint of its TENs by 90%. As shown in the chart below, the campus's peak (winter) heating demand is 42 MW (143 MMBtu/hr), but that only occurs for 1 hour each year. The vast majority of the time (~7,260 hours/yr), the campus needs less than half of that capacity. If you calculate the total annual heating demand that could be serviced from a 21 MW renewable energy system (see area under the green line below), you will find that up to 90% of the campus's annual loads and associated carbon emission could be offset from a heating system that is only half of the campus peak demand! This is important because renewable energy systems tend to be capital intensive, and "right sizing" these systems helps ensure the biggest bang for buck while avoiding diminishing returns which can be the difference between a successful decarbonization project and no project at all.

The "1/2 of peak demand = ~90% of annual consumption" is a good rule of thumb when sizing renewable energy projects to achieve favorable economics across a wide range of thermal energy networks. However, because every system is different, it is recommended to use actual data if available to create unique thermal energy load duration curves for every project.

Creating load duration graphs from your own data is relatively easy. The more data the better, but one year of hourly load data is generally required. In Microsoft Excel^{TM}, use the "Countif" function to return the total number

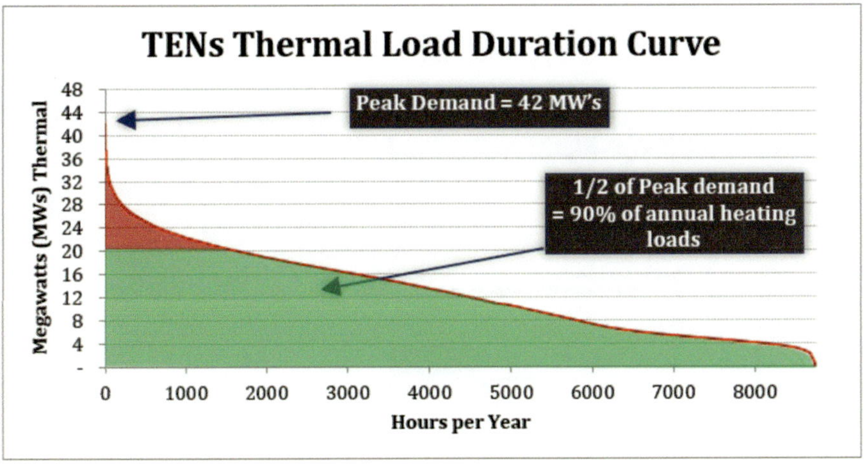

Figure 19.8 TENs thermal load duration curve.

Table 19.2 Example Microsoft Excel formulas for creating load duration curves.

	A	B	C
1	Range (8,760 load data)	Threshold value (MWs)	Total hours per year (Excel formula)
2	30.1	42.00	=COUNTIF(Range,">"&B2)
3	31.2	41.58	=COUNTIF(Range,">"&B3)
4	33.2	41.16	=COUNTIF(Range,">"&B4)
5	33.1	40.74	Etc.

of hours that the system exceeds various thresholds between 0 and systems peak demand (use whatever units preferred: kW, MW, Btu/hr, MBH, tons). See Table 19.2 for the setup and formula.

Next make a graph of the results using the "Scatter Plot with Straight Lines" option as illustrated in Figure 19.9.

19.8.5 In-series configurations

In TENs, renewable energy technologies and heat pumps are often installed in series with more conventional heating and cooling equipment. This simple design ensures prioritization of the most energy efficient equipment while also providing seamless peaking and backup without complicated control sequences. The example diagram in Figure 19.10 shows a heat recovery chiller installed in-series with conventional boilers and chillers in a four-pipe TEN. The heat recovery chiller precools the chilled water before reaching the

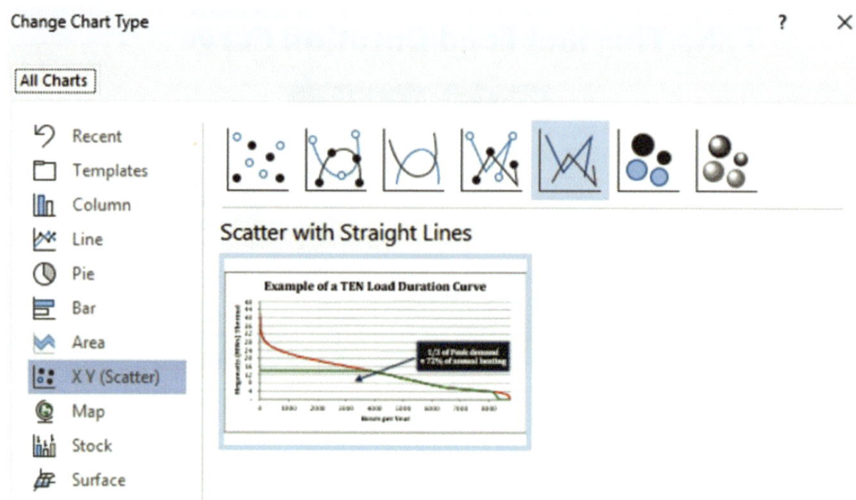

Figure 19.9 Creating a load duration chart in Microsoft Excel.

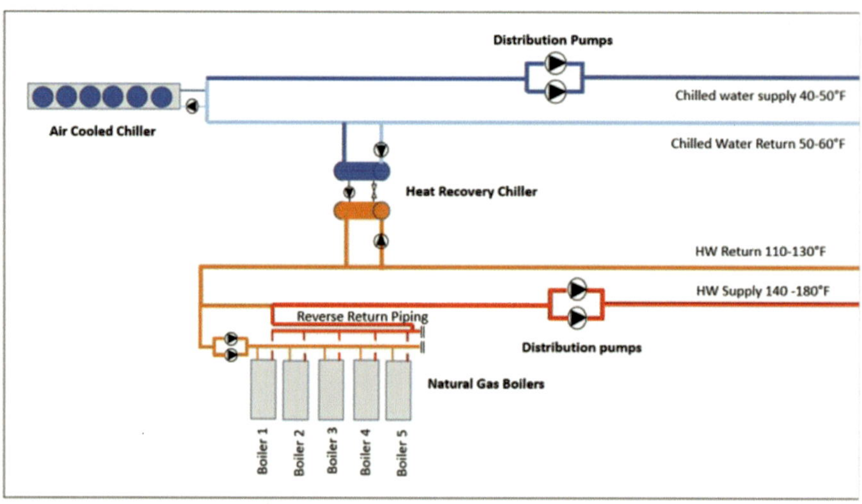

Figure 19.10 In-series configuration of a heat recovery chiller.

main chillers and preheats the hot water before it hits the boilers. During shoulder seasons, when the heating and cooling loads are well balanced, the heat recovery chiller can provide full heating and cooling loads with high COPs. However, when additional heating or cooling is required (as determined by leaving supply water temperature and setpoints), the boilers and chillers will automatically kick in to provide the required peaking and backup.

19.8.6 Thermal energy storage (TES)

Throughout North America, ice and chilled water storages are commonly found in TENs. Hot water storage is less common, but as heating systems electrify, the value of storing heat in water or other materials continues to increase (also see Chapter 17 on energy storage). Like electric batteries, thermal energy storage can leverage off-peak electricity to generate and store heating and cooling energy for dispatch when demands and prices are the highest (also see Chapter 4 on billing rates). Not only does it reduce strain on the electrical grid and reduce utility bills, but it also provides valuable capacity and resiliency to the thermal energy network.

There are many different types of thermal energy storage media including phase change materials, brick, sand batteries, and molten salt. However, for TENs applications, water-based storage tanks are the most common. These tanks range in size and shape but are usually tall and skinny to take advantage of natural thermoclines.

Hourly, daily, weekly, and even seasonal thermal energy storage are possible (see Varanto and Stanford energy storage case studies below).

Geoexchange systems with TENs are gaining in popularity and provide a form of long duration thermal energy storage. Waste heat from summer cooling loads can be stored in the ground and used as a source for heating in the winter. Traditional geoexchange systems operate in the 30–90°F (−1 to 32 °C) range and heat pumps are required to lift the temperature up to useful levels.

19.8.7 Modular design for scalability

All TENs benefit from economies of scale and diversification of loads, but it takes time to reach the critical number of loads to achieve diversification benefits. New TENs must be cost effective from day one while preserving future flexibility to scale. This requires modular thinking, as illustrated in Figure 19.11. Energy centers, in particular, are expensive components and, when oversized, have detrimental effects on the TENs. Modular design allows energy centers to be "right-sized" and installed in a "just in time" manner delaying capital costs until it is needed. Distribution piping systems are somewhat different. The cost of slightly oversized piping systems to

Figure 19.11 Example of a modular TEN made from three shipping containers in Bellingham, Washington.

accommodate for future load growth is minor (up to a point) and should be considered when sizing the distribution piping networks.

19.9 Energy Sources and Technologies

19.9.1 Waste heat – the world's largest untapped energy source

Waste heat is everywhere, from sewers and power stations to grocery stores and data centers. Anytime you see a plume of smoke from a stack or cloud of water vapor rising from a cooling tower, that is wasted heat. According to Danfoss, there is enough excess waste heat in the EU alone to provide almost all of the EU's heating and DHW needs.[8]

Capturing and reusing waste heat provides multiple simultaneous benefits. It is the cheapest form of energy, provides beneficial cooling services, saves water, and reduces emissions at the same time. TENs are uniquely suited to capture waste heat, which often is available away from urban centers, and move the heat to customers who need it.

Waste heat sources can range in temperatures (see also Chapter 20). High grade waste heat from industrial processes can be captured and used directly to heat buildings with only a heat exchanger to provide hydraulic separation. Low grade waste heat from refrigeration or sewers typically requires heat pumps (see Chapter 10) to elevate the temperature to useful levels.

[8]Danfoss a https://www.whyenergyefficiency.com/solutions/allsolutions/the-worlds-largest-untapped-energy-source-excess-heat

19.9.2 Waste to energy

These large industrial plants are built primarily to avoid landfills and generate electricity. Waste heat capture is a tertiary service/benefit made possible with TENs, but not the primary factor.

19.9.3 Bioenergy

Bioenergy including biomass, biogas, and other organic fuels have been successfully used for hundreds to thousands years by humans.[9] The Nordic countries of Europe in particular use a lot of biomass for heating TENs. Byproducts from other industries, such as sustainable forestry management and construction, are the primarily fuel sources and usually considered to be low carbon emissions (see Chapter 2).

19.9.4 Electric boilers

Depending on the "spark spread" (the cost of electricity vs. natural gas) and the carbon emissions intensity of the electrical grid, electric boilers can provide cost effective, low carbon thermal energy and are especially relevant to higher temperature networks.

19.9.5 Geothermal

Geothermal, which refers to thermal energy from the Earth's core (not to be confused with geoexchange systems), provides firm 24/7 heating and sometimes electric power from the Earth's core. Geothermal energy has been utilized around the world for millennia. Iceland in particular leads the world in geothermal energy, providing nearly 90% of the country's heating and DHW from geothermal sources.[10]

19.9.6 Solar thermal

Solar thermal (see Chapter 18) can be incorporated into TENs but is not as common as other energy sources. It is usually most effective when designed in combination with long-term thermal energy storage.

---

[9] https://www.nhm.ac.uk/discover/news/2025/december/earliest-fire-making-dating-back-400-000-years-unearthed-in-suffolk-england.html

[10] https://www.greencitytimes.com/geothermal-district-heating-in-iceland/

19.9.7 Fossil fuels

Fossil fuels, such as natural gas and oil, are typically used for peaking and backup during outages or extreme weather. Fossil fuel boilers are relatively inexpensive to install and provide a high degree of reliability to the network.

19.10 Ownership and Business Models

Throughout North America, TENs are commonly found on university and government campuses. This is likely because campuses tend to be long-term owners of both the TENs and the buildings they service, and are better positioned to take full advantage of the total capital and operating savings.

Municipally owned TENs are also common in many cities in North America. Canada especially has seen a large influx of municipally owned TENs in recent years. This is largely driven by affordability and environmental goals, which are aligned with broader urban planning and climate policies.

Privately owned TENs are another common ownership model throughout North America. They tend to be unregulated legacy steam and chilled water systems that serve downtown cores. Newer TENs owned by regulated utilities have seen rapid growth in recent years, incentivized by government policies.

Hybrid ownership structures or public–private partnerships (P3's) combine public oversight with private investment and operation. These business models leverage the strengths of both sectors – public accountability and private sector efficiency.

19.10.1 Tariff structures

Like other utilities, TENs typically bill customers monthly (see Chapter 4). Bills typically have two components.

Fixed charges. Cover infrastructure costs and ensure a return on capital invested by the utility. This is usually based on the contracted demand for energy services or square footage of the buildings.

Variable charges. This is based on the actual energy consumption as measured at the buildings meters. This pass-through of costs rewards responsible use of energy and promotes energy efficiency.

19.11 TEN Enablers

National, regional, and local codes, policies, and standards play a key role in enabling TENs.

Nationally, highly efficient TENs can complement green building standards such as LEED,[11] Passive House, and Net Zero buildings by eliminating on-site emissions and providing a pathway to credit buildings for the waste heat that is exported.

Additionally federal and state funding programs can create grants, tax credits, and low-interest loans support capital investment in TEN infrastructures.[12]

Regionally, states and provinces can create urban thermal energy network legislation to encourage legacy gas and electric utilities to transition away from fossil fuels. And regional carbon pricing and emissions regulations can create financial incentives for switching to low carbon TENs, and penalize high-emission alternatives.

Locally, municipalities can encourage TENs through zoning and planning bylaws through building height or floor space ratio relaxation, or by designating thermal energy zones and requiring connection to TENs in new developments.

19.12 Case Studies

19.12.1 University of Stanford

A good example of a modern four-pipe TENs with simultaneous heating and cooling and diurnal thermal energy storage can be found at the University of Stanford. In 2015, the campus converted from steam to hot water and installed two 4.3 million gallon chilled water tanks and a 2.3 million gallon hot water tank. These tanks are used in combination with heat recovery chillers to provide 88% of the campus heating demands and 54% of its cooling. This one project reduced the campus's GHG emissions by 68%, potable water by 18% and enabled a resilient pathway to zero carbon emissions in the future. The main system components are illustrated in Figure 19.12.

---

[11]https://www.usgbc.org/credits/new-construction-core-and-shell-schools-new-construction-retail-new-construction-data-89?view=language

[12]https://cib-bic.ca/en/medias/articles/canada-infrastructure-bank-cibc-and-markham-district-energy-close-270-million-district-energy-infrastructure-investment/

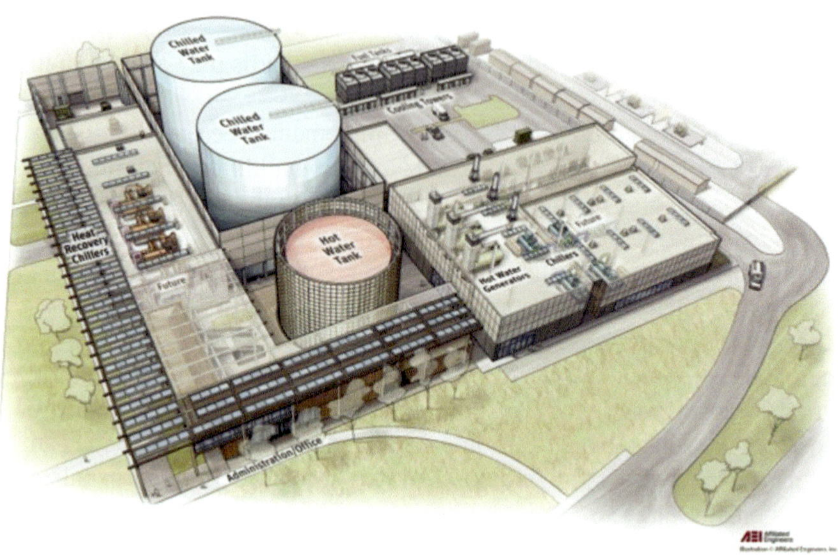

Figure 19.12 University of Stanford Energy Center.

19.12.2 Varanto Cavern Thermal Energy Storage

Not all thermal energy storage uses above-ground tanks. Some of the largest thermal energy storage systems use underground caverns or boreholes. Varanto [13] is the world's largest cavern thermal energy storage system currently in construction in Vantaa Finland and is shown in Figure 19.13. When completed, it will provide 90 GWh of thermal energy storage, which is enough to heat a medium-sized city for a year!

19.12.3 Enwave's Deep Lake Water Cooling

Located in the heart of Toronto, Canada is the world's largest deep lake cooling system shown in Figure 19.14. Using cold 39 °F (4 °C) water from the bottom of Lake Ontario, this system provides cooling to over 100 buildings in downtown Toronto at over eight times the efficiency of traditional cooling systems while avoiding over 60 MW of power demand.[14]

---

[13] Varanto https://www.vantaanenergia.fi/en/about-us/projects/varanto-the-cavern-thermal-energy-storage/

[14]https://www.enwave.com/enwave-commissions-expansion-of-world-renowned-deep-lake-water-cooling-system-internationally-recognized-by-the-united-nations-in-partnership-with-toronto-water

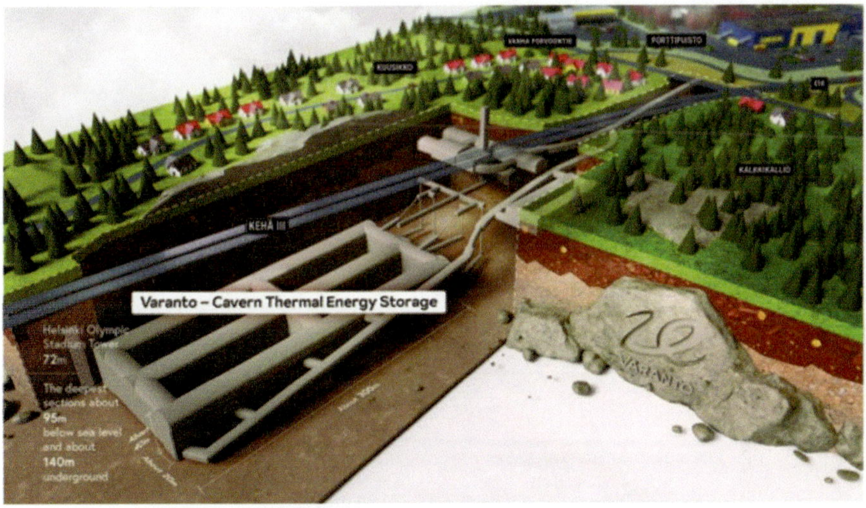

Figure 19.13 Cavern Thermal Energy Storage in Vantaan, Finland.

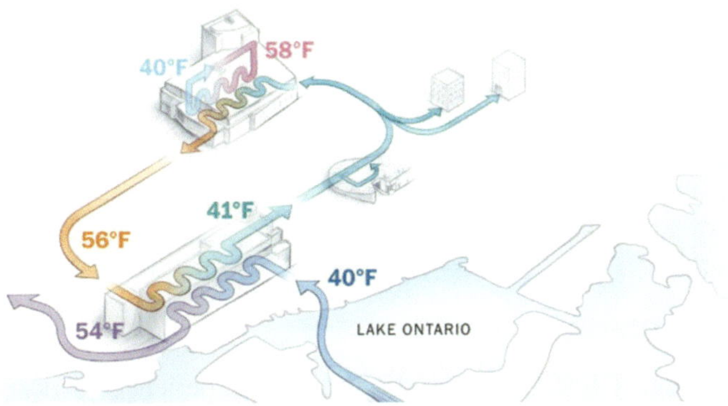

Figure 19.14 Enwave's deep lake cooling system.

19.12.4 South East False Creek NEU

Operational since 2010, Vancouver BC's Southeast False Creek Neighbourhood Energy Utility (see Figure 19.15) provides low carbon heating and domestic hot water to over 6 million square feet of residential and commercial buildings. 70% of the annual heating load is met by sewage heat recovery.[15]

[15]https://vancouver.ca/home-property-development/southeast-false-creek-neighbourhood-energy-utility.aspx

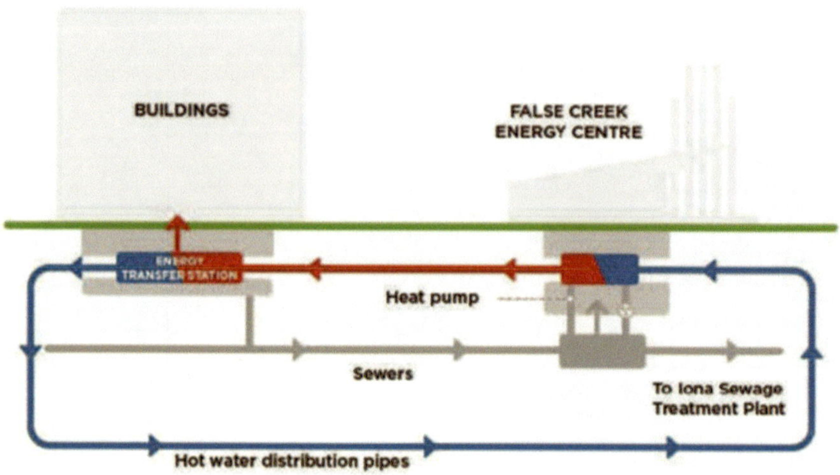

Figure 19.15 South East False Creek Sewer Heat Recovery diagram.

19.13 Conclusion

Thermal energy networks are a transformative solution for sustainable heating and cooling. By leveraging centralized and decentralized energy sources, TENs reduce emissions, improve efficiency, and support climate goals. Their integration with policy, technology, and community engagement makes them a cornerstone of sustainable development.

Author Biography

Jeff Giffin, M.Eng., CEM, has over 17 years of experience in designing, operating, and maintaining innovative low carbon thermal energy networks. From campuses to communities, Jeff is passionate about the potential of TENs to meet our future needs for sustainable heating and cooling. Jeff currently resides in Bellingham, Washington, where he oversees the Bellingham District Energy Utility, a modern TEN that harnesses waste heat from buildings and an adjacent power station. Jeff holds a master's in clean energy engineering from the University of British Columbia and is Certified Energy Manger (CEM).

Acknowledgments

This chapter has benefitted from review and/or specific contributions from Eric Mazzi, Paul Holt, and Gerard MacDonald.

References

ASHRAE Handbook—HVAC Systems and Equipment (2024), Chapter 47 "Heat Exchangers"

ASHRAE Handbook—HVAC Systems and Equipment (2024), Chapter 12 "District heating and cooling"

Ostergaard D. and Svendsen S. (2017) "Space heating with ultra-low temperature district heating – a case study of four single-family houses from the 1980's". Energy Procedia, 116. The 15th International Symposium on District Heating and Cooling.

solution for renewable urban thermal energy supply" Advances in Geoscience, 49.

Buffa S., Cozzini M., D'Antonia M., Baratieri M., and Fedrizzi R. (2019) "5th generation district heating and cooling systems: A review of existing cases in Europe" Renewable and Sustainable Energy Reviews, 104.

Lund H., Werner S., Wiltshire R., Svendsen S., Thorsen J., Hvelplund F., and Vad Mathiesen B. (2014) "4th Generation District Heating (4GDH) Integrating smart thermal grids into future sustainable energy systems" Energy, 68.

IEA-EDGAR CO_2, a component of the EDGAR (Emissions Database for Global Atmospheric Research) Community GHG database version EDGAR_2025_GHG (2025) including or based on data from IEA (2024) Greenhouse Gas Emissions from Energy, www.iea.org/data-and-statistics, as modified by the Joint Research Centre.

International District Energy Association "District Cooling Best Practice Guide: First Edition. https://www.districtenergy.org/districtcooling/best-practice-guide

19.14 Practice Problems

1. Calculate the diversified peak heating load of a 10-building mixed use (residential and commercial) TEN with average building peak loads of 500 kW and a diversification factor of 80%.

2. Estimate the minimum-sized renewable energy system needed to provide 90% of the above heating systems annual energy from renewable sources.
3. Calculate the maximum flow rate in gallons per minute (GPM) that can be delivered through a 12” (1’ ID pipe) with a maximum velocity of 11.5 feet per second. Then calculate potential heating energy that can be delivered through the same pipe with delta Ts (ΔTs) of 10, 20, 40, and 80 °F. Discuss the pros and cons of higher and lower delta Ts.
4. Calculate the distribution losses from 1,000 feet of 12” nominal steel pipe (12.75”OD) with 2”of closed cell polyurethane insulation (16.75” total outer diameter) operating at 158 °F. Assume an average ground temperature of 50°F and a total thermal conductivity k-valve of the piping material (pipe + insulation + outer jacket) of 0.013 Btu/(hr-°F ft2-ft). Now compare these losses to potential thermal energy delivered in Problem 3 with a 40 °F ΔT.
5. Describe a Fourth Generation four-pipe heating and cooling TEN with centralized heat recovery and a two-pipe Ambient TEN with building level heat recovery. List some of the major environmental, economic, and ownership considerations.
6. Describe the process you would take to model a load duration curve of an existing TEN.

20

Industrial and Process Energy Efficiency

Eric Mazzi

Mazzi Consulting, Canada

20.1 Introduction

This chapter provides an introduction to industrial and process energy efficiency. Industrial operations are large consumers of energy and emitters of GHG emissions. As introduced in Chapter 2, industrial GHGs are about 24% of total GHGs globally, 24% in the U.S., and 37% in Canada.[1] If industrial indirect emissions are included such as from purchased electricity (Scope 2 emissions, see Chapter 2), then industry's global share of GHGs is about 33% (Rissman, 2024).[2] As also summarized in Chapter 2, the science of climate change demonstrates that achieving net-zero GHGs globally is required to ultimately stabilize Earth's climate as well as to meet many current policy goals. Therefore, given the industrial sector's one-third contribution to global GHGs, it is clear that industrial efficiency and decarbonization are essential to achieving climate policy goals. Industrial operations are also critical for decarbonization of other economic sectors, because industrial GHG emissions represent the Scope 3 emissions that are reported by other sectors such as buildings and transportation.

This chapter begins with a brief introduction to industrial energy efficiency, followed by several sections on energy efficiency for common industrial and process equipment including: combustion and boilers, fans and pumps, compressed air, heat recovery, and combined heat and power (CHP). While these systems are described mainly as applied to industrial applications, the concepts and methods are equally applicable to buildings.

[1]See Figures 2-2, 2-4, and 2-5 in Chapter 2. These figures exclude agriculture and the electricity production. For Canada, the oil and gas sector is included.

[2]These figures are approximate and there are differences in accounting from different sources. For example, Rissman's summary excludes methane leakage.

Before launching into content on energy-efficient technologies for the rest of this chapter, it is important to emphasize the common organizational priorities for industrial operations. There is typically a ranked order of priorities: safety first, followed by production (product quality and quantity or production rate). After these top two priorities come other issues in various orders depending on the operation: profitability, shareholder value, maintenance, and sustainability (which includes energy efficiency and decarbonization), etc. For professionals working with industrial operations (whether as employees, consultants, contractors, vendors, etc.), an approach to increase the chances of success is to always assume that safety and production are the top priorities. When proposing energy efficiency or low-carbon solutions for an industrial operation, it is more likely that the solution will gain better credibility, be accepted, and put into action if these priorities are explicitly addressed.

20.2 Overview of Industrial Energy Efficiency

Industrial operations span a wide range of processes and applications, which are too numerous to describe in one chapter (or even one book!). Industries include a seemingly infinite list of manufacturing processes and products that most of us use daily. Food and beverage industries, transportation systems (e.g., cars, aircraft), pulp and paper and forest products, mining, chemicals, pharmaceuticals, cement, iron and steel, oil and gas production, plastics and textiles are some of the major segments contributing to industrial energy consumption and emissions. An exercise one can do to gain some appreciation for the number and diversity of industrial processes is to peruse through the U.S. Industrial Training and Assessment Center database of energy assessments at https://itac.university/search/assessments.

Despite the apparent limitless range of industries and complexity of processes involved, there are some common approaches to industrial energy efficiency that are effective. To begin, it is useful to broadly group industrial energy efficiency into two categories: specialized process equipment and ancillary equipment.

20.2.1 Specialized industrial equipment and processes

To illustrate what is meant by specialized industrial equipment, Table 20.1 provides a sample list of specialized industrial technologies or processes. For each technology or process, an example of a real-world energy efficiency measure (EEM) is provided.

Table 20.1 Sample of specialized industrial processes and technologies and EEMs.

Technology or process	Description of technology and EEM
Pulp refiner	Electric power machine that crushes wood chips between metal discs as part of mechanical pulping. More efficient refiner plate design.
Walking beam furnace	Industrial furnace used for metal heating, where metal beams move the metal through different heating zones. Insulation and alter operating procedures to reduce heat loss.
Regenerative thermal oxidize	Pollution control equipment that burns or oxidizes volatile organic compounds to CO_2 and H_2O before releasing to atmosphere, and includes heat recovery.
Yankee dryer	Specialized industrial dryer used for making tissue paper. More efficient dewatering technology.
Weaving machine	Machine used for manufacturing fabric in the textile industry. Energy-efficient looms.
Full air separation process	Process for producing pure liquid nitrogen using air as the feedstock. More efficient process design.
Pulp process lime kiln	Kiln which heats raw materials to produce lime. More efficient dust recycling and pneumatics.
Concrete batching process	Plant which mixes cement, aggregates, water, and admixtures to produce concrete. Variable speed conveyance and regenerative power.
Electric arc furnace	Technology powered by electricity and used to melt scrap metal or reduced iron. Controls optimization to reduce power time, shorten smelting cycle.
Thermal curtain	Engineered screen used for greenhouses to reduce heat losses and control.
Automotive parts	More efficient controls and scheduling.
Cement kiln clinker	More efficient rotary kiln designs with pre-heat and pre-calciners.

It is outside the scope of this book to provide a wide range of energy efficiency solutions for specialized industrial processes. But we can describe the best practice for identifying and implementing energy-efficient solutions. The best practice is actually the same as described in Chapter 3 (i.e., the flow chart in Figure 3-1). What is somewhat unique for specialized industrial equipment is to ensure the project team working on energy efficiency (or decarbonization) solutions includes three key parties:

1. *Plant O&M staff* (engineering, operations, maintenance) have the best knowledge and understanding of the production needs and O&M history for specific plants where they work.

2. *Specialized consultants* can bring unique expertise to the table. Often there are consultants who specialize in specific industries and processes. Consultants can bring valuable experience based on working on many different plants or projects for clients within a specific industry, and from different regions and different companies. The key is finding the consultant with the specific expertise needed for the solution that is needed.
3. *Technology vendors* are essential as the supplier of equipment for new construction or retrofits. Progressive technology suppliers are continuously working on new and improved technologies to serve their customers. Similar to consultants, technology suppliers can bring unique expertise for specific industrial processes based on applications across many different companies and from different regions (perhaps even different continents).

The best practice is for project teams working on identifying and implementing energy efficiency solutions for specialized industrial processes to have all three parties adequately represented, and working together in a cohesive manner.

Some additional guidance on approaching energy efficiency for industrial processes is provided by Doty (2020) in Chapter 4 on manufacturing. Also, Williams (2022) "Industrial Energy Systems Handbook" Chapter 6 describes approaches to energy assessments for industrial operations.

20.2.2 Case study: efficient processes for production of liquid nitrogen

A new industrial facility was designed to produce liquid nitrogen with a production rate of 150 ton/day for 360 days per year. Nitrogen liquefaction processes use air as the feedstock (air is about 79% nitrogen by volume), separate the nitrogen, then cool it to a liquid as shown in Figure 20.1. Two competing process designs were considered and compared based on the efficiency of electricity consumption using a physical-thermodynamic metric (see Chapter 1, Table 20.1–20.5). The baseline process energy intensity is 690 kWh/ton, and was measured using actual data for an existing plant that operates at 180 ton/day using a standard efficiency process. A more efficient plant design is modeled to consume 610 kWh/ton at the new plant production rate of 150 ton/day. The efficiency improvement is due to improved efficiency of the process components, combined with a reduction in the quantity of nitrogen that is vented and not captured as the end product, liquid nitrogen.

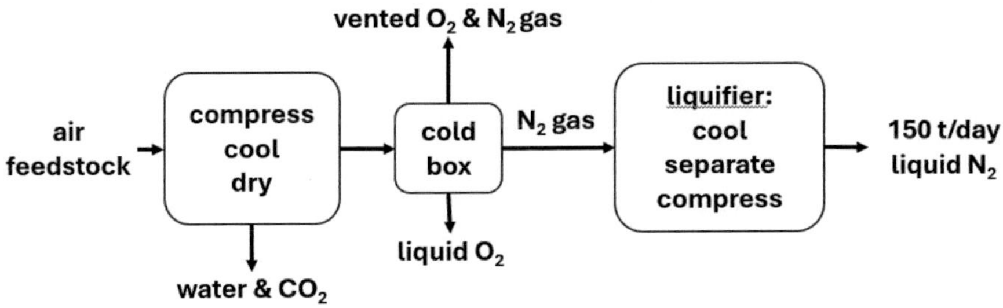

Figure 20.1 Simplified schematic of liquid nitrogen production.

To estimate the electricity savings, it is important to consider that the baseline energy intensity is measured at a different production rate (180 ton/day) than the design production rate for the new facility (150 ton/day). In general, the energy intensity of industrial processes improves (i.e., intensity decreases) as production increases. In other words, higher production rates typically lead to at least modest improvements in energy efficiency. For this reason, in general, it would not be correct to calculate electricity savings using kWh/ton metrics without adjustment, because the baseline and energy-efficient plants are not on an "apples to apples" production basis. However, since the baseline plant data is for a higher production rate, it is likely that the baseline energy intensity would be a lower value than 690 kWh/ton at a production rate of 150 ton/day. To be conservative,[3] assume the baseline energy intensity at 150 ton/day would be 690 kWh/ton and estimate the electricity savings as follows:

Savings = 150 ton/day × 360 days/yr × (690 – 610) kWh/ton = *4,320,000 kWh/yr*

20.2.3 Advanced efficiency and decarbonization for industry

In the big picture, what is necessary to achieve steep improvements in energy efficiency and decarbonization of industrial operations comes down to four major actions according to Rissman (2024). While some of these actions are outside the scope of this book, it is important for professionals to be aware of them.

[3]"Conservative" as commonly defined in M&V such as the IPMVP protocol (www.evo-world.org) means that, when necessary, make assumptions that tend to moderately underestimate energy savings. This common practice in M&V is for business risk management purposes to avoid over-estimating energy savings.

1. **Industrial energy efficiency**. This is the main focus of this chapter and much of this book.
2. **Industrial electrification**. Electrification is introduced in Chapter 2; an introduction to industrial electric heat pumps is provided in Chapter 10; a limited discussion of electric heating equipment is included later in this chapter. As a reminder, electrification decarbonizes energy systems only if the electricity supply is truly low carbon (again, see Chapter 2).
3. **Alternative fuels**. Hydrogen and hydrogen-derived fuels, liquid and gas biofuels (including renewable natural gas, discussed in Chapter 2); biomass (briefly addressed later in this chapter).
4. **Industrial materials** (Allwood & Cullen, 2015; McDonough, 2002). Material efficiency, material substitution, and circular economy are key strategies. Material efficiency is making products while using less material, without compromising product quality. Material substitution is using lower carbon materials to replace higher carbon materials. Circular economy refers to using materials and products for the highest value applications while minimizing the need for mining raw materials and includes maximizing the reuse and repair of products.

Another visionary approach to decarbonizing industrial processes is integrative design, which was developed by Amory Lovins (2013). Integrative design is a method that optimizes an entire system, versus focusing on isolated parts or individual technologies. The 10 design principles for integrative design are directly quoted below (Lovins, 2010):

1. Focusing on the desired end-use places purposes and application before equipment, efficiency before supply, passive before active, simple before complex.
2. Broadening design scope embraces whole systems and sets end-use performance metrics.
3. Designing from scratch, at least initially, creatively harnesses "beginner's mind," spans disciplinary silos, surpasses traditional solutions, and further expands the design space.
4. Analyzing gaps between theoretical minimum requirements and typical usage reveals overlooked opportunities for elegant frugality.
5. Optimizing systems, not isolated parts, lets single expenditures yield multiple benefits.
6. Evidence-based analysis supplants rules of thumb.
7. Measurement and prudence replace mindless oversizing and allow operational risks to be managed explicitly and intelligently.

8. End-use savings multiply upstream energy and capital savings, so efficiency logic is sequenced in the direction opposite to energy flow.
9. Design satisfies rare conditions (making appropriate tradeoffs and engaging end-users), but emphasizes typical conditions to maximize performance integrated over the range.
10. Controls and embedded sensors create intelligence and learning, so design can be optimized in real operation and further improved in future applications."

Elements of integrative design are embedded in the content of this book. For example, the goal to deliver the right level of energy services (not over- or under-deliver), defined by quality and quantity, is aligned with integrative design principles. The emphasis on the sequence of solutions focusing on passive systems first (e.g., Chapter 3, Figure 3-1) is another example. The interdisciplinary content of this book, including fundamentals of rational economics, finance, climate science, and human behavior change (including behavioral economics), is also an attempt to bridge disciplinary silos as encouraged by integrative design. Additional resources on integrative design such as publications, webinars, courses, and case studies can be found at the following Stanford University website: https://integrative-design-for-radical-energy-efficiency.stanford.edu.

20.2.4 Energy efficiency for ancillary industrial systems

The remainder of this chapter is focused on ancillary energy systems. What is meant by ancillary systems is the non-specialized process equipment that is commonly used in industrial plants. This includes: heating technologies, combustion systems, steam systems, fans and pumps, compressed air systems, heat recovery equipment, and combined heat and power (CHP). In addition to the content provided in the remainder of this chapter, it is important to point out that a number of energy efficiency practices and technologies applicable to industrial operations are covered elsewhere in this book:

- Energy management systems (see Chapters 1 and 3)
- Behavior change methods (see Chapter 7)
- HVAC (see Chapter 9)
- Industrial heat pumps (see Chapter 10)
- Industrial refrigeration (see Chapter 10; also see Williams, 2022, Chapter 16)
- Insulation (see Chapter 11)

- Lighting (see Chapter 12)
- Efficient electric systems (see Chapter 13)
- Efficient motors and drives (see Chapter 14)
- Efficient controls (see Chapter 15)
- Process commissioning (see Chapter 16; also see Killcross, 2021)
- Thermal and electric storage (see Chapter 17)
- Solar thermal and solar PV (see Chapter 18)
- Thermal energy networks (see Chapter 19)
- Energy-efficient maintenance practices (see Chapter 21)

20.3 Heating Technology, Combustion, and Steam Systems

In this section, we cover heating technology, combustion processes, and water-based systems (i.e., hot water and steam). We introduce the basics that energy professionals need to know in order to assess and compare technologies, perform practical calculations, estimate performance, and effectively manage heating systems. First, we provide an overview of heating technologies (both fuel-fired and electric), followed by combustion fundamentals, and then the basics of water and steam systems.

20.3.1 Heating technology

Heating equipment is used in almost every home, commercial building, and industrial plant.

As introduced in Chapter 2, there is an urgent need to decarbonize economies globally, and this includes industrial heat. A general approach to designing efficient and low-carbon industrial heating systems should ideally consider whole systems thinking as per the integrative design method introduced above. At a facility or system level, identifying highly efficient, low-carbon industrial heat solutions should employ the approach described in Chapter 3, Figure 3-1. Specifically, it is important to always assess the heating service that is required (this includes both retrofit and new plant designs), defined by quality (e.g., temperature, tolerable contaminants in heating fluids) and quantity (e.g., average and peak demands). Then, before selecting the heating technology, it is useful to carefully assess passive improvements first, before sizing and selecting heating equipment. Additionally, integration of thermal storage may be an economical strategy to "right size" the heating technology (see Chapters 17 and 19). As an example, it would be uneconomical to purchase heat production equipment that serves a process that is uninsulated, operates with a high rate of avoidable fluid leakage, and has a peak demand that could be met more cost-effectively through use of storage.

In assessing the appropriate heating technology, the most critical variable to consider is the heating temperature required for the process. Table 20.2 provides an overview of typical process temperatures required for various industrial activities (Rissman, 2024).

Three common types of heating equipment cover most applications: electric and absorption heat pumps, fuel-fired boilers and heaters, and electric-powered boilers and heaters. Heat pumps are introduced in Chapter 10, including both electric and absorption cycle designs, as well as industrial applications. In this section, we will overview both fuel-fired and electric heating technologies.

The basic design of fuel-fired heating equipment is shown in Figure 20-2. With fuel-fired designs, fuel and air are inputs to the combustion process that releases heat that is transferred to a working fluid, which is usually water in the form of liquid water or steam. Some systems use other fluids such as thermal oils. Combustion products exit the heating equipment as flue gas through an exhaust stack. Section 20.2.2 covers the fundamentals of combustion including estimation of flue gas composition, combustion, and thermal efficiency. Section 20.2.3 covers the basics of water and steam systems.

Common fuels used in combustion processes were introduced in Chapter 1 (see Table 20.1–20.2) and heating values are provided in Appendix B. It is important to distinguish between higher heating value (HHV) and lower heating value (LHV) for fuels, which is covered in the next section on combustion. It is useful to group fuels according to their chemical state:

- **Gaseous fuels**: fossil natural gas, hydrogen, and biofuels such as renewable natural gas (RNG, introduced in Chapter 2) and landfill gas.
- **Liquid fuels**: No. 2 oil (diesel), No. 6 oil (heavy oil which must be heated to liquify it), propane, gasoline, liquified petroleum gas (LPG), and jet fuel are common. Liquid biofuels are also sometimes available.
- **Solid fuels**: coal of different types (see Appendix B), wood or biomass, and municipal solid waste.

Boilers are the most common fuel-fired heating equipment. Boiler is a term that typically refers to any water-based heating equipment, whether the production is in the form of hot (liquid) water or steam. There are two

Table 20.2 Typical process temperature requirements for common industrial activities.

Industrial activity	Temperature range	Example industries
Smelting	430–1,650 °C 800–3,000 °F	Iron and steel, non-ferrous metals
Melting metals	430–1,650 °C 800–3,000 °F	Iron and steel, non-ferrous metals
Melting non-conductive materials	800–1,650 °C 1,470–3,000 °F	Glass, ceramics
Calcining	800–1,100 °C 1,470–2,000 °F	Cement, lime
Space heating	50–100 °C 122–212 °F	All industries
Domestic hot water, dish-washing, laundry	50–90 °C 122–194 °F	All industries
Steam production, boiling, distillation	70–540 °C 180–1,000 °F	Chemicals, food processing
Welding	900–1,500 °C 1,650–2,700 °F	Machinery, vehicles, construction, metal products
Cutting, drilling, boring	120–3,500 °C 250–6,300 °F	Machinery, vehicles, construction, metal products
Heat treating of metals	100–800 °C 212–1,470 °F	Metal products, vehicles
Warming non-conductive, heat-sensitive materials	70–300 °C 160–570 °F	Food processing, plastic products, wood products
Curing adhesives and coatings	100–350 °C 212–660 °F	Vehicles, miscellaneous products
Molding and forming	120–300 °C 250–570 °F	Plastic products
Sterilizing and pasteurizing	100–370 °C 212–700 °F	Food processing
Baking, thawing, drying	100–370 °C 212–700 °F	Food processing, rubber, mineral and construction materials
Exceptionally high-temperature processes	2,000–3,000 °C 3,600–5,400 °F	Synthetic graphite, specialty metals, vapor deposition coatings

common designs: fire tube and water tube. Fire tube boilers are designed so that hot combustion gases are directed through tubes, and transfers heat to the water surrounding the tubes as illustrated in Figure 20.3. In this design, fuel and air are supplied by fuel piping and a combustion air fan, the mixture is ignited, and then the hot gases make multiple loops or passes through the tubes. Hot water or steam is produced and exits the top of the boiler. Fire tube

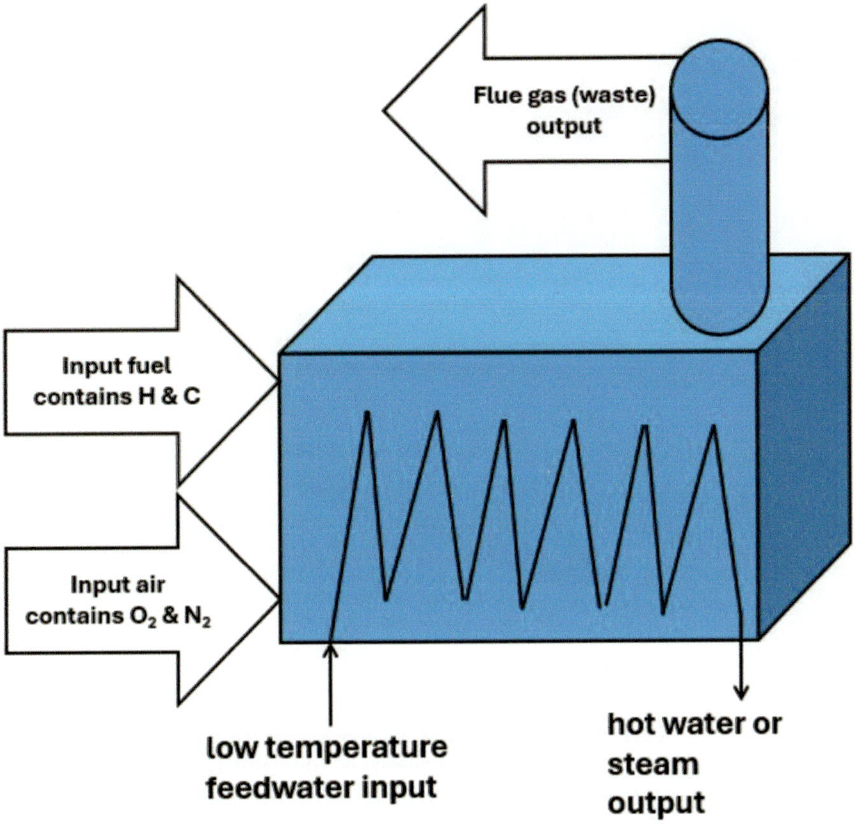

Figure 20.2 Simplified schematic of fuel-fired boilers and heaters.

boilers are usually used for smaller applications, such as building heating systems.

Water tube boilers are the opposite of fire tube, in that water and steam runs through tubes and the tubes make multiple passes through the combustion gases. Water tube boilers tend to be used for larger systems, such as industrial heating and electricity generation. A boiler can be thought of as a large heat exchanger between combustion gases and water. Combustion gases gradually get cooler as heat is transferred to the water side of the system. Traditional systems were designed to ensure the exhaust flue gas temperature remained hot enough to avoid condensation of moisture or acid gases to avoid corrosion. Typically this meant exhaust gas temperatures that exit through a stack are in the range of 340 °F (170 °C) or higher.

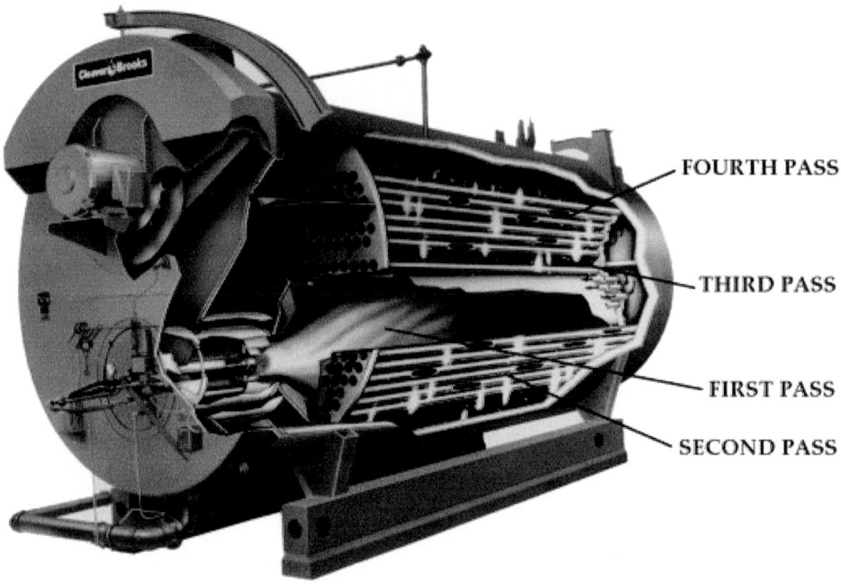

Figure 20.3 Cutaway view of a fire tube boiler (courtesy of Cleaver Brooks; Capehart, 2016).

Condensing boilers, which are further described in the next section, are boiler designs where an extra heat exchanger is added to condense out the water vapor in the exhaust. This heat exchanger is usually designed with copper or stainless steel for corrosion protection. The water vapor that is condensed out of the flue gas is called condensate (not to be confused with condensate return water in a steam distribution system, described in Section 20.2.3). This condensate is corrosive with a low pH (i.e., acidic) and must be chemically neutralized to achieve a pH of around 7. Condensing boilers and heaters can readily be identified during energy audits because there will be a condensate collection system. Figure 20.4 is in image of a pair of condensing hot water heaters in a commercial building. In the figure, condensate treatment system is in the white cylinder at ground level that collects condensate, neutralizes it, and discharges to sanitary sewers, which ultimately go to sewage treatment plants (not storm sewers, which ultimately drain to local water bodies).

Another common fuel-fired heating technology is infrared radiant heating. Radiant heaters are designed for a wide range of temperature applications (ASHRAE Handbook Systems and Equipment, 2024, Chapter 16). This

Figure 20.4 Two condensing hot water heaters in a commercial building.

technology is designed to work with radiation heat transfer in the infrared range (see Chapter 11 for a review of heat transfer modes). One common application is to use infrared heaters for industrial space heating such as the natural gas-fired tubular heater shown in Figure 20.5. Note that these heaters are usually mounted high up on ceilings or walls, and use an aluminum reflector to direct the heat to the intended locations and improve system efficiency.

Example 20-1

During an energy audit, you find an old low pressure (15 psig steam or about 100 kPa gage), watertube steam boiler shown in Figure 20.6. The boiler

Figure 20.5 Gas-fired, infrared tubular heater with an aluminum reflector used for industrial space heating.

nameplate states that it is rated for 100 boiler horsepower output and with an input rating of 4.185 MMBtu/hr. What is the rated thermal efficiency at full load? Estimate the monthly (30 days) fuel consumption in therms of natural gas if the boiler operates at 75% of full load for 24 hours daily. Also estimate the full load fuel consumption in SI units of GJ/hr.

From Appendix A, it can be found that boiler horsepower is a measure of boiler output or production and 1 blhp = 33,475 Btu/hr. Thus, the rated thermal efficiency at full load is:

η = output/input = (100 × 33,475 Btu/hr)/4,185,000 Btu/hr = *0.80 or 80%*

Also from Appendix A, 1 therm = 100,000 Btu. 30-day consumption at 80% of full load is estimated as follows:

Figure 20.6 Water tube boiler installed in the early 1960s for low pressure steam for space heating in an industrial plant.

30 days × 24hr/day × 0.75 × 4,185,000 Btu/hr × (therm/100,000 Btu) = *22,600 therms*

The fuel consumption in SI units is: 4.185 MMBtu/hr × (1.055/GJ) = *4.415 GJ/hr*

In lieu of using fossil fuels, there are two common approaches to decarbonization of industrial heating. One approach is to use biofuels, which, if sourced sustainability and fully replenished in less than 20 years from harvesting, can have low GHG emission factors (see Chapter 2). Combustion of biofuels in heating equipment works the same as fossil fuels. Biofuels are, like fossil fuels, are composed mostly of molecules constructed with carbon and hydrogen and the fundamental combustion process is the same as described in the next section.

The other common approach to decarbonizing industrial heating (and non-industrial as well) is to use electric heating technologies. In Chapter 10, industrial heat pumps are introduced, which are typically the most energy-efficient electric technology. A key limitation of heat pumps is that the process fluid is limited to around 100 °C (212 °F) with current commercial equipment. However, technology is evolving rapidly with designs in the range of 150 °C (300 °F). For higher temperature needs, the following are key industrial electric technologies (Rissman, 2024; Guidehouse, 2025).

Induction heating uses alternating current through an electromagnet that creates electric "eddy currents" in a conductor. With induction, the current is generated directly in the material to be heated, unlike resistance heating, which generates heat that must be then transferred to the material to be heated. Induction heating can convert over 90% of electricity to heat, but because it delivers it directly to the target material, it can be more efficient and faster than resistance heat. Induction heating technologies range from 100 °C (212 °F) to 3,000 °C (5,400 °F). Induction cooking stoves for home and commercial use have increased adoption in recent years.

Electric arc and plasma furnaces use high-voltage electricity that produces plasma, which is an ionized gas. Lightning produced in the atmosphere is an example of an electric arc. Electric arc furnaces are commercially used for steel, other metals, phosphorus, and calcium carbine and can heat these materials to 1,800 °C (3,300 °F).

Infrared heating operates by emitting thermal radiation in the infrared wavelength range. Systems are designed with emitters and sometimes with a reflector. Ceramic emitters are about 96% efficient in converting heat into infrared radiation. Quartz lamp with tungsten filament emitters are about 85% efficient, and tubular emitters are about 56%. Infrared heating is used for industrial space heating and industrial drying. Ceramic emitters can heat materials to about 700 °C (1,300 °F), while quartz can heat to about 1,400 °C (2,600 °F).

Electric resistance heating has electric current passing through a resistor to dissipate heat. Nickel chromium designs can produce temperatures to 1,250 °C (2,280 °F). Tungsten designs can produce up to 3,000 °C (5,400 °F). Nominally, the efficiency of resistance heating is 100%. But ancillary losses may exist and the practical efficiency may be slightly less.

Electric boilers that use resistance heating in various ways are becoming more widely adopted as a low-carbon option, especially in jurisdictions with low carbon electricity supplies. The three most common electric boiler designs are immersion heaters, electrode boilers, and immersed electrode

Figure 20.6a Jet-type electrode boiler for industrial process steam production (courtesy: Thermon Inc.).

boilers.[4] Immersion boilers pass an electric current through a resistor, which then transfers to heat water. Immersion technology can be used to produce hot water or steam directly. Immersion heaters come in 600 V designs for small loads, and 4,260–6,900 V designs for larger systems. Jet-type electrode boilers work by spraying water across electrodes where the water acts as a resistor and generates steam. Immersed electrode boilers work by immersing electrodes in water, and the water resistor produces heat to generate hot water or steam. Figure 20.7 is an example of an industrial jet-type electrode boiler.[5] All electrode boilers require careful control of water quality. Some electrode designs cannot produce hot water directly, and in these cases, a steam to water heat exchanger is required to produce hot water. Electric boiler

[4]See, for example, models offered by Cleaver Brooks at https://cleaverbrooks.com/Catalog/boilers/electric-and-electrode and AERCO www.aerco.com/products/hvac-hot-water-solutions/boilers/sequoia-s

[5]Electrode boiler for steam production in the range: 800-50,000 kW, 2,700-167,000 lb/hr, 100-405 psig, and 4.16-15 kV from www.thermon.com

Figure 20.7 Electric process heater (courtesy of Sigma Thermal Inc.).

products are typically stated to have thermal efficiencies of nearly 100%. However, considering ancillary losses, the overall efficiency may be slightly lower. ASHRAE Handbook (2024), Chapter 32, states that the overall thermal efficiency of electric boilers is 92%–96%.

Electric resistance heating equipment is also used for industrial process fluids, and not only for water or steam. Figure 20.7 is an example of a factory-engineered industrial electric process heater. Applications for this type of heater include heating of asphalt, materials for extruding, and petroleum products to maintain targeted viscosity.

20.3.2 Fuel combustion fundamentals

Fuel combustion is a chemical reaction that releases thermal energy, and it is the heat release that is the key purpose and benefit of combustion systems. Chemical reactions are described with chemical balance equations as illustrated in Figure 20.8, which illustrates a typical combustion process (note: chemical balance multipliers are not shown).

$$\underbrace{C + H + O_2 + N_2 + S + \text{ash}}_{\textit{reactants}} \rightarrow \underbrace{CO_2 + H_2O + N_2 + \text{heat} + \text{excess } O_2 + CO + NO_x + VOC + SO_x + \text{particles}}_{\textit{products}}$$

Figure 20.8 Basic combustion chemical reaction equation.

On the left side, the reactants typically include hydrocarbon fuels and air (which is ~20.9% O_2, 79% N_2, and other trace gases). Depending on the fuel composition, the reactants may also contain contaminants such as sulfur or ash (or even more contaminants, e.g., consider municipal solid waste combustion).

On the right side of the chemical equation, the products include CO_2, H_2O, and N_2. It is important to understand that essentially all carbon in the fuel converts to CO_2, assuming complete combustion, and all hydrogen in the fuel converts to H_2O. Nitrogen is mostly inert and carries through to the products, except for trace levels that form nitrogen oxides (NO_x) or N_2O, which are air pollutants. As shown in Figure 20.8, there are a number of air contaminants that may also be present in the products (i.e., exhaust stack) depending on the fuel properties and combustion conditions. A summary of common air contaminants is provided in Table 20.3, which lists the key impacts of the pollutants and common control methods. As noted in the table, this includes common GHGs, which are covered in Chapter 2. The formation of pollutants is, in part, driven by the "3 T's" of combustion, which are time (i.e., residence time for combustion), temperature (i.e., hotter vs. cooler flames), and turbulence (i.e., how well mixed is the fuel and air). The 3 T's are inherent in the design and operation of the combustion equipment.

The concentration of oxygen (O_2) in the products is critical for efficiency and pollution management. Too little O_2 results in incomplete combustion and formation of CO and other pollutants. Too much O_2 results in more heat pushed out of the exhaust, lowering the thermal efficiency. The ratio of fuel to air is the key variable that determines the quantity of excess O_2 in the exhaust. With the exact amount of air required to combust all carbon in the fuel, the concentration of O_2 in the exhaust is 0% and (ideally) all carbon reacts to form CO_2. This is called the *stoichiometric ratio* and is described as supplying *100% theoretical air* or 0% excess air (or sometimes stated as 0% excess O_2 in the flue gas). Due to imperfect conditions and the 3 T's, all real combustion processes must be supplied with greater than 100% theoretical air to operate safely. Control of excess air (or excess O_2) is critical in combustion equipment design and operation. A common technology that is used is O_2 trim control (see Chapter 15).

Because fuel combustion is a chemical reaction, if the fuel composition is known (at least approximately) and the concentration of excess O_2 is known, it turns out that other combustion parameters can be accurately estimated. Appendix I provides a table based on the combustion of natural gas, assuming a common fuel composition. Use of this table is illustrated by example.

Table 20.3 Common air pollutants, impacts, and control methods.

Chemical or abbreviation	**Pollutant names**	**Formation mechanism**	**Main impacts**	**Common methods to reduce emissions (in addition to improved energy efficiency)**
PM, PM_{10}, $PM_{2.5}$	Particulate matter (solids + aerosols). Aerodynamic diameter $\leq 2.5\ \mu m$ or $10\ \mu m$	Ash and other solids in fuel. Incomplete combustion	Human health (cardiovascular, respiratory), regional visibility	Switch fuel (e.g., coal to natural gas). Post-combustion removal: electrostatic precipitator, baghouse (filter), scrubber
NO_x = NO + NO_2	NO_x is the sum of nitric oxide (NO) plus nitrogen dioxide (NO_2)	Nitrogen in air and fuel oxidize during combustion	Human health: NO_x plus VOC form ozone (O_3) in atmosphere	Switch fuel (e.g., oil to gas). Improve 3T's. Post-combustion removal: ammonia and catalyst, scrubber
CO	Carbon monoxide	Incomplete combustion	Acute toxicity to human health	Improve 3T's. Increase excess air. Catalyst
SO_x	Sulfur oxides such as SO_2 and SO_3	Sulfur in fuel oxidizes	Acid rain (sulfuric acid). Human health	Reduce sulfur in fuel (e.g., ultra-low sulfur diesel). Post-combustion scrubber
VOC's	Volatile organic compounds of hydrogen and carbon	Incomplete combustion	Human health: NOx + VOC form ozone (O_3) in atmosphere	Improve 3T's. Post-combustion removal: scrubber, catalyst
Other toxics	Metals, chlorinated organics, aldehydes, etc.	Beyond the scope of this book		
Principal GHGs CO_2, N_2O, CH_4	Carbon dioxide Nitrous oxide Methane	Combustion and 3 T's	Global climate change. See Chapter 2, including Tables 2-1 and 2-3	Substitute fossil fuel for sustainable biofuels or low-carbon electricity

Example 20-2

Natural gas is combusted in a hot water heater, and the concentration of O_2 in the exhaust is measured at 4.4% on a dry basis (this means the O_2 meter dries the sample prior to the gas sensor). Estimate the following: % theoretical air, air-to-fuel volume ratio, air-to-fuel mass ratio, flue gas % volume of moisture, flue gas % CO_2, flue gas dry % N_2 dry, and the dry density of the flue gas in lbm/ft^3.

Using the natural gas combustion table in Appendix I, the following values are looked up in the row of the table based on the given O_2 value of 4.4% dry (see column F):

- Column A: % theoretical air = 124.0 %
- Column B: air-to-fuel volume ratio = 12.21
- Column C: air-to-fuel mass ratio 20.24
- Column E: % volume of moisture = 15.27%
- Column H: % CO_2 dry = 9.37%
- Column J: % N_2 dry = 86.2%
- Column M: dry density of the flue gas = 0.0783 lbm/ft^3 (1.26 kg/m^3)

Thermal efficiency for combustion equipment is defined as the ratio of heat released by the fuel to heat transferred to the water or steam as shown in Figure 20.9. Combustion efficiency is similar, but only counts energy losses through the exhaust. Using the variables as defined in the figure, the equations for each efficiency are (Cengel et al., 2024):

Thermal Efficiency = E_{H2O}/E_{input}

Combustion Efficiency = $(E_{input} - E_{stack})/E_{input}$

Note: Combustion efficiency is always greater than thermal efficiency

There are two critical variables that determine combustion efficiency: flue gas %O_2 and stack temperature rise (STR). STR is defined as the difference between the flue gas temperature and the temperature of the incoming combustion air. STR and %O_2 are two variables that are readily measured, and then combustion efficiency can be accurately estimated using well-proven industry procedures (Allwood & Cullen, 2015). Using estimates or approximate assumptions for the "other" losses, thermal efficiency can then be estimated as well. Appendix E provides charts of thermal efficiency for different values of STR at different levels of excess O_2 for three different fuels: natural gas, Number 2 oil (diesel fuel), and biomass with 40% moisture content. We will illustrate the use of these charts with an example.

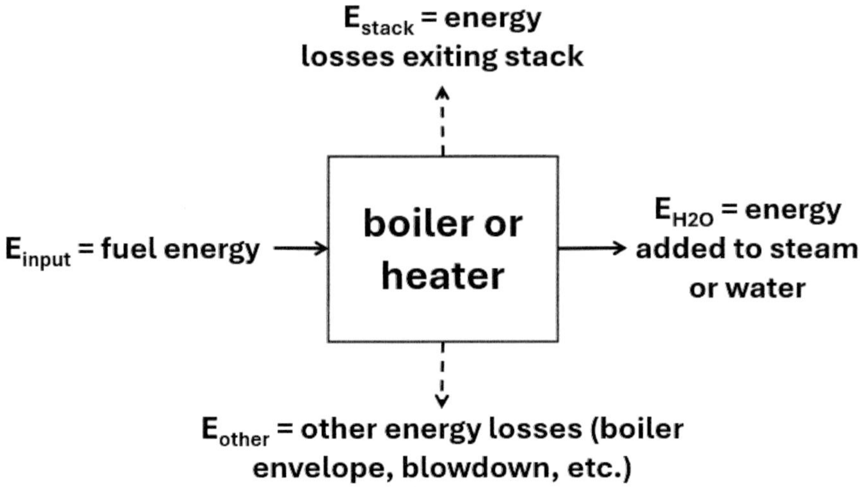

Figure 20.9 Schematic of energy inputs and output for fuel-fired boilers and heaters.

Example 20-3

A pulp mill power boiler is firing biomass with 40% moisture content. The incoming combustion air temperature is 12 °C (54 °F). The flue gas measurements at the boiler exit are 212 °C (414 °F) and 8.0% O_2 dry. Estimate the thermal efficiency for the power boiler.

STR = 212 – 12 = 200 °C

Boiler exit O_2 is stated to be 8.0% dry

Using the chart in Appendix E for biomass with 40% moisture shown in Figure 20.10, the boiler efficiency is found to be 70.0% on an HHV basis.

The thermal efficiency values in the Appendix E charts are stated to be based on the fuel higher heating value (HHV) basis, and not the lower heating value (LHV). What does this mean? The difference between HHV and LHV was introduced in Chapter 1, along with fuel calorific values in Table 20.1–20.2 and Appendix B. The difference between HHV and LHV is that HHV is the useful heat extracted if all the moisture in the exhaust is condensed out to become a liquid. LHV is the heat extracted if all the water in the exhaust remains as a vapor. In other words, HHV is the amount of extractable energy from the fuel assuming the latent heat in water vapor is recovered. Look back at Example 20-1 and note that the volume of moisture in the exhaust is 15.3% by volume for natural gas fuel at the stated conditions. This is a significant

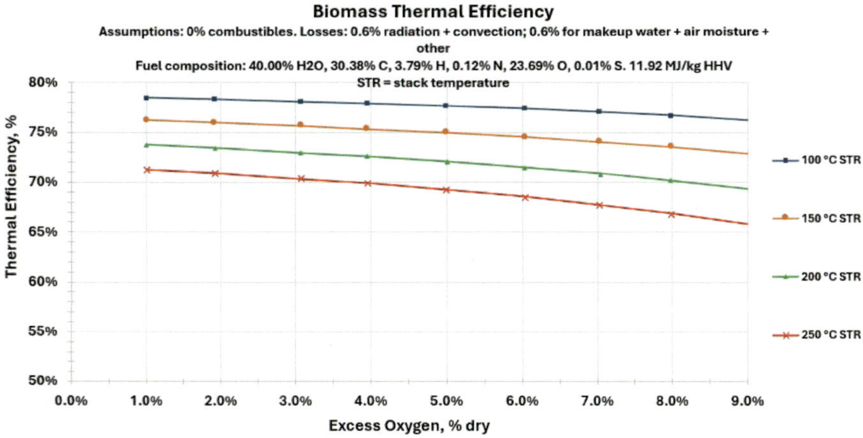

Figure 20.10 Thermal efficiency for 40% moisture biomass vs. STR and excess O_2.

quantity of moisture, and condensing out that moisture means that a significant quantity of heat could be recovered. This is the benefit of "condensing" boilers and heaters (introduced earlier in this chapter). Condensing boilers have an extra heat exchanger to condense out the water vapor in the exhaust, meaning there is a real improvement in thermal efficiency. For the condensing heat exchange process to work effectively, it is critical that the boiler return water temperature be cold enough to condense out the water vapor. The cooler the return water temperature, then condensing boiler efficiency is improved. Figure 20.11 illustrates this relationship for a condensing commercial boiler[6] (both °F and °C units are shown). Note that for condensing boilers, the thermal efficiency improves at reduced load.

A critical issue is to ensure that *when thermal efficiencies are stated, be sure to confirm whether the efficiency is stated on an HHV or LHV basis*. The formulas to convert efficiencies for natural gas and No. 2 oil are[7]:

Natural gas: $\eta_{\mathrm{HHV}} = 0.91 \times \eta_{\mathrm{LHV}}$

No. 2 Fuel oil (diesel fuel): $\eta_{\mathrm{HHV}} = 0.94 \times \eta_{\mathrm{LHV}}$

---

[6]These curves are for a specific commercial boiler. In practice, it is best to use manufacturer curves for a specific boiler.

[7]Technically, the conversion constants in these formulas depend on the specific fuel composition. These values are based on the fuel composition shown in the charts in Appendix J. But these constants do not change much with variations in fuel composition, and these formulas are appropriate for energy assessment purposes.

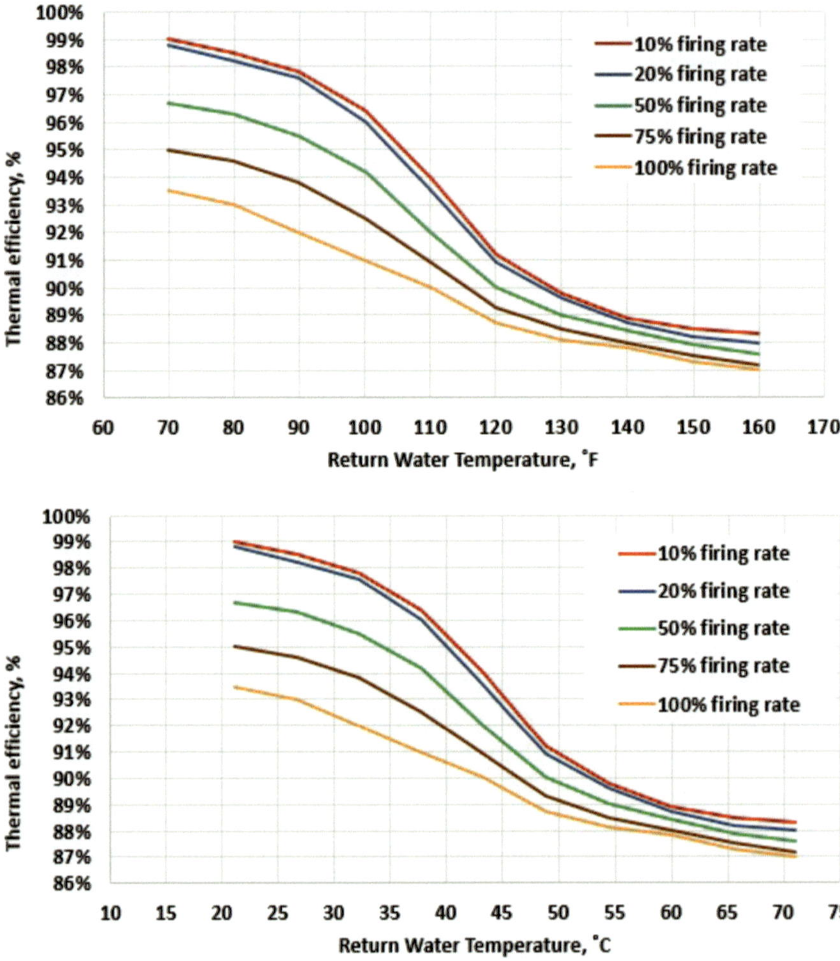

Figure 20.11 Thermal efficiency (HHV) vs. return water temperature for a commercial condensing boiler firing natural gas.

Example 20-4

The nameplate on a fire tube boiler states that the thermal efficiency at full load rated conditions is 88.0% on an HHV basis for Number 2 oil. Estimate thermal efficiency on an LHV basis.

η_{LHV} = 0.88/0.94 = 0.936 or 93.6%.

We conclude this section with a list of common strategies to maintain and improve the efficiency of combustion systems:

- Include key combustion equipment in energy management system activities such as tracking KPIs and training (see Chapters 1 and 7).
- Ensure the scheduling of combustion equipment is optimized to meet the service needs including operating hours and loading (see Chapter 15).
- Optimize fuel and air controls to minimize excess O_2, while maintaining pollutants within acceptable limits and combustion safety (see Chapter 15).
- Employ predictive maintenance practices for combustion equipment, including calibration of instruments, tuning of controls, and regular cleaning of heat transfer tubes and coils (see Chapter 21).
- Install heat recovery (covered later in this chapter). Two common heat recovery options are recovering heat from flue gas to either preheat air or preheat feedwater.
- Assess fan and pump efficiency improvements for combustion air and exhaust or induced draft fans, and feedwater pumps (see later section in this chapter).
- Ensure all hot surfaces and components exposed to the mechanical room or outdoors are insulated (see Chapter 11) to reduce heat loss and for personnel safety.

Additional EEMs can be found in Chapter 9 of the Industrial Energy Systems Handbook (Williams, 2022), Doty's energy audit guide (Doty, 2020), and the DOE steam sourcebook (DOE, 2012). Also, the U.S. DOE's MEASUR tool[8] is an open source app that is useful for identifying and quantifying energy savings opportunities for industrial systems.

20.3.3 Steam systems

Water-based heating and cooling systems are widely used. In this section, we will highlight the benefits of using water as an energy carrier, describe common steam distribution systems, learn how to estimate key properties, and overview water quality control requirements. The overall goal is to be able to estimate physical properties of water and steam and perform practical calculations needed for common energy management projects. This section will focus on steam because there are a lot of similarities between steam and liquid water (hydronic) systems, but steam is more complicated. As well, Chapter 10 (HVAC) and Chapter 19 (Thermal Energy Networks) both provide guidance on hydronic systems.

[8] www.energy.gov/eere/ito/measur

To begin, it is useful to highlight the benefits of using water as an energy carrier:

- Water is widely available and is of low cost
- Water and steam have good heat transfer properties
- Generally safe to use
- Clean, non-toxic
- Good ability to control pressure and temperature
- Easy to transport over long distances

Figure 20.12 is a schematic of a typical steam distribution system. It is useful to view steam systems in terms of supply, distribution, and end uses. Here we will provide a brief overview of steam distribution systems. For more detailed content, DOE's publication "Improving Steam System Performance: A Sourcebook for Industry" (DOE, 2012) and the "Energy Management Handbook" Chapter 14 (Williams, 2025) are recommended.

The supply part of a steam distribution system includes steam generation with combustion boilers (or, increasingly, electric technologies) where steam is discharged to the distribution piping. The combustion side of boiler systems is covered earlier in this chapter. There is also the water side of the boiler system, which includes water quality treatment to remove unwanted

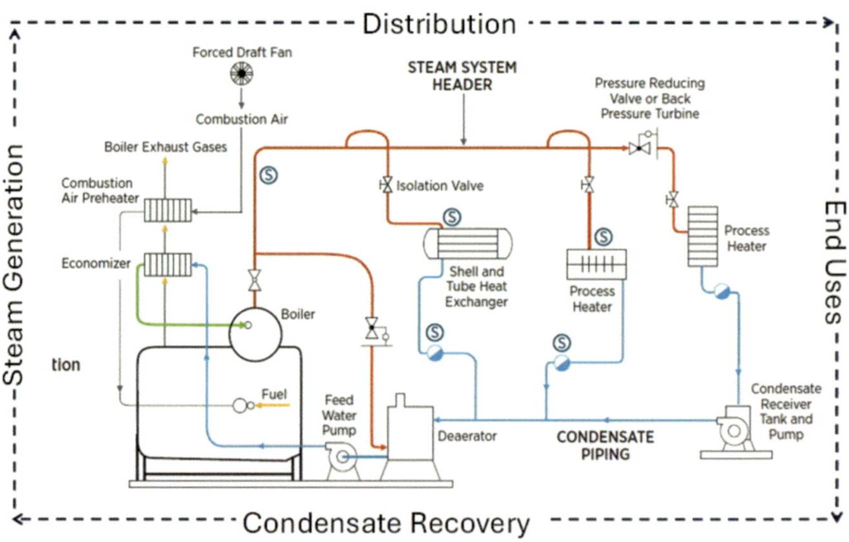

Figure 20.12 Schematic diagram of a typical steam distribution system (source: DOE, 2012).

contaminants. Control of contaminant gases (e.g., air, CO_2) includes deaerators and deaerating feedwater heaters. Equipment to remove solid contaminants (e.g., suspended and dissolved solids, minerals, and ions) includes clarifiers, filters, softeners, dealkalizers, decarbonators, reverse osmosis units, and demineralizers. "Blowdown" of boiler water systems is where a portion of the circulating feedwater water is removed for treatment. Blowdown extraction typically exists from the top of the boiler, called surface blowdown (removes circulating water), and also from the bottom, called bottom blowdown (removes sludge and solids). Control of water quality is critical to maintain efficiency and reliability of boiler and steam systems. Chapter 15 includes a case study illustrating the benefits of blowdown controls. To further demonstrate the link between water treatment and energy efficiency, Figure 20.13 charts the percentage of fuel loss as a function of scale thickness on the water side of the tubes in a boiler.[9] For example, one millimeter scale inside a boiler tube can increase fuel loss from 2.5% to 4% compared to a new system with no scale.

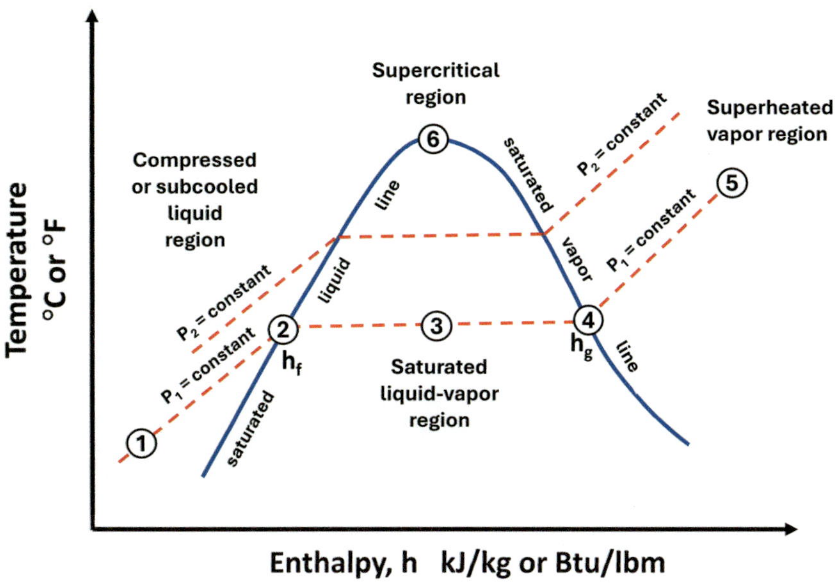

Figure 20.13 Scale thickness on water side of boiler tubes vs. percentage of fuel loss.

[9]Source: www.energy.gov/sites/prod/files/2014/05/f16/steam7_surfaces.pdf

The distribution system is composed of key components including steam supply piping. Steam piping is typically steel pipe and be insulated for energy efficiency and safety (see Chapter 11 for insulation basics). The hot water that remains after heat is removed from the steam and condensed to form a liquid, which is called condensate. The condensate return piping is a critical part of system efficiency. In industrial plants, a high percentage (if not 100%) of condensate is usually returned to the boiler. For district heating systems, especially older systems, it is common that condensate piping is corroded so that only a small (or sometimes zero) amount of condensate is returned. Every unit of condensate that is not returned to the boiler means an added unit of make-up water required that must be purchased, heated, and undergoes water quality treatment.

Most steam systems used for process or space heating are designed to produce saturated (not superheated) steam, as described later in this chapter. With saturated steam systems, heat losses lead to formation of some condensate in the steam piping. Steam traps are devices that perform three functions: (1) remove non-condensible gases (e.g., air), (2) remove liquid condensate, and (3) retain the live, dry steam. Table 20.4 describes three common steam trap designs. A steam trap is "leaking" if it is discharging dry steam to the condensate return system, which is wasted energy. A steam trap is "blocked" if it malfunctions and does not remove the condensate or gases. Buildup of liquid condensate in steam piping is a serious safety hazard that can lead to a phenomenon called water hammers that can cause catastrophic system failures and personnel injuries.

The end-use equipment for steam systems is typically a heat exchanger of which there are various designs. Heat exchangers are discussed later in this chapter in the section on heat recovery.

The properties of steam and water have been carefully measured and were first published in the 1840s. To obtain properties for specific applications, steam tables are commonly used and published in thermodynamic textbooks (e.g., Cengel et al., 2024) and software apps. The properties needed for energy calculations in the scope of this book include[10]:

- *Enthalpy* (use the symbol, h in units kJ/kg or Btu/lbm) which is a measure of the thermal energy.

[10]Entropy is a measure of the quantity of energy per unit temperature (kJ/kg °C or Btu/lbm °F) that is unavailable to do work. Entropy is critical for engineers designing thermal equipment, but is not needed for any of the energy calculations in this book.

Table 20.4 Common steam trap designs (DOE, 2012).

Type	How it works	Example applications	Visual
Mechanical (float, ball, bucket)	Based on density (buoyancy) difference between steam and condensate	Process equipment requiring fast removal of condensate.	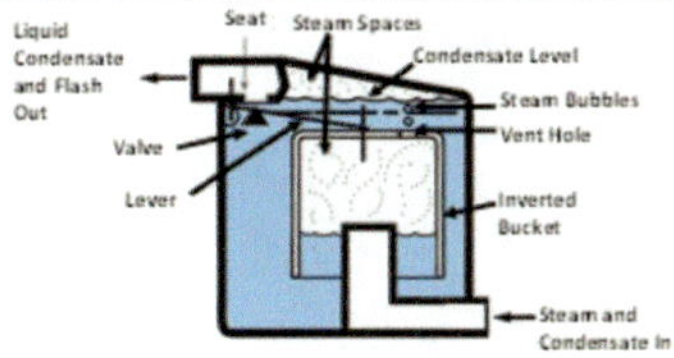 Inverted bucket
Thermodynamic (disc)	Based on velocity and pressure differences between steam and condensate	Steam main lines	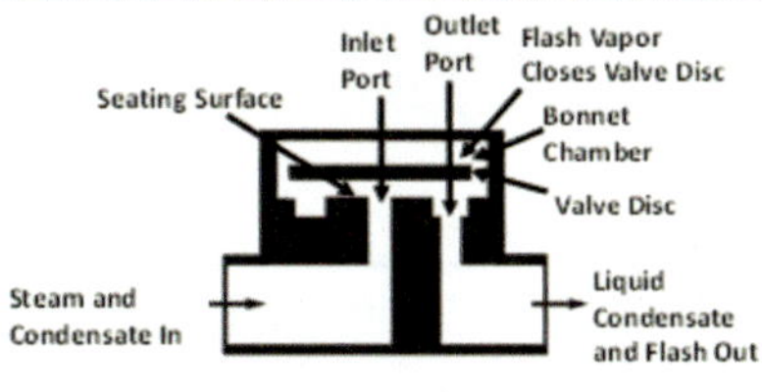 Thermodynamic disc steam trap
Thermostatic (temperature, balanced pressure)	Based on temperature difference between steam and condensate	Steam sterilizers. Heat tracing.	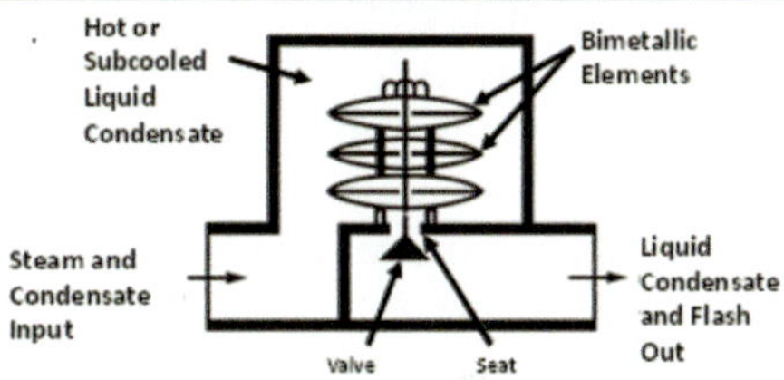 Thermostatic steam trap with a bimetallic element

- *Density* (use the symbol, ρ in units kg/m^3 or lbm/ft^3), or the inverse of density which is called specific volume (units m^3/kg or ft^3/lbm). For most problems though, the density at standard conditions can be used, which is 1.00 kg/L (1,000 kg/m^3) or 62.4 lbm/ft^3 (8.34 lbm/gal).
- *Saturation temperature* (use the symbol, T in units °C or °F), which is the temperature that water boils at a given pressure. Similarly, sometimes the question is reversed where temperature is provided, and we need to find the *saturation pressure, P*, which is the pressure at which water boils for the given temperature. For example, at sea level pressure of 101.3 kPa or 14.7 psia, we know that water boils at 100 °C or 212 °F.

Figure 20.14 is a generic temperature-enthalpy diagram, which is a thermodynamic model that helps us understand how certain thermal processes

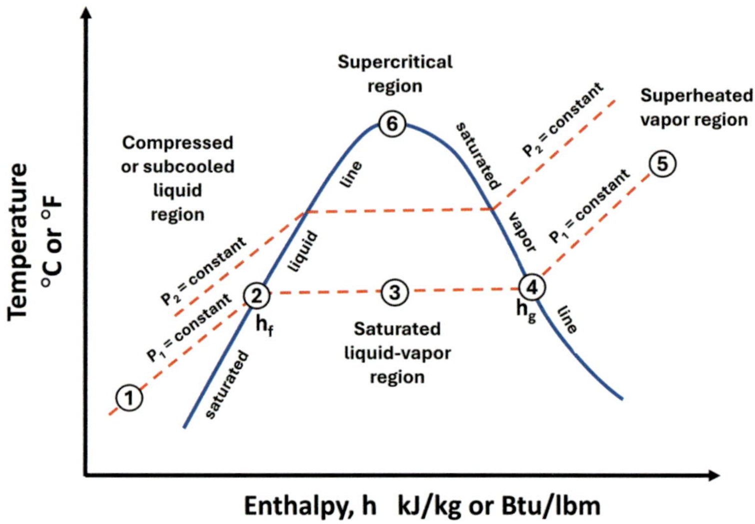

Figure 20.14 Temperature-enthalpy diagram for water.

work[11] and how to look up useful properties for water and steam at different conditions.

If you could heat water while maintaining it at a constant pressure (e.g., imagine water inside a cylinder with a piston on top applying a constant force all the time as it is held over a burner that continuously adds heat), the process would follow one of the dashed lines in the figure. Let us say you are doing your experiment at sea level, with a starting temperature of 20 °C, and following the dashed line from point 1 to point 5. So P_1 = 101.3 kPa and T_1 = 20 °C, and you begin to heat the water *at constant pressure*. If you carefully measure temperature, you will find the temperature rises until it reaches 100 °C when the first drop of water vaporizes, which is point 2 on the diagram. This is the saturation temperature at sea level pressure of 101.3 kPa. We describe the state at point 2 as a "saturated liquid" and use the symbol h_f for the enthalpy of a saturated liquid. Suppose you continue to add heat at constant pressure and you will notice that the temperature stays constant as more liquid water vaporizes. At some point, if you could measure the relative amounts of liquid and vapor, there would be 50% liquid/50% vapor, which is approximately point 3 in the diagram. Keep adding heat until 100%

[11]Temperature vs. enthalpy diagrams are also very useful for understanding refrigerant properties in vapor compression cycles, which are covered in Chapters 10 and 11.

of the water is now a vapor (0% liquid), which is point 4 on the diagram. We describe the state at point 4 as a "saturated vapor" and use the symbol h_g for the enthalpy of a saturated vapor. Then if you continue to add heat, you will notice the temperature starts rising again until point 5. Point 5 is described as superheated steam. If you start the experiment over, but do it at a higher pressure (say at two atmospheres or P_2 = 202.6 kPa), this would be illustrated with the upper dashed line in Figure 20.14. This experiment can be done at many different pressures to build up a database of properties. This hypothetical experiment has actually been done in real laboratories, in order to develop steam tables. The main regions in the diagram represent common states of water or steam:

- *Compressed or sub-cooled liquid region* is where point 1 lies in the diagram, and is the region where water is a liquid at a temperature below the boiling point. If you drink a glass of water at home, you are technically drinking a compressed liquid! The enthalpy of a compressed liquid can be looked up in a compressed liquid table (not contained in this book, because it is not needed), or accurately approximated as the h_f value at the temperature of the water. For liquid water, only temperature (not pressure) is needed to look up the enthalpy.
- *Saturation region* is where points 2, 3, and 4 lie in the diagram, and is the region where water transitions from 100% liquid to a 100% vapor. In between points 2 and 4 is a region where water exists in two phases, part liquid and part vapor. To estimate the enthalpy of a two-phase mixture of liquid water and steam, use the following:
  $h = h_f + X\, h_{fg}$ where:
  X is the saturated steam quality = mass fraction of steam vapor
  $h_{fg} = h_g - h_f$
- *Superheat region* is where water vapor has been heated to a temperature higher than the saturation temperature.
- *Supercritical region* is the region above the hump in the curve of the diagram, and point 6 is the critical point. We will not be covering supercritical steam in this book. It is commonly used for electric power generation with supercritical boilers, is never used in buildings for space heat or hot water, and is very rarely used in industrial processes. See any thermodynamics text to learn more (e.g., Cengel et al., 2024).

Each of these regions is represented in the steam tables that are contained in Appendix F. Steam tables in the appendix include the following:

- Saturated steam, organized by temperature

- Saturated steam, organized by pressure
- Superheated steam

Tables are provided in both English and SI units. Also, steam tables *always use absolute pressure* (not gage pressure). So any gage pressure that is measured or stated must be converted to absolute pressure:

$$P_{abs} = P_{gage} + P_{atm} .$$

We will illustrate how to use the steam tables to estimate enthalpy values by way of an example.

Example 20-5

Feedwater is supplied to a saturated steam boiler at 300 °F and 285.3 psig (gage pressure) in a location located at sea level. Assume boiler pressure losses are negligible and the steam exits the boiler at 285.3 psig and 500 °F. Estimate the following five enthalpy values in Btu/lbm, and convert each value to kJ/kg:

h_1 = Enthalpy of feedwater supplied to the boiler (point 1 in Figure 20.14)
h_2 = Enthalpy of saturated liquid (point 2 in Figure 20.14)
h_3 = Enthalpy of the state when 40% by mass of the feedwater has vaporized and 60% remains as a liquid (point 3 in Figure 20.14)
h_4 = Enthalpy of the saturated vapor (point 4 in Figure 20.14)
h_5 = Enthalpy of the delivered steam (point 5 in Figure 20.14)

1. As stated above, the enthalpy of a compressed liquid (point 1 in the diagram) can be accurately estimated as the h_f value for a given temperature. Since we want to find the saturated liquid enthalpy and we know the temperature, use the saturated steam table organized by temperature in Appendix J. The enthalpy of saturated liquid at 300 °F is 269.8 Btu/lbm. Converting this to SI units: h_1 = 269.8 Btu/lbm × (3,600 kJ/3,412 Btu) × (lbm/0.4536 kg) = 627.6 kJ/kg.
2. To look up the saturated liquid enthalpy at 285.3 psig, we must first calculate the absolute pressure: 285.3 + 14.7 = 300 psia. Since we are working with saturated steam and we know pressure, use the table organized by pressure in Appendix F. The enthalpy of saturated liquid at 300 psia is 394.0 Btu/lbm. Converting this to SI units: h_2 = 394.0 Btu/lbm × (3,600 kJ/3,412 Btu) × (lbm/0.4536 kg) = 916.5 kJ/kg.
   From point 1 to point 2 on the temperature vs. enthalpy diagram, this region is liquid water with no phase change. As such, we should have been able to estimate the change in enthalpy from point 1 to point 2

using the sensible heat approximation, which is (see Appendix A):

$$\Delta h = Cp \times \Delta T.$$

We can look up the temperature for point 2 that is a saturated liquid at 300 psia, and the value from the Appendix J table is 417.4 °F. So the enthalpy h_2 can also be estimated as follows:

$$h_2 = h_1 + \Delta h = 269.8 Btu/lbm + [(1.00 Btu/lbm^\circ F)$$

$$\times (417.4 - -300^\circ F] = 387.2 Btu/lbm.$$

This value differs from the h_2 lookup value of 394.0 Btu/lbm by 6.8 Btu/lbm, or a 1.7% difference. It provides approximately the same answer.

3. h_3 = h_f + X h_{fg} at the saturation pressure of 300 psia (saturation temperature of 417.4 °F). It is given that the steam quality, X = 40% (i.e., 40% of the feedwater has vaporized). From the saturated steam pressure table in Appendix F: h_g = 1,203.0 Btu/lbm, so h_{fg} = 1,203.0 – 394.0 = 809.0 Btu/lbm. Note that h_{fg} can also be looked up directly in the steam table. Then h_3 = 394.0 Btu/lbm + (0.40 × 809.0 Btu/lbm) = 717.6 Btu/lbm or 1,669 kJ/kg.
4. The enthalpy h_4 is h_g at the saturation pressure of 300 psia. We looked up this value in part 3, which is 1,203 Btu/lbm or 2,798 kJ/kg.
5. Part 5 of this problem states the final temperature of the steam is 500 °F. It is not stated whether this is saturated steam or superheated steam. However, given the pressure is 300 psia and we know that the saturation temperature at 300 psia is 417.4 °F, for the steam to be at a temperature greater than the saturation temperature (i.e., 500 °F in this case) means it must be superheated steam. To look up enthalpy values for superheated steam requires both temperature and pressure to be known. Looking up in the superheated steam table at 300 psia and 500 °F, h_5 = 1,257.9 Btu/lbm or 2,926 Btu/lbm.

One last aspect of steam that is important to understand is the phenomenon of flash steam. Flash steam is defined as the steam vapor that is formed when high pressure, saturated liquid condensate drops in pressure (ASHRAE Handbook–HVAC Systems and Equipment 2024, Chapter 11). A common method to drop the pressure is to install a tank (commonly called a "flash tank") where the hot liquid can expand to a lower pressure. The process is illustrated in Figure 20.15. If hot condensate is a saturated liquid at point a in the diagram (at higher pressure P_2), and it undergoes a pressure drop to lower pressure P_1, then the mass fraction of condensate that flashes to become steam is

$$\text{Fraction of flash steam} = (h_{f,\text{high}} - -h_{f,\text{low}})/(h_{fg,\text{low}})$$
$$= (h_a - -h_b)/(h_c - -h_b).$$

The use of this formula is demonstrated with a practice problem at the end of this chapter.

We conclude this section with a list of common strategies to maintain and improve the efficiency of steam systems:

- Include steam equipment in energy management system activities such as tracking KPIs and training (see Chapters 1 and 7).
- Ensure the programmed schedules of steam equipment are optimized to meet the service needs including operating hours and loading (see Chapter 15).
- Optimize water quality controls, including water quality management and boiler blowdown (see Chapter 15).
- Employ predictive maintenance practices for steam equipment, including calibration of instruments, tuning of controls, and regular cleaning of the water-side heat transfer tubes and coils to minimize scale buildup (see Chapter 21). Steam traps are an especially critical piece of equipment requiring regular testing and maintenance.
- Install heat recovery such as for boiler blowdown. Boiler makeup water and feedwater are excellent candidates for capturing waste heat from various sources (see heat recovery later in this chapter).
- Assess pump efficiency improvements for feedwater and hydronic system pumps (see pumps and fans later in this chapter).
- Ensure all hot surfaces and components exposed to the mechanical room or outdoors are insulated (see Chapter 11) to reduce heat loss and for personnel safety. This includes piping, valves, expansion joints, heat exchangers, tanks, and other hot components.

Additional EEMs can be found in Chapter 9 of the Industrial Energy Systems Handbook (Williams, 2022), Doty's energy audit guide (Doty, 2020), and the DOE steam sourcebook (DOE, 2012). Also, the U.S. DOE's MEASUR tool[12] is an open source app that is useful for identifying and quantifying energy savings opportunities for industrial systems.

---

[12]www.energy.gov/eere/ito/measur

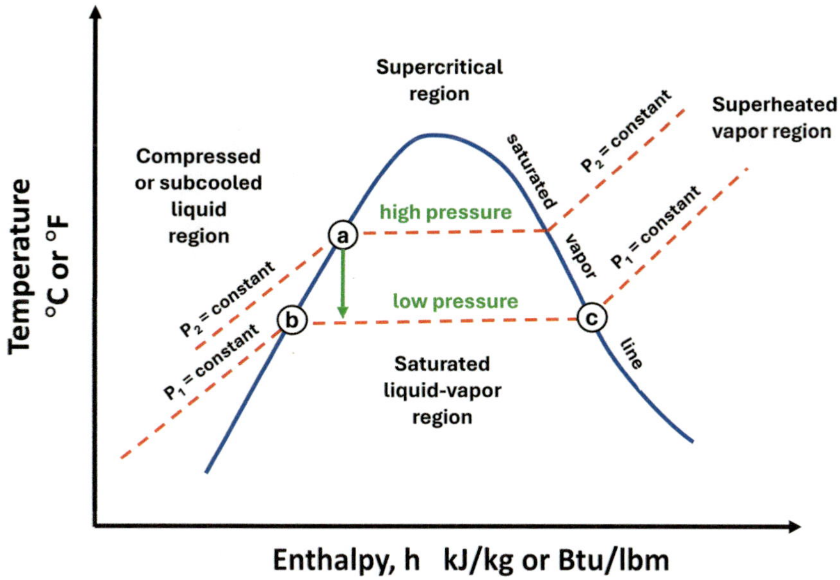

Figure 20.15 Temperature enthalpy diagram for flash steam.

20.4 Fans and Pumps

Pumps and fans are widely used in buildings and industrial processes. On a lifecycle basis, energy is a significant cost as shown in Figure 20.15. This figure shows that energy cost accounts for about 60% of the lifecycle cost for industrial pumps and fans, and this is for a jurisdiction that has low cost electricity.[13] If electricity prices are higher, energy can account for a much larger proportion of lifecycle costs. One source estimates the lifecycle cost for pumps and fans to be 92% energy, 4% operation, and 4% capital (Kaya et al., 2021).

In this section, we will introduce how to perform basic energy calculations for pumps and fans, followed by a detailed description of a 10-step approach to identifying and implementing energy-efficient designs, and then conclude with a few case studies. We will not describe the details of fan and pump hardware options as there are many excellent sources elsewhere:

[13]The cost data are based on industrial pump systems ranging in size from 100 to 900 hp. Fan data are based on a sample of 10 systems ranging in size from 50 to 2,750 hp.

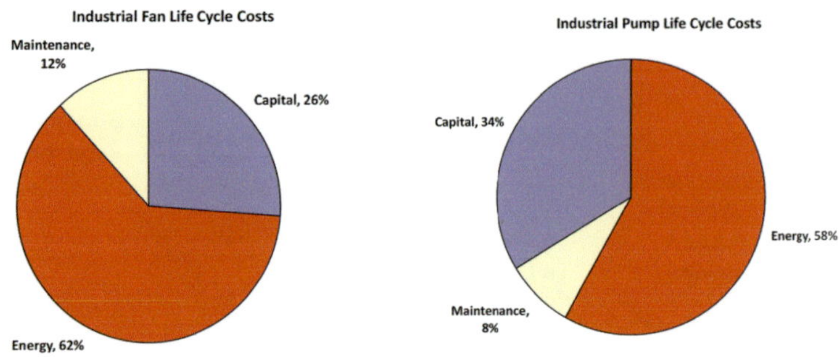

Figure 20.16 Lifecycle costs for industrial pumps and fans in a jurisdiction with low electricity prices.

- ASHRAE Handbook Systems and Equipment 2024, Chapter 21 (Fans) and Chapter 44 (Centrifugal Pumps).
- Energy Management Handbook (Williams, 2025), Chapter 19 (Pumps) and Chapter 20 (Fans).
- DOE fan sourcebook (DOE, 2003a) and pump sourcebook (DOE, 2006).
- Air and Movement Control Association Fans and Systems handbook (AMCA, 2023).
- Hydraulic Institute (www.pumps.org) has many pump publications, a free pump savings calculator, and offers pump training.
- U.S. DOE's free MEASUR app has pump and fan modules (www.energy.gov/eere/ito/measur).

20.4.1 Pump and fan energy calculations

Fundamentally, the purpose of pumps and fans is to increase the pressure of the liquid or gas, in order to move the fluid from one location to another. Fans (which move gases) and pumps (which move liquids) have much in common in terms of performance and energy efficiency opportunities, and thus both are covered together here.

The first thing to understand about energy efficiency of pumps and fans is that this equipment is always part of a system. Figure 20.17 illustrates the typical conversion of energies in a fan or pump system[14] from electric power supply (color coded in red in the figure) to mechanical power (color coded

[14]Same for compressors, conveyers, and other equipment too, although not always converted to fluid power.

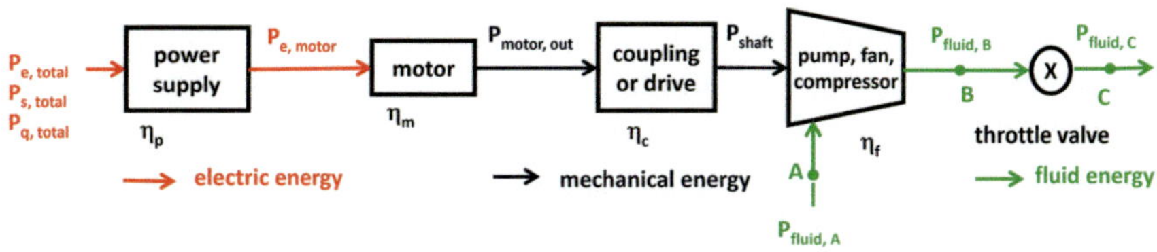

Figure 20.17 Typical energy conversion chain for electric-driven fluid moving equipment.

in blue) to ultimately delivering fluid power to a system (color coded green). The system efficiency will be the product of the four conversion efficiencies shown in the figure (this is review from Chapter 14). Appendix A provides definitions of the variables shown in this figure as well as common formulas.

For pumps and fans, the figure above describes only part of the system of interest for designing energy-efficient pump and fan systems. To make a complete assessment, it is important to analyze the hydraulic system served by the pumps or fans. Figure 20.18 provides an example of a hydraulic process that is powered by pumps. Marked up in red in the figure are some common potential opportunities to improve the system design.

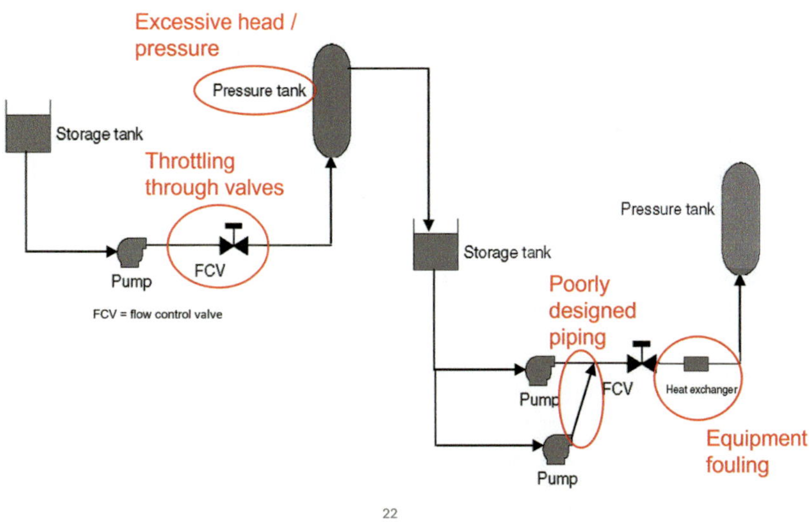

Figure 20.18 Example of a hydraulic system powered by pumps.

The efficiency of pumps and fans is defined as the ratio of output fluid power to input shaft power. The formula sheet in Appendix A provides a summary of key equations and defined variables from fluid mechanics and electric power systems that are needed to perform energy assessments of electric-driven equipment. These equations apply to all types of pumps and fans: rotary, reciprocating, centrifugal, and axial. It should be noted that while electric-driven equipment is the most common design choice, there are cases where mechanically powered systems exist. For example, some older steam plants have fans run by steam turbines. One industrial plant ran a cooling water pump powered by a water turbine which extracted energy from the cooling water return pipe. However, in general, we will assume that pumps and fans are electric driven.

A good place to start is with the definition of pump and fan efficiency, which is

$$\eta_f = P_{\text{fluid}}/P_{\text{shaft}}$$

.

One of the limits of this basic formula is that shaft power is rarely measured. In practice, what is commonly measured is input electric power. If the efficiency of the electric equipment is known (e.g., transformer or variable frequency drive (VFD) – see Chapters 13 and 14), efficiency of the motor is known, and (if applicable) efficiency of a mechanical conversion device (e.g., belt and sheave) can be estimated, then the shaft power can be estimated.

Working the problem from the hydraulic process side, fluid power can be estimated if there are measurements or modeled results for flow and pressure head using the formulas below. The formulas below for fluid power apply only for steady-state flow where compressibility effects are negligible and density is approximately constant. These conditions are valid for a wide range of pump and fan applications moving air, water, or other low pressure gases. For compressed air or gas compression systems and high velocity fluids that approach the speed of sound, the fluid power equations below are not accurate.

Fluid power added by pumps

SI units: $P_{\text{fluid}} = \gamma \dot{V}\ H_{\text{pump}}/1{,}000$

P_{fluid} = fluid power, kW

γ = specific weight, N/m^3 = sg $\times\ \gamma_{water}$ [γ_{water} = 9,810 N/m^3]

sg = specific gravity = $\gamma_{\text{fluid}}/\gamma_{\text{water}}$

$\dot{V}$ = volume flow rate of liquid, m^3/s

H_{pump} = fluid head added by pump, m

English units: $P_{\text{fluid}} = \gamma\ \dot{V}\ H_{\text{pump}}/550$

P_{fluid} = fluid power = hp

γ = lbf/ft^3 = sg $\times$ γ_{water} [γ_{water} = 62.4 lbf/ft^3]

$\dot{V}$ = volume flow rate of liquid, ft^3/s

H_{pump} = fluid head added by pump, ft

Note: when $\dot{V}$ is in U.S. gal/min: P_{fluid} = sg $\dot{V}\ H_p$/3,956

Fluid power added by fans

SI units: $P_{\text{fluid}} = \Delta\text{TP}\ \dot{V}/60$

P_{fluid} = air or gas fluid power, kW

ΔTP = fan total pressure rise, kPa

ΔTP is total fluid pressure added by the fan, approximated by fan static pressure from a fan curve

$\dot{V}$ = actual air or gas flow, m^3/min

60 = seconds per minute

English units: $P_{\text{fluid}} = \Delta\text{TP}\ \dot{V}/6{,}343$

P_{fluid} = air or gas fluid power, hp

ΔTP = fan total pressure rise, inches of water

ΔTP is total fluid pressure added by the fan, approximated by fan static pressure from a fan curve

$\dot{V}$ = actual air or gas flow, ft^3/min

6,343 = unit conversion constant

Example 20-6

To illustrate the usefulness of estimating the fluid power, consider an industrial ventilation system that ventilates warm air during winter at 170 m^3/min for 4,000 hours annually. The system is powered by a centrifugal fan. A heat recovery heat exchanger is to be retrofit, which will have a design pressure drop of 500 Pa (0.500 kPa) at 170 m^3/min. Estimate the added annual electricity consumption by the ventilation fan, assuming fan efficiency of 60% and electric motor efficiency of 95% (there is no VFD and no mechanical losses between the motor and fan shafts).

Assume that pressure losses through the heat exchanger are equal to the incremental static pressure that must be delivered by the fan:

$$\Delta P_{\text{fluid}} = \Delta TP\dot{V}/60 = (0.50{\times}170)/60 = 1.417kW.$$

This means that an extra 1.417 kW of fluid power is required to overcome the pressure losses of the new heat exchanger. Then the change in electric

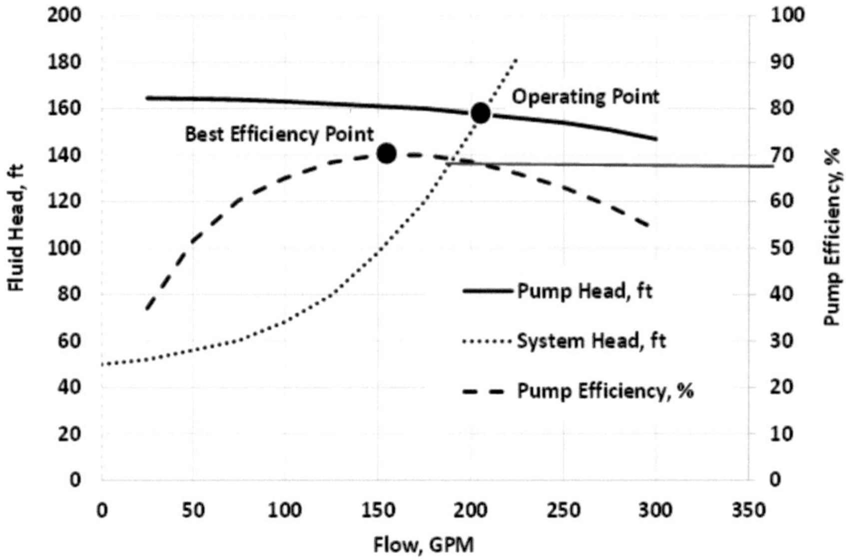

Figure 20.19 Sample performance curve for a water pump.

power consumption is calculated as follows:

$$\Delta P_e = 1.417kW/(0.60\times 0.95) = 2.485kW.$$

Annual increase in electricity consumption = 2.485 kW × 4,000 hr/yr = *9,940 kWh/yr.*

Essential to working with pumps and fans is to understand the manufacturer performance curves. Figure 20.19 is an example of a pump curve for water, and fan curves are very similar. In fluid mechanics, the energy of the fluid is described as fluid "head" in units of length (e.g., feet or meters). Fluid head is defined as the quantity of energy per unit weight of fluid and the units work out to be length (i.e., ft-lbf/lbf = ft or N m/N = m). This pump curve plots the head which the pump adds to the fluid at different flow rates. Pump efficiency is also plotted. Every pump and fan will have a maximum efficiency, described as the "best efficiency point," which is shown in the pump curve. System designers should aim to maximize the operating time for a pump or fan that occurs at or near the best efficiency point. The system that is served by a pump or fan also has a characteristic curve that quantifies how much fluid head is needed to move a specific flow rate, which is also shown in Figure 20.19. *A pump (or fan) will always operate at the intersection of the pump head curve and the system head curve.*

Example 20-7

For the pump curve above, estimate the electric power consumption (kW) at the operating point shown. Assume the motor efficiency is 94%.

From the curve, we see the operating point flow rate is about 210 GPM. At 210 GPM, the pump head delivered is about 159 ft of head and the pump efficiency is about 68%. The fluid power delivered is estimated as follows:

$$P_{\text{fluid}} = sg\dot{V}H_p/3,956 = (1.0\times210\times159)/3,956 = 8.44\text{hp}$$

$$8.44\text{hp}\times0.7457 = 6.294\text{kW shaft power}$$

$$\text{Electric power:} P_e = 6.294kW/0.94 = 6.70kW.$$

20.4.2 General approach to energy-efficient fans and pumps

The best practice for assessing and implementing pump and fan energy improvements is provided in the flow chart in Figure 20.20. This flow chart is designed to apply to both new projects and retrofits.

A brief description and/or rationale for each step in the flow chart follows:

Step 1: Identify the pumps and fans of interest. There are multiple criteria that could be used for selecting pumps or fans for upgrades including: high energy cost, record of poor operation (e.g., turndown, peak capacity, ramp rates), record of excessive maintenance (e.g., failures leading to production downtime, excessive vibration). Experience working with industrial operations reveals that some pumps and fans are considered "bad actors" based on O&M data, and these may be candidates for upgrade that could have a side benefit of saving energy.

Step 2: Review O&M records, confirm service requirements, and conduct baseline energy measurements. It cannot be emphasized often enough to be sure to collect baseline data – energy data and O&M data. Too often this step is skipped or initiated too late to collect a complete set of data.

Step 3: Assess whether pumps or fans can be eliminated in the design. While it is relatively rare, there are examples where this can be accomplished. Two of the case studies at the end of this section involved elimination of pumps.

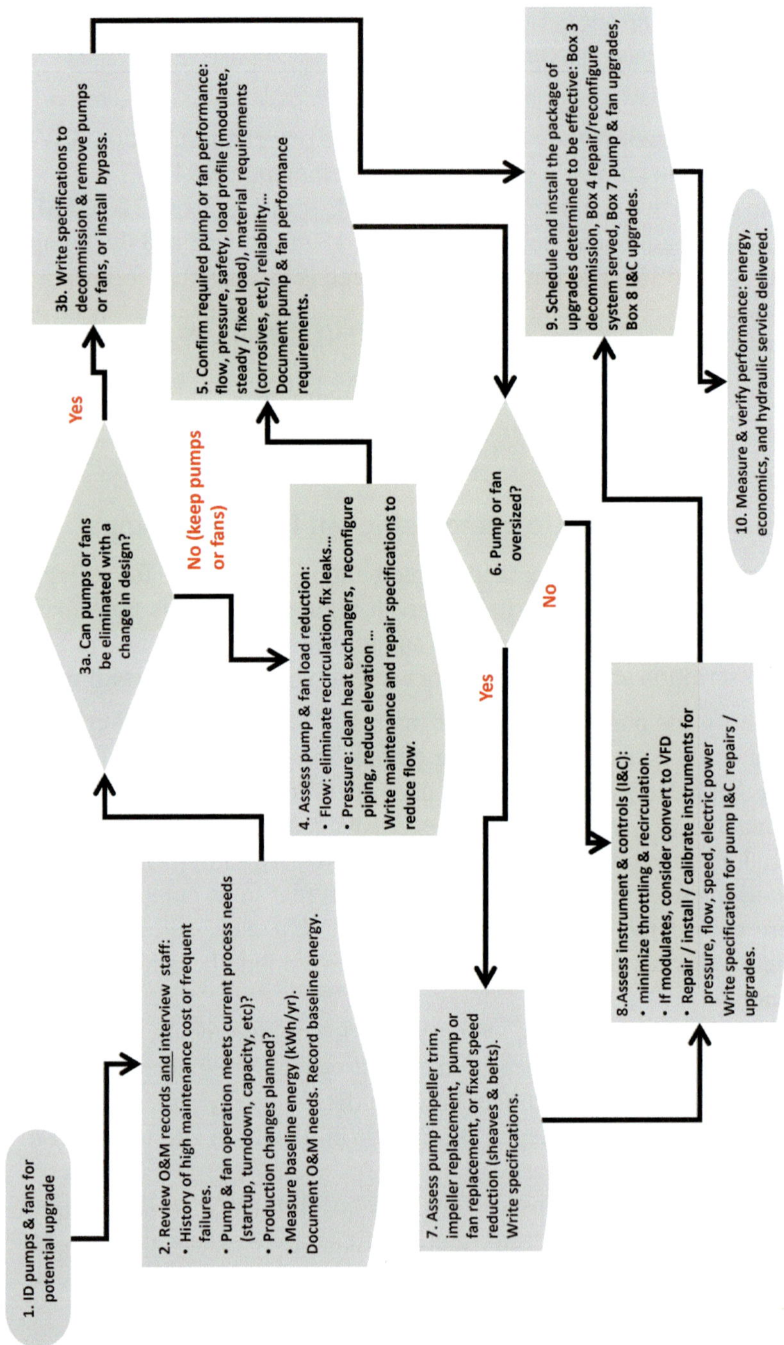

Figure 20.20 Flow chart of 10-step process for identifying and implementing energy-efficient pumps and fan systems.

Step 4: Assess the opportunities to reduce flow requirements by reducing leakage to a practical minimum and/or reducing recirculation. Also assess the opportunities to economically reduce pressure requirements such as: reducing length of pipe runs, eliminating pipe bends, replacing or cleaning components with lower pressure losses (e.g. valves, heat exchangers), and installing larger pipe diameter. In some real-world projects, it has been found that just cleaning (or perhaps refurbishing) old heat exchangers can dramatically reduce pressure losses (see one case study at the end of this section). Fluid engineering science demonstrates that pressure losses through straight pipes are inversely proportional to diameter to the 5th power.[15] Therefore, undersized pipe diameter has a dramatic effect on pressure losses that must be overcome by pumps or fans.

Step 5: Confirm pump or fan fluid performance requirements (flow, pressure head) as well as other important design requirements related to safety, controllability, and maintenance. For retrofits, do not always assume that past or present production requirements will persist indefinitely. Also consider whether production changes are planned for the future that should factor into the pump or fan performance requirements.

Step 6: Assess whether the pump or fan is oversized. For new design projects, the proper size should have been established in Step 5. For retrofits, it might be discovered that existing pumps or fans are oversized (or possibly undersized).

Step 7: If pumps or fans are oversized, assess the most cost-effective method to "right size" the equipment, which could include: impeller change, impeller trim, speed change, or replacement with new pumps or fans.

Step 8: Assess instrument and control upgrades that might be beneficial and economical. For example, energy sub-metering, flow metering, vibration monitoring, additional temperature or pressure sensors, new performance dashboards.

Step 9: Specify, procure, schedule, and install the optimum package of upgrades.

[15] See www.tec-science.com/mechanics/gases-and-liquids/pressure-loss-in-pipe-systems/

Step 10: Perform measurement and verification (see Chapter 1) of energy performance, but not only energy. Confirm that hydraulic performance and other service requirements (safety, maintainability, controllability) are achieved as per the specifications.

This flow chart procedure for pumps and fans is carefully designed and sequenced to achieve the optimum outcome. It is important to emphasize that following the procedure leads to consideration of pump or fan replacement only in the later steps in the process. Novice energy professionals sometimes think of fan or pump replacement first, before: collecting baseline data, conducting an assessment of the system served by the equipment, assessing actual performance requirement needs, and/or reviewing O&M records. Jumping to hasty conclusions about equipment replacement will often lead to higher capital and/or lifecycle costs.

To summarize, Table 20.5 provides an overview of the common types of EEMs for pumps and fans. Chapters 13 and 14 cover EEMs related to electrical systems and variable speed. Additional EEMs can be found in Chapters 19 (pumps) and 20 (fans) of the Energy Management Handbook (Williams, 2025), Doty's energy audit guide (Doty, 2020), the DOE fan sourcebook (DOE, 2003a), and the DOE pump system source book (DOE, 2006). Also, the U.S. DOE's MEASUR tool[16] is an open source app that is useful for identifying and quantifying energy savings opportunities for industrial systems.

20.4.3 Fan and pump energy efficiency case studies

A sample of six case studies is provided below. These case studies were chosen to illustrate the range of solutions for pumps and fans for industrial operations including impeller replacements, process hydraulic load reduction (heat exchanger cleaning), elimination of pumps in process design, and variable speed conversions. All pumps and fans in the case studies below were centrifugal designs. The payback period for the projects below ranged from 2 to 12 years (not including any utility or government subsidies).

Case study 1: Water treatment plant expansion – eliminated need for new pumps

A water treatment plant implemented a plant expansion to add more water treatment modules. The key treatment technology was a treatment membrane that requires pumps to move the water through the membranes. The baseline

---

[16] www.energy.gov/eere/ito/measur

Table 20.5 Overview of common EEMs for pumps and fans. The percentage energy savings are not additive as there are significant interactive effects between measures.

	EEM	Range of energy savings
>Improve fan or pump assembly (fan or pump assembly includes the motor and control device such as a valve, damper, hydraulic drive, or VFD)	Better control (eliminate throttle losses, reduce speed, use it only when you need it). See Chapter 14 for variable speed savings assessment.	5%–50%
	Upgrade or replace with a more efficient fan or pump	10%–50%
	Replace impeller with a more efficient design. Impeller changes are standard for large industrial pumps. For fans, impeller changes are usually custom-designed and usually only economical for large industrial fans (say 200 hp and larger). Impeller trimming is common for large industrial fans and pumps.	10%–40%
	Replace a V-belt with an energy-efficient "synchronous" belt.	3%–5%
	More efficient motor	1%–5%
Improve hydraulic system	Modify the process to reduce flow and/or pressure required	5%–100%

design was to add a new pump skid consuming 66 kW of electricity (pump design efficiency of 75%). In the efficient design, a siphon system was designed so that new treatment modules could be added with no additional pumps. The need for new pumps was completely eliminated. The siphon system incurred extra capital cost compared to baseline, but this incremental cost was offset by the avoided cost of pumps and reduced ongoing energy and maintenance costs. Verified electricity savings were 520,000 kWh/yr.

Case study 2: Industrial dryer fan impeller replacement[17]

A large industrial dryer operates with an exhaust fan equipped with a 4,500 hp motor and discharges to a pollution control system. This large fan was the subject of a study and it was determined that the impeller was not appropriately sized, and that a more efficient impeller design was feasible. Both the baseline and new impellers were radial blade designs, but the baseline fan was tested at 58% fan efficiency vs. 64% average efficiency (78% peak) for the new impeller. The custom-made impeller was retrofit while retaining the original fan housing, motor, and foundation. Verified electricity

[17]Williams (2025) Chapter 14 and AMCA (2024) provide detailed descriptions of radial fan impeller designs vs. other designs such as forward curved, backward curved, and airfoil.

savings were 1,650,000 kWh/yr. It was confirmed that the new impellers did not impose any new maintenance requirements or reliability issues.

Case study 3: Slurry pump impeller replacement

An industrial operation with multiple slurry pumps totaling 2,100 hp of motor capacity was the subject of an energy study. It was concluded that the pumps were oversized with significant throttling. A retrofit was implemented to refit the pumps with new impellers so that throttling would be substantially reduced. The resulting electricity savings were verified to be 1,080,000 kWh/yr. It was also confirmed that the maintenance requirements of the new impellers did not increase pump maintenance costs.

Case study 4: Gas cooling heat exchanger chemical cleaning

An industrial plant has three large cold boxes, which are plate fin heat exchangers, to cool a gas for further processing. Upstream of the cold boxes is a desiccant system to remove moisture. The main compression for the plant is composed of multiple gas compressors, totaling 31,000 hp for the motors. This case study is technically not about a fan or pump, but compressors that are very similar. The plant had been operating for many years, and in spite of regular maintenance on the heat exchangers, material from the desiccant built up on the gas coolers. An assessment of an advanced chemical cleaning was made and implemented, which included some changes to limit future contamination of the cold boxes from the desiccant process. The one-time cleaning substantially reduced pressure losses through the cold boxes. Verified electricity savings during the first 12 months of operation were 3,900,00 kWh/yr.

Case study 5: Pumping system reconfiguration

An industrial plant redesigned a washing system and reconfigured the piping and pump system. As part of the re-design, one pump was eliminated, one was converted to VFD, and two pumps had impellers trimmed. Also, the pipe routing was simplified to reduce system pressure losses. The verified electricity savings were 410,000 kWh/yr, which exceeded pre-project estimates by about 20%.

Case study 6: Convert water pumps to VFD

An industrial plant converted one 600 hp pump and another 700 hp pump from valve control to variable speed, using a VFD. Each pump operated as one of a pair of parallel pumps. Overall, the two pumps were part of a complex system of 11 pumps for two different plant water systems. The control scheme for the retrofit system involved changing the control of other pumps. Total verified electricity savings were 1,920,000 kWh/yr. However, the savings were about 40% lower than pre-project estimates as the estimates of VFD control savings were inaccurate.

20.5 Compressed Air Systems

Compressed air systems are a common ancillary service in industrial operations as well as many buildings and small commercial operations. It is often described as the fourth utility, in addition to electricity, natural gas, and water (CAGI, 2016). Like other equipment covered in this chapter, the energy cost typically comprises a large fraction of the lifecycle cost. Various sources estimate that energy accounts for 60%–90% of lifecycle costs for industrial compressed air systems (Kaya et al., 2021; Saidur et al., 2010).

20.5.1 Compressed air system components

Figure 20.21 illustrates the typical compressed air system. Like many other systems, it is useful to think of compressed air in terms of supply, distribution, and end uses.

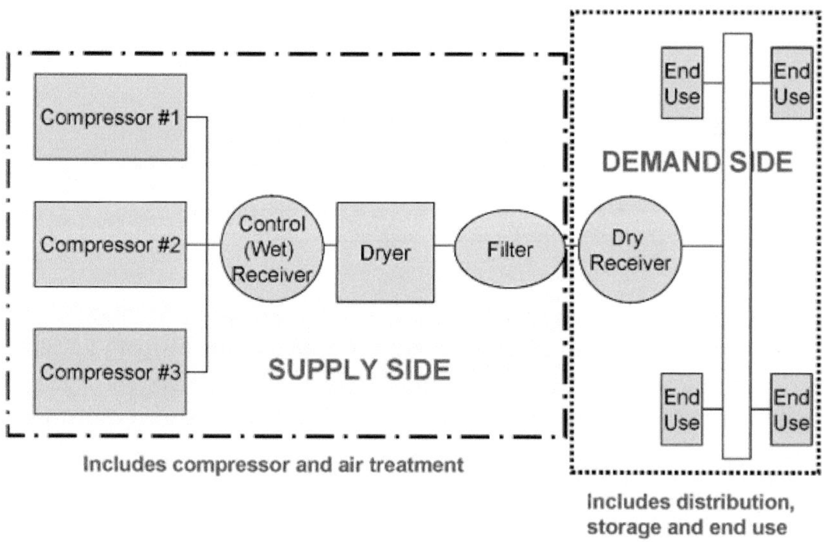

Figure 20.21 Typical compressed air system.

Short descriptions of common components are provided below. Detailed descriptions of compressed air hardware components are widely available in other sources such as DOE's compressed air sourcebook,[18] the Compressed Air and Gas Handbook (CAGI, 2016), and Williams Energy Management Handbook (2025), Chapter 18.

Compressors. One or more supply compressors may be installed and arranged in series or parallel. Common compressor types are in two categories: positive displacement (reciprocating and rotary screw/scroll/sliding vane) and dynamic (centrifugal and axial). Table 20.6 lists some common compressor types along with specific power consumption of electricity (Williams, 2022). Lubrication and cooling components are important design choices affecting the compressor performance and reliability. Compressed air is heated due to the compression process and typically requires intercoolers and aftercoolers installed. Compressors are either water cooled or air cooled.

Table 20.6 Specific power consumption for various compressor types.

Compressor type	kW per m^3/min free air	kW per 100 cfm free air
Reciprocating single acting, 1 stage	7.8–8.5	22–24
Reciprocating single acting, 2 stage	6.4–8.1	18–23
Reciprocating double acting (uses both sides of piston to compress), 1 stage	8.5–10.2	24–29
Reciprocating double acting (uses both sides of piston to compress), 2 stage	5.3–5.7	15–16
Lubricated screw, 1 stage	6.0–7.8	17–22
Lubricated screw, 2 stage	5.7–6.7	16–19
Lubricant-free screw, 2 stage	6.4–7.8	18–22
Centrifugal, 3 stage	5.7–7.1	16–20

Dryers. Compressing and cooling of air leads to moisture formation that must be managed. Separators remove liquids usually downstream of air coolers. There are several drying technologies including refrigerant dryers, regenerative desiccant dryers that use a porous material to adsorb moisture (water adheres to surfaces), and deliquescent dryers that use a medium to absorb moisture (materials chemically react with water). Drying technologies can have substantial impact on the system electricity consumption.

[18]www1.eere.energy.gov/manufacturing/tech_assistance/pdfs/compressed_air_sourcebook.pdf

Air filters. Air filters may be used in different locations in the compressed air system such as inlet air filters for particles, or other locations to remove lubricants or condensate. Some applications, especially instrument air, require a higher level of contaminant removal.

Receivers. Wet (i.e., before moisture removal) and dry (i.e., after moisture removal) receivers are storage components in the system that can have a significant effect on energy consumption by enabling the supply compressors to operate in a more efficient manner with less cycling and/or operation nearer to the best efficiency load.

Process controls. Compressor control strategies include start/stop (i.e., compressor on or off), load/unload where the motor runs continuously but unloads from compressing air for periods of time, modulating controls using inlet throttle valves, or variable speed with a VFD. Sequencing and loading of multiple compressor systems is important for optimizing performance and energy efficiency.

Distribution. Piping and valves that transport compressed air from the source air compressor to the end uses are composed of trunk lines, hoses, and valves. Older compressed air systems can evolve to be inefficient with high pressure losses, unnecessarily long pipe runs, recirculation, extra valves, or other components that dissipate pressure.

End uses. Table 20.7 provides examples of compressed air uses for various applications.

20.5.2 Compressed air system energy calculations

Next we will examine the energy consumption of compressed air systems and look at overall system efficiency. First let us introduce formulas to estimate the fluid power produced in compressed air systems. As mentioned in the section on pumps and fans, the equation for estimating fluid power for fans does not work for compressed air because the assumption of incompressibility does not apply. Based on engineering thermodynamics, compressed air has been found to work as a polytropic process (Cengel et al., 2024). The mechanical shaft energy consumption per unit mass of compressed air flow is estimated with the following equation. Below is an explanation of each

Table 20.7 Examples of compressed air end uses for various applications.

Industry	Example end uses
Food and beverage	Dehydration, bottling, controls and actuators, conveying, spraying coatings, cleaning, vacuum packing
Textiles	Agitating liquids, clamping, conveying, automated equipment, controls and actuators, loom jet weaving, spinning, texturizing
Apparel	Conveying, clamping, tool powering, controls and actuators, automated equipment
Lumber and wood	Sawing, hoisting, clamping, pressure treatment, controls and actuators
Furniture	Air piston powering, tool powering, clamping, spraying, controls and actuators
Pulp and paper	Conveying, controls and actuators
Chemicals	Conveying, controls and actuators
Petroleum and gas production	Process gas compressing, controls and actuators
Rubber and plastics	Tool powering, clamping, controls and actuators, forming, mold press powering, injection molding
F-Stone, clay, and glass	Conveying, blending, mixing, controls and actuators, glass blowing and molding, cooling
Agriculture	Farm equipment, materials handling, spraying crops, dairy machines
Mining	Pneumatic tools, hoists, pumps, controls and actuators
Transportation	Pneumatic tools, hoists, air brakes
Wastewater treatment	Vacuum filters, conveying
Primary metals	Vacuum melting, controls and actuators, hoisting
Metals fabrication	Assembly station powering, tool powering, controls and actuators, injection molding, spraying

variable for the equation, following by a sample calculation:

$$\frac{P_{\text{fluid}}}{\dot{m}} = \left(\frac{n\,R\,T_1}{(n-1)}\right)(P_2/P_1)^{(n-1)/n}$$

$$\frac{P_{\text{shaft}}}{\dot{m}} = \frac{\frac{P_{\text{fluid}}}{\dot{m}}}{\eta_c}$$

$\dot{m}$ = mass flow of air, $\frac{\text{kg}}{\text{s}}$ or lbm/hr

$\frac{P_{\text{fluid}}}{\dot{m}}$ = energy of the compressed air, kJ/kg or Btu/lbm

$\frac{P_{\text{shaft}}}{\dot{m}}$ = mechanical or shaft energy per unit mass, kJ/kg or Btu/lbm

η_c = compressor efficiency, typically 0.7–0.9

n = polytropic compression exponent. n = 1.4 if no intercooling. $1.0 < n < 1.4$ if there is intercooling
T_1 = inlet air temperature in absolute temperature, Kelvin (K) or Rankine (R)
R = ideal gas constant for air = 0.287 kJ/kg K or 53.35 ft-lbf/lbm-R
P_1 = inlet air pressure, typically assumed to be 101.3 kPa or 14.696 lbf/in^2
P_2 = compressor discharge absolute pressure in kPa or lbf/in^2

Example 20-8

Leaks are measured[19] to be 0.10 kg/s for a compressed air system operating at sea level. The inlet air temperature is 18 °C (64 °F). The compressor efficiency is 0.70. There is intercooling, so assume the polytropic exponent is 1.2. The compressor discharge pressure is 7 bar (103 psig) gage pressure. The compressor electric motor is 92% efficient. Estimate the electric power consumption (kW) associated with the leaks, and the system efficiency from electricity to fluid power.

The fluid power that is required to deliver the leaking compressed air is equal to the fluid power per kg of air leakage, multiplied by the mass rate of air leakage. Using the equation above, the value for each variable is
$\dot{m} = 0.30\ \mathrm{kg/s}$
n = 1.2
T_1 = 273 + 18 = 291 K
R = 0.287 kJ/kg K
η_c = 0.70
P_1 = 101.3 kPa
P_2 = 7 bar × (100 kPa/bar) + 101.3 kPa = 801.3 kPa

Now we can estimate the fluid energy per unit mass in kJ/kg:

$$\frac{P_{\text{fluid}}}{\dot{m}} = \left(\frac{1.2 \times 0.287 \frac{\mathrm{kJ}}{\mathrm{kg\ K}} \times 291\ \mathrm{K}}{(1.2-1)} \right) (801.3/101.3)^{(1.2-1)/1.2}$$

$$\frac{P_{\text{fluid}}}{\dot{m}} = (\,501.10\,)\,(7.910)^{(0.16667)}$$

$$\frac{P_{\text{fluid}}}{\dot{m}} = 707.33 kJ/kg.$$

Next divide by the compressor efficiency to estimate the mechanical or shaft energy per kg of air leakage (see Figure 20.16):

[19] Cengel (2024) provides an equation and example calculation to estimate mass flow of air leakage based on the temperature and pressure of the compressed air, and the leakage hole diameter.

$$\frac{P_{\text{shaft}}}{\dot{m}} = 707.33\frac{\text{kJ}}{\text{kg}} \;/\; 0.70 \;=\; 1,010.5\ \text{kJ/kg}.$$

Next multiply by the mass flow leakage rate to estimate the shaft power, then divide my motor efficiency to estimate the electric power:

$$P_{\text{shaft}} = 1,010.5\ kJ/kg \times 0.30\ kg/s = 303.5\ kW$$
$$P_e = 303.15\ kW/0.92 = 329.5\ kW.$$

The system efficiency from electricity to fluid power can now be estimated. The fluid power delivered in kW is: 707.3 kJ/kg × 0.30 kg/s = 212.2 kW. Therefore, the system efficiency is the ratio of output fluid power to input electric power:

212.2 kW/329.5 kW = 0.64 or 64%.

Many sources cite that the total system efficiency of compressed air usage is in the range of 5%–20%, but most sources do not state how efficiency is defined (Nehler, 2018; Saidur, 2010; Williams, 2025). A review of several sources reveals the definition of efficiency to be electricity to shaft power of a pneumatic device (European Commission, 2009). If the efficiency of electricity to fluid power is about 64% (per the example above), and total system efficiency is less than 20%, then there must be at least another 45% energy lost as heat or consumed elsewhere. If the true system efficiency is less than 20%, the other 45% (or more) of energy loss must be due to (DOE, 2003b):

- leakage, which is cited to be as high as 30%;
- over-pressuring of end uses, which is commonly called "artificial demand";
- energy losses due to distribution piping (including valves, bends, filters);
- dryers that can consume more than 10% of compressed air for purging;
- heat removed from the air by cooling of the air compressor, and cooling of the compressed air in the distribution piping;
- electricity consumed by refrigerant dryers, which may also reheat the compressed air after cooling it;
- energy conversion efficiency of end-use devices, such as air motors

So the 5%–20% system efficiency does seem to be realistic, at least for poorly designed and operated systems. In summary, this brief review of system efficiencies confirms that compressed air has a low overall system

Table 20.8 Common EEMs for compressed air systems.[20] Percentage energy savings are not additive as there are significant interactive effects between measures.

	EEM	**Range of energy savings**
Supply	Improved controls: sequencers and partial load control	10%–20%
	Reduce system over-pressure (eliminate artificial demand)	5%–15%
	New energy-efficient compressors	10%–50%
	Heat recovery (saves heating energy for the heat sink system, e.g., space heat or DHW)	10%–30%
	Improved maintenance practices	5%–10%
	Improved controls: VFD	10%–30%
	More efficient motor	1%–5%
Distribution and treatment	Leak detection and repair	10%–30%
	Improved cooling devices	10%–20%
	Resize and/or relocate wet and/or dry receivers	10%–20%
	Air filters with lower pressure losses	5%–15%
	Reduce pressure losses due to piping bends and run lengths, and better designed components	5%–20%
End uses	Substitute inappropriate end uses. Examples: • Open blowing for cleaning or drying → Low-pressure blowers, brooms, efficient nozzles, air knives • Cooling personnel or equipment → Electric fans • Air motors used to mix or agitate → Electric motors • Idle equipment → Use air-stop valve • Abandoned equipment → Disconnect air supply	10%–40%
	Improved maintenance of end-use equipment	5%–15%

efficiency. This leads to the next section, which describes what can be done to improve efficiency.

20.5.3 Compressed air system EEMs

Due to the low system efficiencies, it is consistently found that electricity savings up to 50% for individual measures are realistically achievable (Saidur, 2010; DOE 2003b; Williams, 2025). Table 20.8 lists common EEMs for compressed air systems, categorized into supply, distribution and treatment, and end uses. It is important to emphasize that the percentage of savings of all the measures is not additive, because there are significant interactive

[20] Additional sources: www.compressedairchallenge.org/library and www.bpa.gov/energy-and-services/efficiency/ee-sectors/industrial/compressed-air

effects between measures. Additional details on these EEMs can be found in many sources, including DOE's compressed air sourcebook (DOE, 2003b), the Compressed Air and Gas Handbook (CAGI, 2016), Williams Energy Management Handbook (2025) Chapter 18, and Doty's energy audit book (2020), Chapter 14.

Some additional rule of thumb findings for compressed air systems to improve energy efficiency and/or production capacity follow[21]:

- System pressure: every 2 psig (14 kPa) decrease in system pressure results in 1% energy savings.
- Limit distribution pressure loss to 10% of discharge pressure.
- Limit total leakage to 10% of total compressed air flow demand. Unmanaged leaks will typically increase 5% each year.
- Inlet filters: every 4 inches water column (1 kPa) pressure drop reduces compressor capacity by 1%.
- Inlet air temperature: every 10 °F (6 °C) drop in inlet air temperature saves 1%–3% energy. Higher relative humidity of inlet air will make the compressed air system inefficient and reduces dry air to the plant.
- Oil separators: a dirty oil separator can increase your compressor power by 5%.

20.5.4 Compressed air EEM case studies

The case studies below are all based on real-world projects. The simple payback period for these projects ranged from 1 to 10 years (not including any utility or government subsidies).

Case study 1: Compressed air for a bioenergy boiler

A bioenergy-fired boiler used compressed air for sootblowers. The existing compressor was deemed to be at end of life, so the baseline for the study was not the existing compressor, but a new, low-cost compressor that met project technical requirements. The baseline design was a 150 hp two-stage lubricated screw compressor with load/unload control. The efficient design was a reciprocating design, two-stage oil-free with three-step unloading control (0%/50%/100%). The receiver (and storage volume) remained unchanged. Electricity savings of 125,000 kWh/yr (15% of baseline) and 54 kW peak demand were realized.

[21] www.energy.gov/eere/ito/compressed-air-systems and www.bpa.gov/energy-and-services/efficiency/ee-sectors/industrial/compressed-air and Williams (2022).

Case study 2: Compressed air for pulp mill

One process in a pulp mill was designed with two compressors, sized 450 hp and 800 hp for various processes including instrument air. The 450 hp was operated as a trim compressor with the larger compressor was base loaded. The compressors were two-stage screw designs, oil-free and water cooled. The system was equipped with two wet-side receivers at 1,100 gallons each, pre- and post-drying air filters, and two desiccant dryers. Baseline specific power consumption was measured to be 19 kW per 100 cfm (6.7 kW per m^3/min) and 5,609,000 kWh/yr. Baseline air leakage was assessed to be 11% of total production. The optimum energy efficiency upgrade was determined to replace the smaller compressor with a new two-stage dry screw compressor with VFD drive. Additionally, compressor optimizing sequencing controls were designed into the retrofit. Total electricity savings were determined to be 1,255,000 kWh/yr (22% of baseline) and 146 kW peak demand reduction.

Case study 3: Compressed air replacement at a terminal operation

An industrial operation that stores and transfers agricultural products was the subject of a comprehensive energy study to upgrade the compressed air system. The baseline system was composed of two 225 hp compressors with helical rotors and valve controls, and equipped with refrigerant and desiccant dryers. The baseline electricity consumption was 1,110,000 kWh/yr based on two weeks of detailed performance measurement of electricity consumption, flow, and pressure. Multiple manufacturers and compressor configurations were assessed. The selected solution was to install two rotary screw compressors equipped with VFDs and permanent magnet motors. Each compressor was also designed with one refrigerant and one new desiccant dryer. The estimated energy savings were conservatively estimated at 519,000 kWh/yr (47% of baseline) and 102 kW peak demand reduction.

Case study 4: Compressed air upgrade in a transportation facility

A facility operating in the transportation sector had a comprehensive compressed air study completed. The baseline system was a collection of five small compressors operating independently. Compressor motors were sized at 19–70 kW with a combination of load/unload and modulating controls, 10 air receivers and 4 unheated desiccant dryers (dryer purge air was 18% of total production). Baseline specific power consumption ranged from 17 to 20 KW per 100 cfm (6–7 kW per m^3/min). Five EEMs were implemented resulting in 453,000 kWh/yr savings and 98 kVA reduction in peak demand.

- Replaced air motor end use with an electric motor.
- Replaced unheated desiccant dryers with heated desiccant dryers.
- Removed heat tracing, which was possible due to the retrofit of heated desiccant dryers.
- Reduced header pressure from 115 to 100 psig (highest end use need was 90 psig).
- Piping reconfigured to network multiple compressors together, upgraded control programming, and eliminated one existing compressor.

20.6 Waste Heat Recovery

Industrial waste heat represents a large opportunity for improvements in energy efficiency. It is estimated for the U.S. that 20%–50% of industrial energy input is lost as waste heat (DOE, 2008). This section focuses on utilizing waste heat from industrial liquid and gas sources. There are additional opportunities such as heat radiated from hot surfaces and emitted from industrial products, but we will focus only on fluids that are the most common and economical options. Although focused on industrial applications, the same concepts and methods apply equally to buildings. Additionally, Chapter 9 includes content on heat recovery for HVAC applications.

When assessing waste heat utilization, the first consideration should be to assess whether the source of waste heat can be cost-effectively reduced or eliminated. Very likely that would be the most sustainable solution, if it is feasible. If the waste heat source could be substantially reduced or eliminated, installing waste heat recovery could be viewed as locking in the need for waste heat for a long period of time. Proceeding forward, the discussion of waste heat recovery below is based on the assumption that the waste heat source cannot be cost-effectively eliminated.

In this chapter, the production or availability of waste heat will be referred to as waste heat sources or just "sources," and uses of waste heat will be referred to as "sinks."

20.6.1 Key factors that determine the viability of heat recovery opportunities

The most critical variable that defines the quality of a heat source is temperature. Table 20.9 provides common categories of waste heat based on temperature ranges (CEATI, 2012; DOE, 2008). Common waste heat sources include: combustion exhaust, heat rejected by refrigeration and cooling systems, process off gases, and steam system condensate. Waste heat sinks are any liquid or gas that can absorb waste heat for some useful purpose such as: cold water that needs heated for DHW or industrial hot water circuits, boiler feedwater preheating, combustion air preheating, and ventilation air preheating.

Table 20.9 Waste heat temperature ranges, common sources, and sinks.

Temperature of source	**Heat recovery fluid**	**Common sources**
High $\geq$ 600 °C $\geq$ 1,100 °F	Mostly flue gas	Combustion furnaces, gas turbines, boilers
		Cement kilns
		Incinerators
Medium 200-600 °C 400-1,100 °C	Mostly flue gas	Combustion furnaces, gas turbines, boilers, and reciprocating engines
		Drying processes
		High temperature cooling systems
Low $<$ 200 °C $<$ 400 °F	Gas and liquid effluent	Cooling systems (superheated refrigerant)
		Compressed air
		Process steam or condensate
		Hot liquids from processes

Heat recovery opportunities are generally defined by three factors:

- **Quality of the source**. Higher temperature and minimal contaminants are more likely to be economically viable, while low temperatures and corrosive or particle-laden fluids are less likely to be viable.
- **Proximity of the source to the potential sink**. The closer that sinks and sources are located, the better the opportunity. Longer distances mean there will be longer piping or ductwork runs, larger heat losses, and higher costs.
- **Timing of source availability vs. sink needs**. If the source and sink timing is fully synchronized, this presents the best opportunity. If the source is available in the summer, but the heat is needed in the winter, it is unlikely to be viable. Day vs. night and weekday vs. weekend mismatches in timing are not ideal, but practical solutions may be found by adding thermal storage to the design.

It is important to avoid all-or-nothing thinking in the use of waste heat. In many cases, the source is insufficient to meet the quantity of heat and/or peak temperature required by the sink. In these cases, *preheating* with waste heat may be a viable strategy, and then the remaining need for heat may be supplemented with fossil fuel or electricity heaters.

20.6.2 Four common heat recovery applications

In general, there are four common applications for industrial heat recovery:

1. Passive heat recovery with heat exchangers
2. Active heat recovery with heat pumps
3. Cooling with absorption cooling
4. Electricity generation from waste heat

Passive heat recovery is the most common application where a high temperature source supplies heat to a sink. Heat flows naturally from the higher temperature source to the lower temperature sink. This type of heat recovery is common in HVAC systems, as described in Chapter 9, Section 9.6.2, on air-to-air heat recovery. Common heat exchanger designs are also described in Chapter 9 including: fixed plate, heat wheels, heat pipes, run around loops, and dual core technologies. Additional heat exchanger types include shell and tube and plate and frame (see ASHRAE Handbook Systems and Equipment, 2024, Chapter 47).

In the practice of energy management, the detailed calculations of heat exchanger performance are usually completed by the heat exchanger vendor engineers or professional engineering consultants. Detailed coverage of heat transfer analysis is outside the scope of this book.

However, we will introduce some basic formulas for two of the most common designs of passive heat exchangers: parallel flow and counterflow. Schematic diagrams and temperature charts for these exchangers are illustrated in Figure 20.22. Heat balances for heat exchangers of all types typically assume that energy losses are negligible, so that all the heat lost from the hot fluid is added to the cold fluid. While zero energy loss is technically not possible, this assumption works well in practice for insulated heat exchangers. For exchange of heat that is sensible only (i.e., no phase change is occurring such as condensing vapors to liquids, or evaporating liquids back to vapors), heat flow for each fluid is calculated using the specific heat of the fluid with the equation: $\dot{Q}=\dot{m}\ c_P\Delta T$. For counterflow and parallel flow exchangers, the key equations are shown below (subscript “c” for cold, and “h” for hot). After the equations are introduced, an example problem is provided:

$$\dot{Q} = \dot{m}_h C_{p,h}\left(T_{h,\text{in}} - T_{h,\text{out}}\right) = \dot{m}_c C_{p,c}\left(T_{c,\text{out}} - T_{c,\text{in}}\right) \tag{20.1}$$

- $\dot{Q}$ is the rate of heat exchange in kW or Btu/hr
- $\dot{m}$ is the mass flow rate in kg/s or lbm/hr
- C_p is the specific heat in Btu/lbm °F or kJ/kg °C
- This equation is valid for all heat exchangers operating in steady state, which are well insulated and have sensible heat exchange only (i.e., no latent heat exchange is occurring).

$$\dot{Q}=U\ A\ \Delta T_{lm}$$

- U = overall heat transfer coefficient in Btu/hr ft^2 oF or W / m^2 oC
- A = heat transfer surface area in the heat exchanger design in ft^2 or m^2
- This equation is valid for counterflow and parallel flow only. For other passive heat exchanger designs such as cross flow or multi-pass, there are correction factors applied to this equation. See any engineering heat transfer textbook for further details (e.g., Cengel et al., 2022).

$$\Delta T_{lm}=\frac{\Delta T_1-\Delta T_2}{\ln(\Delta T_1-\Delta T_2)}$$

- ΔT_{lm} is log mean temperature difference, which uses the natural log function.
- Counterflow: $\Delta T_1 = T_{h,in} - T_{c,out}$ and $\Delta T_1 = T_{h,out} - T_{c,in}$
- Parallel flow: $\Delta T_1 = T_{h,in} - T_{c,in}$ and $\Delta T_1 = T_{h,out} - T_{c,out}$
- For other heat exchanger designs, see any engineering heat transfer textbook.

Example 20-9

A counterflow heat exchanger is supplied with hot thermal oil to heat boiler feedwater. Thermal oil mass flow is 3.5 kg/s, the specific heat is 3.0 kJ/kg °C, and the oil inlet temperature is 80 °C. Feedwater flow is 2.5 kg/s, the specific heat is 4.19 kJ/kg °C, the inlet temperature is 10 °C, and the heat exchanger must be sized to deliver feedwater at 55 °C. The overall heat transfer coefficient for the heat exchanger is 850 W/m^2 oC (0.850 kW/m^2 oC). Estimate: (a) the rate of heat flow (kW), (b) the outlet temperature of the hot oil (°C), and (c) the surface area required for the heat exchanger (m^2).

(a) $\dot{Q}= \dot{m}_c\ C_{p,c}\ (T_{c,\text{out}}- T_{c,\text{in}})$
$= 2.5$ kg/s × 4.19 kJ/kg °C × (55 − 10 °C)
$= 471.38$ kJ/s = *471 kW*.

(b) $\dot{Q}= \dot{m}_h\ C_{p,h}\ (T_{h,\text{in}}- T_{h,\text{out}})$
471.38 kJ/s = 3.5 kg/s × 3.0 kJ/kg °C × (80 – $T_{h,out}$)

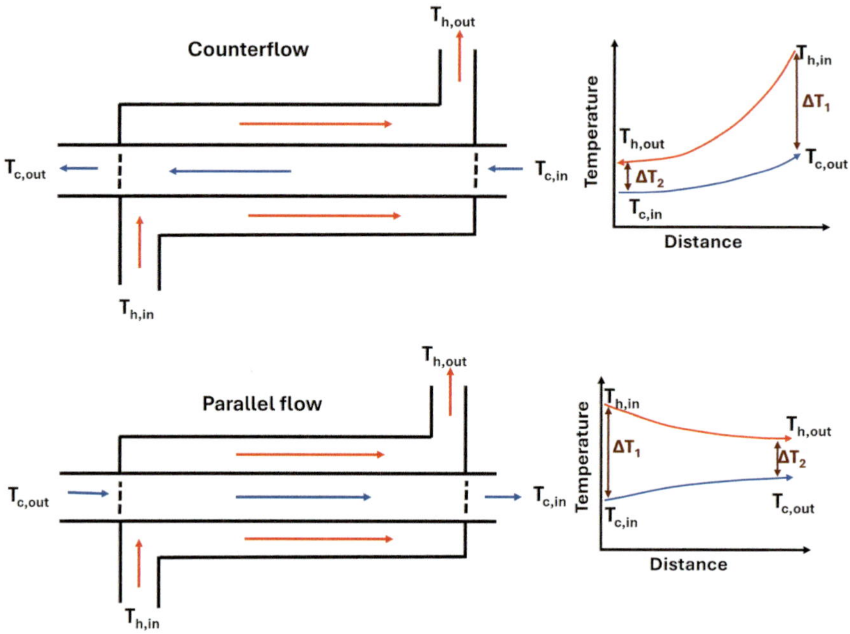

Figure 20.22 Diagrams of counterflow and parallel flow heat recovery heat exchangers.

Solve for $T_{h,\text{out}}$ = *35.11 °C.*

(c) $A = \dot{Q} / (U \;\; \Delta T_{\text{lm}})$

ΔT_{lm} = (80 – 55) – (35.11 – 10)/ln[(80 – 55)/(35.11 – 10)]

= –0.11/–0.0043903

= 25.05 °C

A = 471.38 kJ/s/[(0.850 kW/m^2 ○C) × 25.05 °C]

A = *22.1 m*^{2}.

Active heat recovery with heat pumps is the second common type of waste heat system. Active heat recovery is used when the heat source temperature is lower than the heat sink temperature. This may seem counterintuitive at first, but it is how heat pumps work. Consider that an electric-driven heat pump warms an indoor space in winter by removing heat from cold outdoor air (e.g., –20 °C or 4 °F) and delivers warm air to heat the indoor space. As long as a heat source is above absolute zero temperature (i.e., greater than –273 °C or –460 °F), then there is thermal energy that can potentially be utilized. The same can be done in industrial processes where a low temperature heat source might not be suitable for passive heat recovery, but the use of heat pump technology may be technically and economically viable. Chapter 10 on heat pumps and Chapter 19 on thermal energy networks provide useful content on how heat pumps can be used for heat recovery.

Absorption cooling is the third common application for use of waste heat. If thermal needs are for cooling rather than heating, this is feasible because absorption chillers have been used for over a century to provide industrial cooling. Systems are commercially available today, which utilize medium temperature waste heat (see Table 20.8) including steam and combustion gases from engines. The sample system in Figure 20.23 illustrates one such system using 530 °F or 277 °C waste gas to produce chilled water with a COP of 1.35 to deliver 90 tons of chilled water.[22]

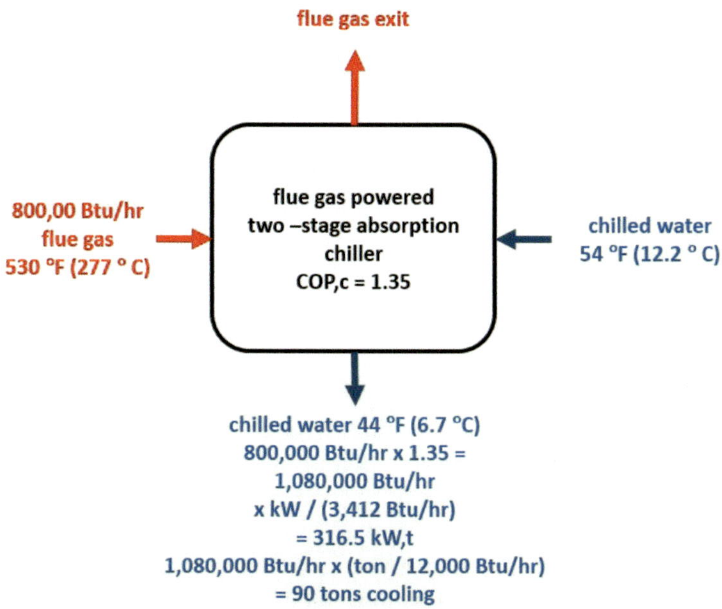

Figure 20.23 Example of an absorption chiller supplied by a medium temperature heat source.

Waste heat to electricity is the fourth and final common application to utilize waste heat. Organic Rankine cycle technology has been commercialized for

[22]Waste heat supply temperature and COP are from: www.energy.gov/sites/prod/files/2017/06/f35/CHP-Absorption%20Chiller-compliant.pdf

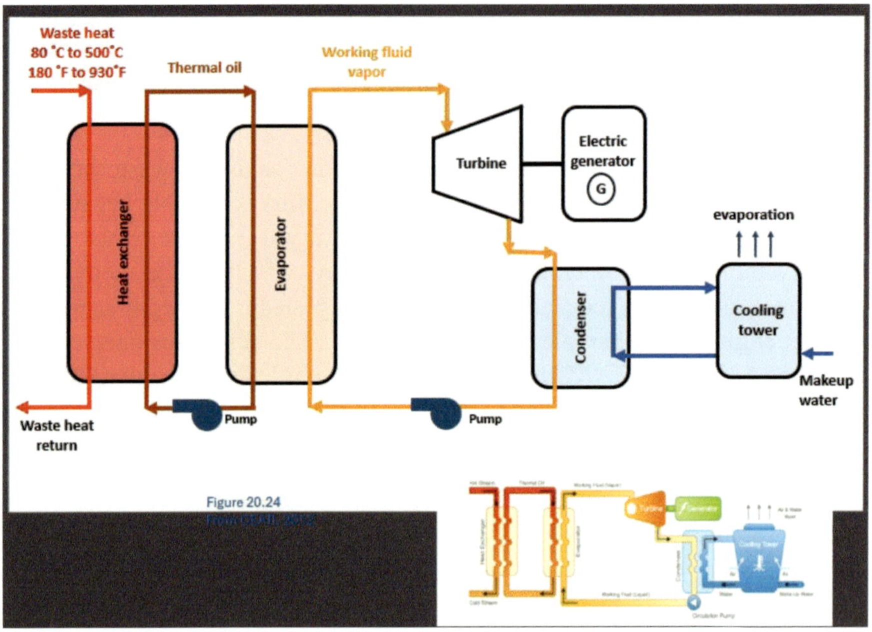

Figure 20.24 Organic Rankine cycle schematic to generate electricity from waste heat.

decades (CEATI, 2012; Mota-Babiloni et al., 2023) and can work with source temperatures in the range of 80–500 °C (180–930 °F). The diagram in Figure 20.24 illustrates the basic components of a common system design (CEATI, 2012). Hot waste gas exchanges heat with a thermal oil, which then evaporates a working fluid that runs through a turbine. The working fluid is an organic substance with low boiling point, typically in the range of 20–100 °C. The working fluid is fully condensed to a liquid downstream of the turbine, and then pumped to higher pressure to complete the cycle. Stated efficiencies range from 8% to 20% (waste heat extraction to electricity, accounting for ancillary loads). A similar technology is the Kalina cycle, which uses two working fluids, usually a mixture of ammonia and water (Jouhara et al., 2018). Kalina cycles are stated to achieve higher efficiencies, in the range of 15%–25%. The next section on combined heat and power includes one case study on the use of organic Rankine cycle technology in a wastewater treatment plant.

20.6.3 Heat recovery case studies

Case study 1: Compressed air heat recovery

A manufacturing plant uses water cooling for a 4,000 hp compressed air system, which produces warm water in the range of 19–22 °C, depending on the production rate. A study identified a heat recovery design using a glycol loop with a VFD-driven pump and a plate and frame heat exchanger to recover heat from the compressor cooling water. The plant uses a large quantity of ventilation air from a natural gas-fired makeup air unit. The existing air equipment lacked available space to integrate heat recovery coils, so the heat recovery design specified a new air handling unit to utilize the waste heat. The new air handling unit was designed with supplementary heat from a gas burner to ensure adequate air temperature could always be achieved. The reduction in natural gas savings was conservatively estimated to be 6,900 GJ/yr (6,540 MMBtu/yr) with a modest increase in electricity consumption of 80,000 kWh/yr. The simple payback period was estimated at 12 years, but with a utility incentive, the project was made economically viable.

Case study 2: Lumber kiln heat recovery to preheat makeup air

Two lumber drying kilns each consume over 100,000 MMBtu/yr (105,500 GJ/yr) of natural gas for drying lumber from about 40% moisture to under 20%. The lumber drying is a batch process. Dimensioned lumber is heated with natural gas burners and air circulating fans. Humidity levels in the kiln are maintained by exhausting moist air from the top of the kiln, and cold outdoor makeup air is introduced to offset the exhausted air. Heat recovery ventilation units (5,000 cfm each) were installed to recover heat from the moist, vented air to preheat the makeup air. Natural gas savings were estimated to be 15,800 MMBtu/yr (16,700 GJ/yr). Offsetting the fuel cost savings was an additional 20,000 kWh/yr of electricity consumption and added maintenance cost for the HRVs of approximately $8,000 annually. The simple payback period was estimated to be 6.5 years without any incentives.

Case study 3: Flash steam heat recovery

A brewery consumes about 20 MMBtu/hr (21,100 GJ/yr) of natural gas to produce 50 psig (~350 kPa) saturated steam. In the baseline, flash steam that is produced in the steam traps is fed back to the feedwater tank where it is vented to outdoors. A condensate recovery system was designed to capture 90% of flash steam and exchange heat to the boiler feedwater with a new heat exchanger. Natural gas savings were estimated to be 14,700 MMBtu/yr (15,500 GJ/yr). The simple payback period was estimated to be 2.1 years.

Case study 4: Heat pump heat recovery for a dairy operation

A dairy operation uses an ammonia-based refrigeration system to cool a glycol loop for process cooling. Refrigeration heat was rejected though the use of cooling towers in the existing (baseline) operation. In addition to cooling, the operation uses a significant quantity of process heat for clean in place, pasteurizing, equipment cleaners, and product heaters. The process temperature requirements for various end uses targeted for heat recovery ranged from 50 to 85 °C (122 to 185 °F). The waste heat source operated at about 60 °C (140 °F), which limited the quantity of recovered heat that could be utilized with passive heat exchangers. A heat recovery system was designed using a single 550 hp heat pump to lift the temperature to the range of 85 °C. The design included a stratified thermal storage tank (45 m^3 or 1,600 ft^3) to balance the waste heat supply with end uses. Although it was estimated that waste heat recovery was sufficient for 100% of the targeted process heat end uses, boiler steam was still available to top up and ensure all process heating needs would be met. Natural gas savings were estimated to be 29,500 GJ/yr (31,150 MMBtu/yr). Increased electricity consumption of 1,900,000 kWh/yr (equivalent energy 6,840 GJ/yr) was required to power the heat recovery heat pump plus ancillary pump loads.

20.7 Combined Heat and Power

Combined heat and power (CHP) is defined as an energy system that simultaneously produces electricity and thermal energy.[23] CHP systems are essentially the same as waste heat recovery covered in the previous section of this chapter, with the stipulation that either: (a) the waste heat source is from electricity generation, or (b) waste heat is used to generate electricity. CHP systems have been in commercial use since the first installation in New York City in 1882.[24] The traditional rationale for CHP is that it has a high system efficiency as illustrated in Figure 20.25. In a traditional

[23]Sources: www.eia.gov/tools/glossary, https://iea-etsap.org/E-TechDS/PDF/E04-CHP-GS-gct_ADfinal.pdf & https://natural-resources.canada.ca/energy-sources/renewable-energy/combined-heat-power. Some sources (e.g., Williams, 2025) broadly define CHP as simultaneous production of either electric or mechanical power, plus thermal power.

[24]www.powermag.com/history-of-power-the-evolution-of-the-electric-generation-industry/

energy system, electricity is supplied from the grid that is dominated by Rankine cycle electric power plants at about 32% efficiency. Then thermal energy is produced at a facility by a non-condensing steam boiler at 80% thermal efficiency. The traditional system efficiency is 50% (80 units of useful energy divided by 160 units of input energy). The CHP system is 80% efficient (80 units output divided by 100 units input). This model for CHP comparison is becoming less relevant over time, because new electricity supplies are increasingly from renewable sources such as wind and solar and less from fossil fuel.[25] Moreover, while existing heating equipment is still dominated by non-condensing natural gas combustion, new installations are increasingly using condensing gas-fired boilers, electric boilers, or heat pumps. Nonetheless, the overall system efficiency of CHP often presents an attractive opportunity to improve efficiency and/or reduce GHG emissions.

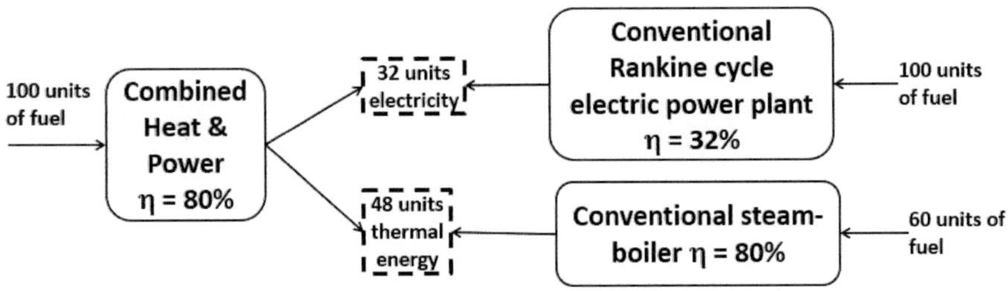

Figure 20.25 Comparison of CHP to supply of heat and electricity from traditional sources.

CHP systems are usually categorized as either topping cycles or bottoming cycles.[26] A topping cycle is where electricity is the primary production output, and residual (lower quality) heat is used for thermal processes as shown in Figure 20.26. A bottoming cycle is where the primary production is thermal output, and then residual heat is of sufficient temperature to produce steam and electricity as shown in Figure 20.27. Another common CHP design is a combined cycle CHP system where a conventional combined cycle electric power plant (which has both a gas turbine generator and steam turbine generator) uses residual heat for thermal end uses as shown in Figure 20.28a.

The main device that converts the fuel into electricity and heat is described as the prime mover. Table 20.10 lists the basic technical and cost performance parameters for CHP systems using prime movers with the

[25] www.iea.org/reports/world-energy-outlook-2025

[26] www.eia.gov/tools/glossary

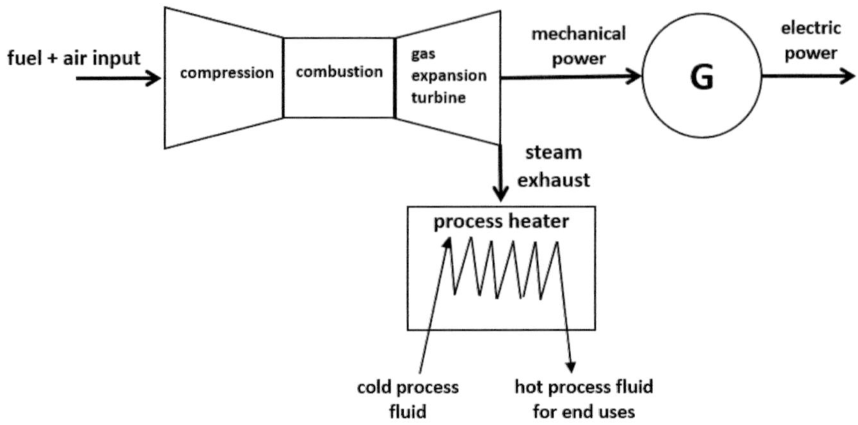

Figure 20.26 CHP topping cycle diagram with a gas turbine.

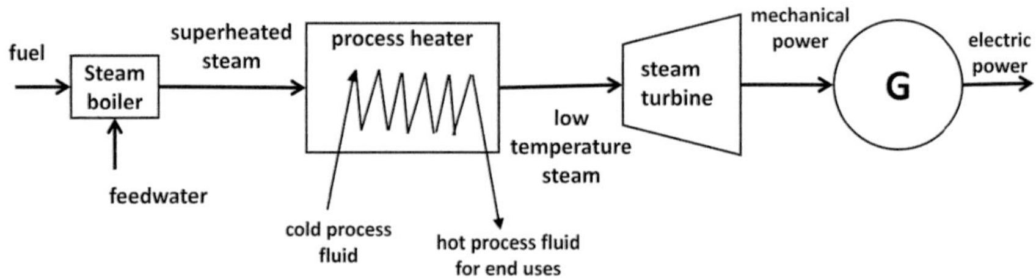

Figure 20.27 CHP bottoming cycle diagram.

highest adoption rates in the U.S. (DOE, 2024). Notice that all of the prime movers utilize gas turbines, microturbines, or reciprocating engines. None of the systems listed in the table are bottoming cycles, because very few bottoming cycles have been installed in recent years in the U.S. The overall efficiency shown in Table 20.10 is the total system efficiency, including electricity and thermal output on a net basis (i.e., ancillary loads have been counted). For all natural gas-fired systems, air emissions of NOx and CO will be the main concerns (see Table 20.3). For oil-fired or biomass-fired systems, in addition to NOx and CO, particulate matter and possibly other air toxic species may be emitted in significant levels. For all systems, CO_2 emissions will be proportional to efficiency. However, emissions of CH_4 and N_2O can vary widely depending on the fuel, the conversion technology, and combustion conditions.

Table 20.10 Technology performance and cost factors for common CHP systems.

Sector	Prime mover & fuel*	Thermal recovery	Electric capacity kW AC	Overall efficiency HHV	Fuel inputBtu/hr HHV	Fuel inputkJ/hr HHV	Useful thermal output Btu/hr	Useful thermal output kJ/hr	Electric heat rate Btu/kWhe	Electric heat rate kJ/kWhe	Capital$ per kW 2022 US$	O&M$ per kWh 2022 US$	GHG** g CO2e/ kWhe	GHG** lbm CO2e/ MWhe
Commercial	Reciprocating engine	Hot water	300	79.7%	3,129	3,301	1,470	1,551	10,431	11,005	$3,750	$0.019	565	124
	Reciprocating engine (No. 2 oil)	Hot water	350	81.3%	3,549	3,744	1,653	1,744	10,139	10,697	$3,100	$0.019	743	164
	Gas turbine	Hot water & steam	3,300	65.8%	45,540	48,040	18,710	19,740	13,801	14,560	$3,900	$0.015	747	165
	Microturbine	Hot water & steam	200	68.8%	2,300	2,427	840	886	11,350	11,974	$3,100	$0.017	614	135
Industrial	Reciprocating engine	Hot water & steam	1,000	78.7%	9,210	9,717	3,890	4,104	9,210	9,717	$3,100	$0.019	498	110
	Reciprocating engine	Hot water & steam	3,000	78.8%	25,250	26,640	9,680	10,210	8,417	8,880	$2,600	$0.016	456	100
	Gas turbine	Process heat, hot water & steam	5,000	68.8%	60,920	64,270	27,420	28,930	12,185	12,855	$2,600	$0.015	659	145
	Gas turbine	Process heat, hot water & steam	10,000	68.4%	120,900	127,500	58,390	61,600	12,091	12,756	$2,000	$0.013	654	144
	Gas turbine	Process heat & steam	25,000	69.3%	243,000	256,400	83,970	88,590	9,720	10,255	$1,700	$0.010	526	116
	Gas turbine	Process heat & steam	40,000	70.3%	384,400	405,500	132,000	139,300	9,611	10,140	$1,400	$0.010	520	115
	Combined cycle	Process heat & steam	100,000	65.5%	674,900	712,000	341,200	360,000	6,749	7,120	$1,600	$0.004	365	81
	Combined cycle	Process heat & steam	375,000	63.3%	2,342,000	2,471,000	683,200	720,800	6,246	6,590	$1,400	$0.003	338	75

* Second row is No. 2 oil. All other rows are natural gas fuel.

** GHG's estimated using 0.0513 gCO2e/GJ for gas, and 0.0695 gCO2e/GJ for oil.

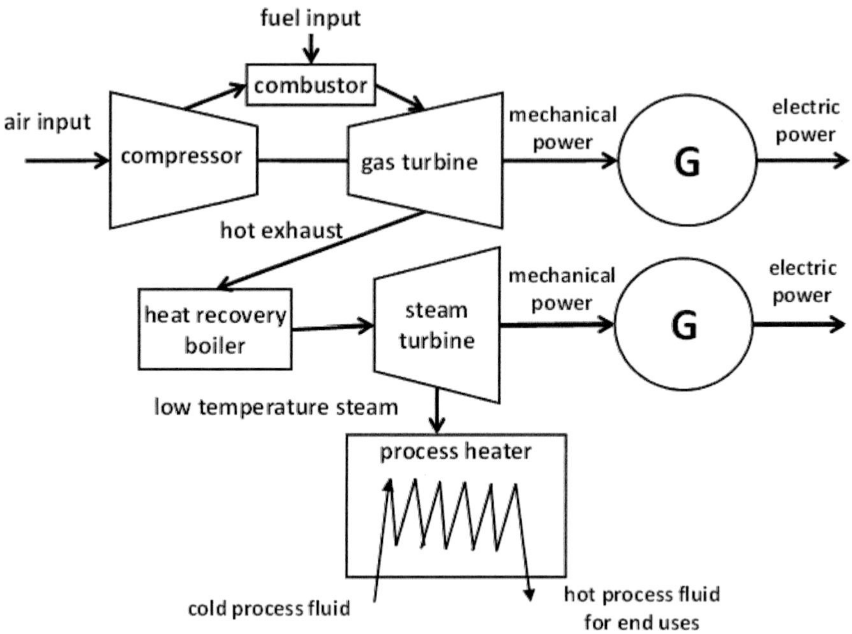

Figure 20.28a Combined cycle CHP system diagram.

To assess CHP for potential site-specific applications, what is needed is the electric and thermal conversion efficiencies of the main prime movers at different system capacities. This information is provided in the following four figures.[27] While these efficiencies are all stated based on the use of natural gas, engines and turbines can be designed to use a wide range of liquid or gaseous fuels such as diesel, biogas, LPG, gasoline, propane, landfill gas, or possibly hydrogen. Fuel cells tend to be limited to gaseous fuels such as natural gas, biogas, or, of course, pure hydrogen if available.

- Natural gas-fired reciprocating engine efficiencies are provided in Figure 20.28a. Over 75% of reciprocating engine CHP systems in the U.S. are under 1,000 kW electric output. These CHP systems tend to have the highest total system efficiency, at about 80% on an HHV basis. Reciprocating engines operate relatively efficiently down to about 25% of full load.

[27] Source for data in all four figures is https://chp.ecatalog.ornl.gov/resources/technologies

- Natural gas-fired gas turbine efficiencies are provided in Figure 20.28b. These systems mostly range in capacity from 2,000 to 25,000 kW electric output and operate around 70% of total system efficiency. Gas turbine designs are characterized by: operating speed ranges 3,000–20,000 rpm, inlet/outlet compression ratio up to 25, and exhaust temperatures over 800 °F (430 °C). Gas turbines are best suited for base load operation, not part load.
- Natural gas-fired microturbine efficiencies are provided in Figure 20.29. Microturbines are small gas turbines. These systems are characterized by: sizes under 1,000 kW, operating speeds of 40,000–100,000 rpm, single-stage compressors with inlet/outlet pressure ratios of 2–5, and exhaust temperatures of 500–650 °F (260–340 °C). Microturbines are suitable for modular designs using multiple units.
- Fuel cell efficiencies are provided in Figure 20.30. These technologies operate with hydrogen, which is rarely available as a pure fuel. Hydrogen is typically produced in the fuel cell package by reforming natural gas where CH_4 is reacted to form $CO_2 + H_2$. Hydrogen is converted electro-chemically in the fuel cell to produce electricity and heat, and the CO_2 is emitted to the atmosphere. The efficiencies shown in the figure are for phosphoric acid and molten carbonate fuel cell designs. Fuel cells tend to operate relatively efficiently down to about 50% part load.

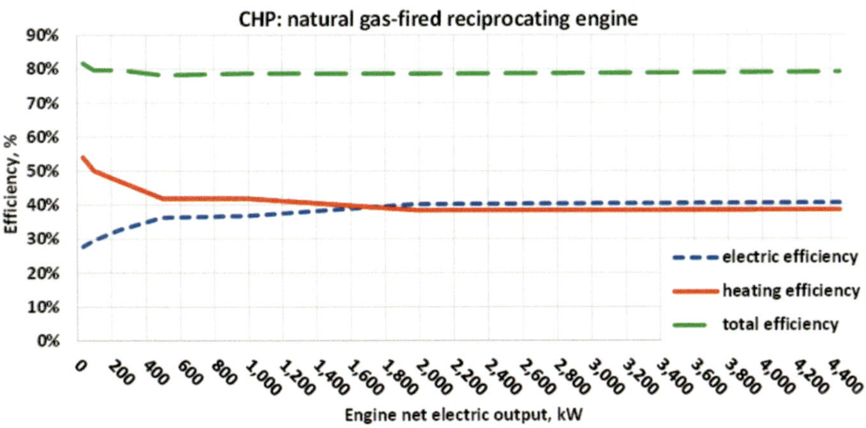

Figure 20.28b CHP electric, thermal, and total efficiency for natural gas-fired reciprocating engines.

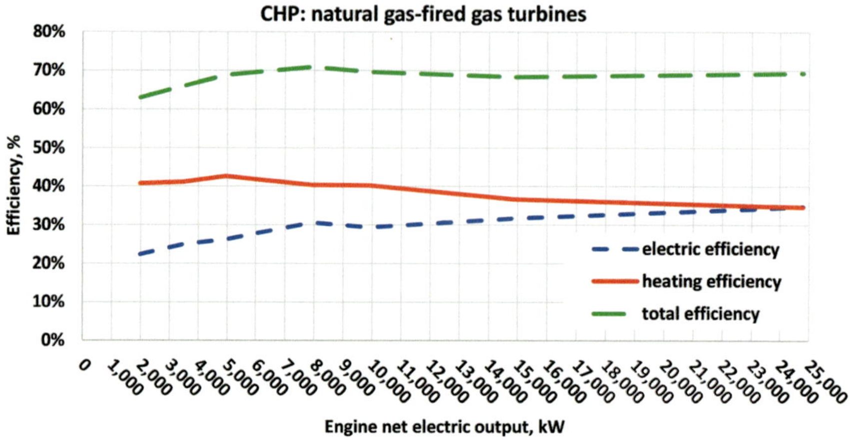

Figure 20.29 CHP electric, thermal, and total efficiency for natural gas-fired gas turbines.

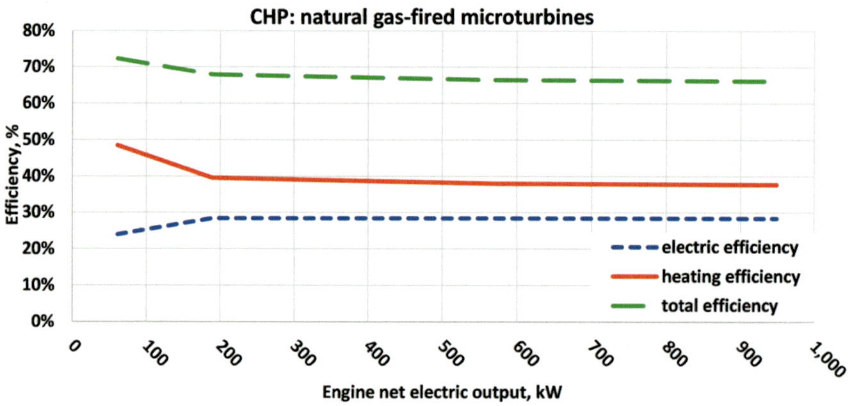

Figure 20.30 CHP electric, thermal, and total efficiency for natural gas-fired microturbines.

There are many factors to consider besides energy efficiency for real-world CHP applications (ASHRAE Handbook Systems and Equipment, 2024, Chapter 7). The following are key factors in system design and assessment[28]:

[28] www.epa.gov/chp/chp-project-development-steps

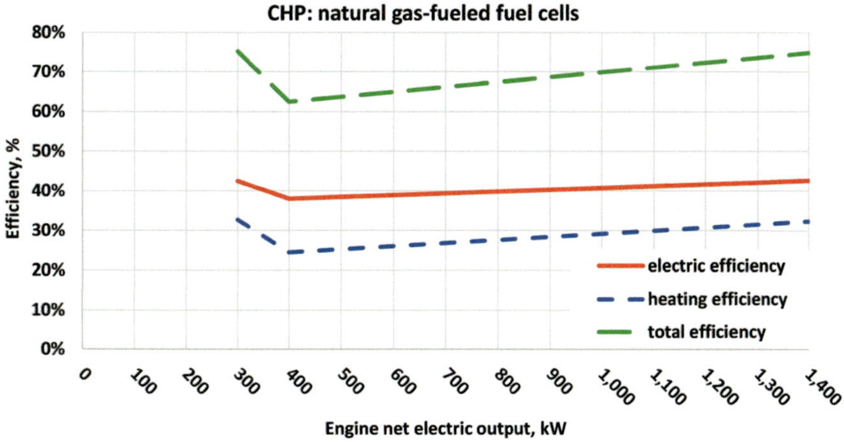

Figure 20.31 CHP electric, thermal, and total efficiency for natural gas-fueled fuel cells.

- Multiple benefits or services can be realized from the CHP system including: electricity demand response services, backup power, and thermal supplies during power outages (resiliency).
- **Thermal and electric load types and profiles.** For the targeted facilities and end uses, both the timing of electric thermal demands and the ratio of thermal-to-electric are important. While CHP has traditionally been thought of as electricity and heat only, the need for cooling services is increasingly critical due to global climate change. CHP systems can readily be designed to deliver cooling services.[29]
- **Utility interconnection.** A critical factor is whether electricity would be exported to the grid, or be consumed within the facility. For grid connection, there are many potential regulations, codes, and rate structures that need to be navigated.
- **Permits and codes.** Many different permits may be required such as building permits, state/provincial environmental permits, and utility permits. Emissions of common air pollutants and GHG emissions need to meet requirements of local bylaws and/or organizational sustainability policies. Land use zoning, building/structural codes, electrical, and fire code requirements need to be addressed.
- **Financing and costs.** Key issues include access to capital, O&M costs, availability of grants or subsidies, utility net metering, or net billing rates

[29] www.energy.gov/sites/prod/files/2017/06/f35/CHP-Absorption%20Chiller-compliant.pdf

(see Chapter 4). Future fuel costs and utility rates significantly affect project economics.

- **Procurement and contracting.** Securing the resources to design, build, commission, operate, and maintain the CHP plant within business case time limits. Use of packaged or modular systems vs. custom engineered systems is a design choice that can affect cost and performance. For some applications, the skilled trades and operators needed for the CHP system may be different from the qualifications of the current facility O&M staff.
- **Fuel source reliability and cost.** All CHP systems utilize a fuel and the cost and availability of the fuel will likely vary over the life of a project. For example, future natural gas prices, biomass availability. Flexibility for more using more than one fuel is an option.

This section concludes with a selection of five CHP case studies.

Case study 1: Organic Rankine cycle electricity generation (Mota-Babiloni, 2023)

A wastewater treatment plant produces biogas from anaerobic digestion of sewage sludge, with gas composition of about 65% methane. The biogas is burned in reciprocating engines to produce 1,600 kW of electricity. An organic Rankine cycle unit is installed on the exhaust which produces additional electricity and 70 °C (158 °F) hot water that is used in the digestion process.

Case study 2: Bioenergy CHP

The university of British Columbia has a bioenergy plant, which includes a nominal 2,000 kW Jenbacher reciprocating engine.[30] The engine was originally designed to burn a syngas produced on site from biomass gasification. It later transitioned to utilize renewable natural gas purchased from the regional gas utility. The engine produces net electricity of about 1,600 kW and is also equipped with heat recovery to produce steam at approximately 2,800 kW thermal. Steam runs through a heat exchanger to produce hot water for the campus district energy system. Every unit of recovered heat displaces hot water production from the campus natural gas-fired heating boilers.

The next three case studies are quoted from the U.S. Department of Energy database of CHP case studies.[31] Cases are selected to illustrate

[30]https://facilities.ubc.ca/infrastructure-and-systems/bioenergy-research-demonstration-facility-brdf/ & https://vpfo-facilities.sites.olt.ubc.ca/files/2025/02/UBC-EN-Outreach-Report.pdf

[31]https://dg.resilienceguide.ornl.gov/case-studies

the diversity of CHP systems and delivered benefits, beyond basic energy efficiency.

Case study 3: Breevort East CHP

"The Breevort East high-rise co-op in Greenwich Village, NYC, installed a 300 kW CHP system in order to reduce electricity and thermal energy costs for the facility. Three natural gas-fired reciprocating engines are used to provide 300 kW of continuous power and 2,100 MBtu/h of thermal output for hot water. The recovered heat is also used to provide cooling via a 90 ton absorption chiller. The Brevoort was one of the only buildings in the lower third of Manhattan that was able to maintain power during the extended grid outages caused by the storm. The CHP system has also helped to provide the building with significant energy cost reductions."

Case study 4: Greenwich Hospital CHP

"The Greenwich Hospital is a 175 bed, 500,000 sq ft medical center located in Greenwich, CT. Its CHP system, installed in 2008, consists of two 1,250 kW natural gas-fired reciprocating engines. The hospital also has a 2,000 kW backup generator. Following Hurricane Sandy, the area surrounding Greenwich Hospital lost power for approximately 7 days. When the hospital lost grid power, it went down for about 7 seconds before the backup generators kicked in and power was restored. The transition from using grid power to operating solely on the CHP system went as planned, with the CHP system shutting down and restarting in island mode, while power was supplied to the hospital by backup generators. The whole transition process takes approximately 5 minutes. Due to its CHP system, Greenwich Hospital was able to continue normal operations throughout the storm."

Case study 5: Sikorsky Aircraft CHP plus solar

"Sikorsky Aircraft Corporation, an aircraft manufacturer based in Stratford, Connecticut, installed a 10.7 MW gas turbine CHP system in 2011 to align with corporate sustainability goals. The CHP system provides the 2,000,000 square foot manufacturing facility with 84% of its electric and 85% of its thermal needs. Due to the reliability of the CHP system, the facility did not experience any interruptions during Superstorm Sandy, and was able to sustain normal operation during storm and aftermath. The facility also provided helicopter transport services for disaster relief personnel and supplies, and the manufacturing campus was opened as a place of refuge (meals, showers, phone charging) for the 6,650 employees and their families."

References

Allwood and Cullen (2015) "Sustainable materials without the hot air: making buildings, vehicles, and products efficiently with less new material" Bloomsbury Publishing.

ASHRAE Handbook—HVAC Systems and Equipment (2024) Chapter 7 "Combined Heat and Power Systems"

ASHRAE Handbook—HVAC Systems and Equipment (2024) Chapter 11 "Steam Systems"

ASHRAE Handbook—HVAC Systems and Equipment (2024) Chapter 16 "Infrared Radiant Heating"

ASHRAE Handbook—HVAC Systems* and Equipment (2024) Chapter 44 "Centrifugal Pumps"

ASHRAE Handbook—HVAC Systems and Equipment (2024) Chapter 26 "Air to Air Energy Recovery Equipment"

ASHRAE Handbook—HVAC Systems and Equipment (2024) Chapter 21 "Fans"

ASHRAE Handbook—HVAC Systems and Equipment (2024) Chapter 32 "Boilers"

ASHRAE Handbook—HVAC Systems and Equipment (2024) Chapter 47 "Heat Exchangers"

ASHRAE Handbook HVAC Applications (2023), Chapter 32 "Ventilation of the Industrial Environment"

ASME Power Test Code 4-2013 (2013) "Fired Steam Generators". American Society of Mechanical Engineers.

CEATI (2012) "Conversion of waste heat to electricity: technology update and assessment of potential applications" Report No. T102700-529. www.ceati.com

Cengel Y., Boles M., Kanoglu M. (2024) "Thermodynamics : an Engineering Approach" 10th edition. McGraw Hill.

Cengel Y., Cimbala J., Ghajar A. (2022) "Fundamentals of Thermal Fluid Sciences" 6th edition. McGraw Hill.

CAGI (2016) "Compressed Air and Gas Handbook" 6th edition. Compressed Air and Gast Institute www.cagi.org

DOE (2003a) "Improving Fan System Performance: A Sourcebook for Industry". U.S. Department of Energy. DOE (2003a) "Improving Fan System Performance: A Sourcebook for Industry". U.S. Department of Energy.

DOE (2003b) "Improving Compressed Air System Performance a sourcebook for industry". U.S. Department of Energy.

DOE (2006) "Improving Pump System Performance: A Sourcebook for Industry". U.S. Department of Energy.

DOE (2008) "Waste Heat Recovery: Technology and Opportunities in the U.S. Industry" prepared by BCS Inc. for the U.S. Department of Energy.

DOE (2012) "Improving Steam System Performance: A Sourcebook for Industry". U.S. Department of Energy.

DOE (2024) "Distributed Generation, Battery Storage, and Combined Heat and Power System Characteristics and Costs in the Buildings and Industrial Sectors" U.S. Energy Information Administration and Department of Energy.

Doty (2020) "Commercial Energy Auditing Reference Handbook" Chapter 4 "Manufacturing and Unit Operations" River Publishers

European Commission (2009) "Reference Document on Best Available Techniques for Energy Efficiency" https://bureau-industrial-transformation.jrc.ec.europa.eu/sites/default/files/2019-11/ENE_Adopted_02-2009.pdf

Guidehouse (2024) "Electrification of High Temperature Heat: Electric Heating Capacity and Revenue Forecasts for High Temperature Industrial Processes, Global Market 2024-2033"

Jouhara H., Khordehgah N., Almahmoud S., Delpech B., Chauhan A., and Tassou S. (2018) "Waste heat recovery technologies and applications" Thermal Science and Engineering Progress 6.

Kaya D., Canka Kilic F., and Huseyin Ozturk H. (2021) "Energy Management and Energy Efficiency in Industry" Springer Nature Switzerland.

Kerry F. (2007) "Industrial gas handbook: gas separation and purification" CRC Press.

Killcross M. (2021) "Chemical and Process Plant Commissioning Handbook" 2nd edition. Butterworth-Heinemann.

Lovins (2010) "Integrative Design: A Disruptive Source of Expanding Returns to Investments in Energy Efficiency" white paper available at www.rmi.org.

Lovins (2013) "Reinventing Fire: bold business solutions for the new energy era" Chelsea Green Publishing.

McDonough (2002) "Cradle to Cradle: Remaking the Way We Make Things"

Mota-Babiloni A., Amat-Albuixech M., Moles-Ribera F., and Joaquin-Esbri J. (2023) "Techno-Economic Feasibility of Organic Rankine Cycles (ORC) for Waste Heat Recovery" chapter in Borge-Diez E. and Rosales-Asensio E. "Heat Energy Recovery for Industrial Processes and Wastes" Springer Green Enegy and Technology.

Rissman (2024) "Zero Carbon Industry" Columbia University Press
Williams, A. (2022) "Industrial Energy Systems Handbook" River Publishers.
Williams, A. (2025) "Energy Management Handbook" River Publishers.

Author Biography

Eric Mazzi is a consultant and educator focusing on technology and policy for demand-side energy systems. He specializes in performing independent technical analysis, M&V, policy studies, and educational services. He advocates and contributes to efforts to reduce energy consumption and achieve net-zero GHG emissions, while maintaining or improving energy services and advancing energy equity. His experience includes air quality consulting, utility maintenance, and five years as full-time engineering faculty at the University of British Columbia. He volunteers as a board member for the U.S. Northwest Energy Coalition, and the International Performance M&V Protocol committee. Eric teaches part-time, delivering university and professional certification courses. He is an adjunct professor of Energy Management at the New York Institute of Technology – Vancouver. He has B.Sc. and M.Sc. degrees in mechanical engineering, and a PhD in environmental policy. He is a registered engineer in California and British Columbia.

Acknowledgments

The chapter benefited from review by Sheila Glessman.

Practice Problems

1. An industrial plant currently is capable of producing up to 5,000 t/yr (tons per year) of a product but only produces 3,000 t/yr. Based on performance tests, the measured baseline electrical energy consumption as a function of annual production is shown in the table below. With a capital upgrade subsidized by the electric utility, the plant can improve the energy efficiency as a function of production as shown in the same table. The energy efficiency measure (EEM) also enabled the plant to increase full production to 6,000 tons/yr. Plant management intends to ramp up production over the next five years as follows:

Year 1 = 3,000 t/yr
Year 2 = 4,000 t/yr
Year 3 = 5,000 t/yr
Year 4 = 5,000 t/yr
Year 5 = 6,000 t/yr

For the purpose of reporting to the electric utility, estimate the total estimated electricity savings (kWh) over the five-year period if the plant accepts the utility subsidy and upgrades the plant to the more efficient condition by the beginning of year 1. Assume the electric utility's program rules do not allow incentives to be paid for increases in production that are enabled by the EEM, which is called a rebound effect. From the utility perspective, the baseline electricity consumption is calculated based on what would have happened if the EEM had not been implemented.

Tons/yr	Baseline kWh/t	Efficiency upgrade kWh/t
0	n/a	n/a
1,000	1,750	1,500
2,000	1,250	1,000
3,000	1,083	833
4,000	1,000	750
5,000	950	700
6,000	n/a	667

2. A condensing boiler firing natural gas is rated at 800 MBH input. The boiler operates at 50% load for 3,000 hours annually. Through design of the hydronic heating system, the average return water temperature can be reduced from 130 to 90 °F. Estimate the annual fuel cost savings assuming natural gas costs $7.00 per MMBtu. Assume the thermal efficiency curves for this boiler are shown in Figure 20.10.
3. Saturated steam is flowing in a main pipeline at 200 °C. 180 kg/hr of condensate is removed with steam traps and discharged to a drain at atmospheric pressure. Estimate the kg/hr of flash steam formed. Assume the condensate pressure (before flashing) is the same as the saturated steam.
4. A commercially available heat recovery unit recovers heat from a residential shower drain, and uses it to pre-heat the incoming cold water. It is expected to raise the incoming water temperature by 21 °C at a flow rate of 15 L/min. Assume:

- 10 minutes per shower
- 4 showers per day
- 350 days per year usage
- Specific heat of water is 4.2 kJ/kg °C
- Water density = 1.00 kg/L

(a) Estimate the annual useful energy delivered by the heat recovery system GJ/yr.

(b) Estimate the annual natural gas cost savings ($/yr) assuming 80% for the hot water heating system efficiency, and $8/GJ gas cost.

5. A compressed air system draws air from a boiler room all year round at an average inlet temperature of 78 °F in winter. The compressor operates at a constant flow of 2,000 scfm (68 °F reference temperature). By moving the compressor air inlet to outdoors, the average inlet temperature in winter is reduced to 48 °F. Estimate the reduction in compressor electric power in kW during winter. Assume compressor efficiency of 80% and motor efficiency of 94%. The operation is located at sea level and delivers compressed air at 7.5 atmospheres. There is no intercooling in the system design. Comment on whether your estimate is in agreement with the rule of thumb of 1%–3% compressor power savings for every 10 °F drop in inlet temperature. If your estimate is not in agreement, suggest reasons why not.
6. A pump delivers 100 kW in fluid power to a water system.

   (a) If the pump is 48% efficient, what is the shaft power required to drive the pump in units of kW and horsepower (hp)?

   (b) If the pump operates 8,700 hours per year, how much electricity does it use (kWh/yr)? Assume the pump has a magnetic coupling with efficiency of 98%, and the motor efficiency is 93.5%.

   (c) How much electricity can be saved annually (kWh/yr) by replacing the 48% efficient pump with an 82% efficient pump?

7. Water flows through the water turbine at a rate of 30 m^3/s as shown. How much power is delivered by the turbine assuming a turbine efficiency of 90%?

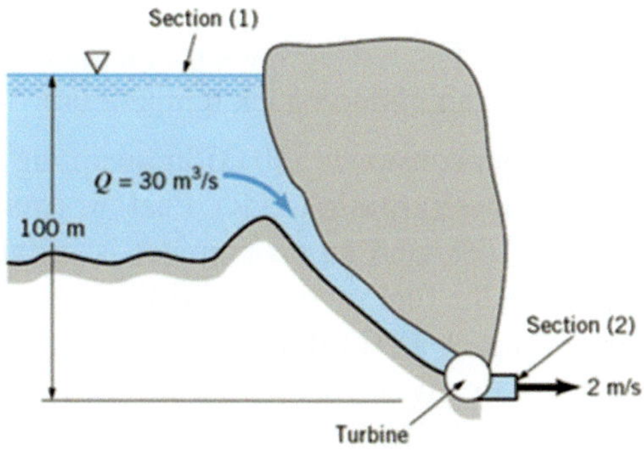

8. A fan with performance curve shown below is connected to a 100 hp motor by a belt drive coupling that is 96% efficient. The curve shows fan performance with no throttling or speed reduction.
   (a) The system head curve (not shown) intersects with the fan head curve at 22,000 acfm. Estimate the fan efficiency.
   (b) Estimate the motor load factor.

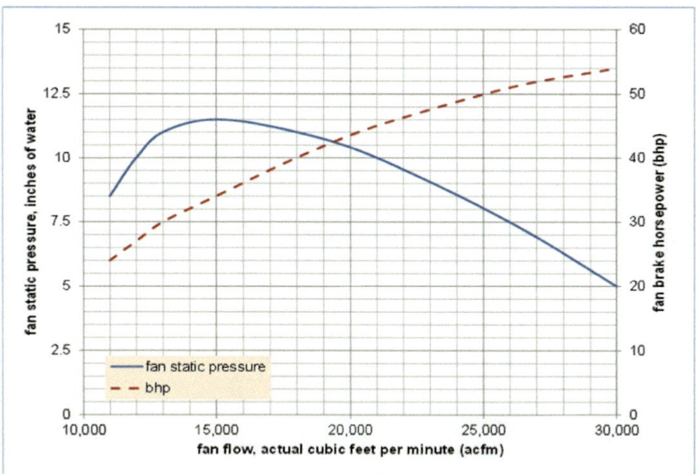

9. Which approach is most likely to provide the greatest improvement in boiler efficiency (assume no change in fuel properties)?
   (a) Increase stack temperature, increase combustion air temperature
   (b) Reduce CO_2 in the exhaust, increase combustion air temperature
   (c) Reduce stack temperature, decrease combustion air temperature

(d) Increase boiler excess O_2, reduce stack temperature
(e) Reduce boiler excess air, reduce stack temperature

10. A No. 2 fuel oil-fired boiler consumes 500 GJ/hr operating at 30% excess air and 300 °C stack temperature rise (STR). Fuel-air controls are tuned to reduce excess air to 10%, and an economizer is installed to reduce STR to 150 °C. Estimate the fuel savings in GJ/hr.
11. A steam boiler operating at 1,500 lbm/hr is supplied with feedwater at 100 °F, and delivers steam at 185.3 psig and 600 °F. The thermal efficiency is 82%, and the boiler operates at sea level. Calculate input rate of natural gas in units of MBH.
12. The manufacturer's specification for a gas turbine firing natural gas is provided below. Estimate the annual natural gas consumption in GJ/yr (HHV basis) if the gas turbine operates at the design conditions with 5,670 kW electric output at the generator terminals for 7,500 hours per year.

Performance

Output Power	5670 kWe
Heat Rate	11 430 kJ/kWe-hr (10,830 Btu/kWe-hr)
Exhaust Flow	78 370 kg/hr (172,770 lb/hr)
Exhaust Temp.	510°C (950°F)

Application Performance

Steam (Unfired)	13.5 tonnes/hr (29,750 lb/hr)
Steam (Fired) 1536°C (2800°F)	58.9 tonnes/hr (129,830 lb/hr)
Chilling (Absorp.)	11 650 kW (3310 refrigeration tons)

Nominal rating – per ISO
At 15°C (59°F), sea level
No inlet/exhaust losses
Relative humidity 60%
Natural gas fuel with
LHV = 35 MJ/Nm³ (940 Btu/scf)
No accessory losses
Engine efficiency: 31.5%
(measured at generator terminals)

13. A natural gas-fired CHP system is to be sized to deliver 100 kW of electricity using a reciprocating engine with exhaust temperature of 600 °F (320 ° C). The system is to use a new absorption heat pump

technology that can produce hot water in heating mode with a COP of 0.85, and chilled water in a cooling mode with a COP of 1.40. Assume that the fuel energy (HHV basis) is converted as follows: 30% to electricity, 50% used by the heat pump, and 20% is exhausted. Estimate (a) the heating production in kW and system efficiency while operating in heating mode, and (b) the cooling production in both kW and tons and system efficiency while operating in cooling mode.

21

Maintenance and Energy Systems

Eric Mazzi

Mazzi Consulting, Canada

21.1 Introduction

This chapter overviews the potential energy benefits of effective facility maintenance practices. Maintenance of facilities is primarily concerned with safety, equipment and process performance, and maintenance cost management. There is ample evidence to support the idea that maintenance and energy management can be synergistic – each can help the other achieve its primary objectives. This chapter provides a general introduction to common maintenance methods and practices, and describes the link between maintenance and energy management. Maintenance practices are summarized and presented from the perspective of energy management. Included are examples of maintenance activities (e.g., leak repairs) to illustrate the potential benefits of diligent maintenance to reduce energy costs.

What is maintenance? A good definition of maintenance is "*the decisions and actions regarding upkeep of property and equipment. These are inclusive, but not limited to, the following: 1) actions focused on scheduling, procedures, and work/systems control and optimization; and 2) performance of routine, preventive, predictive, scheduled and unscheduled actions aimed at preventing equipment failure or decline with the goal of increasing efficiency, reliability, and safety.*" (DOE, 2010).

Maintenance and energy management have many things in common, including:

- Both are more effective if done proactively (not reactively) and using proven methods.
- Table 21.1 provides an overview of key elements of effective maintenance. Comparing this table to Table 1-6 in Chapter 1, it can be seen

that both require similar organizational structures, senior management recognition and support, and similar activities such as training, tools and performance indicators.

- Effective maintenance practices can reduce energy costs, and effective energy management practices can improve equipment reliability and reduce maintenance costs.
- In some cases, maintenance management and energy management employ the same fundamental approaches. Perhaps the best example of this is the plan-do-check-act cycle, which was introduced in Chapter 1 for energy management. ASHRAE Handbook (ASHRAE Handbook, 2023) states that "*A useful concept in operation & maintenance is to improve O&M activities using the Deming Institute's Plan, Do, Check, Act cycle*." So energy and maintenance management often employ the same work process.

In the next sections, we will overview the common approaches or methodologies for maintenance management, and illustrate how maintenance and energy management are interrelated for each.

21.2 Introduction to Maintenance Methodologies: Prerequisites and Barriers

There are many systems and methodologies (or approaches) for maintenance management, and the ways in which maintenance systems are classified varies from one source to another. In this chapter, we focus on three approaches to maintenance: reactive, preventive, and predictive (including reliability-centered). No matter what maintenance method is chosen, the most effective maintenance managers do not treat maintenance as a cost of doing business, but treat it as a system for achieving reliability, safety, economic, and energy performance goals.

Before describing the various approaches to maintenance, it is important to highlight two important precursors to effectively implement any maintenance method: accurate equipment records and inventories, and condition assessments. In practice, accurate equipment records mean having reasonably accurate as-built drawings for systems and equipment, and sufficiently detailed equipment lists (e.g., industrial process equipment, motor size and design types, air handling unit manufacturer and model numbers, types and locations of steam traps, etc.). Ideally, these system drawings and records would be in digital format. Similarly, condition assessments for key equipment are critical to make effective maintenance management

Table 21.1 Key elements of effective maintenance management.

Element	Description
Management commitment	An official, documented recognition of the maintenance management, staff responsibilities, policies, and objectives.
Designated lead person	A designated lead person who is responsible for maintenance management.
Support team	Support teams are critical to carry out the many technical and financial functions.
Cost centers	Effective management of maintenance costs has been found to require dedicated cost centers.
Materials	Material resources are important, such as tools and maintenance equipment, instrumentation and sensors, maintenance data collection and storage, and software for analytics, scheduling, and planning (i.e., computerized maintenance management).
Maintenance assessments	Quantify and analyze maintenance data such as failure rates, safety statistics, labor costs, and material costs.
Performance monitoring	Establish key parameters and key performance indicators (KPIs) for the facility.
Reporting	Systematic reporting of maintenance activities and performance data. Careful consideration of how to visually display actual versus targeted performance is important.
Training	All maintenance team members should be afforded training in skills essential to their role. Trades and hands-on staff should be continuously learning new maintenance practices and techniques. Management and professional staff should be continuously learning maintenance best practices.

decisions (Chapter 3 includes an introduction to facility condition assessment practices). Developing equipment condition assessments requires that maintenance staff perform inspections and (in some cases) measurements on key equipment, and document conditions utilizing coding systems and standard categories to rate equipment condition and defects. Ideally, equipment records and condition assessments would be combined into one digital system (e.g., geographic information system) with live links to condition assessment databases, which are updated periodically. O&M staff should have access to these digital records and databases, and ideally update these records as part of their routine job tasks. Put another way, the lack of accurate records and equipment condition data is a critical barrier to effective maintenance. Attempting to design and implement an effective maintenance program is more or less an impossible task with inaccurate as-built drawings, missing equipment records, and outdated or missing condition assessments.

Other common barriers to effective maintenance include budget constraints, lack of skilled staff, inadequate documentation, and resistance to change (e.g., status quo bias; see Chapter 7). Standardized procedures and training are critical to overcoming these challenges (ASHRAE Handbook, 2023).

21.3 Reactive Maintenance

Reactive maintenance, also known as run-to-failure maintenance, refers to the practice of making repairs that occur only after equipment has failed (DOE, 2010). This approach is generally the least energy efficient and most costly over time, due to unplanned downtime and system inefficiencies. This strategy often leads to compromised operational performance and increased repair costs due to a lack of early problem detection (Petrocelly, 2020). Although reactive maintenance has its drawbacks, it may be justified in cases where equipment failure does not impact energy consumption, safety, or productivity in any significant way. If reactive maintenance is deliberately adopted, it should be done with caution and consideration of hidden costs. As highlighted in one publication (DOE, 2010), cost savings through reactive maintenance may not be realized because "*In reality, during the time we believe we are saving maintenance and capital cost, we are really spending more dollars than we would have under a different maintenance approach.*" For systems that are energy-intensive and critical for the safety and productivity of facilities (e.g., chillers and boilers), a reactive strategy is discouraged.

21.4 Preventive Maintenance

Preventive maintenance is a time-based or usage-based maintenance strategy that aims to prevent system failures through regularly scheduled inspections and servicing. This method typically involves conducting maintenance tasks based on calendar schedules or operating hours. Preventive maintenance is designed with the goal that systems operate within their optimal design parameters, thereby reducing energy waste and enhancing reliability. For common building systems, preventive maintenance has been estimated to save 8% to 12% in maintenance costs versus reactive maintenance (DOE, 2010).

Examples of common tasks include lubricating moving parts, replacing filters, calibrating sensors, and checking refrigerant levels. Table 21.2 is a list of common systems found in buildings and industrial applications for

Table 21.2 Systems for which ASHRAE Standard 180 provides preventive maintenance schedules.

Air distribution systems	Engines, microturbines
Air handlers	Fans (e.g., exhaust, supply, transfer, and return)
Boilers	Fan coils, hot-water, and steam unit heaters
Chillers, absorption	Furnaces, combustion unit heaters
Chillers, air-cooled	HVAC water distribution systems
Chillers, water-cooled	Indoor section duct-free splits
Coils and radiators	Outdoor-air heat-exchanging systems
Condensing units	Package terminal air conditioners/heat pumps
Control systems	Pumps
Cooling towers / evaporative- cooled devices	Rooftop units
Dehumidification and humidification devices	Steam distribution systems
Economizers, air-side	Terminal and control boxes (e.g., VAV, fan-powered, and bypass)
Water-source heat pumps	

which ASHRAE Standard 180 (ASHRAE, 2018) provides more detailed preventive maintenance schedules. To illustrate what maintenance schedules include, Table 21.3 provides a sample of a preventive maintenance schedule for water-source heat pumps (ASHRAE, 2018).

There are many other sources of guidelines for preventive maintenance schedules and activities. First and foremost are the manufacturers of various equipment. All equipment suppliers, whether for standard equipment or custom-designed, will come with vendor-recommended best practices for maintenance. For common energy system equipment (e.g., chillers, air handling units, and compressed air), the U.S. DOE has published detailed guidelines (DOE, 2010). One publication by Pacific Northwest National Laboratory (DOE, 2022) includes links for best maintenance practices for equipment such as ground source heat pumps, onsite wind turbines, solar water heating, and other applications. Furthermore, many industry associations provide maintenance guidelines. For boilers, the American Boiler Manufacturers Association publishes guidelines for boiler maintenance (see www.abma.org). Maintenance practices for common electrical equipment can be obtained from the National Electrical Manufacturers Association (www.nema.org). For lighting systems, best practices for maintenance of

Table 21.3 Sample preventive maintenance schedule for water-source heat pumps (ASHRAE, 2018).

Inspection task	**Maintenance task**	**Frequency**	**Recommended corrective action**
Check for particulate accumulation on filters	Clean or replace as needed to ensure proper operation	Quarterly	Evaluate the frequency of change requirements
Check the ultraviolet lamp	Clean and verify that it is functioning	Quarterly	Replace as needed to ensure proper operation
Check P-trap	Prime as needed to ensure proper operation	Quarterly	Replace the damaged P-trap
Check drain pain, drain line, coil, and other areas of moisture accumulation for visible signs of biological growth	Clean and verify proper operation	Quarterly	Disinfect as needed
Check the control system and devices for evidence of improper operation	Clean, lubricate, and adjust.	Semi-annually	Repair or replace components as needed to ensure proper operation
Check for damage or evidence of leaks on the water-cooled heat exchangers surfaces	Record the location of identified leaks	Semi-annually	Repair or replace components as needed to ensure proper operation
Check VFD for proper operation	Correct as needed. Clean housing and tighten connections. Clean or replace air filter.	Semi-annually	Repair or replace components as needed to ensure proper operation
Check the air filter fit and housing seal integrity	Clean and verify proper fit/finish.	Annually	Repair or replace components as needed to ensure proper operation

lighting systems are provided in the Illuminating Engineering Society Standard 36 (IES, 2024).

One example of the energy benefits of preventive maintenance is provided by a case study for a commercial office building that implemented a calendar-based preventive maintenance plan. Regular filter replacements, belt adjustments, and airflow inspections were carried out quarterly. According to the (ASHRAE Handbook, 2023), this led to a 15% reduction in HVAC-related energy consumption over a two-year period. Equipment lifespan was also extended, reducing replacement costs.

21.5 Predictive and Reliability-centered Maintenance

Predictive maintenance utilizes condition monitoring (ideally in real-time) and diagnostics to anticipate equipment failures before they occur. This approach relies on technologies such as vibration analysis, thermal imaging, and system data analytics. The Department of Energy's "O&M Best Practices Guide" (DOE, 2010) identifies predictive maintenance as a best practice for improving operational efficiency and reducing costs. Predictive programs go further than preventive maintenance in various ways, such as optimizing setpoints, detecting fouling, misalignment, leaks, and control drift before they result in failures. Predictive maintenance potentially reduces equipment downtime, part-load penalties, and distribution losses. (DOE, 2010) cites typical results of well-run predictive maintenance to save 8% to 12% maintenance cost savings versus preventive alone.

Reliability-centered maintenance (RCM) is sometimes described as a type of predictive maintenance (ASHRAE Handbook, 2023) and sometimes described as a stand-alone category of maintenance (DOE, 2022). Regardless of how it is categorized, RCM takes standard predictive maintenance practices further by focusing on critical systems whose failure could result in operational, safety, or environmental consequences.

RCM is a structured process that selects the right maintenance method for each asset based on function, failure modes, consequences, and economics. It relies heavily on predictive maintenance practices, data such as mean time between failures, and root-cause analysis. Predictive and reliability-centered maintenance help achieve energy and cost savings via the following mechanisms (DOE, 2010; DOE, 2022):

1. *Lower parasitic power losses*. Bearings aligned and lubricated properly reduce drive losses; clean heat-exchange surfaces lower approach temperatures and fan or pump power.
2. *Avoiding emergency modes*. Failures push systems into inefficient overrides (manual, full-speed, or bypass). RCM reduces "run-to-failure" and minimizes wasteful hours of operation.
3. *Fewer leaks and losses*. Detect steam, compressed air, and hydronic leaks that result in direct energy waste (see section 21.8)

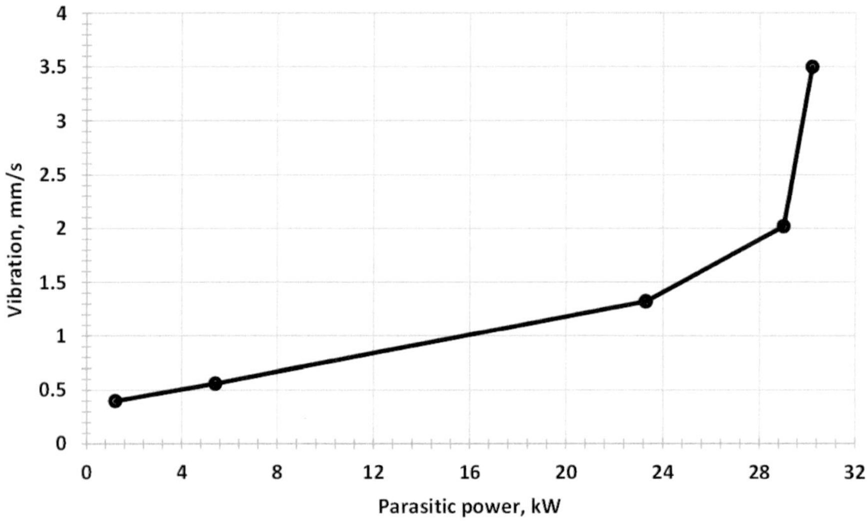

Figure 21.1 Pump parasitic power versus measured vibration (data source: ASME, 2015).

4. *Optimized maintenance intervals.* Prevents both over- and under-maintenance; avoid replacing efficient components early (embodied energy, cost); minimize running hours with inefficient components.

Core technologies that are part of predictive and reliability-centered maintenance methods include:

- *Vibration and motor analysis* identify imbalances, misalignment, and rotor/stator faults that increase kW draw for the same load. To illustrate how vibration analysis can be impactful, Figure 21.1 provides a chart of parasitic power for a pump versus vibration levels (data source: ASME, 2015). Parasitic power is defined as the power imparted on the pump shaft that does not contribute to fluid power (i.e., energy lost as heat).
- *Infrared thermography* (or thermal imaging) is useful for detecting problems such as hot electrical connections (see Figure 21.2), refractory breakdown in boilers, and building envelope insulation failures (see Figure 21.3). These problems contribute to both wasted energy, as well as increased failure rates, loss of thermal comfort, and safety hazards. Furthermore, the visual impact of using thermal imaging in energy studies has also been shown to increase the adoption of energy efficiency measures (Goodhew, 2015).

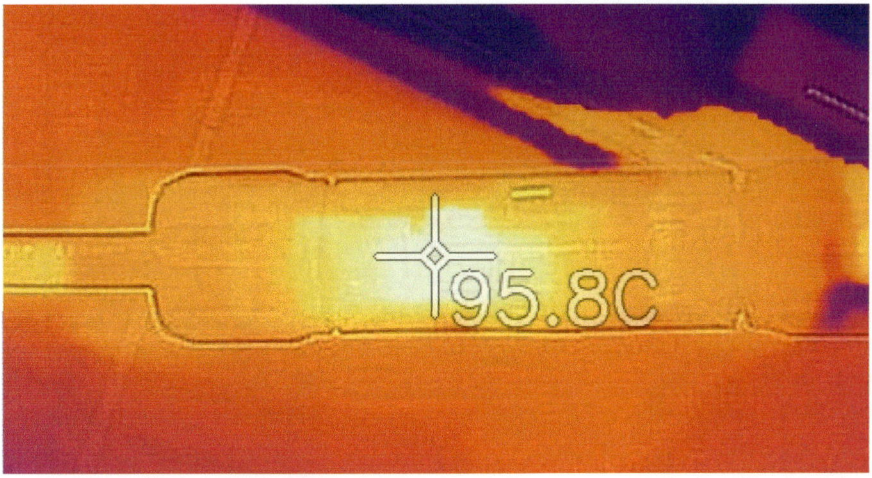

Figure 21.2 Thermal image shows an overheated (95.8 °C or 204 °F) solar PV electrical connection (source: https://energy.sandia.gov).

- *Ultrasonics and lubrication oil analysis* data drive maintenance activities. These data help identify bearing wear and lubrication breakdown that drive friction losses, overheating, and increased failure rates.

One case study of the energy benefits of effective maintenance is provided by a manufacturing plant that adopted an RCM approach for its chiller system (DOE, 2022). Vibration sensors and flow meters were installed and detected early signs of bearing wear and heat exchanger fouling. Maintenance practices were implemented to replace components before energy performance degraded. This data-driven strategy, using practices described in the DOE guide, reduced energy use by 12% and was credited with avoiding an unplanned shutdown.

21.6 Other Approaches to Maintenance

In addition to the standard maintenance practices described so far, there are additional maintenance methodologies available. Two additional methods are described here.

Total productive maintenance (TPM)[1] is described as an approach to maintenance that "*seeks to engage all levels and functions in an organization*

[1] www.epa.gov/sustainability/lean-thinking-and-methods-tpm

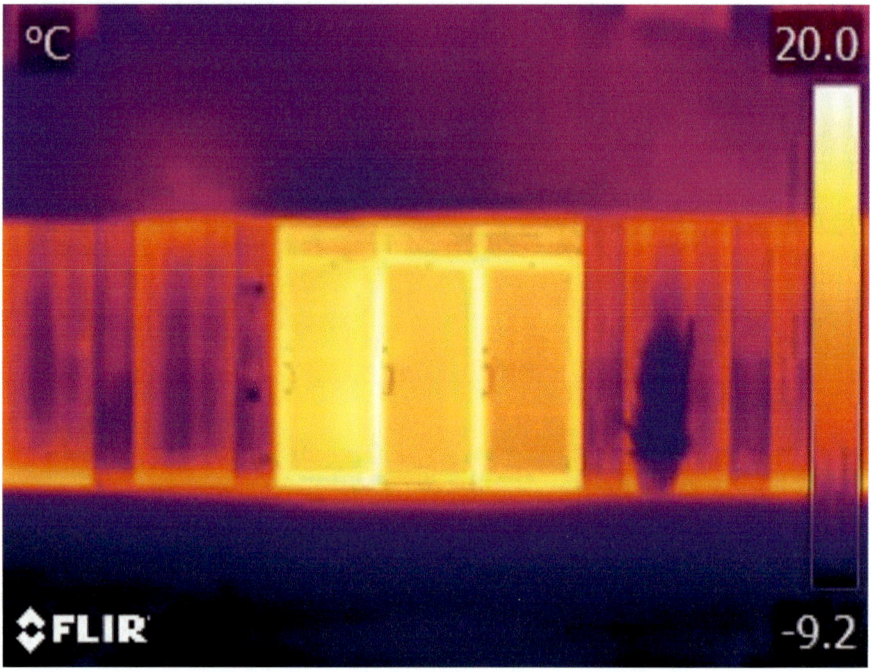

Figure 21.3 Thermal image indicating thermal losses through metal-framed entry doors (source: www.nrel.gov).

to maximize the overall effectiveness of production equipment. This method further tunes up existing processes and equipment by reducing mistakes and accidents. Whereas maintenance departments are the traditional center of preventive maintenance programs, TPM seeks to involve workers in all departments and levels, from the plant-floor to senior executives, to ensure effective equipment operation." TPM also takes the approach to provide front-line O&M staff the tools and autonomy to make O&M (not capital) changes, sometimes rapidly, such as within 72 hours of diagnosing a problem. One approach often used in TPM is the *Kaizen* process, which is a lean production method[2].

Alshakhshir and Howell (2021) provide a maintenance model called *Data Driven Energy Centered Maintenance*. They define energy-centered maintenance "*as a continuous improvement maintenance regime that combines the physical preventive and predictive maintenance tasks with energy-related*

[2] www.epa.gov/sustainability/lean-thinking-and-methods-kaizen

maintenance tasks that maintain the operational parameters of the equipment and its efficiency (i.e., motor current or fan flow rate). The energy-centered maintenance model applies to all energy-critical equipment to extend its life, maintain equipment efficiency, prevent excess energy use, and reduce energy waste." As the name implies, energy-centered maintenance uses minimizing energy consumption as the primary goal in the design and implementation of maintenance tasks. Similar to RCM, energy-centered maintenance heavily emphasizes the use of equipment O&M data as a key activity to achieve energy optimization goals.

21.7 Common Maintenance Tasks to Save Energy

Presented here are some common maintenance tasks and data to illustrate the connection between maintenance and energy management. These tasks would typically be considered to be part of a predictive or reliability-centered maintenance program, but in some cases might be included in a preventive maintenance program.

Detecting and estimating compressed air leaks is a common maintenance task (DOE, 2014). Leaks are typically detected using ultrasonic devices, and then the electricity impacts can be estimated based on the measurement of hole diameter and header pressure using Table 21.4 (Woodruff, 2019). The example below illustrates how to use the table to estimate utility bill savings by eliminating compressed air leaks.

Example 21-1

A 6 mm hole is measured for a compressed air system with a 7 bar (102 psig) header pressure. The compressed air system operates 12 hours per day,

Table 21.4 Compressed air leak table (English and International units). Compressed air system average electricity consumption (kWh) for one year (8,760 hours) based on header pressure.

Hole diameter mm (inches)	4 bar (58 psig)	5 bar (72 psig)	6 bar (87 psig)	7 bar (102 psig)	8 bar (116 psig)	9 bar (131 psig)
1 (0.04)	900	1,300	1,600	2,000	2,400	2,900
2 (0.08)	3,700	5,100	6,500	8,100	9,700	11,500
3 (0.12)	8,400	11,400	14,600	18,200	21,900	25,900
6 (0.24)	33,700	45,500	58,600	72,700	87,700	103,500
9 (0.35)	75,800	102,400	131,800	163,600	197,400	232,800

5 days per week, for 50 weeks annually. Estimate the electricity bill savings if electricity is valued at \$0.12 per kWh.

From Table 21.4, the average electricity consumption associated with this leak is 72,700 kWh for 8,760 hours. The leaking compressed air system operates 3,000 hours per year (60 hr/week × 50 weeks). The utility bill savings are estimated:

(3,000/8,760) × (72,700 kWh) × (\$0.12/kWh) = \$2,988 per year.

Steam leaks are another common occurrence in steam operations (e.g., industrial plants and district energy systems). Thermal losses from steam leaks can be estimated based on the leak diameter and header pressure using the data in Table 21.5 (Woodruff, 2019). The example below illustrates how to use the table.

Example 21-2

A steam trap is leaking live steam through a 3/8 inch (9.53 mm) discharge port. The steam pressure is 100 psig (6.9 bar). The steam system operates 24 hours every day, for 50 weeks annually. Steam is supplied by a natural gas boiler operating at an average efficiency of 83% (HHV basis). Estimate the natural gas bill savings by eliminating this leak, if natural gas is valued \$6.00 per MMBtu.

From Table 21.5, the average thermal loss associated with this leak (0.375 inch) is 2,915 MMBtu/yr for 8,760 hours. The boiler system operates 50 weeks per year (8,400 hr/yr). The utility bill savings are estimated:

Table 21.5 Thermal energy losses for steam leaks in MMBtu/yr (GJ/yr in parentheses) for one year (8,760 hours) based on steam header pressure.

Hole size inches (mm)	50 psig (3.4 bar)	100 psig (6.9 bar)	150 psig (10.3 bar)	200 psig (13.8 bar)
0.125 (3.18)	189 (200)	324 (342)	453 (478)	577 (609)
0.25 (6.35)	756 (798)	1,296 (1,367)	1,811 (1,911)	2,307 (2,434)
0.375 (9.53)	1,702 (1,796)	2,915 (3,076)	4,075 (4,299)	5,192 (5,478)
0.50 (12.7)	3,026 (3,192)	5,182 (5,468)	7,244 (7,643)	9,230 (9,738)
0.75 (19.1)	6,807 (7,182)	11,661 (12,303)	16,299 (17,196)	20,767 (21,910)
1.0 (25.4)	12,102 (12,768)	20,730 (21,871)	28,976 (30,571)	36,919 (38,952)

(8,400/8,760) × (2,915 MMBtu) × ($6.00/MMBtu)/0.83 = $20,206 per year.

Note: Divide by boiler efficiency to convert thermal losses to fuel losses.

Another common deficiency in older steam operations is having uninsulated sections of piping. The data in Table 21.6 can be used to estimate the thermal losses that can be avoided by adding standard thickness insulation to uninsulated piping. The example that follows illustrates how to use the table.

Example 21-3

An energy audit measures 86 meters of uninsulated 102 mm diameter steam pipe in a plant. The steam pressure is 1,030 kPa (1 bar = 100 kPa). The steam system operates 24 hours every day, for 7 months annually. Steam is supplied by a natural gas boiler operating at an average efficiency of 88% (HHV basis). Estimate the natural gas bill savings by insulating the pipe with standard insulation, if natural gas is valued $7.50 per GJ.

From Table 21.6, the average thermal loss associated with this uninsulated pipe (102 mm and 10.3 bar) is 896.8 GJ/yr for 8,760 hours, and for each 30.5 meters of pipe length. The boiler system operates 7 months per year, and the uninsulated pipe length is 86 meters. The utility bill savings are estimated:

(7/12) × 896.8 GJ × (86/30.5) × $7.50/GJ/0.88 = $12,572 per year.

Note: Divide by boiler efficiency to convert thermal losses to fuel losses.

Table 21.6 Thermal energy losses for uninsulated pipe in MMBtu/yr (GJ/yr in parentheses) for one year (8,760 hours) based on header pressure. Losses are provided for unit pipe lengths of 100 ft (30.5 m).

Nominal pipe diameter inches (mm)	15 psig (1.0 bar)	150 psig (10.3 bar)	300 psig (20.7 bar)	600 psig (41.4 bar)
1.0 (25)	140 (147.7)	285 (300.7)	375 (395.6)	495 (522.3)
2.0 (51)	235 (247.9)	480 (506.4)	630 (664.7)	840 (886.2)
4.0 (102)	415 (437.8)	850 (896.8)	1,120 (1,182)	1,500 (1,583)
8.0 (203)	740 (780.7)	1,540 (1,625)	2,030 (2,142)	2,725 (2,875)
12.0 (305)	1,055 (1,113)	2,200 (2,321)	2,910 (3,070)	3,920 (4,136)

21.8 Implementing Advanced Maintenance Practices

A key theme of this chapter is to provide an overview of common maintenance practices and illustrate the potential benefits of transitioning to more modern maintenance practices such as RCM. To summarize, a qualitative comparison of maintenance practices is provided in Table 21.7 (DOE, 2010; DOE, 2022). A quantitative comparison is provided in Figure 21.4, using pumping systems to illustrate the maintenance cost benefits of more modern methods (DOE, 2010). These qualitative and quantitative comparisons illustrate the potential benefits of adopting more advanced maintenance practices. While it is difficult to find comprehensive data on what maintenance practices are in use, the available literature (and the author's experience) suggests that adoption of more advanced practices is underutilized.

For facilities that presently rely mostly on reactive or preventive maintenance practices, it does not necessarily mean that all equipment should be rapidly transitioned to predictive or RCM maintenance. A mix of different maintenance practices may be implemented. For example, DOE's guide (2010) presents a portfolio mix for top performers to be less than 15% reactive, about 30% to 40% preventive, and about 45% to 55% predictive. Similarly, (ASHRAE Handbook, 2023) states that "*optimal maintenance programs generally incorporate all three basic strategies selectively. For example, a program might use a run-to-failure replacement strategy for lighting, windows, and door weatherstripping; a preventive strategy for roughing filter changes, strainer cleaning, and residential water heater replacement; and a predictive strategy for chillers and boilers.*"

To begin adopting more advanced practices, a good start is to pilot test advanced maintenance methods such as increasing the use of thermography, vibration monitoring, and RCM techniques. Piloting should include the development of KPIs such as energy indicators in Table 1-5 in Chapter 1, reliability

Table 21.7 Qualitative comparison of maintenance practices.

Maintenance type	Energy efficiency impact	Initial cost	Life cycle cost	System uptime
Reactive	Low	Low	High	Low
Preventive	Medium	Moderate	Moderate	Moderate to high
Predictive/RCM	High	High	Low	High

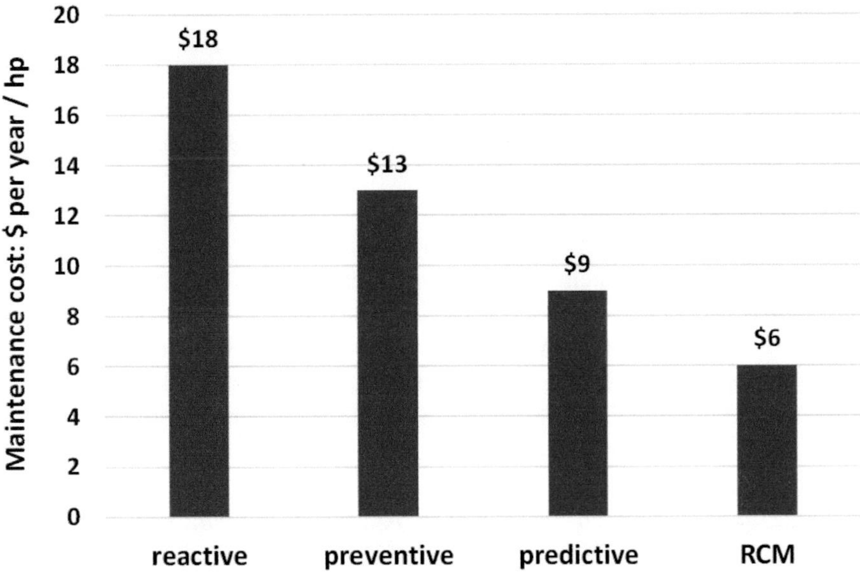

Figure 21.4 Comparing maintenance practices for pumps (dollars per motor horsepower).

indicators, material costs, and maintenance labor. These KPIs should be tracked and ideally be integrated into a modern building management system (see Chapter 15) or computerized maintenance management system.

21.9 Conclusions

Maintenance practices significantly impact energy efficiency, equipment longevity, and operational reliability in residential, commercial, and industrial facilities. Reactive maintenance, while inexpensive in the short term, often leads to higher lifecycle costs and energy inefficiency. Preventive maintenance involves using a calendar or operating time to schedule maintenance and ensures consistent performance through scheduled activities. Predictive and reliability-centered maintenance optimizes maintenance schedules and resources by leveraging condition-based data. Facility managers are encouraged to adopt a layered strategy that transitions from reactive to preventive, and ultimately to predictive maintenance.

References

Alshakhshir F. and Howell M. (2021) "Data Driven Energy Centered Maintenance" e-ISBN: 9788770223560. River Publishers.

ASHRAE Handbook—HVAC Applications SI (2023) "Operation and Maintenance Management" Chapter 40.

ASHRAE Standard 180-2018 – Standard Practice for Inspection and Maintenance of Commercial Building HVAC Systems

ASME (2015) "Pump system energy assessment" ISO/ASME 14414. 1st edition.

DOE (2010) "Operations & Maintenance Best Practices – A Guide to Achieving Operational Efficiency (Release 3.0)". PNNL-19634

DOE (2014) "Improving Compressed Air System Performance: a sourcebook for industry" www.energy.gov/sites/prod/files/2014/05/f16/compressed_air_sourcebook.pdf

DOE (2022) "An Advanced Maintenance Approach: Reliability Centered Maintenance" www.pnnl.gov/projects/om-best-practices/advanced-maintenance-approach-reliability-centered-maintenance#References

Goodhew J., Pahl S., Auburn T., and Goodhew S. (2015) "Making Heat Visible: Promoting Energy Conservation Behaviors Through Thermal Imaging" Environment and Behavior, vol. 47, no. 10.

Illuminating Engineering Society (IES) (2024) "Recommended practice: lighting maintenance. An American National Standard. ANSI/IES/NALMCO RP-36-24.

Petrocelly, K. L. (2020). Physical Plant Equipment Fundamentals. River Publishers.

Woodruff E. and Mazzi E. (2019) "Maintenance Outperforms Wall Street by 10x" www.buildings.com/building-systems-om/article/10185721/maintenance-outperforms-wall-street-by-10-times

Author Biography

Eric Mazzi is a consultant and educator focusing on technology and policy for demand-side energy systems. He specializes in performing independent technical analysis, M&V, policy studies, and educational services. He advocates and contributes to efforts to reduce energy consumption and achieve net zero GHG emissions, while maintaining or improving energy services and advancing energy equity. His experience includes air quality consulting, utility maintenance, and 5 years as full-time engineering faculty at the University of British Columbia. He volunteers as a board member for the U.S. Northwest Energy Coalition and the International Performance M&V Protocol committee. Eric teaches part-time, delivering university and professional certification courses. He is an Adjunct Professor of Energy Management at the New York Institute of Technology – Vancouver. He has B.Sc. and M.Sc. degrees in mechanical engineering and a PhD in environmental policy. He is a registered engineer in California and British Columbia.

Practice questions

1. Considering maintenance cost, reliability, and energy performance, list at least 5 KPI's for HVAC systems for an institutional building that has both heating and cooling. Hint: use ASHRAE Handbook, 2023, Chapter 40 or other cited references.
2. Consider the maintenance tasks that would be performed for a common cooling tower.
   - 2a. List at least 10 preventive maintenance tasks that should be performed, and the recommended frequency for each task. Cite the source(s) used.
   - 2b. Describe at least five maintenance activities that would be part of predictive maintenance.
   - 2c. Describe three activities that would be best described as reliability-centered maintenance.
3. Define mean time between failure (MTBF), describe how MTBF could be measured for common equipment such as boilers and chillers, and describe how to use MTBF in a reliability-centered maintenance program.
4. For the pump plotted in Figure 21.1, the average vibration level can be reduced from 2.0 to 1.0 mm/s through improved maintenance practices. Estimate the annual electricity bill savings if the pump operates 5,000 hours per year, the average motor efficiency is 95%, and electricity is valued at $0.12/kWh.
5. In Chapter 12, group relamping is introduced as a lighting maintenance strategy. A common strategy is to perform group relamping when the lamps have operated 70%–80% of their rated life. Assuming lamps are replaced at 75% of rated life, calculate the relamping interval in calendar years for lamps with a rated life of 30,000 hours, and which operate 4,600 hours annually.

6. For lighting systems, list and describe three factors that cause lumen depreciation (hint: see Chapter 12, and Illuminating Engineering Society standard IES RP-36-24).
7. The two critical precursors to implementing an effective maintenance management program are identified in this chapter as: a) equipment inventory and records, and b) equipment condition assessment. State why each is important in general. Consider a natural gas-fired hot water boiler and hydronic heating system for a commercial building. Describe what would be the key requirements for: a) equipment inventory and records, and b) condition assessment in order to implement a predictive maintenance program.

Appendix A

Part A: Units and Conversions

International Units

mass	kg	kilogram	1 tonne = 1,000 kg
length	m	meter	
time	s	second	
force	N	Newton	$N = kg\ m/s^2$
energy	J	Joule	J = N m
power	W	Watt	W = J/s
temperature	°C or K	Celsius or Kelvin	Kelvin is absolute temperature
pressure	Pa	Pascal	$Pa = N/m^2$

International Unit Conversions and Constants

1 GJ = 277.78 kWh	1 Wh = 3.6 kJ	$k = 10^3$
1 kWh = 3,600 kJ = 3.6 MJ	1 kcal = 4.184 kJ	$M = 10^6$
1 kW = 3,600 kJ/hr = 3.6 MJ/hr	$1{,}000\ L = 1\ m^3$	$G = 10^9$

English Units

mass	lbm	pounds mass	
length	ft	foot	
time	s	second	
force	lbf	pounds force	$F_{gravity} = m\ g/g_c$
energy	Btu	British thermal units	
power	Btu/hr	Btu per hour	
temperature	°F or R	Fahrenheit or Rankine	Rankine is absolute temperature
pressure	lbf/ft^2	pounds per square foot	$1\ psi = 1\ lbf/in^2 = 144\ lbf/ft^2$ $= 27.68\ inch\ H_2O$

English Unit Conversions and Constants

1 therm = 100,000 Btu	1 dekatherm (dth) = 1 MMBtu
1 ton = 2,000 lbm	1 ft^3 = 1,728 in^3
1 ton of thermal energy = 12,000 Btu/hr	g_c = 32.2 ft lbm / lbf s^2
1 ft^3 = 7.4805 US gallon	M = 1,000 = 10^3
1 ft^2 = 144 in^2	MM = 1,000,000 = 10^6
1 hp = 33,000 ft lbf / min	MBH = 1,000 Btu/hr
1 ft^3/s = 448.83 gpm [U.S. gallons per minute]	
Boiler Horsepower = 34.5 lbm/hr saturated steam @ 14.696 psia = 34.5 × 970.3 = 33,475 Btu/hr [output]	

Useful English-to-International Unit Conversions

1 hp = 0.7457 kW
1 lbm = 0.4536 kg
1 lbf = 4.448 N
1 m = 3.2808 ft
1 m^2 = 10.76 ft^2
1 m^3 = 35.315 ft^3
1 kWh = 3,600 kJ = 3,412 Btu
1.055 GJ = 1 MMBtu = 1 × 10^6 Btu
°C = (°F – 32) / 1.8
1 W/m^2 °C = 5.676 Btu/hr ft^2 °F or RSI = R / 5.676

Part B: Thermodynamics

Temperature: $\mathrm{K} = {}^\circ\mathrm{C} + 273.15$ [International units: Kelvin is absolute temperature scale]

Temperature: $\mathrm{R} = {}^\circ\mathrm{F} + 459.67$ [English units: Rankine is absolute temperature scale]

Pressure: $P_{gage} = P_{abs} - P_{atm}$
International: kPag is gage pressure; kPa is absolute pressure
P_{atm} = 101.325 kPa = 101.325 kN/m^2 at sea level
P_{atm} = 0 kPa gage pressure at any elevation
1 bar = 100 kPa is *approximately* equal to one atmosphere of pressure at sea level

English: psig is gage pressure; psia is absolute pressure
Sea level: P_{atm} = 14.696 lbf/in^2 = 2,116 lbf/ft^2 = 406.78 inch H_2O = 14.696 psia

P_{atm} = 0.00 psig at any elevation

Water and Air at Standard Conditions
International: sea level pressure, temperature $\sim$ 20 °C
Water specific heat, C_p = 4.184 kJ / kg °C
Water density, ρ = 1.00 kg/L = 1,000 kg/m^3
Air density, ρ = 1.204 kg/m^3= 0.001204 kg/L
Air specific heat, C_p = 1.006 kJ / kg °C

English: sea level pressure, temperature $\sim$ 68 °F
Water specific heat, C_p = 1.00 Btu / lbm °F
Water density, ρ = 8.34 lbm / gal = 62.4 lbm/ft^3
Air density, ρ = 0.0752 lbm/ft^3
Air specific heat, C_p = 0.2403 Btu/ lbm °F

Quality of a pure substance (temperature-enthalpy or temperature specific volume charts)
x= quality = mass fraction of vapor for a pure substance, two phase mixture of liquid and vapor

$$\mathrm{h} = \mathrm{h_f} + x\mathrm{h_{fg}}$$

Enthalpy definition: $\mathrm{h} = \mathrm{u} + \mathrm{P}v$. From thermodynamics, u is the internal energy of a substance.

Conservation of Mass (steady state)
$\dot{\mathrm{m}} = \rho\dot{\mathcal{V}} = \rho \mathrm{VA} = \frac{\mathrm{VA}}{v}$ and $\dot{\mathcal{V}} = \mathrm{VA}$
steady state: $\dot{\mathrm{m}}_{in} = \dot{\mathrm{m}}_{out} = \dot{\mathrm{m}}$
A = cross section area, ft^2.
$A_{square\ or\ rectangular}$ = (height)(width)
$A_{circular}$ = (π/4)(diameter)^2

International :
$\dot{\mathrm{m}}$ = mass flow rate, kg/s
$\mathcal{V}$= volume, m^3
$\dot{\mathcal{V}}$= volume flow rates, m^3/s
v= specific volume, m^3/kg
ρ= density $= 1/v$, kg/m^3
V = velocity, m/s

English :
$\dot{\mathrm{m}}$ = mass flow rate, lbm/s
$\mathcal{V}$= volume, ft^3
$\dot{\mathcal{V}}$= volume flow rates, ft^3/s

v= specific volume, ft^3/lbm
ρ= density = $1/v$, lbm/ft^3
V = velocity, ft/s

Ideal gas law for air
$p = \rho RT$

International:
p = pressure, kPa
R = gas constant = 0.2870 kJ / kg K = 0.2870 kPa m^3 / kg K for air
T = absolute temperature, K
ρ = density, kg/m^3

Example for air at 1 atm & 20°C:
ρ=p/RT = 101.325 kPa/[(0.2870 kPa m^3/kg K)(273.15+20)K] = 1.204 kg/ m^3

English:
p = pressure, lbf/ft^2
R = gas constant = 53.35 ft-lbf / lbm-R for air
T = absolute temperature, R
ρ = density, lbm/ft^3

Example for air at 1 atm & 68°F:
ρ=p/RT =2,116 lbf/ft^2 / [(53.35 ft-lbf/lbm-R)(459.67+68)R] = 0.0752 lbm/ ft^3

First Law of Thermodynamics (steady state conservation of energy)

- Energy is neither created nor destroyed. It only changes into different forms.
- In thermodynamics, thermal energy or heat is defined as energy that moves due to a temperature difference. Work is defined as all other forms of energy, other than heat.
- Suggested text: Cengel & Boles, "Thermodynamics: An Engineering Approach" (10th ed, 2023) McGraw Hill

First law of thermodynamics for steady state, steady flow control volume (open system):

$$\dot{Q}_{in} + \dot{W}_{in} + \sum_{in} \dot{m}_{in}\left(h_{in} + \frac{V_{in}^2}{2} + gz_{in}\right)$$

$$= \dot{Q}_{out} + \dot{W}_{out} + \sum_{out} \dot{m}_{out}\left(h_{out} + \frac{V_{out}^2}{2} + gz_{out}\right)$$

Steady state heating or cooling of air or water flow with no work, negligible change in elevation, and negligible change in velocity is a common situation. The first law simplifies to:

$$\dot{Q}_{in} - \dot{Q}_{out} = \dot{Q} = \dot{m}\ (h_{out} - h_{in}) = \dot{m}\ \Delta h$$

If there is <u>only sensible heating or cooling</u> (no latent heat involved: no moisture added, no moisture removed):

$$\Delta h = c_P\ \Delta T \text{ and } \dot{Q}_{in} - \dot{Q}_{out} = \mathbf{\dot{Q} = \dot{m}\ c_P\ \Delta T}$$

Steady state change in thermal energy (sensible + latent) of air and water vapour mixtures (psychrometrics)

$$\dot{Q} = \dot{m}\ \Delta h = \dot{\mathcal{V}}\ \rho\ 60\ \Delta h$$

International:
- $\dot{Q}$ = thermal energy loss or gain, kJ/hr
- $\dot{\mathcal{V}}$ = actual dry air flow, m^3 / min
- ρ = density of dry air, kg / m^3 [psychrometric chart provides specific volume, $v = 1 / \rho$]
- 60 = minutes per hour
- Δh = change in enthalpy, kJ / kg of dry air [from psychrometric chart]

English:
- $\dot{Q}$ = thermal energy loss or gain, Btu/hr
- $\dot{\mathcal{V}}$ = actual dry air flow, ft^3 / min
- ρ = density of dry air, lbm / ft^3 [psychrometric chart provides specific volume, $v = 1 / \rho$]
- 60 = minutes per hour
- Δh = change in enthalpy, Btu/lbm of dry air [from psychrometric chart]

<u>Air</u> sensible only (no latent) thermal energy change

$$\dot{Q} = \dot{\mathcal{V}}\ \rho\ C_p\ \Delta T$$

International :
- $\dot{Q}$ = sensible energy change, kW
- $\dot{\mathcal{V}}$ = volume rate of air flow, m^3/s [note: 1 m^3/s = 1000 L/s]
- ρ = 1.204 kg/m^3 = density of air at standard conditions
- C_p = 1.006 kJ / kg °C = specific heat of air at standard conditions
- ΔT = temperature difference, °C

$\dot{Q} = 1.21\ \dot{V}\ \Delta T$

English :

$\dot{Q}$ = sensible energy change, Btu/hr

$\dot{V}$ = volume rate of air flow, ft^3/hr [note: ft^3/hr = ft^3/min × 60 min/hr]

ρ = 0.0752 lbm/ft^3 = density of air at standard conditions

C_p = 0.2403 Btu / lbm °F = specific heat of air at standard conditions

ΔT = temperature difference, °F

$\dot{Q} = 0.0180\ \dot{V}\ \Delta T$

If flow in ft^3/min: $\dot{Q} = 1.08\ \dot{V}\ \Delta T$

Water sensible only (no latent) thermal energy change

$$\dot{Q} = \dot{V}\ \rho\ C_p\ \Delta T$$

International:

$\dot{Q}$ = sensible energy change, kW

$\dot{V}$ = volume rate of water flow, L/s [note: 1 m^3/s = 1000 L/s]

ρ = 1.00 kg/L = density of water at standard conditions

C_p = 4.2 kJ / kg °C = specific heat of water at standard conditions

ΔT = temperature difference, °C

$\dot{Q} = 4.2\ \dot{V}\ \Delta T$

English:

$\dot{Q}$ = sensible energy change, Btu/hr

$\dot{V}$ = volume rate of water flow, ft^3/hr

note: ft^3/hr = ft^3/min × 60 min/hr and U.S. gal/min = ft^3/min × 7.48 gal/ft^3

ρ = 62.4 lbm/ft^3 = density of water at standard conditions

C_p = 1.00 Btu / lbm °F = specific heat of water at standard conditions

ΔT = temperature difference, °F

$\dot{Q} = 62.4\ \dot{V}\ \Delta T$

If flow in U.S. gal/min: $\dot{Q} = 500.5\ \dot{V}\ \Delta T$

Heating and Cooling Degree Days

International:

- DD units are: °C day / unit time [unit of time is often 1 yr, but not always]
- Reference temperature is 18.3 °C for ***heating*** or ***cooling*** unless otherwise specified
- Example: if average outdoor is 10 °C for 1 week then HDD = (18.3 - 10) × 7 = 58.1 °C day / week

English:

- DD units are: °F day / unit time [unit of time is often 1 yr, but not always]
- Reference temperature is 65 °F for ***heating*** and ***cooling*** unless otherwise specified
- Example: if average outdoor is 45 °F for 1 week then HDD = (65 - 45) × 7 = 140 °F day / week

Annual energy loss (heating) or gain (cooling) due to thermal energy transfer through envelope

$$\dot{Q} = U \ A \ DD \ 24$$

International:

$\dot{Q}$ = useful energy loss or gain, Wh / yr [note this is Wh, not kWh]
U = overall thermal conductance, W / m^2 °C
A = heat transfer area, m^2
DD = annual heating or cooling degree-days, °C day / yr
24 = hr/day

English:

$\dot{Q}$ = useful energy loss or gain, Btu/yr
U = overall thermal conductance, Btu / ft^2 hr °F
A = heat transfer area, ft^2
DD = annual heating or cooling degree-days, °F day / yr
24 = hr/day

Energy loss or gain due to air mass loss or gain through a hole or crack (*infiltration or exfiltration*)

$$\dot{Q} = \dot{\mathcal{V}} \ 1440 \ \rho \ C_p \ (HDD + CDD)$$

International:

$\dot{Q}$ = useful energy loss or gain, kJ/yr
$\dot{\mathcal{V}}$ = actual m^3 / min of dry air flow
1440 = number of minutes per day
ρ = density of air at standard conditions = 1.204 kg/ m^3
C_p = specific heat of air at standard conditions = 1.006 kJ / kg °C
HDD = annual heating degree-days, °C day / yr
CDD = annual cooling degree-days, °C day / yr

English:

$\dot{Q}$ = useful energy loss or gain, Btu/yr
$\dot{V}$ = actual ft^3 / min of dry air flow
1440 = number of minutes per day
ρ = density of air at standard conditions = 0.0752 lbm/ ft^3
C_p = specific heat of air at standard conditions = 0.2403 Btu / lbm °F
HDD = annual heating degree-days, °F day / yr
CDD = annual cooling degree-days, °F day / yr

Efficiency of Fuel Combustion:
Natural gas: $\eta_{HHV} = 0.91 \times \eta_{LHV}$
No. 2 Fuel Oil (diesel fuel): $\eta_{HHV} = 0.93 \times \eta_{LHV}$

Performance of vapour compression and other HVAC equipment

International:

COP = useful thermal energy delivered / input energy [dimensionless]
air conditioner or heat pump in cooling mode: $COP_{max} = T_{cold} / (T_{hot} - T_{cold})$
heat pump in heating mode: $COP_{max} = T_{hot} / (T_{hot} - T_{cold})$
note: must use absolute temperatures (Kelvin) in the COP_{max} formulas

$$\frac{kW_{input}}{kW_{thermal}} = \frac{kWh_{input}}{kWh_{thermal}} = \frac{1}{COP}$$

English:

EER = energy efficiency ratio
= useful thermal energy in Btu delivered per Wh input energy [Btu per Watt-hour]

COP is unitless. See International unit formulas above.

$$\frac{kW_{input}}{ton_{thermal}} = \frac{kWh_{input}}{ton.h_{thermal}} = \frac{12}{EER} = \frac{3.517}{COP}$$

HSPF = heating season performance factor = season heat output divided by Wh of energy input
HSPF = Btu of useful energy delivered per Wh input energy [Btu / Wh]
Seasonal average COP = HSPF / 3.412 = HSPF × 0.2931

Per 2020 ASHRAE Handbook—HVAC Systems and Equipment (SI), HSPF and EER are only defined in English units.

"ton" of thermal energy = 12,000 Btu/hr = energy to melt one ton of ice (2,000 lbm @ 32°F) in 24 hr

- ice heat of fusion @ 32°F = 144 Btu/lbm (335 kJ/kg @ 0 °C)

- 2,000 lbm × 144 Btu/lbm = 288,000 Btu
- 288,000 Btu / 24 hr = 12,000 Btu/hr

Part C: Fluid Mechanics

- Liquids and gases are described as "fluids"
- Basic 'law' of fluid mechanics: **fluids flow from high pressure to low pressure**

Energy Balance Equation for Fluids

- Fluid energy includes three terms: pressure, gravity, and kinetic energy
- Fluid energy is expressed in "head" which is energy per unit weight. English units are ft-lbf/lbf = ft. International units are N-m/N = m.
- Key assumptions: incompressible flow, steady state, neglect heat & other energy forms

Bernoulli's Energy Balance Equation (using words)

(initial fluid energy) – (pipe or duct friction losses) – (minor losses such as valves & bends)
+ (energy added by pumps) – (energy extracted by turbines) = (final fluid energy)

Bernoulli's Energy Balance Equation (using math)

$$\frac{p_1}{\gamma} + z_1 + \frac{V_1^2}{2g} - f\frac{L}{D}\frac{V^2}{2g} - K_{\mathrm{L}}\frac{V^2}{2g} + H_{\text{pump}} - H_{\text{turbine}} = \frac{p_2}{\gamma} + z_2 + \frac{V_2^2}{2g}$$

Subscript 1 = initial condition (upstream location along a streamline)
Subscript 2 = final condition (downstream location along a streamline)

International units:

p = pressure, kPa
z = height of fluid, m
g = gravity constant, 9.81 m/s^2
V = average velocity of fluid, m/s
f = dimensionless friction factor from Moody diagram (see any fluid mechanics text book)
L = length of pipe, m
D = internal diameter of pipe, m
K_L = dimensionless minor loss factor; lookup in table (see any fluid mechanics text book)

H_{pump} = head added by pump, m
$H_{turbine}$ = head extracted by turbine, m
γ = the specific weight of the fluid, N/m^3. Standard conditions:
γ_{water} = 9,810 N/m^3
$\gamma = sg\ \gamma_{water}$
sg = specific gravity for a liquid = γ of actual liquid / γ_{water} = dimensionless

English units:
p = pressure, lbf/ft^2
z = height of fluid, ft
g = gravity constant, 32.2 ft/s^2
V = average velocity of fluid, ft/s
f = dimensionless friction factor from Moody diagram (see any fluid mechanics text book)
L = length of pipe, ft
D = internal diameter of pipe, ft
K_L = dimensionless minor loss factor; lookup in table (see any fluid mechanics text book)
H_{pump} = head added by pump, ft
$H_{turbine}$ = head extracted by turbine, ft
γ = the specific weight of the fluid, lbf/ft^3. Standard conditions:
γ_{water} = 62.4 lbf/ft^3
$\gamma = sg\ \gamma_{water}$
sg = specific gravity for a liquid = γ of actual liquid / γ_{water} = dimensionless

Fluid power added by pumps or extracted by turbines

International units:
pumps: $P_{fluid} = \gamma\ \dot{V}\ H_{pump} / 1000$ is fluid power *added* by pump
turbines: $P_{fluid} = \gamma\ \dot{V}\ H_{turbine} / 1000$ is fluid power *extracted* by a turbine
P_{fluid} = fluid power, kW
γ = specific weight, N/m^3 = sg × γ_{water} [γ_{water} = 9,810 N/m^3]
$\dot{V}$ = volume flow rate of liquid, m^3/s
H_{pump} = fluid head added by pump, m
$H_{turbine}$ = fluid head extracted by turbine, m

English units:
pumps: $P_{fluid} = \gamma\ \dot{V}\ H_{pump} / 550$ is fluid power *added* by pump
turbines: $P_{fluid} = \gamma\ \dot{V}\ H_{turbine} / 550$ is fluid power *extracted* by a turbine

P_{fluid} = fluid power = hp
γ = lbf/ft^3 = sg $\times$ γ_{water} [γ_{water} = 62.4 lbf/ft^3]
$\dot{V}$ = volume flow rate of liquid, ft^3/s
H_{pump} = fluid head added by pump, ft
$H_{turbine}$ = fluid head extracted by turbine, ft
Common form of the above formula when $\dot{V}$ is in U.S. gal/min:
$P_{fluid} = sg\ \dot{V}\ H_p / 3{,}956$

Fluid power added by fans

International units:
$P_{fluid} = \Delta TP\ \dot{V} / 60$
P_{fluid} = air or gas fluid power, kW
ΔTP = fan total pressure rise, kPa
ΔTP is total fluid pressure added by the fan, approximated by fan static pressure from a fan curve
$\dot{V}$ = actual air or gas flow, m^3/min
60 = seconds per minute

English units:
$P_{fluid} = \Delta TP\ \dot{V} / 6{,}343$
P_{fluid} = air or gas fluid power, hp
ΔTP = fan total pressure rise, inches of water
ΔTP is total fluid pressure added by the fan, approximated by fan static pressure from a fan curve
$\dot{V}$ = actual air or gas flow, ft^3/min
6,343 = unit conversion constant

Part D: Heat Transfer

- Heat or thermal energy is energy in nature that flows due to a temperature difference.
- Basic 'law' of heat transfer: **heat (thermal energy) flows from high temperature to low temperature.**

Convection mode:

$$\dot{Q}_{conv} = h_{conv}\ A\ (T_h - T_c) = \frac{A}{R_{conv}}(T_h - T_c)$$

International units:
$\dot{Q}_{conv}$ = convective heat flow, W

h_{conv} = convective or film coefficient, W / m^2 °C
[note: h is also used for enthalpy]
A = surface area of solid surface to fluid interface, m^2
T_h = hot temperature, °C or K
T_c = cold temperature, °C or K
R_{conv} = 1/h_{conv} = convective film resistance, m^2 °C / W

English units:
$\dot{Q}_{conv}$ = convective heat flow, Btu/hr
h_{conv} = convective or film coefficient, Btu / hr ft^2 °F
[note: h is also used for enthalpy]
A = surface area of solid surface to fluid interface, ft^2
T_h = hot temperature, °F or R
T_c = cold temperature, °F or R
R_{conv} = 1/h_{conv} = convective film resistance, hr ft^2 °F / Btu

Conduction mode:

$$\dot{Q}_{cond} = \frac{k\,A}{t}\,(T_h - T_c) = \frac{A}{R_{cond}}\,(T_h - T_c)$$

International units:
$\dot{Q}_{cond}$ = conductive heat flow, W
k = thermal conductivity, W / m °C
L = material thickness, m [note: 1000 mm = 100 cm = 1 m]
A = cross section area of heat flow, m^2
T_h = hot temperature, °C or K
T_c = cold temperature, °C or K
R_{cond} = L / k = conductive resistance, m^2 °C / W
Metric conduction resistance is usually described as RSI.
C = 1/R_{cond} is thermal conductance for solids, W / m^2 °C

English units:
$\dot{Q}_{cond}$ = conductive heat flow, Btu/hr
k = thermal conductivity, Btu / hr ft °F
L = material thickness, ft
A = cross section area of heat flow, ft^2
T_h = hot temperature, °F or R
T_c = cold temperature, °F or R
R_{cond} = L / k = conductive resistance, hr ft^2 °F / Btu
C = 1/R_{cond} is thermal conductance for solids, Btu / hr ft^2 °F

Radiation mode:

$$\dot{Q}_{rad} = A \; (T_h^4 - T_c^4) = \frac{A}{R_{rad}} (T_h - T_c)$$

International units:

$\dot{Q}_{rad}$ = radiative heat flow, W

ϵ = thermal emissivity, dimensionless $0 \le \epsilon \le 1$

σ = Stefan-Boltzmann constant = 5.67×10^{-8} W / m^2 K^4

A = cross section area of heat flow, m^2

T_h = hot temperature, K. **Must use absolute temperature for radiation.** Do not use °C

T_c = cold temperature, K. **Must use absolute temperature for radiation.** Do not use °C

Radiation resistance is seldom used due to the 4th power mathematics. But if temperatures are known, Radiative resistance equation is:

$R_{rad} = \frac{1}{(T_h^2 + T_c^2)(T_h + T_c)}$ m^2 K / W

English units:

$\dot{Q}_{rad}$ = radiative heat flow, Btu/hr

ϵ = thermal emissivity, dimensionless $0 \le \epsilon \le 1$

σ = Stefan-Boltzmann constant = 0.1713×10^{-8} Btu / hr ft^2 R^4

A = cross section area of heat flow, ft^2

T_h = hot temperature, R. **Must use absolute temperature for radiation.** Do not use °F

T_c = cold temperature, R. **Must use absolute temperature for radiation.** Do not use °F

Radiation resistance is seldom used due to the 4th power mathematics. But if temperatures are known, Radiative resistance equation is:[1]

$R_{rad} = \frac{1}{(T_h^2 + T_c^2)(T_h + T_c)}$ hr ft^2 R / Btu

Multi-mode thermal resistor model for heat transfer:

$$\dot{Q}_{total} = \frac{A}{\sum R} (T_h - T_c) = U \, A \, (T_h - T_c)$$

International units:

$\sum R$ = sum of all heat transfer resistance values, m^2 °C / W

[1] R_{rad} is derived using the math relation: $(T_h^4 - T_c^4) = ((T_h^2 + T_c^2)(T_h + T_c))(T_h - T_c)$

$U = \frac{1}{\sum R}$ is defined as *overall thermal conductance*, W / m^2 °C

English units:

$\sum R$ = sum of all heat transfer resistance values, hr ft^2 °F / Btu

$U = \frac{1}{\sum R}$ is defined as *overall thermal conductance*, Btu / hr ft^2 °F

Heat exchangers:

Heat exchanger effectiveness = $\dot{Q}_{actual}$ / $\dot{Q}_{max}$

$\dot{Q}_{actual}$ = actual heat transfer

$\dot{Q}_{max}$ = maximum heat transfer

Part E: Electric power and equipment

This part provides the common, approximate formulas for estimating power quantities in fluid moving systems (pumps, fans, compressors) powered by electric motors (applicable to Y or delta alternating current designs). The simplified diagram illustrates the typical arrangement.

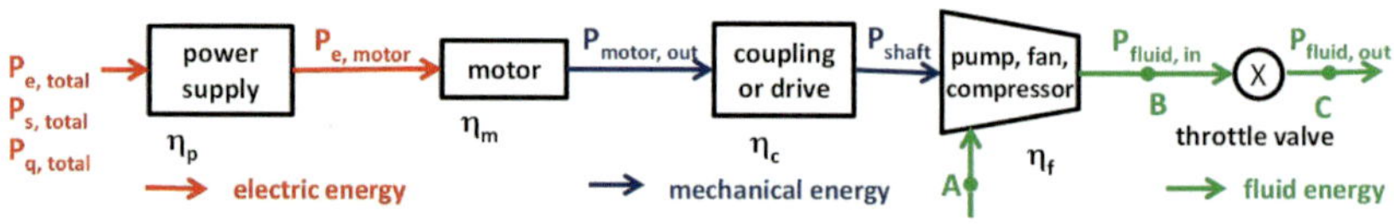

Variables (International units only):

P_e = real or active power kW. Real kWh used for billing purposes by the utility

P_s = total or apparent power, kVA. Peak kVA often used for demand charges by the utility

P_q = reactive power, kVAr

PF = power factor [usually lagging power factor; PF $<$ 1 for inductive loads; PF = 1 for resistive loads]

$P_{e,\ motor}$ = real electric power input to the motor, kW

$P_{motor,\ out}$ = mechanical power out of the motor, kW note: **motors are always rated as output power**

$T_{motor,\ out}$ = torque delivered by the motor, Nm

$P_{motor,\ nameplate}$ = nominal, full load[2] output power, kW. Nameplate is usually stated in hp; convert to kW.

[2] "Full load" here is defined as motor loaded at its nameplate value. This does not necessarily equal full load for pumps, fans, or compressors.

P_{shaft} = power input to pump, fan, or compressor, kW
P_{fluid} = power added to increase pressure and move liquids and gases[3], kW.
$I_{nameplate}$ = full load motor active or "absorbed" current, from nameplate, amps
I_L = measured line current input to the system, amps
V_L = measured line voltage input to the system, Volts
ω = motor or fluid mover shaft rotational speed, rev/min or RPM
MLF = motor load factor = operating load / rated load
η_p = electric power supply efficiency [e.g. transformer or variable frequency drive]
η_m = motor efficiency [varies substantially with operating conditions such as motor load]
η_c = coupling or drive efficiency [e.g. belt drive]
η_f = fluid mover efficiency[4] [compressor, pump or fan; efficiency varies substantially based on design and operating conditions]

Electric and Mechanical Power Equations

1. $P_s = (3)^{1/2}\ I_L\ V_L / 1000$ for 3-phase AC systems
2. $P_s = I_L\ V_L / 1000$ for 1-phase AC systems
3. $P_e = I_L\ V_L / 1000$ for DC systems
4. $P_s^2 = P_e^2 + P_q^2$
5. $PF = P_e / P_s$
6. $P_{motor,\ out} = (\omega\ T_{motor,\ out}) / 9{,}550$
   In English units: $P_{motor,\ out} = (\omega\ T_{motor,\ out}) / 5{,}245$ in units of hp, with torque in units of ft-lbf
7. $P_{e,\ total} = P_{e,motor} / \eta_p$
8. $P_{e,motor} = P_{motor,\ out} / \eta_m$
9. $P_{motor,\ out} = P_{shaft} / \eta_c$
10. $P_{shaft} = P_{fluid} / \eta_f$
11. $MLF = P_{motor,out} / P_{motor,\ nameplate}$
12. $P_{motor,\ out} = T_{motor,\ out}\ \omega_{meas} / 9550$

---

[3] Pfluid can also be calculated using fluid head term, Hpump or Hfan, from the Bernoulli equation: Pfluid = (specific weight) x (flow rate) x (fluid head). Pump manufacturers usually exclude valve losses from pump curves, and include valve losses in system curves. Fan manufacturers usually include inlet vanes or control damper losses in fan curves, and exclude vane or control damper losses from system curves.

[4] This is the conservation of energy law from physics, ratio of output to input energy. It is not the same as isentropic efficiency or other ways of quantifying efficiency for pumps, fans, or compressors.

13. $P_{shaft2} = P_{shaft1} (\omega_2/\omega_1)^n$
    n = 3 is affinity "law" for centrifugal systems
    is only true when static head or gravity forces are negligible (i.e. kinetic energy dominates)
14. $P_{e2,\ total} = P_{e1,\ total} (\omega_2/\omega_1)^n$
    n $\approx$ 1.8 to 2.8 when static head or gravity dominate for centrifugal systems
15. MLF can be determined using three different measurements (current, shaft speed, or electric power)
    $\approx (I_{L,motor} / I_{nameplate})\ (V_{L,motor} / V_{nameplate})$ is less accurate and should only use **when it is known that MLF > 50%**
    $\approx (\omega_{sync} - \omega_{meas})/(\omega_{sync} - \omega_{nameplate})$
    $\approx (P_{motor,\ in})/(P_{motor,in\ @full\ load})$

Part F: Solar Energy

Solar PV

NOCT = normal operating cell temperature
P_{mp} = maximum power point (point on I-V curve where load is drawing maximum power), Watts
P_{DC} = direct current power, Watts (W_{DC})
P_{AC} = alternating current power, Watts (W_{AC})
I_{SC} = short circuit current, amps
I_{mp} = current which corresponds to P_{mp}, amps
I_{DC} = direct current, amps
I_{AC} = alternating current, amps
V_{OC} = open circuit voltage, Volts
V_{DC} = direct current voltage, Volts
V_{AC} = alternating current voltage, Volts
$T_{ambient}$ = environment or air temperature surrounding the PV module
T_{cell} = PV cell temperature
S = solar radiation, kWh/m^2

$$T_{cell} = T_{ambient} + \frac{NOCT - 20}{0.80} S$$

Solar Thermal

$$\dot{Q}_{useful\ heat} = \dot{Q}_{absorbed\ solar} - \dot{Q}_{lost\ to\ ambient}$$
$$= \alpha\tau AI - AU\,(T_{collector} - T_{ambient})$$

α = absorptivity coefficient
τ = transmissivity coefficient
A = area, m^2
I = solar irradiation, W/m^2
U = heat loss coefficient, $W/m^2\ ^\circ C$
$T_{collector}$ = temperature of the collector, °C
$T_{ambient}$ = temperature of the ambient, °C
$\dot{Q}_{useful\ heat}$ = rate of useful heat available from the collector, W
$\dot{Q}_{absorbed\ solar}$ = rate of heat absorption by the collector, W
$\dot{Q}_{lost\ to\ ambient}$ = rate of heat loss to the ambient, W

$$\dot{Q}_{useful\ heat} = \dot{m} c_p \left(T_{fluid\ out} - T_{fluid\ in}\right)$$

$\dot{m}$ = mass flow rate, kg/s
c_p = fluid heat capacity, kJ/kg°C
$T_{fluid\ out}$ = temperature output fluid
$T_{fluid\ in}$ = temperature input fluid

$$\eta_{collector} = \frac{\dot{Q}_{useful\ heat}}{AI} = \alpha\tau - U\frac{T_{collector} - T_{ambient}}{I}$$

$\eta_{collector}$ = collector's efficiency

$$\eta_{collector} = F_R\alpha\tau - F_R U\frac{T_{fluid\ in} - T_{ambient}}{I}$$

F_R = heat removal factors

Appendix B

State	Fuel	HHV MJ/m3	HHV MJ/kg	HHV Btu/ft3	HHV Btu/gal	HHV Btu/lbm	LHV MJ/m3	LHV MJ/kg	LHV Btu/ft3	LHV Btu/gal	LHV Btu/lbm
gas	natural gas	38.5	54.2	1,034		23,310	35.1	49.3	941		21,210
gas	renewable natural gas	39.8	55.5	1,069		23,860	35.9	50.0	964		21,500
gas	methane	39.8	55.5	1,069		23,860	35.9	50.0	964		21,500
gas	hydrogen	12.7	141.7	341		60,920	10.8	120.0	290		51,590
gas	landfill gas	19.3		517			17.5		471		
liquid	#6 oil (residual fuel oil)	40,900	41.8	1,099,000	146,900	17,970	38,230	39.0	1,026,000	137,200	16,770
liquid	#2 oil (diesel)	38,600	45.6	1,035,000	138,400	19,600	36,040	42.6	967,300	129,300	18,310
liquid	propane	25,100	50.4	673,000	90,000	21,670	23,080	46.4	619,500	82,800	19,950
liquid	crude oil	38,000	45.9	1,019,000	136,200	19,730	36,140	42.8	970,000	129,700	18,400
liquid	gasoline	34,200	46.4	917,800	122,700	19,950	31,990	43.4	858,500	114,800	18,660
liquid	jet kerosene	37,900					35,300				
liquid	LPG	26,500	49.3	710,500	95,000	21,190	24,440	45.5	655,800	87,700	19,560
liquid	liquified natural gas (-160 °C; 1 atm)	23,300	54.3	625,327	83,600	23,340	20,520	49.4	550,650	73,600	21,240
solid	coal: lignite		14.0			6,020					
solid	coal: subbituminous		24.4			10,490					
solid	coal: bituminous		30.2			12,980		29.0			12,470
solid	coal: anthracite		32.6			14,020					
solid	dry wood, low estimate		16.2			6,960		15.4			6,620
solid	dry wood, high estimate		21.0			9,030		20.0			8,580
solid	dry wood, average		18.6			8,000		17.7			7,600
solid	wet wood (20% H2O wet basis), low estimate		13.0			5,570					
solid	wet wood (20% H2O wet basis), high estimate		16.8			7,220					
solid	wet wood (20% H2O wet basis), average		14.9			6,400					
solid	municipal solid waste		11.8			5,070					

HHV = higher heating value; is sometimes described as GCV = gross calorific value

HHV or GCV both refer to energy released from combustion where H2O in the exhaust is condensed to a liquid.

LHV = lower heating value; is sometimes described as NCV = net calorific value

LHV or NCV both refer to energy released from combustion where H2O is exhausted as a vapour.

Appendix C

1.0% = i, interest rate						
number of time periods, n	F/P(i%,n)	P/F(i%,n)	F/A(i%,n)	A/F(i%,n)	P/A(i%,n)	A/P(i%,n)
1	1.01000	0.99010	1.00000	1.00000	0.99010	1.01000
2	1.02010	0.98030	2.01000	0.49751	1.97040	0.50751
3	1.03030	0.97059	3.03010	0.33002	2.94099	0.34002
4	1.04060	0.96098	4.06040	0.24628	3.90197	0.25628
5	1.05101	0.95147	5.10101	0.19604	4.85343	0.20604
6	1.06152	0.94205	6.15202	0.16255	5.79548	0.17255
7	1.07214	0.93272	7.21354	0.13863	6.72819	0.14863
8	1.08286	0.92348	8.28567	0.12069	7.65168	0.13069
9	1.09369	0.91434	9.36853	0.10674	8.56602	0.11674
10	1.10462	0.90529	10.46221	0.09558	9.47130	0.10558
11	1.11567	0.89632	11.56683	0.08645	10.36763	0.09645
12	1.12683	0.88745	12.68250	0.07885	11.25508	0.08885
13	1.13809	0.87866	13.80933	0.07241	12.13374	0.08241
14	1.14947	0.86996	14.94742	0.06690	13.00370	0.07690
15	1.16097	0.86135	16.09690	0.06212	13.86505	0.07212
16	1.17258	0.85282	17.25786	0.05794	14.71787	0.06794
17	1.18430	0.84438	18.43044	0.05426	15.56225	0.06426
18	1.19615	0.83602	19.61475	0.05098	16.39827	0.06098
19	1.20811	0.82774	20.81090	0.04805	17.22601	0.05805
20	1.22019	0.81954	22.01900	0.04542	18.04555	0.05542
21	1.23239	0.81143	23.23919	0.04303	18.85698	0.05303
22	1.24472	0.80340	24.47159	0.04086	19.66038	0.05086
23	1.25716	0.79544	25.71630	0.03889	20.45582	0.04889
24	1.26973	0.78757	26.97346	0.03707	21.24339	0.04707
25	1.28243	0.77977	28.24320	0.03541	22.02316	0.04541
26	1.29526	0.77205	29.52563	0.03387	22.79520	0.04387
27	1.30821	0.76440	30.82089	0.03245	23.55961	0.04245
28	1.32129	0.75684	32.12910	0.03112	24.31644	0.04112
29	1.33450	0.74934	33.45039	0.02990	25.06579	0.03990
30	1.34785	0.74192	34.78489	0.02875	25.80771	0.03875
31	1.36133	0.73458	36.13274	0.02768	26.54229	0.03768
32	1.37494	0.72730	37.49407	0.02667	27.26959	0.03667
33	1.38869	0.72010	38.86901	0.02573	27.98969	0.03573
34	1.40258	0.71297	40.25770	0.02484	28.70267	0.03484
35	1.41660	0.70591	41.66028	0.02400	29.40858	0.03400
36	1.43077	0.69892	43.07688	0.02321	30.10751	0.03321
37	1.44508	0.69200	44.50765	0.02247	30.79951	0.03247
38	1.45953	0.68515	45.95272	0.02176	31.48466	0.03176
39	1.47412	0.67837	47.41225	0.02109	32.16303	0.03109
40	1.48886	0.67165	48.88637	0.02046	32.83469	0.03046
41	1.50375	0.66500	50.37524	0.01985	33.49969	0.02985
42	1.51879	0.65842	51.87899	0.01928	34.15811	0.02928
43	1.53398	0.65190	53.39778	0.01873	34.81001	0.02873
44	1.54932	0.64545	54.93176	0.01820	35.45545	0.02820
45	1.56481	0.63905	56.48107	0.01771	36.09451	0.02771
46	1.58046	0.63273	58.04589	0.01723	36.72724	0.02723
47	1.59626	0.62646	59.62634	0.01677	37.35370	0.02677
48	1.61223	0.62026	61.22261	0.01633	37.97396	0.02633
49	1.62835	0.61412	62.83483	0.01591	38.58808	0.02591
50	1.64463	0.60804	64.46318	0.01551	39.19612	0.02551
55	1.72852	0.57853	72.85246	0.01373	42.14719	0.02373
60	1.81670	0.55045	81.66967	0.01224	44.95504	0.02224
65	1.90937	0.52373	90.93665	0.01100	47.62661	0.02100
70	2.00676	0.49831	100.67634	0.00993	50.16851	0.01993
75	2.10913	0.47413	110.91285	0.00902	52.58705	0.01902
80	2.21672	0.45112	121.67152	0.00822	54.88821	0.01822
85	2.32979	0.42922	132.97900	0.00752	57.07768	0.01752
90	2.44863	0.40839	144.86327	0.00690	59.16088	0.01690
95	2.57354	0.38857	157.35376	0.00636	61.14298	0.01636
100	2.70481	0.36971	170.48138	0.00587	63.02888	0.01587

2.0% = i, interest rate							
number of time periods, n	F/P(i%,n)	P/F(i%,n)	F/A(i%,n)	A/F(i%,n)	P/A(i%,n)	A/P(i%,n)
1	1.02000	0.98039	1.00000	1.00000	0.98039	1.02000
2	1.04040	0.96117	2.02000	0.49505	1.94156	0.51505
3	1.06121	0.94232	3.06040	0.32675	2.88388	0.34675
4	1.08243	0.92385	4.12161	0.24262	3.80773	0.26262
5	1.10408	0.90573	5.20404	0.19216	4.71346	0.21216
6	1.12616	0.88797	6.30812	0.15853	5.60143	0.17853
7	1.14869	0.87056	7.43428	0.13451	6.47199	0.15451
8	1.17166	0.85349	8.58297	0.11651	7.32548	0.13651
9	1.19509	0.83676	9.75463	0.10252	8.16224	0.12252
10	1.21899	0.82035	10.94972	0.09133	8.98259	0.11133
11	1.24337	0.80426	12.16872	0.08218	9.78685	0.10218
12	1.26824	0.78849	13.41209	0.07456	10.57534	0.09456
13	1.29361	0.77303	14.68033	0.06812	11.34837	0.08812
14	1.31948	0.75788	15.97394	0.06260	12.10625	0.08260
15	1.34587	0.74301	17.29342	0.05783	12.84926	0.07783
16	1.37279	0.72845	18.63929	0.05365	13.57771	0.07365
17	1.40024	0.71416	20.01207	0.04997	14.29187	0.06997
18	1.42825	0.70016	21.41231	0.04670	14.99203	0.06670
19	1.45681	0.68643	22.84056	0.04378	15.67846	0.06378
20	1.48595	0.67297	24.29737	0.04116	16.35143	0.06116
21	1.51567	0.65978	25.78332	0.03878	17.01121	0.05878
22	1.54598	0.64684	27.29898	0.03663	17.65805	0.05663
23	1.57690	0.63416	28.84496	0.03467	18.29220	0.05467
24	1.60844	0.62172	30.42186	0.03287	18.91393	0.05287
25	1.64061	0.60953	32.03030	0.03122	19.52346	0.05122
26	1.67342	0.59758	33.67091	0.02970	20.12104	0.04970
27	1.70689	0.58586	35.34432	0.02829	20.70690	0.04829
28	1.74102	0.57437	37.05121	0.02699	21.28127	0.04699
29	1.77584	0.56311	38.79223	0.02578	21.84438	0.04578
30	1.81136	0.55207	40.56808	0.02465	22.39646	0.04465
31	1.84759	0.54125	42.37944	0.02360	22.93770	0.04360
32	1.88454	0.53063	44.22703	0.02261	23.46833	0.04261
33	1.92223	0.52023	46.11157	0.02169	23.98856	0.04169
34	1.96068	0.51003	48.03380	0.02082	24.49859	0.04082
35	1.99989	0.50003	49.99448	0.02000	24.99862	0.04000
36	2.03989	0.49022	51.99437	0.01923	25.48884	0.03923
37	2.08069	0.48061	54.03425	0.01851	25.96945	0.03851
38	2.12230	0.47119	56.11494	0.01782	26.44064	0.03782
39	2.16474	0.46195	58.23724	0.01717	26.90259	0.03717
40	2.20804	0.45289	60.40198	0.01656	27.35548	0.03656
41	2.25220	0.44401	62.61002	0.01597	27.79949	0.03597
42	2.29724	0.43530	64.86222	0.01542	28.23479	0.03542
43	2.34319	0.42677	67.15947	0.01489	28.66156	0.03489
44	2.39005	0.41840	69.50266	0.01439	29.07996	0.03439
45	2.43785	0.41020	71.89271	0.01391	29.49016	0.03391
46	2.48661	0.40215	74.33056	0.01345	29.89231	0.03345
47	2.53634	0.39427	76.81718	0.01302	30.28658	0.03302
48	2.58707	0.38654	79.35352	0.01260	30.67312	0.03260
49	2.63881	0.37896	81.94059	0.01220	31.05208	0.03220
50	2.69159	0.37153	84.57940	0.01182	31.42361	0.03182
55	2.97173	0.33650	98.58653	0.01014	33.17479	0.03014
60	3.28103	0.30478	114.05154	0.00877	34.76089	0.02877
65	3.62252	0.27605	131.12616	0.00763	36.19747	0.02763
70	3.99956	0.25003	149.97791	0.00667	37.49862	0.02667
75	4.41584	0.22646	170.79177	0.00586	38.67711	0.02586
80	4.87544	0.20511	193.77196	0.00516	39.74451	0.02516
85	5.38288	0.18577	219.14394	0.00456	40.71129	0.02456
90	5.94313	0.16826	247.15666	0.00405	41.58693	0.02405
95	6.56170	0.15240	278.08496	0.00360	42.38002	0.02360
100	7.24465	0.13803	312.23231	0.00320	43.09835	0.02320

3.0% = i, interest rate

number of time periods, n	F/P(i%,n)	P/F(i%,n)	F/A(i%,n)	A/F(i%,n)	P/A(i%,n)	A/P(i%,n)
1	1.03000	0.97087	1.00000	1.00000	0.97087	1.03000
2	1.06090	0.94260	2.03000	0.49261	1.91347	0.52261
3	1.09273	0.91514	3.09090	0.32353	2.82861	0.35353
4	1.12551	0.88849	4.18363	0.23903	3.71710	0.26903
5	1.15927	0.86261	5.30914	0.18835	4.57971	0.21835
6	1.19405	0.83748	6.46841	0.15460	5.41719	0.18460
7	1.22987	0.81309	7.66246	0.13051	6.23028	0.16051
8	1.26677	0.78941	8.89234	0.11246	7.01969	0.14246
9	1.30477	0.76642	10.15911	0.09843	7.78611	0.12843
10	1.34392	0.74409	11.46388	0.08723	8.53020	0.11723
11	1.38423	0.72242	12.80780	0.07808	9.25262	0.10808
12	1.42576	0.70138	14.19203	0.07046	9.95400	0.10046
13	1.46853	0.68095	15.61779	0.06403	10.63496	0.09403
14	1.51259	0.66112	17.08632	0.05853	11.29607	0.08853
15	1.55797	0.64186	18.59891	0.05377	11.93794	0.08377
16	1.60471	0.62317	20.15688	0.04961	12.56110	0.07961
17	1.65285	0.60502	21.76159	0.04595	13.16612	0.07595
18	1.70243	0.58739	23.41444	0.04271	13.75351	0.07271
19	1.75351	0.57029	25.11687	0.03981	14.32380	0.06981
20	1.80611	0.55368	26.87037	0.03722	14.87747	0.06722
21	1.86029	0.53755	28.67649	0.03487	15.41502	0.06487
22	1.91610	0.52189	30.53678	0.03275	15.93692	0.06275
23	1.97359	0.50669	32.45288	0.03081	16.44361	0.06081
24	2.03279	0.49193	34.42647	0.02905	16.93554	0.05905
25	2.09378	0.47761	36.45926	0.02743	17.41315	0.05743
26	2.15659	0.46369	38.55304	0.02594	17.87684	0.05594
27	2.22129	0.45019	40.70963	0.02456	18.32703	0.05456
28	2.28793	0.43708	42.93092	0.02329	18.76411	0.05329
29	2.35657	0.42435	45.21885	0.02211	19.18845	0.05211
30	2.42726	0.41199	47.57542	0.02102	19.60044	0.05102
31	2.50008	0.39999	50.00268	0.02000	20.00043	0.05000
32	2.57508	0.38834	52.50276	0.01905	20.38877	0.04905
33	2.65234	0.37703	55.07784	0.01816	20.76579	0.04816
34	2.73191	0.36604	57.73018	0.01732	21.13184	0.04732
35	2.81386	0.35538	60.46208	0.01654	21.48722	0.04654
36	2.89828	0.34503	63.27594	0.01580	21.83225	0.04580
37	2.98523	0.33498	66.17422	0.01511	22.16724	0.04511
38	3.07478	0.32523	69.15945	0.01446	22.49246	0.04446
39	3.16703	0.31575	72.23423	0.01384	22.80822	0.04384
40	3.26204	0.30656	75.40126	0.01326	23.11477	0.04326
41	3.35990	0.29763	78.66330	0.01271	23.41240	0.04271
42	3.46070	0.28896	82.02320	0.01219	23.70136	0.04219
43	3.56452	0.28054	85.48389	0.01170	23.98190	0.04170
44	3.67145	0.27237	89.04841	0.01123	24.25427	0.04123
45	3.78160	0.26444	92.71986	0.01079	24.51871	0.04079
46	3.89504	0.25674	96.50146	0.01036	24.77545	0.04036
47	4.01190	0.24926	100.39650	0.00996	25.02471	0.03996
48	4.13225	0.24200	104.40840	0.00958	25.26671	0.03958
49	4.25622	0.23495	108.54065	0.00921	25.50166	0.03921
50	4.38391	0.22811	112.79687	0.00887	25.72976	0.03887
55	5.08215	0.19677	136.07162	0.00735	26.77443	0.03735
60	5.89160	0.16973	163.05344	0.00613	27.67556	0.03613
65	6.82998	0.14641	194.33276	0.00515	28.45289	0.03515
70	7.91782	0.12630	230.59406	0.00434	29.12342	0.03434
75	9.17893	0.10895	272.63086	0.00367	29.70183	0.03367
80	10.64089	0.09398	321.36302	0.00311	30.20076	0.03311
85	12.33571	0.08107	377.85695	0.00265	30.63115	0.03265
90	14.30047	0.06993	443.34890	0.00226	31.00241	0.03226
95	16.57816	0.06032	519.27203	0.00193	31.32266	0.03193
100	19.21863	0.05203	607.28773	0.00165	31.59891	0.03165

4.0% = i, interest rate						
number of time periods, n	**F/P(i%,n)**	**P/F(i%,n)**	**F/A(i%,n)**	**A/F(i%,n)**	**P/A(i%,n)**	**A/P(i%,n)**
1	1.04000	0.96154	1.00000	1.00000	0.96154	1.04000
2	1.08160	0.92456	2.04000	0.49020	1.88609	0.53020
3	1.12486	0.88900	3.12160	0.32035	2.77509	0.36035
4	1.16986	0.85480	4.24646	0.23549	3.62990	0.27549
5	1.21665	0.82193	5.41632	0.18463	4.45182	0.22463
6	1.26532	0.79031	6.63298	0.15076	5.24214	0.19076
7	1.31593	0.75992	7.89829	0.12661	6.00205	0.16661
8	1.36857	0.73069	9.21423	0.10853	6.73274	0.14853
9	1.42331	0.70259	10.58280	0.09449	7.43533	0.13449
10	1.48024	0.67556	12.00611	0.08329	8.11090	0.12329
11	1.53945	0.64958	13.48635	0.07415	8.76048	0.11415
12	1.60103	0.62460	15.02581	0.06655	9.38507	0.10655
13	1.66507	0.60057	16.62684	0.06014	9.98565	0.10014
14	1.73168	0.57748	18.29191	0.05467	10.56312	0.09467
15	1.80094	0.55526	20.02359	0.04994	11.11839	0.08994
16	1.87298	0.53391	21.82453	0.04582	11.65230	0.08582
17	1.94790	0.51337	23.69751	0.04220	12.16567	0.08220
18	2.02582	0.49363	25.64541	0.03899	12.65930	0.07899
19	2.10685	0.47464	27.67123	0.03614	13.13394	0.07614
20	2.19112	0.45639	29.77808	0.03358	13.59033	0.07358
21	2.27877	0.43883	31.96920	0.03128	14.02916	0.07128
22	2.36992	0.42196	34.24797	0.02920	14.45112	0.06920
23	2.46472	0.40573	36.61789	0.02731	14.85684	0.06731
24	2.56330	0.39012	39.08260	0.02559	15.24696	0.06559
25	2.66584	0.37512	41.64591	0.02401	15.62208	0.06401
26	2.77247	0.36069	44.31174	0.02257	15.98277	0.06257
27	2.88337	0.34682	47.08421	0.02124	16.32959	0.06124
28	2.99870	0.33348	49.96758	0.02001	16.66306	0.06001
29	3.11865	0.32065	52.96629	0.01888	16.98371	0.05888
30	3.24340	0.30832	56.08494	0.01783	17.29203	0.05783
31	3.37313	0.29646	59.32834	0.01686	17.58849	0.05686
32	3.50806	0.28506	62.70147	0.01595	17.87355	0.05595
33	3.64838	0.27409	66.20953	0.01510	18.14765	0.05510
34	3.79432	0.26355	69.85791	0.01431	18.41120	0.05431
35	3.94609	0.25342	73.65222	0.01358	18.66461	0.05358
36	4.10393	0.24367	77.59831	0.01289	18.90828	0.05289
37	4.26809	0.23430	81.70225	0.01224	19.14258	0.05224
38	4.43881	0.22529	85.97034	0.01163	19.36786	0.05163
39	4.61637	0.21662	90.40915	0.01106	19.58448	0.05106
40	4.80102	0.20829	95.02552	0.01052	19.79277	0.05052
41	4.99306	0.20028	99.82654	0.01002	19.99305	0.05002
42	5.19278	0.19257	104.81960	0.00954	20.18563	0.04954
43	5.40050	0.18517	110.01238	0.00909	20.37079	0.04909
44	5.61652	0.17805	115.41288	0.00866	20.54884	0.04866
45	5.84118	0.17120	121.02939	0.00826	20.72004	0.04826
46	6.07482	0.16461	126.87057	0.00788	20.88465	0.04788
47	6.31782	0.15828	132.94539	0.00752	21.04294	0.04752
48	6.57053	0.15219	139.26321	0.00718	21.19513	0.04718
49	6.83335	0.14634	145.83373	0.00686	21.34147	0.04686
50	7.10668	0.14071	152.66708	0.00655	21.48218	0.04655
55	8.64637	0.11566	191.15917	0.00523	22.10861	0.04523
60	10.51963	0.09506	237.99069	0.00420	22.62349	0.04420
65	12.79874	0.07813	294.96838	0.00339	23.04668	0.04339
70	15.57162	0.06422	364.29046	0.00275	23.39451	0.04275
75	18.94525	0.05278	448.63137	0.00223	23.68041	0.04223
80	23.04980	0.04338	551.24498	0.00181	23.91539	0.04181
85	28.04360	0.03566	676.09012	0.00148	24.10853	0.04148
90	34.11933	0.02931	827.98333	0.00121	24.26728	0.04121
95	41.51139	0.02409	1,012.78465	0.00099	24.39776	0.04099
100	50.50495	0.01980	1,237.62370	0.00081	24.50500	0.04081

5.0% = i, interest rate

number of time periods, n	F/P(i%,n)	P/F(i%,n)	F/A(i%,n)	A/F(i%,n)	P/A(i%,n)	A/P(i%,n)
1	1.05000	0.95238	1.00000	1.00000	0.95238	1.05000
2	1.10250	0.90703	2.05000	0.48780	1.85941	0.53780
3	1.15763	0.86384	3.15250	0.31721	2.72325	0.36721
4	1.21551	0.82270	4.31013	0.23201	3.54595	0.28201
5	1.27628	0.78353	5.52563	0.18097	4.32948	0.23097
6	1.34010	0.74622	6.80191	0.14702	5.07569	0.19702
7	1.40710	0.71068	8.14201	0.12282	5.78637	0.17282
8	1.47746	0.67684	9.54911	0.10472	6.46321	0.15472
9	1.55133	0.64461	11.02656	0.09069	7.10782	0.14069
10	1.62889	0.61391	12.57789	0.07950	7.72173	0.12950
11	1.71034	0.58468	14.20679	0.07039	8.30641	0.12039
12	1.79586	0.55684	15.91713	0.06283	8.86325	0.11283
13	1.88565	0.53032	17.71298	0.05646	9.39357	0.10646
14	1.97993	0.50507	19.59863	0.05102	9.89864	0.10102
15	2.07893	0.48102	21.57856	0.04634	10.37966	0.09634
16	2.18287	0.45811	23.65749	0.04227	10.83777	0.09227
17	2.29202	0.43630	25.84037	0.03870	11.27407	0.08870
18	2.40662	0.41552	28.13238	0.03555	11.68959	0.08555
19	2.52695	0.39573	30.53900	0.03275	12.08532	0.08275
20	2.65330	0.37689	33.06595	0.03024	12.46221	0.08024
21	2.78596	0.35894	35.71925	0.02800	12.82115	0.07800
22	2.92526	0.34185	38.50521	0.02597	13.16300	0.07597
23	3.07152	0.32557	41.43048	0.02414	13.48857	0.07414
24	3.22510	0.31007	44.50200	0.02247	13.79864	0.07247
25	3.38635	0.29530	47.72710	0.02095	14.09394	0.07095
26	3.55567	0.28124	51.11345	0.01956	14.37519	0.06956
27	3.73346	0.26785	54.66913	0.01829	14.64303	0.06829
28	3.92013	0.25509	58.40258	0.01712	14.89813	0.06712
29	4.11614	0.24295	62.32271	0.01605	15.14107	0.06605
30	4.32194	0.23138	66.43885	0.01505	15.37245	0.06505
31	4.53804	0.22036	70.76079	0.01413	15.59281	0.06413
32	4.76494	0.20987	75.29883	0.01328	15.80268	0.06328
33	5.00319	0.19987	80.06377	0.01249	16.00255	0.06249
34	5.25335	0.19035	85.06696	0.01176	16.19290	0.06176
35	5.51602	0.18129	90.32031	0.01107	16.37419	0.06107
36	5.79182	0.17266	95.83632	0.01043	16.54685	0.06043
37	6.08141	0.16444	101.62814	0.00984	16.71129	0.05984
38	6.38548	0.15661	107.70955	0.00928	16.86789	0.05928
39	6.70475	0.14915	114.09502	0.00876	17.01704	0.05876
40	7.03999	0.14205	120.79977	0.00828	17.15909	0.05828
41	7.39199	0.13528	127.83976	0.00782	17.29437	0.05782
42	7.76159	0.12884	135.23175	0.00739	17.42321	0.05739
43	8.14967	0.12270	142.99334	0.00699	17.54591	0.05699
44	8.55715	0.11686	151.14301	0.00662	17.66277	0.05662
45	8.98501	0.11130	159.70016	0.00626	17.77407	0.05626
46	9.43426	0.10600	168.68516	0.00593	17.88007	0.05593
47	9.90597	0.10095	178.11942	0.00561	17.98102	0.05561
48	10.40127	0.09614	188.02539	0.00532	18.07716	0.05532
49	10.92133	0.09156	198.42666	0.00504	18.16872	0.05504
50	11.46740	0.08720	209.34800	0.00478	18.25593	0.05478
55	14.63563	0.06833	272.71262	0.00367	18.63347	0.05367
60	18.67919	0.05354	353.58372	0.00283	18.92929	0.05283
65	23.83990	0.04195	456.79801	0.00219	19.16107	0.05219
70	30.42643	0.03287	588.52851	0.00170	19.34268	0.05170
75	38.83269	0.02575	756.65372	0.00132	19.48497	0.05132
80	49.56144	0.02018	971.22882	0.00103	19.59646	0.05103
85	63.25435	0.01581	1,245.08707	0.00080	19.68382	0.05080
90	80.73037	0.01239	1,594.60730	0.00063	19.75226	0.05063
95	103.03468	0.00971	2,040.69353	0.00049	19.80589	0.05049
100	131.50126	0.00760	2,610.02516	0.00038	19.84791	0.05038

6.0% = i, interest rate						
number of time periods, n	F/P(i%,n)	P/F(i%,n)	F/A(i%,n)	A/F(i%,n)	P/A(i%,n)	A/P(i%,n)
1	1.06000	0.94340	1.00000	1.00000	0.94340	1.06000
2	1.12360	0.89000	2.06000	0.48544	1.83339	0.54544
3	1.19102	0.83962	3.18360	0.31411	2.67301	0.37411
4	1.26248	0.79209	4.37462	0.22859	3.46511	0.28859
5	1.33823	0.74726	5.63709	0.17740	4.21236	0.23740
6	1.41852	0.70496	6.97532	0.14336	4.91732	0.20336
7	1.50363	0.66506	8.39384	0.11914	5.58238	0.17914
8	1.59385	0.62741	9.89747	0.10104	6.20979	0.16104
9	1.68948	0.59190	11.49132	0.08702	6.80169	0.14702
10	1.79085	0.55839	13.18079	0.07587	7.36009	0.13587
11	1.89830	0.52679	14.97164	0.06679	7.88687	0.12679
12	2.01220	0.49697	16.86994	0.05928	8.38384	0.11928
13	2.13293	0.46884	18.88214	0.05296	8.85268	0.11296
14	2.26090	0.44230	21.01507	0.04758	9.29498	0.10758
15	2.39656	0.41727	23.27597	0.04296	9.71225	0.10296
16	2.54035	0.39365	25.67253	0.03895	10.10590	0.09895
17	2.69277	0.37136	28.21288	0.03544	10.47726	0.09544
18	2.85434	0.35034	30.90565	0.03236	10.82760	0.09236
19	3.02560	0.33051	33.75999	0.02962	11.15812	0.08962
20	3.20714	0.31180	36.78559	0.02718	11.46992	0.08718
21	3.39956	0.29416	39.99273	0.02500	11.76408	0.08500
22	3.60354	0.27751	43.39229	0.02305	12.04158	0.08305
23	3.81975	0.26180	46.99583	0.02128	12.30338	0.08128
24	4.04893	0.24698	50.81558	0.01968	12.55036	0.07968
25	4.29187	0.23300	54.86451	0.01823	12.78336	0.07823
26	4.54938	0.21981	59.15638	0.01690	13.00317	0.07690
27	4.82235	0.20737	63.70577	0.01570	13.21053	0.07570
28	5.11169	0.19563	68.52811	0.01459	13.40616	0.07459
29	5.41839	0.18456	73.63980	0.01358	13.59072	0.07358
30	5.74349	0.17411	79.05819	0.01265	13.76483	0.07265
31	6.08810	0.16425	84.80168	0.01179	13.92909	0.07179
32	6.45339	0.15496	90.88978	0.01100	14.08404	0.07100
33	6.84059	0.14619	97.34316	0.01027	14.23023	0.07027
34	7.25103	0.13791	104.18375	0.00960	14.36814	0.06960
35	7.68609	0.13011	111.43478	0.00897	14.49825	0.06897
36	8.14725	0.12274	119.12087	0.00839	14.62099	0.06839
37	8.63609	0.11579	127.26812	0.00786	14.73678	0.06786
38	9.15425	0.10924	135.90421	0.00736	14.84602	0.06736
39	9.70351	0.10306	145.05846	0.00689	14.94907	0.06689
40	10.28572	0.09722	154.76197	0.00646	15.04630	0.06646
41	10.90286	0.09172	165.04768	0.00606	15.13802	0.06606
42	11.55703	0.08653	175.95054	0.00568	15.22454	0.06568
43	12.25045	0.08163	187.50758	0.00533	15.30617	0.06533
44	12.98548	0.07701	199.75803	0.00501	15.38318	0.06501
45	13.76461	0.07265	212.74351	0.00470	15.45583	0.06470
46	14.59049	0.06854	226.50812	0.00441	15.52437	0.06441
47	15.46592	0.06466	241.09861	0.00415	15.58903	0.06415
48	16.39387	0.06100	256.56453	0.00390	15.65003	0.06390
49	17.37750	0.05755	272.95840	0.00366	15.70757	0.06366
50	18.42015	0.05429	290.33590	0.00344	15.76186	0.06344
55	24.65032	0.04057	394.17203	0.00254	15.99054	0.06254
60	32.98769	0.03031	533.12818	0.00188	16.16143	0.06188
65	44.14497	0.02265	719.08286	0.00139	16.28912	0.06139
70	59.07593	0.01693	967.93217	0.00103	16.38454	0.06103
75	79.05692	0.01265	1,300.94868	0.00077	16.45585	0.06077
80	105.79599	0.00945	1,746.59989	0.00057	16.50913	0.06057
85	141.57890	0.00706	2,342.98174	0.00043	16.54895	0.06043
90	189.46451	0.00528	3,141.07519	0.00032	16.57870	0.06032
95	253.54625	0.00394	4,209.10425	0.00024	16.60093	0.06024
100	339.30208	0.00295	5,638.36806	0.00018	16.61755	0.06018

8.0% = i, interest rate

number of time periods, n	F/P(i%,n)	P/F(i%,n)	F/A(i%,n)	A/F(i%,n)	P/A(i%,n)	A/P(i%,n)
1	1.08000	0.92593	1.00000	1.00000	0.92593	1.08000
2	1.16640	0.85734	2.08000	0.48077	1.78326	0.56077
3	1.25971	0.79383	3.24640	0.30803	2.57710	0.38803
4	1.36049	0.73503	4.50611	0.22192	3.31213	0.30192
5	1.46933	0.68058	5.86660	0.17046	3.99271	0.25046
6	1.58687	0.63017	7.33593	0.13632	4.62288	0.21632
7	1.71382	0.58349	8.92280	0.11207	5.20637	0.19207
8	1.85093	0.54027	10.63663	0.09401	5.74664	0.17401
9	1.99900	0.50025	12.48756	0.08008	6.24689	0.16008
10	2.15892	0.46319	14.48656	0.06903	6.71008	0.14903
11	2.33164	0.42888	16.64549	0.06008	7.13896	0.14008
12	2.51817	0.39711	18.97713	0.05270	7.53608	0.13270
13	2.71962	0.36770	21.49530	0.04652	7.90378	0.12652
14	2.93719	0.34046	24.21492	0.04130	8.24424	0.12130
15	3.17217	0.31524	27.15211	0.03683	8.55948	0.11683
16	3.42594	0.29189	30.32428	0.03298	8.85137	0.11298
17	3.70002	0.27027	33.75023	0.02963	9.12164	0.10963
18	3.99602	0.25025	37.45024	0.02670	9.37189	0.10670
19	4.31570	0.23171	41.44626	0.02413	9.60360	0.10413
20	4.66096	0.21455	45.76196	0.02185	9.81815	0.10185
21	5.03383	0.19866	50.42292	0.01983	10.01680	0.09983
22	5.43654	0.18394	55.45676	0.01803	10.20074	0.09803
23	5.87146	0.17032	60.89330	0.01642	10.37106	0.09642
24	6.34118	0.15770	66.76476	0.01498	10.52876	0.09498
25	6.84848	0.14602	73.10594	0.01368	10.67478	0.09368
26	7.39635	0.13520	79.95442	0.01251	10.80998	0.09251
27	7.98806	0.12519	87.35077	0.01145	10.93516	0.09145
28	8.62711	0.11591	95.33883	0.01049	11.05108	0.09049
29	9.31727	0.10733	103.96594	0.00962	11.15841	0.08962
30	10.06266	0.09938	113.28321	0.00883	11.25778	0.08883
31	10.86767	0.09202	123.34587	0.00811	11.34980	0.08811
32	11.73708	0.08520	134.21354	0.00745	11.43500	0.08745
33	12.67605	0.07889	145.95062	0.00685	11.51389	0.08685
34	13.69013	0.07305	158.62667	0.00630	11.58693	0.08630
35	14.78534	0.06763	172.31680	0.00580	11.65457	0.08580
36	15.96817	0.06262	187.10215	0.00534	11.71719	0.08534
37	17.24563	0.05799	203.07032	0.00492	11.77518	0.08492
38	18.62528	0.05369	220.31595	0.00454	11.82887	0.08454
39	20.11530	0.04971	238.94122	0.00419	11.87858	0.08419
40	21.72452	0.04603	259.05652	0.00386	11.92461	0.08386
41	23.46248	0.04262	280.78104	0.00356	11.96723	0.08356
42	25.33948	0.03946	304.24352	0.00329	12.00670	0.08329
43	27.36664	0.03654	329.58301	0.00303	12.04324	0.08303
44	29.55597	0.03383	356.94965	0.00280	12.07707	0.08280
45	31.92045	0.03133	386.50562	0.00259	12.10840	0.08259
46	34.47409	0.02901	418.42607	0.00239	12.13741	0.08239
47	37.23201	0.02686	452.90015	0.00221	12.16427	0.08221
48	40.21057	0.02487	490.13216	0.00204	12.18914	0.08204
49	43.42742	0.02303	530.34274	0.00189	12.21216	0.08189
50	46.90161	0.02132	573.77016	0.00174	12.23348	0.08174
55	68.91386	0.01451	848.92320	0.00118	12.31861	0.08118
60	101.25706	0.00988	1,253.21330	0.00080	12.37655	0.08080
65	148.77985	0.00672	1,847.24808	0.00054	12.41598	0.08054
70	218.60641	0.00457	2,720.08007	0.00037	12.44282	0.08037
75	321.20453	0.00311	4,002.55662	0.00025	12.46108	0.08025
80	4.71955E+02	2.11885E-03	5.88694E+03	1.69868E-04	12.47351	0.08017
85	6.93456E+02	1.44205E-03	8.65571E+03	1.15531E-04	12.48197	0.08012
90	1.01892E+03	9.81436E-04	1.27239E+04	7.85920E-05	12.48773	0.08008
95	1.49712E+03	6.67949E-04	1.87015E+04	5.34716E-05	12.49165	0.08005
100	2.19976E+03	4.54595E-04	2.74845E+04	3.63841E-05	12.49432	0.08004

10.0% = i, interest rate number of time periods, n	F/P(i%,n)	P/F(i%,n)	F/A(i%,n)	A/F(i%,n)	P/A(i%,n)	A/P(i%,n)
1	1.10000	0.90909	1.00000	1.00000	0.90909	1.10000
2	1.21000	0.82645	2.10000	0.47619	1.73554	0.57619
3	1.33100	0.75131	3.31000	0.30211	2.48685	0.40211
4	1.46410	0.68301	4.64100	0.21547	3.16987	0.31547
5	1.61051	0.62092	6.10510	0.16380	3.79079	0.26380
6	1.77156	0.56447	7.71561	0.12961	4.35526	0.22961
7	1.94872	0.51316	9.48717	0.10541	4.86842	0.20541
8	2.14359	0.46651	11.43589	0.08744	5.33493	0.18744
9	2.35795	0.42410	13.57948	0.07364	5.75902	0.17364
10	2.59374	0.38554	15.93742	0.06275	6.14457	0.16275
11	2.85312	0.35049	18.53117	0.05396	6.49506	0.15396
12	3.13843	0.31863	21.38428	0.04676	6.81369	0.14676
13	3.45227	0.28966	24.52271	0.04078	7.10336	0.14078
14	3.79750	0.26333	27.97498	0.03575	7.36669	0.13575
15	4.17725	0.23939	31.77248	0.03147	7.60608	0.13147
16	4.59497	0.21763	35.94973	0.02782	7.82371	0.12782
17	5.05447	0.19784	40.54470	0.02466	8.02155	0.12466
18	5.55992	0.17986	45.59917	0.02193	8.20141	0.12193
19	6.11591	0.16351	51.15909	0.01955	8.36492	0.11955
20	6.72750	0.14864	57.27500	0.01746	8.51356	0.11746
21	7.40025	0.13513	64.00250	0.01562	8.64869	0.11562
22	8.14027	0.12285	71.40275	0.01401	8.77154	0.11401
23	8.95430	0.11168	79.54302	0.01257	8.88322	0.11257
24	9.84973	0.10153	88.49733	0.01130	8.98474	0.11130
25	10.83471	0.09230	98.34706	0.01017	9.07704	0.11017
26	11.91818	0.08391	109.18177	0.00916	9.16095	0.10916
27	13.10999	0.07628	121.09994	0.00826	9.23722	0.10826
28	14.42099	0.06934	134.20994	0.00745	9.30657	0.10745
29	15.86309	0.06304	148.63093	0.00673	9.36961	0.10673
30	17.44940	0.05731	164.49402	0.00608	9.42691	0.10608
31	19.19434	0.05210	181.94342	0.00550	9.47901	0.10550
32	21.11378	0.04736	201.13777	0.00497	9.52638	0.10497
33	23.22515	0.04306	222.25154	0.00450	9.56943	0.10450
34	25.54767	0.03914	245.47670	0.00407	9.60857	0.10407
35	28.10244	0.03558	271.02437	0.00369	9.64416	0.10369
36	30.91268	0.03235	299.12681	0.00334	9.67651	0.10334
37	34.00395	0.02941	330.03949	0.00303	9.70592	0.10303
38	37.40434	0.02673	364.04343	0.00275	9.73265	0.10275
39	41.14478	0.02430	401.44778	0.00249	9.75696	0.10249
40	45.25926	0.02209	442.59256	0.00226	9.77905	0.10226
41	49.78518	0.02009	487.85181	0.00205	9.79914	0.10205
42	54.76370	0.01826	537.63699	0.00186	9.81740	0.10186
43	60.24007	0.01660	592.40069	0.00169	9.83400	0.10169
44	66.26408	0.01509	652.64076	0.00153	9.84909	0.10153
45	72.89048	0.01372	718.90484	0.00139	9.86281	0.10139
46	80.17953	0.01247	791.79532	0.00126	9.87528	0.10126
47	88.19749	0.01134	871.97485	0.00115	9.88662	0.10115
48	97.01723	0.01031	960.17234	0.00104	9.89693	0.10104
49	106.71896	0.00937	1,057.18957	0.00095	9.90630	0.10095
50	117.39085	0.00852	1,163.90853	0.00086	9.91481	0.10086
55	189.05914	0.00529	1,880.59142	0.00053	9.94711	0.10053
60	304.48164	0.00328	3,034.81640	0.00033	9.96716	0.10033
65	490.37073	0.00204	4,893.70725	0.00020	9.97961	0.10020
70	789.74696	0.00127	7,887.46957	0.00013	9.98734	0.10013
75	1,271.89537	0.00079	12,708.95371	0.00008	9.99214	0.10008
80	2.04840E+03	4.88186E-04	2.04740E+04	4.88424E-05	9.99512	0.10005
85	3.29897E+03	3.03125E-04	3.29797E+04	3.03217E-05	9.99697	0.10003
90	5.31302E+03	1.88217E-04	5.31202E+04	1.88252E-05	9.99812	0.10002
95	8.55668E+03	1.16868E-04	8.55568E+04	1.16881E-05	9.99883	0.10001
100	1.37806E+04	7.25657E-05	1.37796E+05	7.25710E-06	9.99927	0.10001

12.0% = i, interest rate

number of time periods, n	F/P(i%,n)	P/F(i%,n)	F/A(i%,n)	A/F(i%,n)	P/A(i%,n)	A/P(i%,n)
1	1.12000	0.89286	1.00000	1.00000	0.89286	1.12000
2	1.25440	0.79719	2.12000	0.47170	1.69005	0.59170
3	1.40493	0.71178	3.37440	0.29635	2.40183	0.41635
4	1.57352	0.63552	4.77933	0.20923	3.03735	0.32923
5	1.76234	0.56743	6.35285	0.15741	3.60478	0.27741
6	1.97382	0.50663	8.11519	0.12323	4.11141	0.24323
7	2.21068	0.45235	10.08901	0.09912	4.56376	0.21912
8	2.47596	0.40388	12.29969	0.08130	4.96764	0.20130
9	2.77308	0.36061	14.77566	0.06768	5.32825	0.18768
10	3.10585	0.32197	17.54874	0.05698	5.65022	0.17698
11	3.47855	0.28748	20.65458	0.04842	5.93770	0.16842
12	3.89598	0.25668	24.13313	0.04144	6.19437	0.16144
13	4.36349	0.22917	28.02911	0.03568	6.42355	0.15568
14	4.88711	0.20462	32.39260	0.03087	6.62817	0.15087
15	5.47357	0.18270	37.27971	0.02682	6.81086	0.14682
16	6.13039	0.16312	42.75328	0.02339	6.97399	0.14339
17	6.86604	0.14564	48.88367	0.02046	7.11963	0.14046
18	7.68997	0.13004	55.74971	0.01794	7.24967	0.13794
19	8.61276	0.11611	63.43968	0.01576	7.36578	0.13576
20	9.64629	0.10367	72.05244	0.01388	7.46944	0.13388
21	10.80385	0.09256	81.69874	0.01224	7.56200	0.13224
22	12.10031	0.08264	92.50258	0.01081	7.64465	0.13081
23	13.55235	0.07379	104.60289	0.00956	7.71843	0.12956
24	15.17863	0.06588	118.15524	0.00846	7.78432	0.12846
25	17.00006	0.05882	133.33387	0.00750	7.84314	0.12750
26	19.04007	0.05252	150.33393	0.00665	7.89566	0.12665
27	21.32488	0.04689	169.37401	0.00590	7.94255	0.12590
28	23.88387	0.04187	190.69889	0.00524	7.98442	0.12524
29	26.74993	0.03738	214.58275	0.00466	8.02181	0.12466
30	29.95992	0.03338	241.33268	0.00414	8.05518	0.12414
31	33.55511	0.02980	271.29261	0.00369	8.08499	0.12369
32	37.58173	0.02661	304.84772	0.00328	8.11159	0.12328
33	42.09153	0.02376	342.42945	0.00292	8.13535	0.12292
34	47.14252	0.02121	384.52098	0.00260	8.15656	0.12260
35	52.79962	0.01894	431.66350	0.00232	8.17550	0.12232
36	59.13557	0.01691	484.46312	0.00206	8.19241	0.12206
37	66.23184	0.01510	543.59869	0.00184	8.20751	0.12184
38	74.17966	0.01348	609.83053	0.00164	8.22099	0.12164
39	83.08122	0.01204	684.01020	0.00146	8.23303	0.12146
40	93.05097	0.01075	767.09142	0.00130	8.24378	0.12130
41	104.21709	0.00960	860.14239	0.00116	8.25337	0.12116
42	116.72314	0.00857	964.35948	0.00104	8.26194	0.12104
43	130.72991	0.00765	1,081.08262	0.00092	8.26959	0.12092
44	146.41750	0.00683	1,211.81253	0.00083	8.27642	0.12083
45	163.98760	0.00610	1,358.23003	0.00074	8.28252	0.12074
46	183.66612	0.00544	1,522.21764	0.00066	8.28796	0.12066
47	205.70605	0.00486	1,705.88375	0.00059	8.29282	0.12059
48	230.39078	0.00434	1,911.58980	0.00052	8.29716	0.12052
49	258.03767	0.00388	2,141.98058	0.00047	8.30104	0.12047
50	289.00219	0.00346	2,400.01825	0.00042	8.30450	0.12042
55	509.32061	0.00196	4,236.00505	2.36071E-04	8.31697	0.12024
60	897.59693	0.00111	7,471.64111	1.33839E-04	8.32405	0.12013
65	1,581.87249	0.00063	13,173.93742	7.59075E-05	8.32807	0.12008
70	2,787.79983	0.00036	23,223.33190	4.30601E-05	8.33034	0.12004
75	4,913.05584	0.00020	40,933.79867	2.44297E-05	8.33164	0.12002
80	8.65848E+03	1.15494E-04	7.21457E+04	1.38608E-05	8.33237	0.12001
85	1.52592E+04	6.55342E-05	1.27152E+05	7.86462E-06	8.33279	0.12001
90	2.68919E+04	3.71859E-05	2.24091E+05	4.46247E-06	8.33302	0.12000
95	4.73928E+04	2.11003E-05	3.94931E+05	2.53208E-06	8.33316	0.12000
100	8.35223E+04	1.19729E-05	6.96011E+05	1.43676E-06	8.33323	0.12000

15.0% = i, interest rate						
number of time periods, n	F/P(i%,n)	P/F(i%,n)	F/A(i%,n)	A/F(i%,n)	P/A(i%,n)	A/P(i%,n)
1	1.15000	0.86957	1.00000	1.00000	0.86957	1.15000
2	1.32250	0.75614	2.15000	0.46512	1.62571	0.61512
3	1.52088	0.65752	3.47250	0.28798	2.28323	0.43798
4	1.74901	0.57175	4.99338	0.20027	2.85498	0.35027
5	2.01136	0.49718	6.74238	0.14832	3.35216	0.29832
6	2.31306	0.43233	8.75374	0.11424	3.78448	0.26424
7	2.66002	0.37594	11.06680	0.09036	4.16042	0.24036
8	3.05902	0.32690	13.72682	0.07285	4.48732	0.22285
9	3.51788	0.28426	16.78584	0.05957	4.77158	0.20957
10	4.04556	0.24718	20.30372	0.04925	5.01877	0.19925
11	4.65239	0.21494	24.34928	0.04107	5.23371	0.19107
12	5.35025	0.18691	29.00167	0.03448	5.42062	0.18448
13	6.15279	0.16253	34.35192	0.02911	5.58315	0.17911
14	7.07571	0.14133	40.50471	0.02469	5.72448	0.17469
15	8.13706	0.12289	47.58041	0.02102	5.84737	0.17102
16	9.35762	0.10686	55.71747	0.01795	5.95423	0.16795
17	10.76126	0.09293	65.07509	0.01537	6.04716	0.16537
18	12.37545	0.08081	75.83636	0.01319	6.12797	0.16319
19	14.23177	0.07027	88.21181	0.01134	6.19823	0.16134
20	16.36654	0.06110	102.44358	0.00976	6.25933	0.15976
21	18.82152	0.05313	118.81012	0.00842	6.31246	0.15842
22	21.64475	0.04620	137.63164	0.00727	6.35866	0.15727
23	24.89146	0.04017	159.27638	0.00628	6.39884	0.15628
24	28.62518	0.03493	184.16784	0.00543	6.43377	0.15543
25	32.91895	0.03038	212.79302	0.00470	6.46415	0.15470
26	37.85680	0.02642	245.71197	0.00407	6.49056	0.15407
27	43.53531	0.02297	283.56877	0.00353	6.51353	0.15353
28	50.06561	0.01997	327.10408	0.00306	6.53351	0.15306
29	57.57545	0.01737	377.16969	0.00265	6.55088	0.15265
30	66.21177	0.01510	434.74515	0.00230	6.56598	0.15230
31	76.14354	0.01313	500.95692	0.00200	6.57911	0.15200
32	87.56507	0.01142	577.10046	0.00173	6.59053	0.15173
33	100.69983	0.00993	664.66552	0.00150	6.60046	0.15150
34	115.80480	0.00864	765.36535	0.00131	6.60910	0.15131
35	133.17552	0.00751	881.17016	0.00113	6.61661	0.15113
36	153.15185	0.00653	1,014.34568	0.00099	6.62314	0.15099
37	176.12463	0.00568	1,167.49753	0.00086	6.62881	0.15086
38	202.54332	0.00494	1,343.62216	0.00074	6.63375	0.15074
39	232.92482	0.00429	1,546.16549	0.00065	6.63805	0.15065
40	267.86355	0.00373	1,779.09031	0.00056	6.64178	0.15056
41	308.04308	0.00325	2,046.95385	0.00049	6.64502	0.15049
42	354.24954	0.00282	2,354.99693	0.00042	6.64785	0.15042
43	407.38697	0.00245	2,709.24647	0.00037	6.65030	0.15037
44	468.49502	0.00213	3,116.63344	0.00032	6.65244	0.15032
45	538.76927	0.00186	3,585.12846	0.00028	6.65429	0.15028
46	619.58466	0.00161	4,123.89773	0.00024	6.65591	0.15024
47	712.52236	0.00140	4,743.48239	0.00021	6.65731	0.15021
48	819.40071	0.00122	5,456.00475	0.00018	6.65853	0.15018
49	942.31082	0.00106	6,275.40546	0.00016	6.65959	0.15016
50	1,083.65744	0.00092	7,217.71628	0.00014	6.66051	0.15014
55	2.17962E+03	4.58795E-04	1.45241E+04	6.88509E-05	6.66361	0.15007
60	4.38400E+03	2.28102E-04	2.92200E+04	3.42231E-05	6.66515	0.15003
65	8.81779E+03	1.13407E-04	5.87786E+04	1.70130E-05	6.66591	0.15002
70	1.77357E+04	5.63834E-05	1.18231E+05	8.45799E-06	6.66629	0.15001
75	3.56729E+04	2.80325E-05	2.37812E+05	4.20499E-06	6.66648	0.15000
80	7.17509E+04	1.39371E-05	4.78333E+05	2.09060E-06	6.66657	0.15000
85	1.44317E+05	6.92921E-06	9.62104E+05	1.03939E-06	6.66662	0.15000
90	2.90272E+05	3.44504E-06	1.93514E+06	5.16758E-07	6.66664	0.15000
95	5.83841E+05	1.71279E-06	3.89227E+06	2.56920E-07	6.66666	0.15000
100	1.17431E+06	8.51561E-07	7.82875E+06	1.27734E-07	6.66666	0.15000

20.0% = i, interest rate

number of time periods, n	F/P(i%,n)	P/F(i%,n)	F/A(i%,n)	A/F(i%,n)	P/A(i%,n)	A/P(i%,n)
1	1.20000	0.83333	1.00000	1.00000	0.83333	1.20000
2	1.44000	0.69444	2.20000	0.45455	1.52778	0.65455
3	1.72800	0.57870	3.64000	0.27473	2.10648	0.47473
4	2.07360	0.48225	5.36800	0.18629	2.58873	0.38629
5	2.48832	0.40188	7.44160	0.13438	2.99061	0.33438
6	2.98598	0.33490	9.92992	0.10071	3.32551	0.30071
7	3.58318	0.27908	12.91590	0.07742	3.60459	0.27742
8	4.29982	0.23257	16.49908	0.06061	3.83716	0.26061
9	5.15978	0.19381	20.79890	0.04808	4.03097	0.24808
10	6.19174	0.16151	25.95868	0.03852	4.19247	0.23852
11	7.43008	0.13459	32.15042	0.03110	4.32706	0.23110
12	8.91610	0.11216	39.58050	0.02526	4.43922	0.22526
13	10.69932	0.09346	48.49660	0.02062	4.53268	0.22062
14	12.83918	0.07789	59.19592	0.01689	4.61057	0.21689
15	15.40702	0.06491	72.03511	0.01388	4.67547	0.21388
16	18.48843	0.05409	87.44213	0.01144	4.72956	0.21144
17	22.18611	0.04507	105.93056	0.00944	4.77463	0.20944
18	26.62333	0.03756	128.11667	0.00781	4.81219	0.20781
19	31.94800	0.03130	154.74000	0.00646	4.84350	0.20646
20	38.33760	0.02608	186.68800	0.00536	4.86958	0.20536
21	46.00512	0.02174	225.02560	0.00444	4.89132	0.20444
22	55.20614	0.01811	271.03072	0.00369	4.90943	0.20369
23	66.24737	0.01509	326.23686	0.00307	4.92453	0.20307
24	79.49685	0.01258	392.48424	0.00255	4.93710	0.20255
25	95.39622	0.01048	471.98108	0.00212	4.94759	0.20212
26	114.47546	0.00874	567.37730	0.00176	4.95632	0.20176
27	137.37055	0.00728	681.85276	0.00147	4.96360	0.20147
28	164.84466	0.00607	819.22331	0.00122	4.96967	0.20122
29	197.81359	0.00506	984.06797	0.00102	4.97472	0.20102
30	237.37631	0.00421	1,181.88157	0.00085	4.97894	0.20085
31	284.85158	0.00351	1,419.25788	0.00070	4.98245	0.20070
32	341.82189	0.00293	1,704.10946	0.00059	4.98537	0.20059
33	410.18627	0.00244	2,045.93135	0.00049	4.98781	0.20049
34	492.22352	0.00203	2,456.11762	0.00041	4.98984	0.20041
35	590.66823	0.00169	2,948.34115	0.00034	4.99154	0.20034
36	708.80187	0.00141	3,539.00937	0.00028	4.99295	0.20028
37	850.56225	0.00118	4,247.81125	0.00024	4.99412	0.20024
38	1,020.67470	0.00098	5,098.37350	0.00020	4.99510	0.20020
39	1,224.80964	0.00082	6,119.04820	0.00016	4.99592	0.20016
40	1,469.77157	0.00068	7,343.85784	0.00014	4.99660	0.20014
41	1,763.72588	0.00057	8,813.62941	1.13461E-04	4.99717	0.20011
42	2,116.47106	0.00047	10,577.35529	9.45416E-05	4.99764	0.20009
43	2,539.76527	0.00039	12,693.82635	7.87785E-05	4.99803	0.20008
44	3,047.71832	0.00033	15,233.59162	6.56444E-05	4.99836	0.20007
45	3,657.26199	0.00027	18,281.30994	5.47007E-05	4.99863	0.20005
46	4.38871E+03	2.27857E-04	2.19386E+04	4.55818E-05	4.99886	0.20005
47	5.26646E+03	1.89881E-04	2.63273E+04	3.79834E-05	4.99905	0.20004
48	6.31975E+03	1.58234E-04	3.15937E+04	3.16518E-05	4.99921	0.20003
49	7.58370E+03	1.31862E-04	3.79135E+04	2.63758E-05	4.99934	0.20003
50	9.10044E+03	1.09885E-04	4.54972E+04	2.19794E-05	4.99945	0.20002
55	2.26448E+04	4.41602E-05	1.13219E+05	8.83244E-06	4.99978	0.20001
60	5.63475E+04	1.77470E-05	2.81733E+05	3.54947E-06	4.99991	0.20000
65	1.40211E+05	7.13213E-06	7.01048E+05	1.42644E-06	4.99996	0.20000
70	3.48889E+05	2.86624E-06	1.74444E+06	5.73250E-07	4.99999	0.20000
75	8.68147E+05	1.15188E-06	4.34073E+06	2.30376E-07	4.99999	0.20000
80	2.16023E+06	4.62914E-07	1.08011E+07	9.25828E-08	5.00000	0.20000
85	5.37534E+06	1.86035E-07	2.68767E+07	3.72070E-08	5.00000	0.20000
90	1.33756E+07	7.47632E-08	6.68778E+07	1.49526E-08	5.00000	0.20000
95	3.32827E+07	3.00457E-08	1.66413E+08	6.00913E-09	5.00000	0.20000
100	8.28180E+07	1.20747E-08	4.14090E+08	2.41493E-09	5.00000	0.20000

25.0% = i, interest rate

number of time periods, n	F/P(i%,n)	P/F(i%,n)	F/A(i%,n)	A/F(i%,n)	P/A(i%,n)	A/P(i%,n)
1	1.25000	0.80000	1.00000	1.00000	0.80000	1.25000
2	1.56250	0.64000	2.25000	0.44444	1.44000	0.69444
3	1.95313	0.51200	3.81250	0.26230	1.95200	0.51230
4	2.44141	0.40960	5.76563	0.17344	2.36160	0.42344
5	3.05176	0.32768	8.20703	0.12185	2.68928	0.37185
6	3.81470	0.26214	11.25879	0.08882	2.95142	0.33882
7	4.76837	0.20972	15.07349	0.06634	3.16114	0.31634
8	5.96046	0.16777	19.84186	0.05040	3.32891	0.30040
9	7.45058	0.13422	25.80232	0.03876	3.46313	0.28876
10	9.31323	0.10737	33.25290	0.03007	3.57050	0.28007
11	11.64153	0.08590	42.56613	0.02349	3.65640	0.27349
12	14.55192	0.06872	54.20766	0.01845	3.72512	0.26845
13	18.18989	0.05498	68.75958	0.01454	3.78010	0.26454
14	22.73737	0.04398	86.94947	0.01150	3.82408	0.26150
15	28.42171	0.03518	109.68684	0.00912	3.85926	0.25912
16	35.52714	0.02815	138.10855	0.00724	3.88741	0.25724
17	44.40892	0.02252	173.63568	0.00576	3.90993	0.25576
18	55.51115	0.01801	218.04460	0.00459	3.92794	0.25459
19	69.38894	0.01441	273.55576	0.00366	3.94235	0.25366
20	86.73617	0.01153	342.94470	0.00292	3.95388	0.25292
21	108.42022	0.00922	429.68087	0.00233	3.96311	0.25233
22	135.52527	0.00738	538.10109	0.00186	3.97049	0.25186
23	169.40659	0.00590	673.62636	0.00148	3.97639	0.25148
24	211.75824	0.00472	843.03295	0.00119	3.98111	0.25119
25	264.69780	0.00378	1,054.79118	0.00095	3.98489	0.25095
26	330.87225	0.00302	1,319.48898	0.00076	3.98791	0.25076
27	413.59031	0.00242	1,650.36123	0.00061	3.99033	0.25061
28	516.98788	0.00193	2,063.95153	0.00048	3.99226	0.25048
29	646.23485	0.00155	2,580.93941	0.00039	3.99381	0.25039
30	807.79357	0.00124	3,227.17427	0.00031	3.99505	0.25031
31	1,009.74196	0.00099	4,034.96783	0.00025	3.99604	0.25025
32	1,262.17745	0.00079	5,044.70979	0.00020	3.99683	0.25020
33	1,577.72181	0.00063	6,306.88724	0.00016	3.99746	0.25016
34	1,972.15226	0.00051	7,884.60905	0.00013	3.99797	0.25013
35	2,465.19033	0.00041	9,856.76132	0.00010	3.99838	0.25010
36	3,081.48791	0.00032	12,321.95164	8.11560E-05	3.99870	0.25008
37	3,851.85989	0.00026	15,403.43956	6.49206E-05	3.99896	0.25006
38	4,814.82486	0.00021	19,255.29944	5.19338E-05	3.99917	0.25005
39	6,018.53108	0.00017	24,070.12430	4.15453E-05	3.99934	0.25004
40	7,523.16385	0.00013	30,088.65538	3.32351E-05	3.99947	0.25003
41	9.40395E+03	1.06338E-04	3.76118E+04	2.65874E-05	3.99957	0.25003
42	1.17549E+04	8.50706E-05	4.70158E+04	2.12695E-05	3.99966	0.25002
43	1.46937E+04	6.80565E-05	5.87707E+04	1.70153E-05	3.99973	0.25002
44	1.83671E+04	5.44452E-05	7.34644E+04	1.36120E-05	3.99978	0.25001
45	2.29589E+04	4.35561E-05	9.18315E+04	1.08895E-05	3.99983	0.25001
46	2.86986E+04	3.48449E-05	1.14790E+05	8.71153E-06	3.99986	0.25001
47	3.58732E+04	2.78759E-05	1.43489E+05	6.96918E-06	3.99989	0.25001
48	4.48416E+04	2.23007E-05	1.79362E+05	5.57531E-06	3.99991	0.25001
49	5.60519E+04	1.78406E-05	2.24204E+05	4.46023E-06	3.99993	0.25000
50	7.00649E+04	1.42725E-05	2.80256E+05	3.56817E-06	3.99994	0.25000
55	2.13821E+05	4.67681E-06	8.55281E+05	1.16921E-06	3.99998	0.25000
60	6.52530E+05	1.53250E-06	2.61012E+06	3.83124E-07	3.99999	0.25000
65	1.99136E+06	5.02168E-07	7.96546E+06	1.25542E-07	4.00000	0.25000
70	6.07716E+06	1.64550E-07	2.43086E+07	4.11376E-08	4.00000	0.25000
75	1.85460E+07	5.39199E-08	7.41841E+07	1.34800E-08	4.00000	0.25000
80	5.65980E+07	1.76685E-08	2.26392E+08	4.41712E-09	4.00000	0.25000
85	1.72723E+08	5.78960E-09	6.90893E+08	1.44740E-09	4.00000	0.25000
90	5.27110E+08	1.89714E-09	2.10844E+09	4.74284E-10	4.00000	0.25000
95	1.60861E+09	6.21654E-10	6.43445E+09	1.55414E-10	4.00000	0.25000
100	4.90909E+09	2.03704E-10	1.96364E+10	5.09259E-11	4.00000	0.25000

30.0% = i, interest rate						
number of time periods, n	F/P(i%,n)	P/F(i%,n)	F/A(i%,n)	A/F(i%,n)	P/A(i%,n)	A/P(i%,n)
1	1.30000	0.76923	1.00000	1.00000	0.76923	1.30000
2	1.69000	0.59172	2.30000	0.43478	1.36095	0.73478
3	2.19700	0.45517	3.99000	0.25063	1.81611	0.55063
4	2.85610	0.35013	6.18700	0.16163	2.16624	0.46163
5	3.71293	0.26933	9.04310	0.11058	2.43557	0.41058
6	4.82681	0.20718	12.75603	0.07839	2.64275	0.37839
7	6.27485	0.15937	17.58284	0.05687	2.80211	0.35687
8	8.15731	0.12259	23.85769	0.04192	2.92470	0.34192
9	10.60450	0.09430	32.01500	0.03124	3.01900	0.33124
10	13.78585	0.07254	42.61950	0.02346	3.09154	0.32346
11	17.92160	0.05580	56.40535	0.01773	3.14734	0.31773
12	23.29809	0.04292	74.32695	0.01345	3.19026	0.31345
13	30.28751	0.03302	97.62504	0.01024	3.22328	0.31024
14	39.37376	0.02540	127.91255	0.00782	3.24867	0.30782
15	51.18589	0.01954	167.28631	0.00598	3.26821	0.30598
16	66.54166	0.01503	218.47220	0.00458	3.28324	0.30458
17	86.50416	0.01156	285.01386	0.00351	3.29480	0.30351
18	112.45541	0.00889	371.51802	0.00269	3.30369	0.30269
19	146.19203	0.00684	483.97343	0.00207	3.31053	0.30207
20	190.04964	0.00526	630.16546	0.00159	3.31579	0.30159
21	247.06453	0.00405	820.21510	0.00122	3.31984	0.30122
22	321.18389	0.00311	1,067.27963	0.00094	3.32296	0.30094
23	417.53905	0.00239	1,388.46351	0.00072	3.32535	0.30072
24	542.80077	0.00184	1,806.00257	0.00055	3.32719	0.30055
25	705.64100	0.00142	2,348.80334	0.00043	3.32861	0.30043
26	917.33330	0.00109	3,054.44434	0.00033	3.32970	0.30033
27	1,192.53329	0.00084	3,971.77764	0.00025	3.33054	0.30025
28	1,550.29328	0.00065	5,164.31093	0.00019	3.33118	0.30019
29	2,015.38126	0.00050	6,714.60421	0.00015	3.33168	0.30015
30	2,619.99564	0.00038	8,729.98548	0.00011	3.33206	0.30011
31	3.40599E+03	2.93600E-04	1.13500E+04	8.81059E-05	3.33235	0.30009
32	4.42779E+03	2.25846E-04	1.47560E+04	6.77692E-05	3.33258	0.30007
33	5.75613E+03	1.73728E-04	1.91838E+04	5.21274E-05	3.33275	0.30005
34	7.48297E+03	1.33637E-04	2.49399E+04	4.00964E-05	3.33289	0.30004
35	9.72786E+03	1.02798E-04	3.24229E+04	3.08424E-05	3.33299	0.30003
36	1.26462E+04	7.90750E-05	4.21507E+04	2.37244E-05	3.33307	0.30002
37	1.64401E+04	6.08269E-05	5.47969E+04	1.82492E-05	3.33313	0.30002
38	2.13721E+04	4.67900E-05	7.12370E+04	1.40376E-05	3.33318	0.30001
39	2.77837E+04	3.59923E-05	9.26091E+04	1.07981E-05	3.33321	0.30001
40	3.61189E+04	2.76864E-05	1.20393E+05	8.30614E-06	3.33324	0.30001
41	4.69545E+04	2.12972E-05	1.56512E+05	6.38930E-06	3.33326	0.30001
42	6.10409E+04	1.63825E-05	2.03466E+05	4.91482E-06	3.33328	0.30000
43	7.93531E+04	1.26019E-05	2.64507E+05	3.78062E-06	3.33329	0.30000
44	1.03159E+05	9.69377E-06	3.43860E+05	2.90816E-06	3.33330	0.30000
45	1.34107E+05	7.45674E-06	4.47019E+05	2.23704E-06	3.33331	0.30000
46	1.74339E+05	5.73596E-06	5.81126E+05	1.72080E-06	3.33331	0.30000
47	2.26641E+05	4.41227E-06	7.55465E+05	1.32369E-06	3.33332	0.30000
48	2.94633E+05	3.39406E-06	9.82106E+05	1.01822E-06	3.33332	0.30000
49	3.83022E+05	2.61081E-06	1.27674E+06	7.83246E-07	3.33332	0.30000
50	4.97929E+05	2.00832E-06	1.65976E+06	6.02496E-07	3.33333	0.30000
55	1.84878E+06	5.40898E-07	6.16258E+06	1.62270E-07	3.33333	0.30000
60	6.86438E+06	1.45680E-07	2.28813E+07	4.37039E-08	3.33333	0.30000
65	2.54870E+07	3.92358E-08	8.49565E+07	1.17707E-08	3.33333	0.30000
70	9.46313E+07	1.05673E-08	3.15438E+08	3.17020E-09	3.33333	0.30000
75	3.51359E+08	2.84609E-09	1.17120E+09	8.53827E-10	3.33333	0.30000
80	1.30457E+09	7.66535E-10	4.34857E+09	2.29960E-10	3.33333	0.30000
85	4.84379E+09	2.06450E-10	1.61460E+10	6.19350E-11	3.33333	0.30000
90	1.79846E+10	5.56030E-11	5.99488E+10	1.66809E-11	3.33333	0.30000
95	6.67757E+10	1.49755E-11	2.22586E+11	4.49265E-12	3.33333	0.30000
100	2.47934E+11	4.03334E-12	8.26445E+11	1.21000E-12	3.33333	0.30000

40.0% = i, interest rate						
number of time periods, n	F/P(i%,n)	P/F(i%,n)	F/A(i%,n)	A/F(i%,n)	P/A(i%,n)	A/P(i%,n)
1	1.40000	0.71429	1.00000	1.00000	0.71429	1.40000
2	1.96000	0.51020	2.40000	0.41667	1.22449	0.81667
3	2.74400	0.36443	4.36000	0.22936	1.58892	0.62936
4	3.84160	0.26031	7.10400	0.14077	1.84923	0.54077
5	5.37824	0.18593	10.94560	0.09136	2.03516	0.49136
6	7.52954	0.13281	16.32384	0.06126	2.16797	0.46126
7	10.54135	0.09486	23.85338	0.04192	2.26284	0.44192
8	14.75789	0.06776	34.39473	0.02907	2.33060	0.42907
9	20.66105	0.04840	49.15262	0.02034	2.37900	0.42034
10	28.92547	0.03457	69.81366	0.01432	2.41357	0.41432
11	40.49565	0.02469	98.73913	0.01013	2.43826	0.41013
12	56.69391	0.01764	139.23478	0.00718	2.45590	0.40718
13	79.37148	0.01260	195.92869	0.00510	2.46850	0.40510
14	111.12007	0.00900	275.30017	0.00363	2.47750	0.40363
15	155.56810	0.00643	386.42024	0.00259	2.48393	0.40259
16	217.79533	0.00459	541.98833	0.00185	2.48852	0.40185
17	304.91347	0.00328	759.78367	0.00132	2.49180	0.40132
18	426.87885	0.00234	1,064.69714	0.00094	2.49414	0.40094
19	597.63040	0.00167	1,491.57599	0.00067	2.49582	0.40067
20	836.68255	0.00120	2,089.20639	0.00048	2.49701	0.40048
21	1,171.35558	0.00085	2,925.88894	3.41776E-04	2.49787	0.40034
22	1,639.89781	0.00061	4,097.24452	2.44066E-04	2.49848	0.40024
23	2,295.85693	0.00044	5,737.14232	1.74303E-04	2.49891	0.40017
24	3,214.19970	0.00031	8,032.99925	1.24487E-04	2.49922	0.40012
25	4,499.87958	0.00022	11,247.19895	8.89110E-05	2.49944	0.40009
26	6.29983E+03	1.58734E-04	1.57471E+04	6.35038E-05	2.49960	0.40006
27	8.81976E+03	1.13382E-04	2.20469E+04	4.53578E-05	2.49972	0.40005
28	1.23477E+04	8.09869E-05	3.08667E+04	3.23974E-05	2.49980	0.40003
29	1.72867E+04	5.78478E-05	4.32143E+04	2.31405E-05	2.49986	0.40002
30	2.42014E+04	4.13199E-05	6.05011E+04	1.65286E-05	2.49990	0.40002
31	3.38820E+04	2.95142E-05	8.47025E+04	1.18060E-05	2.49993	0.40001
32	4.74348E+04	2.10816E-05	1.18585E+05	8.43280E-06	2.49995	0.40001
33	6.64087E+04	1.50583E-05	1.66019E+05	6.02340E-06	2.49996	0.40001
34	9.29722E+04	1.07559E-05	2.32428E+05	4.30241E-06	2.49997	0.40000
35	1.30161E+05	7.68279E-06	3.25400E+05	3.07314E-06	2.49998	0.40000
36	1.82226E+05	5.48770E-06	4.55561E+05	2.19509E-06	2.49999	0.40000
37	2.55116E+05	3.91979E-06	6.37787E+05	1.56792E-06	2.49999	0.40000
38	3.57162E+05	2.79985E-06	8.92903E+05	1.11994E-06	2.49999	0.40000
39	5.00027E+05	1.99989E-06	1.25006E+06	7.99959E-07	2.50000	0.40000
40	7.00038E+05	1.42849E-06	1.75009E+06	5.71399E-07	2.50000	0.40000
41	9.80053E+05	1.02035E-06	2.45013E+06	4.08142E-07	2.50000	0.40000
42	1.37207E+06	7.28824E-07	3.43018E+06	2.91530E-07	2.50000	0.40000
43	1.92090E+06	5.20588E-07	4.80226E+06	2.08235E-07	2.50000	0.40000
44	2.68926E+06	3.71849E-07	6.72316E+06	1.48740E-07	2.50000	0.40000
45	3.76497E+06	2.65606E-07	9.41242E+06	1.06243E-07	2.50000	0.40000
46	5.27096E+06	1.89719E-07	1.31774E+07	7.58875E-08	2.50000	0.40000
47	7.37934E+06	1.35513E-07	1.84484E+07	5.42054E-08	2.50000	0.40000
48	1.03311E+07	9.67953E-08	2.58277E+07	3.87181E-08	2.50000	0.40000
49	1.44635E+07	6.91395E-08	3.61588E+07	2.76558E-08	2.50000	0.40000
50	2.02489E+07	4.93854E-08	5.06223E+07	1.97541E-08	2.50000	0.40000
55	1.08904E+08	9.18244E-09	2.72259E+08	3.67298E-09	2.50000	0.40000
60	5.85709E+08	1.70733E-09	1.46427E+09	6.82933E-10	2.50000	0.40000
65	3.15009E+09	3.17452E-10	7.87521E+09	1.26981E-10	2.50000	0.40000
70	1.69419E+10	5.90252E-11	4.23548E+10	2.36101E-11	2.50000	0.40000
75	9.11177E+10	1.09748E-11	2.27794E+11	4.38993E-12	2.50000	0.40000
80	4.90053E+11	2.04060E-12	1.22513E+12	8.16239E-13	2.50000	0.40000
85	2.63562E+12	3.79417E-13	6.58905E+12	1.51767E-13	2.50000	0.40000
90	1.41750E+13	7.05467E-14	3.54375E+13	2.82187E-14	2.50000	0.40000
95	7.62366E+13	1.31171E-14	1.90591E+14	5.24683E-15	2.50000	0.40000
100	4.10019E+14	2.43891E-15	1.02505E+15	9.75565E-16	2.50000	0.40000

50.0% = i, interest rate

number of time periods, n	F/P(i%,n)	P/F(i%,n)	F/A(i%,n)	A/F(i%,n)	P/A(i%,n)	A/P(i%,n)
1	1.50000	0.66667	1.00000	1.00000	0.66667	1.50000
2	2.25000	0.44444	2.50000	0.40000	1.11111	0.90000
3	3.37500	0.29630	4.75000	0.21053	1.40741	0.71053
4	5.06250	0.19753	8.12500	0.12308	1.60494	0.62308
5	7.59375	0.13169	13.18750	0.07583	1.73663	0.57583
6	11.39063	0.08779	20.78125	0.04812	1.82442	0.54812
7	17.08594	0.05853	32.17188	0.03108	1.88294	0.53108
8	25.62891	0.03902	49.25781	0.02030	1.92196	0.52030
9	38.44336	0.02601	74.88672	0.01335	1.94798	0.51335
10	57.66504	0.01734	113.33008	0.00882	1.96532	0.50882
11	86.49756	0.01156	170.99512	0.00585	1.97688	0.50585
12	129.74634	0.00771	257.49268	0.00388	1.98459	0.50388
13	194.61951	0.00514	387.23901	0.00258	1.98972	0.50258
14	291.92926	0.00343	581.85852	0.00172	1.99315	0.50172
15	437.89389	0.00228	873.78778	0.00114	1.99543	0.50114
16	656.84084	0.00152	1,311.68167	0.00076	1.99696	0.50076
17	985.26125	0.00101	1,968.52251	0.00051	1.99797	0.50051
18	1,477.89188	0.00068	2,953.78376	0.00034	1.99865	0.50034
19	2,216.83782	0.00045	4,431.67564	0.00023	1.99910	0.50023
20	3,325.25673	0.00030	6,648.51346	0.00015	1.99940	0.50015
21	4,987.88510	2.00486E-04	9,973.77019	1.00263E-04	1.99960	0.50010
22	7,481.82764	1.33657E-04	14,961.65529	6.68375E-05	1.99973	0.50007
23	11,222.74146	8.91048E-05	22,443.48293	4.45564E-05	1.99982	0.50004
24	16,834.11220	5.94032E-05	33,666.22439	2.97034E-05	1.99988	0.50003
25	25,251.16829	3.96021E-05	50,500.33659	1.98018E-05	1.99992	0.50002
26	3.78768E+04	2.64014E-05	7.57515E+04	1.32011E-05	1.99995	0.50001
27	5.68151E+04	1.76009E-05	1.13628E+05	8.80063E-06	1.99996	0.50001
28	8.52227E+04	1.17340E-05	1.70443E+05	5.86705E-06	1.99998	0.50001
29	1.27834E+05	7.82264E-06	2.55666E+05	3.91135E-06	1.99998	0.50000
30	1.91751E+05	5.21510E-06	3.83500E+05	2.60756E-06	1.99999	0.50000
31	2.87627E+05	3.47673E-06	5.75251E+05	1.73837E-06	1.99999	0.50000
32	4.31440E+05	2.31782E-06	8.62878E+05	1.15891E-06	2.00000	0.50000
33	6.47160E+05	1.54521E-06	1.29432E+06	7.72608E-07	2.00000	0.50000
34	9.70740E+05	1.03014E-06	1.94148E+06	5.15072E-07	2.00000	0.50000
35	1.45611E+06	6.86761E-07	2.91222E+06	3.43381E-07	2.00000	0.50000
36	2.18416E+06	4.57841E-07	4.36833E+06	2.28921E-07	2.00000	0.50000
37	3.27625E+06	3.05227E-07	6.55249E+06	1.52614E-07	2.00000	0.50000
38	4.91437E+06	2.03485E-07	9.82874E+06	1.01742E-07	2.00000	0.50000
39	7.37155E+06	1.35657E-07	1.47431E+07	6.78283E-08	2.00000	0.50000
40	1.10573E+07	9.04377E-08	2.21147E+07	4.52189E-08	2.00000	0.50000
41	1.65860E+07	6.02918E-08	3.31720E+07	3.01459E-08	2.00000	0.50000
42	2.48790E+07	4.01945E-08	4.97580E+07	2.00973E-08	2.00000	0.50000
43	3.73185E+07	2.67964E-08	7.46370E+07	1.33982E-08	2.00000	0.50000
44	5.59777E+07	1.78642E-08	1.11955E+08	8.93212E-09	2.00000	0.50000
45	8.39666E+07	1.19095E-08	1.67933E+08	5.95475E-09	2.00000	0.50000
46	1.25950E+08	7.93966E-09	2.51900E+08	3.96983E-09	2.00000	0.50000
47	1.88925E+08	5.29311E-09	3.77850E+08	2.64655E-09	2.00000	0.50000
48	2.83387E+08	3.52874E-09	5.66775E+08	1.76437E-09	2.00000	0.50000
49	4.25081E+08	2.35249E-09	8.50162E+08	1.17625E-09	2.00000	0.50000
50	6.37622E+08	1.56833E-09	1.27524E+09	7.84164E-10	2.00000	0.50000
55	4.84194E+09	2.06529E-10	9.68388E+09	1.03264E-10	2.00000	0.50000
60	3.67685E+10	2.71972E-11	7.35369E+10	1.35986E-11	2.00000	0.50000
65	2.79211E+11	3.58153E-12	5.58421E+11	1.79076E-12	2.00000	0.50000
70	2.12026E+12	4.71641E-13	4.24051E+12	2.35821E-13	2.00000	0.50000
75	1.61007E+13	6.21091E-14	3.22014E+13	3.10546E-14	2.00000	0.50000
80	1.22265E+14	8.17898E-15	2.44529E+14	4.08949E-15	2.00000	0.50000
85	9.28447E+14	1.07707E-15	1.85689E+15	5.38534E-16	2.00000	0.50000
90	7.05039E+15	1.41836E-16	1.41008E+16	7.09180E-17	2.00000	0.50000
95	5.35389E+16	1.86780E-17	1.07078E+17	9.33900E-18	2.00000	0.50000
100	4.06561E+17	2.45965E-18	8.13122E+17	1.22983E-18	2.00000	0.50000

100.0% = i, interest rate

number of time periods, n	F/P(i%,n)	P/F(i%,n)	F/A(i%,n)	A/F(i%,n)	P/A(i%,n)	A/P(i%,n)
1	2.00000	0.50000	1.00000	1.00000	0.50000	2.00000
2	4.00000	0.25000	3.00000	0.33333	0.75000	1.33333
3	8.00000	0.12500	7.00000	0.14286	0.87500	1.14286
4	16.00000	0.06250	15.00000	0.06667	0.93750	1.06667
5	32.00000	0.03125	31.00000	0.03226	0.96875	1.03226
6	64.00000	0.01563	63.00000	0.01587	0.98438	1.01587
7	128.00000	0.00781	127.00000	0.00787	0.99219	1.00787
8	256.00000	0.00391	255.00000	0.00392	0.99609	1.00392
9	512.00000	0.00195	511.00000	0.00196	0.99805	1.00196
10	1,024.00000	0.00098	1,023.00000	0.00098	0.99902	1.00098
11	2,048.00000	4.88281E-04	2,047.00000	4.88520E-04	0.99951	1.00049
12	4,096.00000	2.44141E-04	4,095.00000	2.44200E-04	0.99976	1.00024
13	8,192.00000	1.22070E-04	8,191.00000	1.22085E-04	0.99988	1.00012
14	16,384.00000	6.10352E-05	16,383.00000	6.10389E-05	0.99994	1.00006
15	32,768.00000	3.05176E-05	32,767.00000	3.05185E-05	0.99997	1.00003
16	65,536.00000	1.52588E-05	65,535.00000	1.52590E-05	0.99998	1.00002
17	131,072.00000	7.62939E-06	131,071.00000	7.62945E-06	0.99999	1.00001
18	262,144.00000	3.81470E-06	262,143.00000	3.81471E-06	1.00000	1.00000
19	524,288.00000	1.90735E-06	524,287.00000	1.90735E-06	1.00000	1.00000
20	1,048,576.00000	9.53674E-07	1,048,575.00000	9.53675E-07	1.00000	1.00000
21	2.09715E+06	4.76837E-07	2.09715E+06	4.76837E-07	1.00000	1.00000
22	4.19430E+06	2.38419E-07	4.19430E+06	2.38419E-07	1.00000	1.00000
23	8.38861E+06	1.19209E-07	8.38861E+06	1.19209E-07	1.00000	1.00000
24	1.67772E+07	5.96046E-08	1.67772E+07	5.96046E-08	1.00000	1.00000
25	3.35544E+07	2.98023E-08	3.35544E+07	2.98023E-08	1.00000	1.00000
26	6.71089E+07	1.49012E-08	6.71089E+07	1.49012E-08	1.00000	1.00000
27	1.34218E+08	7.45058E-09	1.34218E+08	7.45058E-09	1.00000	1.00000
28	2.68435E+08	3.72529E-09	2.68435E+08	3.72529E-09	1.00000	1.00000
29	5.36871E+08	1.86265E-09	5.36871E+08	1.86265E-09	1.00000	1.00000
30	1.07374E+09	9.31323E-10	1.07374E+09	9.31323E-10	1.00000	1.00000
31	2.14748E+09	4.65661E-10	2.14748E+09	4.65661E-10	1.00000	1.00000
32	4.29497E+09	2.32831E-10	4.29497E+09	2.32831E-10	1.00000	1.00000
33	8.58993E+09	1.16415E-10	8.58993E+09	1.16415E-10	1.00000	1.00000
34	1.71799E+10	5.82077E-11	1.71799E+10	5.82077E-11	1.00000	1.00000
35	3.43597E+10	2.91038E-11	3.43597E+10	2.91038E-11	1.00000	1.00000
36	6.87195E+10	1.45519E-11	6.87195E+10	1.45519E-11	1.00000	1.00000
37	1.37439E+11	7.27596E-12	1.37439E+11	7.27596E-12	1.00000	1.00000
38	2.74878E+11	3.63798E-12	2.74878E+11	3.63798E-12	1.00000	1.00000
39	5.49756E+11	1.81899E-12	5.49756E+11	1.81899E-12	1.00000	1.00000
40	1.09951E+12	9.09495E-13	1.09951E+12	9.09495E-13	1.00000	1.00000
41	2.19902E+12	4.54747E-13	2.19902E+12	4.54747E-13	1.00000	1.00000
42	4.39805E+12	2.27374E-13	4.39805E+12	2.27374E-13	1.00000	1.00000
43	8.79609E+12	1.13687E-13	8.79609E+12	1.13687E-13	1.00000	1.00000
44	1.75922E+13	5.68434E-14	1.75922E+13	5.68434E-14	1.00000	1.00000
45	3.51844E+13	2.84217E-14	3.51844E+13	2.84217E-14	1.00000	1.00000
46	7.03687E+13	1.42109E-14	7.03687E+13	1.42109E-14	1.00000	1.00000
47	1.40737E+14	7.10543E-15	1.40737E+14	7.10543E-15	1.00000	1.00000
48	2.81475E+14	3.55271E-15	2.81475E+14	3.55271E-15	1.00000	1.00000
49	5.62950E+14	1.77636E-15	5.62950E+14	1.77636E-15	1.00000	1.00000
50	1.12590E+15	8.88178E-16	1.12590E+15	8.88178E-16	1.00000	1.00000
55	3.60288E+16	2.77556E-17	3.60288E+16	2.77556E-17	1.00000	1.00000
60	1.15292E+18	8.67362E-19	1.15292E+18	8.67362E-19	1.00000	1.00000
65	3.68935E+19	2.71051E-20	3.68935E+19	2.71051E-20	1.00000	1.00000
70	1.18059E+21	8.47033E-22	1.18059E+21	8.47033E-22	1.00000	1.00000
75	3.77789E+22	2.64698E-23	3.77789E+22	2.64698E-23	1.00000	1.00000
80	1.20893E+24	8.27181E-25	1.20893E+24	8.27181E-25	1.00000	1.00000
85	3.86856E+25	2.58494E-26	3.86856E+25	2.58494E-26	1.00000	1.00000
90	1.23794E+27	8.07794E-28	1.23794E+27	8.07794E-28	1.00000	1.00000
95	3.96141E+28	2.52435E-29	3.96141E+28	2.52435E-29	1.00000	1.00000
100	1.26765E+30	7.88861E-31	1.26765E+30	7.88861E-31	1.00000	1.00000

Appendix D
Lighting Tables

D.1 Legacy fluorescent lighting

Generic designation NEMA	Lamp watts	Bulb type	Base	Rated life[1] (h)	Maximum overall length (mm)	Maximum overall length (in.)	Lamp current (A)	Lamp voltage (V)	Approx. Initial Lumens[2]	Lumens Per Watt	Color[3] temperature and / or CRI
Twin tube 2700K, 3500K, 4100K, 5000K[4], CRI 80 +											
CFT5W / G23	5	T-4	G23	7500	105	4-1/4	0.180	38	250	50	82
CFT7W / G23	7	T-4	G23	10,000	135	5-7/16	0.180	45	400	51	82
CFT9W / G23	9	T-4	G23	10,000	167	6-9/16	0.180	59	600	67	82
CFT13W / GX23	13	T-4	GX23	10,000	191	7-1/2	0.285	60	888	68	82
CFT5W / 2G7	5	T-4	2G7	10,000	85	3-11/32	0.180	35	250	50	82
FT7W / 2G7	7	T-4	2G7	10,000	115	4-17/32	0.180	45	400	57	82
CFT7W / 2G7[5,6]	9	T-4	2G7	10,000	145	5-22/32	0.180	59	600	67	82
CFT13W / 2GX7	13	T-4	2GX7	10,000	175	6-29/32	0.285	59	900	69	82
FT18W / 2G11[7]	18	T-5	2G11	12,000	229	9	0.375	60	1250	69	82
FT18W / 2G11RS	18	T-5	2G11	20,000	267	10-1/2	0.250	76	1250	69	82
FT24W / 2G11	24-27	T-5	2G11	12,000	328	12-29/32	0.340	91	1800	69	82
FT36W / 2G11	36-39	T-5	2G11	12,000	422	10-5/8	0.430	111	2900	76	82
FT40W / 2G11	40	T-5	2G11	20,000	574	22-19/32	0.270	169	1350	79	82
FT50W / 2G11	50	T-5	2G11	14,000	574	22-19/32	0.43	147	4000	80	—
Quad 2700K, 3500K, 4100K, 5000K; CRI 80 +											
CFQ9W / G23	9	T-4	G23-2	10,000	111	4-3/8	0.180	15	575	21	82
CFQ13W / GX23	13	T-4	GX23-2	10,000	125	4-29/32	0.285	15	860	59	82
CFQ10W / G24d	10	T-4	G24d-1	10,000	118	46	0.140	64	600	60	82
CFQ13W / G24d	13	T-4	G24d-1	10,000	152	6	0.170	91	900	69	82
CFQ18W / G24d	18	T-4	G24d-2	10,000	175	6-29/32	0.220	100	1250	69	82
CFQ28W / G24d	26	T-4	G24d-3	10,000	196	7-23/32	0.315	105	1800	69	82
CFQ15W / GX32d	15	T-5	GX32d-1	10,000	140	5-1/2	—	60	900	60	—
CFQ22W / GX32d	20	T-5	GX32d-2	10,000	151	5-15/16	—	53	1200	60	—
CFQ26W / GX32d	27	T-5	GX32d-3	10,000	173	6-13/16	—	54	1600	59	—
CFQ10W / G24q	10	T-4	G24q-1	10,000	117	4-5/8	0.190	64	600	60	82
CFQ13W / G24q	13	T-4	G24q-1	10,000	152	6	0.170	77	900	69	82
CFQ18W / G24q	18	T-4	G24q-2	10,000	173	6-13/16	0.110	80	1250	69	82
CFQ26W / G24q	26	T-4	G24q-3	10,000	194	7-5/8	0.158	80	1800	69	82

Generic designation	Ballast type	Incandescent equivalent (W)	Lamp watts	Bulb type	Base	Related life[1] (h)	Maximum overall length[9] (mm)	Maximum overall length[9] (in.)	Lamp current (A)	Lamp voltage (V)	Approx. Initial Lumens[2] (Candle-power)	Lumens Per watt	Color[3] temperature and / or CRI
7 W[10]	Electronic	25	7	T-4	Med. Screw	10,000	140	5-1 / 2	0.140	120	400	57	—
11 W[10]	Electronic	40	11	T-4	Mod. Screw	10,000	140	5-1 / 2	0.170	120	600	55	—
11 W reflector[10]	Electronic	50W .R30	11	P-35	Med. Screw	10,000	148	5-13 / 16	0.170	120	(315)	—	—
11 W globe[10]	Electronic	30	11	G-32	Med. Screw	10,000	168	6-9 / 16	0.170	120	450	41	—
15 W[5,7,10]	Electronic	60	15	T-4	Mod. Screw	10,000	172	6-25 / 32	0.240	120	900	60	81
15 W reflector[10]	Electronic	75W .R40	15	RSB	Med. Screw	10,000	183	7-7 / 32	0.240	120	(1335)	—	—
15 W globe[10]	Electronic	50	15	G-38	Med. Screw	10,000	188	7-13 / 32	0.240	120	700	47	—
15 W globe[7]	Magnetic	40-60	18	G-30	Med. Screw	9,000	160	6-5 / 16	—	120	700	47	82
17 W decorative diffuser[5]	Electronic	60	17	T-24	Med. Screw	10,000	149	5-7 / 8	0.265	120	950	57	—
18 W decorative diffuser[7]	Electronic	75	18	T-24	Med. Screw	10,000	183	7-7 / 32	0.240	120	1100	61	—
18 W reflector	Electronic	75W .R40	18	R-40	Mod. Screw	10,000	142	5-19 / 32	0.240	120	800	44	—
18 W[11]	Electronic	75	18	T-4	Med. Screw	10,000	175	6-7 / 8	—	120	1100	61	—
20 W[5,7,10]	Electronic	75	20	T-4	Med. Screw	10,000	203	8	0.265[12]	120	1200	60	—
22 W circular lamp[7]	Magnetic	75	22	T-9	Med. Screw	10,000	130	5-1 / 8	—	120	1200	54	52
23 W[5,10]	Electronic	90	23	T-4	Med. Screw	10,000	176	6-5 / 16	0.325	120	1550	67	—
26 W[7]	Electronic	90	26	Y-5	Med. Screw	10,000	203	8	0.430	120	1500	58	—
Square 2700 K and 3500 K[7]													
F102D / 827 / 4P		10		2D	GR10Q 4PI	8000	94	3-23 / 32	0.18	72	650	65	82
F162D / 827 / 4P		16		2D	GR10Q 4PI	8000	140	5-1 / 2	0.195	103	1050	66	82
F212D / 827 / 4P		21		2D	GR10Q 4PI	8000	140	5-1 / 2	0.26	101	1350	64	82
F282D / 827 / 4P		28		2D	GR10Q 4PI	10,000	203	8	0.32	107	2050	73	82
F382D / 827 / 4P[13,14]		38		2D	GR10Q 4PI	10,000	203	8	0.43	110	2850	79	82

1. At 3 hours per start.
2. Values for reflector lamps are in beam candlepower.
3. These values are the averages of several manufacturers' data.
4. Not all color temperatures available from each manufacturer.
5. Philips
6. Also available in red, blue, green.
7. General Electric
8. Screw base adapters with integral ballasts are available for retrofitting twin and quad tube lamps into incandescent fixtures.
9. Some lamps are available in shorter MOLS.
10. Osram
11. Sylvania
12. Lamps are available with higher power factors and lower line currents.
13. Add 2 Watts for cathodes when operating on rapid-start circuits.
14. Rapid start life is estimated as 12,000 hours at 3 or more hours per start.

D.2: Checklist of Energy Saving Guidelines for Lighting Ideas

Lighting Needs		
☐	Visual tasks: specification	Identify specific visual tasks and locations to determine recommended illuminances for tasks and for surrounding areas.
☐	Safety and esthetics	Review lighting requirements forgiven applications to satisfy safety and esthetic criteria.
☐	Over-lighted application	In existing spaces, identify applications where maintained illumination is greater than recommended. Reduce energy by adjusting illuminance to meet recommended levels.
☐	Groupings: similar visual tasks	Group visual tasks having the same illuminance requirements, and avoid widely separated workstations.
☐	Task lighting	Illuminate work surfaces with luminaires properly located in or on furniture; provide lower ambient levels.
☐	Luminance ratios	Use wall washing and lighting of decorative objects to balance brightness.
Space Design and Utilization		
☐	Space plan	When possible, arrange for occupants working after hours to work in close proximity to one another.
☐	Room surfaces	Use light colors for walls, floors, ceilings, and furniture to increase utilization of light, and reduce connected lighting power to achieve required illumination. Avoid glossy finishes on room and work surfaces to limit reflected glare.
☐	Space utilization: branch circuit wiring	Use modular brank circuit wiring to allow for flexibility in moving, relocating, or adding luminaires to suit changing space configurations.
☐	Space utilization: occupancy	Light building for occupied periods only, and when required for security or cleaning purposes.
Daylighting		
☐	Daylight compensation	If daylighting can be used to replace some electric lighting near fenestration during substantial periods of the day, lighting in those areas should be circuited so that it may be controlled manually or automatically by switching or dimming.
☐	Daylight sensing	Daylight sensors and dimming systems can reduce electric lighting energy.
☐	Daylight control	Maximize the effectiveness of existing fenestration-shading controls (interior and exterior) or replace with proper devices or shielding media.

☐	Space utilization	Use daylighting in transition zones, in lounge and recreational areas, and for functions where the variation in color, intensity and direction may be desirable. Consider applications where daylight can be utilized as ambient lighting, supplemented by local task lights.
Lighting sources: Lamps and Ballasts		
☐	Source efficacy	Install light sources with the highest efficacies to provide the desired light source color and distribution requirements.
☐	Fluorescent lamps	Use T8 and/or T5 fluorescent for improved source efficacy and color quality.
☐	Ballasts	Use electronic or energy efficient ballasts with fluorescent and HID lamps.
☐	System change	Substitute solid state (i.e. LED) lighting systems for existing incandescent, fluorescent, mercury vapor metal halide and high-pressure sodium lighting systems.
☐	LED luminaires and retrofit kits	Use LED luminaires or retrofit kits, where possible, to replace incandescent, fluorescent and HID luminaires.
☐	LED lamps replacements	If keeping existing luminaires, use LED lamps with good lumen maintenance and colour rendering to replace existing incandescent, fluorescent and HID lamps. Follow the manufacturers instructions as technically required to keep, remove or replace existing ballasts.
☐	Control compatibility	If a control system is used, check compatibility of light sources with the control device.
Luminaires		
☐	Maintain efficiency	Select luminaires which do not collect dirt rapidly and which can be easily cleaned.
☐	Improved maintenance	Improved maintenance procedures may enable a lighting system with reduced wattage to provide adequate illumination throughout system or component life.
☐	Luminaire efficiency: replacement or relocation	Check luminaire effectiveness for task lighting and for overall efficiency. If ineffective or inefficient, consider replacement or relocation.
☐	Heat removal	When luminaire temperatures exceed optimal system operating limits, consider using heat removal luminaires to improve lamp performance and reduce heat gain to the space. The decrease in lamp temperature may, however, actually increase power consumption.
☐	Maintained efficiency	Select lamp replacement schedule for all light sources, to more accurately predict light loss factors and possibly decrease the number of luminaires required.
Lighting Controls		
☐	Switching: local control	Install switches for local and convenient control of lighting by occupants. This should be in combination with a building-wide system to turn lights off when the building is occupied.

☐	Selective switching	Install selective switching of luminaires according to groupings of working tasks and different working hours.
☐	Low-voltage switching systems	Use low-voltage switching systems to obtain maximum switching capability.
☐	Master control system	Use a programmable low-voltage master switching system for the entire building to turn lights on and off automatically as needed, with overrides at individual areas.
☐	Multipurpose spaces	Install multi-circuit switching or preset dimming controls to provide flexibility when spaces are used for multiple purposes and require different ranges of illuminance for various activities. Clearly label the control cover plates.
☐	"Tuning" illuminance	Use switching and dimming systems as a means of adjusting illuminance for variable lighting requirements.
☐	Scheduling	Operate lighting according to predetermined schedule.
☐	Occupant/motion sensors	Use occupants/motion sensors for unpredictable patterns of occupancy.
☐	Lumen maintenance	Fluorescent dimming systems may be utilized to maintain illuminance throughout lamp life, thereby saving energy by compensating for lamp-lumen depreciation and other light loss factors.
☐	Ballast switching	Use multi-level ballasts and local inboard-outboard lamp switching where a reduction in illuminances is sometimes desired.
Operation and Maintenance		
☐	Education	Analyze lighting used during working and building cleaning periods, and institute an education program to have personnel turn off lighting when in use (fluorescent lamps if not used for 5 minutes or longer; HID lamps if space will not be used 30 minutes or longer).
☐	Parking	Restrict parking after hours to specific lots so lighting can be reduced to minimum security requirements in unused parking areas.
☐	Custodial service	Schedule routine building cleaning during occupied hours.
☐	Reduced illuminance	Reduce illuminance during building cleaning periods if building is not otherwise occupied.
☐	Cleaning schedules	Adjust cleaning schedules to minimize time of operation, by concentrating cleaning activities in fewer spaces at the same time and by turning off lights in unoccupied areas.
☐	Program evaluation	Evaluate the present lighting maintenance program, and revise it as necessary to provide the most efficient use of the lighting system

☐	Cleaning and maintenance	Clean luminaires and replace lamps on a regular maintenance schedule to ensure proper illumination levels are maintained.
☐	Regular system checks	Check to see if all components are in good working condition. Transmitting or diffusing media should be examined, and badly discolored or deteriorated media placed to improve efficiency.
☐	Renovation of luminaires	Replace outdated or damages luminaires with modern ones which have good cleaning capabilities and which use lamps with higher efficacy and good lumen maintenance characteristics.
☐	Area maintenance	Trim trees and bushes that may be obstructing outdoor luminaire distribution and creating unwanted shadows.

Appendix E

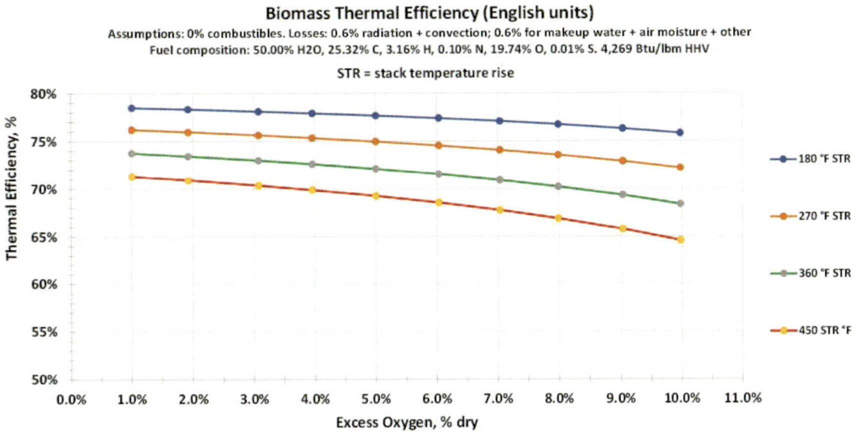

Biomass Thermal Efficiency (English units)
Assumptions: 0% combustibles. Losses: 0.6% radiation + convection; 0.6% for makeup water + air moisture + other
Fuel composition: 50.00% H2O, 25.32% C, 3.16% H, 0.10% N, 19.74% O, 0.01% S. 4,269 Btu/lbm HHV
STR = stack temperature rise
Thermal Efficiency, %
80%
75%
70%
65%
60%
55%
50%
0.0%
1.0%
2.0%
3.0%
4.0%
5.0%
6.0%
7.0%
8.0%
9.0%
10.0%
11.0%
Excess Oxygen, % dry
180 °F STR
270 °F STR
360 °F STR
450 STR °F

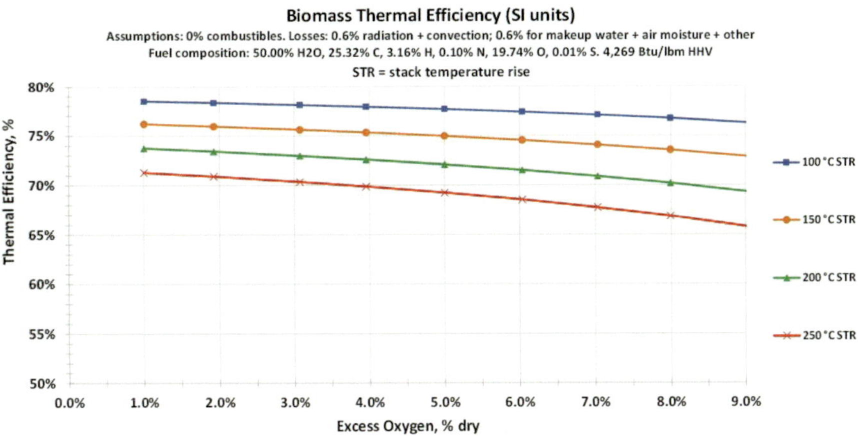

Biomass Thermal Efficiency (SI units)
Assumptions: 0% combustibles. Losses: 0.6% radiation + convection; 0.6% for makeup water + air moisture + other
Fuel composition: 50.00% H2O, 25.32% C, 3.16% H, 0.10% N, 19.74% O, 0.01% S. 4,269 Btu/lbm HHV
STR = stack temperature rise
Thermal Efficiency, %
80%
75%
70%
65%
60%
55%
50%
0.0%
1.0%
2.0%
3.0%
4.0%
5.0%
6.0%
7.0%
8.0%
9.0%
Excess Oxygen, % dry
100 °C STR
150 °C STR
200 °C STR
250 °C STR

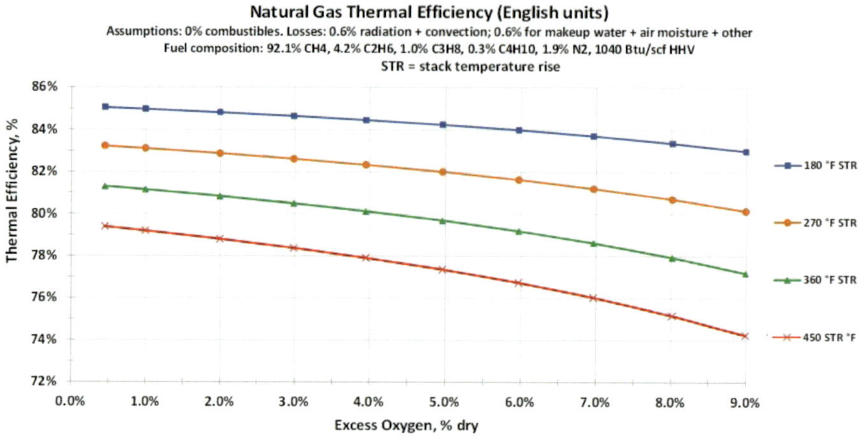
Natural Gas Thermal Efficiency (English units)
Assumptions: 0% combustibles. Losses: 0.6% radiation + convection; 0.6% for makeup water + air moisture + other
Fuel composition: 92.1% CH4, 4.2% C2H6, 1.0% C3H8, 0.3% C4H10, 1.9% N2, 1040 Btu/scf HHV
STR = stack temperature rise
Thermal Efficiency, %
86%
84%
82%
80%
78%
76%
74%
72%
0.0%
1.0%
2.0%
3.0%
4.0%
5.0%
6.0%
7.0%
8.0%
9.0%
Excess Oxygen, % dry
180 °F STR
270 °F STR
360 °F STR
450 STR °F

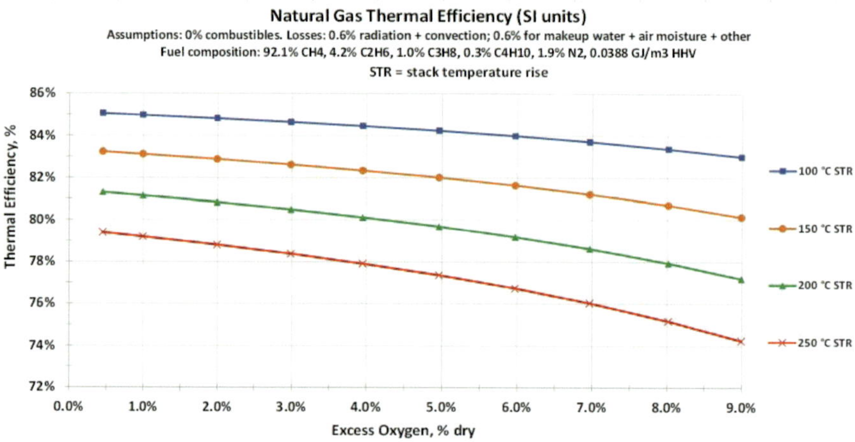
Natural Gas Thermal Efficiency (SI units)
Assumptions: 0% combustibles. Losses: 0.6% radiation + convection; 0.6% for makeup water + air moisture + other
Fuel composition: 92.1% CH4, 4.2% C2H6, 1.0% C3H8, 0.3% C4H10, 1.9% N2, 0.0388 GJ/m3 HHV
STR = stack temperature rise
Thermal Efficiency, %
86%
84%
82%
80%
78%
76%
74%
72%
0.0%
1.0%
2.0%
3.0%
4.0%
5.0%
6.0%
7.0%
8.0%
9.0%
Excess Oxygen, % dry
100 °C STR
150 °C STR
200 °C STR
250 °C STR

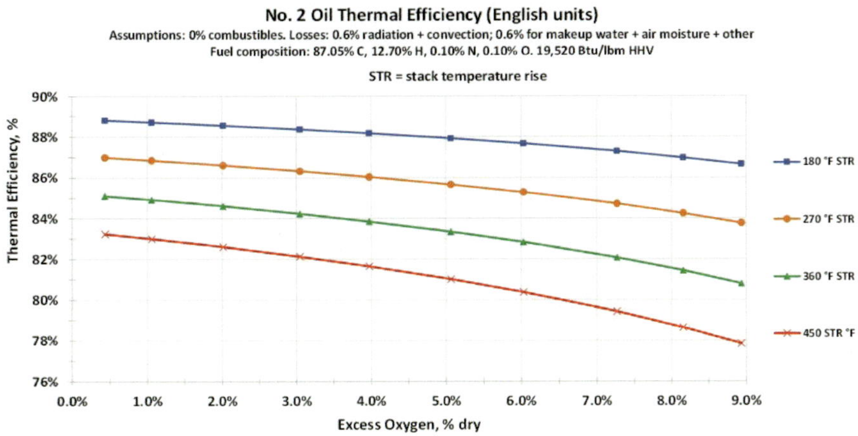
No. 2 Oil Thermal Efficiency (English units)
Assumptions: 0% combustibles. Losses: 0.6% radiation + convection; 0.6% for makeup water + air moisture + other
Fuel composition: 87.05% C, 12.70% H, 0.10% N, 0.10% O. 19,520 Btu/lbm HHV
STR = stack temperature rise
Thermal Efficiency, %
90%
88%
86%
84%
82%
80%
78%
76%
0.0%
1.0%
2.0%
3.0%
4.0%
5.0%
6.0%
7.0%
8.0%
9.0%
Excess Oxygen, % dry
180 °F STR
270 °F STR
360 °F STR
450 STR °F

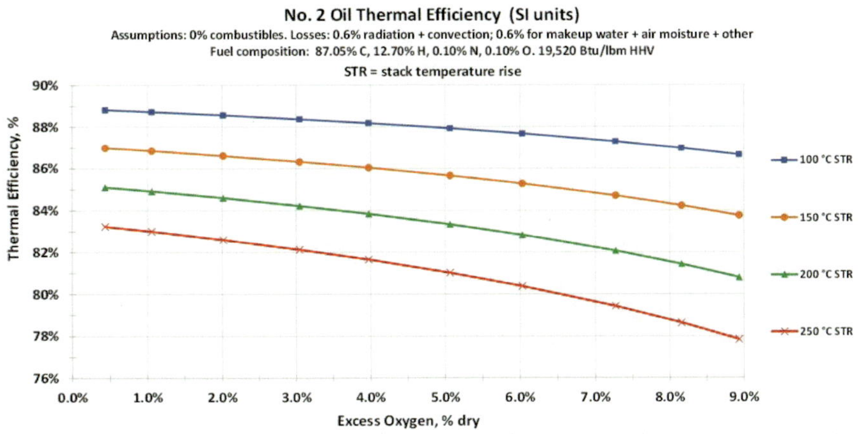
No. 2 Oil Thermal Efficiency (SI units)
Assumptions: 0% combustibles. Losses: 0.6% radiation + convection; 0.6% for makeup water + air moisture + other
Fuel composition: 87.05% C, 12.70% H, 0.10% N, 0.10% O. 19,520 Btu/lbm HHV
STR = stack temperature rise
Thermal Efficiency, %
90%
88%
86%
84%
82%
80%
78%
76%
0.0%
1.0%
2.0%
3.0%
4.0%
5.0%
6.0%
7.0%
8.0%
9.0%
Excess Oxygen, % dry
100 °C STR
150 °C STR
200 °C STR
250 °C STR

Appendix F

Saturated steam - pressure: English units
Saturated steam - temperature: English units
Superheated steam: English units
Saturated steam - pressure: international units
Saturated steam - temperature: international units
Superheated steam: international units
Source: ASME steam tables compact edition (2006). Some values were interpolated where not provided in the ASME tables

Table F.1 Saturated steam: pressure table (English units)

p psia	T °F	v_f ft³/lbm	v_g ft³/lbm	h_f Btu/lbm	h_{fg} Btu/lb	h_g Btu/lb	s_f Btu/(lbm °R)	s_g Btu/(lbm °R)
0.1	35.0	0.01602	2945	3.01	1074.0	1077	0.0061	2.176
0.15	45.4	0.01602	2004	13.48	1067.5	1081	0.0271	2.141
0.3	64.5	0.01604	1039	32.53	1056.5	1089	0.0641	2.080
0.6	85.2	0.01609	539.9	53.24	1044.8	1098	0.1028	2.021
1	101.7	0.01614	333.5	69.73	1035.3	1105	0.1326	1.978
1.5	115.6	0.01619	227.7	83.65	1027.4	1111	0.1571	1.944
3	141.4	0.01630	118.7	109.40	1012.6	1122	0.2009	1.886
6	170.0	0.01645	61.98	138.00	996.0	1134	0.2474	1.829
10	193.2	0.01659	38.42	161.20	981.8	1143	0.2836	1.788
15	213.0	0.01672	26.3	181.20	969.8	1151	0.3137	1.755
30	250.3	0.01700	13.75	218.90	945.1	1164	0.3682	1.700
60	292.7	0.01738	7.176	262.20	915.8	1178	0.4273	1.644
100	327.8	0.01774	4.432	298.60	888.4	1187	0.4744	1.603
200	381.8	0.01839	2.288	355.50	843.5	1199	0.5439	1.546
300	417.4	0.01890	1.543	394.00	809.0	1203	0.5883	1.511
400	444.6	0.01934	1.162	424.20	780.8	1205	0.6217	1.485
500	467.0	0.01975	0.9282	449.50	755.5	1205	0.6490	1.464
600	486.2	0.02014	0.7702	471.70	732.3	1204	0.6723	1.446
700	503.1	0.02051	0.6559	491.60	710.4	1202	0.6928	1.431
850	525.3	0.02106	0.5332	518.40	679.6	1198	0.7197	1.409
1000	544.7	0.02160	0.4461	542.60	650.4	1193	0.7434	1.391
1150	561.9	0.02214	0.3808	564.80	621.2	1186	0.7647	1.373
1300	577.5	0.02270	0.3299	585.50	593.5	1179	0.7843	1.357
1450	591.8	0.02327	0.289	605.20	566.8	1172	0.8025	1.341
1600	604.9	0.02386	0.2553	624.10	538.9	1163	0.8197	1.326
1750	617.2	0.02449	0.2269	642.30	511.7	1154	0.8361	1.311
1900	628.6	0.02516	0.2026	660.10	483.9	1144	0.8519	1.296
2050	639.4	0.02588	0.1814	677.60	455.4	1133	0.8673	1.281
2200	649.5	0.02668	0.1627	695.10	424.9	1120	0.8825	1.266
2350	659.1	0.02757	0.1459	712.70	394.3	1107	0.8976	1.250
2500	668.2	0.02859	0.1307	730.90	361.1	1092	0.9131	1.230
2650	676.8	0.02982	0.1166	749.80	324.2	1074	0.9290	1.210
2800	685.0	0.03136	0.1031	770.30	283.7	1054	0.9463	1.190
2950	692.9	0.03345	0.0894	793.70	234.3	1028	0.9659	1.170
3100	700.3	0.03687	0.074	824.00	165.0	989	0.9914	1.130
3200.11	705.1	0.04975	0.04975	897.50	0.0	898	1.0540	1.054

Table F.2 Saturated steam: temperature table (English units)

T °F	p psia	v_f ft^3/lbm	v_g ft^3/lbm	h_f Btu/lbm	h_{fg} Btu/lb	h_g Btu/lb	s_f Btu/(lbm °R)	s_g Btu/(lbm °R)
32	0.09	0.01602	3,302	-0.0179	1,075	1,075	-3.7E-05	2.187
40	0.12	0.01602	2,443	8.032	1,071	1,079	0.0162	2.159
60	0.26	0.01603	1,206	28.0800	1,059	1,087	0.05554	2.094
80	0.51	0.01607	632	48.07	1,048	1,096	0.0933	2.035
100	0.95	0.01613	350	68.0400	1,037	1,105	0.1296	1.982
120	1.69	0.0162	203	88	1,025	1,113	0.1647	1.933
140	2.89	0.01629	123	108.0000	1,014	1,122	0.1985	1.889
160	4.75	0.01639	77	128	1,002	1,130	0.2313	1.848
180	7.52	0.01651	50	148.0000	990	1,138	0.2631	1.811
200	11.54	0.01663	34	168.1	978	1,146	0.2940	1.776
220	17.20	0.01677	23	188.2000	965	1,153	0.3241	1.744
240	24.99	0.01692	16	208.5	953	1,161	0.3534	1.714
260	35.45	0.01708	12	228.8000	938	1,167	0.382	1.686
280	49.22	0.01726	9	249.2	925	1,174	0.4099	1.660
300	67.02	0.01745	6	269.8000	910	1,180	0.4372	1.635
320	89.65	0.01765	5	290.4	895	1,185	0.4640	1.612
340	118.00	0.01787	4	311.3000	879	1,190	0.4903	1.590
360	153.00	0.01811	3	332.3	863	1,195	0.5162	1.568
380	195.71	0.01836	2	353.6000	845	1,199	0.5416	1.548
400	247.22	0.01864	2	375.1	826	1,201	0.5667	1.528
420	308.71	0.01894	2	396.9000	807	1,204	0.5915	1.509
440	381.44	0.01926	1	419	786	1,205	0.6161	1.490
460	466.71	0.01962	1	441.5000	764	1,205	0.6405	1.471
480	565.92	0.02001	1	464.4	740	1,204	0.6648	1.452
500	680.53	0.02044	1	487.9000	714	1,202	0.689	1.433
520	812.08	0.02092	1	511.9	687	1,199	0.7133	1.415
540	962.23	0.02146	0	536.7000	657	1,194	0.7377	1.395
560	1132.70	0.02208	0	562.3	625	1,187	0.7624	1.375
580	1325.40	0.02279	0	589.0000	589	1,178	0.7875	1.354
600	1542.50	0.02363	0	616.9	550	1,167	0.8133	1.332
620	1786.10	0.02465	0	646.6000	505	1,152	0.84	1.308
640	2059.20	0.02593	0	678.7	453	1,132	0.8683	1.280
660	2364.80	0.02766	0	714.5000	391	1,105	0.8991	1.248
680	2707.30	0.03036	0	757.4	309	1,066	0.9355	1.207
700	3092.90	0.03666	0	822.5000	169	991	0.99	1.135
705.103	3200.10	0.04975	0	897.5	0	898	1.0540	1.054

Table F.3 Superheated steam (English units)

psia Tsat,°F			200 °F	250 °F	300 °F	350 °F	400 °F	450 °F	500 °F	550 °F	600 °F	650 °F	700 °F	750 °F	800 °F	900 °F	1000 °F	1100 °F	1200 °F	1300 °F	1400 °F	1500 °F
1 psia	v	ft3/lbm	392.53	422.42	452.28	482.11	511.93	541.74	571.55	601.35	631.15	660.95	690.74	720.53	750.32	809.91	869.48	929.06	988.64			
101.69 °F	h	Btu/lbm	1,150.1	1,172.8	1,195.7	1218,6	1,241.8	1,265.1	1,288.6	1,312.4	1,336.2	1,360.40	1,384.6	1,409.25	1,433.9	1,484.1	1,535.1	1,587.6	1,640.0			
	s	Btu/lbm R	2.0510	2.0842	2.1152	2.1445	2.1723	2.1986	2.2238	2.2474	2.2710	2.29	2.3146	2.34	2.3554	2.3937	2.4299	2.4636	2.4973			
5 psia	v	ft3/lbm	78.16	84.22	90.25	96.25	102.25	108.23	114.21	120.18	126.15	132.12	138.09	144.06	150.02	161.94	173.87	185.79	197.71			
162.18 °F	h	Btu/lbm	1,148.5	1,171.7	1,194.8	1,218.0	1,241.3	1,264.7	1,288.2	1,312.1	1,335.9	1,360.15	1,384.4	1,409.05	1,433.7	1,483.9	1,535.0	1,587.5	1,640.0			
	s	Btu/lbm R	1.8716	1.9055	1.9370	1.9665	1.9944	2.0209	2.0461	2.0698	2.0934	2.1153	2.1371	2.1575	2.1779	2.2162	2.2525	2.2862	2.3198			
10 psia	v	ft3/lbm	38.85	41.94	44.99	48.02	51.04	54.04	57.40	60.22	63.03	66.02	69.01	71.99	74.98	80.95	86.92	92.88	98.84			
193.16 °F	h	Btu/lbm	1,146.4	1,170.2	1,193.8	1,217.2	1,240.6	1,264.1	1,287.8	1,311.7	1,335.6	1,359.90	1,384.2	1,408.85	1,433.5	1,483.8	1,534.9	1,587.4	1,639.9			
	s	Btu/lbm R	1.7926	1.8275	1.8595	1.8893	1.9174	1.9440	1.9693	1.9930	2.0167	2.01	2.0065	2.05	2.1013	2.1397	2.1760	2.2097	2.2434			
20 psia	v	ft3/lbm		20.796	22.362	23.903	25.430	26.947	28.458	29.963	31.467	32.967	34.467	35.964	37.461	40.451	43.438	46.423	49.408			
227.92 °F	h	Btu/lbm		1,167.2	1191.6·	1,215.5	1,239.3	1,263.0	1,286.9	1,310.9	1,334.9	1,359.25	1,383.6	1,408.35	1,433.1	1,483.4	1,534.6	1,587.2	1,639.7			
	s	Btu/lbm R		1.7477	1.7808	1.8113	1.8398	1.8667	1.8922	1.9160	1.9398	1.9618	1.9838	2.0043	2.0247	2.1)631	2.0994	2.1332	2.1669			
40 psia	v	ft3/lbm			11.041	11.841	12.625	13.398	14.165	14.926	15.686	16.442	17.197	17.950	18.702	20.202	21.700	23.196	24.691			
267.22 °F	h	Btu/lbm			1,186.9	1,212.0	1,236.5	1,260.8	1,285.0	1,309.3	1,333.6	1,358.10	1,382.6	1,407.45	1,432.3	1,482.7	1,534.0	1,586.7	1,639.3			
	s	Btu/lbm R			1.6996	1.7316	1.7610	1.7885	1.8144	1.8385	1.8625	1.88	1.9067	1.93	1.9478	1.9863	2.0227	2.0565	2.0903			
60 psia	v	ft3/lbm			7.260	7.817	8.355	8.881	9.400	9.913	10.425	10.933	11.440	11.944	12.448	13.453	14.454	15.454	16.453			
292.69 °F	h	Btu/lbm			1,181.9	1,208.4	1,233.7	1,258.6	1,283.2	1,307.7	1,332.2	1,356.85	1,381.5	1,406.45	1,431.4	1,482.1	1,533.5	1,586.2	1,638.9			
	s	Btu/lbm R			1.6496	1.6834	1.7138	1.7419	1.7682	1.7925	1.8168	1.8391	1.8613	1.8820	1.9026	1.9413	1.9777	2.0116	2.0454			
80 psia	v	ft3/lbm				5.803	6.219	6.622	7.018	7.408	7.795	8.178	8.561	8.942	9.322	10.078	10.831	11.583	12.333	13.082	13.831	
312.03 °F	h	Btu/lbm				1,204.5	1,230.8	1,256.2	1,281.3	1,306.1	1,330.8	1,355.65	1,380.5	1,405.55	1,430.6	1,481.4	1,532.9	1,585.3	1,638.5	1,693.0	1747.5	
	s	Btu/lbm R				1.6480	1.6795	1.7082	1.7350	1.7602	1.7842	1.81	1.8289	1.85	1.8704	1.9092	1.9457	1.9804	2.0135	2.0445	2.0755	
100 psia	v	ft3/lbm				4.592	4.936	5.266	5.588	5.904	6.217	6.525	6.834	7.140	7.446	8.053	8.658	9.260	9.862	10.461	11.061	
327.82 °F	h	Btu/lbm				1,200.4	1,227.8	1,253.9	1,279.3	1304,5	1,329.5	1,354.45	1,379.4	1,404.60	1,429.8	1,480.7	1,532.3	1,584.8	1,638.1	1,692.7	1747.2	
	s	Btu/lbm R				1.6194	1.6521	1.6816	1.7089	1.7344	1.7586	1.7812	1.8037	1.8245	1.8453	1.8842	1.9209	1.9556	1.9887	2.0198	2.0508	
150 psia	v	ft3/lbm					3.222	3.456	3.680	3.898	4.112	4.321	4.531	4.738	4.944	5.353	5.759	6.163	6.566	6.967	7.369	
358.43 °F	h	Btu/lbm					1,219.7	1,247.6	1,274.3	1,300.3	1,325.9	1,351.35	1,376.8	1,402.25	1,427.7	1,479.0	1,530.9	1,583.6	1,637.1	1,691.8	1746.5	
	s	Btu/lbm R					1.6001	1.6317	1.6602	1.6866	1.7114	1.73	1.7573	1.78	1.7994	1.8386	1.8754	1.9103	1.9435	1.9747	2.0058	
200 psia	v	ft3/lbm					2.361	2.549	2.725	2.894	3.059	3.219	3.379	3.536	3.693	4.003	4.310	4.615	4.918	5.220	5.522	
381.81 °F	h	Btu/lbm					1,210.9	1,241.0	1,269.1	1,296.0	1,322.3	1,348.20	1,374.1	1,399.80	1,425.5	1,477.3	1,529.5	1,582.4	1,636.1	1,691.0	1745.8	
	s	Btu/lbm R					1.5602	1.5943	1.6243	1.6517	1.6771	1.7005	1.7238	1.7451	1.7664	1.8059	1.8430	1.8780	1.9114	1.9426	1.9738	

psia Tsat,°F			200 °F	250 °F	300 °F	350 °F	400 °F	450 °F	500 °F	550 °F	600 °F	650 °F	700 °F	750 °F	800 °F	900 °F	1000 °F	1100 °F	1200 °F	1300 °F	1400 °F	1500 °F
300 psia	v	ft3/lbm						1.637	1.767	1.888	2.005	2.116	2.227	2.335	2.442	2.653	2.861	3.066	3.270	3.474	3.676	3.878
417.37 °F	h	Btu/lbm						1,226.4	1,257.9	1,287.0	1,314.8	1,341.65	1,368.5	1,394.85	1,421.2	1,473.8	1,526.7	1,580.0	1,634.0	1,688.8	1744.3	1800.6
	s	Btu/lbm R						1.5370	1.5706	1.6002	1.6271	1.65	1.6756	1.70	1.7191	1.7593	1.7968	1.8322	1.8657	1.8978	1.9284	1.9579
400 psia	v	ft3/lbm						1.175	1.285	1.384	1.477	1.564	1.651	1.734	1.817	1.978	2.136	2.292	2.446	2.600	2.753	2.905
444.63 °F	h	Btu/lbm						1,209.5	1,245.6	1,277.3	1,306.9	1,334.90	1,362.9	1,389.90	1,416.9	1,470.3	1,523.8	1,577.7	1,632.0	1,687.0	1742.8	1799.3
	s	Btu/lbm R						1.4901	1.5288	1.5611	1.5897	1.6150	1.6402	1.6625	1.6848	1.7257	1.7636	1.7993	1.8331	1.8653	1.8961	1.9257
500 psia	v	ft3/lbm							0.886	0.969	1.043	1.107	1.170	1.237	1.304	1.425	1.543	1.658	1.772	1.885	1.997	2.109
467.05 °F	h	Btu/lbm							1,224.5	1,261.5	1,294.3	1,324.15	1,354.0	1,382.10	1,410.2	1,465.0	1,519.5	1,574.0	1,629.0	1,684.4	1740.5	1797.3
	s	Btu/lbm R							1.4760	1.5137	1.5454	1.57	1.5993	1.62	1.6457	1.6876	1.7263	1.7624	1.7966	1.8290	1.86	1.8898
600 psia	v	ft3/lbm							0.795	0.875	0.946	1.010	1.073	1.132	1.190	1.302	1.411	1.518	1.623	1.726	1.830	1.932
486.25 °F	h	Btu/lbm							1,216.5	1,255.8	1,289.9	1,320.45	1,351.0	1,379.45	1,407.9	1,463.2	1,518.0	1,572.8	1,628.0	1,683.6	1739.8	1796.7
	s	Btu/lbm R							1.4597	1.4996	1.5325	1.5601	1.5877	1.6113	1.6348	1.6770	1.7159	1.7523	1.7865	1.8190	1.8501	1.8799
800 psia	v	ft3/lbm								0.616	0.678	0.733	0.783	0.831	0.877	0.964	1.048	1.130	1.211	1.290	1.368	1.446
518.27 °F	h	Btu/lbm								1,230.5	1,270.8	1,306.00	1,338.4	1,369.00	1,398.6	1,456.0	1,512.1	1,568.0	1,623.9	1,680.1	1736.8	1794
	s	Btu/lbm R								1.4476	1.4866	1.52	1.5476	1.57	1.5975	1.6413	1.6812	1.7182	1.7529	1.7858	1.8171	1.8471
1000 psia	v	ft3/lbm								0.454	0.514	0.564	0.609	0.650	0.688	0.761	0.831	0.898	0.963	1.028	1.091	1.154
544.65 °F	h	Btu/lbm								1,199.1	1,249.3	1,289.40	1,324.9	1,357.80	1,389.0	1,448.5	1,506.2	1,563.0	1,619.7	1,676.6	1733.7	1791.4
	s	Btu/lbm R								1.3971	1.4457	1.4827	1.5140	1.5418	1.5670	1.6125	1.6535	1.6911	1.7264	1.7596	1.7912	1.8214
1200 psia	v	ft3/lbm									0.402	0.450	0.491	0.528	0.562	0.626	0.686	0.743	0.798	0.853	0.906	0.959
567.26 °F	h	Btu/lbm									1,224.1	1,271.10	1,310.5	1,345.90	1,378.9	1,440.9	1,500.1	1,558.1	1,615.6	1,673.0	1730.7	1788.8
	s	Btu/lbm R									1.4061	1.45	1.4842	1.51	1.5408	1.5882	1.6302	1.6686	1.7043	1.7379	1.7698	1.8002
1400 psia	v	ft3/lbm									0.318	0.367	0.400	0.441	0.472	0.529	0.582	0.632	0.681	0.728	0.774	0.820
587.14 °F	h	Btu/lbm									1,193.7	1,250.80	1,295.0	1,333.40	1,368.5	1,433.1	1,493.9	1,553.1	1,611.4	1,669.5	1727.7	1786.1
	s	Btu/lbm R									1.3649	1.4175	1.4566	1.4800	1.5174	1.5668	1.6100	1.6491	1.6854	1.7194	1.7515	1.7821
1600 psia	v	ft3/lbm										0.303	0.342	0.375	0.404	0.456	0.504	0.549	0.592	0.634	0.676	0.716
604.93 °F	h	Btu/lbm										1,227.70	1,278.3	1,320.20	1,357.6	1,425.1	1,487.7	1,548.0	1,607.2	1,665.9	1724.6	1783.5
	s	Btu/lbm R										1.39	1.4302	1.47	1.4959	1.54-75	1.5920	1.6319	1.6687	1.7031	1.7355	1.7663
1800 psia	v	ft3/lbm										0.251	0.291	0.323	0.351	0.399	0.443	0.484	0.524	0.562	0.599	0.635
621.07 °F	h	Btu/lbm										1,200.60	1,259.9	1,306.20	1,346.2	1,416.9	1,481.3	1,542.8	1,602.9	1,662.3	1721.5	1780.8
	s	Btu/lbm R										1.3520	1.4043	1.4434	1.4758	1.5299	1.5756	1.6164	1.6537	1.6885	1.7212	1.7522
2000 psia	v	ft3/lbm										0.206	0.249	0.281	0.308	0.354	0.395	0.433	0.469	0.503	0.537	0.570
635.85 °F	h	Btu/lbm										1,167.50	1,239.7	1,291.20	1,334.3	1,408.5	1,474.9	1,537.6	1,598.6	1,658.7	1718.4	1778.1
	s	Btu/lbm R										1.31	1.3783	1.42	1.4567	1.5134	1.5606	1.6022	1.6401	1.6752	1.7083	1.7395

Table F.4 Saturated steam: pressure table (International units)

kPa	MPa	°C	v_f m^3/kg	v_g m^3/kg	h_f kJ/kg	h_{fg} kJ/kg	h_g kJ/kg	s_f kJ/kg-K	s_g kJ/kg-K
1.0	0.001	6.97	0.0010001	129.18	29.298	2484.4	2513.7	0.1059	8.9749
2.0	0.002	17.5	0.0010014	66.99	73.435	2459.5	2532.9	0.2606	8.7227
4.0	0.004	28.96	0.0010041	34.792	121.4	2432.3	2553.7	0.4224	8.4735
6.0	0.006	36.16	0.0010064	23.734	151.49	2415.2	2566.7	0.5209	8.3291
8.0	0.008	41.51	0.0010085	18.099	173.85	2402.4	2576.2	0.5925	8.2274
10.0	0.010	45.81	0.0010103	14.671	191.81	2392.1	2583.9	0.6492	8.1489
12.0	0.012	49.42	0.0010119	12.359	206.91	2383.4	2590.3	0.6963	8.085
14.0	0.014	52.55	0.0010133	10.691	219.99	2375.8	2595.8	0.7366	8.0312
16.0	0.016	55.31	0.0010147	9.4309	231.55	2369.2	2600.7	0.772	7.9847
18.0	0.018	57.8	0.001016	8.4433	241.95	2363.1	2605	0.8035	7.9437
20.0	0.020	60.06	0.0010171	7.6482	251.4	2357.5	2608.9	0.832	7.9072
25.0	0.025	64.96	0.0010198	6.2034	271.93	2345.5	2617.4	0.8931	7.8302
30.0	0.030	69.1	0.0010222	5.2286	289.23	2335.4	2624.6	0.9439	7.7675
35.0	0.035	72.68	0.0010244	4.5252	304.25	2326.5	2630.7	0.9876	7.7146
40.0	0.040	75.86	0.0010264	3.9931	317.57	2318.5	2636.1	1.0259	7.669
45.0	0.045	78.71	0.0010282	3.5761	329.55	2311.4	2640.9	1.0601	7.6288
50.0	0.050	81.32	0.0010299	3.2401	340.48	2304.7	2645.2	1.091	7.593
60.0	0.060	85.93	0.0010331	2.7318	359.84	2293.1	2652.9	1.1452	7.5311
70.0	0.070	89.93	0.0010359	2.3649	376.68	2282.7	2659.4	1.1919	7.479
80.0	0.080	93.49	0.0010385	2.0872	391.64	2273.6	2665.2	1.2328	7.4339
90.0	0.090	96.69	0.0010409	1.8695	405.13	2265.2	2670.3	1.2694	7.3942
100.0	0.100	99.61	0.0010431	1.694	417.44	2257.5	2674.9	1.3026	7.3588
120.0	0.120	104.78	0.0010473	1.4284	439.3	2243.8	2683.1	1.3608	7.2976
140.0	0.140	109.29	0.001051	1.2366	458.37	2231.6	2690	1.4109	7.246
160.0	0.160	113.3	0.0010544	1.0914	475.34	2220.7	2696	1.4549	7.2014
180.0	0.180	116.91	0.0010576	0.97753	490.67	2210.7	2701.4	1.4944	7.162
200.0	0.200	120.21	0.0010605	0.88574	504.68	2201.5	2706.2	1.5301	7.1269
250.0	0.250	127.41	0.0010672	0.7187	535.35	2181.2	2716.5	1.6072	7.0524
300.0	0.300	133.53	0.0010732	0.60579	561.46	2163.4	2724.9	1.6718	6.9916
350.0	0.350	138.86	0.0010786	0.5242	584.31	2147.7	2732	1.7275	6.9401
400.0	0.400	143.61	0.0010836	0.46239	604.72	2133.4	2738.1	1.7766	6.8954
450.0	0.450	147.91	0.0010882	0.4139	623.22	2120.2	2743.4	1.8206	6.856
500.0	0.500	151.84	0.0010926	0.3748	640.19	2107.9	2748.1	1.8606	6.8206
600.0	0.600	158.83	0.0011006	0.31558	670.5	2085.6	2756.1	1.9311	6.7592
700.0	0.700	164.95	0.001108	0.27276	697.14	2065.6	2762.7	1.9921	6.707
800.0	0.800	170.41	0.0011148	0.24033	721.02	2047.3	2768.3	2.046	6.6615
900.0	0.900	175.36	0.0011212	0.21487	742.72	2030.3	2773	2.0944	6.6212
1,000.0	1.000	179.89	0.0011272	0.19435	762.68	2014.4	2777.1	2.1384	6.585
1,200.0	1.200	187.96	0.001139	0.16325	798.5	1985.3	2783.8	2.2163	6.5217

kPa	MPa	°C	v_f m^3/kg	v_g m^3/kg	h_f kJ/kg	h_{fg} kJ/kg	h_g kJ/kg	s_f kJ/kg-K	s_g kJ/kg-K
1,400.0	1.400	195.05	0.001149	0.14077	830.13	1958.8	2788.9	2.2839	6.4675
1,600.0	1.600	201.38	0.001159	0.12373	858.61	1934.3	2792.9	2.3438	6.42
1,800.0	1.800	207.12	0.001168	0.11036	884.61	1911.4	2796	2.3978	6.3776
2,000.0	2.000	212.38	0.001177	0.099581	908.62	1889.8	2798.4	2.447	6.3392
2,200.0	2.200	217.26	0.001185	0.090695	930.98	1869.2	2800.2	2.4924	6.304
2,400.0	2.400	221.8	0.001193	0.083242	951.95	1849.6	2801.5	2.5344	6.2714
2,600.0	2.600	226.05	0.001201	0.076897	971.74	1830.8	2802.5	2.5738	6.2411
2,800.0	2.800	230.06	0.001209	0.071428	990.5	1812.5	2803	2.6107	6.2126
3,000.0	3.000	233.86	0.001217	0.066664	1008.4	1794.9	2803.3	2.6456	6.1858
3,200.0	3.200	237.46	0.001224	0.062475	1025.5	1777.7	2803.2	2.6787	6.1604
3,400.0	3.400	240.9	0.001231	0.058761	1041.8	1761.2	2803	2.7102	6.1362
3,600.0	3.600	244.19	0.001239	0.055446	1057.6	1744.9	2802.5	2.7403	6.1131
3,800.0	3.800	247.33	0.001246	0.052468	1072.8	1729.0	2801.8	2.769	6.091
4,000.0	4.000	250.36	0.001253	0.049777	1087.4	1713.5	2800.9	2.7967	6.0697
4,200.0	4.200	253.27	0.001259	0.047333	1101.6	1698.3	2799.9	2.8232	6.0492
4,400.0	4.400	256.07	0.001266	0.045103	1115.4	1683.3	2798.7	2.8488	6.0294
4,600.0	4.600	258.78	0.001273	0.04306	1128.8	1668.5	2797.3	2.8736	6.0103
4,800.0	4.800	261.4	0.00128	0.041181	1141.8	1654.0	2795.8	2.8975	5.9917
5,000.0	5.000	263.94	0.001286	0.039446	1154.5	1639.7	2794.2	2.9207	5.9737
5,500.0	5.500	269.97	0.001303	0.035642	1184.9	1604.8	2789.7	2.9759	5.9307
6,000.0	6.000	275.59	0.001319	0.032449	1213.7	1570.9	2784.6	3.0274	5.8901
6,500.0	6.500	280.86	0.001336	0.029728	1241.2	1537.6	2778.8	3.076	5.8515
7,000.0	7.000	285.83	0.001352	0.02738	1267.4	1505.2	2772.6	3.122	5.8146
7,500.0	7.500	290.54	0.001368	0.025331	1292.7	1473.1	2765.8	3.1658	5.7792
8,000.0	8.000	295.01	0.001385	0.023528	1317.1	1441.5	2758.6	3.2077	5.7448
8,500.0	8.500	299.27	0.001401	0.021926	1340.7	1410.3	2751	3.2478	5.7115
9,000.0	9.000	303.35	0.001418	0.020493	1363.7	1379.2	2742.9	3.2866	5.679
9,500.0	9.500	307.25	0.001435	0.019203	1386	1348.4	2734.4	3.324	5.6472
10,000.0	10.000	311	0.001453	0.018034	1407.9	1317.6	2725.5	3.3603	5.6159
11,000.0	11.000	318.08	0.001489	0.015994	1450.3	1256.1	2706.4	3.43	5.5545
12,000.0	12.000	324.68	0.001526	0.014269	1491.3	1194.3	2685.6	3.4965	5.4941
13,000.0	13.000	330.86	0.001566	0.012785	1531.4	1131.5	2662.9	3.5606	5.4339
14,000.0	14.000	336.67	0.00161	0.011489	1570.9	1067.2	2638.1	3.623	5.373
15,000.0	15.000	342.16	0.001657	0.01034	1610.2	1000.7	2610.9	3.6844	5.3108
16,000.0	16.000	347.36	0.00171	0.009308	1649.7	931.1	2580.8	3.7457	5.2463
17,000.0	17.000	352.29	0.001769	0.008369	1690	857.4	2547.4	3.8077	5.1785
18,000.0	18.000	356.99	0.001839	0.007499	1732	777.5	2509.5	3.8717	5.1055
19,000.0	19.000	361.47	0.001925	0.006673	1776.9	688.5	2465.4	3.9396	5.0246
20,000.0	20.000	365.75	0.002039	0.005858	1827.1	584.3	2411.4	4.0154	4.9299
21,000.0	21.000	369.83	0.002212	0.004988	1889.4	448.1	2337.5	4.1093	4.8062
22,000.0	22.000	373.71	0.00275	0.003577	2021.9	142.3	2164.2	4.3109	4.5308
critical	critical	373.946	0.003106	0.003106	2087.5	0.0	2087.5	4.412	4.412

Table F.5 Saturated steam: temperature table (International units)

°C	kPa	MPa	v_f m³/kg	v_g m³/kg	h_f kJ/kg	h_{fg} kJ/kg	h_g kJ/kg	s_f kJ/kg-K	s_g kJ/kg-K
0.01	0.6	0.0006	0.0010002	206	0.001	2500.9	2500.9	0	9.1555
5	0.9	0.0009	0.0010001	147.02	21.019	2489.1	2510.1	0.0763	9.0249
10	1.2	0.0012	0.0010003	106.31	42.021	2477.2	2519.2	0.1511	8.8998
15	1.7	0.0017	0.0010009	77.881	62.984	2465.4	2528.4	0.2245	8.7804
20	2.3	0.0023	0.0010018	57.761	83.92	2453.6	2537.5	0.2965	8.6661
25	3.2	0.0032	0.001003	43.341	104.84	2441.7	2546.5	0.3673	8.5568
30	4.2	0.0042	0.0010044	32.882	125.75	2429.9	2555.6	0.4368	8.4521
35	5.6	0.0056	0.001006	25.208	146.64	2418.0	2564.6	0.5052	8.3518
40	7.4	0.0074	0.0010079	19.517	167.54	2406.0	2573.5	0.5724	8.2557
45	9.6	0.0096	0.0010099	15.253	188.44	2394.1	2582.5	0.6386	8.1634
50	12.4	0.0124	0.0010121	12.028	209.34	2382.0	2591.3	0.7038	8.0749
55	15.8	0.0158	0.0010145	9.5649	230.24	2369.9	2600.1	0.768	7.9899
60	19.9	0.0199	0.0010171	7.6677	251.15	2357.7	2608.8	0.8312	7.9082
65	25.0	0.0250	0.0010199	6.1938	272.08	2345.4	2617.5	0.8935	7.8296
70	31.2	0.0312	0.0010228	5.0397	293.02	2333.1	2626.1	0.955	7.754
75	38.6	0.0386	0.0010258	4.1291	313.97	2320.6	2634.6	1.0156	7.6812
80	47.4	0.0474	0.001029	3.4053	334.95	2308.1	2643	1.0754	7.611
85	57.9	0.0579	0.0010324	2.8259	355.95	2295.4	2651.3	1.1344	7.5434
90	70.2	0.0702	0.0010359	2.3591	376.97	2282.5	2659.5	1.1927	7.4781
95	84.6	0.0846	0.0010396	1.9806	398.02	2269.6	2667.6	1.2502	7.415
100	**101.3**	**0.1013**	**0.0010435**	**1.6719**	**419.1**	**2256.5**	**2675.6**	**1.307**	**7.3541**
105	120.9	0.1209	0.0010474	1.4185	440.21	2243.2	2683.4	1.3632	7.2951
110	143.4	0.1434	0.0010516	1.2094	461.36	2229.7	2691.1	1.4187	7.238
115	169.2	0.1692	0.0010559	1.0359	482.55	2216.1	2698.6	1.4735	7.1827
120	198.7	0.1987	0.0010603	0.8913	503.78	2202.1	2705.9	1.5278	7.1291
125	232.2	0.2322	0.0010649	0.77011	525.06	2188.0	2713.1	1.5815	7.077
130	270.3	0.2703	0.0010697	0.66808	546.39	2173.7	2720.1	1.6346	7.0264
135	313.2	0.3132	0.0010747	0.5818	567.77	2159.1	2726.9	1.6872	6.9772
140	361.5	0.3615	0.0010798	0.50852	589.2	2144.2	2733.4	1.7393	6.9293
145	415.6	0.4156	0.001085	0.44602	610.69	2129.1	2739.8	1.7909	6.8826
150	476.1	0.4761	0.0010905	0.3925	632.25	2113.7	2745.9	1.842	6.837
155	543.4	0.5434	0.0010962	0.3465	653.88	2097.9	2751.8	1.8926	6.7926
160	618.1	0.6181	0.001102	0.30682	675.57	2081.8	2757.4	1.9428	6.7491
165	700.8	0.7008	0.001108	0.27246	697.35	2065.5	2762.8	1.9926	6.7066
170	792.1	0.7921	0.0011143	0.24262	719.21	2048.7	2767.9	2.0419	6.6649
175	892.5	0.8925	0.0011207	0.2166	741.15	2031.6	2772.7	2.0909	6.6241

°C	kPa	MPa	v_f m^3/kg	v_g m^3/kg	h_f kJ/kg	h_{fg} kJ/kg	h_g kJ/kg	s_f kJ/kg-K	s_g kJ/kg-K
180	1,002.6	1.0026	0.0011274	0.19386	763.19	2014.0	2777.2	2.1395	6.5841
185	1,123.3	1.1233	0.0011343	0.17392	785.32	1996.1	2781.4	2.1878	6.5447
190	1,255.0	1.2550	0.0011414	0.15638	807.57	1977.7	2785.3	2.2358	6.506
195	1,398.6	1.3986	0.0011488	0.14091	829.92	1959.0	2788.9	2.2834	6.4679
200	1,554.7	1.5547	0.0011565	0.12722	852.39	1939.7	2792.1	2.3308	6.4303
205	1,724.0	1.7240	0.0011645	0.11509	874.99	1919.9	2794.9	2.3779	6.3932
210	1,907.4	1.9074	0.0011727	0.1043	897.73	1899.7	2797.4	2.4248	6.3565
215	2,105.5	2.1055	0.0011813	0.094689	920.61	1878.8	2799.4	2.4714	6.3202
220	2,319.3	2.3193	0.0011902	0.086101	943.64	1857.5	2801.1	2.5178	6.2842
225	2,549.4	2.5494	0.001199	0.078411	966.84	1835.5	2802.3	2.5641	6.2485
230	2,796.8	2.7968	0.001209	0.07151	990.21	1812.8	2803	2.6102	6.2131
235	3,062.2	3.0622	0.001219	0.065304	1013.8	1789.5	2803.3	2.6561	6.1777
240	3,346.7	3.3467	0.001229	0.05971	1037.5	1765.6	2803.1	2.7019	6.1425
245	3,650.9	3.6509	0.00124	0.054658	1061.5	1740.8	2802.3	2.7477	6.1074
250	3,975.9	3.9759	0.001252	0.050087	1085.7	1715.3	2801	2.7934	6.0722
255	4,322.7	4.3227	0.001264	0.045941	1110.1	1689.0	2799.1	2.8391	6.037
260	4,692.1	4.6921	0.001276	0.042175	1134.8	1661.8	2796.6	2.8847	6.0017
265	5,085.1	5.0851	0.001289	0.038748	1159.8	1633.7	2793.5	2.9304	5.9662
270	5,502.8	5.5028	0.001303	0.035622	1185.1	1604.6	2789.7	2.9762	5.9304
275	5,946.3	5.9463	0.001318	0.032767	1210.7	1574.4	2785.1	3.0221	5.8943
280	6,416.5	6.4165	0.001333	0.030154	1236.7	1543.1	2779.8	3.0681	5.8578
285	6,914.5	6.9145	0.001349	0.027758	1263	1510.7	2773.7	3.1143	5.8208
290	7,441.6	7.4416	0.001366	0.025557	1289.8	1476.8	2766.6	3.1608	5.7832
295	7,999.0	7.9990	0.001385	0.023531	1317	1441.6	2758.6	3.2076	5.7449
300	8,587.7	8.5877	0.001404	0.021663	1344.8	1404.8	2749.6	3.2547	5.7058
305	9,209.2	9.2092	0.001425	0.019937	1373.1	1366.3	2739.4	3.3024	5.6656
310	9,864.7	9.8647	0.001448	0.018339	1402	1325.9	2727.9	3.3506	5.6243
315	10,556.0	10.5560	0.001472	0.016856	1431.6	1283.5	2715.1	3.3994	5.5816
320	11,284.0	11.2840	0.001499	0.015476	1462.1	1238.6	2700.7	3.4491	5.5373
325	12,051.0	12.0510	0.001528	0.014189	1493.4	1191.1	2684.5	3.4997	5.4911
330	12,858.0	12.8580	0.001561	0.012984	1525.7	1140.5	2666.2	3.5516	5.4425
335	13,707.0	13.7070	0.001597	0.011852	1559.3	1086.3	2645.6	3.6048	5.391
340	14,600.0	14.6000	0.001638	0.010784	1594.4	1027.7	2622.1	3.6599	5.3359
345	15,540.0	15.5400	0.001685	0.00977	1631.4	963.6	2595	3.7175	5.2763
350	16,529.0	16.5290	0.00174	0.008801	1670.9	892.7	2563.6	3.7783	5.2109
355	17,570.0	17.5700	0.001808	0.007866	1713.7	812.7	2526.4	3.8438	5.1377
360	18,666.0	18.6660	0.001895	0.006945	1761.5	719.5	2481	3.9164	5.0527
365	19,822.0	19.8220	0.002016	0.006004	1817.6	604.4	2422	4.0011	4.9482
370	21,043.0	21.0430	0.002222	0.004946	1892.6	440.9	2333.5	4.1142	4.7996
critical	22,064.0	22.0640	0.003106	0.003106	2087.5	0.0	2087.5	4.412	4.412

Table F.6 Superheated steam: temperature table (International units)

kPa Tsat,°c		50 °C	100 °C	150 °C	200 °C	225 °C	250 °C	300 °C	325 °C	350 °C	375 °C	400 °C	425 °C	450 °C	475 °C	500 °C	550 °C	600 °C	650 °C	700 °C	750 °C	800 °C
5 kPa	v	29.7820	34.4190	39.0430	43.6630	45.9720	48.2810	52.8980	55.2065	57.5150	59.8230	62.1310	64.4390	66.7470	69.0550	71.3630	75.9790	80.5940	85.2100	89.8260		
32.9 °C	h	2593.4	2688.0	2783.4	2879.8	2928.7	2977.6	3076.9	3127.3	3177.6	3228.8	3280.0	3332.0	3384.0	3436.9	3489.7	3597.1	3706.3	3818.1	3929.9		
	s	8.4976	8.7700	9.0097	9.2251	9.3234	9.4216	9.6027	9.6870	9.7713	9.8503	9.9293	10.0037	10.0780	10.1490	10.2200	10.3540	10.4830	10.6040	10.7250		
10 kPa	v	14.8670	17.1970	19.5140	21.8260	22.9810	24.1360	26.4460	27.6005	28.7550	29.9095	31.0640	32.2180	33.3720	34.5260	35.6800	37.9880	40.2960	42.6040	44.9120		
45.8 °C	h	2592.0	2687.4	2783.0	2879.6	2928.5	2977.4	3076.7	3127.1	3177.5	3228.7	3279.9	3332.0	3384.0	3436.9	3489.7	3597.1	3706.3	3818.1	3929.9		
	s	8.1741	8.4488	8.6892	8.9048	9.0031	9.1014	9.2827	9.3670	9.4513	9.5303	9.6093	9.6839	9.7584	9.8291	9.8997	10.0340	10.1630	10.2840	10.4050		
20 kPa	v		8.5857	9.7488	10.9070	11.4855	12.0640	13.2200	13.7975	14.3750	14.9525	15.5300	16.1070	16.6840	17.2615	17.8390	18.9930	20.1470	21.3010	22.4550		
60.1 °C	h		2686.2	2782.3	2879.1	2928.1	2977.1	3076.5	3127.0	3177.4	3228.6	3279.8	3331.8	3383.8	3436.7	3489.6	3597.0	3706.2	3818.0	3929.8		
	s		8.1262	8.3680	8.5842	8.6827	8.7811	8.9624	9.0468	9.1311	9.2102	9.2892	9.3638	9.4383	9.5090	9.5797	9.7143	9.8431	9.9646	10.0860		
50 kPa	v		3.4188	3.8899	4.3563	4.5885	4.8207	5.2841	5.5156	5.7470	5.9783	6.2095	6.4407	6.6718	6.9029	7.1339	7.5959	8.0578	8.5196	8.9814		
81.3 °C	h		2682.4	2780.2	2877.8	2927.0	2976.2	3075.8	3126.3	3176.8	3228.1	3279.3	3331.4	3383.5	3436.4	3489.2	3596.7	3706.0	3817.9	3929.7		
	s		7.6952	7.9412	8.1591	8.2580	8.3568	8.5386	8.6231	8.7076	8.7867	8.8658	8.9404	9.0150	9.0858	9.1565	9.2912	9.4200	9.5413	9.6625		
100 kPa	v		1.6960	1.9367	2.1725	2.2894	2.4062	2.6389	2.7550	2.8710	2.9869	3.1027	3.2185	3.3342	3.4499	3.5656	3.7968	4.0279	4.2590	4.4900		
99.6 °C	h		2675.8	2776.6	2875.5	2925.0	2974.5	3074.5	3125.2	3175.8	3227.2	3278.5	3330.7	3382.8	3435.8	3488.7	3596.3	3705.6	3817.5	3929.4		
	s		7.3610	7.6147	7.8356	7.9351	8.0346	8.2171	8.3018	8.3865	8.4658	8.5451	8.6198	8.6945	8.7653	8.8361	8.9709	9.0998	9.2211	9.3424		
150 kPa	v			1.2856	1.4445	1.5229	1.6013	1.7571	1.8347	1.9123	1.9897	2.0671	2.1444	2.2217	2.2990	2.3762	2.5305	2.6847	2.8388	2.9929		
111.4 °C	h			2772.9	2873.1	2923.0	2972.9	3073.3	3124.1	3174.9	3226.4	3277.8	3330.0	3382.2	3435.2	3488.2	3595.8	3705.2	3817.2	3929.1		
	s			7.4207	7.6447	7.7449	7.8451	8.0284	8.1134	8.1983	8.2777	8.3571	8.4319	8.5067	8.5776	8.6484	8.7833	8.9123	9.0337	9.1550		
200 kPa	v			0.9599	1.0805	1.1397	1.1989	1.3162	1.3746	1.4330	1.4912	1.5493	1.6074	1.6655	1.7235	1.7814	1.8973	2.0130	2.1287	2.2444		
120.2 °C	h			2769.1	2870.8	2921.1	2971.3	3072.1	3123.0	3173.9	3225.5	3277.0	3329.3	3381.5	3434.6	3487.6	3595.4	3704.8	3816.8	3928.8		
	s			7.2809	7.5081	7.6091	7.7100	7.8940	7.9792	8.0643	8.1439	8.2235	8.2984	8.3733	8.4442	8.5151	8.6501	8.7792	8.9006	9.0220		
300 kPa	v			0.6340	0.7164	0.7565	0.7965	0.8753	0.9145	0.9536	0.9926	1.0315	1.0704	1.1092	1.1480	1.1867	1.2641	1.3414	1.4186	1.4958		
133.5 °C	h			2761.2	2866.0	2917.0	2967.9	3069.6	3120.8	3172.0	3223.7	3275.4	3327.8	3380.2	3433.4	3486.6	3594.5	3704.0	3816.1	3928.2		
	s			7.0791	7.3132	7.4157	7.5181	7.7037	7.7893	7.8749	7.9548	8.0346	8.1097	8.1848	8.2559	8.3269	8.4622	8.5914	8.7129	8.8344		
400 kPa	v			0.4709	0.5343	0.5648	0.5952	0.6549	0.6844	0.7139	0.7433	0.7726	0.8019	0.8311	0.8603	0.8894	0.9475	1.0056	1.0636	1.1215		
143.6 °C	h			2752.8	2861.0	2912.8	2964.6	3067.1	3118.6	3170.0	3222.0	3273.9	3326.5	3379.0	3432.3	3485.5	3593.6	3703.2	3815.4	3927.6		
	s			6.9305	7.1724	7.2765	7.3805	7.5677	7.6538	7.7398	7.8200	7.9001	7.9754	8.0507	8.1219	8.1931	8.3286	8.4579	8.5796	8.7012		
500 kPa	v				0.4250	0.4497	0.4744	0.5226	0.5464	0.5701	0.5937	0.6173	0.6408	0.6642	0.6876	0.7109	0.7576	0.8041	0.8506	0.8970		
151.8 °C	h				2855.9	2908.5	2961.1	3064.6	3116.4	3168.1	3220.2	3272.3	3325.0	3377.7	3431.1	3484.4	3592.6	3702.5	3814.8	3927.0		
	s				7.0611	7.1669	7.2726	7.4614	7.5480	7.6345	7.7150	7.7954	7.8709	7.9464	8.0178	8.0891	8.2247	8.3543	8.4760	8.5977		
600 kPa	v				0.3521	0.3730	0.3939	0.4344	0.4544	0.4743	0.4940	0.5137	0.5334	0.5530	0.5725	0.5920	0.6309	0.6698	0.7086	0.7473		
158.8 °C	h				2850.7	2904.2	2957.7	3062.1	3114.1	3166.1	3218.4	3270.7	3323.6	3376.4	3429.9	3483.3	3591.7	3701.7	3814.1	3926.5		
	s				6.9684	7.0759	7.1834	7.3740	7.4610	7.5480	7.6288	7.7095	7.7852	7.8609	7.9324	8.0039	8.1398	8.2694	8.3913	8.5131		
800 kPa	v				0.2609	0.2771	0.2932	0.3242	0.3393	0.3544	0.3694	0.3843	0.3991	0.4139	0.4286	0.4433	0.4726	0.5019	0.5310	0.5601	0.5892	0.6182
170.4 °C	h				2839.8	2895.2	2950.5	3056.9	3109.6	3162.2	3214.9	3267.6	3320.7	3373.8	3427.5	3481.2	3589.9	3700.1	3811.9	3925.3	4040.3	4157.1
	s				6.8176	6.9290	7.0403	7.2345	7.3226	7.4106	7.4920	7.5733	7.6494	7.7255	7.7973	7.8690	8.0053	8.1353	8.2598	8.3794	8.4947	8.606
1,000 kPa	v				0.2060	0.2194	0.2327	0.2580	0.2703	0.2825	0.2946	0.3066	0.3185	0.3304	0.3423	0.3541	0.3777	0.4011	0.4245	0.4478	0.4711	0.4944
179.9 °C	h				2828.3	2885.8	2943.2	3051.7	3105.0	3158.2	3211.3	3264.4	3317.8	3371.2	3425.1	3479.0	3588.1	3698.6	3810.5	3924.1	4039.3	4156.2
	s				6.6955	6.8111	6.9266	7.1247	7.2138	7.3028	7.3848	7.4668	7.5433	7.6198	7.6919	7.7640	7.9007	8.0309	8.1557	8.2755	8.3909	8.5024
1,500 kPa	v				0.1324	0.1422	0.1520	0.1697	0.1782	0.1866	0.1948	0.2030	0.2111	0.2192	0.2272	0.2352	0.2510	0.2668	0.2825	0.2981	0.3137	0.3293
198.3 °C	h				2796.0	2860.0	2924.0	3038.3	3093.2	3148.0	3202.2	3256.4	3310.6	3364.7	3419.2	3473.6	3583.5	3694.6	3807.2	3921.2	4036.7	4153.9
	s				6.4537	6.5824	6.7111	6.9199	7.0117	7.1035	7.1872	7.2708	7.3484	7.4259	7.4988	7.5716	7.7093	7.8404	7.9657	8.0860	8.2018	8.3135

kPa Tsat,°c		50 °C	100 °C	150 °C	200 °C	225 °C	250 °C	300 °C	325 °C	350 °C	375 °C	400 °C	425 °C	450 °C	475 °C	500 °C	550 °C	600 °C	650 °C	700 °C	750 °C	800 °C
2,000 kPa	v						0.1115	0.1255	0.1321	0.1386	0.1449	0.1512	0.1574	0.1635	0.1696	0.1757	0.1877	0.1996	0.2115	0.2233	0.2350	0.2467
212.4 °C	h						2903.2	3024.3	3081.0	3137.6	3192.9	3248.2	3303.2	3358.1	3413.1	3468.1	3578.9	3690.7	3803.8	3918.2	4034.2	4151.6
	s						6.5474	6.7685	6.8634	6.9582	7.0436	7.1290	7.2077	7.2863	7.3599	7.4335	7.5723	7.7042	7.8301	7.9509	8.067	8.1791
3,000 kPa	v						0.0706	0.0812	0.0859	0.0906	0.0950	0.0994	0.1037	0.1079	0.1121	0.1162	0.1244	0.1324	0.1405	0.1484	0.1563	0.1642
233.9 °C	h						2856.5	2994.3	3055.2	3116.1	3173.9	3231.6	3288.2	3344.7	3400.9	3457.0	3569.6	3682.8	3797.0	3912.3	4029.0	4147.0
	s						6.2893	6.5412	6.6431	6.7449	6.8341	6.9233	7.0043	7.0853	7.1605	7.2356	7.3767	7.5102	7.6373	7.7590	7.8759	7.9885
4,000 kPa	v							0.0589	0.0627	0.0665	0.0700	0.0734	0.0767	0.0800	0.0832	0.0864	0.0927	0.0989	0.1049	0.1110	0.1170	0.1229
250.4 °C	h							2961.7	3027.5	3093.3	3153.9	3214.4	3272.7	3331.0	3388.4	3445.8	3560.2	3674.8	3790.2	3906.4	4023.8	4142.5
	s							6.3638	6.4741	6.5843	6.6778	6.7712	6.8548	6.9383	7.0151	7.0919	7.2353	7.3704	7.4989	7.6215	7.7391	7.8523
5,000 kPa	v							0.0453	0.0487	0.0520	0.0549	0.0578	0.0606	0.0633	0.0660	0.0686	0.0737	0.0787	0.0836	0.0885	0.0933	0.0982
263.9 °C	h							2925.6	2997.5	3069.3	3133.0	3196.6	3256.8	3317.0	3375.8	3434.5	3550.8	3666.8	3783.3	3900.5	4018.6	4137.9
	s							6.2109	6.3312	6.4515	6.5498	6.6481	6.7345	6.8208	6.8993	6.9778	7.1235	7.2604	7.3901	7.5137	7.6321	7.7459
6,000 kPa	v							0.0362	0.0393	0.0423	0.0449	0.0474	0.0498	0.0522	0.0545	0.0567	0.0610	0.0653	0.0694	0.0735	0.0776	0.0816
275.6 °C	h							2885.5	2964.7	3043.9	3111.1	3178.2	3240.5	3302.8	3362.9	3422.9	3541.2	3658.8	3776.4	3894.5	4013.4	4133.3
	s							6.0702	6.2029	6.3356	6.4394	6.5431	6.6324	6.7216	6.8020	6.8824	7.0306	7.1692	7.3002	7.4248	7.5439	7.6583
7,000 kPa	v							0.0295	0.0326	0.0353	0.0377	0.0400	0.0421	0.0442	0.0462	0.0482	0.0520	0.0557	0.0593	0.0628	0.0664	0.0698
285.8 °C	h							2839.8	2935.5	3016.8	3090.4	3159.1	3223.7	3288.2	3349.8	3411.3	3531.5	3650.6	3769.4	3888.5	4008.1	4128.6
	s							5.9335	6.0970	6.2303	6.3460	6.4501	6.5426	6.6351	6.7174	6.7997	6.9505	7.0909	7.2232	7.3488	7.4687	7.5837
8,000 kPa	v							0.0243	0.0274	0.0300	0.0323	0.0343	0.0363	0.0382	0.0400	0.0418	0.0452	0.0485	0.0517	0.0548	0.0579	0.0610
295.0 °C	h							2786.4	2898.3	2988.1	3066.9	3139.3	3206.3	3273.2	3336.3	3399.4	3521.8	3642.4	3762.4	3882.4	4002.9	4124.0
	s							5.7935	5.9849	6.1319	6.2560	6.3657	6.4617	6.5577	6.6421	6.7264	6.8798	7.0221	7.1557	7.2823	7.403	7.5186
9,000 kPa	v								0.0233	0.0258	0.0280	0.0300	0.0318	0.0335	0.0352	0.0368	0.0399	0.0429	0.0458	0.0486	0.0514	0.0541
303.4 °C	h								2857.0	2957.2	3042.2	3118.8	3188.4	3257.9	3322.6	3387.3	3511.9	3634.2	3755.4	3876.4	3997.6	4119.4
	s								5.8736	6.0378	6.1716	6.2875	6.3873	6.4871	6.5736	6.6601	6.8163	6.9605	7.0955	7.2231	7.3446	7.4608
10,000 kPa	v								0.0199	0.0224	0.0246	0.0264	0.0281	0.0298	0.0313	0.0328	0.0357	0.0384	0.0410	0.0436	0.0461	0.0486
311.0 °C	h								2810.2	2924.0	3016.2	3097.4	3169.9	3242.3	3308.7	3375.1	3501.9	3625.8	3748.3	3870.3	3992.3	4114.7
	s								5.7593	5.9458	6.0910	6.2139	6.3178	6.4217	6.5105	6.5993	6.7584	6.9045	7.0409	7.1696	7.2918	7.4086
11,000 kPa	v								0.0170	0.0196	0.0217	0.0235	0.0251	0.0267	0.0282	0.0296	0.0322	0.0347	0.0371	0.0395	0.0418	0.0441
318.1 °C	h								2755.6	2887.8	2988.7	3075.1	3150.7	3226.2	3294.4	3362.6	3491.9	3617.5	3741.2	3864.2	3987.0	4110.1
	s								5.6373	5.8541	6.0129	6.1438	6.2522	6.3605	6.4518	6.5430	6.7050	6.8531	6.9910	7.1207	7.2437	7.3612
12,000 kPa	v								0.0143	0.0172	0.0193	0.0211	0.0227	0.0242	0.0255	0.0268	0.0293	0.0317	0.0339	0.0361	0.0383	0.0404
324.7 °C	h								2688.4	2848.0	2959.5	3051.9	3130.9	3209.8	3279.9	3350.0	3481.7	3609.0	3734.1	3858.0	3981.6	4105.4
	s								5.4988	5.7607	5.9362	6.0762	6.1895	6.3027	6.3965	6.4902	6.6553	6.8055	6.9448	7.0756	7.1994	7.3175
13,000 kPa	v									0.0151	0.0173	0.0190	0.0205	0.0220	0.0233	0.0245	0.0269	0.0291	0.0312	0.0332	0.0352	0.0372
330.9 °C	h									2803.6	2928.3	3027.6	3110.3	3192.9	3265.0	3337.1	3471.4	3600.5	3726.9	3851.9	3976.3	4100.7
	s									5.6635	5.8600	6.0104	6.1290	6.2475	6.3440	6.4404	6.6087	6.7610	6.9018	7.0336	7.1583	7.2771
14,000 kPa	v									0.0132	0.0155	0.0172	0.0187	0.0201	0.0213	0.0225	0.0248	0.0268	0.0288	0.0308	0.0326	0.0345
336.7 °C	h									2752.9	2894.9	3002.2	3088.9	3175.6	3249.9	3324.1	3461.0	3591.9	3719.7	3845.7	3970.9	4096.0
	s									5.5595	5.7832	5.9457	6.0701	6.1945	6.2938	6.3931	6.5648	6.7192	6.8615	6.9944	7.12	7.2393
15,000 kPa	v									0.0115	0.0139	0.0157	0.0171	0.0185	0.0197	0.0208	0.0229	0.0249	0.0268	0.0286	0.0304	0.0321
342.2 °C	h									2693.0	2858.9	2975.5	3066.7	3157.8	3234.3	3310.8	3450.5	3583.3	3712.4	3839.5	3965.6	4091.3
	s									5.4435	5.7049	5.8817	6.0125	6.1433	6.2456	6.3479	6.5230	6.6797	6.8235	6.9576	7.0839	7.2039

Appendix G

Psychrometric Chart

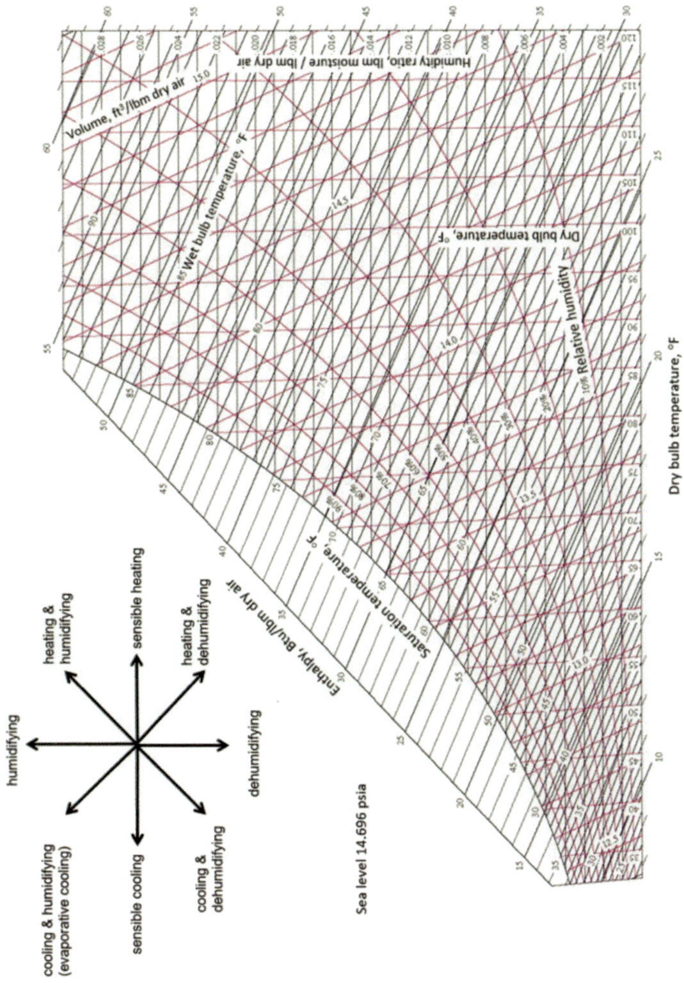

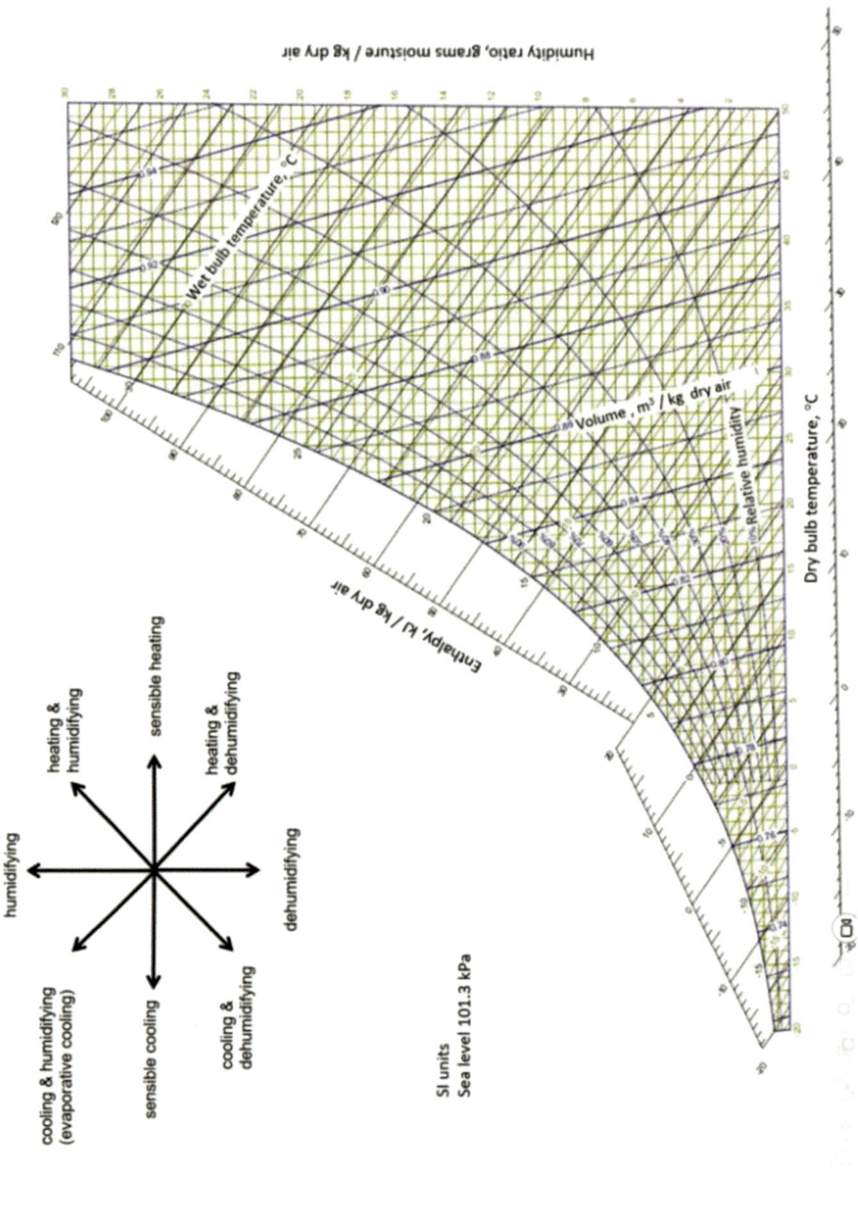

Humidity ratio, grams moisture / kg dry air
Wet bulb temperature, °C
Volume, m³ / kg dry air
Relative humidity
Dry bulb temperature, °C
Enthalpy, kJ / kg dry air
SI units
Sea level 101.3 kPa
humidifying
heating & humidifying
sensible heating
heating & dehumidifying
dehumidifying
cooling & dehumidifying
sensible cooling
cooling & humidifying (evaporative cooling)

Appendix H

Sample survey for building users or occupants

The **Facility Owner/Management** is implementing various measures to save energy, energy costs, and reduce greenhouse gas emissions. At the same time, our goal is to help maintain and improve building services such as thermal comfort, lighting quality, and hot water delivery. As part of our assessment, we would like to learn from building users how they find the building systems so that we can deliver more effective solutions. This short questionnaire is specifically for users of the **BUILDING ABC123.**

As a user of this facility, we would appreciate if you could take 10 minutes to complete this survey using the online link provided. **This survey will close at 8:00 pm on September XX, 2024.**

If you have any questions about facility energy and GHG upgrades, please contact xxxx@ABC123.ca

1. Approximately how many years have you been a user of the **ABC123** facility?

   (a) Less than 1 year
   (b) 1 to 5 years
   (c) More than 5 years

2. What days of the week and hours of the day do you currently use **ABC123**? (for example: Monday through Friday 9am to 5pm, weekends only, 2 days in the office weekly, etc.)

(user entries)

3. What type of activity do you engage in at the **ABC123** facility?

   (a) Mainly office work, using a computer, and/or meetings
   (b) Activity involving significant physical labour such as maintenance work, cleaning, operating equipment.
   (c) Other

4. How satisfied are you with the facility lighting at **ABC123**? Consider lighting levels, brightness, glare, and color rendering.
   (a) Very unsatisfied
   (b) Somewhat unsatisfied
   (c) Neither satisfied or dissatisfied
   (d) Somewhat satisfied
   (e) Very satisfied
5. How satisfied are you with the **ABC123** facility heating (temperature and humidity) during the heating season?
   (a) Very unsatisfied
   (b) Somewhat unsatisfied
   (c) Neither satisfied or dissatisfied
   (d) Somewhat satisfied
   (e) Very satisfied
   (f) I rarely use the facility during Winter or times with cold outdoor air temperature
6. How satisfied are you with the **ABC123** facility cooling (temperature and humidity) during the cooling season?
   (a) Very unsatisfied
   (b) Somewhat unsatisfied
   (c) Neither satisfied or dissatisfied
   (d) Somewhat satisfied
   (e) Very satisfied
   (f) I rarely use the facility during Summer or times with hot outdoor air temperature
7. How satisfied are you with the **ABC123** facility domestic hot water (such as hand washing, showers, kitchen use, laundry)?
   (a) Very unsatisfied
   (b) Somewhat unsatisfied
   (c) Neither satisfied or dissatisfied
   (d) Somewhat satisfied
   (e) Very satisfied
   (f) I rarely use the facility hot water system

8. How satisfied are you with the **ABC123** facility indoor air and environmental quality? Consider stuffy or stale air, nuisance noise, odours, draftiness, or dust.
   (a) Very unsatisfied
   (b) Somewhat unsatisfied
   (c) Neither satisfied or dissatisfied
   (d) Somewhat satisfied
   (e) Very satisfied
9. Do you have any recommendations to improve the **ABC123** facility services such as heating, cooling, lighting, air quality, hot water, power supplies for computers and other plug loads, or noise reduction?

(user entries)

10. Do you have any recommendations to save energy and reduce utility bill costs for facility **ABC123**? Examples may include scheduling, HVAC control changes, equipment repairs such as leaks, or equipment replacement.

(user entries)

Index

About the Editor

Eric Mazzi is a consultant and educator focusing on technology and policy for demand-side energy systems. He specializes in performing independent technical analysis, measurement and verification (M&V), policy studies, and educational services. He advocates and contributes to efforts to reduce energy consumption and achieve net zero GHG emissions, while maintaining or improving energy services and advancing energy equity. His experience includes energy and environmental consulting, utility maintenance, and 5 years as full-time engineering faculty at the University of British Columbia. He volunteers as a board member for the U.S. Northwest Energy Coalition, and the International Performance M&V Protocol committee. Eric teaches part-time, delivering university and professional certification courses. He has B.Sc. and M.Sc. degrees in mechanical engineering, and a Ph.D. in environmental policy. He is a registered engineer in California and British Columbia.